Lecture Notes in Computational Science and Engineering

Volume 137

This series contains monographs of lecture notes type, lecture course material, and high-quality proceedings on topics described by the term "computational science and engineering". This includes theoretical aspects of scientific computing such as mathematical modeling, optimization methods, discretization techniques, multiscale approaches, fast solution algorithms, parallelization, and visualization methods as well as the application of these approaches throughout the disciplines of biology, chemistry, physics, engineering, earth sciences, and economics.

Jens M. Melenk • Ilaria Perugia •
Joachim Schöberl • Christoph Schwab
Editors

Spectral and High Order Methods for Partial Differential Equations ICOSAHOM 2020+1

Selected Papers from the ICOSAHOM Conference, Vienna, Austria, July 12-16, 2021

Springer

Editors
Jens M. Melenk
TU Wien
Wien, Austria

Ilaria Perugia
Universität Wien
Wien, Austria

Joachim Schöberl
TU Wien
Wien, Austria

Christoph Schwab
ETH Zürich
Zürich, Switzerland

ISSN 1439-7358 ISSN 2197-7100 (electronic)
Lecture Notes in Computational Science and Engineering
ISBN 978-3-031-20431-9 ISBN 978-3-031-20432-6 (eBook)
https://doi.org/10.1007/978-3-031-20432-6

Mathematics Subject Classification: 65M70, 65N35, 65N30, 74S25, 76M10, 76M22, 78M10, 78M22

Cover illustration: "Magnetic flux induced by an electric current through a coil with the color indicating its strength. The simulation was performed with the open source finite element library NGSolve (https://urldefense.com/v3/__http://www.ngsolve.org__;!!NLFGqXoFfo8MMQ!oF0w1iiz98O5s65n4FBNy07HnMyrsf5V2AB_nl-Un5rpBo41jcXxtY51fKJ5ffIIJSF86ai-LyiwKAk1oCtyrg$) using high order H(curl)-conforming finite elements."

This Springer imprint is published by the registered company Springer Nature Switzerland AG
The registered company address is: Gewerbestrasse 11, 6330 Cham, Switzerland

Preface

This volume comprises selected papers from the thirteenth International Conference on Spectral and High-Order Methods ("ICOSAHOM 2020+1"). The 2020+1 edition of ICOSAHOM was originally scheduled to be held in Vienna, Austria, July 6–10th, 2020. Due to the Covid-19 pandemic, it had to be postponed to July 12–16th, 2021, and was held as a virtual conference. The ICOSAHOM conference series is an established regular forum and global meeting place for researchers with interests in the theoretical, applied, and computational aspects of higher-order discretization methods for partial differential equations. In spite of being an online conference, ICOSAHOM 2020+1 was one of the largest one in this series with about 380 registered users and 220 accepted presentations. The week-long conference included 8 plenary talks by internationally leading researchers, 35 minisymposia focusing on a variety of current topics in the field, and was rounded out by about 80 contributed presentations.

The ICOSAHOM conference series looks back upon a long tradition with previous meetings in Como, Italy (1989), Montpellier, France (1992), Houston, TX, USA (1995), Tel Aviv, Israel (1998), Uppsala, Sweden (2001), Providence, RI, USA (2004), Beijing, China (2007), Trondheim, Norway (2009), Gammarth, Tunisia (2012), Salt Lake City, UT, USA (2014), Rio de Janeiro, Brazil (2016), and London, United Kingdom (2018).

The content of these proceedings is organized as follows. First, contributions from the invited speakers are included. The remainder of the volume consists of selected papers highlighting the broad spectrum of topics presented at ICOSAHOM 2020+1. These papers were carefully refereed by internationally renowned experts.

The success of ICOSAHOM 2020+1 was aided through financial support of sponsors, in particular the Austrian Science Fund (FWF) and TU Wien. We would like to give special thanks to our local organizing committee, the local staff, and the many student helpers of the TU Wien for their contribution to the smooth running of the conference. Particular thanks go to the many reviewers of the proceedings submissions.

We acknowledge the publisher, Springer Verlag, for the professional handling of administrative matters related to the publishing of these proceedings.

Vienna, Austria — Jens M. Melenk
Vienna, Austria — Ilaria Perugia
Vienna, Austria — Joachim Schöberl
Zürich, Switzerland — Christoph Schwab

Contents

Part I
Invited Papers

On Arbitrarily Regular Conforming Virtual Element Methods for Elliptic Partial Differential Equations

P. F. Antonietti, G. Manzini, S. Scacchi, and M. Verani

1 Introduction

The conforming finite element method is based on the construction of a finite dimensional approximation spaces that are typically only C^0-continuous [41] on the meshes covering the computational domain. The construction of such approximation spaces with higher regularity is normally deemed a difficult task because it requires a set of basis functions with such global regularity. Examples in this direction can be found all along the history of finite elements from the oldest works in the sixties of the last century, e.g., [12, 28, 42] to the most recent attempts in [46, 47, 63, 64]. Despite its intrinsic difficulty, designing approximations with global C^1- or higher regularity is still a major research topic. Such kind of approximations have indeed a natural application in the numerical treatment of problems involving high-order differential operators.

The Virtual Element Method (VEM) [16] does not require the explicit knowledge of the basis functions spanning the approximation spaces in its formulation and implementation. The crucial idea behind the VEM is that the elemental approximation spaces, which are globally "glued" in a highly regular conforming way, are defined elementwise as the solutions of a partial differential equation. The functions

P. F. Antonietti · M. Verani (✉)
MOX, Dipartimento di Matematica, Politecnico di Milano, Milan, Italy
e-mail: paola.antonietti@polimi.it; marco.verani@polimi.it

G. Manzini
Theoretical Division, Los Alamos National Laboratory, Los Alamos, NM, USA
e-mail: gmanzini@lanl.gov

S. Scacchi
Dipartimento di Matematica, Università degli Studi di Milano, Milan, Italy
e-mail: simone.scacchi@unimi.it

J. M. Melenk et al. (eds.), *Spectral and High Order Methods for Partial Differential Equations ICOSAHOM 2020+1*, Lecture Notes in Computational Science and Engineering 137, https://doi.org/10.1007/978-3-031-20432-6_1

that belong to such approximation spaces are dubbed as "virtual" as they are never really computed, with the noteworthy exception of a subspace of polynomials that are indeed used in the formulation of the method. The virtual element functions are uniquely characterized by a set of values, the *degrees of freedom*, that are actually solved for in the method.

The virtual element "paradigm" thus provides a major breakthrough in obtaining highly regular Galerkin methods as it allows the construction of numerical approximation of any order of accuracy and global conforming regularity that work on unstructured two-dimensional and three-dimensional meshes with very general polytopal elements.

Roughly speaking, the VEM is a Galerkin-type projection method that generalize the finite element method, which was originally designed for simplicial and quadrilateral/hexahedral meshes, to polytopal meshes. Other important families of methods that are suited to polytopal meshes are the polygonal/polyhedral finite element method [61] the mimetic finite difference method [18] the discontinuous Galerkin method on polygonal/polyhedral grids [3, 36]; the hybrid discontinuous Galerkin method [43]; and the hybrid high-order method [44].

The conforming VEM was first developed for second-order elliptic problems in primal formulation [16, 22], and then in mixed formulation [21, 35] and nonconforming formulation [15]. Despite its relative youthness (the first paper was published in 2013), the VEM has been very successful in a wide range of scientific and engineering applications. A non-exhaustive list of applications includes, for example, the works of References [2, 5, 8, 10, 13, 14, 29, 30, 37, 39, 52, 58–60, 62]. Virtual element spaces forming de Rham complexes for the Stokes, Navier-Stokes and Maxwell equations were proposed in [20, 24, 25]. A VEM for Helmholtz problems based on non-conforming approximation spaces of Trefftz functions, i.e., functions that belong to the kernel of the Helmholtz operator, is found [53].

The first works using a C^1-regular conforming VEM addressed the classical plate bending problems [34, 40], second-order elliptic problems [17, 19], and the nonlinear Cahn-Hilliard equation [4]. More recently, highly regular virtual element spaces were considered for the von Kármán equation modelling the deformation of very thin plates [51], geostrophic equations [55] and fourth-order subdiffusion equations [50], two-dimensional plate vibration problem of Kirchhoff plates [54], the transmission eigenvalue problems [56] the fourth-order plate buckling eigenvalue problem [57]. In [7], we proposed the highly-regular conforming VEM for the two-dimensional polyharmonic problem $(-\Delta)^{p_1}u = f$, $p_1 \geq 1$. The VEM is based on an approximation space that locally contains polynomials of degree $r \geq 2p_1-1$ and has a global H^{p_1} regularity. In [11], we extended this formulation to a virtual element space that can have arbitrary regularity $p_2 \geq p_1 \geq 1$ and contains polynomials of degree $r \geq p_2$. This VEM is a generalization of the VEMs for second- and fourth-order problems since the approximation space for $p_2 = p_1 = 1$ coincides with the conforming virtual element spaces for the Poisson equation of Reference [16] and the approximation space for $p_2 = p_1 = 2$ coincides with the conforming virtual element spaces for the biharmonic equation of Reference [34]. VEMs for three-dimensional problems are also available for the fourth-order linear elliptic equation [27] (see also [33]), and highly-regular conforming VEM in any dimension has been proposed in [49]. Numerical experiments for the conforming approximation of the three-dimensional biharmonic problem can be found in Reference [27], while the two dimensional case is treated, e.g., in References [11, 40]. For numerical tests related to the conforming approximation of a second-order elliptic problem employing higher-regular virtual element spaces, we refer to References [11, 17, 19].

Generally speaking, high regularity of the numerical approximation is of primary importance when dealing with high-order differential problems. Moreover, global smoothness can be useful to directly compute physical quantities (such as fluxes, strains, stresses) without resorting to post-processing as in classical C^0-continuous Finite Element Method (FEM), to develop anisotropic error estimators based on the Hessian or to devise better eigenvalue approximation (studies in isogeometric analysis have shown that highly regular discrete spaces may give a better approximation of the high end of the spectrum).

In this paper, we review the detailed construction of the virtual element spaces with arbitrary order of accuracy and regularity for the numerical approximation of two-dimensional problems involving the polyharmonic operator of degree p_1. Such a construction follows the standard guidelines of the VEM, which we briefly summarize here. As the VEM is a conforming Galerkin variational method, its formulation requires the definition of a suitable finite dimensional approximation space, which is obtained by combining in a conforming way local (elemental) finite dimensional spaces. The local virtual element spaces are defined in every mesh element by all the solutions of a specific polyharmonic problem of degree $p_2 \geq p_1$. The loading terms of the partial differential equations defining the elemental virtual element spaces can be all the polynomials of degree (up to) $r - 2p_1$, where the integer number $r \geq p_2$ is the order of the virtual element space. The traces of the virtual element functions and all its normal derivatives of order j from one to $p_2 - 1$ on the elemental edges are univariate polynomials of degree at least $r - j$ (in some cases the polynomial degree can be a little higher than $r - j$). From the definition, it also follows the fundamental property that the polynomials of degree up to r inside all elements are a linear subspace of the virtual element space of degree r. Then, the elemental spaces are "glued" together to form a global space with H^{p_2}-regularity. A very careful choice of the degrees of freedom, which as usual are nodal values associated with the mesh vertices or polynomial moments associated with edges and elements, makes the elliptic projection onto the polynomials of degree r computable. An L^2-orthogonal projection onto the polynomials of degree $r - p_1$ in every mesh element is also computable in the "modified" (or "enhanced") formulation of the virtual element method. In this work, we also present a detailed discussion of the enhanced formulation and its major properties. The enhanced formulation is obtained by extending the similar construction for the Poisson equation presenting in the pioneering paper [1] to our case. These polynomial projection operators are finally used to construct the discrete approximation of the bilinear form and the right-hand side that are used in the virtual element approximation. An abstract convergence result holds, that can be proved by assuming only a few fundamental properties of the virtual element formulation.

The remaining part of the manuscript is organized as follows. In Sect. 2 we introduce the continuous polyharmonic problem and its weak formulation. In Sect. 3 we introduce the virtual element discretization and recall the main abstract convergence result. In Sect. 4 we present the formulation of the conforming virtual element approximation with higher-order regularity. Finally, in Sect. 5 we draw our conclusions.

2 The Continuous Problem

Let $\Omega \subset \mathbb{R}^2$ be an open, bounded, convex domain with polygonal boundary Γ. For any integer $p_1 \geq 1$, we consider the polyharmonic problem

$$(-\Delta)^{p_1} u = f \qquad \text{in } \Omega, \tag{1a}$$

$$\partial_n^j u = 0 \qquad \text{for } j = 0, \ldots, p_1 - 1 \text{ on } \Gamma, \tag{1b}$$

where $\partial_n u = \mathbf{n} \cdot \nabla u$ is the normal derivative of u and $\partial_n^j u$ is the normal derivative applied j times to u with the useful convention that $\partial_n^0 u = u$ for $j = 0$. Let

$$V := H_0^{p_1}(\Omega) = \left\{ v \in H^{p_1}(\Omega) : \partial_n^j v = 0 \text{ on } \Gamma,\ j = 0, \ldots, p_1 - 1 \right\}.$$

Denoting the duality pairing between V and its dual V' by $\langle \cdot, \cdot \rangle$, the variational formulation of the polyharmonic problem (1) reads as

$$\textit{Find } u \in V \textit{ such that}: \quad a_{p_1}(u, v) = \langle f, v \rangle \quad \forall v \in V, \tag{2}$$

where, for any nonnegative integer ℓ, the bilinear form $a_{p_1}(\cdot, \cdot) : V \times V \to \mathbb{R}$ is given by

$$a_{p_1}(u, v) := \begin{cases} \displaystyle\int_\Omega \nabla \Delta^\ell u \cdot \nabla \Delta^\ell v \, d\mathbf{x} & \text{for } p_1 = 2\ell + 1, \ell \geq 0, \\ \displaystyle\int_\Omega \Delta^\ell u \, \Delta^\ell v \, d\mathbf{x} & \text{for } p_1 = 2\ell, \ell \geq 1. \end{cases} \tag{3}$$

If $f \in L^2(\Omega)$ we have

$$\langle f, v \rangle := (f, v) = \int_\Omega f v \, d\mathbf{x}, \tag{4}$$

where $(\cdot,\cdot)$ denotes the L^2-inner product. The bilinear form $a_{p_1}(\cdot,\cdot)$ is coercive and continuous with respect to $||u||_V := (a_{p_1}(u,u))^{1/2}$, which is a norm on $H_0^{p_1}(\Omega)$. The coercivity and continuity constants are respectively denoted by α and M, and their value depends on the regularity of Ω and its boundary Γ. Coercivity and continuity implies existence and uniqueness of the solution to (2) from an application of the Lax-Milgram theorem [32, Theorem 2.7.7]. About the regularity of the solution to (2), it is worth mentioning the result in [45, Corollary 2.21]. Accordingly, if the domain boundary $\partial\Omega$ is C^k-regular for $k \geq 2p_1$ and $f \in H^{k-2p_1}(\Omega)$, then $u \in H^k(\Omega) \cap H_0^{p_1}(\Omega)$ and it holds that $||u||_k \leq C||f||_{k-2p_1}$. As pointed out in [11], the regularity of u for domains with irregular boundaries is still an open issue. However, we know that a similar result holds for the biharmonic problem, i.e., $p_1 = 2$, if Ω is a bounded, convex, polygonal domain see [31].

3 The Discrete Problem and an Abstract Convergence Result

Let r and p_2 be two integer numbers such that $r \geq p_2 \geq p_1 \geq 1$. The virtual element approximation to the variational problem (2) reads as

$$\textit{Find } u_h \in V_{h,r}^{p_2,p_1} \textit{ such that}: \quad a_h(u_h, v_h) = \langle f_h, v_h \rangle \quad \forall v_h \in V_{h,r}^{p_2,p_1}, \tag{5}$$

where the virtual element space $V_{h,r}^{p_2,p_1}$ is a finite-dimensional conforming subspace of V; $a_h(\cdot,\cdot) : V_{h,r}^{p_2,p_1} \times V_{h,r}^{p_2,p_1} \to \mathbb{R}$ is the virtual element bilinear form that approximates the bilinear form (3); $\langle f_h, \cdot \rangle : V_{h,r}^{p_2,p_1} \to \mathbb{R}$ is the continuous linear functional that approximates (4) through an element f_h of the dual space $(V_{h,r}^{p_2,p_1})^*$ of $V_{h,r}^{p_2,p_1}$. The formal definition and properties of $V_{h,r}^{p_2,p_1}$, $a_h(\cdot,\cdot)$ and f_h are discussed in the next section.

3.1 Mesh Notation, Mesh Regularity and Some Basic Definitions

The virtual element method is formulated on the mesh family $\{\Omega_h\}_h$, where each mesh Ω_h is a partition of the computational domain Ω into nonoverlapping polygonal elements P and is labeled by the mesh size parameter h that is defined below. A polygonal element P is a compact subset of $\mathbb{R}^2$ with boundary $\partial\mathsf{P}$, area $|\mathsf{P}|$, center $\mathbf{x}_\mathsf{P}$, and diameter $h_\mathsf{P} = \sup_{\mathbf{x},\mathbf{y}\in\mathsf{P}} |\mathbf{x}-\mathbf{y}|$. The mesh elements of Ω_h form a finite cover of Ω such that $\overline{\Omega} = \cup_{\mathsf{P}\in\Omega_h}\mathsf{P}$ and the mesh size labeling each mesh Ω_h is defined by $h = \max_{\mathsf{P}\in\Omega_h} h_\mathsf{P}$. A mesh edge e has center $\mathbf{x}_e$ and length h_e and we denote the set of mesh edges by $\mathcal{E}_h$. A mesh vertex v has position vector $\mathbf{x}_\mathsf{v}$ and we denote the set of mesh vertices by $\mathcal{V}_h$. Moreover, in the definition of the degrees of freedom of the next section, we associate every vertex V with a characteristic length h_{V}. This characteristic length h_{V} can be the average of the diameters of the polygons sharing V.

For any integer number $\ell \geq 0$, we let $\mathbb{P}_\ell(\mathsf{P})$ and $\mathbb{P}_\ell(e)$ denote the space of polynomials defined on P and e, respectively, and $\mathbb{P}_\ell(\Omega_h)$ denotes the space of piecewise polynomials of degree ℓ on the mesh Ω_h. Accordingly, if $q \in \mathbb{P}_\ell(\Omega_h)$ then it holds that $q_{|\mathsf{P}} \in \mathbb{P}_\ell(\mathsf{P})$ for all $\mathsf{P} \in \Omega_h$. Finally, we define the (broken) seminorm of a function $v \in \prod_{\mathsf{P}\in\Omega_h} H^{p_1}(\mathsf{P})$ by

$$||v||_h^2 = \sum_{\mathsf{P}\in\Omega_h} a_{p_1}^{\mathsf{P}}(v, v)$$

where $a_{p_1}^{\mathsf{P}}(\cdot,\cdot)$ denotes the restriction to P of the bilinear form $a_{p_1}(\cdot,\cdot)$.

Throughout the paper, we use the multi-index notation, so that $\nu = (\nu_1, \nu_2)$ is a two-dimensional index defined by the two integer numbers $\nu_1, \nu_2 \geq 0$. Moreover, $D^\nu w = \partial^{|\nu|}w/\partial x^{\nu_1}\partial y^{\nu_2}$ denotes the partial derivative of order $|\nu| = \nu_1 + \nu_2 > 0$ of a given bivariate function $w(x, y)$, and we use the conventional notation that $D^{(0,0)}w = w$ for $\nu = (0, 0)$. We denote the partial derivatives of w versus x and y by the shortcuts $\partial_x w$, $\partial_y w$, $\partial_{xx} w$, $\partial_{xy} w$, $\partial_{yy} w$, etc. We denote the normal and tangential derivatives with respect to a given edge and their mixed combination by $\partial_n w$, $\partial_t w$, $\partial_{tt} w$, $\partial_{nt} w$, $\partial_{nn} w$, etc, and use the shorter notation $\partial_t^j w$ and $\partial_n^j w$ for the tangential and normal derivatives of w of order j.

3.2 *Abstract Convergence Theorem*

For the mathematical formulation of the virtual element approximation (5), we require the two following assumptions on the virtual element space $V_{h,r}^{p_2,p_1}$ and the bilinear form $a_h(\cdot,\cdot)$:

(H1). For all $h > 0$, the *global* virtual element space $V_{h,r}^{p_2,p_1}$ is a conforming, finite-dimensional subspace of $V = H_0^{p_1}(\Omega) \cap H^{p_2}(\Omega)$ such that for all elements P of all mesh partitions Ω_h it holds that

- $V_{h,r}^{p_2,p_1}(\mathsf{P})$, the *local* (elemental) virtual element space that is defined as the restriction of $V_{h,r}^{p_2,p_1}$ to the element P is a finite-dimensional subspace of $H^{p_2}(\mathsf{P})$;
- $\mathbb{P}_r(\mathsf{P})$, the space of polynomials of degree up to r defined on P is a subspace of $V_{h,r}^{p_2,p_1}(\mathsf{P})$.

The two above conditions can be summarized as

$$\mathbb{P}_r(P) \subseteq V_{h,r}^{p_2,p_1}(P) \subset H^{p_2}(P) \subseteq H^{p_1}(P), \quad p_2 \geq p_1.$$

(H2). The bilinear form $a_h(\cdot,\cdot) : V_{h,r}^{p_2,p_1} \times V_{h,r}^{p_2,p_1} \to \mathbb{R}$ admits the elementwise decomposition

$$a_h(u_h, v_h) = \sum_{\mathsf{P}\in\Omega_h} a_h^{\mathsf{P}}(u_h, v_h) \quad \forall u_h,\ v_h \in V_{h,r}^{p_2,p_1},$$

where for all element P the local bilinear form $a_h^{\mathsf{P}}(\cdot,\cdot)$ is symmetric and such that

(*r*-**Consistency**): for every polynomial $q \in \mathbb{P}_r(\mathsf{P})$ and every virtual element function $v_h \in V_{h,r}^{p_2,p_1}(\mathsf{P})$ it holds that

$$a_h^{\mathsf{P}}(v_h, q) = a_{p_1}^{\mathsf{P}}(v_h, q); \tag{6}$$

(**Stability**): there exist two positive constants α_*, α^* independent of h and P such that for every $v_h \in V_{h,r}^{p_2,p_1}(\mathsf{P})$ it holds that

$$\alpha_* a_{p_1}^{\mathsf{P}}(v_h, v_h) \le a_h^{\mathsf{P}}(v_h, v_h) \le \alpha^* a_{p_1}^{\mathsf{P}}(v_h, v_h). \tag{7}$$

The stability constant α_* and α^* may depend on the polynomial approximation degree r, see, e.g., [6] for the case $p_1 = 1$.

Assumption **(H2)** implies that the symmetric bilinear form $a_h(\cdot,\cdot)$ is coercive and continuous, so that the existence and uniqueness of the solution u_h follows from an application of the Lax-Milgram theorem [32, Theorem 2.7.7]. Under these assumptions we can prove this abstract convergence result.

Theorem 3.1 *Let $u \in V$ be the solution to the variational problem (1) and $u_h \in V_{h,r}^{p_2,p_1}$, $r \ge p_2 \ge p_1 \ge 1$, the solution to the virtual element approximation (5) under assumptions **(H1)–(H2)**. Then, there exists a constant C independent of h such that*

$$||u - u_h||_V \le C\Big(||u - u_I||_V + ||u - u_\pi||_h + ||f_h - f||_{(V_{h,r}^{p_2,p_1})^*}\Big), \tag{8}$$

for every virtual element approximation u_I in $V_{h,r}^{p_2,p_1}$ and any piecewise polynomial approximation $u_\pi \in \mathbb{P}_r(\Omega_h)$ of u. The constant C is proportional to $(M/\alpha)\,(\alpha^*/\alpha_*)$.

The proof of this theorem was first published in [7] for the case with $p_2 = p_1$ and then extended to the case for $p_2 \ge p_1$ in [11].

4 The Virtual Element Spaces of Higher-Order Continuity

In this section, we present the formulation of the virtual element method of Eq. (5). To this end, we first introduce the local virtual element spaces, the degrees of freedom, and the global virtual element space, which is obtained by "gluing" in a conforming way the local spaces. Then, we discuss the computability of the elliptic projection operator, and the enhancement of the local spaces, which allows us to compute the orthogonal projection operators onto the subspace of polynomials of degree up to $p_1 - 1$. Finally, we discuss the construction of the bilinear form $a_h(\cdot, \cdot)$ and the load term $\langle f_h, \cdot \rangle$.

4.1 Local Space Definitions

Let P be a mesh element. For $p_2 \leq r \leq 2p_2 - 2$, we consider the local virtual element space defined as

$$V_{h,r}^{p_2,p_1}(\mathsf{P}) = \Big\{ v_h \in H^{p_2}(\mathsf{P}) : \Delta^{p_2} v_h \in \mathbb{P}_{r-2p_1}(\mathsf{P}),\ \partial_n^j v_h \in \mathbb{P}_{\alpha_j(p_2,r)}(e), \quad j = 0, \ldots, p_2 - 1\ \forall e \in \partial \mathsf{P} \Big\}, \tag{9}$$

where $\alpha_j(p_2, r) = \max\{2(p_2 - j) - 1, r - j\}$. For $r \geq 2p_2 - 1$ it holds that $\alpha_j = r - j$ for all $j = 0, \ldots, p_2 - 1$ and the definition of the local virtual element space on the element P takes the simpler form

$$V_{h,r}^{p_2,p_1}(\mathsf{P}) = \Big\{ v_h \in H^{p_2}(\mathsf{P}) : \Delta^{p_2} v_h \in \mathbb{P}_{r-2p_1}(\mathsf{P}),\ \partial_n^j v_h \in \mathbb{P}_{r-j}(e), \quad j = 0, \ldots, p_2 - 1\ \forall e \in \partial \mathsf{P} \Big\}. \tag{10}$$

In both definitions (9) and (10) we use the conventional notation that $\mathbb{P}_r(\mathsf{P}) = \{0\}$ if $r < 0$.

Remark 4.1 The space of polynomials $\mathbb{P}_r(\mathsf{P})$ is a subspace of $V_{h,r}^{p_2,p_1}(\mathsf{P})$ for both definitions (9) and (10).

Remark 4.2 Let $\#(\mathcal{P})$ denote the cardinality of a (finite dimensional) space $\mathcal{P}$ and $N_{\mathsf{P}}^{\mathcal{E}}$ and $N_{\mathsf{P}}^{\mathcal{V}}$ the number of edges and vertices of element P. The dimension of the local virtual element space (9) is given by

$$\dim V_{h,r}^{p_2,p_1}(\mathsf{P}) = \#\big(\mathbb{P}_{r-2p_1}(\mathsf{P})\big) + \sum_{e \in \partial \mathsf{P}} \sum_{j=0}^{p_2-1} \#\big(\mathbb{P}_{\alpha_j(p_2,r)}(e)\big) - N_{\mathsf{P}}^{\mathcal{V}} \frac{(p_2+1)p_2}{2}$$

$$= \frac{(r-2p_1+1)(r-2p_1+2)}{2} + N_{\mathsf{P}}^{\mathcal{E}} \sum_{j=0}^{p_2-1} \big(\alpha_j(p_2,r)+1\big)$$

$$- N_{\mathsf{P}}^{\mathcal{V}} \frac{(p_2+1)p_2}{2}. \tag{11}$$

The dimension of the local virtual element space (10) is given by

$$\dim V_{h,r}^{p_2,p_1}(\mathsf{P}) = \#\big(\mathbb{P}_{r-2p_1}(\mathsf{P})\big) + \sum_{e\in\partial\mathsf{P}} \sum_{j=0}^{p_2-1} \#\big(\mathbb{P}_{r-j}(e)\big) - N_{\mathsf{P}}^{\mathcal{V}} \frac{(p_2+1)p_2}{2}$$

$$= \frac{(r-2p_1+1)(r-2p_1+2)}{2} + N_{\mathsf{P}}^{\mathcal{E}} \frac{p_2\big(2r+3-p_2\big)}{2}$$

$$- N_{\mathsf{P}}^{\mathcal{V}} \frac{(p_2+1)p_2}{2}. \tag{12}$$

In both Eqs. (11) and (12), the last term of the right-hand side, i.e., $N_{\mathsf{P}}^{\mathcal{V}} p_2(p_2+1)/2$, is subtracted to take into account the C^{p_2-1}-regularity of v_h at the elemental vertices.

4.2 Local Degrees of Freedom

Let $\beta_j = \alpha_j - \min\{2(p_2-j)-1, r-j\} - 1$. For $r = 2p_2 - 1 - k$ with $k = 1, \dots, p_2 - 1$, the virtual element functions in the elemental space (9) are uniquely identified by the following degrees of freedom:

(D1) $h_{\mathsf{V}}^{|\nu|} D^{\nu} v_h(\mathsf{V})$, $|\nu| \le p_2 - 1$ for any vertex V of $\partial\mathsf{P}$;

(D2) $h_e^{-1+j} \int_e q\, \partial_n^j v_h\, ds$ for any $q \in \mathbb{P}_{\beta_j}(e)$ and edge e of $\partial\mathsf{P}$, $j = k+1, \dots, p_2 - 1$;

(D3) $h_{\mathsf{P}}^{-2} \int_{\mathsf{P}} q v_h\, d\mathbf{x}$ for any $q \in \mathbb{P}_{r-2p_1}(\mathsf{P})$.

For $r \ge 2p_2 - 1$, we note that $\beta_j = r - (2p_2 - j)$ and we consider the polynomial edge moments in **(D2)** for $j = 0, \dots, p_2 - 1$.

Remark 4.3 The L^2-projection operator $\Pi_{r-2p_1}^{0,\mathsf{P}}$ onto the polynomial space $\mathbb{P}_{r-2p_1}(\mathsf{P})$ is computable from the degrees of freedom **(D3)**.

A virtual element function in $v_h \in V_{h,r}^{p_2,p_1}(\mathsf{P})$ has the regularity property that $(v_h)_{|\partial\mathsf{P}} \in C^{p_2-1}(\partial\mathsf{P})$. This regularity is reflected by the choice of the degrees of freedom, and is, indeed, provided by the vertex degrees of freedom of **(D1)**. Furthermore, the traces of v_h ($j = 0$) and the j-th normal derivatives (up to order

$j = p_2-1$) on every edge $e \in \partial \mathsf{P}$ are univariate polynomials of degree at least $r-j$. The information provided by (**D1**) makes it possible to build polynomial traces of degree higher than $r - j$ if r is equal to p_2 (or not "too bigger" than p_2 as shown in the following examples). In such a case, we can compute the edge traces $(v_h)_{|e}$ and $(\partial_n^j v_h)_{|e}$ by solving the interpolation problem that uses the vertex values of v_h and its partial derivatives.

In the next three examples, we discuss the trace interpolation problem for $r \geq p_2$ and $j = 0$, $j = 1$, $j \geq 2$. This process is also shown in Table 2 and Fig. 1.

Example 4.4 ($j = 0$) We derive the higher-order *tangential* derivatives of v_h by repetitively applying the differential operator $\mathbf{t} \cdot \nabla = \left(t_x \partial_x + t_y \partial_y\right)$ to the univariate polynomial trace of v_h, i.e., $\partial_t^\ell v_h$ along every elemental edge (recall that $\partial_t^0 v_h = v_h$ for $\ell = 0$). For example, for $\ell = 1, 2, 3, 4$, the tangential derivatives $\partial_t^\ell v_h$ are given by

$$\begin{aligned}
\partial_t v_h(\mathsf{v}_i) &= t_x \partial_x (v_h)(\mathsf{v}_i) + t_y \partial_y (v_h)(\mathsf{v}_i),\\
\partial_t^2 v_h(\mathsf{v}_i) &= \left(t_x \partial_x + t_y \partial_y\right)\left(\partial_t v_h\right)(\mathsf{v}_i)\\
&= t_x \partial_x \left(\partial_t v_h\right)(\mathsf{v}_i) + t_y \partial_y \left(\partial_t v_h\right)(\mathsf{v}_i)\\
&= t_x t_x \partial_{xx} v_h(\mathsf{v}_i) + 2 t_x t_y \partial_{xy} v_h(\mathsf{v}_i) + t_y t_y \partial_{yy} v_h(\mathsf{v}_i),\\
\partial_t^3 v_h(\mathsf{v}_i) &= \left(t_x \partial_x + t_y \partial_y\right)\left(\partial_t^2 v_h\right)(\mathsf{v}_i)\\
&= t_x \partial_x \left(\partial_t^2 v_h\right)(\mathsf{v}_i) + t_y \partial_y \left(\partial_t^2 v_h\right)(\mathsf{v}_i)\\
&= t_x t_x t_x \partial_{xxx} v_h(\mathsf{v}_i) + 3 t_x t_x t_y \partial_{xxy} v_h(\mathsf{v}_i) + 3 t_x t_y t_y \partial_{xyy} v_h(\mathsf{v}_i)\\
&\quad + t_y t_y t_y \partial_{yyy} v_h(\mathsf{v}_i),\\
\partial_t^4 v_h(\mathsf{v}_i) &= \left(t_x \partial_x + t_y \partial_y\right)\left(\partial_t^3 v_h\right)(\mathsf{v}_i)\\
&= t_x \partial_x \left(\partial_t^3 v_h\right)(\mathsf{v}_i) + t_y \partial_y \left(\partial_t^3 v_h\right)(\mathsf{v}_i)\\
&= t_x t_x t_x t_x \partial_{xxxx} v_h(\mathsf{v}_i) + 4 t_x t_x t_x t_y \partial_{xxxy} v_h(\mathsf{v}_i) + 6 t_x t_x t_y t_y \partial_{xxyy} v_h(\mathsf{v}_i)\\
&\quad + 4 t_x t_y t_y t_y \partial_{xyyy} v_h(\mathsf{v}_i) + t_y t_y t_y t_y \partial_{yyyy} v_h(\mathsf{v}_i).
\end{aligned}$$

It is easy to recognize the pattern of the combinatorial coefficients in these expansions. According to the third column ($j = 0$) of Table 1, the degrees of freedom (**D1**) at vertex v_i for a given regularity index p_2 yield p_2 pieces of information, $v_h(\mathsf{v}_i)$, $\partial_t v_h(\mathsf{v}_i)$, $\partial_t^2 v_h(\mathsf{v}_i)$, $\ldots \partial_t^{p_2-1} v_h(\mathsf{v}_i)$. Since each edge has two vertices, we have $2p_2$ pieces of information and we can interpolate the edge trace of v_h as a univariate polynomial of degree $2p_2 - 1$. Such polynomial degree is

Table 1 Vertex degrees of freedom for the trace interpolation process on the elemental edges. The first column shows the value of $\max\{|\nu|\} = p_2 - 1$ that we use to define the degrees of freedom (**D1**). The second column shows the degrees of freedom at vertex v_i corresponding to $p_2 - 1$ in the first column. The remaining columns shows the quantities that we can compute using the degrees of freedom listed in the second column. Recalling that $|\nu| \leq p_2 - 1$ and $0 \leq j \leq p_2 - 1$, on the columns for $j = 0, \dots, 4$ we read the pieces of information that are available for the interpolation of $\partial_n^j v_h$ (with $\partial_n^0 v_h = v_h$ for $j = 0$). For example, if $p_2 = 3$, we can use only the objects of the first three table rows (i.e., $|\nu| = 0, 1, 2$), and each column for $j = 0, 1, 2$ lists the pieces of information that are available to construct the polynomial trace of v_h, $\partial_n v_h$ and $\partial_n^2 v_h$ on each edge. In such a case, the vertex degrees of freedom allows us to interpolate the trace of v_h as a polynomial of degree 5, the trace of $\partial_n v_h$ as a polynomial of degree 3, and the trace of $\partial_n^2 v_h$ as a polynomial of degree 1. These trace interpolations are consistent with $r = p_2$ and a global virtual element space with C^{p_2-1}-regularity on Ω

$p_2 - 1$	Degrees of freedom (**D1**)	$j = 0$	$j = 1$	$j = 2$	$j = 3$	$j = 4$
0	$v_h(\mathsf{v}_i)$	v_h	–	–	–	–
1	$\partial_x v_h(\mathsf{v}_i)$, $\partial_y v_h(\mathsf{v}_i)$	$\partial_t v_h$	$\partial_n v_h$	–	–	–
2	$\partial_{xx} v_h(\mathsf{v}_i)$, $\partial_{xy} v_h(\mathsf{v}_i)$, $\partial_{yy} v_h(\mathsf{v}_i)$	$\partial_t^2 v_h$	$\partial_t \partial_n v_h$	$\partial_n^2 v_h$	–	–
3	$\partial_{xxx} v_h(\mathsf{v}_i)$, $\partial_{xxy} v_h(\mathsf{v}_i)$, $\partial_{xyy} v_h(\mathsf{v}_i)$, $\partial_{yyy} v_h(\mathsf{v}_i)$	$\partial_t^3 v_h$	$\partial_t^2 \partial_n v_h$	$\partial_t \partial_n^2 v_h$	$\partial_n^3 v_h$	–
4	$\partial_{xxxx} v_h(\mathsf{v}_i)$, $\partial_{xxxy} v_h(\mathsf{v}_i)$, $\partial_{xxyy} v_h(\mathsf{v}_i)$, $\partial_{xyyy} v_h(\mathsf{v}_i)$, $\partial_{yyyy} v_h(\mathsf{v}_i)$	$\partial_t^4 v_h$	$\partial_t^3 \partial_n v_h$	$\partial_t^2 \partial_n^2 v_h$	$\partial_t \partial_n^3 v_h$	$\partial_n^4 v_h$

clearly bigger than r if we choose r such that $p_2 \leq r < 2p_2 - 1$. This fact is not in conflict with the property that the virtual element space contains the subspace of polynomials of degree r.

The polynomial degrees of the edge trace of v_h that we can interpolate from the degrees of freedom (**D1**)–(**D2**) are illustrated in Table 2 by the rows for $j = 0$ and different values of p_2. In this table, the values of r such that $r = 2p_2 - 1$ are reported in bold, and the ones for $r < 2p_2 - 1$ are those preceding the bold ones on the same row. For these values of r the trace of v_h can be interpolated from the information provided by (**D1**). However, if we increase the polynomial degree r so that $r > 2p_2 - 1$, the degrees of freedom (**D1**) are no longer enough to solve the interpolation problem. In such a case, we need the additional degrees of freedom of (**D2**), i.e., the moments of v_h against a (basis of) polynomials of degree $r-(2p_2-1)$ defined on e. $\square$

Example 4.5 ($j = 1$) As for the case $j = 0$, we derive the higher-order *tangential* derivatives of $\partial_n v_h = n_x \partial_x v_h + n_y \partial_y v_h$ by repetitively applying the differential operator $\mathbf{t} \cdot \nabla = (t_x \partial_x + t_y \partial_y)$ to the univariate polynomial trace of $\partial_n v_h$, i.e, $\partial_t^\ell \partial_n v_h$

Table 2 Polynomial orders of the edge traces of v_h and its normal derivatives $\partial_n^j v_h$ for $p_2 = 1, 2, 3, 4$ and $r = p_2, \ldots 7$ (we recall that $r \geq p_2$). The first column on the left reports the value of p_2 and $j = 0, \ldots, p_2 - 1$; the second column reports the value of $2(p_2 - j) - 1$, which is a threshold value, and the remaining columns the possible values of $r - j$ (remember that on each edge $\partial_n^j v_h \in \mathbb{P}_{\alpha_j}$ with $\alpha_j = \max\{2(p_2 - j) - 1, r - j\}$). The values of the polynomial degree $r - 1$ such that $r - j = 2(p_2 - j) - 1$ (or, equivalently, that $r = 2p_2 - j - 1$) are reported in bold font. The polynomial traces with degree equal or higher than this bold value, which are above it in every column ad correspond to the smaller order j of derivation, can be interpolated using only the vertex degrees of freedom (**D1**). To interpolate the remaining edge traces we need the additional information provided by (**D2**)

		$r - j$							
$\mathbf{p_2 = 1}$	$\mathbf{2(p_2 - j) - 1}$	$\mathbf{r = 1}$	$\mathbf{r = 2}$	$\mathbf{r = 3}$	$\mathbf{r = 4}$	$\mathbf{r = 5}$	$\mathbf{r = 6}$	$\mathbf{r = 7}$	...
$j = 0$	**1**	**1**	2	3	4	5	6	7	...
$\mathbf{p_2 = 2}$	$\mathbf{2(p_2 - j) - 1}$	$\mathbf{r = 1}$	$\mathbf{r = 2}$	$\mathbf{r = 3}$	$\mathbf{r = 4}$	$\mathbf{r = 5}$	$\mathbf{r = 6}$	$\mathbf{r = 7}$	...
$j = 0$	**3**	–	2	**3**	4	5	6	7	...
$j = 1$	**1**	–	**1**	2	3	4	5	6	...
$\mathbf{p_2 = 3}$	$\mathbf{2(p_2 - j) - 1}$	$\mathbf{r = 1}$	$\mathbf{r = 2}$	$\mathbf{r = 3}$	$\mathbf{r = 4}$	$\mathbf{r = 5}$	$\mathbf{r = 6}$	$\mathbf{r = 7}$	...
$j = 0$	**5**	–	–	3	4	**5**	6	7	...
$j = 1$	**3**	–	–	2	**3**	4	5	6	...
$j = 2$	**1**	–	–	**1**	2	3	4	5	...
$\mathbf{p_2 = 4}$	$\mathbf{2(p_2 - j) - 1}$	$\mathbf{r = 1}$	$\mathbf{r = 2}$	$\mathbf{r = 3}$	$\mathbf{r = 4}$	$\mathbf{r = 5}$	$\mathbf{r = 6}$	$\mathbf{r = 7}$	...
$j = 0$	**7**	–	–	–	4	5	6	**7**	...
$j = 1$	**5**	–	–	–	3	4	**5**	6	...
$j = 2$	**3**	–	–	–	2	**3**	4	5	...
$j = 3$	**1**	–	–	–	**1**	2	3	4	...

along every elemental edge (recall that $\partial_t^0 \partial_n v_h = \partial_n v_h$ for $\ell = 0$). For example, for $\ell = 1, 2, 3$ we find that

$$
\begin{aligned}
\partial_t \partial_n v_h(\mathsf{V}_i) &= \big(t_x \partial_x + t_y \partial_y\big)\big(\partial_n v_h\big)(\mathsf{V}_i) \\
&= t_x \partial_x \big(\partial_n v_h\big)(\mathsf{V}_i) + t_y \partial_y \big(\partial_n v_h\big)(\mathsf{V}_i) \\
&= t_x n_x \partial_{xx} v_h(\mathsf{V}_i) + (t_x n_y + t_y n_x)\partial_{xy} v_h(\mathsf{V}_i) + t_y t_y \partial_{yy} v_h(\mathsf{V}_i), \\
\partial_t^2 \partial_n v_h(\mathsf{V}_i) &= \big(t_x \partial_x + t_y \partial_y\big)\big(\partial_t \partial_n v_h\big)(\mathsf{V}_i) \\
&= t_x \partial_x \big(\partial_t \partial_n v_h\big)(\mathsf{V}_i) + t_y \partial_y \big(\partial_t \partial_n v_h\big)(\mathsf{V}_i) \\
&= t_x t_x n_x \partial_{xxx} v_h(\mathsf{V}_i) + \big(t_x\big(t_x n_y + t_y n_x\big) + t_y\big(t_x n_y + t_y n_x\big)\big)\partial_{xxy} v_h(\mathsf{V}_i) \\
&\quad + t_x t_y \big(n_x + n_y\big)\partial_{xyy} v_h(\mathsf{V}_i) + t_y t_y n_y \partial_{yyy} v_h(\mathsf{V}_i), \\
\partial_t^3 \partial_n v_h(\mathsf{V}_i) &= \big(t_x \partial_x + t_y \partial_y\big)\big(\partial_t^2 \partial_n v_h\big)(\mathsf{V}_i)
\end{aligned}
$$

$$= \big(t_x\partial_x + t_y\partial_y\big)\big(\partial_n^2 v_h\big)(\mathsf{v}_i)$$

$$= t_x\partial_x\big(\partial_t^2\partial_n v_h(\mathsf{v}_i)\big)(\mathsf{v}_i) + t_y\partial_y\big(\partial_t^2\partial_n v_h(\mathsf{v}_i)\big)(\mathsf{v}_i),$$

$$= \dots$$

According to the fourth column ($j = 1$) of Table 1, the degrees of freedom (**D1**) at vertex v_i for a given p_2 yield $p_2 - 1$ pieces of information, $\partial_n v_h(\mathsf{v}_i)$, $\partial_t\partial_n v_h(\mathsf{v}_i)$, $\partial_t^2\partial_n v_h(\mathsf{v}_i), \dots \partial_t^{p_2-2}\partial_n v_h(\mathsf{v}_i)$. Since each edge has two vertices, we have $2(p_2 - 1)$ pieces of information that we can interpolate as a polynomial of degree $2p_2 - 3$. Such polynomial degree is clearly bigger than $r - 1$ if we choose r such that $p_2 \leq r < 2(p_2 - 1)$.

The polynomial degrees of the edge trace of $\partial_n v_h$ that we can interpolate from the degrees of freedom (**D1**)–(**D2**) are illustrated in Table 2 by the rows for $j = 1$ and different values of p_2. In this table, the values of r such that $r-1 = 2(p_2-1)-1$ are reported in bold, and the ones for $r-1 < 2(p_2-1)-1$ are those preceding the bold ones on the same row. For these values of r the trace of $\partial_n v_h$ can be interpolated from the information provided by (**D1**). However, if we increase the polynomial degree r so that $r - 1 > 2(p_2 - 1) - 1$, the degrees of freedom (**D1**) are no longer enough to solve the interpolation problem. In such a case, we need the additional degrees of freedom of (**D2**), i.e., the moments of $\partial_n v_h$ against a (basis of) polynomials of degree $r - 2(p_2 - 1)$ defined on e. □

Example 4.6 ($j \geq 2$) As for the cases $j = 0$ and $j = 1$, we derive the higher-order tangential derivatives of $\partial_n^j v_h$ by repetitively applying the differential operator $\mathbf{t}\cdot\nabla = \big(t_x\partial_x + t_y\partial_y\big)$ to the univariate polynomial trace of $\partial_n^j v_h$, i.e., $\partial_t^\ell\partial_n^j v_h$ along every elemental edge. For example, for $j = 2$, since $\partial_n^2 v_h = n_x n_x\partial_{xx} v_h + 2n_x n_y\partial_{xy} v_h + t_y t_y\partial_{yy} v_h$, we find that

$$\partial_t\partial_n^2 v_h(\mathsf{v}_i) = \big(t_x\partial_x + t_y\partial_y\big)\big(\partial_n^2 v_h\big)(\mathsf{v}_i)$$

$$= t_x\partial_x\big(\partial_n^2 v_h\big)(\mathsf{v}_i) + t_y\partial_y\big(\partial_n^2 v_h\big)(\mathsf{v}_i)$$

$$= t_x n_x n_x\partial_{xxx} v_h + (2t_x t_x n_x + t_y n_x n_x)\partial_{xxy} v_h$$

$$+ (t_x n_y n_y + 2t_y n_x n_y)\partial_{xyy} v_h + t_y n_y n_y\partial_{yyy} v_h.$$

For $j \geq 0$, each edge vertex v_i, $i = 1, 2$, provides the values of $\partial_n^j v_h$ and its first $p_2 - j - 1$ tangential derivatives. Hence, there are $2(p_2 - j)$ pieces of information available on each edge and we can interpolate the edge trace of $\partial_n^j v_h$ as a univariate polynomial of degree $2(p_2 - j) - 1$. Such polynomial degree is clearly bigger than $r - j$ if we choose r such that $p_2 \leq r < 2p_2 - j - 1$.

The polynomial degrees of the edge trace of $\partial_n^j v_h$ that we can interpolate from the degrees of freedom (**D1**)–(**D2**) are illustrated in Table 2 by the rows for $j \geq 0$

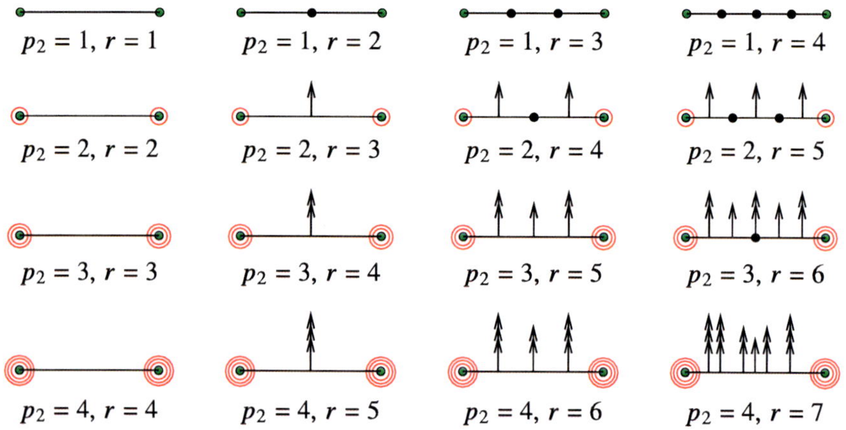

Fig. 1 Edge degrees of freedom of the virtual element space $V_{h,r}^{p_2,p_1}$ with regularity index $p_2 = 1$ (Laplace operator), $p_2 = 2$ (bi-harmonic operator), $p_2 = 3$ (tri-harmonic operator), $p_2 = 4$, and polynomial degree r such that $p_2 \leq r \leq p_2 + 3$. The (green) dots at the vertices represent the vertex values and each (red) vertex circle represents an order of derivation. The (black) dots on the edge represent the polynomial moments of the trace $v_{h|e}$; the arrows represent the polynomial moments of $\partial_n v_{h|e}$; the double arrows represent the polynomial moments of $\partial_n^2 v_{h|e}$

and different values of p_2. In this table, the values of r such that $r = 2p_2 - j - 1$ are reported in bold, and the ones for $r < 2p_2 - j - 1$ are those preceding the bold ones on the same row. For these values of r the trace of $\partial_n^j v_h$ can be interpolated from the information provided by (**D1**). However, if we increase the polynomial degree r so that $r > 2p_2 - j - 1$, the degrees of freedom (**D1**) are no longer enough to solve the interpolation problem. In such a case, we need the additional degrees of freedom of (**D2**), i.e., the moments of $\partial_n^j v_h$ against a (basis of) polynomials of degree $r - (2p_2 - j - 1)$ defined on e. It is worth noting that for increasing values of r, we need to supplement this information starting from the higher-order normal derivatives (see Fig. 1). □

Lemma 4.7 *The degrees of freedom* (**D1**)–(**D3**) *are unisolvent in the virtual element space* $V_{h,r}^{p_2,p_1}(\mathsf{P})$.

Proof Let P be a polygonal element. First, a counting argument shows that the number of degrees of freedom (**D1**)–(**D3**) is equal to the dimension of $V_{h,r}^{p_2,p_1}(\mathsf{P})$ (see Remark 4.2). Then, we prove that a virtual element function v_h is necessarily zero if all its degrees of freedom (**D1**)–(**D3**) are zero. In particular, assuming that the degrees of freedom (**D1**) and (**D2**) are zero implies that the polynomial traces of v_h and its normal derivatives of order up to $p_2 - 1$ are identically zero on all edges of $\partial\mathsf{P}$, and so are their tangential derivatives of any order. Likewise, assuming that the degrees of freedom (**D3**) are zero implies that the elemental moments of v_h against the polynomials of degree up to $r - 2p_1$ (for $r \geq 2p_1$) are zero.

Consider separately the case of odd and even values of p_2. If $p_2 = 2\ell + 1$ with $\ell \geq 0$, a repeated application of the integration by parts formula yields

$$\int_{\mathsf{P}} \left|\nabla \Delta^{\ell} v_h\right|^2 \, d\mathbf{x} = -\int_{\mathsf{P}} \left(\Delta^{p_2} v_h\right) v_h \, d\mathbf{x} + \int_{\partial\mathsf{P}} \left(\partial_n \Delta^{\ell} v_h\right) \Delta^{\ell} v_h \, ds$$
$$+ \sum_{i=1}^{\ell} \left(\int_{\partial\mathsf{P}} \left(\partial_n \Delta^{p_2 - i} v_h\right) \Delta^{i-1} v_h \, ds - \int_{\partial\mathsf{P}} \left(\Delta^{p_2-i} v_h\right) \partial_n \Delta^{i-1} v_h \, ds \right). \tag{13}$$

Similarly, if $p_2 = 2\ell$ with $\ell \geq 1$, we find that

$$\int_{\mathsf{P}} \left|\Delta^{\ell} v_h\right|^2 \, d\mathbf{x} = \int_{\mathsf{P}} \left(\Delta^{p_2} v_h\right) v_h \, d\mathbf{x}$$
$$- \sum_{i=1}^{\ell} \left(\int_{\partial\mathsf{P}} \left(\partial_n \Delta^{p_2 - i} v_h\right) \Delta^{i-1} v_h \, ds - \int_{\partial\mathsf{P}} \left(\Delta^{p_2-i} v_h\right) \partial_n \Delta^{i-1} v_h \, ds \right). \tag{14}$$

Since $\Delta^{p_2} v_h$ is a polynomial of degree $r - 2p_1$ according to the definition of the virtual element space, the volume integral in the right-hand sides of (13) and (14) is an elemental moment of v_h. This integral must be zero since we assumed that the degrees of freedom **(D3)** of v_h are zero.

To prove that the edge integrals in (13) and (14) are zero, we first note that such integrals contain the edge trace of $\Delta^{\mu} v_h$ for $\mu = 0, \ldots, p_2 - 1$ and its normal and tangential derivatives. Since $\Delta v_{h|e} = \partial_t^2 v_h + \partial_n^2 v_h$, it holds that

$$\Delta^{\mu} v_h = \left(\partial_t^2 + \partial_n^2\right)^{\mu} v_h = \sum_{\nu=0}^{\mu} C_{\mu,\nu} \, \partial_t^{2(\mu-\nu)} \, \partial_n^{2\nu} v_h, \tag{15}$$

where $C_{\mu,\nu}$ denote the ν-th combinatorial coefficient of the μ-th power expansion. Therefore, all the edge integrals either contain the normal derivatives $\partial_n^{\ell} v_h$ for some integer $\ell = 0, \ldots, p_2 - 1$, or the tangential derivatives of these quantities. As noted at the beginning of this proof, all these quantities are zero since we assumed that the degrees of freedom **(D1)**–**(D2)** of v_h are zero.

Finally, we note that a function $v_h \in V_{h,r}^{p_2,p_1}(\mathsf{P})$ with all zero degrees of freedom also belongs to $H_0^{p_2}(\mathsf{P}) = \{v \in H^{p_2}(\mathsf{P}) : \partial^j v_{|\partial\mathsf{P}} = 0 \, \forall j = 0, \ldots, p_2 - 1\}$. Since both left-hand sides of (13) and (14) are a norm on $H_0^{p_2}(\mathsf{P})$, it follows that $v_h = 0$.

□

4.3 Global Virtual Element Spaces

Building upon the local virtual element spaces, the *global* conforming virtual element space $V_{h,r}^{p_2,p_1}$ is defined on Ω as

$$V_{h,r}^{p_2,p_1} = \Big\{ v_h \in H_0^{p_1}(\Omega) \cap H^{p_2}(\Omega) \,:\, v_{h|\mathsf{P}} \in V_{h,r}^{p_2,p_1}(\mathsf{P}) \ \forall \mathsf{P} \in \Omega_h \Big\}, \tag{16}$$

where $V_{h,r}^{p_2,p_1}(\mathsf{P})$ is the local space defined in (9) if $p_2 \leq r \leq 2p_2 - 2$ and the local space defined in (10) if $r \geq 2p_2 - 1$.

Remark 4.8 Let $N^{\mathcal{P}}$, $N^{\mathcal{E}}$ and $N^{\mathcal{V}}$ denote the number of element, edges and vertices of Ω_h. The dimension of the global virtual element space built upon (9) is given by

$$\dim V_{h,r}^{p_2,p_1} = N^{\mathcal{P}} \frac{(r-2p_1+1)(r-2p_1+2)}{2} + N^{\mathcal{E}} \sum_{j=0}^{p_2-1} \big(\alpha_j(p_2,r)+1\big) - N^{\mathcal{V}} \frac{(p_2+1)p_2}{2}.$$

The dimension of the global virtual element space built upon (10) is given by

$$\dim V_{h,r}^{p_2,p_1} = N^{\mathcal{P}} \frac{(r-2p_1+1)(r-2p_1+2)}{2} + N^{\mathcal{E}} \frac{(p_2-1)\big(2(r+1)-(p_2-1)\big)}{2} - N^{\mathcal{V}} \frac{(p_2+1)p_2}{2}.$$

The set of global degrees of freedom are inherited from the local degrees of freedom of Sect. 4.2. Therefore, we consider

(D1) $h_{\mathsf{V}}^{|\nu|} D^{\nu} v_h(\mathsf{V})$, $|\nu| \leq p_2 - 1$ for every vertex V of $\mathcal{V}_h$;

(D2) $h_e^{-1+j} \displaystyle\int_e q \partial_n^j v_h \, ds$ for any $q \in \mathbb{P}_{\beta_j}(e)$ and every edge e of $\mathcal{E}_h$, $j = k+1, \dots, p_2 - 1$;

(D3) $h_{\mathsf{P}}^{-2} \displaystyle\int_{\mathsf{P}} q v_h \, d\mathbf{x}$ for any $q \in \mathbb{P}_{r-2p_1}(\mathsf{P})$ and every element P of Ω_h,

where, again, $\beta_j = \alpha_j - \min\{2(p_2-j)-1, r-j\} - 1$ and $\alpha_j(p_2,r) = \max\{2(p_2-j)-1, r-j\}$, $j = 0, \dots, p_2-1$. For $r \geq 2p_2 - 1$, these degrees of freedom become

(D1) $h_{\mathsf{V}}^{|\nu|} D^{\nu} v_h(\mathsf{V})$, $|\nu| \leq p_2 - 1$ for every interior vertex V of $\mathcal{V}_h$;

(D2) $h_e^{-1+j} \displaystyle\int_e q \partial_n^j v_h \, ds$ for any $q \in \mathbb{P}_{r-2p_2+j}(e)$ $j = 0, \dots, p_2 - 1$ and every interior edge e of $\mathcal{E}_h$;

(D3) $h_{\mathsf{P}}^{-2}\int_{\mathsf{P}} q v_h \, d\mathbf{x}$ for any $q \in \mathbb{P}_{r-2p_1}(\mathsf{P})$ and every element P of Ω_h.

We remark that the associated global space is made of $H^{p_2}(\Omega)$ functions. Indeed, the restriction of a virtual element function v_h to each element P belongs to $H^{p_2}(\mathsf{P})$ and glues with C^{p_2-1}-regularity across the internal mesh faces.

Finally, the unisolvence of these degrees of freedom is an immediate consequence of the unisolvence of the elementwise degrees of freedom in any elemental space $V_{h,r}^{p_2,p_1}(\mathsf{P})$, cf. Lemma 4.7.

4.4 Elliptic Projection Operator

The elliptic projection operator $\Pi_r^{p_1,\mathsf{P}} : V_{h,r}^{p_2,p_1}(\mathsf{P}) \to \mathbb{P}_r(\mathsf{P})$ is such that for all $v_h \in V_{h,r}^{p_2,p_1}(\mathsf{P})$, the projection $\Pi_r^{p_1,\mathsf{P}} v_h$ is the solution of the finite dimensional variational problem

$$a_{p_1}^{\mathsf{P}}(\Pi_r^{p_1,\mathsf{P}} v_h - v_h, q) = 0 \quad \forall q \in \mathbb{P}_r(\mathsf{P}), \tag{17}$$

$$\int_{\partial\mathsf{P}} \left(\Pi_r^{p_1,\mathsf{P}} v_h - v_h\right) q \, ds = 0 \quad \forall q \in \mathbb{P}_{p_1-1}(\mathsf{P}). \tag{18}$$

Condition (18) allows us to fix the nontrivial kernel of $a_{p_1}^{\mathsf{P}}(\cdot,\cdot)$, which is the subspace of polynomials of degree (up to) $p_1 - 1$.

Remark 4.9 Instead of (18), we can consider the alternative condition [7]

$$\widehat{\Pi}^{\mathsf{P}} D^{\nu} \Pi_r^{p_1,\mathsf{P}} v_h = \widehat{\Pi}^{\mathsf{P}} D^{\nu} v_h \quad \text{with } |\nu| \le p_1 - 1,$$

by using the *vertex average projection* $\widehat{\Pi}^{\mathsf{P}} : C(\mathsf{P}) \to \mathbb{P}_0(\mathsf{P})$, which is such that

$$\widehat{\Pi}^{\mathsf{P}} \psi = \frac{1}{N^{\mathcal{P}}} \sum_{\mathsf{v} \in \partial\mathsf{P}} \psi(\mathsf{v}), \tag{19}$$

for all continuous function ψ.

Lemma 4.10 *The elliptic projection operator $\Pi_r^{p_1,\mathsf{P}}$ is polynomial preserving in the sense that $\Pi_r^{p_1,\mathsf{P}} q = q$ for every $q \in \mathbb{P}_r(\mathsf{P})$.*

Proof Let $\mathbb{P}_r(\mathsf{P}) \setminus \mathbb{P}_{p_1-1}(\mathsf{P})$ denote the linear space of polynomials of degrees s such that $p_1 \le s \le r$, and consider the decomposition

$$\mathbb{P}_r(\mathsf{P}) = \mathbb{P}_r(\mathsf{P}) \setminus \mathbb{P}_{p_1-1}(\mathsf{P}) \oplus \mathbb{P}_{p_1-1}(\mathsf{P}). \tag{20}$$

We expand the polynomial $q \in \mathbb{P}_r(\mathsf{P})$ and its projection $\Pi_r^{p_1,\mathsf{P}} q$ as follows

$$q = \sum_{\ell'} c_{\ell'}(q)\mu_{\ell'} + \sum_{\ell'} \widetilde{c}_{\ell'}(q)\widetilde{\mu}_{\ell'}, \tag{21}$$

$$\Pi_r^{p_1,\mathsf{P}} q = \sum_{\ell'} c_{\ell'}(\Pi_r^{p_1,\mathsf{P}} q)\mu_{\ell'} + \sum_{\ell'} \widetilde{c}(\Pi_r^{p_1,\mathsf{P}} q)\widetilde{\mu}_{\ell'}, \tag{22}$$

where $\{\mu_{\ell'}\}$ is a basis of $\mathbb{P}_r(\mathsf{P}) \setminus \mathbb{P}_{p_1-1}(\mathsf{P})$, $\{\widetilde{\mu}_{\ell'}\}$ is a basis of $\mathbb{P}_{p_1-1}(\mathsf{P})$, and $c_{\ell'}(q)$, $\widetilde{c}_{\ell'}(q)$, $c_{\ell'}(\Pi_r^{p_1,\mathsf{P}} q)$, and $\widetilde{c}(\Pi_r^{p_1,\mathsf{P}} q)$ are the coefficients of such expansions. The range of the summation index ℓ', which is not explicitly indicated in (21) and (22), is consistent with the dimensions of $\mathbb{P}_r(\mathsf{P}) \setminus \mathbb{P}_{p_1-1}(\mathsf{P})$ and $\mathbb{P}_{p_1-1}(\mathsf{P})$. We assume that the polynomials $\mu_{\ell'}$ are orthogonal with respect to the semi-inner product $a_{p_1}^{\mathsf{P}}(\cdot,\cdot)$, which is the restriction of $a_{p_1}^{\mathsf{P}}(\cdot,\cdot)$ to a polygonal element P, so that $a_{p_1}^{\mathsf{P}}(\mu_{\ell'},\mu_{\ell}) = |\mathsf{P}|\,\delta_{\ell',\ell}$. Since the polynomials $\widetilde{\mu}_{\ell'}$ belong to the kernel of $a_{p_1}^{\mathsf{P}}(\cdot,\cdot)$, we substitute the expansions (21) and (22) in (17) (with $v_h = q$ and $q = \mu_{\ell}$) and we find that

$$\begin{aligned} 0 = a_{p_1}^{\mathsf{P}}(\Pi_r^{p_1,\mathsf{P}} q - q, \mu_{\ell}) &= \sum_{\ell'} \big(c_{\ell'}(\Pi_r^{p_1,\mathsf{P}} q) - c_{\ell'}(q)\big) a_{p_1}^{\mathsf{P}}(\mu_{\ell},\mu_{\ell'}) \\ &= |\mathsf{P}| \sum_{\ell'} \big(c_{\ell'}(\Pi_r^{p_1,\mathsf{P}} q) - c_{\ell'}(q)\big)\delta_{\ell,\ell'} = |\mathsf{P}|\,\big(c_{\ell}(\Pi_r^{p_1,\mathsf{P}} q) - c_{\ell}(q)\big), \end{aligned}$$

which holds for all possible integers ℓ. This relation implies that

$$\Pi_r^{p_1,\mathsf{P}} q - q = \sum_{\ell'} \big(\widetilde{c}_{\ell'}(\Pi_r^{p_1,\mathsf{P}} q) - \widetilde{c}_{\ell'}(q)\big)\widetilde{\mu}_{\ell'} \in \mathbb{P}_{p_1-1}(\mathsf{P}). \tag{23}$$

Then, we assume that the polynomials $\widetilde{\mu}_{\ell'}$ are orthogonal with respect to the inner product $(v,u)_{\partial\mathsf{P}} = \int_{\partial\mathsf{P}} vu\,ds$, so that $(\widetilde{\mu}_{\ell'},\widetilde{\mu}_{\ell})_{\partial\mathsf{P}} = |\partial\mathsf{P}|\,\delta_{\ell',\ell}$, where $|\partial\mathsf{P}|$ is the perimeter of $\partial\mathsf{P}$. We substitute (23) in (18) (with $v_h = q$ and $q = \widetilde{\mu}_{\ell}$) and we find that

$$\begin{aligned} 0 = \int_{\partial\mathsf{P}} (\Pi_r^{p_1,\mathsf{P}} q - q)\widetilde{\mu}_{\ell}\,ds &= \sum_{\ell'} \big(\widetilde{c}_{\ell'}(\Pi_r^{p_1,\mathsf{P}} q) - \widetilde{c}_{\ell'}(q)\big) \int_{\partial\mathsf{P}} \widetilde{\mu}_{\ell'}\widetilde{\mu}_{\ell}\,ds \\ &= |\partial\mathsf{P}| \sum_{\ell'} \big(\widetilde{c}_{\ell'}(\Pi_r^{p_1,\mathsf{P}} q) - \widetilde{c}_{\ell'}(q)\big)\delta_{\ell',\ell} = |\partial\mathsf{P}|\,\big(\widetilde{c}_{\ell}(\Pi_r^{p_1,\mathsf{P}} q) - \widetilde{c}_{\ell}(q)\big) \end{aligned}$$

which holds for all possible integers ℓ. This implies that $\Pi_r^{p_1,\mathsf{P}} q - q = 0$, which is the assertion of the lemma. □

Lemma 4.11 *The polynomial projection* $\Pi_r^{p_1,\mathsf{P}} v_h$ *is* computable *using only the degrees of freedom* **(D1)**–**(D3)** *of* $v_h \in V_{h,r}^{p_2,p_1}(\mathsf{P})$.

Proof To prove the assertion of the lemma, we only need to prove that $a^{\mathsf{P}}_{p_1}(v_h, q)$ and $(v, u)_{\partial\mathsf{P}} = \int_{\partial\mathsf{P}} v_h q\, ds$ are computable for all $v_h \in V^{p_2,p_1}_{h,r}(\mathsf{P})$ and scalar polynomial $q \in \mathbb{P}_r(\mathsf{P})$. To this end, we integrate by parts $a^{\mathsf{P}}_{p_1}(v_h, q)$. For an odd p_1, i.e., $p_1 = 2\ell + 1$, we find that

$$a^{\mathsf{P}}_{p_1}(v_h, q) = -\int_{\mathsf{P}} (\Delta^{p_1} v_h)\, q\; d\mathbf{x} + \int_{\partial\mathsf{P}} (\partial_n \Delta^{\ell} v_h)\, \Delta^{\ell} q\; ds$$
$$+ \sum_{i=1}^{\ell} \left(\int_{\partial\mathsf{P}} (\partial_n \Delta^{p_1 - i} v_h)\, \Delta^{i-1} q\; ds - \int_{\partial\mathsf{P}} (\Delta^{p_1 - i} v_h)\, \partial_n \Delta^{i-1} q\; ds \right). \tag{24}$$

For an even p_1, i.e., $p_1 = 2\ell$, we find that

$$a^{\mathsf{P}}_{p_1}(v_h, q) = \int_{\mathsf{P}} (\Delta^{p_1} v_h)\, q\; d\mathbf{x}$$
$$- \sum_{i=1}^{\ell} \left(\int_{\partial\mathsf{P}} (\partial_n \Delta^{p_1 - i} v_h)\, \Delta^{i-1} q\; ds - \int_{\partial\mathsf{P}} (\Delta^{p_1 - i} v_h)\, \partial_n \Delta^{i-1} q\; ds \right). \tag{25}$$

The first integral of the right-hand side of both formulas (24) and (25) is computable from the degrees of freedom (**D3**). In turn, all the edge integrals are computable since we can expand the trace of $\Delta^{\mu} v_h$ in terms of $\partial_t^{2(\mu-\nu)} \partial_n^{2\mu} v_h$ and use the same argument of the proof of Lemma 4.7. Since the edge traces of v_h and its normal derivatives (and all their tangential derivatives) are computable from the degrees of freedom of (**D1**)–(**D2**) through a polynomial interpolation, we deduce that all the edge integrals for both odd and even p_1 and the boundary integral $(v, u)_{\partial\mathsf{P}}$ are computable. □

4.5 Enhancement

As noted in Remark 4.3, the L^2-projection operator $\Pi^{0,\mathsf{P}}_{r-2p_1} : V^{p_2,p_1}_{h,r}(\mathsf{P}) \to \mathbb{P}_r(\mathsf{P})$ is computable from the degrees of freedom (**D3**). Instead, to compute the orthogonal projection onto the polynomial subspace $\mathbb{P}_{r-p_1}(\mathsf{P})$, we need to modify the space definition as follows thus obtaining the so called "*enhanced*" virtual element space. Our construction follows the guidelines in [1]. First, we consider the mesh element

P and the "*extended*" virtual element space for $r \geq 2p_2 - 1$ (recall that $p_2 \geq p_1$) defined as

$$\widetilde{V}_{h,r}^{p_2,p_1}(\mathsf{P}) := \Big\{ v_h \in H^{p_2}(\mathsf{P}) : \Delta^{p_2} v_h \in \mathbb{P}_{r-p_1}(\mathsf{P}),\ \partial_n^j v_h \in \mathbb{P}_{r-j}(e),$$
$$j = 0, \ldots, p_2 - 1\ \forall e \in \partial\mathsf{P} \Big\}. \quad (26)$$

Then, we define the enhanced virtual element space as

$$W_{h,r}^{p_2,p_1}(\mathsf{P}) := \Big\{ v_h \in \widetilde{V}_{h,r}^{p_2,p_1}(\mathsf{P}) : \int_{\mathsf{P}} v_h q\, d\mathbf{x} = \int_{\mathsf{P}} \Pi_{r-p_1}^{p_1,\mathsf{P}} v_h q\, d\mathbf{x}$$
$$\forall q \in \mathbb{P}_{r-p_1} \setminus \mathbb{P}_{r-2p_1}(\mathsf{P}) \Big\}. \quad (27)$$

The polynomial space $\mathbb{P}_r(\mathsf{P})$ is a subspace of $W_{h,r}^{p_2,p_1}(\mathsf{P})$ and, thus, of $\widetilde{V}_{h,r}^{p_2,p_1}(\mathsf{P})$, and the elliptic projection $\Pi_{r-p_1}^{p_1,\mathsf{P}} : \widetilde{V}_{h,r}^{p_2,p_1}(\mathsf{P}) \to \mathbb{P}_{r-p_1}(\mathsf{P})$ that is defined in (17)–(18) is still computable and only depends on the degrees of freedom **(D1)**, **(D2)** and **(D3)**. This assertion can easily be proved by repeating the argument of Lemma 4.10.

The virtual element functions of the space $\widetilde{V}_{h,r}^{p_2,p_1}(\mathsf{P})$ are uniquely characterized by the set of degrees of freedom **(D1)**, **(D2)**, **(D3)** and the set of additional degrees of freedom $(\widetilde{\mathbf{D3}})$

$(\widetilde{\mathbf{D3}})$ $\quad h_{\mathsf{P}}^{-2} \int_{\mathsf{P}} q v_h\, d\mathbf{x}$ for any $q \in \mathbb{P}_{r-p_1}(\mathsf{P}) \setminus \mathbb{P}_{r-2p_1}(\mathsf{P})$.

We state the unisolvence of these degrees of freedom in the following lemma. The proof is equal to the proof of Lemma 4.7 (consider the degrees of freedom (**(D3)**, $(\widetilde{\mathbf{D3}})$) instead of **(D3)**) and is omitted.

Lemma 4.12 *The degrees of freedom* **(D1)**, **(D2)**, **(D3)**, $(\widetilde{\mathbf{D3}})$ *are unisolvent in the virtual element space* $\widetilde{V}_{h,r}^{p_2,p_1}(\mathsf{P})$.

Remark 4.13 According to Lemma 4.12, the dimension of $\widetilde{V}_{h,r}^{p_2,p_1}(\mathsf{P})$ must be equal to the cardinality of the set of the degrees of freedom **(D1)**, **(D2)**, **(D3)**, $(\widetilde{\mathbf{D3}})$. This statement can also be proved by a counting argument.

Next, we want to prove that the degrees of freedom **(D1)**, **(D2)** and **(D3)** are unisolvent in the enhanced space $W_{h,r}^{p_2,p_1}(\mathsf{P})$. To this end, we first need to establish a technical result. Consider the set of linear, bounded functionals $\lambda_{\ell_1}^{(\mathbf{D1})}, \lambda_{\ell_2}^{(\mathbf{D2})}, \lambda_{\ell_3}^{(\mathbf{D3})} : \widetilde{V}_{h,r}^{p_2,p_1}(\mathsf{P}) \to \mathbb{R}$, which respectively return the degrees of freedom **(D1)**, **(D2)** and **(D3)** when applied to a virtual element function $v_h \in \widetilde{V}_{h,r}^{p_2,p_1}(\mathsf{P})$. The indices ℓ_1, ℓ_2, and ℓ_3 run from 1 to #**(D1)**, #**(D2)** and #**(D3)**, respectively, where $\#(\mathcal{D})$ denotes the cardinality of the discrete set $\mathcal{D}$. Renumbering ℓ_2 and ℓ_3 may require the introduction of suitable sets of basis functions for the polynomial spaces $\mathbb{P}_{r-j}(e)$ and $\mathbb{P}_{r-2p_1}(\mathsf{P})$ in **(D2)** and **(D3)**, respectively. We left this aspect undefined as this technicality is not crucial in this presentation, although important in the practical implementations of the method. We introduce the additional set of linear functionals $\lambda_{\ell}^{(\widetilde{\mathbf{D3}})}$ that are such that

$$\lambda_{\widetilde{\ell}_3}^{(\widetilde{\mathbf{D3}})}(v_h) = h_{\mathsf{P}}^{-2} \int_{\mathsf{P}} q_{\widetilde{\ell}_3} \big(\Pi_{r-p_1}^{p_1,\mathsf{P}} - \Pi_{r-p_1}^{0,\mathsf{P}}\big) v_h \, d\mathbf{x} \qquad \widetilde{\ell}_3 = 1,\dots,\#(\widetilde{\mathbf{D3}})$$

where $\{q_{\widetilde{\ell}_3}\}_{\widetilde{\ell}_3}$ is a basis of $\mathbb{P}_{r-p_1}(\mathsf{P}) \setminus \mathbb{P}_{r-2p_1}(\mathsf{P})$, and the index $\widetilde{\ell}_3$ runs from 1 to $\#(\widetilde{\mathbf{D3}})$, the number of degrees of freedom of $(\widetilde{\mathbf{D3}})$.

Then, we collect these different types of functionals in the functional set

$$\Lambda = \big(\lambda_\ell\big) = \Big(\lambda_{\ell_1}^{(\mathbf{D1})}, \lambda_{\ell_2}^{(\mathbf{D2})}, \lambda_{\ell_3}^{(\mathbf{D3})}, \lambda_{\widetilde{\ell}_3}^{(\widetilde{\mathbf{D3}})}\Big). \tag{28}$$

We assume that the integer index ℓ is consistent with a suitable renumbering of such degrees of freedom; so that ℓ runs form 1 to $m' = m + \#(\widetilde{\mathbf{D3}}) = \dim \widetilde{V}_{h,r}^{p_2,p_1}(\mathsf{P})$ and $m = \#(\mathbf{D1}) + \#(\mathbf{D2}) + \#(\mathbf{D3})$. These functionals satisfy the property stated in the following lemma.

Lemma 4.14 *The linear functionals* Λ *are linearly independent.*

Proof Let $v_h \in \widetilde{V}_{h,r}^{p_2,p_1}(\mathsf{P})$ such that $\lambda_\ell(v_h) = 0$ for all $\ell = 1,\dots,m'$. Now, the degrees of freedom $(\mathbf{D1})$, $(\mathbf{D2})$ and $(\mathbf{D3})$ of v_h are (obviously) zero as they are the values of the functionals $\lambda_{\ell_1}^{(\mathbf{D1})}(v_h)$, $\lambda_{\ell_2}^{(\mathbf{D2})}(v_h)$ and $\lambda_{\ell_3}^{(\mathbf{D3})}(v_h)$. Moreover, it holds that $\Pi_r^{p_1,\mathsf{P}} v_h = 0$, and, hence, $\Pi_{r-p_1}^{p_1,\mathsf{P}} v_h = 0$, as these projections only depend on the degrees of freedom $(\mathbf{D1})$, $(\mathbf{D2})$ and $(\mathbf{D3})$, cf. Lemma 4.10. Then, we observe that the definition of the orthogonal projection $\Pi_{r-p_1}^{0,\mathsf{P}} v_h$ and the facts that $\lambda_{\widetilde{\ell}_3}^{(\widetilde{\mathbf{D3}})}(v_h) = 0$ and $\Pi_{r-p_1}^{p_1,\mathsf{P}} v_h = 0$ imply that

$$\int_{\mathsf{P}} q v_h \, d\mathbf{x} = \int_{\mathsf{P}} q \Pi_{r-p_1}^{0,\mathsf{P}} v_h \, d\mathbf{x} = \int_{\mathsf{P}} q \Pi_{r-p_1}^{p_1,\mathsf{P}} v_h \, d\mathbf{x} = 0 \tag{29}$$

for all $q \in \mathbb{P}_{r-p_1}(\mathsf{P}) \setminus \mathbb{P}_{r-2p_1}(\mathsf{P})$. Therefore, the degrees of freedom $(\widetilde{\mathbf{D3}})$ of v_h are equal to zero and finally $v_h = 0$ because the degrees of freedom $(\mathbf{D1})$, $(\mathbf{D2})$, $(\mathbf{D3})$ and $(\widetilde{\mathbf{D3}})$ are unisolvent in $\widetilde{V}_{h,r}^{p_2,p_1}(\mathsf{P})$. This argument proves that the intersection of the kernels of all the linear functionals λ_ℓ contains only the virtual element function that is identically zero over P, so that these linear functionals are necessarily linearly independent. □

Using the linear functionals Λ, we reformulate the definition of space $W_{h,r}^{p_2,p_1}(\mathsf{P})$ in the following equivalent way:

$$W_{h,r}^{p_2,p_1}(\mathsf{P}) := \Big\{ v_h \in \widetilde{V}_{h,r}^{p_2,p_1}(\mathsf{P}) : \lambda_{\widetilde{\ell}_3}^{(\widetilde{\mathbf{D3}})}(v_h) = 0 \quad \forall \widetilde{\ell}_3 = m+1,\dots,m' \Big\}. \tag{30}$$

In other words, $W_{h,r}^{p_2,p_1}(\mathsf{P})$ belongs to the intersection of the kernels of all the additional linear functionals λ_ℓ with $\ell = m+1,\dots,m'$. The space $W_{h,r}^{p_2,p_1}(\mathsf{P})$ has the two important properties that are stated in the following lemma.

Lemma 4.15 *The virtual element space $W_{h,r}^{p_2,p_1}(\mathsf{P})$ has the same dimension of the "regular" space $V_{h,r}^{p_2,p_1}(\mathsf{P})$ and the set of degrees of freedom* **(D1)**, **(D2)** *and* **(D3)** *are unisolvent in $W_{h,r}^{p_2,p_1}(\mathsf{P})$.*

Proof In view of Lemma 4.14 the linear functionals in Λ are linearly independent and the cardinality of Λ, i.e., $m' = \#(\Lambda)$, is equal to the dimension of $\widetilde{V}_{h,r}^{p_2,p_1}(\mathsf{P})$. Therefore, $\big(\mathsf{P}, \mathbb{P}_r(\mathsf{P}), \Lambda\big)$ is a *finite element* in the sense of Ciarlet, cf. [41]. So, there exists a set of m' dual basis functions ψ_ℓ such that

$$\lambda_\ell(\psi_{\ell'}) = \delta_{\ell,\ell'} \qquad \ell,\ell' = 1,\dots,m'.$$

Now, it holds that $\lambda_{\widetilde{\ell}_3}^{(\widetilde{\mathbf{D3}})}(\psi_{\ell'}) = \lambda_\ell(\psi_{\ell'}) = 0$ for $\ell' = 1,\dots,m$, $\ell = m+1,\dots,m'$ (and the corresponding values of the index $\widetilde{\ell}_3$). This fact has the following consequences. The first m linearly independent functions $\psi_{\ell'}$, $\ell' = 1,\dots,m$, belong to $W_{h,r}^{p_2,p_1}(\mathsf{P})$, cf. formulation (30), since $\lambda_{\widetilde{\ell}_3}^{(\widetilde{\mathbf{D3}})}(\psi_{\ell'}) = \lambda_\ell(\psi_{\ell'}) = 0$ for $\ell = m+1,\dots,m'$ (and corresponding indices $\widetilde{\ell}_3$). This implies that $\dim W_{h,r}^{p_2,p_1}(\mathsf{P}) \geq m$. Furthermore, according to the space definition (30) *all* virtual element functions $w_h \in W_{h,r}^{p_2,p_1}(\mathsf{P})$ are such that $\lambda_{\widetilde{\ell}_3}^{(\widetilde{\mathbf{D3}})}(w_h) = \lambda_\ell(w_h) = 0$ for $\ell = m+1,\dots,m'$. Therefore, such functions can be written as a linear combination of *only* the first m basis functions $\psi_{\ell'}$, $\ell' = 1,\dots,m$, are thus identified by the values of the linear functionals $\lambda_{\ell_1}^{(\mathbf{D1})} v_h$, $\lambda_{\ell_2}^{(\mathbf{D2})} v_h$ and $\lambda_{\ell_3}^{(\mathbf{D3})} v_h$. Consequently, all virtual element functions of $W_{h,r}^{p_2,p_1}(\mathsf{P})$ are uniquely identified by the degrees of freedom **(D1)**, **(D2)**, **(D3)** and, consequently, $\dim W_{h,r}^{p_2,p_1}(\mathsf{P}) = m$. □

In view of Lemma 4.15, the orthogonal projection operator $\Pi_{r-p_1}^{0,\mathsf{P}} : W_{h,r}^{p_2,p_1}(\mathsf{P}) \to \mathbb{P}_{r-p_1}(\mathsf{P})$ is computable from the degrees of freedom **(D1)**, **(D2)** and **(D3)**.

Finally, we collect the local virtual element spaces into a *global* conforming virtual element space $W_{h,r}^{p_2,p_1}$ defined on Ω as

$$W_{h,r}^{p_2,p_1} = \Big\{ w_h \in H_0^{p_1}(\Omega) \cap H^{p_2}(\Omega) \,:\, w_{h|\mathsf{P}} \in W_{h,r}^{p_2,p_1}(\mathsf{P}) \;\forall \mathsf{P} \in \Omega_h \Big\}, \tag{31}$$

where $W_{h,r}^{p_2,p_1}(\mathsf{P})$ is the local space defined above.

4.6 The Virtual Element Bilinear Form $a_h(\cdot,\cdot)$

We discuss the definition of the bilinear form $a_h(\cdot,\cdot)$ that approximates the bilinear form $a(\cdot,\cdot)$ in the virtual element discretization (5). This construction is the same for the "regular" virtual element spaces (9) and (10) and the "enhanced" space (27). In this section we use the symbol $V_{h,r}^{p_2,p_1}(\mathsf{P})$ to denote both choices of the spaces. However, such construction holds also for the enhanced virtual element space (31). The symmetric bilinear form $a_h : V_{h,r}^{p_2,p_1} \times V_{h,r}^{p_2,p_1} \to \mathbb{R}$, is written as the sum of local terms

$$a_h(u_h, v_h) = \sum_{\mathsf{P}\in\Omega_h} a_h^{\mathsf{P}}(u_h, v_h), \tag{32}$$

where each local term $a_h^{\mathsf{P}} : V_{h,r}^{p_2,p_1}(\mathsf{P}) \times V_{h,r}^{p_2,p_1}(\mathsf{P}) \to \mathbb{R}$ is a symmetric bilinear form. We set

$$a_h^{\mathsf{P}}(u_h, v_h) = a_{p_1}^{\mathsf{P}}(\Pi_r^{p_1,\mathsf{P}} u_h, \Pi_r^{p_1,\mathsf{P}} v_h) + S^{\mathsf{P}}(u_h - \Pi_r^{p_1,\mathsf{P}} u_h, v_h - \Pi_r^{p_1,\mathsf{P}} v_h), \tag{33}$$

where $S^{\mathsf{P}} : V_{h,r}^{p_2,p_1}(\mathsf{P}) \times V_{h,r}^{p_2,p_1}(\mathsf{P}) \to \mathbb{R}$ provides the stabilization term. The stabilization form $S^{\mathsf{P}}(\cdot,\cdot)$ is a symmetric, positive definite bilinear form for which there exist two positive constants σ_* and σ^* such that

$$\sigma_* a_{p_1}^{\mathsf{P}}(v_h, v_h) \leq S^{\mathsf{P}}(v_h, v_h) \leq \sigma^* a_{p_1}^{\mathsf{P}}(v_h, v_h) \quad \forall v_h \in V_{h,r}^{p_2,p_1}(\mathsf{P}) \text{ with } \Pi_r^{p_1,\mathsf{P}} v_h = 0. \tag{34}$$

The constants σ_*, σ^* are independent of h (and P). A possible proof of the validity of (34) for the so called "dofi-dofi" stabilization in the context of arbitrarily regular conforming VEM can be found in [49] (for the case $p_1 = 2$ see also [23, 38]). This construction has the *r-consistency* and *stability* properties stated in (6) and (7),

4.7 The Virtual Element Approximation of the Load Term

To approximate the right-hand side term of (5) we first assume the elemental decomposition

$$\langle f_h, v_h \rangle = \sum_{\mathsf{P}\in\Omega_h} \int_{\mathsf{P}} f_h v_h \, d\mathbf{x}. \tag{35}$$

In Eq. (35), the elemental term $f_{h|\mathsf{P}}$ is defined as

$$f_{h|\mathsf{P}} = \begin{cases} \Pi^{0,\mathsf{P}}_{r-2p_1} f, & (a) \text{ if } p_2 + 2p_1 - 1 \leq r, \\ \Pi^{0,\mathsf{P}}_{r-p_1} f, & (b) \text{ if } p_2 \leq r \leq p_2 + 2p_1 - 2. \end{cases} \tag{36}$$

We discuss the two definitions of f_h given above separately.

Remark 4.16 The right-hand side of (35) is fully computable by using only the degrees of freedom (**D3**) if $r \geq 2p_1$ and we choose f_h as the piecewise polynomial approximation of f on Ω_h in accordance with (a). In such a case, we do not need to resort to the enhanced virtual element space defined in (27).

Now, consider decomposition (35) and definition (a). Since $p_2 \geq p_1$, it holds that $r - 2p_1 \geq p_2 - 1$ (equivalently, $r \geq 3p_1 - 1$). Thus, using the definition of the L^2-orthogonal projection, from (35), we find that

$$\langle f_h, v_h \rangle = \sum_{\mathsf{P}\in\Omega_h} \int_{\mathsf{P}} \Pi^{0,\mathsf{P}}_{r-2p_1} f \, v_h \, d\mathbf{x} = \sum_{\mathsf{P}\in\Omega_h} \int_{\mathsf{P}} \Pi^{0,\mathsf{P}}_{r-2p_1} f \Pi^{0,\mathsf{P}}_{p_1-1} v_h \, d\mathbf{x}. \tag{37}$$

Applying standard approximation results to (37) and recalling that $v_h \in V^{p_2,p_1}_{h,r} \subset H^{p_1}_0(\Omega) \cap H^{p_2}(\Omega)$ yield the following estimate

$$\langle f - f_h, v_h \rangle \leq Ch^{r-p_1+1} |v_h|_{p_1} |f|_{r-2p_1+1},$$

for some positive constant C that is independent of h. In particular, for $p_1 = p_2 \geq 2$ it is enough to choose $r \geq 2p_2 + 1$ (the case $p_1 = p_2 = 2$ and $r \geq 5$ has been originally treated in [34]). Note that for fixed values of p_1, larger values of the regularity parameter p_2 ensure higher convergence rate for the approximation of the right-hand side. This is a specific attractive feature of arbitrarily regular conforming VEM (which can not be exploited, e.g., in the nonconforming setting).

Now, consider decomposition (35) and definition (b). Similarly to the previous case, using the definition of the L^2-orthogonal projection yields

$$\langle f_h, v_h \rangle = \sum_{\mathsf{P}\in\Omega_h} \int_{\mathsf{P}} \Pi^{0,\mathsf{P}}_{r-p_1} f \, v_h \, d\mathbf{x} = \sum_{\mathsf{P}\in\Omega_h} \int_{\mathsf{P}} \Pi^{0,\mathsf{P}}_{r-p_1} f \, \Pi^{0,\mathsf{P}}_{0} v_h \, d\mathbf{x}. \tag{38}$$

Applying again standard approximation results to (38) we we find that

$$\langle f - f_h, v_h \rangle \leq Ch^{r-p_1+2} |v_h|_{p_1} |f|_{r-p_1+1}.$$

For arbitrary values of p_1 and p_2, the use of the enhancement approach might be avoided using arguments similar to those employed in [48].

4.8 Error Analysis

In this section, we briefly recall a convergence result in the energy norm [7] (see also [9, 49]) for the approximation of (1a)–(1b). In particular, employing Theorem 3.1 together with standard results of approximation (see, e.g., Reference [22, 26, 49]) and the approximation properties of the right-hand side contained in Sect. 4.7.

Theorem 4.17 *Let $u \in H_0^{p_1}(\Omega) \cap H^{r+1}(\Omega)$ be the solution of the polyharmonic problem* (1a)–(1b) *and let $u_h \in V_{h,r}^{p_2,p_1}$ be the solution of the discrete problem* (5). *Assume that f is sufficiently regular. Then, there exists a positive constant C independent of h such that*

$$\|u - u_h\|_V \leq C h^{r-(p_1-1)}. \tag{39}$$

Remark 4.18 Convergence estimates in lower order norms can be established provided that classical duality arguments can be used and that the polynomial approximation order r is sufficient large[7, 9, 40].

5 Conclusion

We reviewed the construction of highly regular virtual element spaces for the conforming approximations in two spatial dimensions of elliptic problems of order $2p_1$, $p_1 \geq 1$. The resulting finite dimensional virtual spaces are subspaces of $H^{p_2}(\Omega)$, $p_2 \geq p_1$. We presented an abstract convergence result in a suitably defined energy norm. Moreover, after discussing the construction of the approximation spaces and major aspects such as the choice and unisolvence of the degrees of freedom, we provided specific examples of highly regular virtual spaces, corresponding to various practical cases. Finally, a detailed discussion of the properties of the "enhanced" formulation of the virtual element spaces is provided.

Acknowledgments PFA and MV acknowledge the financial support of PRIN research grant number 201744KLJL "*Virtual Element Methods: Analysis and Applications*" and PRIN research grant number 20204LN5N5 "*Advanced polyhedral discretisations of heterogeneous PDEs for multiphysics problems*" funded by MIUR. GM acknowledges the financial support of the LDRD program of Los Alamos National Laboratory under project number 20220129ER. Los Alamos National Laboratory is operated by Triad National Security, LLC, for the National Nuclear Security Administration of U.S. Department of Energy (Contract No. 89233218CNA000001). The Authors are affiliated to GNCS-INdAM (Italy).

References

1. Ahmad, B., Alsaedi, A., Brezzi, F., Marini, L.D., Russo, A.: Equivalent projectors for virtual element methods. Comput. Math. Appl. **66**(3), 376–391 (2013)
2. Aldakheel, F., Hudobivnik, B., Hussein, A., Wriggers, P.: Phase-field modeling of brittle fracture using an efficient virtual element scheme. Comput. Methods Appl. Mech. Engrg. **341**, 443–466 (2018)
3. Antonietti, P.F., Giani, S., Houston, P.: hp-version composite discontinuous Galerkin methods for elliptic problems on complicated domains. SIAM J. Sci. Comput. **35**(3), A1417–A1439 (2013)
4. Antonietti, P.F., Beirão da Veiga, L., Scacchi, S., Verani, M.: A C^1 virtual element method for the Cahn-Hilliard equation with polygonal meshes. SIAM J. Numer. Anal. **54**(1), 34–56 (2016)
5. Antonietti, P.F., Bruggi, M., Scacchi, S., Verani, M.: On the virtual element method for topology optimization on polygonal meshes: a numerical study. Comput. Math. Appl. **74**(5), 1091–1109 (2017)
6. Antonietti, P.F., Mascotto, L., Verani, M.: A multigrid algorithm for the p-version of the virtual element method. ESAIM Math. Model. Numer. Anal. **52**(1), 337–364 (2018)
7. Antonietti, P.F., Manzini, G., Verani, M.: The conforming virtual element method for polyharmonic problems. Comput. Math. Appl. **79**(7), 2021–2034 (2020)
8. Antonietti, P.F., Bertoluzza, S., Prada, D., Verani, M.: The virtual element method for a minimal surface problem. Calcolo **57**(4), Paper No. 39, 21 (2020)
9. Antonietti, P.F., Berrone, S., Borio, A., D'Auria, A., Verani, M., Weisser, S.: Anisotropic a posteriori error estimate for the virtual element method. IMA J. Numer. Anal. (2021). https://doi.org/10.1093/imanum/drab001
10. Antonietti, P.F., Manzini, G., Mazzieri, I., Mourad, H., Verani, M.: The arbitrary-order virtual element method for linear elastodynamics models: Convergence, stability and dispersion-dissipation analysis. Int. J. Numer. Methods Eng. **122**, 934–971 (2021)
11. Antonietti, P.F., Manzini, G., Scacchi, S., Verani, M.: A review on arbitrarily regular conforming virtual element methods for second- and higher-order elliptic partial differential equations. Math. Models Methods Appl. Sci. **31**(14), 2825–2853 (2021)
12. Argyris, J.H., Fried, I., Scharpf, D.W.: The TUBA family of plate elements for the matrix displacement method. Aeronaut. J. R. Aeronaut. Soc. **72**, 701–709 (1968)
13. Artioli, E., de Miranda, S., Lovadina, C., Patruno, L.: A stress/displacement virtual element method for plane elasticity problems. Comput. Methods Appl. Mech. Eng. **325**, 155–174 (2017)
14. Artioli, E., de Miranda, S., Lovadina, C., Patruno, L.: A family of virtual element methods for plane elasticity problems based on the Hellinger-Reissner principle. Comput. Methods Appl. Mech. Eng. **340**, 978–999 (2018)
15. Ayuso de Dios, B., Lipnikov, K., Manzini, G.: The non-conforming virtual element method. ESAIM Math. Model. Numer. **50**(3), 879–904 (2016)
16. Beirão da Veiga, L., Brezzi, F., Cangiani, A., Manzini, G., Marini, L.D., Russo, A.: Basic principles of virtual element methods. Math. Models Methods Appl. Sci. **23**(1), 199–214 (2013)
17. Beirão da Veiga, L., Manzini, G.: A virtual element method with arbitrary regularity. IMA J. Numer. Anal. **34**(2), 782–799 (2014)
18. Beirão da Veiga, L., Lipnikov, K., Manzini, G.: The mimetic finite difference method for elliptic problems. Modeling, Simulations and Applications, vol. 11, I edn. Springer, Berlin (2014)
19. Beirão da Veiga, L., Manzini, G.: Residual a posteriori error estimation for the virtual element method for elliptic problems. ESAIM Math. Model. Numer. Anal. **49**(2), 577–599 (2015)
20. Beirão da Veiga, L., Brezzi, F., Marini, L.D., Russo, A.: H(div) and H(curl)-conforming VEM. Numer. Math. **133**(2), 303–332 (2016)

21. Beirão da Veiga, L., Brezzi, F., Marini, L.D., Russo, A.: Mixed virtual element methods for general second order elliptic problems on polygonal meshes. ESAIM: Math. Model. Numer. Anal. **50**(3), 727–747 (2016)
22. Beirão da Veiga, L., Brezzi, F., Marini, L.D., Russo, A.: Virtual element methods for general second order elliptic problems on polygonal meshes. Math. Models Methods Appl. Sci. **26**(4), 729–750 (2016)
23. Beirão da Veiga, L., Lovadina, C., Russo, A.: Stability analysis for the virtual element method. Math. Models Methods Appl. Sci. **27**(13), 2557–2594 (2017)
24. Beirão da Veiga, L., Lovadina, C., Vacca, G.: Divergence free virtual elements for the Stokes problem on polygonal meshes. ESAIM: Math. Model. Numer. Anal. **51**(2), 509–535 (2017)
25. Beirão da Veiga, L., Lovadina, C., Vacca, G.: Virtual elements for the Navier-Stokes problem on polygonal meshes. SIAM J. Numer. Anal. **56**(3), 1210–1242 (2018)
26. Beirão da Veiga, L., Mora, D., Rivera, G.: Virtual elements for a shear-deflection formulation of Reissner-Mindlin plates. Math. Comput. **88**(315), 149–178 (2019)
27. Beirão da Veiga, L., Dassi, F., Russo, A.: A C^1 virtual element method on polyhedral meshes. Comput. Math. Appl. **79**(7), 1936–1955 (2020)
28. Bell, K.: A refined triangular plate bending finite element. Int. J. Numer. Meth. Eng. **1**(1), 101–122 (1969).
29. Benedetto, M.F., Berrone, S., Borio, A.: The virtual element method for underground flow situations in fractured data. In: Advances in Discretization Methods. SEMA SIMAI Springer Series, vol. 12, pp. 167–186. Springer, Cham (2016)
30. Benvenuti, E., Chiozzi, A., Manzini, G., Sukumar, N.: Extended virtual element method for the Laplace problem with singularities and discontinuities. Comput. Methods Appl. Mech. Eng. **356**, 571–597 (2019)
31. Blum, H., Rannacher, R.: On the boundary value problem of the biharmonic operator on domains with angular corners. Math. Methods Appl. Sci. **2**(4), 556–581 (1980)
32. Brenner, S.C., Scott, R.: The Mathematical Theory of Finite Element Methods, vol. 15. Springer Science & Business Media, New York (2008)
33. Brenner, S.C., Sung, L.Y.: Virtual enriching operators. Calcolo **56**(4), 1–25 (2019)
34. Brezzi, F., Marini, L.D.: Virtual element methods for plate bending problems. Comput. Methods Appl. Mech. Eng. **253**, 455–462 (2013)
35. Brezzi, F., Falk, R.S., Marini, L.D.: Basic principles of mixed virtual element methods. ESAIM Math. Model. Numer. Anal. **48**(4), 1227–1240 (2014)
36. Cangiani, A., Dong, Z., Georgoulis, E.H., Houston, P.: *hp*-Version Discontinuous Galerkin Methods on Polygonal and Polyhedral Meshes. Springer Briefs in Mathematics. Springer, Cham (2017)
37. Certik, O., Gardini, F., Manzini, G., Vacca, G.: The virtual element method for eigenvalue problems with potential terms on polytopic meshes. Appl. Math. **63**(3), 333–365 (2018)
38. Chen, L., Huang, J.: Some error analysis on virtual element methods. Calcolo **55**(1), Paper No. 5, 23 (2018)
39. Chi, H., Pereira, A., Menezes, I.F., Paulino, G.H.: Virtual element method (VEM)-based topology optimization: an integrated framework. Struct. Multidiscip. Optim. **62**(3), 1089–1114 (2020)
40. Chinosi, C., Marini, L.D.: Virtual element method for fourth order problems: L^2-estimates. Comput. Math. Appl. **72**(8), 1959–1967 (2016)
41. Ciarlet, P.G.: The finite element method for elliptic problems. Classics Appl. Math. **40**, 1–511 (2002)
42. Clough, R.W., Tocher, J.L. (eds.): Finite element stiffness matrices for analysis of plates in bending. In: Proceedings of the Conference on Matrix Methods in Structural Mechanics (1965)
43. Cockburn, B., Dong, B., Guzmán, J.: A superconvergent LDG-hybridizable Galerkin method for second-order elliptic problems. Math. Comput. **77**(264), 1887–1916 (2008)
44. Di Pietro, D.A., Droniou, J.: The Hybrid High-Order Method for Polytopal Meshes: Design, Analysis, and Applications. Modeling, Simulations and Applications. Springer, Berlin (2020)

45. Gazzola, F., Grunau, H.C., Sweers, G.: Polyharmonic Boundary Value Problems. Lecture Notes in Mathematics, vol. 1991. Springer, Berlin (2010)
46. Hu, J., Zhang, S.: The minimal conforming H^k finite element spaces on $\mathbb{R}^n$ rectangular grids. Math. Comput. **84**(292), 563–579 (2015)
47. Hu, J., Lin, T., Wu, Q.: A construction of C^r conforming finite element spaces in any dimension (2021). arXiv:2103.14924
48. Huang, X.: Nonconforming virtual element method for $2m$th order partial differential equations in $\mathbb{R}^n$ with $m > n$. Calcolo **57**(4), Paper No. 42, 38 (2020)
49. Chen, C., Huang, X., Wei, H.: H^m-conforming virtual elements in arbitrary dimension. SIAM J. Numer. Anal. **60**(6), 3099–3123 (2022). https://doi.org/10.1137/21M1440323
50. Li, M., Zhao, J., Huang, C., Chen, S.: Conforming and nonconforming vems for the fourth-order reaction–subdiffusion equation: a unified framework. IMA J. Numer. Anal. (2021)
51. Lovadina, C., Mora, D., Velásquez, I.: A virtual element method for the von Kármán equations. ESAIM Math. Model. Numer. Anal. **55**(2), 533–560 (2021)
52. Mascotto, L., Perugia, I., Pichler, A.: A nonconforming Trefftz virtual element method for the Helmholtz problem. Math. Models Methods Appl. Sci. **29**(9), 1619–1656 (2019)
53. Mascotto, L., Perugia, I., Pichler, A.: A nonconforming Trefftz virtual element method for the Helmholtz problem: numerical aspects. Comput. Methods Appl. Mech. Eng. **347**, 445–476 (2019)
54. Mora, D., Rivera, G., Velásquez, I.: A virtual element method for the vibration problem of Kirchhoff plates. ESAIM Math. Model. Numer. Anal. **52**(4), 1437–1456 (2018)
55. Mora, D., Silgado, A.: A C^1 virtual element method for the stationary quasi-geostrophic equations of the ocean. Comput. Math. Appl. (2021)
56. Mora, D., Velásquez, I.: A virtual element method for the transmission eigenvalue problem. Math. Models Methods Appl. Sci. **28**(14), 2803–2831 (2018)
57. Mora, D., Velásquez, I.: Virtual element for the buckling problem of Kirchhoff-Love plates. Comput. Methods Appl. Mech. Eng. **360**, 112687, 22 (2020)
58. Park, K., Chi, H., Paulino, G.H.: On nonconvex meshes for elastodynamics using virtual element methods with explicit time integration. Comput. Methods Appl. Mech. Eng. **356**, 669–684 (2019)
59. Park, K., Chi, H., Paulino, G.H.: Numerical recipes for elastodynamic virtual element methods with explicit time integration. Int. J. Numer. Methods Eng. **121**(1), 1–31 (2020)
60. Perugia, I., Pietra, P., Russo, A.: A plane wave virtual element method for the Helmholtz problem. ESAIM Math. Model. Num. **50**(3), 783–808 (2016)
61. Sukumar, N., Tabarraei, A.: Conforming polygonal finite elements. Int. J. Numer. Methods Eng. **61**(12), 2045–2066 (2004)
62. Wriggers, P., Rust, W.T., Reddy, B.D.: A virtual element method for contact. Comput. Mech. **58**(6), 1039–1050 (2016)
63. Zhang, S.: A family of 3D continuously differentiable finite elements on tetrahedral grids. Appl. Numer. Math. **59**(1), 219–233 (2009)
64. Zhang, S.: A family of differentiable finite elements on simplicial grids in four space dimensions. Math. Numer. Sin. **38**(3), 309–324 (2016)

Broadband Recursive Skeletonization

Abinand Gopal and Per-Gunnar Martinsson

1 Introduction

When numerically solving scattering problems, it is often convenient to work with integral equation formulations. For problems with homogeneous media, which are the foci of this paper, this has the advantage of reducing the dimension of the region that must be discretized. Moreover, the free space fundamental solution can be incorporated into the ansatz for the scattered field, allowing for radiation boundary conditions to be easily enforced and the mitigation of dispersion errors upon discretization. Finally, the resulting equations can often be formulated as second kind Fredholm equations on L^2, allowing for high-order discretizations without spurious ill-conditioning when appropriate quadratures are used [56].

A challenge with integral equation formulations is that the coefficient matrices that arise upon discretization are dense, and, as such, efficient inversion in even two dimensions is not straightforward. In some environments, an iterative solver, such as GMRES, paired with a fast-matrix vector multiplication algorithm, such as the fast multipole method or the FFT, is sufficient to obtain a fast solver. However, in near-resonant environments or environments where multiple incident fields are of interest, iterative solvers may require a prohibitively large number of iterations. In such regimes, it is typically advantageous to instead use direct solvers.

The past couple of decades have seen many advances in direct solvers for integral equations. There have been several different lines of work on this subject, such as

A. Gopal
Yale University, New Haven, CT, USA
e-mail: abinand.gopal@yale.edu

P.-G. Martinsson (✉)
University of Texas at Austin, Austin, TX, USA
e-mail: pgm@oden.utexas.edu

J. M. Melenk et al. (eds.), *Spectral and High Order Methods for Partial Differential Equations ICOSAHOM 2020+1*, Lecture Notes in Computational Science and Engineering 137, https://doi.org/10.1007/978-3-031-20432-6_2

HSS methods (e.g., [19–21, 79]), $\mathcal{H}$-matrix methods (e.g., [6, 7, 9–12, 37, 44–47]), and recursive skeletonization-based methods (e.g, [2, 8, 14, 22, 23, 34, 36, 40, 42, 51–53, 55, 58, 59, 62, 63, 69]). While the techniques discussed in this manuscript can likely be extended to other types of solvers, we will focus on solvers based on recursive skeletonization (RS) and refer the reader to the textbook [57] and monographs [6, 40] for discussion on how the other methods relate.

Direct solvers using RS typically proceed in three stages. First, there is a *compression stage* where low rank approximations to the off-diagonal blocks (i.e., blocks corresponding to degrees of freedom that are separated based on some hierarchical division of physical space) of the coefficient matrix are computed. Then, there is an *inversion stage* where the rank-structured approximation is inverted using, for example, hierarchical factorizations (e.g., [53, 63]), sparse embeddings (e.g., [51]), variants of the Woodbury formula (e.g., [34, 55]), or scattering matrices (e.g., [14]). Finally, there is a *solve stage* where the computed inverse is rapidly applied to the right-hand side. While there now exist solvers for which all three stages are asymptotically quasilinear in both two and three dimensions, the constants in the scaling can vary drastically. While the exact relations depend on the specifics of the solver, we find empirically that for a broad class of solvers in two dimensions, based on so-called *weak admissibility*, the compression stage is around an order of magnitude slower than the inversion stage, which in turn is about an order of magnitude slower than the solve stage.

In applications where multiple incident fields must be considered, this breakdown of costs is rather convenient since both the compression and inversion stages must be done only once. In broadband environments, where one is interested in solving problems at many different frequencies, all three stages must be repeated at each wavenumber, which gets expensive. Such environments arise across a broad range of applications, such as in computing sonar/radar cross sections, transmission problems, eigenvalue and resonance problems (e.g., [80]), and Fourier transform-based solvers for the wave equation (e.g., [3]). In this manuscript, we introduce a new technique for accelerating the application of recursive skeletonization-based direct solvers in this setting.

Specifically, we accelerate the compression stage of the direct solvers. Instead of computing low-rank approximations from scratch at each wavenumber of interest, we compute *universal bases*, which are valid on an entire interval of wavenumbers, for the column and row space of each block that must be compressed in a rank-structured representation. Then, to obtain a rank-structured approximation to the coefficient matrix at any wavenumber, all that must be done is a projection of each block of the coefficient matrix onto the computed bases. When the interpolative decomposition is used, this effectively amounts to a series of kernel evaluations, which can be executed efficiently. We refer to this procedure as *broadband recursive skeletonization*. While the ranks produced in the resulting rank-structured approximation are necessarily suboptimal, we find that in most cases of interest, the ranks are only slightly higher than the ranks at the highest frequency of interest, and the resulting acceleration in the compression stage more than compensates for the minor increase in cost in the inversion and solve stages. The result is that the overall computation can often be accelerated by a factor of between 2 and 10.

The manuscript is structured as follows: In Sect. 2, we survey some preliminaries pertaining to the direct solution of the integral equations of scattering theory. This is followed by Sect. 3 and 4 where we formulate the main theoretical observations that underpin our work. In Sect. 5, we present algorithms for computing universal bases. How these algorithms can be incorporated into recursive skeletonization is explained in Sect. 6. This is applied in Sect. 7, where we demonstrate our method in a few different environments. Finally, we briefly summarize this manuscript and outline some directions for future work in Sect. 8.

Remark 1.1 (High Frequency Breakdown) All techniques described in this manuscript rely on low rank approximation of the off-diagonal blocks of the coefficient matrix. This property is known to break down in the limit as the wavelength goes to zero. Alternative representations have been proposed to overcome this challenge [43], but such techniques are outside the scope of the present work. Practically speaking, we expect the techniques described here to be viable for problems up to a few hundred wavelengths in diameter in two dimensions.

2 Preliminaries

The key contribution of the manuscript is a technique for rapidly constructing low-rank factorizations to off-diagonal blocks of the dense matrices that arise upon the discretization of boundary integral operators (BIEs). In this section, we introduce a representative BIE for an external scattering problem, briefly describe how the BIE can be discretized, discuss how the "interpolative decomposition" or "skeletonization" is used to factorize off-diagonal blocks, and then briefly sketch how this can be used to develop fast direct solvers.

2.1 *Integral Equations of Scattering Theory*

Let $\Omega \subset \mathbb{R}^d$ for $d = 2$ or 3 be a scatterer with boundary Γ and let κ be the wavenumber. Then, it is well-known that the scattered field for a standard scattering problem can be represented by

$$u(x) = \int_\Gamma \left[\frac{\partial G_\kappa(x, y)}{\partial n(y)} - i\eta G_\kappa(x, y) \right] \sigma(y)\, ds(y), \quad x \in \mathbb{R}^d \backslash \overline{\Omega} \tag{1}$$

where $\eta > 0$, $n(y)$ is the outward normal of Ω at y, and G_κ is the free space fundamental solution of the Helmholtz equation given by

$$G_\kappa(x, y) = \begin{cases} \frac{i}{4} H_0(\kappa \|x - y\|), & d = 2, \\ \frac{1}{4\pi} \frac{e^{i\kappa\|x-y\|}}{\|x - y\|}, & d = 3 \end{cases} \tag{2}$$

where H_0 is the first-kind Hankel function of order zero. For any $\sigma \in L^2(\Gamma)$, (1) satisfies the Helmholtz equation in the exterior of $\overline{\Omega}$ and the Sommerfeld radiation condition. Therefore, σ is chosen so that u will satisfy the necessary boundary condition on Γ. This can be done by solving an integral equation on Γ that is derived by taking the limit of (1) to the boundary. For example, for the sound-hard (Dirichlet) scattering problem, σ must satisfy the equation

$$\sigma(x) + 2 \int_\Gamma \left[\frac{\partial G_\kappa(x, y)}{\partial n(y)} - i\eta G_\kappa(x, y) \right] \sigma(y)\, ds(y) = -2u^{\text{inc}}(x), \quad x \in \Gamma \tag{3}$$

where u^{inc} is the incident field. Obviously, (3) is a Fredholm equation of the second kind on $L^2(\Gamma)$.

Similar integral equations can be derived for other boundary conditions and interior problems. For more details, we refer the reader to the books [26] and [27].

2.2 *Nyström Discretization*

Consider an integral operator of the form $T : L^2(\Gamma_\text{s}) \to L^2(\Gamma_\text{t})$ where

$$[T\sigma](x) = \int_{\Gamma_\text{s}} K(x, y)\sigma(y)\, ds(y), \quad x \in \Gamma_t \tag{4}$$

where Γ_s and Γ_t are curves or surfaces in $\mathbb{R}^d$. Assume that K is smooth on $\Gamma_\text{t} \times \Gamma_\text{s}$. Given suitable quadrature rules $\{(v_i, x_i)\}_{i=1}^n$ and $\{(w_i, y_i)\}_{i=1}^n$ on Γ_t and Γ_s, respectively, we can discretize the operator T to obtain

$$\boldsymbol{u} = \boldsymbol{A}\boldsymbol{\sigma}, \tag{5}$$

where

$$\boldsymbol{A}(i, j) = \sqrt{v_i}\sqrt{w_j} K(x_i, y_j), \text{ and } \boldsymbol{\sigma}(i) = \sqrt{w_j}\sigma(y_j) \tag{6}$$

for $i, j = 1, 2, \ldots, n$. The expectation is that for sufficiently large n, $\boldsymbol{u}(i) \approx \sqrt{v_i}[T\sigma](x_i)$. Note that the scaling in (6) was chosen so that the discrete inner products approximate the continuous inner products, which can be important for conditioning reasons [13].

For a second kind integral equation of the form

$$\sigma(x) + \int_\Gamma K(x, y)\sigma(y)\, ds(y) = f(x), \quad x \in \Gamma, \tag{7}$$

we can choose the collocation points to coincide with the quadrature nodes. This corresponds to what is known as a *Nyström discretization* and yields a linear system of the form

$$(\boldsymbol{I} + \boldsymbol{A})\boldsymbol{\sigma} = \boldsymbol{f}, \tag{8}$$

where $\boldsymbol{I}$ is the $n \times n$ identity,

$$\boldsymbol{A}(i, j) = \sqrt{w_i}\sqrt{w_j}K(x_i, x_j), \ \boldsymbol{\sigma}(i) = \sqrt{w_i}\sigma_i, \text{ and } \boldsymbol{f}(i) = \sqrt{w_i}f(x_i), \tag{9}$$

for $i, j = 1, 2, \ldots, n$. We of course expect $\sigma_i \approx \sqrt{w_i}\sigma(x_i)$ for sufficiently large n, and indeed, under some mild assumptions on the regularity of K, f, and Γ, it is known that the solution of a Nyström discretization converges to the values of the true solution at the same order to which the integral is discretized. For more details, we refer the reader to the books [4] and [56].

The above discussion was restricted to the case of smooth kernels. When the kernel has a singularity such as (2), it is typically not possible to find a single quadrature rule which will accurately discretize the integral in (7) at each collocation node. Fortunately, it is typically possible to account for the singularity using a few local corrections (e.g., for each collocation point x_i we adjust the quadrature weights corresponding to nodes x_j for x_j sufficiently close to x_i). This corresponds to modifying some of the entries near the diagonal of $\boldsymbol{A}$ in (8), and so most of the rank structure in the off-diagonal blocks is preserved. We refer the reader to the survey paper [49] for more details.

2.3 Interpolative Decomposition

The *row interpolative decomposition* (ID) of a rank-k matrix $\boldsymbol{A} \in \mathbb{C}^{m\times n}$ is a factorization of the form

$$\boldsymbol{A} = \boldsymbol{U}\boldsymbol{A}(I, :) \tag{10}$$

where $I \subset [1, 2, \ldots, n]$ is a k-element index vector and $\boldsymbol{U}$ is an $m \times k$ matrix for which $\boldsymbol{U}(I, :)$ is the $k \times k$ identity matrix. The elements of I and $[1, 2, \ldots, n]\backslash I$ are typically called the *skeleton indices* and *residual indices*, respectively. In other words, a row ID is a factorization where a subset of the rows of $\boldsymbol{A}$ is used to span the row space of $\boldsymbol{A}$. Of course, in practice the matrices we work with are only of low *numerical* rank. As a result, (10) only holds to some tolerance, which we would like to be close to the optimal truncation error given by the singular value decomposition. Moreover, for the purposes of numerical stability, it is desirable for the magnitude of the entries of $\boldsymbol{U}$ to be small (ideally not much larger than 1 in modulus). A column interpolative decomposition can be similarly defined.

A key advantage of the ID is that it allows for efficient projection. Suppose that $\boldsymbol{U}$ is an $m \times k$ matrix such that $\boldsymbol{U}(I, :)$ is the $k \times k$ identity matrix for an index vector I with $|I| = k$, and $\boldsymbol{A}$ is an $m \times n$ matrix. Then, letting $\|\cdot\|$ denote the spectral norm, we have the relation

$$\begin{aligned}
\|\boldsymbol{U}\boldsymbol{A}(I, :) - \boldsymbol{A}\| &\leq \|\boldsymbol{U}\boldsymbol{A}(I, :) - \boldsymbol{U}\boldsymbol{U}^\dagger\boldsymbol{A}\| + \|\boldsymbol{U}\boldsymbol{U}^\dagger\boldsymbol{A} - \boldsymbol{A}\| && (11)\\
&\leq \|\boldsymbol{U}\|\|\boldsymbol{A}(I, :) - \boldsymbol{U}^\dagger\boldsymbol{A}\| + \|\boldsymbol{U}\boldsymbol{U}^\dagger\boldsymbol{A} - \boldsymbol{A}\| && (12)\\
&= \|\boldsymbol{U}\|\|\boldsymbol{A}(I, :) - \boldsymbol{U}(I, :)\boldsymbol{U}^\dagger\boldsymbol{A}\| + \|\boldsymbol{U}\boldsymbol{U}^\dagger\boldsymbol{A} - \boldsymbol{A}\| && (13)\\
&\leq (1 + \|\boldsymbol{U}\|)\|\boldsymbol{U}\boldsymbol{U}^\dagger\boldsymbol{A} - \boldsymbol{A}\|, && (14)
\end{aligned}$$

where in (13) we used the fact that $\boldsymbol{U}(I, :)$ is the $k \times k$ identity matrix. As a result, if $\boldsymbol{U}$ is well-conditioned we can use $\boldsymbol{U}\boldsymbol{A}(I, :)$ as a rank-k approximation to $\boldsymbol{A}$ in place of $\boldsymbol{U}\boldsymbol{U}^\dagger\boldsymbol{A}$ without a significant loss of accuracy. Typically, the former is significantly cheaper to compute, especially when the entries of $\boldsymbol{A}$ are given by evaluations of a kernel function, which is the case in Nyström discretizations of integral equations.

The above observation enables the *method of proxy sources* to be used efficiently, which is crucial for achieving quasilinear complexity recursive skeletonization algorithms (see [23, Sec. 5.2] and [57, Sec. 17.2]). Using the CPQR factorization, a rank-k ID can be computed in $O(mnk)$ floating point operations; for details on this and more about the ID in general, we refer the reader to the article [23]. Using randomized algorithms, this cost can be reduced to $O(mn \log k + k^2 n)$ [74].

2.4 Recursive Skeletonization

Due to space constraints, we do not provide a full description of the recursive skeletonization procedure here. Instead, we sketch the main idea and refer the reader to the textbook [57], survey article [34], and research papers [51, 58].

The linear systems that arise upon discretizing integral equations such as (3) are dense, and so for a problem with n degrees of freedom, straightforward Gaussian elimination would take $O(n^3)$ floating point operations, which is prohibitive when n is large. Fortunately, the coefficient matrix has additional structure that we can exploit. In particular, while the matrix is dense, many of its off-diagonal blocks will be of low numerical rank. This is due to the fact that most of the entries are given by a scaling of the kernel that is smooth "away from the diagonal", cf. (9). Specifically, when an off-diagonal block encodes interactions between sources and targets that are physically separated in space, it will typically have low numerical rank, due to the smoothness of the kernel. This is essentially the same idea as in the fast multipole method, where interactions between particles that are well-separated are efficiently encoded by multipole expansions [41].

In recursive skeletonization, custom expansions produced by the ID are used in lieu of multipole expansions. This leads to lower ranks and also enables the compression of blocks that are immediately adjacent in physical space [57, p. 94]. In physical terms, using the ID corresponds to viewing each degree of freedom in a Nyström discretization as a point charge, partitioning the point charges based on their location in physical space, and then selecting a subset of representative charges in each group (which we call *skeletons*) that are used to encode all interactions between groups. The grouping of point charges for a smooth five-pointed star contour is illustrated in Fig. 1a, where each color represents a group. The interactions between groups can then be compressed and represented as interactions between skeletons which are plotted in Fig. 1b. After one step of compression is done, the skeletons are then merged to form new groups. This is illustrated in Fig. 1c, where each group is denoted by a different color. The interactions between these groups are then compressed, as can be seen in Fig. 1d. This procedure of grouping and compression is repeated until after there are only two groups left and the interactions between them have been compressed (see Fig. 1e and f).

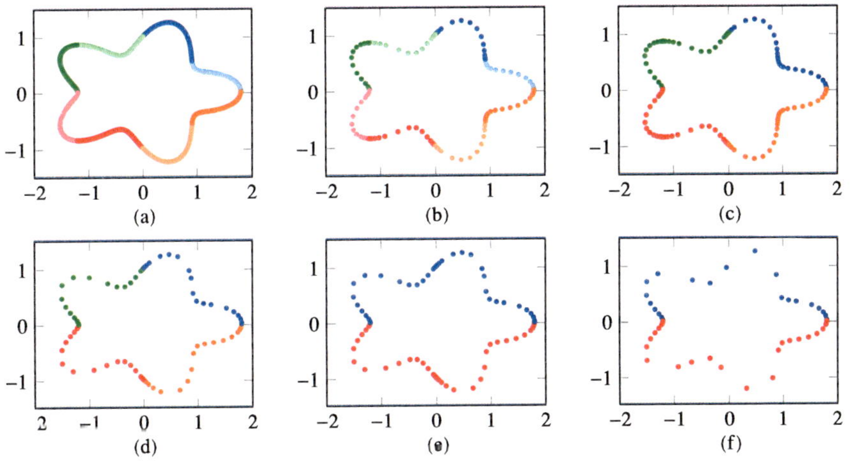

Fig. 1 Illustration of four-level recursive skeletonization for a contour in 2D. Each piece of the contour corresponding to a different leaf is shown in a different color in (**a**). After the interactions between the leaves are compressed, the skeleton points that remain are shown in (**b**). These skeleton points are then grouped by sibling pairs to introduce new low-rank structure in (**c**). The interactions between these new groups are again compressed to obtain the skeleton points in (**d**). This procedure is then repeated in (**e**) and (**f**)

After recursive skeletonization has been performed, various linear algebra operations, such as matrix-vector products and inversion can be done efficiently. We refer the reader to the textbook [57] for more details.

3 Theoretical Foundations

The main discovery reported is that when using fast algorithms for scattering problems that are based on low-rank approximation of interactions between disjoint subdomains, it is possible to construct local bases that are valid for a whole interval $I = [\kappa_{\min}, \kappa_{\max}]$ of frequencies. In this section, we make precise what we mean by this claim, we justify it analytically for the case of a circular domain, and then substantiate it numerically for a wide collection of general domains. To simplify slightly, the punchline is that the rank of interaction for the whole interval I is in many geometries basically the same as the rank required in order to compress only the largest wavenumber $\kappa_{\max}$. For the particular case of a curve segment in the plane, this rank turns out to be almost exactly that dictated by the Shannon–Nyquist theorem, requiring two points per wavelength along the curve.

3.1 Universal Bases

Let $\boldsymbol{A}_\kappa$ be a $2n \times 2n$ coefficient matrix corresponding to the discretization of an integral equation formulation of a scattering problem with wavenumber κ. For a single-level scheme, we could partition $\boldsymbol{A}_\kappa$ into a 2×2 block matrix of the form

$$\boldsymbol{A}_\kappa = \begin{bmatrix} \boldsymbol{A}_{11}^{(\kappa)} & \boldsymbol{A}_{12}^{(\kappa)} \\ \boldsymbol{A}_{21}^{(\kappa)} & \boldsymbol{A}_{22}^{(\kappa)} \end{bmatrix}, \tag{15}$$

where all of the matrices in the right-hand side are $n \times n$. We would then like to compute low-rank factorizations of the off-diagonal blocks $\boldsymbol{A}_{12}^{(\kappa)}$ and $\boldsymbol{A}_{21}^{(\kappa)}$ by computing $n \times l^{(\kappa)}$ matrices $\boldsymbol{U}_1^{(\kappa)}, \boldsymbol{U}_2^{(\kappa)}, \boldsymbol{V}_1^{(\kappa)}$, and $\boldsymbol{V}_2^{(\kappa)}$, as well as $l^{(\kappa)} \times l^{(\kappa)}$ matrices $\widetilde{\boldsymbol{A}}_{12}^{(\kappa)}$ and $\widetilde{\boldsymbol{A}}_{21}^{(\kappa)}$ with $l^{(\kappa)} < n$ such that

$$\boldsymbol{A}_\kappa \approx \begin{bmatrix} \boldsymbol{A}_{11}^{(\kappa)} & \boldsymbol{U}_1^{(\kappa)} \widetilde{\boldsymbol{A}}_{12}^{(\kappa)} (\boldsymbol{V}_2^{(\kappa)})^* \\ \boldsymbol{U}_2^{(\kappa)} \widetilde{\boldsymbol{A}}_{21}^{(\kappa)} (\boldsymbol{V}_1^{(\kappa)})^* & \boldsymbol{A}_{22}^{(\kappa)} \end{bmatrix}. \tag{16}$$

After a factorization of the form (16) is obtained, the structure can be exploited for efficient matrix algebra; for example, it is obvious that (16) can be written as a block diagonal plus low-rank matrix, which is amenable to inversion by the Woodbury formula. Typically, the matrices $\boldsymbol{A}_{11}^{(\kappa)}$ and $\boldsymbol{A}_{22}^{(\kappa)}$ themselves have off-diagonal blocks that are numerically of low rank, so the idea can be applied recursively. Frequently, algorithms with a complexity that is linear or a small power in the number of degrees of freedom are attainable in both 2D and 3D.

The cost of compressing the operator relative to the costs of doing matrix algebra depends on the specifics of the direct solver used and the problem. As aforementioned, the cost of compression is typically significantly larger than the cost of doing inversion, which is an order of magnitude larger than the cost of doing a matrix-vector multiplication or applying the inverse to a vector. In many applications, the compression and inversion stages must only be done once, so this is not problematic. However, in applications where multiple frequencies must be considered, this is rather inconvenient since the costly compression stage must be repeated at each wavenumber.

The key idea of this manuscript is that instead of computing factorizations of the form (16) for each $\kappa \in [\kappa_{\min}, \kappa_{\max}]$, we could compute a set of $n \times l$ matrices $\boldsymbol{U}_1$, $\boldsymbol{U}_2$, $\boldsymbol{V}_1$, and $\boldsymbol{V}_2$ such that there exist $l \times l$ matrices $\widetilde{\boldsymbol{A}}_{12}^{(\kappa)}$ and $\widetilde{\boldsymbol{A}}_{21}^{(\kappa)}$ for which

$$\boldsymbol{A}_\kappa \approx \begin{bmatrix} \boldsymbol{A}_{11}^{(\kappa)} & \boldsymbol{U}_1 \widetilde{\boldsymbol{A}}_{12}^{(\kappa)} (\boldsymbol{V}_2)^* \\ \boldsymbol{U}_2 \widetilde{\boldsymbol{A}}_{21}^{(\kappa)} (\boldsymbol{V}_1)^* & \boldsymbol{A}_{22}^{(\kappa)} \end{bmatrix}, \quad \text{for all } \kappa \in [\kappa_{\min}, \kappa_{\max}]. \tag{17}$$

In other words, the matrices used to span the row and column spaces of the off-diagonal blocks of the coefficient matrix are independent of the wavenumber. Thus, we refer to the columns of the matrices $\boldsymbol{U}_1$, $\boldsymbol{U}_2$, $\boldsymbol{V}_1$, and $\boldsymbol{V}_2$ as *universal bases*. At first glance, it is not clear that using (17) is actually more efficient than (16) since the matrices $\widetilde{\boldsymbol{A}}_{12}^{(\kappa)}$ and $\widetilde{\boldsymbol{A}}_{21}^{(\kappa)}$ must still be computed at each wavenumber. A crucial observation is that when the ID is used for the low-rank factorizations, these matrices can be directly extracted as submatrices of $\boldsymbol{A}_\kappa$ (cf. Sect. 2.3). As such, all that must be done at each wavenumber is a series of kernel evaluations. The hierarchical extension of these ideas is straightforward.

There are two disadvantages to this approach. First, the size of the discretization must be kept constant for all wavenumbers. Since most direct solvers that are applied scale either linearly or almost linearly with the total number of degrees of freedom, the associated computational trade-off is rather minor. The second and more significant issue is that the ranks of the resulting rank-structured approximation must be kept constant across $[\kappa_{\min}, \kappa_{\max}]$. In the above discussion, this corresponds to the fact that the parameter l does not depend on κ. Obviously, for (17) to hold to some tolerance, l must be taken to be greater than that which $l^{(\kappa)}$ must be taken for (16) for the same tolerance for all $\kappa \in [\kappa_{\min}, \kappa_{\max}]$. It turns out that in many cases, the ranks when the universal bases are used are not much greater than the ranks required at the highest frequency. That is, there are many circumstances where $l \approx l^{(\kappa_{\max})}$. The goal of the next few sections is to formalize this observation to a degree.

3.2 Broadband Analysis on Concentric Circles

When collocation discretizations are used, the off-diagonal blocks of the coefficient matrix can typically be interpreted as approximations to certain integral operators mapping charges on one portion of the boundary of the scatterer to targets on another. Thus, it is instructive to look at the rank behavior of such an operator in the broadband environment. Suppose $\Gamma_\mathrm{s} \subset \mathbb{R}^2$ and $\Gamma_\mathrm{t} \subset \mathbb{R}^2$ are disjoint curves in two dimensions. For simplicity, we will consider the single layer operator $S_\kappa : L^2(\Gamma_\mathrm{s}) \to L^2(\Gamma_\mathrm{t})$ given by

$$[S_\kappa \sigma](x) = \int_{\Gamma_\mathrm{s}} G_\kappa(x, y)\sigma(y)\, ds(y), \qquad x \in \Gamma_\mathrm{t}. \tag{18}$$

In certain very special cases, we can work out a low-rank approximations to the continuum operator analytically. Let $\Gamma_\mathrm{s} \subset \mathbb{R}^2$ and $\Gamma_\mathrm{t} \subset \mathbb{R}^2$ be concentric circles around the origin with radii R_s and R_t, respectively, with $R_\mathrm{s} > 2R_\mathrm{t}$. Then, by Graf's addition theorem (see [1, (9.1.79)]), we see that for any $\kappa > 0$, (18) is equivalent in this case to

$$[S_\kappa \sigma](R_\mathrm{t} e^{i\theta}) = \frac{iR_\mathrm{s}}{4} \sum_{j=-\infty}^{\infty} H_j(\kappa R_\mathrm{s}) J_j(\kappa R_\mathrm{t}) e^{ij\theta} \int_0^{2\pi} e^{-ij\varphi} \sigma(R_\mathrm{s} e^{ij\varphi})\, d\varphi \tag{19}$$

$$= \sum_{j=-\infty}^{\infty} u_j(\theta) \int_0^{2\pi} v_j^{(\kappa)}(\varphi) \sigma(R_\mathrm{s} e^{ij\varphi})\, d\varphi \tag{20}$$

where

$$u_j(\theta) = e^{ij\theta} \text{ and } v_j^{(\kappa)}(\varphi) = \frac{iR_\mathrm{s}}{4} H_j(\kappa R_\mathrm{s}) J_j(\kappa R_\mathrm{t}) e^{-ij\varphi}, \qquad j \in \mathbb{Z}. \tag{21}$$

In particular, we see that we can span the range of S_κ for any $\kappa \in [\kappa_{\min}, \kappa_{\max}]$ using the same set of functions, namely the trigonometric polynomials. Moreover, it follows from the properties of Bessel and Hankel functions that the norm of the terms in (20) will start to superalgebraically decay after approximately $|j| \approx 2\kappa R_\mathrm{t}$ [28]. Thus, to span the ranges of all S_κ for $\kappa \in [\kappa_{\min}, \kappa_{\max}]$, one would only need approximately $2\kappa_{\max} R_\mathrm{t}$ functions, which is what is needed to approximate the range of $S_{\kappa_{\max}}$ alone.

3.3 Broadband Analysis in 2D

For more general geometries, there is in general no analytic formula for universal bases available. Instead, we numerically construct them. To this end given a source curve $\Gamma_\mathrm{s} \subset \mathbb{R}^2$ and a target curve $\Gamma_\mathrm{t} \subset \mathbb{R}^2$, we let $\boldsymbol{A}_\kappa$ denote an $m \times n$ discretization of the single layer operator (18) obtained in the manner described in Sect. 2.2. Given a wavenumber interval $[\kappa_{\min}, \kappa_{\max}]$ and an integer $N > 1$, we let $\{\kappa_j^{(N)}\}$ denote the nodes of the N-point second kind Chebyshev rule in $[\kappa_{\min}, \kappa_{\max}]$, and define the block matrix

$$A^{(N)}_{[\kappa_{\min},\kappa_{\max}]} = \left[A_{\kappa_1^{(N)}}\ A_{\kappa_2^{(N)}}\ \cdots\ A_{\kappa_N^{(N)}}\right] \in \mathbb{C}^{m\times Nn}. \tag{22}$$

For notational convenience, we also define

$$A^{(1)}_{[\kappa_{\min},\kappa_{\max}]} = A_{\kappa_{\max}}. \tag{23}$$

Clearly, if an $m \times l^{(N)}$ matrix $U^{(N)}$ satisfies

$$\left\| A^{(N)}_{[\kappa_{\min},\kappa_{\max}]} - U^{(N)}(U^{(N)})^{\dagger} A^{(N)}_{[\kappa_{\min},\kappa_{\max}]} \right\| < \varepsilon \tag{24}$$

then it satisfies

$$\left\| A_{\kappa_j^{(N)}} - U^{(N)}(U^{(N)})^{\dagger} A_{\kappa_j^{(N)}} \right\| < \varepsilon, \quad j = 1, 2, \ldots, N. \tag{25}$$

Since the entries of the A_κ are smooth with the wavenumber κ, we might hope that (24) for sufficiently large N also implies a bound such as

$$\left\| A_\kappa - U^{(N)}(U^{(N)})^{\dagger} A_\kappa \right\| < 2\varepsilon, \quad \kappa \in [\kappa_{\min}, \kappa_{\max}]. \tag{26}$$

We first consider the question of how large $l^{(N)}$ must be in order to satisfy (24). In other words, we examine the ε-ranks of the $A^{(N)}_{[\kappa_{\min},\kappa_{\max}]}$. After all, if $U^{(N)}$ in (24) had to be $m \times m$, then this entire discussion would be vacuous. We consider the source and target curves given by Fig. 2a and the case where $\kappa_{\min} = 1$ and

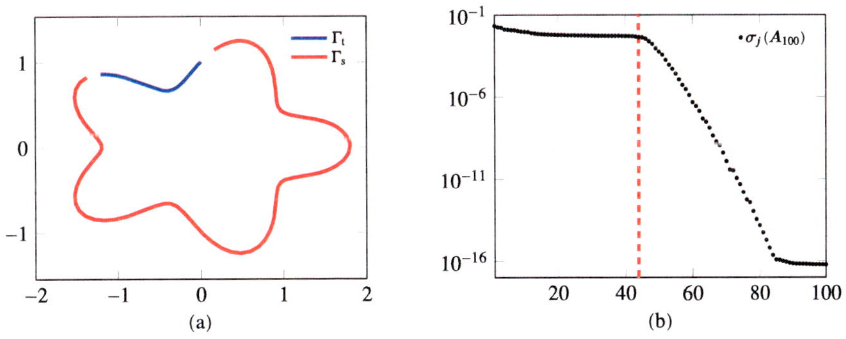

Fig. 2 (**a**) Geometry of the model problem considered in Sect. 3.3. (**b**) First 100 singular values of A_{100} computed in double precision arithmetic. The red, dashed line demarcates the values $\kappa|\Gamma_t|/\pi \approx 43.80$

$\kappa_{\max} = 100$. The singular values of A_{100} computed in double precision are plotted in Fig. 2b, where the red dashed line denotes the value $\kappa|\Gamma_t|/\pi$. The behavior of the singular values of A_{100} in Fig. 2b will come as no surprise to researchers in scattering theory. By essentially the Shannon–Nyquist theorem, a certain number of modes are required to resolve the oscillations of the fields in the range of the operator, and the corresponding singular values are all roughly equal. In this case, about $100|\Gamma_t|/\pi$ such modes are required, which corresponds to roughly two times the arc length of the target curve measured in wavelengths. After the oscillations are resolved, the remaining singular values superalgebraically decay.

The singular values of $A^{(N)}_{[1,100]}$ for this geometry for $N = 3, 5, 7$ are plotted in Fig. 3a. In every case, the singular values behave similarly to those of A_{100}. That is, there are roughly $100|\Gamma_t|/\pi$ singular values all of the same order, and the remainder superalgebraically decay. As such, it should come as no surprise that the ε-ranks in all four cases are similar. For example, for $\varepsilon = 10^{-10}$, the ε-ranks of A_{100}, $A^{(3)}_{[1,100]}$, $A^{(5)}_{[1,100]}$, and $A^{(7)}_{[1,100]}$ are 70, 72, 73, and 73, respectively.

We now turn our attention to the relation between (24) and (26). Again, keeping $\varepsilon = 10^{-10}$, we let $U^{(N)}$ denote the matrix containing the ε-rank leading singular vectors of $A^{(N)}_{[1,100]}$ and plot the projection error

$$E^{(N)}_{\kappa} = \|A_{\kappa} - U^{(N)}(U^{(N)})^{\dagger}A_{\kappa}\| \tag{27}$$

as a function of κ in Fig. 3b. As is clear from the figure, $U^{(1)}$ forms a basis for the range of A_{100} up to the desired accuracy, but this is not true in general for A_{κ} with $\kappa < 100$. Indeed, there are wavenumbers for which up to four digits are lost. The situation is substantially improved when looking at projection onto the columns of $U^{(3)}$. Now an error of less than 10^{-10} is obtained near $\kappa = 100$ and also for all $\kappa \in [1, 52]$. Moreover, no more than two digits are lost across the entire interval.

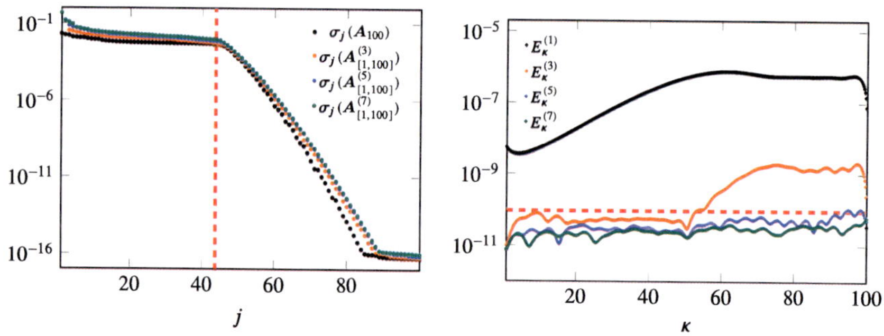

(a) (b)

Fig. 3 (**a**) First 100 singular values of $A^{(N)}_{[1,100]}$ for $N = 1, 3, 5, 7$ computed in double precision. The red, dashed line demarcates the value $100|\Gamma_t|/\pi \approx 43.80$. (**b**) Errors in the spectral norm when A_{κ} is projected onto the columns of $U^{(N)}_{\varepsilon}$ with $\varepsilon = 10^{-10}$ and $N = 1, 3, 5, 7$. The red, dashed line demarcates the value 10^{-10}

The situation is improved further with $\boldsymbol{U}^{(5)}$. Now the errors are below 10^{-10} on the majority of the interval, and we lose at most one digit. Finally, with $\boldsymbol{U}^{(7)}$ the errors are below 10^{-10} for all $\kappa \in [1, 100]$.

The upshot is that for this geometry a universal basis for the approximate column space of $\boldsymbol{A}_\kappa$ for all $\kappa \in [1, 100]$ can be computed and the size of this basis is not much larger than what would be required to approximate the column space of $\boldsymbol{A}_{100}$. This is somewhat intuitive, since the column space of $\boldsymbol{A}_\kappa$ essentially consists of "bandlimited" fields on $\Gamma_{\rm t}$. Therefore, we expect the numerical rank to be controlled by the highest frequency. While we have illustrated this here for the specific geometry in Fig. 2a, this observation holds more generally. The left images in Fig. 4 show four more configurations of source and target curves in red and blue, respectively. For each geometry, we consider the wavenumbers in the interval $[1, \kappa_{\max}]$, where the values of $\kappa_{\max}$ and $|\Gamma_{\rm t}|$ are shown in the top right corner of each plot. The images in the right column show the singular values of the corresponding matrices $\boldsymbol{A}_{\kappa_{\max}}$ and $\boldsymbol{A}_{[\kappa_{\min}, \kappa_{\max}]}^{(N)}$, where N is chosen sufficiently large so that taking the columns of $\boldsymbol{U}^{(N)}$ to be the ε-rank leading left singular vectors of $\boldsymbol{A}_{[\kappa_{\min}, \kappa_{\max}]}^{(N)}$ suffices for (26) hold with $\varepsilon = 10^{-10}$. In all cases, the singular values of $\boldsymbol{A}_{[\kappa_{\min}, \kappa_{\max}]}^{(N)}$ start to superalgebraically decay after there are approximately $\kappa_{\max}|\Gamma_{\rm t}|/\pi$ modes.

Establishing a rigorous result about this is outside of the scope of this work, so we summarize the above rather informally in the following observation:

Observation 3.1 *Let $\Gamma_{\rm s}$ and $\Gamma_{\rm t}$ be two curves in $\mathbb{R}^2$ that are sufficiently separated. Let $\boldsymbol{A}_\kappa$ denote an $m \times n$ collocation discretization obtained in the fashion described in Sect. 2.2. Then, there exists an $m \times m$ matrix $\boldsymbol{U}$ such that the projection error*

$$\sup_{\kappa \in [\kappa_{\min}, \kappa_{\max}]} \|\boldsymbol{A}_\kappa - \boldsymbol{U}(:, 1:l)\boldsymbol{U}(:, 1:l)^\dagger \boldsymbol{A}_\kappa\| \tag{28}$$

starts to decay superalgebraically as a function of l when

$$l \approx \frac{\kappa_{\max}|\Gamma_{\rm t}|}{\pi}. \tag{29}$$

We have been intentionally vague about what constitutes sufficient separation in Observation 3.1. Empirically, we find that it suffices for the sources and targets to be separated by a little over a wavelength (at the highest frequency) except in small neighborhoods around isolated points. For example, $\Gamma_{\rm s}$ and $\Gamma_{\rm t}$ may be components of the same contour in 2D with touching endpoints. Of course, if $|\Gamma_{\rm s}| < |\Gamma_{\rm t}|$, we could instead build a universal basis for the row space of the $\boldsymbol{A}_\kappa$.

When there are symmetries, Observation 3.1 can be rather pessimistic. This is illustrated in Fig. 5 which considers a case where the source and target curves correspond to two halves of an ellipse with semi-major axis 5 and semi-minor axis 1 (see Fig. 5a) and $\kappa_{\max} = 50$. Due to the symmetry in the geometry, the point at which the singular values of $\boldsymbol{A}_{[1,50]}^{(N)}$ start to decay is about half of what is

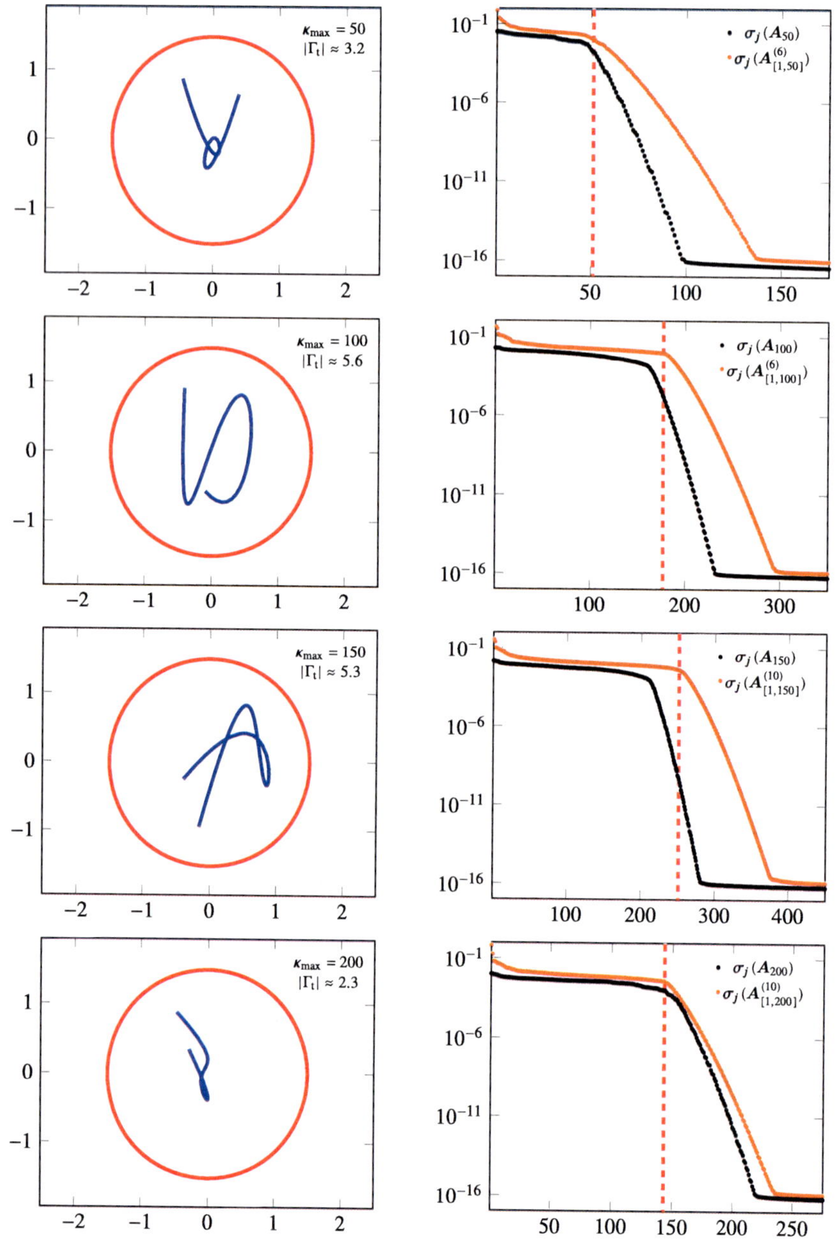

Fig. 4 The left column shows the four geometries with Γ_s denoted in red and Γ_t denoted in blue. The right column shows the singular values of $A_{\kappa_{max}}$ and $A^{(N)}_{[\kappa_{min},\kappa_{max}]}$ for the smallest N sufficient for (26) to hold with $\varepsilon = 10^{-10}$ for all $\kappa \in [\kappa_{min}, \kappa_{max}]$. The red, dashed line demarcates the value $\kappa_{max}|\Gamma_t|/\pi$

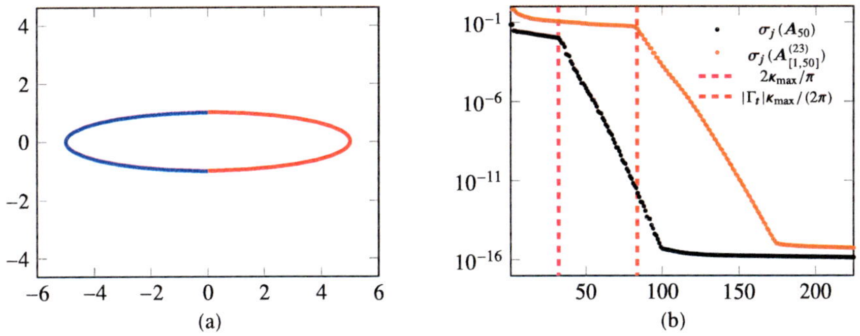

Fig. 5 (**a**) Geometry of a problem where Γ_s (denoted in red) and Γ_t (denoted in blue) correspond to halves of an ellipse. (**b**) Corresponding singular values computed for this problem when $\kappa_{\max} = 50$. The red and magenta, dashed lines correspond to the values $|\Gamma_t|\kappa_{\max}/(2\pi) \approx 83.60$ and $2\kappa_{\max}/\pi \approx 31.83$, respectively

predicted by Observation 3.1. This example also illustrates that while the size of the universal basis depends on the arc length of the source and target curves measured in wavelengths, this is not true for the single frequency case. Due to the elongated nature of the domain, a better predictor for the point where the singular values of $\boldsymbol{A}_{50}$ start to superalgebraically decay is given by two times the length of the minor axis of the ellipse measured in wavelengths. In general, there will be a discrepancy when the problem is elongated or has a strong directionality to it. The ranks in the single frequency case in such settings have been studied extensively (see e.g., [15, 16, 24, 31–33, 65, 70, 73]). Essentially, this amounts to the absence of Green's theorem in the multifrequency regime. In our experience, unless one is specifically studying special geometries, such as elongated scatterers, such cases where there is a significant discrepancy between the ranks in the multifrequency case and single frequency case at the highest frequency rarely arise.

Remark 3.2 Readers familiar with tensor decompositions will likely notice that there are close connections between tensors and the ideas in this section. Indeed, the $m \times Nn$ matrices $\boldsymbol{A}^{(N)}_{[\kappa_{\min},\kappa_{\max}]}$ can be viewed as matricizations of third-order tensors $\mathfrak{A}_{[\kappa_{\min},\kappa_{\max}]}^{(N)}$ with dimensions $m \times n \times N$, where the first, second, and third dimensions correspond to targets, sources, and frequencies, respectively. The problem of finding a universal basis then amounts to the problem of computing a low-rank Tucker decomposition. In particular, an $m \times l$ matrix $\boldsymbol{U}$ and $l \times n \times N$ tensor $\widetilde{\mathfrak{A}}_{[\kappa_{\min},\kappa_{\max}]}^{(N)}$ are computed, such that the contraction of these two yields a good approximation to $\mathfrak{A}_{[\kappa_{\min},\kappa_{\max}]}^{(N)}$. We refer the reader to the survey article [54] for more information on tensors and the Tucker decomposition.

3.4 Broadband Analysis in 3D

The idea of universal bases extends naturally to three dimensions. Here, two modes per wavelength will be required in each direction, for a total of four modes per unit area. This is illustrated in Fig. 6, where the red dashed line is now on

$$\frac{(\kappa_{\max})^2|\Gamma_t|}{\pi^2}. \tag{30}$$

We summarize this in Observation 3.3.

Observation 3.3 *Let Γ_s and Γ_t be well-behaved surfaces in $\mathbb{R}^3$ that are sufficiently separated. Let $\boldsymbol{A}_\kappa$ denote an $m \times n$ collocation discretization obtained in the fashion described in Sect. 2.2. Then, there exists an $m \times m$ matrix $\boldsymbol{U}$ such that the projection error*

$$\sup_{\kappa \in [\kappa_{\min}, \kappa_{\max}]} \|\boldsymbol{A}_\kappa - \boldsymbol{U}(:, 1:l)\boldsymbol{U}(:, 1:l)^\dagger \boldsymbol{A}_\kappa\| \tag{31}$$

starts to decay superalgebraically as a function of l when

$$l \approx \frac{(\kappa_{\max})^2|\Gamma_t|}{\pi^2}. \tag{32}$$

What is meant by well-behaved in Observation 3.3 is left intentionally vague, as there are more pathological cases in 3D than in 2D. For example, in the event that Γ_t is long and thin, then the operator behaves more like the 2D problem and a

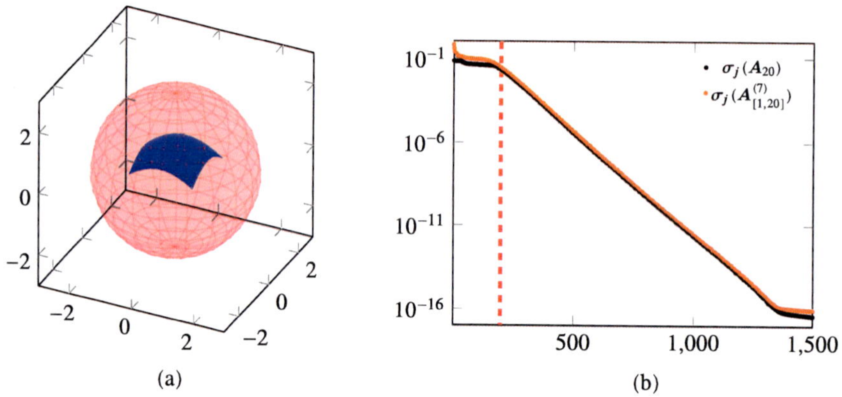

Fig. 6 **(a)** Geometry of a 3D problem where Γ_t is a surface (denoted in blue) and Γ_s is a sphere (denoted in red). **(b)** Singular values of $\boldsymbol{A}_{20}$ and $\boldsymbol{A}^{(7)}_{[1,20]}$ corresponding to this geometry. The red, dashed line is located at $\kappa^2_{\max}|\Gamma_t|/\pi^2 \approx 196.98$

better predictor for where the singular values start to decay is given by two times the supremum of the lengths of the minimal geodesics on Γ_t measured in wavelengths.

4 Connection with Fourier Operators

Unsurprisingly, the ideas in Sect. 3 are closely related to ideas from the theory of bandlimited functions. For a given wavenumber κ in 2D, it is well-known that when the sources are sufficiently separated from the targets, fields generated by operators such as (18) can be well-approximated by functions of the form

$$u(x) = \int_S e^{i\kappa x\cdot\omega}\rho(\omega)\,ds(\omega), \quad x \in \Gamma_t \tag{33}$$

where S is the unit circle in 2D and $\rho : S \to \mathbb{C}$, which is to say that the fields are bandlimited functions. In consequence, such fields can be well-approximated as linear combinations of the leading left singular vectors of the operator F_κ : $L^2(S) \to L^2(\Gamma_t)$ given by

$$[F_\kappa\rho](x) = \int_S e^{i\kappa x\cdot\omega}\,ds(\omega), \quad x \in \Gamma_t \tag{34}$$

where S is the unit circle.

In the broadband regime, the relevant Fourier operator is instead $\widetilde{F}_{\kappa_{\max}}$: $L^2(D) \to L^2(\Gamma_t)$ where

$$[\widetilde{F}_{\kappa_{\max}}\rho](x) = \int_D e^{i\kappa_{\max}x\cdot\omega}\,d\omega, \quad x \in \Gamma_t \tag{35}$$

where D is the unit disk. Indeed, we can take the universal bases in the previous section to be discretizations of the leading left singular functions of (35) as its columns, and the question of how many vectors are necessary in the universal bases can be reframed in terms of the behavior of the singular values of the operator in (35). Numerical experiments shows that in most circumstances, the operator in (35) will have about $\kappa_{\max}|\Gamma_t|/\pi$ order one singular values and the remaining will superalgebraically decay. This is illustrated in Fig. 7a which shows the singular values of (35) with $\kappa_{\max} = 150$ and Γ_t taken to be the blue curve in the third image of Fig. 4. Figure 7b demonstrates that the left singular functions of (35) can indeed be used as a universal bases by plotting

$$E_j = \left\| \boldsymbol{A}^{(10)}_{[1,150]} - \boldsymbol{U}(:, 1:j)(\boldsymbol{U}(:, 1:j))^* \boldsymbol{A}^{(10)}_{[1,150]} \right\| \tag{36}$$

where $\boldsymbol{U}$ is the matrix containing the singular vectors of the discretization of (35) with the usual ordering.

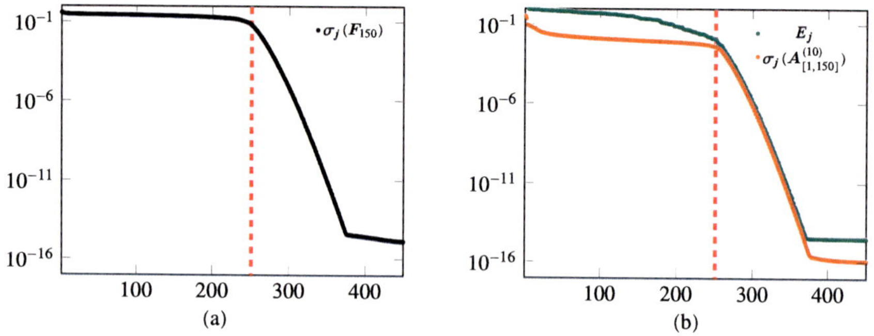

Fig. 7 (**a**) The plot of the singular values of a discretization of (35) computed in double precision arithmetic with $\kappa_{\max} = 150$ and Γ_t given by the blue curve in the third row of Fig. 4. (**b**) The singular values of $A^{(10)}_{[1,150]}$, and E_j as defined in (36). The red, dashed lines demarcate the value $\kappa_{\max}|\Gamma_t|/\pi$

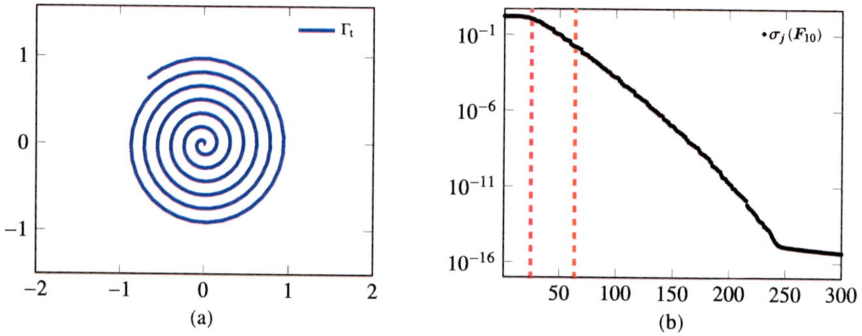

Fig. 8 (**a**) Spiral geometry. Note that $|\Gamma_t| = 10.06$, whereas diam $\Gamma_t < 1$. (**b**) Singular values of F_{10}, which is a discretization of (35) for this geometry. The red and magenta, dashed lines denote the value $\kappa_{\max}|\Gamma_t|/\pi \approx 63.86$ and $(\kappa_{\max})^2/4 = 25$

It is interesting to consider the spectral behavior of the operator in (35) in the case where Γ_t is approximately space filling. Such a case is illustrated in Fig. 8 where Γ_t is a spiral with $|\Gamma_t| \approx 10.06$ (as shown in Fig. 8a) and $\kappa_{\max} = 10$. As can be seen from this picture, the singular values start to decay significantly before $\kappa_{\max}|\Gamma_t|/\pi$. Indeed, a better approximation for the point of decay is given by $(\kappa_{\max})^2/4 = 25$, which is well-known to be the point at which the singular values of the restricted Fourier operator from the disk to the disk in 2D start to decay [38, 66]. Naturally, the entire discussion here is closely-related to the study of prolate spheroidal wave functions and generalized prolate spheroidal functions [64, 67, 68].

5 Algorithms

This section describes practical techniques for actually building the universal bases. To be precise suppose that $\kappa_{\min}$, $\kappa_{\max}$, Γ_s, and Γ_t are given, and let $\boldsymbol{A}_\kappa \in \mathbb{C}^{m\times n}$ for any $\kappa > 0$ denote a discretization of the single, double, or combined layer operator that maps sources on Γ_s to targets on Γ_t with wavenumber κ. Given $\varepsilon > 0$ our goal is to construct an ID with interpolation matrix $\boldsymbol{U}$ and skeleton indices I such that

$$\|\boldsymbol{A}_\kappa - \boldsymbol{U}\boldsymbol{A}_\kappa(I,:)\| < \varepsilon \tag{37}$$

for all $\kappa \in [\kappa_{\min}, \kappa_{\max}]$.

We take the approach of Sect. 3.1 which constructed a universal basis by concatenating coefficient matrices with different wavenumbers. We again use the notation $\boldsymbol{A}^{(N)}_{[\kappa_{\min},\kappa_{\max}]}$ to denote the matrix (22). For shorthand we will suppress the subscript and refer to $\boldsymbol{A}^{(N)}_{[\kappa_{\min},\kappa_{\max}]}$ simply as $\boldsymbol{A}^{(N)}$. For all odd N we also introduce the notation

$$\widetilde{\boldsymbol{A}}^{(N)} = \left[\boldsymbol{A}_{\kappa_2^{(N)}}\ \boldsymbol{A}_{\kappa_4^{(N)}}\ \cdots\ \boldsymbol{A}_{\kappa_{N-1}^{(N)}}\right], \tag{38}$$

so that $\widetilde{\boldsymbol{A}}^{(N)}$ is a submatrix of $\boldsymbol{A}^{(N)}$ that omits every other block.

The value of N needed depends non-trivially on the geometry and wavenumber interval, so in practice it is best determined adaptively. We start by setting the parameter p equal to 1, assemble the matrix $\boldsymbol{A}^{(2^p+1)}$, and compute a matrix $\boldsymbol{U}^{(p)}$ and index vector $I^{(p)}$ such that

$$\left\|\boldsymbol{A}^{(2^p+1)} - \boldsymbol{U}^{(p)}\boldsymbol{A}^{(2^p+1)}(I^{(p)},:)\right\| < \varepsilon. \tag{39}$$

This of course guarantees

$$\left\|\boldsymbol{A}_{\kappa_j^{(2^p+1)}} - \boldsymbol{U}^{(p)}\boldsymbol{A}_{\kappa_j^{(2^p+1)}}(I^{(p)},:)\right\| < \varepsilon, \quad j = 1, 2, \ldots, 2^p+1. \tag{40}$$

In order to check that we have sampled at enough wavenumbers, we check the error on a finer grid of wavenumbers consisting of $\{\kappa_j^{(2^{p+1}+1)}\}_{j=1}^{2^{p+1}+1}$. Due to our choice of Chebyshev points, $\{\kappa_j^{(2^p+1)}\}_{j=1}^{2^p+1} = \{\kappa_{2j+1}^{(2^{p+1}+1)}\}_{j=0}^{2^p}$, so it suffices to simply check that

$$\left\|\boldsymbol{A}_{\kappa_{2j}^{(2^{p+1}+1)}} - \boldsymbol{U}^{(p)}\boldsymbol{A}_{\kappa_{2j}^{(2^{p+1}+1)}}(I^{(p)},:)\right\| < \varepsilon, \quad j = 1, 2, \ldots, 2^p. \tag{41}$$

If (41) is not met, then we increment the value of p by 1, build new $\boldsymbol{U}^{(p)}$ and $I^{(p)}$ satisfying (40), and check the error (41). Note that in the process of evaluating the left-hand side of (41) in the previous iteration, we have already evaluated all of the entries of the matrix $\boldsymbol{A}^{(2^p+1)}$ in the current iteration. This may be looped with p incremented by 1 at each iteration until (41) is satisfied. This algorithm is summarized in Algorithm 1.

Algorithm 1 Build universal bases

Input: Wavenumber interval $[\kappa_{\min}, \kappa_{\max}]$, compression tolerance $\varepsilon > 0$, and a function handle that given $\kappa \in [\kappa_{\min}, \kappa_{\max}]$ returns $\boldsymbol{A}_\kappa$
Output: Interpolation matrix $\boldsymbol{U}$ and skeleton indices I

1: **for all** $p = 1, 2, \ldots$ **do**
2: Form $\boldsymbol{A}^{(2^p+1)}$ in (22)
3: Compute $\boldsymbol{U}^{(p)}$ and $I^{(p)}$ such that (39) holds
4: **if** $\max_{j=1,2,\ldots,2^p} \left\| \boldsymbol{A}_{\kappa_{2j}^{(2^{p+1}+1)}} - \boldsymbol{U}^{(p)} \boldsymbol{A}_{\kappa_{2j}^{(2^{p+1}+1)}}(I^{(p)}, :) \right\| < \varepsilon$ **then**
5: **return** $\boldsymbol{U} = \boldsymbol{U}^{(p)}$ and $I = I^{(p)}$
6: **end if**
7: **end for**

In practice we find that the termination criterion (41) is a good proxy for the quality of the universal bases throughout the entire interval $[\kappa_{\min}, \kappa_{\max}]$. Indeed, if (41) is met for some ε with $\boldsymbol{U}^{(p)}$ and $I^{(p)}$ chosen from (40), we typically find that (37) also holds and almost never see the left-hand side of (37) exceeding 2ε. Furthermore, we observe that in the majority of cases only a few iterations are required. In the numerical examples of the next section, typically only one to three iterations are required, and the number of iterations does not exceed six in any of the examples, despite considering intervals of wavenumbers corresponding to geometries differing in diameters by over a thousand wavelengths.

At first glance, it is rather unsatisfactory that (40) requires us to compute the ID of a matrix with a number of columns that grows exponentially with the iteration count ($\boldsymbol{A}^{(2^p+1)}$). Fortunately, this can be avoided when implementing Algorithm 1, by using ideas from streaming algorithms. The key idea here is that instead of working with $\boldsymbol{A}^{(2^p+1)}$ at each iteration, we work primarily with a random sketch. In the first iteration, we compute

$$\boldsymbol{Y}^{(1)} = \boldsymbol{A}^{(3)} \boldsymbol{\Omega}^{(1)} \in \mathbb{C}^{m \times \widetilde{n}} \tag{42}$$

where $\boldsymbol{\Omega}^{(1)} \in \mathbb{C}^{3n \times \widetilde{n}}$ with $\widetilde{n} < 3n$ is either a normalized Gaussian or a sparse sign matrix (see [25, 30, 60, 71] and the references therein for discussion of sparse sign matrices). If $\widetilde{n}$ is chosen appropriately (in practice we find $\widetilde{n} \approx 2n$ is a good choice), then the leading left singular vectors of $\boldsymbol{Y}^{(1)}$ align well with those of $\boldsymbol{A}^{(3)}$ with high probability. We can therefore compute factors $\boldsymbol{U}^{(1)}$ and $I^{(1)}$ such that (40) likely holds by instead computing an ID of the smaller matrix $\boldsymbol{Y}^{(1)}$. In subsequent iterations, we can form a sketch of $\boldsymbol{A}^{(2^{p+1}+1)}$ given $\boldsymbol{Y}^{(p)} \in \mathbb{C}^{m\times k}$, a sketch of $\boldsymbol{A}^{(2^p+1)}$, by evaluating

$$\boldsymbol{Y}^{(p+1)} = \boldsymbol{Y}^{(p)} + \widetilde{\boldsymbol{A}}^{(2^{p+1}+1)}\boldsymbol{\Omega}^{(p+1)} \in \mathbb{C}^{m\times\widetilde{n}} \tag{43}$$

where $\boldsymbol{\Omega}^{(p+1)} \in \mathbb{C}^{2^p n\times\widetilde{n}}$ is another randomized embedding with columns drawn in the same fashion as those of $\boldsymbol{\Omega}^{(1)}$. When sparse sign matrices are used, (43) can be evaluated in $O(m2^{p+1}n)$ floating point operations. Now $\boldsymbol{Y}^{(p+1)}$ may be used as a proxy for $\boldsymbol{A}^{(2^{p+1}+1)}$ in all subsequent steps of the algorithm, so the sizes of the matrices for which we must compute the ID remain constant. For example, to get an ID for $\boldsymbol{A}^{(2^{p+1}+1)}$, we can just do CPQR on $\boldsymbol{Y}^{(p+1)}$, which is what we do in our numerics. If the parameters are chosen appropriately, then there exist theoretical guarantees that state that using the sketches introduces negligible errors with high probability [30, 48, 60, 61, 72, 74]. Moreover, one can cheaply compute a "certificate of accuracy" using a small Gaussian test matrix at each iteration, which improves robustness [60, Section 12.2].

Algorithm 1 is still rather suboptimal since at each iteration the new information added consists of entire coefficient matrices $\boldsymbol{A}_\kappa$ for varying values of κ, which we already know are of low numerical rank. Therefore, at each iteration a lot of redundant information is introduced. A faster algorithm may be obtained through randomized "coordinate-wise" sampling of the wavenumbers. In this case, we fix a value of N such that $N = 2^P + 1$ for some $P \in \mathbb{N}$, let $p = 1$, and set $\boldsymbol{B}^{(1)} = \sqrt{N/3}\boldsymbol{A}^{(3)} \in \mathbb{C}^{m\times 3n}$. An ID of $\boldsymbol{B}^{(p)}$ is computed, and $\boldsymbol{U}^{(p)}$ and $\boldsymbol{I}^{(p)}$ satisfying

$$\left\|\boldsymbol{B}^{(p)} - \boldsymbol{U}^{(p)}\boldsymbol{B}^{(p)}(\boldsymbol{I}^{(p)}, :)\right\| < \varepsilon \tag{44}$$

are produced. We now draw the matrix $\boldsymbol{C}^{(p)} \in \mathbb{C}^{m\times\widetilde{n}}$ (in practice we find that $\widetilde{n} \approx 2n$ works well) where the columns of $\boldsymbol{C}^{(p)}$ are chosen from the columns of $\boldsymbol{A}^{(N)}$ selected with uniform probability and without replacement, and scaled by a factor of $\sqrt{Nn/\widetilde{n}}$. In other words, $\boldsymbol{C}^{(p)}$ is given by

$$\boldsymbol{C}^{(p)} = \sqrt{\frac{Nn}{\widetilde{n}}}\left[\boldsymbol{A}_{\kappa_1'}(:, i_1)\ \boldsymbol{A}_{\kappa_2'}(:, i_2)\ \ldots\ \boldsymbol{A}_{\kappa_{\widetilde{n}}'}(:, i_{\widetilde{n}})\right] \tag{45}$$

where $\{i_j\}_{j=1}^{\widetilde{n}}$ is randomly drawn uniformly from $\{1, 2, \ldots, Nn\}$ and $\{\kappa_j'\}_{j=1}^{\widetilde{n}}$ is randomly selected from $\{\kappa_j^{(N)}\}_{j=1}^{N}$. For our termination criterion, we check the condition

$$\max_{1\le j\le\widetilde{n}}\left\|\boldsymbol{C}^{(p)}(:, j) - \boldsymbol{U}^{(p)}\boldsymbol{C}^{(p)}(\boldsymbol{I}^{(p)}, j)\right\| < \varepsilon. \tag{46}$$

If this condition is not satisfied, we increment p, and let

$$\boldsymbol{B}^{(p)} = \frac{1}{\sqrt{3n + p\tilde{n}}} \left[\sqrt{3n + (p-1)\tilde{n}}\, \boldsymbol{B}^{(p-1)} \;\; \sqrt{\tilde{n}}\, \boldsymbol{C}^{(p-1)} \right]. \tag{47}$$

The process is then repeated with $\boldsymbol{U}^{(p)}$ and $I^{(p)}$ produced by computing an ID of $\boldsymbol{B}^{(p)}$, a new matrix $\boldsymbol{C}^{(p)}$ drawn in an identical manner as $\boldsymbol{C}^{(1)}$, and the condition (46) checked. Again, this procedure is iterated until (46) is met, and sketching may be used, in a manner analogous to how Algorithm 1 was accelerated, to keep the size of $\boldsymbol{B}^{(p)}$ effectively constant across all iterations. This is summarized in Algorithm 2.

Algorithm 2 Build universal bases (accelerated)

Input: Wavenumber interval $[\kappa_{\min}, \kappa_{\max}]$, compression tolerance $\varepsilon > 0$, and a function handle that given $\kappa \in [\kappa_{\min}, \kappa_{\max}]$ returns $\boldsymbol{A}_\kappa$
Output: Interpolation matrix $\boldsymbol{U}$ and skeleton indices I

```
B^(1) = sqrt(N/3) A^(3)
for all p = 1, 2, ... do
    Compute U^(p) and I^(p) such that (44) holds
    Form C^(p) with columns given by (45)
    if max_{j=1,2,...,ñ} ‖C^(p)(:, j) − U^(p) C^(p)(I^(p), j)‖ < ε then
        return U = U^(p) and I = I^(p)
    else
        Form B^(p+1) in (47)
    end if
end for
```

Since this algorithm depends on coordinate-wise sampling, its success is related to the concept of coherence in the matrix completion and randomized algorithm literature [17, 18, 60]. We are not aware of any results establishing coherence bounds on matrices like $\boldsymbol{A}^{(N)}$ whose entries arise from evaluating free space fundamental solutions of the Helmholtz equation. We find that in our context this algorithm is robust in practice and convergence is usually achieved in merely a couple of iterations. Therefore, we typically prefer Algorithm 2 to Algorithm 1, since it is typically faster and capable of achieving the same level of accuracy, despite the fact that the latter is on slightly better theoretical foundations.

Remark 5.1 As mentioned in Theorem 3.2, the computation of universal bases is intimately related to the low-rank approximation of certain third-order tensors in the Tucker decomposition, and in fact, Algorithms 1 and 2 can be viewed as algorithms for computing such approximations in our very specific context (cf. [54, Section 4.2]). It is also worth noting that similar ideas to those in Algorithm 1 were applied in [29, 50] in the context of function approximation.

6 Broadband Recursive Skeletonization

Applying the algorithms in Sect. 5 in recursive skeletonization is relatively straightforward. Given an interval of wavenumbers $[\kappa_{\min}, \kappa_{\max}]$ and a tree structure, we simply compute universal bases for each block that must be compressed. The method of proxy sources can of course still be used. After this precomputation, the compression stage for any wavenumber in the interval at which we wish to do RS consists purely of assembling the sibling-, self-, and near- (if applicable) interactions. In a skeletonization based-scheme, this assembly consists simply of forming submatrices of the coefficient matrix corresponding to the wavenumber. We note that since each node in the hierarchical tree can be processed independently of the others, this step is highly parallelizable. After this, any rank-structured matrix arithmetic (e.g., inversion or matrix-vector multiplication) can be done as usual. The trade-offs here are that the size of the discretization and the ranks in the rank-structured format are kept constant across $[\kappa_{\min}, \kappa_{\max}]$. In spite of this, we still achieve considerable accelerations in many regimes as illustrated in Sect. 7. We refer to the variant of RS where universal bases are computed and used, as broadband recursive skeletonization or BRS for short.

In the language of the FMM, the precomputation of the universal bases can be viewed as constructing multipole-to-multipole and local-to-local translation operators that are valid for all κ in $[\kappa_{\min}, \kappa_{\max}]$. At each κ, the multipole-to-local translation operators must, however, be computed. The trade-offs are now that the number of particles and the lengths of the multipole and local expansions must be kept constant across the interval of wavenumbers.

7 Applications and Numerical Experiments

We now illustrate BRS in various settings. The timings reported are based on experiments on a workstation with 2 Intel Xeon Gold 6254 processors (18 cores at 3.1 GHz base frequency) with 768 GB of RAM, with all codes implemented in MATLAB 2019b. We use the FMMLIB2D [39] and FMM3D [35] libraries for 2D and 3D FMMs, respectively. The zeta quadratures are computed using the codes at [75] and [76].

7.1 2D Boundary Integral Equation

For our first numerical example, we consider a standard combined field integral operator of the form (3) when the domain is the smooth, five-pointed star pictured in Fig. 9a and parametrized by $\gamma = (\gamma_1, \gamma_2) : [0, 2\pi) \to \mathbb{R}^2$ where

$$\begin{aligned} \gamma_1(t) &= 1.5\cos(t) + 0.15\cos(6t) + 0.15\cos(4t) \\ \gamma_2(t) &= \sin(t) + 0.15\sin(6t) - 0.15\sin(4t). \end{aligned} \tag{48}$$

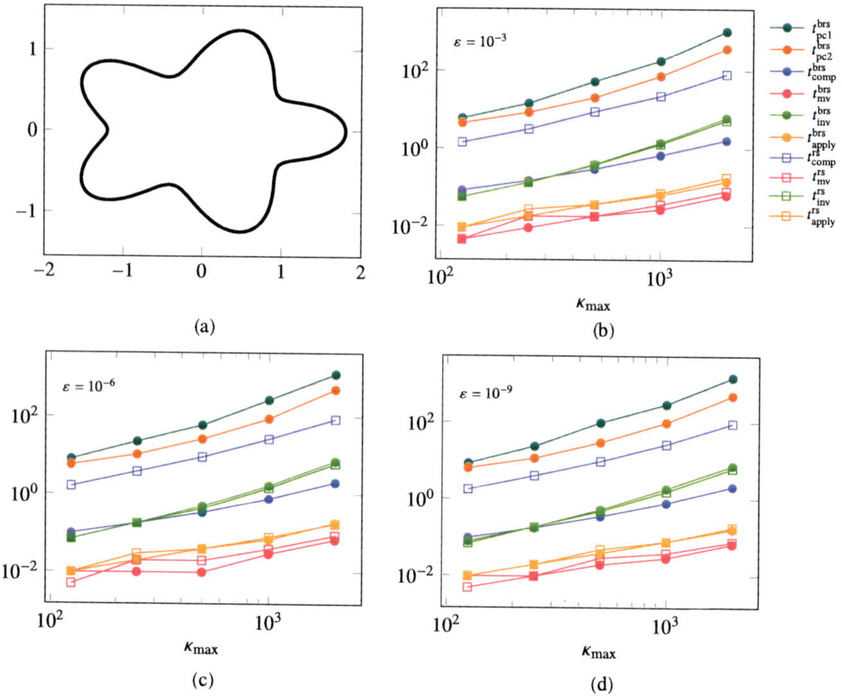

Fig. 9 (**a**) Smooth, five-pointed star geometry as considered in Sect. 7.1. The arc length of the contour is about 10.27 and its diameter is about 3.39. (**b**)–(**d**) Timing data comparing rank-structured matrix algebra when BRS is used for an interval $[1, \kappa_{\max}]$ compared to when RS is naïvely applied at just $\kappa_{\max}$ when the off-diagonal blocks are compressed to tolerances 10^{-3}, 10^{-6}, and 10^{-9}, respectively

We first study the performance and accuracy of BRS on intervals of wavenumbers $[1, \kappa_{\max}]$ for varying values of $\kappa_{\max}$. We will use 10th-order zeta quadratures [78] and take the number of degrees of freedom in our discretization for each interval to be

$$400 + \left\lceil \frac{10|\Gamma_t|\kappa_{\max}}{2\pi} \right\rceil \approx 400 + \lceil 16.34\kappa_{\max} \rceil, \tag{49}$$

which corresponds to taking 400 plus 10 points times the length of the contour measured in wavelengths at the highest frequency. The values of $\kappa_{\max}$ range from 125 to 2000. The largest interval of wavenumbers, when $\kappa_{\max} = 2000$, corresponds to domains ranging from under a wavelength to over a thousand wavelengths in diameter. For each interval, we first do a one-time computation of the universal bases. Then for each wavenumber, we obtain an HBS approximation to the coefficient matrix by simply evaluating the sibling-sibling and self-interaction matrices. After this, we use standard algorithms for fast matrix-vector products, inversion, and application of the inverse. In our experiments, we use a scattering matrix-based solver for inversion (see, for example [14]). Theoretically, in two dimensions these algorithms all scale linearly with the number of degrees of freedom, whereas inverting the rank-structured matrix scales cubically and matrix-vector products and applying the inverse scale quadratically with the wavenumber. These complexities all depend on the ranks of the various off-diagonal blocks which, since universal bases are used, stay constant across all wavenumbers. To gauge accuracy we would like to measure the error

$$\frac{\|\boldsymbol{M}_\kappa - \boldsymbol{M}_\kappa^\varepsilon\|}{\|\boldsymbol{M}_\kappa\|} \tag{50}$$

where $\boldsymbol{M}_\kappa$ is the true coefficient matrix and $\boldsymbol{M}_\kappa^\varepsilon$ is the HBS approximation when the off-diagonal blocks are compressed to tolerance ε in the spectral norm. Since computing (50) directly is prohibitively expensive, we instead use a power iteration where a high-accuracy FMM is used to compute matrix-vector products with $\boldsymbol{M}_\kappa$ and $\boldsymbol{M}_\kappa^*$. We approximate (50) in this fashion at $K = \lceil \kappa \rceil$ equispaced wavenumbers in the interval and take the maximum.

The following quantities are reported in Tables 1, 2, and 3, which consider the cases when the compression tolerance is 10^{-3}, 10^{-6}, and 10^{-9}, respectively:

$t_{\text{pc1}}^{\text{brs}}$	Time required for one-time computation of the universal bases for HBS matrices when Algorithm 1 is used
$t_{\text{pc2}}^{\text{brs}}$	Time required for one-time computation of the universal bases for HBS matrices when Algorithm 2 is used
$t_{\text{comp}}^{\text{brs}}$	Time required to obtain an HBS approximation to the coefficient matrix at each wavenumber after the universal bases are computed
$t_{\text{mv}}^{\text{brs}}$	Time required for a single matrix-vector product at each wavenumber with universal bases
$t_{\text{inv}}^{\text{brs}}$	Time required to build a rank-structured inverse at each wavenumber with universal bases
$t_{\text{apply}}^{\text{brs}}$	Time required to apply the inverse, after it is computed, at each wavenumber with universal bases
E_1^{brs}	Error metric defined in (50) when Algorithm 1 is used to compute the universal bases
E_2^{brs}	Error metric defined in (50) when Algorithm 2 is used to compute the universal bases

The data in these tables are summarized in Fig. 9b–d. In all three cases, we find roughly the same behavior. The computation of the universal bases is significantly more expensive than the other steps by roughly a couple orders of magnitude and using Algorithm 2 for this step yields a factor of 2–3 speedup over using Algorithm 1. Fortunately, this step must only be done once for the entire interval of wavenumbers. As $\kappa_{\max}$ doubles, the cost of each computation increases by a factor of 2–6, slightly better than the theoretical prediction of 8 for the computation of the universal bases and the inversion stage. Our error metric is also typically within an order of magnitude of the compression tolerance, which is standard. It is important to note that the accuracy lost is not due to our algorithms failing to obtain accurate

Table 1 Data for Sect. 7.1 when BRS is used and the compression tolerance $\varepsilon = 10^{-3}$

$[\kappa_{\min}, \kappa_{\max}]$	$t^{\text{brs}}_{\text{pc1}}$	$t^{\text{brs}}_{\text{pc2}}$	$t^{\text{brs}}_{\text{comp}}$	$t^{\text{brs}}_{\text{mv}}$	$t^{\text{brs}}_{\text{inv}}$	$t^{\text{brs}}_{\text{apply}}$	E^{brs}_1	E^{brs}_2
[1, 125]	6.18	4.82	0.09	0.00	0.06	0.01	6.38×10^{-4}	7.27×10^{-4}
[1, 250]	15.05	9.11	0.16	0.01	0.14	0.02	1.30×10^{-3}	1.73×10^{-3}
[1, 500]	55.71	22.00	0.32	0.02	0.42	0.04	2.07×10^{-3}	3.55×10^{-3}
[1, 1000]	190.94	79.78	0.73	0.03	1.51	0.07	5.06×10^{-3}	5.60×10^{-3}
[1, 2000]	1110.31	404.08	1.97	0.07	6.59	0.16	9.48×10^{-3}	2.40×10^{-2}

Table 2 Data for Sect. 7.1 when BRS is used and the compression tolerance $\varepsilon = 10^{-6}$

$[\kappa_{\min}, \kappa_{\max}]$	$t^{\text{brs}}_{\text{pc1}}$	$t^{\text{brs}}_{\text{pc2}}$	$t^{\text{brs}}_{\text{comp}}$	$t^{\text{brs}}_{\text{mv}}$	$t^{\text{brs}}_{\text{inv}}$	$t^{\text{brs}}_{\text{apply}}$	E^{brs}_1	E^{brs}_2
[1, 125]	8.04	5.89	0.10	0.01	0.07	0.01	9.92×10^{-7}	7.36×10^{-7}
[1, 250]	22.70	10.78	0.18	0.01	0.18	0.02	1.42×10^{-6}	2.23×10^{-6}
[1, 500]	60.37	27.28	0.34	0.01	0.50	0.04	2.73×10^{-6}	1.59×10^{-6}
[1, 1000]	271.37	92.12	0.78	0.03	1.69	0.07	5.67×10^{-6}	6.39×10^{-6}
[1, 2000]	1264.17	527.51	2.09	0.07	7.31	0.19	8.36×10^{-6}	1.10×10^{-5}

Table 3 Data for Sect. 7.1 when BRS is used and the compression tolerance $\varepsilon = 10^{-9}$

$[\kappa_{\min}, \kappa_{\max}]$	$t^{\text{brs}}_{\text{pc1}}$	$t^{\text{brs}}_{\text{pc2}}$	$t^{\text{brs}}_{\text{comp}}$	$t^{\text{brs}}_{\text{mv}}$	$t^{\text{brs}}_{\text{inv}}$	$t^{\text{brs}}_{\text{apply}}$	E^{brs}_1	E^{brs}_2
[1, 125]	8.65	6.66	0.10	0.01	0.08	0.01	1.39×10^{-9}	1.54×10^{-9}
[1, 250]	24.20	12.12	0.18	0.01	0.19	0.02	3.24×10^{-9}	1.98×10^{-9}
[1, 500]	100.62	31.44	0.36	0.02	0.54	0.04	2.41×10^{-9}	4.51×10^{-9}
[1, 1000]	303.56	103.99	0.81	0.03	1.89	0.08	6.84×10^{-9}	4.89×10^{-9}
[1, 2000]	1525.61	527.21	2.16	0.07	7.59	0.17	6.70×10^{-9}	4.20×10^{-9}

universal bases but rather the accumulation of errors that would occur even at a single wavenumber. This being the case, the error grows with the number of levels in the tree, which is why the errors increase with $\kappa_{\max}$. There also does not seem to be a substantial difference in accuracy between when Algorithms 1 and 2 are used. In subsequent experiments we will only use Algorithm 2 since it is usually faster. The dependence of the timings on the compression tolerance ε is fairly weak with the cost of each computation growing substantially slower than $\log(1/\varepsilon)$.

To put these numbers into perspective, we compare them with the analogous values when regular RS is done at just the highest wavenumber. This is the focus of Tables 4, 5, and 6, which tabulate the quantities:

$t^{\text{rs}}_{\text{comp}}$	Time required to obtain an HBS approximation to the coefficient matrix without universal bases
$t^{\text{rs}}_{\text{mv}}$	Time required for a single matrix-vector product without universal bases
$t^{\text{rs}}_{\text{inv}}$	Time required to build a rank-structured inverse without universal bases
$t^{\text{rs}}_{\text{apply}}$	Time required to apply the inverse after it is computed without universal bases
E^{rs}	Error metric defined in (50) without universal bases

Table 4 Data for Sect. 7.1 when regular RS compression is done at a single wavenumber and with compression tolerance $\varepsilon = 10^{-3}$

κ	t^{rs}_{comp}	t^{rs}_{mv}	t^{rs}_{inv}	t^{rs}_{apply}	E^{rs}
125	1.50	0.00	0.06	0.01	1.44×10^{-3}
250	3.32	0.02	0.14	0.03	1.18×10^{-3}
500	9.35	0.02	0.40	0.04	2.92×10^{-3}
1000	23.85	0.04	1.38	0.08	2.89×10^{-3}
2000	87.29	0.09	5.64	0.20	4.53×10^{-3}

Table 5 Data for Sect. 7.1 when regular RS compression is done at a single wavenumber and with compression tolerance $\varepsilon = 10^{-6}$

κ	t^{rs}_{comp}	t^{rs}_{mv}	t^{rs}_{inv}	t^{rs}_{apply}	E^{rs}
125	1.62	0.00	0.07	0.01	2.73×10^{-6}
250	3.86	0.02	0.18	0.03	2.19×10^{-6}
500	9.22	0.02	0.43	0.04	9.06×10^{-6}
1000	26.94	0.04	1.47	0.08	5.21×10^{-6}
2000	87.68	0.09	6.23	0.18	1.05×10^{-5}

Table 6 Data for Sect. 7.1 when regular RS compression is done at a single wavenumber and with compression tolerance $\varepsilon = 10^{-9}$

κ	t^{rs}_{comp}	t^{rs}_{mv}	t^{rs}_{inv}	t^{rs}_{apply}	E^{rs}
125	1.80	0.00	0.07	0.01	1.14×10^{-9}
250	4.13	0.01	0.19	0.02	2.24×10^{-9}
500	9.93	0.03	0.49	0.05	3.90×10^{-9}
1000	27.77	0.04	1.58	0.08	4.29×10^{-9}
2000	97.74	0.08	6.37	0.19	1.67×10^{-8}

The data in these tables are also summarized in Fig. 9b–d. We first examine the cost of inverting the HBS matrices. Across the board, this is about 20% cheaper than with BRS. This is, of course, to be expected; the universal bases computed in BRS are valid at $\kappa_{\max}$, so the ranks when compression is done at just $\kappa_{\max}$ are necessarily lower and therefore we should expect inversion to be faster. The same reasoning applies to the cost of matrix-vector products and applying the inverse, but since these computations are light on arithmetic, simply consisting of a series of matrix-vector products, communication costs are more pronounced. As a result, these costs are observed to be essentially the same as in BRS.

It is important to compare the compression cost in regular RS with the corresponding costs in BRS. In all of these experiments, the compression cost in regular RS is substantially less than the cost of computing the universal bases for BRS. Even when the accelerated algorithm is used to compute the universal bases, this difference ranges from a factor of 4–6. However, if the universal bases have already been computed, then forming an HBS approximation to the coefficient matrix using BRS is about a factor of 10–40 faster than the cost of computing it from scratch. Since the compression cost dominates the inversion cost, this suggests that BRS can yield a significant speedup in the per wavenumber cost in multi-frequency applications.

Of course, when regular RS is used at each wavenumber the rank-structured matrix algebra costs depend on the wavenumber, and this dependence is theoretically quadratic or cubic. This is in contrast to BRS where these costs are effectively constant for an entire interval. To gain a practical speedup, the constants in the complexities are important. We will more closely study this trade-off by focusing on the case when $\kappa_{\max} = 1000$ and examining how the costs associated with BRS (displayed in the fourth row of Tables 1, 2, and 3) compare to the corresponding *average costs* of dense linear algebra and regular RS in various subintervals of $[1, 1000]$. For this the size of the discretization should be chosen based on the wavenumber, so for $\kappa \in [1, 1000]$ we will always take $400 + \lceil 16.34\kappa \rceil$ degrees of freedom. In Tables 7 and 8, we tabulate the following quantities:

$t_{\text{form}}^{\text{dense}}$	Average time required to assemble the full dense coefficient matrix for $\kappa \in [\kappa_{\min}, \kappa_{\max}]$
$t_{\text{mv}}^{\text{dense}}$	Average time required for a single matrix-vector product after dense coefficient matrix is formed for $\kappa \in [\kappa_{\min}, \kappa_{\max}]$
$t_{\text{LU}}^{\text{dense}}$	Average time required to compute a dense LU factorization of the coefficient matrix for $\kappa \in [\kappa_{\min}, \kappa_{\max}]$
$t_{\text{apply}}^{\text{dense}}$	Average time required to do a single backward and forward substitution after the dense LU factors have been computed for $\kappa \in [\kappa_{\min}, \kappa_{\max}]$
$t_{\text{comp}}^{\text{rs}}$	Average time required to compute an HBS approximation of the coefficient matrix for $\kappa \in [\kappa_{\min}, \kappa_{\max}]$ without universal bases
$t_{\text{mv}}^{\text{rs}}$	Average time required to do a single matrix-vector product with the HBS matrix for $\kappa \in [\kappa_{\min}, \kappa_{\max}]$ without universal bases
$t_{\text{inv}}^{\text{rs}}$	Average time required to build an inverse of an HBS matrix for $\kappa \in [\kappa_{\min}, \kappa_{\max}]$
$t_{\text{apply}}^{\text{rs}}$	Average time required to apply the inverse, after it is computed, for $\kappa \in [\kappa_{\min}, \kappa_{\max}]$ without universal bases

Let us consider a couple of concrete cases to interpret this data. Suppose we are interested in solving a single integral equation at K different wavenumbers in the interval $[1, 1000]$ where the wavenumbers are chosen fairly uniformly. If we were to use dense linear algebra, then we would expect the total cost to be approximately K times $t_{\text{form}}^{\text{dense}} + t_{\text{LU}}^{\text{dense}} + t_{\text{LU}}^{\text{dense}} \approx 8.57$ s. On the other hand, if we were to do regular RS with $\varepsilon = 10^{-9}$ at each iteration, then we would expect the cost to be approximately K times $t_{\text{comp}}^{\text{rs}} + t_{\text{inv}}^{\text{rs}} + t_{\text{apply}}^{\text{rs}} \approx 11.16$ s. From this it follows that in this case, there would be no point in doing regular RS over dense linear algebra. For BRS with

Table 7 Data for Sect. 7.1 showing average times required by dense linear algebra and regular RS ($\varepsilon = 10^{-3}$) are done for various intervals of wavenumbers

	Dense				RS ($\varepsilon = 10^{-3}$)			
$[\kappa_{\min}, \kappa_{\max}]$	$t_{\text{form}}^{\text{dense}}$	$t_{\text{mv}}^{\text{dense}}$	$t_{\text{LU}}^{\text{dense}}$	$t_{\text{apply}}^{\text{dense}}$	$t_{\text{comp}}^{\text{rs}}$	$t_{\text{mv}}^{\text{rs}}$	$t_{\text{inv}}^{\text{rs}}$	$t_{\text{apply}}^{\text{rs}}$
[1, 1000]	3.84	0.02	4.41	0.32	9.71	0.01	0.52	0.03
[500, 1000]	6.36	0.04	6.71	0.50	14.96	0.02	0.82	0.05
[750, 1000]	9.35	0.04	10.30	0.70	17.71	0.03	1.07	0.06
[875, 1000]	10.31	0.06	12.18	0.78	19.07	0.03	1.17	0.06
[999, 1000]	11.17	0.06	14.24	0.80	23.73	0.03	1.38	0.06

Table 8 Data for Sect. 7.1 showing average times required by regular RS ($\varepsilon = 10^{-6}$ and $\varepsilon = 10^{-9}$) are done for various intervals of wavenumbers

	RS ($\varepsilon = 10^{-6}$)				RS ($\varepsilon = 10^{-9}$)			
$[\kappa_{\min}, \kappa_{\max}]$	$t_{\text{comp}}^{\text{rs}}$	$t_{\text{mv}}^{\text{rs}}$	$t_{\text{inv}}^{\text{rs}}$	$t_{\text{apply}}^{\text{rs}}$	$t_{\text{comp}}^{\text{rs}}$	$t_{\text{mv}}^{\text{rs}}$	$t_{\text{inv}}^{\text{rs}}$	$t_{\text{apply}}^{\text{rs}}$
[1, 1000]	10.28	0.02	0.56	0.03	10.50	0.02	0.62	0.04
[500, 1000]	15.70	0.02	0.90	0.05	16.20	0.02	0.98	0.05
[750, 1000]	18.88	0.03	1.18	0.07	19.43	0.03	1.27	0.07
[875, 1000]	20.74	0.03	1.33	0.07	22.61	0.03	1.45	0.07
[999, 1000]	24.18	0.03	1.48	0.07	27.57	0.03	1.61	0.08

$\varepsilon = 10^{-9}$ the total cost would be the time it takes to compute the universal bases (103.99 s) plus K times $t_{\text{comp}}^{\text{brs}} + t_{\text{inv}}^{\text{brs}} + t_{\text{apply}}^{\text{brs}} \approx 2.78$ s. Per wavenumber, BRS is about 3 times faster than dense linear algebra and 4 times faster than regular RS, and for sufficiently large K (in this case, $K \geq 18$), BRS yields a lower total running time. In cases where we restrict ourselves to a narrower interval around $\kappa_{\max} = 1000$ or where the wavenumbers are drawn more densely at higher wavenumbers, the advantages of BRS grow more pronounced. For example, if we were to stay in the same setting but instead were to consider the wavenumber interval [999, 1000], the average per wavenumber costs of dense linear algebra and regular RS would change to 26.21 s and 29.26 s, respectively, but the per wavenumber cost associated with BRS would essentially remain the same. The average per wavenumber speedup with BRS would then be a factor of about 10 when compared with both methods and the break-even point for the universal bases computation would be at about $K = 5$.

Remark 7.1 While the above analysis has shown that the strategy of applying BRS for a large interval of wavenumbers such as [1, 1000] can still yield practical acceleration, this is a rather crude strategy. Better practical performance could be achieved by partitioning $[\kappa_{\min}, \kappa_{\max}]$ into a series of subintervals and then applying BRS to each interval. This would lower the time required to compute the universal bases as well. How this partitioning should be done is very application dependent, so we do not explore this matter further here.

7.2 2D Eigenvalue Problems

Closely related to scattering problems is the problem of finding the eigenvalues of the Laplacian. One approach to this is to formulate the problem as a standard boundary integral equation for an interior Helmholtz equation and solve the resulting nonlinear eigenvalue problem to determine the wavenumbers at which the integral operator is singular. These wavenumbers are called the *eigenfrequencies* and are the square roots of the eigenvalues. In this section, we will consider the problem of finding the eigenfrequencies in given intervals $[\kappa_{\min}, \kappa_{\max}]$ of the drum domain

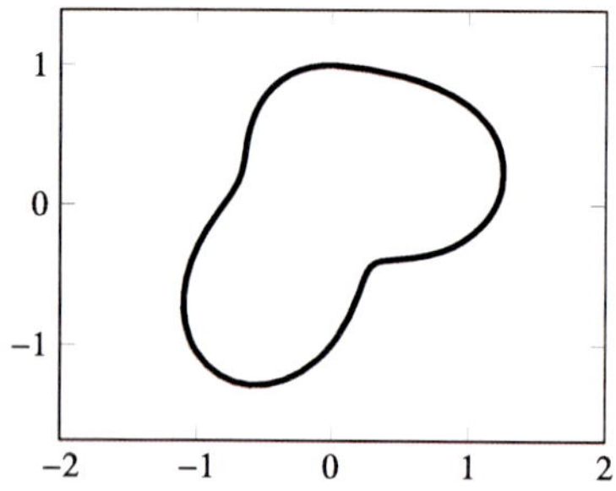

(a)

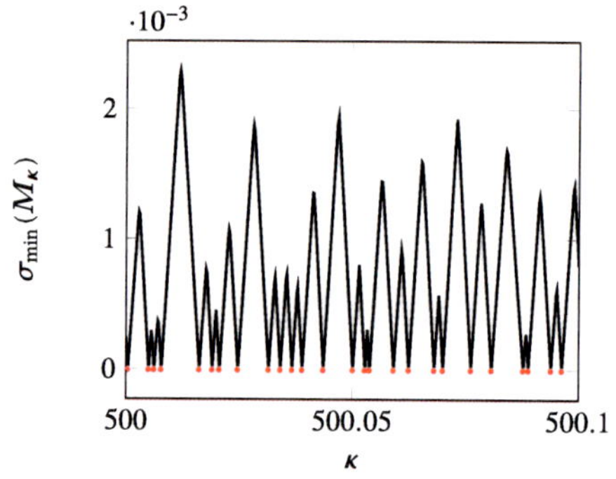

(b)

Fig. 10 (**a**) Drum domain. The diameter of the domain is about 2.69 and the arc length of its boundary is approximately 7.29. (**b**) Plot of the smallest singular value of a discretization of the operator (51) for $\kappa \in [500, 500.1]$. The red dots denote the computed eigenfrequencies, which are the squares of the eigenvalues

introduced in [80], pictured in Fig. 10a, and whose boundary is parametrized by $r(\theta) = 1 + 0.2 \cos 3\theta + 0.3 \sin 2\theta$ for $\theta \in [0, 2\pi)$.

We shall accomplish this by finding κ such that the operator $A_\kappa : L^2(\Gamma) \to L^2(\Gamma)$ given by

$$[A_\kappa \sigma](x) = \frac{1}{2}\sigma(x) - \int_\Gamma \frac{\partial G_\kappa}{\partial n(y)}(x, y)\sigma(y)\, ds(y) \tag{51}$$

is singular. There exist many different methods of solving this nonlinear eigenvalue problem, and here we will take one of the simplest approaches, which was used in [58, Section 6.3]. To this end let $\boldsymbol{M}_\kappa$ denote a discretization of A_κ in the manner described in Sect. 2.2. We use 16th-order zeta quadratures [78], and for a given interval $[\kappa_{\min}, \kappa_{\max}]$, we will take the number of points in our discretization to be $\lceil 10|\Gamma|\kappa_{\max}/2\pi \rceil \approx 11.60\kappa_{\max}$. We will find the eigenfrequencies by finding the values of κ where $\boldsymbol{M}_\kappa$ becomes singular. This in turn will be accomplished by root-finding on the function $f(\kappa) = \sigma_{\min}(\boldsymbol{M}_\kappa)$.

When $\kappa_{\max}$ is small, it is efficient to simply form the dense matrix $\boldsymbol{M}_\kappa$ and to compute its smallest singular value using bidiagonalization and the QR algorithm. For larger values of $\kappa_{\max}$, it can be faster to utilize direct solver techniques. We shall do this by computing an HBS approximation to $\boldsymbol{M}_\kappa$ for each wavenumber, inverting it using the HBS-LU/RSF algorithm [53, 57], and then using a power iteration. Naturally, BRS can be used to accelerate this first step. We apply dense linear algebra, regular RS, and BRS to the problem of finding the eigenfrequencies of the drum in the intervals [500, 500.1] and [1000, 1000.1]. For regular RS and BRS, we set the compression tolerance to 10^{-10}. The root-finding is done by initially evaluating f on a coarse grid in $[\kappa_{\min}, \kappa_{\max}]$ and finding local minima, followed by a refinement step, which uses a quasi-Newton algorithm as implemented in

Table 9 Data for Sect. 7.2

	$t_{\text{pc2}}^{\text{brs}}$	$t_{\text{sv}}^{\text{brs}}$	$t_{\text{sv}}^{\text{rs}}$	$t_{\text{sv}}^{\text{dense}}$
[500, 500.1] (26 eigenvalues)	12.35	1.35	7.10	8.64
[1000, 1000.1] (52 eigenvalues)	30.33	3.01	18.39	42.13

MATLAB's `fminunc` algorithm. The resulting solver finds all eigenfrequencies in our two intervals to six correct digits after the decimal in about 72 function evaluations per root. The performance of the various methods is summarized in Table 9, which tabulates the following quantities:

- $t_{\text{sv}}^{\text{brs}}$ Average time required to compute the smallest singular value by forming an HBS matrix using the universal bases, inverting the rank-structured matrix using an HBS-LU factorization, and then a power iteration in the interval
- $t_{\text{sv}}^{\text{rs}}$ Average time required to compute the smallest singular value by forming an HBS matrix from scratch at each wavenumber, inverting the rank-structured matrix using an HBS-LU factorization, and then a power iteration in the interval
- $t_{\text{sv}}^{\text{dense}}$ Average time required to compute the smallest singular value by forming the dense coefficient matrix, and then bidiagonalization and QR in the interval

We see that each evaluation of f when BRS is used is about a factor of 5 on the interval [500, 500.1] and a factor of 6 on the interval [1000, 1000.1] faster than when regular RS is used, which is in turn faster than dense linear algebra. Also, the one-time cost of computing the universal bases is negligible in light of the hundreds of function evaluations we are doing. The eigenfrequencies and a plot of the function f on the interval [500, 500.1] are shown in Fig. 10b.

Remark 7.2 The optimization strategy employed here to solve the nonlinear eigenvalue problem is rather primitive. A better approach would be to follow the approach developed in [80] and extended in [5] and compute the Fredholm determinant of $2A_\kappa$ where A_κ is given in (51). This is the subject of ongoing work.

7.3 3D Boundary Integral Equation

Our last example will be a simple extension of the experiments in Sect. 7.1 to three dimensions. We consider a torus pictured in Fig. 11a and which can be parametrized by $\gamma = (\gamma_1, \gamma_2, \gamma_3) : [0, 2\pi) \times [0, 2\pi) \to \mathbb{R}^3$ where

$$\begin{aligned} \gamma_1(\theta, \phi) &= (2 + \cos(\theta)) \cos(\phi) \\ \gamma_2(\theta, \phi) &= (2 + \cos(\theta)) \sin(\phi) \\ \gamma_2(\theta, \phi) &= \sin(\theta). \end{aligned} \tag{52}$$

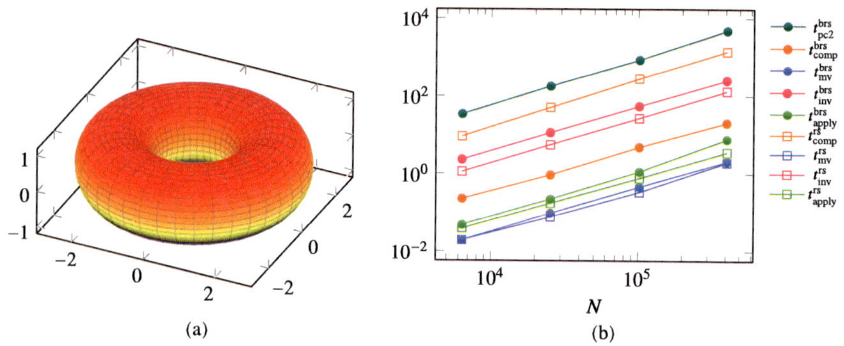

Fig. 11 (**a**) Torus geometry. The diameter of the domain is about 6. (**b**) Timing data for Sect. 7.3

Table 10 Data for Sect. 7.3 when BRS is used

n	$t_{\text{pc2}}^{\text{brs}}$	$t_{\text{comp}}^{\text{brs}}$	$t_{\text{mv}}^{\text{brs}}$	$t_{\text{inv}}^{\text{brs}}$	$t_{\text{apply}}^{\text{brs}}$	E^{brs}
6400	35.10	0.24	0.02	2.45	0.05	7.77×10^{-6}
25,600	184.64	0.99	0.10	12.00	0.23	1.38×10^{-5}
102,400	877.95	5.15	0.48	57.98	1.18	3.21×10^{-5}
409,600	4980.86	21.54	2.18	268.97	8.05	2.00×10^{-3}

Table 11 Data for Sect. 7.3 when regular RS is used

n	$t_{\text{comp}}^{\text{rs}}$	$t_{\text{mv}}^{\text{rs}}$	$t_{\text{inv}}^{\text{rs}}$	$t_{\text{apply}}^{\text{rs}}$	E^{rs}
6400	9.65	0.02	1.19	0.04	3.53×10^{-5}
25,600	53.40	0.08	5.90	0.18	1.98×10^{-4}
102,400	294.41	0.35	28.37	0.80	1.10×10^{-3}
409,600	1489.03	2.04	140.83	3.71	5.33×10^{-3}

Note that the torus is about 6 units in diameter. We use 3rd-order zeta quadratures [77] for simplicity since these are diagonal, but the timings would be similar for higher-order quadratures. We will compare BRS with regular RS as the number of degrees of freedom increases when HBS matrices are used with compression tolerance $\varepsilon = 10^{-5}$, $\kappa_{\text{min}} = 1$, and $\kappa_{\text{max}} = 10$. The timings for BRS are shown in Table 10 and regular RS in Table 11, and the timings are illustrated in Fig. 11b.

We see that the compression and inversion costs of both regular RS and BRS are slightly better than the theoretical asymptotic $O(n^{3/2})$ complexity. In every case the cost of regular RS compression is about a factor of 4 less than the cost of computing the universal bases in BRS. However, after the universal bases have been computed, the cost of the compression step at each wavenumber in BRS is about 40–70 times cheaper than that of regular RS. Given the generally higher ranks in 3D, we see that now there is a substantial increase in the subsequent matrix algebra costs in BRS over regular RS. Inversion, for example, is consistently about a factor of 2 slower. Nonetheless, this trade-off will still yield practical advantages to using BRS in a broad range of applications. Restricting ourselves to the case where there are 102,400 degrees of freedom, the cost per wavenumber to do a single compression, inversion, and application of the inverse using BRS is approximately $t^{\mathrm{brs}}_{\mathrm{comp}} + t^{\mathrm{brs}}_{\mathrm{inv}} + t^{\mathrm{brs}}_{\mathrm{apply}} \approx 64.30$ s. The analogous cost for regular RS is $t^{\mathrm{rs}}_{\mathrm{comp}} + t^{\mathrm{rs}}_{\mathrm{inv}} + t^{\mathrm{rs}}_{\mathrm{apply}} \approx 323.58$ s, which is about a factor of 5 greater. Moreover, the cost of computing the universal bases is compensated for after only about 4 wavenumbers.

8 Conclusions and Future Work

We presented a new technique for accelerating the compression stage of direct solvers based on the recursive skeletonization technique in broadband applications. The technique is based on the observation that universal bases, which are valid for an entire interval of wavenumbers, can be computed for all necessary blocks in a rank-structured format, often with only a minor increase in the ranks. Efforts to improve the algorithms in Sect. 5 using ideas in the tensor decomposition literature as well as to develop an adaptive wavenumber partitioning strategy (cf. Theorem 7.1) are currently underway. Furthermore, the application of this technique to time-dependent wave equation solvers, eigenvalue finders, and transmission problems is also currently being pursued.

Acknowledgments The authors would like to thank Alex Barnett and Vladimir Rokhlin for valuable discussions about the topics under consideration. The work reported was supported by the Office of Naval Research (N00014-18-1-2354), by the National Science Foundation (DMS-1952735 and DMS-2012606), and by the Department of Energy ASCR (DE-SC0022251).

References

1. Abramowitz, M., Stegun, I.A.: Handbook of Mathematical Functions with Formulas, Graphs, and Mathematical Tables, vol. 55. US Government Printing Office, Washington (1964)
2. Ambikasaran, S., Borges, C., Imbert-Gerard, L.M., Greengard, L.: Fast, adaptive, high-order accurate discretization of the Lippmann–Schwinger equation in two dimensions. SIAM J. Sci. Comput. **38**(3), A1770–A1787 (2016)
3. Anderson, T.G., Bruno, O.P., Lyon, M.: High-order, dispersionless "fast-hybrid" wave equation solver. Part I: $O(1)$ sampling cost via incident-field windowing and recentering. SIAM J. Sci. Comput. **42**(2), A1348–A1379 (2020)
4. Anselone, P.: Collectively Compact Operator Approximation Theory. Prentice Hall Series in Automatic Computation. Prentice Hall, Hoboken (1971)
5. Askham, T., Rachh, M.: A boundary integral equation approach to computing eigenvalues of the Stokes operator. Adv. Comput. Math. **46**(2), 1–42 (2020)
6. Ballani, J., Kressner, D.: Matrices with hierarchical low-rank structures. In: Exploiting Hidden Structure in Matrix Computations: Algorithms and Applications, pp. 161–209. Springer, Berlin (2016)
7. Bebendorf, M.: Hierarchical Matrices. Springer, Berlin/Heidelberg (2008)

8. Borges, C., Gillman, A., Greengard, L.: High resolution inverse scattering in two dimensions using recursive linearization. SIAM J. Imag. Sci. **10**(2), 641–664 (2017)
9. Börm, S.: $\mathcal{H}^2$-matrix arithmetics in linear complexity. Computing **77**(1), 1–28 (2006)
10. Börm, S.: Construction of data-sparse $\mathcal{H}^2$-matrices by hierarchical compression. SIAM J. Sci. Comput. **31**(3), 1820–1839 (2009)
11. Börm, S.: Efficient Numerical Methods for Non-local Operators: $\mathcal{H}^2$-matrix Compression, Algorithms and Analysis. Tracts in Mathematics, vol. 14. European Mathematical Society, Madralin (2010)
12. Börm, S., Grasedyck, L., Hackbusch, W.: Hierarchical Matrices. Lecture Notes, Max Planck Institute for Mathematics in the Sciences, vol. 21 (2003)
13. Bremer, J.: On the Nyström discretization of integral equations on planar curves with corners. Appl. Comput. Harmonic Anal. **32**(1), 45–64 (2012)
14. Bremer, J., Gillman, A., Martinsson, P.G.: A high-order accurate accelerated direct solver for acoustic scattering from surfaces. BIT Numer. Math. **55**(2), 367–397 (2015)
15. Bucci, O.M., Franceschetti, G.: On the spatial bandwidth of scattered fields. IEEE Trans. Antennas Propag. **35**(12), 1445–1455 (1987)
16. Bucci, O.M., Franceschetti, G.: On the degrees of freedom of scattered fields. IEEE Trans. Antennas Propag. **37**(7), 918–926 (1989)
17. Candès, E.J., Tao, T.: The power of convex relaxation: near-optimal matrix completion. IEEE Trans. Inf. Theory **56**(5), 2053–2080 (2010)
18. Candès, E.J., Li, X., Ma, Y., Wright, J.: Robust principal component analysis? J. ACM **58**(3), 1–37 (2011)
19. Chandrasekaran, S., Gu, M., Lyons, W.: A fast adaptive solver for hierarchically semiseparable representations. Calcolo **42**(3), 171–185 (2005)
20. Chandrasekaran, S., Gu, M., Pals, T.: A fast ULV decomposition solver for hierarchically semiseparable representations. SIAM J. Matrix Anal. Appl. **28**(3), 603–622 (2006)
21. Chandrasekaran, S., Dewilde, P., Gu, M., Lyons, W., Pals, T.: A fast solver for HSS representations via sparse matrices. SIAM J. Matrix Anal. Appl. **29**(1), 67–81 (2006)
22. Chen, Y.: A fast, direct algorithm for the Lippmann–Schwinger integral equation in two dimensions. Adv. Comput. Math. **16**(2), 175–190 (2002)
23. Cheng, H., Gimbutas, Z., Martinsson, P.G., Rokhlin, V.: On the compression of low rank matrices. SIAM J. Sci. Comput. **26**(4), 1389–1404 (2005)
24. Chew, W.C., Jin, J.M., Michielssen, E.: Fast and Efficient Algorithms in Computational Electromagnetics. Artech House, Norwood (2001)
25. Cohen, M.B.: Nearly tight oblivious subspace embeddings by trace inequalities. In: Proceedings of the Twenty-Seventh Annual ACM-SIAM Symposium on Discrete Algorithms, pp. 278–287. SIAM, Philadelphia (2016)
26. Colton, D., Kress, R.: Integral Equation Methods in Scattering Theory. Classics in Applied Mathematics. SIAM, Philadelphia (2013)
27. Colton, D., Kress, R.: Inverse Acoustic and Electromagnetic Scattering Theory, vol. 93, 4th edn. Springer, New York (2019)
28. Crutchfield, W., Gimbutas, Z., Greengard, L., Huang, J., Rokhlin, V., Yarvin, N., Zhao, J.: Remarks on the implementation of wideband FMM for the Helmholtz equation in two dimensions. Contemp. Math. **408**(01) (2006)
29. Dolgov, S., Kressner, D., Strossner, C.: Functional Tucker approximation using Chebyshev interpolation. SIAM J. Sci. Comput. **43**(3), A2190–A2210 (2021)
30. Dong, Y., Martinsson, P.G.: Simpler is better: a comparative study of randomized algorithms for computing the CUR decomposition (2021). Preprint, arXiv:2104.05877
31. Engquist, B., Ying, L.: Fast directional multilevel algorithms for oscillatory kernels. SIAM J. Sci. Comput. **29**(4), 1710–1737 (2007)
32. Engquist, B., Ying, L., et al.: A fast directional algorithm for high frequency acoustic scattering in two dimensions. Commun. Math. Sci. **7**(2), 327–345 (2009)
33. Engquist, B., Zhao, H.: Approximate separability of the Green's function of the Helmholtz equation in the high frequency limit. Commun. Pure Appl. Math. **71**(11), 2220–2274 (2018)

34. Gillman, A., Young, P., Martinsson, P.G.: A direct solver with $O(N)$ complexity for integral equations on one-dimensional domains. Front. Math. China **7**(2), 217–247 (2012)
35. Gimbutas, Z., Greengard, L., Lu, L., Jeremy Magland, D.M., O'Neil, M., Rachh, M., Rokhlin, V.: FMM3D. Release 1.0.0, https://github.com/flatironinstitute/FMM3D
36. Gopal, A., Martinsson, P.G.: An accelerated, high-order accurate direct solver for the Lippmann-Schwinger equation for acoustic scattering in the plane. Adv. Comput. Math. **48**(4), 1–31 (2022)
37. Grasedyck, L., Hackbusch, W.: Construction and arithmetics of $\mathcal{H}$-matrices. Computing **70**(4), 295–334 (2003)
38. Greengard, P.: On generalized prolate spheroidal functions. Ph.D. Thesis, Yale University (2019)
39. Greengard, L., Gimbutas, Z.: FMMLIB2D (2021). https://github.com/zgimbutas/fmmlib2d, Version 1.2.3
40. Greengard, L., Gueyffier, D., Martinsson, P.G., Rokhlin, V.: Fast direct solvers for integral equations in complex three-dimensional domains. Acta Numer. 1–33 (2009)
41. Greengard, L., Rokhlin, V.: A fast algorithm for particle simulations. J. Comput. Phys. **73**(2), 325–348 (1987)
42. Greengard, L., Rokhlin, V.: On the numerical solution of two-point boundary value problems. Commun. Pure Appl. Math. **44**(4), 419–452 (1991)
43. Guo, H., Liu, Y., Hu, J., Michielssen, E.: A butterfly-based direct integral-equation solver using hierarchical LU factorization for analyzing scattering from electrically large conducting objects. IEEE Trans. Antennas Propag. **65**(9), 4742–4750 (2017)
44. Hackbusch, W.: A sparse matrix arithmetic based on $\mathcal{H}$-matrices. Part I: introduction to $\mathcal{H}$-matrices. Computing **62**(2), 89–108 (1999)
45. Hackbusch, W.: Hierarchical Matrices: Algorithms and Analysis, vol. 49. Springer, Berlin/Heidelberg (2015)
46. Hackbusch, W., Khoromskij, B.N.: A sparse $\mathcal{H}$-matrix arithmetic. Part II: application to multi-dimensional problems. Computing **64**(1), 21–47 (2000)
47. Hackbusch, W., Khoromskij, B., Sauter, S.A.: On $\mathcal{H}^2$-Matrices. Lectures on Applied Mathematics, pp. 9–29. Springer, Berlin/Heidelberg (2000)
48. Halko, N., Martinsson, P.G., Tropp, J.A.: Finding structure with randomness: probabilistic algorithms for constructing approximate matrix decompositions. SIAM Rev. **53**(2), 217–288 (2011)
49. Hao, S., Barnett, A.H., Martinsson, P.G., Young, P.: High-order accurate methods for Nyström discretization of integral equations on smooth curves in the plane. Adv. Comput. Math. **40**(1), 245–272 (2014)
50. Hashemi, B., Trefethen, L.N.: Chebfun in three dimensions. SIAM J. Sci. Comput. **39**(5), C341–C363 (2017)
51. Ho, K., Greengard, L.: A fast direct solver for structured linear systems by recursive skeletonization. SIAM J. Sci. Comput. **34**(5), A2507–A2532 (2012)
52. Ho, K., Greengard, L.: A fast semi-direct least squares algorithm for hierarchically block separable matrices. SIAM J. Matrix Anal. Appl. **35**(2), 725–748 (2014)
53. Ho, K., Ying, L.: Hierarchical interpolative factorization for elliptic operators: integral equations. Commun. Pure Appl. Math. **69**, 1314–1353 (2016)
54. Kolda, T.G., Bader, B.W.: Tensor decompositions and applications. SIAM Rev. **51**(3), 455–500 (2009)
55. Kong, W.Y., Bremer, J., Rokhlin, V.: An adaptive fast direct solver for boundary integral equations in two dimensions. Appl. Comput. Harmon. Anal. **31**(3), 346–369 (2011)
56. Kress, R.: Linear Integral Equations. Applied Mathematical Sciences, vol. 82, 3rd edn. Springer, New York (2014)
57. Martinsson, P.G.: Fast Direct Solvers for Elliptic PDEs. CBMS–NSF Conference Series, vol. 96. SIAM (2019)
58. Martinsson, P.G., Rokhlin, V.: A fast direct solver for boundary integral equations in two dimensions. J. Comput. Phys. **205**(1), 1–23 (2005)

59. Martinsson, P.G., Rokhlin, V.: A fast direct solver for scattering problems involving elongated structures. J. Comput. Phys. **221**(1), 288–302 (2007)
60. Martinsson, P.G., Tropp, J.A.: Randomized numerical linear algebra: foundations and algorithms. Acta Numer. **29**, 403–572 (2020). https://doi.org/10.1017/S0962492920000021
61. Martinsson, P.G., Rokhlin, V., Tygert, M.: A randomized algorithm for the decomposition of matrices. Appl. Comput. Harmon. Anal. **30**(1), 47–68 (2011)
62. Michielssen, E., Boag, A., Chew, W.C.: Scattering from elongated objects: direct solution in $O(N \log^2 N)$ operations. IEE Proc. Microwaves Antennas Propag. **143**(4), 277–283 (1996)
63. Minden, V., Ho, K., Damle, A., Ying, L.: A recursive skeletonization factorization based on strong admissibility. Multiscale Model. Simul. **15**(2), 768–796 (2017)
64. Osipov, A., Rokhlin, V., Xiao, H., et al.: Prolate Spheroidal Wave Functions of Order Zero. Springer Series in Applied Mathematical Science, vol. 187 (2013)
65. Rokhlin, V.: Sparse diagonal forms for translation operators for the Helmholtz equation in two dimensions. Appl. Comput. Harmon. Anal.**5**(1), 36–67 (1998)
66. Shkolnisky, Y.: Prolate spheroidal wave functions on a disc——integration and approximation of two-dimensional bandlimited functions. Appl. Comput. Harmon. Anal. **22**(2), 235–256 (2007)
67. Slepian, D.: Prolate spheroidal wave functions, fourier analysis and uncertainty IV. Bell Syst. Tech. J. **43**(6), 3009–3057 (1964)
68. Slepian, D., Pollak, H.O.: Prolate spheroidal wave functions, fourier analysis and uncertainty I. Bell Syst. Tech. J. **40**(1), 43–63 (1961)
69. Starr, P., Rokhlin, V.: On the numerical solution of two-point boundary value problems II. Commun. Pure Appl. Math. **47**(8), 1117–1159 (1994)
70. Toraldo Di Francia, G.: Degrees of freedom of an image. J. Opt. Soc. Am. **59**(7), 799–804 (1969)
71. Tropp, J.A., Yurtsever, A., Udell, M., Cevher, V.: Fixed-rank approximation of a positive-semidefinite matrix from streaming data. Adv. Neural Inf. Proces. Syst. **30** (2017)
72. Voronin, S., Martinsson, P.G.: Efficient algorithms for CUR and interpolative matrix decompositions. Adv. Comput. Math. **43**(3), 495–516 (2017)
73. Wiscombe, W.J.: Improved Mie scattering algorithms. Appl. Opt. **19**(9), 1505–1509 (1980)
74. Woolfe, F., Liberty, E., Rokhlin, V., Tygert, M.: A fast randomized algorithm for the approximation of matrices. Appl. Comput. Harmon. Anal. **25**(3), 335–366 (2008)
75. Wu, B.: ZetaTrap2D. https://github.com/bobbielf2/ZetaTrap2D (2021)
76. Wu, B.: ZetaTrap3D. https://github.com/bobbielf2/ZetaTrap3D (2022)
77. Wu, B., Martinsson, P.G.: Corrected trapezoidal rules for boundary integral equations in three dimensions. Numer. Math. **149**(4), 1025–1071 (2021)
78. Wu, B., Martinsson, P.G.: Zeta correction: a new approach to constructing corrected trapezoidal quadrature rules for singular integral operators. Adv. Comput. Math. **47**(3), 1–21 (2021)
79. Xia, J., Chandrasekaran, S., Gu, M., Li, X.S.: Fast algorithms for hierarchically semiseparable matrices. Numer. Linear Algebra Appl. **17**(6), 953–976 (2010)
80. Zhao, L., Barnett, A.: Robust and efficient solution of the drum problem via Nyström approximation of the Fredholm determinant. SIAM J. Numer. Anal. **53**(4), 1984–2007 (2015)

A Novel Spectral Method for the Subdiffusion Equation

Chuanju Xu and Wei Zeng

1 Introduction

Fractional partial differential equations (FPDEs) appear in the investigation of transport dynamics in complex systems which are characterized by the anomalous diffusion and nonexponential relaxation patterns. Related equations of importance are the space/time fractional diffusion equations, the fractional advection-diffusion equation for anomalous diffusion with sources and sinks, and the fractional Fokker-Planck equation [9, 23, 24, 41] for anomalous diffusion in an external field, etc. In fact, it has been found that anomalous diffusion is ubiquitous in physical and biological systems where trapping and binding of particles can occur [5, 10, 17, 21, 40, 42, 43, 48–50, 53]. Anomalous diffusion deviates from the standard Fichean description of Brownian motion, the main character of which is that its mean squared displacement is a nonlinear growth with respect to time, such as $\langle x^2(t)\rangle \sim t^\alpha$.

The universality of anomalous diffusion phenomenon in physical and biological experiments has led to an intensive investigation on the fractional differential equations in recent years. The time fractional diffusion equation (TFDE) considered in this paper is of interest not only in its own right, but also in that it constitutes the basic part in solving many other FPDEs. The TFDE and related equations have been investigated in analytical and numerical frames by a large number of authors, see, e.g., [1, 7, 8, 18–20, 27, 36, 38, 39, 55, 56, 60–62]. Spectral methods have also been applied in solving the TFDE. As is well known, any discretization including low-order approaches of a fractional derivative has to take into account its non-

C. Xu (✉) · W. Zeng
School of Mathematical Sciences and Fujian Provincial Key Laboratory of Mathematical Modeling and High Performance Scientific Computing, Xiamen University, Xiamen, China
e-mail: cjxu@xmu.edu.cn; zengwei@xmu.edu.cn

J. M. Melenk et al. (eds.), *Spectral and High Order Methods for Partial Differential Equations ICOSAHOM 2020+1*, Lecture Notes in Computational Science and Engineering 137, https://doi.org/10.1007/978-3-031-20432-6_3

local structure, which results in a full linear system and a high storage requirement. Therefore it is very natural to consider a global method, such as the spectral method, since the high accuracy of spectral methods may significantly reduce the storage requirements. The first attempt in this direction was made by Li and Xu in [31, 32]. It was proven that the exponential convergence of the proposed method is attainable for smooth solutions.

A main difficulty in numerically solving the TFDE comes from the fact that the solutions of the TFDE are usually of low regularity, which lowers the accuracy of the above mentioned methods. Some efforts have been made in developing and analyzing numerical methods for solutions of low regularity. Among them, the modified time-stepping schemes are prominent, which can be roughly divided into two categories, i.e., piecewise polynomial interpolation based on a class of nonuniform grids and convolution quadrature with initial correction. Stynes et al. [54] proposed to use graded meshes in L1 scheme for a reaction-diffusion problem, and an error analysis was given taking the starting time singularity into consideration. Later on, Liao et al. [34, 35] gave a more general error analysis of L1 formula on nonuniform grids based on a discrete fractional Grönwall inequality. Some researchers [14, 28, 29, 37] achieved optimal convergence rate by correcting the first several time steps. Several other works focused on spectral methods for non-smooth solutions of some related fractional equations, using polyfractonomials [58, 59], generalized Jacobi functions [12], Müntz Jacobi polynomials [25, 26], and log orthogonal functions/generalized log orthogonal functions [11, 13]. Numerical experiments or theoretical analysis presented therein have shown exponential convergence for non-smooth solutions having specific singularity.

Unlike these existing numerical approaches, in this paper we propose to first re-scale the time-fractional problems, then use the traditional approximations to the re-scaled problems. Li et al. [33] has tried this idea using a specific scaling function and proposed two finite difference schemes based on the linear interpolation and quadratic interpolation. The advantage of this approach is that the regularity of the re-scaled fractional operator can be much higher than that of the original operator, which is more conducive to construct high-order schemes. Below, we describe the main contributions of the paper and how the paper is organized.

Our first contribution is the development of the ψ-fractional Sobolev spaces presented in Sect. 2, which lays the foundation for the establishment of a new variational framework in Sect. 3. In detail, we introduce the concept of ψ-fractional operators, and propose ψ-fractional Sobolev spaces on this basis and prove the equivalence of related norms of ψ-fractional Sobolev spaces.

The second contribution is to propose a new Galerkin spectral method based on the generalized Jacobi polynomials (GJPs) under the new variational framework introduced in Sect. 3. The well-posedness of the weak problem is proved in the ψ-fractional Sobolev spaces, together with error estimation established in the non-uniform Jacobi-weighted norm. Moreover, it's shown that the new approach is as efficient as the Müntz spectral method [25, 26] by using suitable scaling functions. The novel approach not only provides a theoretical support of the Müntz spectral method but also gives a guideline for the selection of the scaling parameters.

Finally, the proposed approach is applied to the time fractional subdiffusion equations in Sect. 4. A space-time Galerkin spectral method is developed based on the re-scaled weak formulation and a combination of temporal GJPs and spatial Legendre polynomials. In Sect. 5, we present some numerical tests to confirm the theoretical findings. Some concluding remarks are given in Sect. 6.

2 Functional Spaces

In order to develop the re-scaling method for fractional differential equations, we need some preparations, mainly including an introduction of the ψ-fractional Sobolev spaces and establishment of the associated variational framework. Throughout this paper, let c stand for a generic positive constant independent of any functions and of any discretization parameters. In what follows, we use the expression $A \lesssim B$ (respectively, $A \gtrsim B$) to mean that $A \leqslant cB$ (respectively, $A \geqslant cB$).

The first part of this section is devoted to introducing the ψ-fractional integrals, derivatives and a crucial variable transformation.

2.1 ψ-Fractional Operators and Variable Transformation

We recall some definitions of ψ-fractional integrals and ψ-fractional derivatives; see Kilbas et al. [30, Sect. 2.5] or Samko et al. [47, Sect. 18.2]. Let $\Gamma(\cdot)$ denote the Gamma function. For any positive integer n and real number δ, $n-1 \leq \delta < n$, v is an integrable function in the bounded interval $[a_\psi, b_\psi]$ with respect to the function $\psi : [a_\psi, b_\psi] \to \mathbb{R}$ that is increasing and differentiable such that $\psi'(t) \neq 0$. The ψ-fractional integral, ψ-Caputo derivative, and ψ-Riemann–Liouville derivative of order δ of v are respectively defined as follows: $\forall\, t \in [a_\psi, b_\psi]$,

$$\text{left } \psi\text{-fractional integral:} \quad I_t^{\delta,\psi} v(t) := \frac{1}{\Gamma(\delta)} \int_{a_\psi}^{t} \psi'(z)(\psi(t) - \psi(z))^{\delta-1} v(z) dz, \tag{I1}$$

$$\text{right } \psi\text{-fractional integral:} \quad {}_tI^{\delta,\psi} v(t) := \frac{1}{\Gamma(\delta)} \int_{t}^{b_\psi} \psi'(z)(\psi(z) - \psi(t))^{\delta-1} v(z) dz, \tag{I2}$$

$$\text{left } \psi\text{-Caputo derivative:} \quad {}^CD_t^{\delta,\psi} v(t) := I_t^{n-\delta,\psi} \left(\frac{1}{\psi'(t)} \frac{d}{dt} \right)^n v(t), \tag{D1}$$

right ψ-Caption derivative: $${}_{t}^{C}D^{\delta,\psi}v(t) := {}_{t}I^{n-\delta,\psi}\left(-\frac{1}{\psi'(t)}\frac{d}{dt}\right)^{n} v(t), \tag{D2}$$

left ψ-Riemann–Liouville derivative: $$D_{t}^{\delta,\psi}v(t) = \left(\frac{1}{\psi'(t)}\frac{d}{dt}\right)^{n} I_{t}^{n-\delta,\psi}v(t), \tag{D3}$$

right ψ-Riemann–Liouville derivative: $${}_{t}D^{\delta,\psi}v(t) := \left(-\frac{1}{\psi'(t)}\frac{d}{dt}\right)^{n} {}_{t}I^{n-\delta,\psi}v(t). \tag{D4}$$

When $\psi(t) = t$, the above definitions degenerate into the classical fractional integral, Caputo derivative and Riemann–Liouville derivative; see [44, 45]. In particular, when $\delta \in (0,1)$, ψ-Caputo fractional derivative ${}^{C}D_{t}^{\delta,\psi}v(t)$ becomes ${}^{C}D_{t}^{\delta}v(t)$, where

$${}^{C}D_{t}^{\delta}v(t) := \frac{1}{\Gamma(1-\delta)}\int_{0}^{t}(t-z)^{-\delta}v'(z)dz. \tag{1}$$

In contrast, by a change of variable $\psi(t)$, the classical Caputo fractional derivative can be turned into a class of ψ-Caputo fractional derivative. For example, a direct calculation gives

$$\begin{aligned} {}^{C}D_{s}^{\delta}u(s) &= \frac{1}{\Gamma(1-\delta)}\int_{0}^{s}(s-x)^{-\delta}u'(x)dx, \quad s \in (0,T] \\ &\overset{s=t^{1/\gamma}}{=\!=\!=} \frac{1}{\Gamma(1-\delta)}\int_{0}^{t^{1/\gamma}}\left(t^{1/\gamma}-x\right)^{-\delta}u'(x)dx, \quad t \in (0,T^{\gamma}] \\ &\overset{x=z^{1/\gamma}}{=\!=\!=} \frac{1}{\Gamma(1-\delta)}\int_{0}^{t}\left(t^{1/\gamma}-z^{1/\gamma}\right)^{-\delta}du(z^{1/\gamma}), \quad t \in (0,T^{\gamma}]. \end{aligned} \tag{2}$$

Let $\psi(t) = t^{1/\gamma}$, $v(t) := u(\psi(t))$ for $t \in (0,T^{\gamma}]$. Then the new fractional derivative ${}_{*}^{C}D_{t}^{\delta}v(t)$, defined by

$${}_{*}^{C}D_{t}^{\delta}v(t) := \frac{1}{\Gamma(1-\delta)}\int_{0}^{t}\left(t^{1/\gamma}-z^{1/\gamma}\right)^{-\delta}v'(z)dz, \quad t \in (0,T^{\gamma}] \tag{3}$$

can be regarded as a class of ψ-Caputo fractional derivative of $v(t)$ with $\psi(t) = t^{1/\gamma}$.

The following results about ψ-fractional operators are frequently used; see [2, 3].

It is noted that the ψ-Riemann–Liouville fractional derivative and ψ-Caputo fractional derivative of v have the following relationship

$${}^{C}D_{t}^{\delta,\psi}v(t) = D_{t}^{\delta,\psi}[v(t) - v(a_{\psi})], \quad \delta \in (0,1). \tag{4}$$

And left ψ-Riemann–Liouville fractional derivative and integral of order δ satisfy

$$D_t^{\delta,\psi} I_t^{\delta,\psi} v(t) = v(t). \tag{5}$$

With the above notations and properties, we are in a position to introduce the ψ-fractional Sobolev spaces.

2.2 ψ-Fractional Sobolev Spaces

We begin with some additional notations. Let $0 < \gamma \leq 1, I = (a, b)$. The function $\psi^{-1}(\cdot)$ denotes the inverse function of $\psi(\cdot)$. Let $\Lambda = (a_\psi, b_\psi) := (\psi^{-1}(a), \psi^{-1}(b))$. Thus if $s \in I$, then $t = \psi^{-1}(s) \in \Lambda$.

Define the space

$$L_\psi^2(\Lambda) = \left\{ v : \Lambda \to \mathbb{R} \middle| v \text{ is measurable and } \int_\Lambda |v(t)|^2 \psi'(t) dt < \infty \right\}.$$

It can be easily seen that $L_\psi^2(\Lambda)$ is a Hilbert space with respect to the scalar product

$$(v, w)_{L_\psi^2(\Lambda)} = \int_\Lambda v(t) w(t) \psi'(t) dt, \quad \forall v, w \in L_\psi^2(\Lambda). \tag{6}$$

The norm in $L_\psi^2(\Lambda)$ induced by the scalar product $(\cdot, \cdot)_{L_\psi^2(\Lambda)}$ is defined by

$$\|v\|_{L_\psi^2(\Lambda)} = \sqrt{(v, v)_{L_\psi^2(\Lambda)}} = \left(\int_\Lambda |v(t)|^2 \psi'(t) dt \right)^{1/2}, \quad \forall v \in L_\psi^2(\Lambda).$$

In particular, for $\psi(t) = t$, the space $L_\psi^2(\Lambda)$ is reduced to the classical space $L^2(\Lambda)$. Let us denote by $(\cdot, \cdot)_{L^2(\Lambda)}$ and $\|\cdot, \cdot\|_{L^2(\Lambda)}$ the inner product and norm in $L^2(\Lambda)$, respectively.

We now introduce the ψ-fractional Sobolev spaces. Let $\mathcal{F}(v)$ denote the Fourier transform of v, $\tilde{v}(\cdot) := v(\psi^{-1}(\cdot))$. Define the space

$$H^{\delta,\psi}(\mathbb{R}) := \left\{ v \middle| v \in L_\psi^2(\mathbb{R});\ (1 + |\xi|^\delta)\mathcal{F}(\tilde{v})(\xi) \in L^2(\mathbb{R}) \right\}, \ \delta \geq 0, \tag{7}$$

endowed with the semi-norm and norm

$$|v|_{H^{\delta,\psi}(\mathbb{R})} = \left\| |\xi|^\delta \mathcal{F}(\tilde{v})(\xi) \right\|_{L^2(\mathbb{R})},$$

$$\|v\|_{H^{\delta,\psi}(\mathbb{R})} = \left(\|v\|^2_{L_\psi^2(\mathbb{R})} + |v|^2_{H^{\delta,\psi}(\mathbb{R})} \right)^{1/2},$$

respectively. Note that $\mathcal{F}(\tilde{v})$ rather than $\mathcal{F}(v)$ was used in the definition (7).

The ψ-fractional Sobolev space for the bounded domain Λ is defined by

$$H^{\delta,\psi}(\Lambda) := \left\{ v \in L^2_\psi(\Lambda) \middle| \exists\, v_e \in H^{\delta,\psi}(\mathbb{R}) \text{ such that } v_e|_\Lambda = v \right\},$$

equipped with the norm

$$\|v\|_{H^{\delta,\psi}(\Lambda)} = \inf_{v_e \in H^{\delta,\psi}(\mathbb{R}),\ v_e|_\Lambda = v} \|v_e\|_{H^{\delta,\psi}(\mathbb{R})}.$$

It is readily seen that $H^{\delta,\psi}(\Lambda)$ degenerates into the classic Sobolev space $H^\delta(\Lambda)$ when $\psi(t) = t$.

We define

$$^{L}H^{\delta,\psi}(\Lambda) := \left\{ v \middle| \|v\|_{^{L}H^{\delta,\psi}(\Lambda)} < \infty \right\},$$

where $|\cdot|_{^{L}H^{\delta,\psi}(\Lambda)}$ is the norm:

$$\|v\|_{^{L}H^{\delta,\psi}(\Lambda)} = \left(\|v\|^2_{L^2_\psi(\Lambda)} + |v|^2_{^{L}H^{\delta,\psi}(\Lambda)} \right)^{1/2}, \quad |v|_{^{L}H^{\delta,\psi}(\Lambda)} = \left\| D_t^{\delta,\psi} v \right\|_{L^2_\psi(\Lambda)}.$$

Similarly, we define

$$^{R}H^{\delta,\psi}(\Lambda) := \left\{ v \middle| \|v\|_{^{R}H^{\delta,\psi}(\Lambda)} < \infty \right\},$$

with

$$\|v\|_{^{R}H^{\delta,\psi}(\Lambda)} = \left(\|v\|^2_{L^2_\psi(\Lambda)} + |v|^2_{^{R}H^{\delta,\psi}(\Lambda)} \right)^{1/2}, \quad |v|_{^{R}H^{\delta,\psi}(\Lambda)} = \left\| {}_tD^{\delta,\psi} v \right\|_{L^2_\psi(\Lambda)};$$

and

$$^{c}H^{\delta,\psi}(\Lambda) := \left\{ v \middle| \|v\|_{^{c}H^{\delta,\psi}(\Lambda)} < \infty \right\},$$

with

$$\|v\|_{^{c}H^{\delta,\psi}(\Lambda)} = \left(\|v\|^2_{L^2_\psi(\Lambda)} + |v|^2_{^{c}H^{\delta,\psi}(\Lambda)} \right)^{1/2}, \quad |v|_{^{c}H^{\delta,\psi}(\Lambda)} = \left| (D_t^{\delta,\psi} v, {}_tD^{\delta,\psi} v)_{L^2_\psi(\Lambda)} \right|^{1/2}. \tag{8}$$

Let $C_0^\infty(\Lambda)$ is the space of smooth functions with compact support in Λ. Let $^{L}H_0^{\delta,\psi}(\Lambda)$, $^{R}H_0^{\delta,\psi}(\Lambda)$, $^{c}H_0^{\delta,\psi}(\Lambda)$, and $H_0^{\delta,\psi}(\Lambda)$ be the closures of $C_0^\infty(\Lambda)$ with respect to the norms $\|v\|_{^{L}H^{\delta,\psi}(\Lambda)}$, $\|v\|_{^{R}H^{\delta,\psi}(\Lambda)}$, $\|v\|_{^{c}H^{\delta,\psi}(\Lambda)}$ and $\|v\|_{H^{\delta,\psi}(\Lambda)}$, respectively. Besides, let ${}_0H^{\delta,\psi}(\Lambda)$ denote the closure of ${}_0C^\infty(\Lambda)$ with respect to

$\|\cdot\|_{H^{\delta,\psi}(\Lambda)}$, where ${}_0C^\infty(\Lambda)$ is the space of smooth functions with compact support in $(a_\psi, b_\psi]$.

Next we give some crucial lemmas, especially the equivalence results of the related norms of the ψ-fractional Sobolev spaces. These results play a key role in the subsequent analysis, including the well-posedness analysis and error estimation of the numerical methods to be constructed.

2.3 *Some Useful Lemmas*

Define the convolution of the functions $h_1(t)$ and $h_2(t)$ as follows:

$$h_1(t) * h_2(t) := \int_{-\infty}^{+\infty} h_1(t-\tau)h_2(\tau)d\tau = \int_{-\infty}^{+\infty} h_1(\tau)h_2(t-\tau)d\tau,$$

where $h_1(t), h_2(t) \in (-\infty, +\infty)$.

It is known [45] that if the Fourier transform of $h_1(t)$ and $h_2(t)$ exists, then

$$\mathcal{F}\{h_1(t) * h_2(t);\ \xi\} = H_1(\xi)H_2(\xi), \tag{9}$$

where $H_1(\xi) = \mathcal{F}\{h_1(t);\ \xi\}$, $H_2(\xi) = \mathcal{F}\{h_2(t);\ \xi\}$.

Then we can define Fourier transform of ψ- fractional derivatives on the above basis.

Lemma 2.1 (Fourier Transform of ψ-Fractional Derivatives) *Let $v \in C_0^\infty(\mathbb{R})$, $0 < \delta < 1$. Assume $\psi(\infty) = \infty$. Then*

$$\mathcal{F}({}_{-\infty}D_t^{\delta,\psi} v(t)) = (i\xi)^\delta \mathcal{F}(\tilde{v})(\xi), \quad \mathcal{F}({}_tD_{+\infty}^{\delta,\psi} v(t)) = (-i\xi)^\delta \mathcal{F}(\tilde{v})(\xi). \tag{10}$$

Proof We first evaluate the Fourier transform of the ψ-fractional integral ${}_{-\infty}I_t^{\delta,\psi} v$. The Laplace transform $\mathcal{L}$ of $s^{\delta-1}$ reads

$$\mathcal{L}\{s^{\delta-1};\ \tau\} = \int_0^\infty s^{\delta-1}e^{-\tau s}ds = \Gamma(\delta)\tau^{-\delta}. \tag{11}$$

Note that the above integral makes sense for all $\delta > 0$ by the Dirichlet theorem. Let $h(s)$ be the function

$$h(s) = \begin{cases} \frac{s^{\delta-1}}{\Gamma(\delta)}, & s > 0, \\ 0, & s \le 0. \end{cases}$$

Then a direct calculation using (11) shows

$$\mathcal{F}\{h(s);\xi\} = \int_{-\infty}^{\infty} h(s)e^{-i\xi s}ds = \frac{1}{\Gamma(\delta)}\int_0^{\infty} s^{\delta-1}e^{-i\xi s}ds = \frac{\mathcal{L}\{s^{\delta-1};i\xi\}}{\Gamma(\delta)} = (i\xi)^{-\delta}. \tag{12}$$

Let $s = \psi(t)$, then the ψ-fractional integral ${}_{-\infty}I_t^{\delta,\psi}v$ can be expressed as a convolution of the functions $h(\psi(t))$ and $\tilde{v}(\psi(t))$, i.e.,

$$\begin{aligned} {}_{-\infty}I_t^{\delta,\psi}v(t) &= \frac{1}{\Gamma(\delta)}\int_{-\infty}^{\psi(t)} (\psi(t)-\psi(\tau))^{\delta-1}\tilde{v}(\psi(\tau))d\psi(\tau) \\ &= h(s) * \tilde{v}(s), \end{aligned}$$

where $\tilde{v}(\cdot) := v(\psi^{-1}(\cdot))$. Thus, it follows from (9) and (12):

$$\mathcal{F}\{{}_{-\infty}I_t^{\delta,\psi}v(t);\xi\} = (i\xi)^{-\delta}\mathcal{F}(\tilde{v})(\xi).$$

Next, we calculate the Fourier transform of the ψ-fractional derivatives. Note that

$${}_{-\infty}D_t^{\delta,\psi}v(t) = {}_{-\infty}I_t^{n-\delta,\psi}\left(\frac{1}{\psi'(t)}\frac{d}{dt}\right)^n v(t),$$

we have

$$\begin{aligned} \mathcal{F}\{{}_{-\infty}D_t^{\delta,\psi}v(t)\} &= (i\xi)^{\delta-n}\mathcal{F}\left\{\left(\frac{d}{d\psi}\right)^n \tilde{v}(\psi(t));\xi\right\} \\ &= (i\xi)^{\delta-n}(i\xi)^n\mathcal{F}(\tilde{v})(\xi) \\ &= (i\xi)^{\delta}\mathcal{F}(\tilde{v})(\xi). \end{aligned}$$

The second equality in (10) can be proved in a similar way. □

With the help of the Fourier transform of the ψ-fractional derivatives, we can derive the following equivalence result for the ψ-fractional Sobolev spaces on the whole line $\mathbb{R}$.

Lemma 2.2 *Let $\delta > 0$, $\delta \neq n - 1/2$, $n \in \mathbb{N}$. Then the spaces ${}^{L}H^{\delta,\psi}(\mathbb{R})$, ${}^{R}H^{\delta,\psi}(\mathbb{R})$, ${}^{c}H^{\delta,\psi}(\mathbb{R})$, and $H^{\delta,\psi}(\mathbb{R})$ are equal to each other with equivalent semi-norms and norms.*

Proof The proof will be divided into three steps.

Step 1: the equivalence of the spaces ${}^{L}H^{\delta,\psi}(\mathbb{R})$ and $H^{\delta,\psi}(\mathbb{R})$.

For a function $v \in {}^{L}H^{\delta,\psi}(\mathbb{R})$, we have $D_t^{\delta,\psi}v \in L^2_{\psi}(\mathbb{R})$. Using Lemma 2.1 and Plancherel's theorem gives

$$\int_{\mathbb{R}} |\xi|^{2\delta}\,|\mathcal{F}(\tilde{v})(\xi)|^2\,d\xi = \int_{\mathbb{R}} \left|D_t^{\delta,\psi}v\right|^2 \psi'(t)dt.$$

Thus,

$$\left\| |\xi|^\delta \mathcal{F}(\tilde{v})(\xi) \right\|_{L^2(\mathbb{R})} = |v|_{{}^L H^{\delta,\psi}(\mathbb{R})} .$$

The desired result follows immediately from the above equality and the definition of $H^{\delta,\psi}(\mathbb{R})$.

Step 2: the equivalence of the spaces ${}^L H^{\delta,\psi}(\mathbb{R})$ and ${}^R H^{\delta,\psi}(\mathbb{R})$.

Again, using the results of Lemma 2.1 and Plancherel's theorem, we have

$$|v|^2_{{}^L H^{\delta,\psi}(\mathbb{R})} = \int_{\mathbb{R}} \left| (i\xi)^\delta \mathcal{F}(\tilde{v})(\xi) \right|^2 d\xi. \tag{13}$$

Similarly,

$$|v|^2_{{}^R H^{\delta,\psi}(\mathbb{R})} = \int_{\mathbb{R}} \left| (-i\xi)^\delta \mathcal{F}(\tilde{v})(\xi) \right|^2 d\xi. \tag{14}$$

Note that $|(i\xi)^\delta| = |(-i\xi)^\delta|$. Thus the semi-norms $|v|_{{}^L H^{\delta,\psi}(\mathbb{R})}$ and $|v|_{{}^R H^{\delta,\psi}(\mathbb{R})}$, consequently the norms $\|v\|_{{}^L H^{\delta,\psi}(\mathbb{R})}$ and $\|v\|_{{}^R H^{\delta,\psi}(\mathbb{R})}$, are equivalent.

Step 3: the equivalence of the spaces ${}^c H^{\delta,\psi}(\mathbb{R})$ and ${}^L H^{\delta,\psi}(\mathbb{R})$.

Analogous to [15, Lemma 2.4], with the help of some related properties of the Fourier transform, we obtain

$$\left(D_t^{\delta,\psi} v,\ {}_t D^{\delta,\psi} v \right)_{L^2_\psi(\mathbb{R})} = \cos(\pi\delta) \left\| D_t^{\delta,\psi} v \right\|^2_{L^2_\psi(\mathbb{R})} . \tag{15}$$

That is,

$$|v|^2_{{}^c H^{\delta,\psi}(\mathbb{R})} = |\cos(\pi\delta)|\, |v|^2_{{}^L H^{\delta,\psi}(\mathbb{R})} .$$

Thus the semi-norms of ${}^c H^{\delta,\psi}(\mathbb{R})$ and ${}^L H^{\delta,\psi}(\mathbb{R})$ are equivalent. So are their norms, which implies the equivalence of ${}^c H^{\delta,\psi}(\mathbb{R})$ and ${}^L H^{\delta,\psi}(\mathbb{R})$.

We conclude by combining Step 1–Step 3.

□

The equivalence of different ψ-fractional spaces on the bounded interval Λ are established below.

Lemma 2.3 *Let $\delta > 0$, $\delta \neq n - 1/2$, $n \in \mathbb{N}$. Then the spaces ${}^L H_0^{\delta,\psi}(\Lambda)$, ${}^R H_0^{\delta,\psi}(\Lambda)$, ${}^c H_0^{\delta,\psi}(\Lambda)$, and $H_0^{\delta,\psi}(\Lambda)$ are equal to each other with equivalent semi-norms and norms.*

Proof The proof is split into two steps.

Step 1: the equivalence of the spaces ${}^c H_0^{\delta,\psi}(\Lambda)$ and $H_0^{\delta,\psi}(\Lambda)$.

For $v \in C_0^\infty(\Lambda)$, let v_e be the extension of v by zero outside of Λ. Then

$$\begin{aligned}&\operatorname{supp}(v_e) \subset \Lambda,\\&\operatorname{supp}\left(D_t^{\delta,\psi} v_e\right) \subset (a_\psi, \infty),\\&\operatorname{supp}\left({}_tD^{\delta,\psi} v_e\right) \subset (-\infty, b_\psi).\end{aligned}$$

Thus,

$$\operatorname{supp}\left(D_t^{\delta,\psi} v_e \, {}_tD^{\delta,\psi} v_e\right) \subset \Lambda,$$

from which it follows

$$|v|_{{}^cH_0^{\delta,\psi}(\Lambda)} = |v_e|_{{}^cH_0^{\delta,\psi}(\mathbb{R})}.$$

On the other side, we have

$$|v|_{H_0^{\delta,\psi}(\Lambda)} = |v_e|_{H_0^{\delta,\psi}(\mathbb{R})}.$$

Then the semi-norm equivalence of ${}^cH_0^{\delta,\psi}(\mathbb{R})$ and $H_0^{\delta,\psi}(\mathbb{R})$, proved in Lemma 2.2, gives

$$|v|_{{}^cH_0^{\delta,\psi}(\Lambda)} = |v|_{H_0^{\delta,\psi}(\Lambda)}.$$

Thus the spaces ${}^cH_0^{\delta,\psi}(\Lambda)$ and $H_0^{\delta,\psi}(\Lambda)$ are equal with equivalent norms.

Step 2: the equivalence of the spaces ${}^LH_0^{\delta,\psi}(\Lambda)$, ${}^RH_0^{\delta,\psi}(\Lambda)$, and $H_0^{\delta,\psi}(\Lambda)$.

It follows from (13) and the definition of $|\cdot|_{H_0^{\delta,\psi}(\Lambda)}$:

$$\left\|D_t^{\delta,\psi} v\right\|_{L^2_\psi(\Lambda)} = |v|_{{}^LH^{\delta,\psi}(\Lambda)} \le |v_e|_{{}^LH^{\delta,\psi}(\mathbb{R})} = |v_e|_{H^{\delta,\psi}(\mathbb{R})} = |v|_{H_0^{\delta,\psi}(\Lambda)}.$$

This gives

$$H_0^{\delta,\psi}(\Lambda) \subseteq {}^LH_0^{\delta,\psi}(\Lambda).$$

Combining the result proved in Step 1 and Young's inequality, we obtain

$$\begin{aligned}|v|^2_{H_0^{\delta,\psi}(\Lambda)} &\le c\,|v|^2_{{}^cH_0^{\delta,\psi}(\Lambda)}\\&= c\left|\int_\Lambda D_t^{\delta,\psi} v \, {}_tD^{\delta,\psi} v \psi'(t)dt\right|\\&\le \frac{c}{4\epsilon}\left\|D_t^{\delta,\psi} v\right\|^2_{L^2_\psi(\Lambda)} + \epsilon c \left\|{}_tD^{\delta,\psi} v\right\|^2_{L^2_\psi(\Lambda)}.\end{aligned}$$

Furthermore, it follows from (14) and the definition of $|\cdot|_{H_0^{\delta,\psi}(\Lambda)}$:

$$\left\|{}_tD^{\delta,\psi}v\right\|_{L^2_\psi(\Lambda)} = |v|_{{}^RH^{\delta,\psi}(\Lambda)} \leq |v_e|_{{}^RH^{\delta,\psi}(\mathbb{R})} = |v_e|_{H^{\delta,\psi}(\mathbb{R})} = |v|_{H_0^{\delta,\psi}(\Lambda)}.$$

Combining the last two inequalities gives

$$|v|^2_{H_0^{\delta,\psi}(\Lambda)} \leq \frac{c}{2\epsilon}|v|^2_{{}^LH_0^{\delta,\psi}(\Lambda)} + \frac{\epsilon c}{2}|v|^2_{H_0^{\delta,\psi}(\Lambda)}.$$

Taking $\epsilon = 1/c$ in the above inequality yields

$$|v|^2_{H_0^{\delta,\psi}(\Lambda)} \leq c^2|v|^2_{{}^LH_0^{\delta,\psi}(\Lambda)}.$$

This gives

$${}^LH_0^{\delta,\psi}(\Lambda) \subseteq H_0^{\delta,\psi}(\Lambda).$$

This ends the proof of the semi-norm equivalence of the spaces ${}^LH_0^{\delta,\psi}(\Lambda)$ and $H_0^{\delta,\psi}(\Lambda)$, and thus the equivalence of the spaces themselves. In a similar way, we can prove the equivalence of the spaces ${}^RH_0^{\delta,\psi}(\Lambda)$ and $H_0^{\delta,\psi}(\Lambda)$. The proof is completed.

□

Now we turn to derive some Poincaré-Friedrichs-type inequalities for the functions in ψ-fractional spaces. The following mapping properties are useful.

Lemma 2.4 (Mapping Properties) *All the following mappings are bounded linear operator.*

(i) $I_t^{\delta,\psi} : L^2_\psi(\Lambda) \to L^2_\psi(\Lambda)$.
(ii) $I_t^{\delta,\psi} : L^2_\psi(\Lambda) \to {}^LH^{\delta,\psi}(\Lambda)$.
(iii) $D_t^{\delta,\psi} : {}^LH^{\delta,\psi}(\Lambda) \to L^2_\psi(\Lambda)$.
(iv) ${}_tI^{\delta,\psi} : L^2_\psi(\Lambda) \to L^2_\psi(\Lambda)$.
(v) ${}_tI^{\delta,\psi} : L^2_\psi(\Lambda) \to {}^RH^{\delta,\psi}(\Lambda)$.
(vi) ${}_tD^{\delta,\psi} : {}^RH^{\delta,\psi}(\Lambda) \to L^2_\psi(\Lambda)$.

Proof

(i) By using [4, Lemma 2.2], we have

$$\left\|I_t^{\delta,\psi}v\right\|_{L^2_\psi(\Lambda)} \leq \frac{1}{\Gamma(\delta)\sqrt{2\delta-1}}\|v\|_{L^2_\psi(\Lambda)},$$

which means $I_t^{\delta,\psi}$ is a bounded linear operator from $L^2_\psi(\Lambda)$ to $L^2_\psi(\Lambda)$.

(ii) Combining (5) and the definition of ${}^{L}H^{\delta,\psi}(\Lambda)$, then using (i), one obtains

$$\left\| I_t^{\delta,\psi} v \right\|_{{}^{L}H^{\delta,\psi}(\Lambda)} = \left(\left\| I_t^{\delta,\psi} v \right\|^2_{L^2_\psi(\Lambda)} + \left\| D_t^{\delta,\psi} I_t^{\delta,\psi} v \right\|^2_{L^2_\psi(\Lambda)} \right)^{1/2}$$
$$\lesssim \|v\|_{L^2_\psi(\Lambda)} .$$

This proves that $I_t^{\delta,\psi}$ is a bounded linear operator from $L^2_\psi(\Lambda)$ to ${}^{L}H^{\delta,\psi}(\Lambda)$.

(iii) It follows from the definition of the norm $\|\cdot\|_{{}^{L}H^{\delta,\psi}(\Lambda)}$:

$$\left\| D_t^{\delta,\psi} v \right\|_{L^2_\psi(\Lambda)} \leq \left(\|v\|^2_{L^2_\psi(\Lambda)} + \left\| D_t^{\delta,\psi} v \right\|^2_{L^2_\psi(\Lambda)} \right)^{1/2}$$
$$=: \|v\|_{{}^{L}H^{\delta,\psi}(\Lambda)} .$$

This shows that $D_t^{\delta,\psi}$ is a bounded linear operator from ${}^{L}H^{\delta,\psi}(\Lambda)$ to $L^2_\psi(\Lambda)$.

(iv)–(vi) can be proved similarly. □

Lemma 2.5 (ψ-Fractional Poincaré-Friedrichs Inequalities) *The following two Poincaré-Friedrichs-type inequalities hold*

$$\|v\|_{L^2_\psi(\Lambda)} \lesssim |v|_{{}^{L}H^{\delta,\psi}(\Lambda)}, \quad \forall v \in {}^{L}H^{\delta,\psi}(\Lambda).$$

$$\|v\|_{L^2_\psi(\Lambda)} \lesssim |v|_{{}^{R}H^{\delta,\psi}(\Lambda)}, \quad \forall v \in {}^{R}H^{\delta,\psi}(\Lambda).$$

Proof For all $v \in {}^{L}H^{\delta,\psi}(\Lambda)$, it follows from (5) and Lemma 2.4 that

$$\|v\|_{L^2_\psi(\Lambda)} = \|I_t^{\delta,\psi} D_t^{\delta,\psi} v\|_{L^2_\psi(\Lambda)} \lesssim \|D_t^{\delta,\psi} v\|_{L^2_\psi(\Lambda)} = |v|_{{}^{L}H^{\delta,\psi}(\Lambda)}.$$

This proves the first inequality. The second inequality can be derived similarly. □

One of the remarkable properties of the ψ-Riemann–Liouville fractional derivative is given in the following lemma.

Lemma 2.6 *For all* $0 < \delta < 1$, *if* $v \in H^{\delta,\psi}(\Lambda)$, $w \in C_0^\infty(\Lambda)$, *then*

$$\left(D_t^{\delta,\psi} v(t),\ w(t) \right)_{L^2_\psi(\Lambda)} = \left(v(t),\ {}_tD^{\delta,\psi} w(t) \right)_{L^2_\psi(\Lambda)} . \tag{16}$$

Proof By using integration by parts, we have

$$
\begin{aligned}
&\left(D_t^{\delta,\psi} v(t),\ w(t)\right)_{L^2_\psi(\Lambda)}\\
=&\frac{1}{\Gamma(1-\delta)}\int_\Lambda \frac{1}{\psi'(t)}\frac{d}{dt}\int_{a_\psi}^t \psi'(s)(\psi(t)-\psi(s))^{-\delta}v(s)ds w(t)\psi'(t)dt\\
=&\frac{1}{\Gamma(1-\delta)}\Big[w(t)\int_{a_\psi}^t \psi'(s)(\psi(t)-\psi(s))^{-\delta}v(s)ds\big|_{\partial\Lambda}\\
&-\int_\Lambda\int_{a_\psi}^t \psi'(s)(\psi(t)-\psi(s))^{-\delta}v(s)dsdw(t)\Big]\\
=&\frac{-1}{\Gamma(1-\delta)}\int_\Lambda\int_{a_\psi}^t \psi'(s)(\psi(t)-\psi(s))^{-\delta}v(s)dsw'(t)dt\\
=&\frac{-1}{\Gamma(1-\delta)}\int_\Lambda\int_s^{b_\psi} [\psi(t)-\psi(s)]^{-\delta}\, w'(t)dtv(s)\psi'(s)ds.
\end{aligned}
\tag{17}
$$

Furthermore, a direct calculation gives

$$
\begin{aligned}
\frac{d}{ds}\int_s^{b_\psi} \psi'(t)\,[\psi(t)-\psi(s)]^{-\delta}\, w(t)dt =&\frac{d}{ds}\Big[w(t)\frac{[\psi(t)-\psi(s)]^{1-\delta}}{1-\delta}\Big|_s^{b_\psi}\\
&-\int_s^{b_\psi}\frac{[\psi(t)-\psi(s)]^{1-\delta}}{1-\delta}dw(t)\Big]\\
=&-\frac{d}{ds}\int_s^{b_\psi}\frac{[\psi(t)-\psi(s)]^{1-\delta}}{1-\delta}dw(t)\\
=&\psi'(s)\int_s^{b_\psi} [\psi(t)-\psi(s)]^{-\delta}\, w'(t)dt.
\end{aligned}
$$

Thus,

$$
\begin{aligned}
&\left(v(s),\ {}_sD^{\delta,\psi} w(s)\right)_{L^2_\psi(\Lambda)}\\
=&\frac{-1}{\Gamma(1-\delta)}\int_\Lambda \frac{1}{\psi'(s)}\frac{d}{ds}\int_s^{b_\psi} \psi'(t)\,[\psi(t)-\psi(s)]^{-\delta}\, w(t)dtv(s)\psi'(s)ds\\
=&\frac{-1}{\Gamma(1-\delta)}\int_\Lambda \frac{1}{\psi'(s)}\psi'(s)\int_s^{b_\psi} [\psi(t)-\psi(s)]^{-\delta}\, w'(t)dtv(s)\psi'(s)ds\\
=&\frac{-1}{\Gamma(1-\delta)}\int_\Lambda\int_s^{b_\psi} [\psi(t)-\psi(s)]^{-\delta}\, w'(t)dtv(s)\psi'(s)ds\\
=&\left(D_t^{\delta,\psi} v(t),\ w(t)\right)_{L^2_\psi(\Lambda)}.
\end{aligned}
$$

This completes the proof. □

Based on a similar idea introduced in [32], the ψ-fractional derivative can be generalized as a distribution to any $L^2_\psi(\Lambda)$ functions by using the integration by parts (16). That is, for $v \in L^2_\psi(\Lambda)$, the ψ-fractional derivative of v in the distribution sense is defined as the linear functional through

$$\left\langle D_t^{\delta,\psi} v(t),\ w(t)\right\rangle_{L^2_\psi(\Lambda)} = \left(v(t),\ {}_tD^{\delta,\psi} w(t)\right)_{L^2_\psi(\Lambda)}, \quad \forall w \in C_0^\infty(\Lambda).$$

With this convention, we are able to derive, by following the same lines as in [32], a key result which is crucial for the proof of well-posedness of the variational problem. That is, for all $0 < \delta < 1$, if $v, w \in H^{\delta/2,\psi}(\Lambda)$, then

$$\left\langle D_t^{\delta,\psi} v(t),\ w(t)\right\rangle_{L^2_\psi(\Lambda)} = \left(D_t^{\delta/2,\psi} v(t),\ {}_tD^{\delta/2,\psi} w(t)\right)_{L^2_\psi(\Lambda)}. \tag{18}$$

Remark 2.7 It is worth noting that the ψ-fractional variational framework established in this section is valid for quite general function $\psi(t)$. The only assumption on $\psi(t)$ is its increasing differentiability and $\psi'(t) \neq 0$. In what follows we will consider a special case $\psi(t) = t^{1/\gamma}$ to demonstrate how this variational framework can be used to capture some singular solutions of fractional differential equations.

3 A Spectral Method for Fractional Ordinary Differential Equations

As a simple application example, we consider in this section the following initial value problem

$$\begin{cases} {}^CD_s^\delta u(s) + \lambda u(s) = g(s), \ \lambda > 0, \ s \in I, \\ u(a) = \phi. \end{cases} \tag{19}$$

Here $0 < \delta < 1$, ${}^CD_s^\delta$ denotes the classical left-sided Caputo fractional operator defined in (1). This model problem frequently appears in the investigation of the TFDE [16, 46]:

$${}^CD_s^\delta u(x,s) = \Delta u(x,s) + g(x,s), \ x \in \Omega, \ s \in I,$$

where Ω is a spatial domain. The solution of the TFDE can be expended in the space direction by using the eigenfunctions of the Laplacian operator $-\Delta$, resulting in the Eq. (19) with λ being an eigenvalue of $-\Delta$. It is seen that the model problem (19) reflects the main difficulty of solving the TFDE, i.e., singularity feature of the solution in the time direction.

Without loss of generality, we consider the homogeneous initial condition, i.e., $\phi \equiv 0$. The case of non-homogeneous initial condition can be handled by standard homogenization. With $\phi \equiv 0$, the problem (19) can be equivalently written as [57]

$$\begin{cases} D_s^\delta u(s) + \lambda u(s) = g(s), \ \lambda > 0, \ s \in I, \\ I_s^{1-\delta} u(a) = 0, \end{cases} \tag{20}$$

where the operators D_s^δ and $I_s^{1-\delta}$ are defined in (D3) and (I1) with $\psi(s) = s$.

By the change of variable $s = \psi(t)$, and denoting $v(t) = u(\psi(t))$, $f(t) = g(\psi(t))$, the problem (20) can be transformed into the following problem

$$\begin{cases} D_t^{\delta,\psi} v(t) + \lambda v(t) = f(t), \ t \in \Lambda, \\ I_t^{1-\delta,\psi} v(a_\psi) = 0. \end{cases} \tag{21}$$

We propose and analyze below a spectral Galerkin method to solve the transformed problem (21) expressed in the weak form. We first introduce the GJPs (see [22, 52]). Define the shifted GJPs

$$j_n^{\alpha,-1}(t) := (1 + x(t)) J_{n-1}^{\alpha,1}(x(t)), \ \ \alpha > -1, \ t \in \Lambda, \ n = 1, 2, \ldots, \tag{22}$$

where $x(t) = \frac{2t-(a_\psi+b_\psi)}{b_\psi - a_\psi}$, $J_n^{\alpha,\beta}(x)$ are the classical n-th Jacobi polynomials, i.e., orthogonal polynomials with respect to the weight function $\omega^{\alpha,\beta}(x) := (1-x)^\alpha (1+x)^\beta$, $\alpha, \beta > -1$, $n = 0, 1, \ldots$.

It can be checked that

$$\frac{d}{dt} j_n^{\alpha,-1}(t) = \frac{2n}{b_\psi - a_\psi} J_{n-1}^{\alpha+1,0}(x(t)), \ \ t \in \Lambda. \tag{23}$$

Let P_N be the standard polynomial space defined by

$$P_N := \text{span}\{1, t, t^2, \ldots, t^N\}.$$

Set the shifted polynomials space

$$V_N := \text{span}\{v \in P_N | v(0) = 0\} = \text{span}\left\{ j_n^{\alpha,-1}(t), \ \ t \in \Lambda, \ n = 1, \ldots, N \right\}.$$

Define the $L^2_{\omega^{\alpha,-1}}$-orthogonal projection $\pi_N^{\alpha,-1}$: $L^2_{\omega^{\alpha,-1}}(\Lambda) \to V_N$, such that for all $u \in L^2_{\omega^{\alpha,-1}}(\Lambda)$, $\pi_N^{\alpha,-1} u \in V_N$ satisfies

$$\left(\pi_N^{\alpha,-1} u - u, w\right)_{L^2_{\omega^{\alpha,-1}}(\Lambda)} = 0, \ \ \forall w \in V_N.$$

Define the non-uniform Jacobi-weighted Sobolev spaces as follows:

$$B^m_{\omega^{\alpha,-1}}(\Lambda) := \left\{ v : \partial_t^k v \in L^2_{\omega^{\alpha+k,-1+k}}(\Lambda), 0 \le k \le m \right\}.$$

An approximation result of this projection operator is given in the following lemma.

Lemma 3.1 *For any $u \in B^m_{\omega^{\alpha,-1}}(\Lambda)$, $m \in \mathbb{N}$, $m \ge 1$, and $0 \le \mu \le m$, we have*

$$\left\| \pi_N^{\alpha,-1} u - u \right\|_{B^{\mu}_{\omega^{\alpha,-1}}} \lesssim N^{\mu-m} \left\| \partial_t^m u \right\|_{L^2_{\omega^{\alpha+m,-1+m}}}. \tag{24}$$

Proof This approximation result can be proved in the same way as for the projector $\pi_N^{-1,\beta}$ given in [22]. We omit the details in order to limit the length of the paper. □

The spectral approximation we propose for (21) reads: Find $v_N \in V_N$ such that

$$\mathcal{A}(v_N, w_N) = F(w_N), \quad \forall\, w_N \in V_N, \tag{25}$$

where

$$\mathcal{A}(v_N, w_N) = \left(D_t^{\delta,\psi} v_N, w_N \right)_{L^2_\psi(\Lambda)} + \lambda \left(v_N, w_N \right)_{L^2_\psi(\Lambda)},$$
$$F(w_N) = (f, w_N)_{L^2_\psi(\Lambda)},$$

with $(\cdot,\cdot)_{L^2_\psi(\Lambda)}$ being defined in (6).

3.1 Well-Posedness

Theorem 3.2 *For any f satisfying $I_t^{\delta/2,\psi} f \in L^2_\psi(\Lambda)$, $0 < \delta < 1$, the spectral approximation problem (25) is well-posed. Moreover, if v_N is the solution of (25), then it holds*

$$\|v_N\|_{H^{\delta/2,\psi}(\Lambda)} \lesssim \|I_t^{\delta/2,\psi} f\|_{L^2_\psi(\Lambda)}. \tag{26}$$

Proof The proof makes use the classical Lax-Milgram Theorem, which consists in verifying the coercivity and continuity of the bilinear form $\mathcal{A}(\cdot,\cdot)$.

Combining (18) with the definition of $\|\cdot\|_{{}^cH^{\delta,\psi}(\Lambda)}$ gives: for all $v_N \in V_N$,

$$\begin{aligned}
\mathcal{A}(v_N, v_N) &= \left(D_t^{\delta,\psi} v_N, v_N \right)_{L^2_\psi(\Lambda)} + \lambda \left(v_N, v_N \right)_{L^2_\psi(\Lambda)} \\
&= \left(D_t^{\delta/2,\psi} v_N, {}_tD^{\delta/2,\psi} v_N \right)_{L^2_\psi(\Lambda)} + \lambda \left(v_N, v_N \right)_{L^2_\psi(\Lambda)} \\
&\gtrsim |v_N|^2_{{}^cH^{\delta/2,\psi}(\Lambda)}.
\end{aligned}$$

Furthermore, the norm equivalence proved in Lemma 2.3 yields

$$\mathcal{A}(v_N,\ v_N) \gtrsim |v_N|^2_{H^{\delta/2,\psi}(\Lambda)}.$$

Then it follows from the fractional Poincaré-Friedrichs inequality in Lemma 2.5:

$$\mathcal{A}(v_N,\ v_N) \gtrsim \|v_N\|^2_{H^{\delta/2,\psi}(\Lambda)}.$$

By applying (18) again, and using Cauchy-Schwarz inequality, we obtain for all $v_N, w_N \in V_N$,

$$\begin{aligned}
|\mathcal{A}(v_N,\ w_N)| &\leq \left|\left(D_t^{\delta,\psi} v_N,\ w_N\right)_{L^2_\psi(\Lambda)}\right| + \lambda \left|(v_N,\ w_N)_{L^2_\psi(\Lambda)}\right| \\
&\leq \left|\left(D_t^{\delta/2,\psi} v_N,\ {}_tD^{\delta/2,\psi} w_N\right)_{L^2_\psi(\Lambda)}\right| + \lambda \left|(v_N,\ w_N)_{L^2_\psi(\Lambda)}\right| \\
&\leq |v_N|_{{}^LH^{\delta/2,\psi}(\Lambda)} |w_N|_{{}^RH^{\delta/2,\psi}(\Lambda)} + \lambda \|v_N\|_{L^2_\psi(\Lambda)} \|w_N\|_{L^2_\psi(\Lambda)}.
\end{aligned}$$

Finally, we derive from the norm equivalence and Lemma 2.5:

$$\begin{aligned}
|\mathcal{A}(v_N,\ w_N)| &\lesssim |v_N|_{H^{\delta/2,\psi}(\Lambda)} |w_N|_{H^{\delta/2,\psi}(\Lambda)} \\
&\lesssim \|v_N\|_{H^{\delta/2,\psi}(\Lambda)} \|w_N\|_{H^{\delta/2,\psi}(\Lambda)}.
\end{aligned}$$

The well-posedness of (25) is thus proved.

The stability inequality (26) can be derived by using (5), Lemma 2.3, Lemma 2.6, and the Cauchy-Schwarz inequality as follows:

$$\begin{aligned}
\|v_N\|^2_{H^{\delta/2,\psi}(\Lambda)} &\lesssim \mathcal{A}(v_N,\ v_N) = \mathcal{F}(v_N) \\
&= \left(D_t^{\delta/2,\psi} I_t^{\delta/2,\psi} f,\ v_N\right)_{L^2_\psi(\Lambda)} = \left(I_t^{\delta/2,\psi} f,\ {}_tD^{\delta/2,\psi} v_N\right)_{L^2_\psi(\Lambda)} \\
&\leq \left\| I_t^{\delta/2,\psi} f \right\|_{L^2_\psi(\Lambda)} \|{}_tD^{\delta/2,\psi} v_N\|_{L^2_\psi(\Lambda)} \\
&\leq \left\| I_t^{\delta/2,\psi} f \right\|_{L^2_\psi(\Lambda)} \|v_N\|_{H^{\delta/2,\psi}(\Lambda)}.
\end{aligned}$$

This ends the proof of the theorem. □

3.2 Error Estimate

In this subsection we present an error estimate for a specific transformation function, i.e., $\psi(t) = t^{1/\gamma}$. Although this is the only case for which we derive the error estimate here, we are going to see that this specific transformation can well smooth the time fractional diffusion equation, therefore is a good fit for use of the spectral method.

Before carrying out the error analysis, we recall the following definition and lemma from [6]. Define the integral operator $(P_\delta)v(t) = \int_{-\infty}^{t} P_\delta(t,z)v(z)dz$, where $P_\delta(t,z) \geq 0$ is not increasing in t and not decreasing in z in $\{(t,z) \in \mathbb{R}^2 : z < t\}$. For two nonnegative functions ω_1 and ω_2, we set

$$A_{\delta,p,q}(t) = \left(\int_t^{\infty} \left[P_\delta(y,t)^{1/2}\omega_1(y)\right]^q dy\right)^{1/q} \left(\int_{-\infty}^{t} \left[P_\delta(t,y)^{-1/2}\omega_2(y)\right]^{-p'} dy\right)^{1/p'}. \tag{27}$$

Lemma 3.3 (Andersen and Heinig [6]) *Let $\omega_1(t) \geq 0$, $\omega_2(t) \geq 0$, $1/p + 1/p' = 1$. If there exists a constant C such that $A_{\delta,p,q}(t) \leq C, \forall t \in \mathbb{R}$. Then*

$$\left(\int_{-\infty}^{\infty} |\omega_1(t)(P_\delta)v(t)|^q \, dt\right)^{1/q} \leq A_1 C \left(\int_{-\infty}^{\infty} |\omega_2(t)v(t)|^p \, dt\right)^{1/p},$$

where $A_1 = ((p'+q)/q)^{1/p'}((p'+q)/p')^{1/q}$ if $1 < p \leq q < \infty$, and $A_1 = 1$ otherwise.

Next we prove two lemmas which are useful for the error estimation.

Lemma 3.4 *Let $P_\delta(t, z) = \frac{1}{\Gamma(1-\delta)}(t^{1/\gamma} - z^{1/\gamma})^{-\delta}$, $\delta < \alpha < 1$, $\Lambda_1 = (0,1)$,*

$$\omega_1(t) = \begin{cases} (1-t)^{\alpha/2} t^{1/\gamma - 1/2}, & t \in \Lambda_1, \\ 0, & t \in \mathbb{R} \setminus \Lambda_1, \end{cases} \qquad \omega_2(t) = \begin{cases} (1-t)^{(1-\alpha)/2}, & t \in \Lambda_1, \\ 0, & t \in \mathbb{R} \setminus \Lambda_1. \end{cases}$$

For any differentiable function $v(t)$ defined in Λ_1, it holds

$$\int_{\Lambda_1} \left|\omega_1(t)(P_\delta v')(t)\right|^2 dt \lesssim \int_{\Lambda_1} \left|\omega_2(t)(v')(t)\right|^2 dt.$$

Proof First, it is easy to check that $P_\delta(t,z) \geq 0$ is not increasing in t and not decreasing in z. We extend $v(t)$ with zero outside of Λ_1. Taking $p = q = 2$ in (27) gives

$$
\begin{aligned}
& A_{\delta,2,2}(t) \\
=& \left(\int_t^1 \left[(P_\delta(y,t))^{1/2}\omega_1(y)\right]^2 dy\right)^{1/2} \left(\int_0^t \left[(P_\delta(t,y))^{-1/2}\omega_2(y)\right]^{-2} dy\right)^{1/2} \\
=& \left(\int_t^1 (y^{1/\gamma}-t^{1/\gamma})^{-\delta}(1-y)^\alpha y^{2/\gamma-1} dy\right)^{1/2} \left(\int_0^t (t^{1/\gamma}-y^{1/\gamma})^{-\delta}(1-y)^{\alpha-1)} dy\right)^{1/2}.
\end{aligned}
$$

A direct calculation shows

$$
\begin{aligned}
& \int_t^1 \left(y^{1/\gamma}-t^{1/\gamma}\right)^{-\delta}(1-y)^\alpha y^{2/\gamma-1} dy \\
\leq & \int_t^1 \left(y^{1/\gamma}-t^{1/\gamma}\right)^{-\delta} y^{2/\gamma-1} dy \\
=& \gamma \int_t^1 (y-t)^{-\delta}\, y dy \\
\leq & \frac{\gamma}{1-\delta}.
\end{aligned}
$$

Notice $1-y^\gamma \geq \gamma(1-y)$ for $\gamma \in (0,1]$, $y \in (0,1)$, we have

$$
\begin{aligned}
& \int_0^t \left(t^{1/\gamma}-y^{1/\gamma}\right)^{-\delta}(1-y)^{\alpha-1} dy \\
=& \gamma \int_0^t (t-y)^{-\delta}(1-y^\gamma)^{\alpha-1} y^{\gamma-1} dy \\
\leq & \gamma^\alpha \int_0^t (t-y)^{-\delta}(1-y)^{\alpha-1} y^{\gamma-1} dy \\
\leq & \gamma^\alpha B(\alpha-\delta,\gamma) t^{\alpha-\delta+\gamma-1}.
\end{aligned}
$$

Combining the above estimates yields

$$
A_{\delta,2,2}(t) \leq C < \infty.
$$

Then we conclude by using Lemma 3.3. □

Lemma 3.5 *Assume* $v \in {}_0H^{\delta/2,\psi}(\Lambda_1) \cap B^m_{\omega^{-\alpha,-1}}(\Lambda_1)$, *where* $0 < \delta < 1$, $\psi(t) = t^{1/\gamma}$, $m \geq 1$, $\delta < \alpha < 1$. *Then we have*

1. $\|v\|_{L^2_\psi(\Lambda_1)} \lesssim \|v\|_{L^2_{\omega^{-\alpha,-1}}(\Lambda_1)}$,
2. $\left({}^C D_t^{\delta,\psi} v,\, v\right)_{L^2_\psi(\Lambda_1)} \lesssim \|v'\|_{L^2_{\omega^{1-\alpha,0}}(\Lambda_1)} \|v\|_{L^2_{\omega^{-\alpha,-1}}(\Lambda_1)}$.

Proof

1. Noticing $1/\gamma t^{1/\gamma-1} \leq (1-t)^{-\alpha} t^{-1}$, $\forall t \in \Lambda_1$, we have

$$\|v\|^2_{L^2_\psi(\Lambda_1)} = 1/\gamma \int_{\Lambda_1} v^2 t^{1/\gamma-1} dt \leq \int_{\Lambda_1} v^2 (1-t)^{-\alpha} t^{-1} dt = \|v\|_{L^2_{\omega^{-\alpha,-1}}(\Lambda_1)}.$$

2. By a direct computation, we get

$$\begin{aligned}
\left({}^C D_t^{\delta,\psi} v,\, v\right)_{L^2_\psi(\Lambda_1)} =& \frac{1}{\Gamma(1-\delta)} \int_{\Lambda_1} \int_0^t \left(t^{1/\gamma} - z^{1/\gamma}\right)^{-\delta} v'(z) dz v(t) \psi'(t) dt \\
=& 1/\gamma \int_{\Lambda_1} (P_\delta v')(t) \omega_1(t) v(t) (1-t)^{-\alpha/2} t^{-1/2} dt \\
\leq& 1/\gamma \left[\int_{\Lambda_1} \left|\omega_1(t)(P_\delta v')(t)\right|^2 dt\right]^{1/2} \|v\|_{L^2_{\omega^{-\alpha,-1}}(\Lambda_1)} \\
\lesssim& \left(\int_{\Lambda_1} \left|\omega_2(t) v'(t)\right|^2 dt\right)^{1/2} \|v\|_{L^2_{\omega^{-\alpha,-1}}(\Lambda_1)} \\
=& \|v'\|_{L^2_{\omega^{1-\alpha,0}}(\Lambda_1)} \|v\|_{L^2_{\omega^{-\alpha,-1}}(\Lambda_1)}.
\end{aligned}$$

Here, the Cauchy–Schwarz inequality and Lemma 3.4 have been used.

□

With the above preparation, we are now in a position to derive the error estimate.

Theorem 3.6 *Let* v *be the solution of problem* (21), v_N *is the solution of the problem* (25). *Suppose* $v \in {}_0H^{\delta/2,\psi}(\Lambda) \cap B^m_{\omega^{-\alpha,-1}}(\Lambda)$, *where* $0 < \delta < 1$, $\psi(t) = t^{1/\gamma}$, $m \geq 1$, $\delta < \alpha < 1$. *Then the following error estimate holds:*

$$\|v - v_N\|_{H^{\delta/2,\psi}(\Lambda)} \lesssim N^{1/2-m} \left\|\partial_t^m v\right\|_{L^2_{\omega^{-\alpha+m,-1+m}}(\Lambda)} + N^{-m} \|\partial_t^m v\|_{L^2_{\omega^{-\alpha+m,-1+m}}(\Lambda)}. \tag{28}$$

Proof It follows from (21), (25), and Céa lemma that

$$\|v - v_N\|_{H^{\delta/2,\psi}(\Lambda)} \leq \inf_{w_N \in V_N} \|v - w_N\|_{H^{\delta/2,\psi}(\Lambda)} \leq \|v - \pi_N^{-\alpha,-1} v\|_{H^{\delta/2,\psi}(\Lambda)}.$$

Furthermore it is not difficult to derive

$$\|v - \pi_N^{-\alpha,-1} v\|_{H^{\delta/2,\psi}(\Lambda)} \lesssim \|v - \pi_N^{-\alpha,-1} v\|_{H^{\delta/2,\psi}(\Lambda_1)}. \tag{29}$$

Then, it follows from the definition of the norm $\|\cdot\|_{H^{\delta/2,\psi}(\Lambda)}$, the equivalence of ψ-fractional norms, and the relationship (4):

$$\begin{aligned}
&\|v-\pi_N^{-\alpha,-1}v\|_{H^{\delta/2,\psi}(\Lambda_1)}\\
&\lesssim \left|\left(D_t^{\delta,\psi}(v-\pi_N^{-\alpha,-1}v),\ v-\pi_N^{-\alpha,-1}v\right)_{L^2_\psi(\Lambda_1)}\right|^{1/2}+\left\|v-\pi_N^{-\alpha,-1}v\right\|_{L^2_\psi(\Lambda_1)}\\
&=\left|\left({}^C D_t^{\delta,\psi}(v-\pi_N^{-\alpha,-1}v),\ v-\pi_N^{-\alpha,-1}v\right)_{L^2_\psi(\Lambda_1)}\right|^{1/2}+\left\|v-\pi_N^{-\alpha,-1}v\right\|_{L^2_\psi(\Lambda_1)}.
\end{aligned}$$

Using Lemma 3.5 gives

$$\begin{aligned}
&\|v-\pi_N^{-\alpha,-1}v\|_{H^{\delta/2,\psi}(\Lambda_1)}\\
&\lesssim \left\|(v-\pi_N^{-\alpha,-1}v)'\right\|^{1/2}_{L^2_{\omega^{1-\alpha,0}}(\Lambda_1)}\left\|v-\pi_N^{-\alpha,-1}v\right\|^{1/2}_{L^2_{\omega^{-\alpha,-1}}(\Lambda_1)}+\left\|v-\pi_N^{-\alpha,-1}v\right\|_{L^2_{\omega^{-\alpha,-1}}(\Lambda_1)}.
\end{aligned} \tag{30}$$

Finally, the desired estimate follows from combining (29), (30), and Lemma 3.1. □

3.3 *Implementation*

We discuss the implementation issue of the spectral approximation (25). The key is find efficient way to form the stiffness matrix **S**, those entries are

$$\mathbf{S}_{mn} := \left(D_t^{\delta,\psi} j_n^{\alpha,-1}(t),\ j_m^{\alpha,-1}(t)\right)_{L^2_\psi(\Lambda)} = \left({}^C D_t^{\delta,\psi} j_n^{\alpha,-1}(t),\ j_m^{\alpha,-1}(t)\right)_{L^2_\psi(\Lambda)}$$

for $m,\ n = 1, 2, \cdots, N$. We compute the entries $\mathbf{S}_{mn}$ by using (22) and (23) as follows:

$$\begin{aligned}
&\left({}^C D_t^{\delta,\psi} j_n^{\alpha,-1}(t),\ j_m^{\alpha,-1}(t)\right)_{L^2_\psi(\Lambda)} \\
&=\left(\frac{1}{\Gamma(1-\delta)}\int_0^s \left(s^{1/\gamma}-z^{1/\gamma}\right)^{-\delta}\frac{d}{dz}j_n^{\alpha,-1}(z)dz,\ j_m^{\alpha,-1}(s)\right)_{L^2_\psi(\Lambda)} \\
&=\frac{4nT^{1-\delta}}{\gamma\Gamma(1-\delta)}\int_0^1\int_0^1(1-\tau)^{-\delta}\Big(\frac{1-\tau^{1/\gamma}}{1-\tau}\Big)^{-\delta}J_n^{\alpha+1,0}(2\tau s-1)d\tau\cdot s^{(1-\delta)/\gamma+1}J_m^{\alpha,1}(2s-1)ds \\
&=\frac{4nT^{1-\delta}}{\gamma\Gamma(1-\delta)}\sum_{i=0}^{N}\sum_{j=0}^{N}\left(\frac{1-\hat{\eta}_j^{1/\gamma}}{1-\hat{\eta}_j}\right)^{-\delta}J_n^{\alpha+1,0}(2\eta_i\hat{\eta}_j-1)\hat{\omega}_j J_m^{\alpha,1}(2\eta_i-1)\omega_i,
\end{aligned} \tag{31}$$

where the Gauss quadrature point sets $\{\eta_i\}_{i=0}^N$ and $\{\hat{\eta}_j\}_{j=0}^N$ are zeros of the shifted Jacobi polynomials $J_N^{0,\,(1-\delta)/\gamma+1}(x(t))$, $J_N^{-\delta,\,0}(x(t))$, respectively, and $\{\omega_i\}_{i=0}^N$, $\{\hat{\omega}_j\}_{j=0}^N$ are the associated weights. Note that in our calculation, γ is set to be $1/r$, with r being the positive integers so that $\frac{1-\tau^{1/\gamma}}{1-\tau} = \sum_{k=0}^{r-1}\tau^k$, which hasn't a singularity. The singular parts $(1-\tau)^{-\delta}$ and $s^{(1-\delta)/\gamma+1}$ do not appear in the numerical quadrature since they are treated as the associated weights of the Jacobi polynomials. Denote

$$f_m = (f, j_m^{\alpha,-1}(x)),\quad \mathbf{f} = (f_1,\dots,f_N)^T;$$

$$v_N = \sum_{n=1}^{N}\tilde{v}_n j_n^{\alpha,-1}(x),\quad \mathbf{v} = (\tilde{v}_1,\dots,\tilde{v}_N)^T;$$

$$\mathbf{M} = (\mathbf{M}_{mn})_{1\le m,n\le N} \text{ with } \mathbf{M}_{mn} = \left(j_n^{\alpha,-1}(t),\ j_m^{\alpha,-1}(t)\right)_{L^2_\psi(\Lambda)}.$$

Then the matrix form of the problem (25) reads:

$$(\mathbf{S}+\lambda\mathbf{M})\mathbf{v} = \mathbf{f}.$$

Remark 3.7 In [26], the authors proposed a Müntz spectral method based on the Müntz polynomial space span$\{1, t^\gamma, t^{2\gamma}, \dots, t^{N\gamma}\}$ for the fractional differential equation. It can be verified that, with the particular choice of the transformation $\psi(t) = t^{1/\gamma}$, the current method is equivalent to the one in [26] in the sense that the solution $u_N(s)$ computed from the Müntz spectral method is linked to the solution $v_N(t)$ of the ψ-spectral method through $v_N(t) = u_N(\psi(t))$. However, it is worth to note that the numerical analysis of the two methods was conducted using two quite different frameworks. The new approach in the current work not only provides an alternative tool for numerical analysis of the Müntz spectral methods proposed in [25, 26], but also provide a guideline for the selection of parameter γ. The main goal is to choose a suitable transformation function ψ such that $v(\cdot) = u(\psi(\cdot))$ is as smooth as possible.

4 Application to the Time Fractional Subdiffusion Equations

Let $\Omega := (-1, 1)^d$, $d = 1, 2, 3$. Consider the following time fractional diffusion equation:

$$ {}^C D_s^\delta u(x, s) = \Delta u(x, s) - u(x, s) + g(x, s), \ x \in \Omega, \ s \in I \tag{32}$$

subject to the initial and boundary conditions

$$u(x, 0) = 0, \ x \in \Omega, \tag{33}$$

$$u(x, s) = 0, \ x \in \partial\Omega, \ s \in I. \tag{34}$$

We obtain the following transformed equation by applying the transformation $s = \psi(t)$ in the time direction:

$$D_t^{\delta,\psi} v(x, t) = \Delta v(x, t) - v(x, t) + f(x, t), \ x \in \Omega, \ t \in \Lambda. \tag{35}$$

For the Sobolev space X with norm $\|\cdot\|_X$, let

$$H^{\delta,\psi}(\Lambda, X) := \left\{v; \ \|v(\cdot, t)\|_X \in H^{\delta,\psi}(\Lambda)\right\}, \ \delta \geq 0$$

endowed with the norm

$$\|v\|_{H^{\delta,\psi}(\Lambda,X)} := \| \|v(\cdot, t)\|_X \|_{\delta,\Lambda} \, .$$

Let $\mathcal{O} = \Lambda \times \Omega$,

$$\mathcal{H}^{\delta,\psi}(\mathcal{O}) := {}_0H^{\delta,\psi}(\Lambda, L^2(\Omega)) \cap L^2_\psi(\Lambda, H_0^1(\Omega)),$$

$$\mathcal{B}^m_{\omega^{\alpha,\beta}}(\mathcal{O}) := B^m_{\omega^{\alpha,\beta}}(\Lambda, L^2(\Omega)) \cap L^2_{\omega^{\alpha,\beta}}(\Lambda, H_0^1(\Omega)),$$

equipped respectively with the norms

$$\|v\|_{\mathcal{H}^{\delta,\psi}(\mathcal{O})} := \left(\|v\|^2_{H^{\delta,\psi}(\Lambda,L^2(\Omega))} + \|v\|^2_{L^2_\psi(\Lambda,H_0^1(\Omega))} \right)^{1/2},$$

$$\|v\|_{\mathcal{B}^m_{\omega^{\alpha,\beta}}(\mathcal{O})} := \left(\|v\|^2_{B^m_{\omega^{\alpha,\beta}}(\Lambda,L^2(\Omega))} + \|v\|^2_{L^2_{\omega^{\alpha,\beta}}(\Lambda,H_0^1(\Omega))} \right)^{1/2}.$$

Consider the following variational formulation of (35):

$$\mathcal{A}(v, w) = \mathcal{F}(w), \quad \forall w \in \mathcal{H}^{\delta/2,\psi}(O), \tag{36}$$

where

$$\mathcal{A}(v, w) := (D_t^{\delta,\psi} v, w)_O + (\nabla v, \nabla w)_O + (v, w)_O, \ \mathcal{F}(w) := (f, w)_O$$

with $(v, w)_O := \int_\Lambda \int_\Omega v w \psi'(t) dx dt$.

Similar to Theorem 3.2, we can establish the coercivity and continuity of the bilinear form $\mathcal{A}(v, w)$ in the space $\mathcal{H}^{\delta/2,\psi}(O) \times \mathcal{H}^{\delta/2,\psi}(O)$, and therefore the well-posedness of the weak problem (36) for any $f \in \mathcal{H}^{\delta/2,\psi}(O)'$ (the dual space of

$\mathcal{H}^{\delta/2,\psi}(O))$, together with the stability estimate

$$\|v\|_{\mathcal{H}^{\delta/2,\psi}(O)} \lesssim \|f\|_{\mathcal{H}^{\delta/2,\psi}(O)'}.$$

We now propose a space-time Galerkin spectral method to discretize (36). For the time variable, we follow the approach of the previous section. For the space variable, we use standard Legendre polynomials. Let

$$\begin{aligned} &\phi_k(x) = c_k(L_k(x) - L_{k+2}(x)), \quad c_k = \frac{1}{\sqrt{4k+6}}, \\ &a_{jk} = (\partial_x \phi_k(x), \partial_x \phi_j(x)), \quad b_{jk} = (\phi_k(x), \phi_j(x)), \end{aligned} \tag{37}$$

where $L_n(x)$ is the n-th degree Legendre polynomial. Then [51]

$$a_{jk} = \begin{cases} 1, & k = j \\ 0, & k \neq j \end{cases}, \quad b_{jk} = b_{kj} = \begin{cases} c_k c_j\left(\frac{2}{2j+1} + \frac{2}{2j+5}\right), & k = j \\ -c_k c_j \frac{2}{2k+1}, & k = j+2, \\ 0, & \text{otherwise} \end{cases} \tag{38}$$

Set the polynomial space

$$\mathcal{P}_M = \operatorname{span}\{\phi_0(x),\ \phi_1(x),\ \ldots,\ \phi_{M-2}(x)\}.$$

The space-time Galerkin spectral method for (36) is to seek $v_L(x,t) \in \mathcal{P}_M \otimes V_N$ such that

$$\mathcal{A}(v_L, w) = \mathcal{F}(w), \quad \forall\, w \in \mathcal{P}_M \otimes V_N. \tag{39}$$

The error estimate is given in the following theorem without proof.

Theorem 4.1 *Let v be the solution of problem (36), v_L is the solution of problem (39). Suppose $v \in \mathcal{H}^{\delta/2,\psi}(O) \cap B^m_{\omega^{-\alpha,-1}}(\Lambda, H^\sigma(\Omega))$, where $0 < \delta < 1$, $\psi(t) = t^{1/\gamma}$, $m \geq 1$, $\delta < \alpha < 1$. Then the following error estimate holds:*

$$\begin{aligned} \|v - v_N\|_{\mathcal{H}^{\delta/2,\psi}(O)} \lesssim & N^{1/2-m}\left\|\partial_t^m v\right\|_{L^2_{\omega^{-\alpha+m,-1+m}}(\Lambda, L^2(\Omega))} \\ & + N^{1/2-m} M^{-\sigma}\left\|\partial_t^m v\right\|_{L^2_{\omega^{-\alpha+m,-1+m}}(\Lambda, H^\sigma(\Omega))} \\ & + M^{-\sigma}\left(\left\|D_t^{\delta/2,\psi} v\right\|_{L^2_{\omega^{-\alpha,-1}}(\Lambda, H^\sigma(\Omega))} + \|v\|_{L^2_{\omega^{-\alpha,-1}}(\Lambda, H^\sigma(\Omega))}\right) \\ & + M^{1-\sigma}\|v\|_{L^2_{\omega^{-\alpha,-1}}(\Lambda, H^\sigma(\Omega))} + N^{-m}\left\|\partial_t^m v\right\|_{L^2_{\omega^{-\alpha+m,-1+m}}(\Lambda, H_0^1(\Omega))}. \end{aligned}$$

In the implementation, we express the solution v_L of Eq. (39) using the modal basis as follows:

$$v_L(x,t) = \sum_{m=0}^{M-2}\sum_{n=1}^{N} \hat{v}_{nm}\phi_m(x) j_n^{\alpha,-1}(t).$$

Substituting this expression into (39), and taking $w = \phi_p(x) j_q^{\alpha,-1}(t)$, we obtain

$$\sum_{m=0}^{M-2}\sum_{n=1}^{N} \hat{v}_{nm}\Big\{ (\phi_m, \phi_p)_{L^2(\Omega)} \left(D_t^{\delta,\psi} j_n^{\alpha,-1}, j_q^{\alpha,-1}\right)_{L^2_\psi(\Lambda)} + \left(\phi'_m, \phi'_p\right)_{L^2(\Omega)} \left(j_n^{\alpha,-1}, j_q^{\alpha,-1}\right)_{L^2_\psi(\Lambda)}$$
$$+ (\phi_m, \phi_p)_{L^2(\Omega)} \left(j_n^{\alpha,-1}, j_q^{\alpha,-1}\right)_{L^2_\psi(\Lambda)} \Big\} = \left(f, \phi_p j_q^{\alpha,-1}\right)_O.$$

Denote

$$\mathbf{B} = (b_{jk})_{0\le j,k\le M-2},$$
$$f_{nm} = \left(f, \phi_m(x) j_n^{\alpha,-1}(t)\right)_O, \quad \mathbf{F} = (f_{nm})_{1\le n\le N,\ 0\le m\le M-2},$$
$$\mathbf{S} = \left(\mathbf{S}_{nq}\right)_{1\le n,q\le N}, \quad \mathbf{M} = \left(\mathbf{M}_{nq}\right)_{1\le n,q\le N}, \quad \mathbf{V} = \left(\hat{v}_{nm}\right)_{1\le n\le N,\ 0\le m\le M-2}.$$

Using the above notations, (39) can be written under the following matrix form:

$$\mathbf{SVB} + \mathbf{MV} + \mathbf{MVB} = \mathbf{F}.$$

5 Numerical Examples

In this section, we present some numerical examples to illustrate the high accuracy of the proposed method based on GJPs in solving problem (19) with smooth and nonsmooth solutions. In particular, we test the accuracy of the proposed method when the exact solution is unknown. The space-time spectral method based on GJPs and Legendre polynomials presented in Sect. 4 will also be tested for the two-dimensional time fractional subdiffusion equation. The time interval is set to $[a_\psi, b_\psi] := [0, 2^\gamma]$. Note that $\psi(t) = t^{1/\gamma}$ $(0 < \gamma \le 1)$ in the following examples.

Example 1 (Smooth Solution) In this test, we choose the fabricated exact solution $u(s) = s^2$. Naturally, in this case, we take $\gamma = 1$.

Table 1 (Example 1) L^{∞}- and L^2-errors versus N and different δ

	$\delta = 0.1$		$\delta = 0.5$		$\delta = 0.9$	
N	$\|v - v_N\|_{L^\infty}$	$\|v - v_N\|_0$	$\|v - v_N\|_{L^\infty}$	$\|v - v_N\|_0$	$\|v - v_N\|_{L^\infty}$	$\|v - v_N\|_0$
2	3.3307e−16	4.0030e−16	1.1102e−16	1.3878e−16	1.4433e−15	1.4647e−15
4	1.3323e−15	1.4767e−15	1.5543e−15	1.6812e−15	1.1990e−14	1.4989e−14

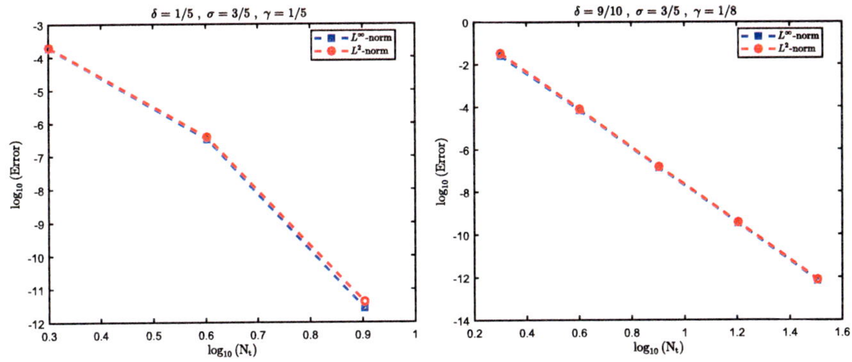

Fig. 1 (Example 2) L^{∞}- and L^2-errors in log scale versus the polynomial degree for fractional power solution

The main purpose of this example is to check the high accuracy of the proposed Galerkin spectral scheme (25) for smooth solutions. The computed results are presented in Table 1, from which we observe that the numerical solutions for some different δ reach the machine accuracy with small polynomial degree N.

Example 2 (Nonsmooth Solution) Consider problem (19) with the fabricated exact solution $u(s) = s^{\sigma}$ for two values of σ : $3/5, \sqrt{2}/2$.

We want to use this example to test the accuracy of the spectral method for nonsmooth solutions. For the fractional $\sigma = 3/5$, we take $\gamma = 1/5$ or $1/8$. The numerical errors versus the polynomial degree N_t for several δ is plotted in Fig. 1. It is observed from this figure that the errors decay exponentially as the polynomial degree increases. For the irrational number $\sigma = \sqrt{2}/2$, we take $\gamma = 1/7$. The obtained result is given in Fig. 2, from which we also observe the spectral convergence.

Example 3 (Unknown Solution) Consider problem (19) with a given source function $f(s) = \sin(s)$. In this case the exact solution and its singularity structure are unknown.

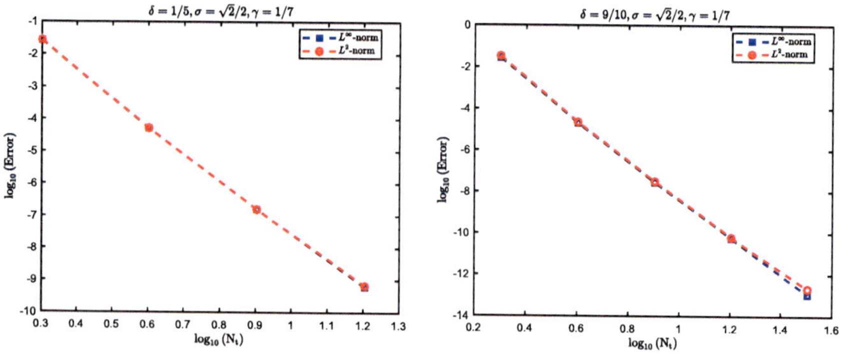

Fig. 2 (Example 2) L^∞- and L^2-errors in log scale versus the polynomial degree for irrational power solution

Since the exact solution is unknown, a numerical solution computed with very fine resolution is served as the reference solution. The solution qualities are compared for different δ by two approaches, i.e., our method and usual spectral method, by plotting the errors versus the polynomial degrees in Fig. 3. We see that more accurate solutions are obtained by using $\gamma = 1/5$ or $1/6$, compared to the classical spectral method, i.e., $\gamma = 1$.

Example 4 (2D Time Fractional Subdiffusion Equation) Consider the 2D subdiffusion Eqs. (32)–(34) with the fabricated exact solution

$$u(x, y, s) = \sin(\pi x)\sin(\pi y)s^{3/5}.$$

In Fig. 4, we depict the exact solution, numerical solution and error at the final time computed with the polynomial degree 20 in both directions. As shown in this figure, a very accurate solution is obtained with pointwise error reaching as small as $O(10^{-13})$. The error history as a function of the polynomial degrees M or N, shown in Fig. 5, confirms the spectral convergence of the used method.

Remark 5.1 For the selection of parameter γ, our fundamental principle is to make $v(\cdot) = u((\cdot)^{1/\gamma})$ sufficiently smooth which can be made according to the following strategy:

Case I: if the solution u is smooth, the optimal value is $\gamma = 1$;

Case II: if the source term $g(\cdot)$ is smooth, then (a) when δ is a rational number p/q, the best choice is $\gamma = 1/q$. Theoretically $\gamma = 1/nq$ $(n = 1, 2, \cdots)$ works

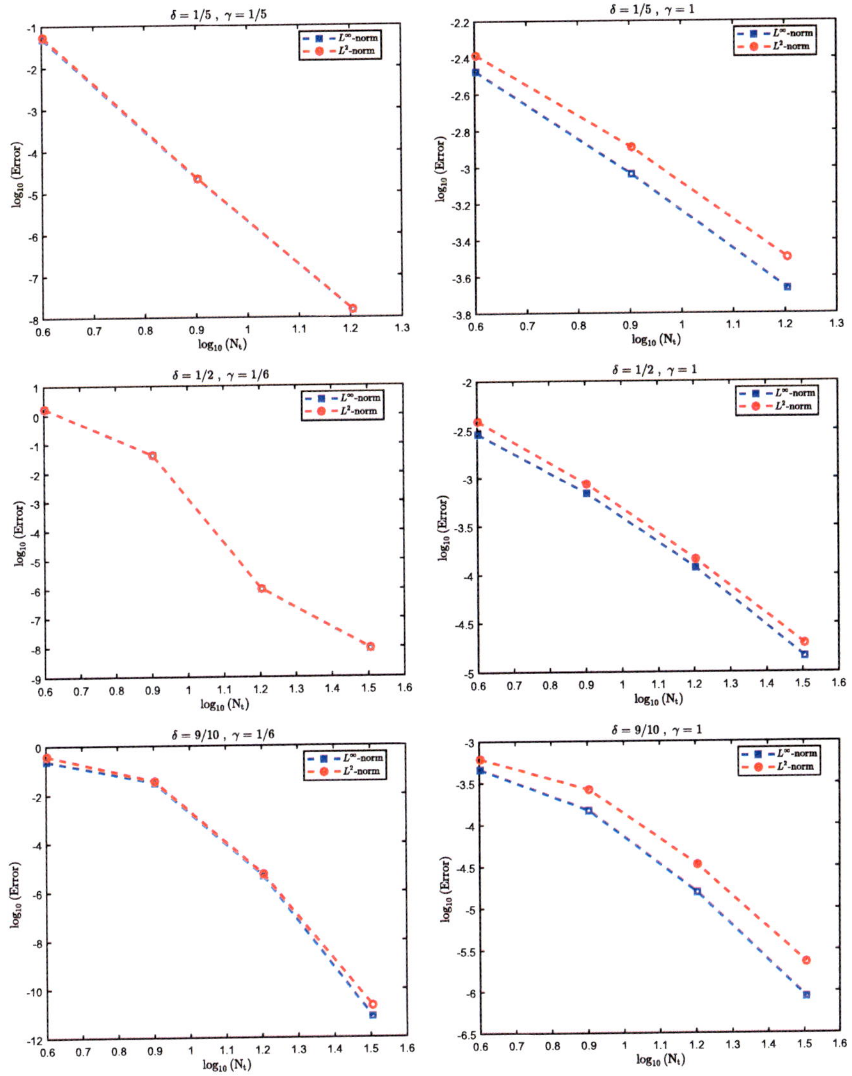

Fig. 3 (Example 3) L^∞- and L^2-errors in log scale versus the polynomial degree N_t for different δ and γ

too, but larger n leads to larger amount of calculation; (*b*) when δ is an irrational number, there is no suitable value of γ to make $u((\cdot)^{1/\gamma})$ smooth. In this case, we can take $\gamma = 1/q$ with a reasonably large q such that $u((\cdot)^{1/\gamma})$ is smooth enough.

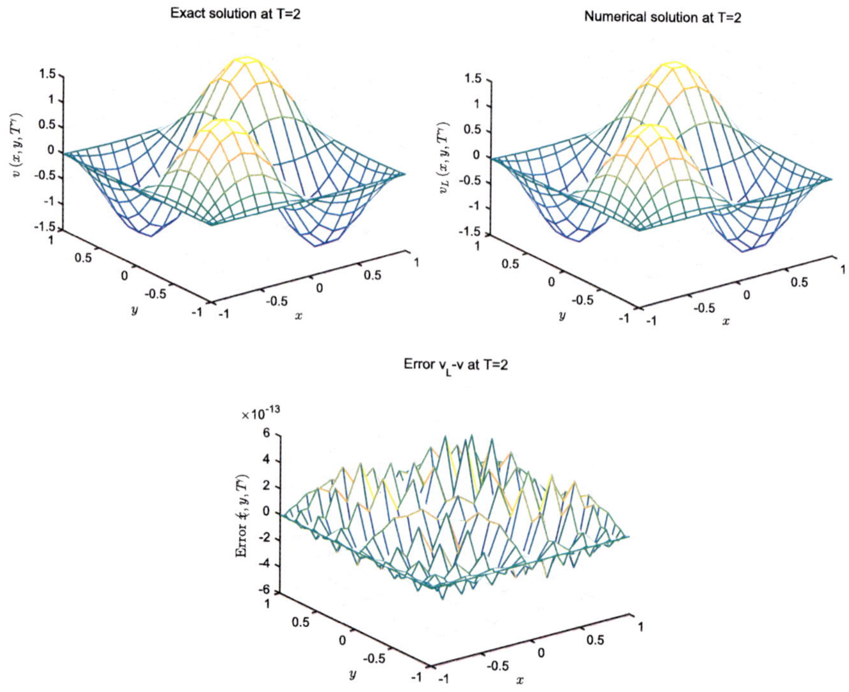

Fig. 4 (Example 4) The exact solution v, numerical solution v_L, and error $v_L - v$ at $T = 2$ with $\gamma = 1/5$

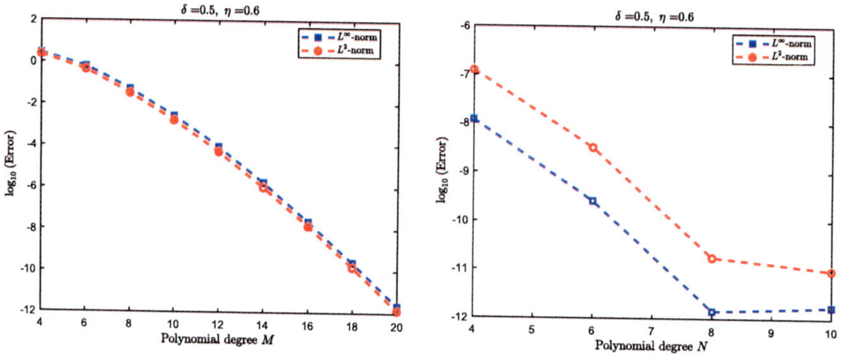

Fig. 5 (Example 4) L^∞- and L^2-errors in semi-log scale versus M or N with $\delta = 0.5$, $\gamma = 1/5$

6 Concluding Remarks

A novel spectral method has been proposed and analyzed for the subdiffusion equation. The main novelty of the proposed method is its variational framework based on fractional Sobolev spaces. The idea was to first apply suitable variable transformation to re-scale the underlying equation, then construct spectral methods for the re-scaled equation. This is particularly useful in numerical solutions of fractional differential equations, to which the solution is often singular and can be smoothed by using appropriate transformation. For this purpose, a new variational framework was established based on the fractional Sobolev spaces, which allows constructing and analyzing numerical methods following the standard Galerkin approach. Our theoretical and numerical investigation showed that the proposed method using suitable transformation is exponentially convergent for general right hand side functions, even though the exact solution has limited regularity. Implementation details was also provided, along with a series of numerical examples to demonstrate the efficiency of the proposed method.

It is worth mentioning here a number of points: First, with some specific choices of the transformation function, the new method can be proved to be equivalent to the Müntz spectral method, recently proposed in a series of papers [25, 26]. The latter was based on the Müntz polynomial approximation to the original equation; Secondly, although the error analysis was carried out only for a particular transformation, it seems extendable to some other choices; Finally, compared to the Müntz spectral method, the main benefit of the current method may be its flexibility in choosing the transformation function. This makes the new method applicable to a larger class of problems.

Acknowledgments This research is supported by the NSFC grant 11971408.

References

1. Alikhanov, A.: A new difference scheme for the time fractional diffusion equation. J. Comput. Phys. **280**(C), 424–438 (2015)
2. Almeida, R.: A Caputo fractional derivative of a function with respect to another function. Commun. Nonlinear Sci Numer Simulat. **44**, 460–481 (2017)
3. Almeida, R., Malinowska, A., Monteiro, M.: Fractional differential equations with a Caputo derivative with respect to a kernel function and their applications. Math. Meth. Appl. Sci. **41**, 336–352 (2018)
4. Almeida, R., Jleli, M., Samet, B.: A numerical study of fractional relaxation-oscillation equations involving ψ-caputo fractional derivative. RACSAM **113**, 1873–1891 (2019)
5. Amblard, F., Maggs, A.C., Yurke, B., Pargellis, A.N., Leibler, S.: Subdiffusion and anomalous local viscoelasticity in actin networks. Phys. Rev. Lett. **77**, 4470 (1996)
6. Andersen, K., Heinig, H.: Weighted norm inequalities for certain integral operators. SIAM J. Numer. Anal. **14**(4), 834–844 (1983)
7. Baffet, D., Hesthaven, J.: High-order accurate adaptive kernel compression time-stepping schemes for fractional differential equations. J. Sci. Comput. **72**(3), 1169–1195 (2017)
8. Baffet, D., Hesthaven, J.: A kernel compression scheme for fractional differential equations. SIAM J. Numer. Anal. **55**(2), 496–520 (2017)
9. Barkai, E., Metzler, R., Klafter, J.: From continuous time random walks to the fractional Fokker-Planck equation. Phys. Rev. E **61**, 132–138 (2000)

10. Brown, E., Wu, E., Zipfel, W., Webb, W.: Measurement of molecular diffusion in solution by multiphoton fluorescence photobleaching recovery. Biophys. J. **77**, 2837–2849 (1999)
11. Chen, S., Shen, J.: Log orthogonal functions: approximation properties and applications. IMA J. Numer. Anal. **00**, 1–32 (2020)
12. Chen, S., Shen, J., Wang, L.: Generalized Jacobi functions and their applications to fractional differential equations. Math. Comput. **85**, 1603–1638 (2016)
13. Chen, S., Shen, J., Zhang, Z., Zhou, Z.: A spectrally accurate approximation to subdiffusion equations using the log orthogonal functions. SIAM J. Sci. Comput. **42**, A849–A877 (2020)
14. Cuesta, E., Lubich, C., Palencia, C.: Convolution quadrature time discretization of fractional diffusion-wave equations. Math. Comput. **75**(254), 673–696 (2006)
15. Ervin, V., Roop, J.: Variational formulation for the stationary fractional advection dispersion equation. Numer. Methods Partial Differ. Equ. **22**(3), 558–576 (2006)
16. Fan, B., Xu, C.J.: Identifying source term in the subdiffusion equation with l2-tv regularization. Inverse Prob. **37**, 105008 (2021)
17. Feder, T., Brust-Mascher, I., Slattery, J., Baird, B., Webb, W.: Constrained diffusion or immobile fraction on cell surfaces: a new interpretation. Biophys. J. **70**, 2767–2773 (1996)
18. Ford, N., Yan, Y.: An approach to construct higher order time discretisation schemes for time fractional partial differential equations with nonsmooth data. Fract. Calc. Appl. Anal. **20**(5), 1076–1105 (2017)
19. Gao, G., Sun, Z., Zhang, Y.: A finite difference scheme for fractional sub-diffusion equations on an unbounded domain using artificial boundary conditions. J. Comput. Phys. **231**(7), 2865–2879 (2012)
20. Gao, G., Sun, Z., Zhang, H.: A new fractional numerical differentiation formula to approximate the caputo fractional derivative and its applications. J. Comput. Phys. **259**(2), 33–50 (2014)
21. Ghosh, R., Webb, W.: Automated detection and tracking of individual and clustered cell low density lipoprotein receptor molecules. Biophys. J. **68**, 766–778 (1994)
22. Guo, B., Shen, J., Wang, L.: Generalized Jacobi polynomials/functions and their applications. Appl. Numer. Math. **59**(5), 1011–1028 (2009)
23. Henry, B., Wearne, S.: Fractional reaction-diffusion. Physica A **276**, 448–455 (2000)
24. Henry, B., Wearne, S.: Existence of turing instabilities in a two-species fractional reaction-diffusion system. SIAM J. Appl. Math. **62**(3), 870–887 (2002)
25. Hou, D., Xu, C.: A fractional spectral method with applications to some singular problems. Adv. Comput. Math. **43**(5), 911–944 (2017)
26. Hou, D., Hasan, M., Xu, C.: Müntz spectral methods for the time-fractional diffusion equation. Comput. Methods Appl. Math. **18**(1), 43–62 (2018)
27. Jiang, S., Zhang, J., Zhang, Q., Zhang, Z.: Fast evaluation of the caputo fractional derivative and its applications to fractional diffusion equations. Commun. Comput. Phys. **21**(3), 650–678 (2017)
28. Jin, B., Lazarov, R., Zhou, Z.: Two fully discrete schemes for fractional diffusion and diffusion-wave equations with nonsmooth data. J. Sci. Comput. **38**(1), A146–A170 (2016)
29. Jin, B., Li, B., Zhou, Z.: Correction of high-order BDF convolution quadrature for fractional evolution equations. SIAM J. Sci. Comput. **39**(6), A3129–A3152 (2017)
30. Kilbas, A., Srivastava, H., Trujillo, J.: Theory and Applications of Fractional Differential Equations. Elsevier, San Diego (2006)
31. Li, X., Xu, C.: A space-time spectral method for the time fractional diffusion equation. SIAM J. Numer. Anal. **47**(3), 2108–2131 (2009)
32. Li, X., Xu, C.: Existence and uniqueness of the weak solution of the space-time fractional diffusion equation and a spectral method approximation. Commun. Comput. Phys. **8**(5), 1016–1051 (2010)
33. Li, D., Sun, W., Wu, C.: A novel numerical approach to time-fractional parabolic equations with nonsmooth solutions. Numer. Math. Theor. Meth. Appl. **14**(2), 355–376 (2021)
34. Liao, H., Li, D., Zhang, J.: Sharp error estimate of the nonuniform L1 formula for linear reaction-subdiffusion equations. SIAM J. Numer. Anal. **56**(2), 1112–1133 (2018)

35. Liao, H., McLean, W., Zhang, J.: A discrete Grönwall inequality with applications to numerical schemes for subdiffusion problems. SIAM J. Numer. Anal. **57**(1), 218–237 (2019)
36. Lin, Y., Xu, C.: Finite difference/spectral approximations for the time-fractional diffusion equation. J. Comput. Phys. **225**(2), 1533–1552 (2007)
37. Lubich, C., Sloan, I., Thomée, V.: Nonsmooth data error estimates for approximations of an evolution equation with a positive-type memory term. Math. Comput. **65**(213), 1–17 (1996)
38. Lv, C., Xu, C.: Improved error estimates of a finite difference/spectral method for time-fractional diffusion equations. Int. J. Numer. Anal. Mod. **12**(2), 384–400 (2015)
39. Lv, C., Xu., C.: Error analysis of a high order method for time-fractional diffusion equation. SIAM J. Sci. Comput. **38**(5), A2699–A2724 (2016)
40. Mainardi, F.: Fractional diffusive waves in viscoelastic solids. In: Wegner, J.L., Norwood, F.R. (eds.) Nonlinear Waves in Solids, ASME Book No. AMR 137, Fairfield, pp. 93–97 (1995)
41. Metzler, R., Klafter, J.: The random walk's guide to anomalous diffusion: a fractional dynamics approach. Phys. Rep. **339**, 1–77 (2000)
42. Müller, H.P., Kimmich, R., Weis, J.: NMR flow velocity mapping in random percolation model objects: evidence for a power-law dependence of the volume-averaged velocity on the probe-volume radius. Phys. Rev. E **54**, 5278–5285 (1996)
43. Nigmatullin, R.R.: Realization of the generalized transfer equation in a medium with fractal geometry. Physica B **133**, 425–430 (1986)
44. Oldham, K., Spanier, J.: The Fractional Caculus. SIAM, Philadelphia (1974)
45. Podlubny, I.: Fractional Difierential Equations. Academic, New York (1999)
46. Sakamoto, K., Yamamoto, M.: Initial value/boundary value problems for fractional diffusion-wave equations and applications to some inverse problems. J. Math. Anal. Appl. **382**(1), 426–447 (2011)
47. Samko, S., Kilbas, A., Marichev, O.: Fractional Integrals and Derivatives: Theory and Applications. Gordon and Breach Science Publishers, Switzerland (1993)
48. Scher, H., Lax, M.: Stochastic transport in a disordered solid. Phys. Rev. B **7**, 4491–4502 (1973)
49. Scher, H., Montroll, E.: Anomalous transit-time dispersion in amorphous solids. Phys. Rev. B **12**, 2455–2477 (1975)
50. Sheets, E., Lee, G., Simson, R., Jacobson, K.: Transient confinement of a glycosylphosphatidylinositol-anchored protein in the plasma membrane. Biochemistry **36**, 12449–12458 (1997)
51. Shen, J.: Efficient spectral-Galerkin method I: direct solvers of second and fourth-order equations using Legendre polynomials. SIAM J. Sci. Comput. **15**(6), 1489–1505 (1994)
52. Shen, J., Tang, T., Wang, L.: Spectral Methods: Algorithms, Analysis and Applications. Series in Computational Mathematics, vol. 41. Springer-Verlag, Berlin/Heidelberg (2011)
53. Smith, P., Morrison, I., Wilson, K., Fernandez, N., Cherry, R.: Anomalous diffusion of major histocompatability complex class i molecules on hela cells determined by single particle tracking. Biophys. J. **76**, 3331–3344 (1999)
54. Stynes, M., O'Riordan, E., Gracia, J.: Error analysis of a finite difference method on graded meshes for a time-fractional diffusion equation. SIAM J. Numer. Anal. **55**(2), 1057–1079 (2017)
55. Sun, Z., Wu, X.: A fully discrete difference scheme for a diffusion-wave system. Appl. Numer. Math. **56**(2), 193–209 (2006)
56. Yan, Y., Sun, Z., Zhang, J.: Fast evaluation of the caputo fractional derivative and its applications to fractional diffusion equations: A second-order scheme. Commun. Comput. Phys. **22**(4), 1028–1048 (2017)
57. Ye, X., Xu, C.: A posteriori error estimates of spectral method for the fractional optimal control problems with non-homogeneous initial conditions. AIMS Math. **6**(11), 12028–12050 (2021)
58. Zayernouri, M., Karniadakis, G.E.: Fractional Sturm-Liouville eigen-problems: theory and numerical approximation. J. Comput. Phys. **252**, 495–517 (2014)
59. Zayernouri, M., Ainsworth, M., Karniadakis, G.E.: A unified Petrov-Galerkin spectral method for fractional PDEs. Comput. Method. Appl. M. **283**(1), 1545–1569 (2015)

60. Zeng, F., Zhang, Z., Karniadakis, G.: Second-order numerical methods for multi-term fractional differential equations: smooth and non-smooth solutions. Comput. Methods Appl. Mech. Eng. **327**, 478–502 (2017)
61. Zeng, F., Turner, I., Burrage, K.: A stable fast time-stepping method for fractional integral and derivative operators. J. Sci. Comput. **77**(1), 283–307 (2018)
62. Zhang, Q., Zhang, J., Jiang, S., Zhang, Z.: Numerical solution to a linearized time fractional KdV equation on unbounded domains. Math. Comput. **87**(310) (2018)

Part II
Contributed Papers

A Face-Based Eight-Order Scheme for Convection-Diffusion Problems with Polyhedral Unstructured Grids

Duarte M. S. Albuquerque and Filipe J. M. Diogo

1 Introduction

High-order accurate methods for unstructured grids have historically been focused on hyperbolic equations. Barth and Frederickson, [1] developed a third-order Finite Volume Methods (FVM) for the resolution of the Euler equations. Olliver-Gooch was a pioneer in the application of finite volume method for the convection-diffusion equation [2]. He later coupled the Euler system with the viscous term, which is discretizated as a diffusion flux [3].

Several polynomial reconstruction techniques applied to FVM can be highlighted: the fourth-order methods of Ollivier-Gooch et al. [4], Cueto-Felgueroso et al. [5], and Nogueira et al. [6], also sixth-order results have been reported by Clain et al. [7]. More recently, Vasconcelos et al. [8] has created an eight-order elliptical operator that can be applied to polyhedral unstructured grids. The current work is an extension of such operator for convection-diffusion problems. More recently, three dimensional results have been reported by R. Costa et al. [9] and a novel unsteady algorithm for convection-diffusion has been proposed by P. Costa et al. [10].

Other authors have also used the weighted least-squares for other numerical applications. Albuquerque et al. [11] has implemented absolute error estimators for second-order with unstructured grids. Martins et al. [12] has created a third-order interpolation method with a divergence free constraint. It was used in immersed boundary applications in conjunction with polyhedral unstructured grids.

D. M. S. Albuquerque (✉) · F. J. M. Diogo
IDMEC, Instituto Superior Técnico, Universidade de Lisboa, Lisbon, Portugal
e-mail: DuarteAlbuquerque@tecnico.ulisboa.pt

J. M. Melenk et al. (eds.), *Spectral and High Order Methods for Partial Differential Equations ICOSAHOM 2020+1*, Lecture Notes in Computational Science and Engineering 137, https://doi.org/10.1007/978-3-031-20432-6_4

2 Finite Volume Discretization in 1D Grids

The convection-diffusion equation is composed by a diffusion term, a convective term and an optional source term. This last one is required, when using manufactured analytical solutions. This general formulation is written in Eq. (1), where ϕ is the variable being studied (for example: velocity, temperature, etc.), $\mathbf{V}$ stands for the velocity vector (associated with the convective field), Γ is the diffusion coefficient and φ_ϕ is the source term.

$$\nabla.(\mathbf{V}\phi) - \nabla.(\Gamma\nabla\phi) = \varphi_\phi \tag{1}$$

by integrating the previous equation over a control volume (CV), Eq. (2) is obtained. This is the first step of the finite volume method (FVM).

$$\int_{CV} \nabla.(\mathbf{V}\phi)dV - \int_{CV} \nabla.(\Gamma\nabla\phi)dV = \int_{CV} \varphi_\phi dV \tag{2}$$

The Gauss Divergence Theorem states that the outward flux property in a close surface is equal to the volume integral of its divergence. In this way, the CV integral from the previous equation can be rearranged into a surface (S) one, with $\mathbf{n}$ being the outward face normal.

$$\int_S (\phi\mathbf{V}).\mathbf{n}dS - \int_S \Gamma(\nabla\phi).\mathbf{n}dS = \int_{CV} \varphi_\phi dV \tag{3}$$

The FMV discretizes the governing equations in a integral form. The integrals over the CV faces, make the method inherently conservative, since the flux that leaves a CV will be the same that enters the adjacent one. This is an advantage of the FVM, considering that most physical properties studied in CFD are conservative (i.e., mass and energy) [13]. The source term integration is done with the classical Gauss quadrature rules.

2.1 *Discretization of the Face Value and Gradient*

The previous Eq. (3) must be discretized in the domain to compute the numerical solution. Figure 1 depicts a 1D domain, with the east (e) and west (w) faces of cell i and their respective face normal vectors. Considering this 1D domain, Eq. (3) can be re-arranged as the following:

$$V\phi_e - V\phi_w + \Gamma(\nabla\phi)_w - \Gamma(\nabla\phi)_e = \int_{CV} \varphi_\phi dV \tag{4}$$

Determining $\nabla\phi_e$, $\nabla\phi_w$, ϕ_e and ϕ_w at the faces with high-order accuracy can be done by using a combination of multiple Taylor series. The Taylor series, written in Eq. (5), is an expansion of a function near a given point. In Eq. (5), x_f is the position of the face f, where the series is being reconstructed and x is the location where it is necessary to estimate the function value.

$$\phi(x) = \sum_{n=0}^{\infty}(x - x_f)^n \frac{\phi^{(n)}(x_f)}{n!} \tag{5}$$

$$\phi(x) = \phi(x_f)+(x-x_f)\frac{\partial\phi(x_f)}{\partial x}+\frac{1}{2!}(x-x_f)^2\frac{\partial^2\phi(x_f)}{\partial x^2}+\frac{1}{3!}(x-x_f)^3\frac{\partial^3\phi(x_f)}{\partial x^3}+\dots$$

The sum of multiple Taylor series, all centered around face f can be used to estimate the values of the neighboring cells. The coefficients that multiply each series can be computed by solving a linear system. This forces the accuracy to be as high as possible, it is necessary that the coefficients combinations cancel the low order terms. Below is an example of the linear system obtained for a fourth order gradient estimation on the west face of Fig. 1, by using cells $i+1$, i, $i-1$ and $i-2$. Assuming that each Taylor series is multiplied by the constants A, B, C and D, respectively, the system will be similar to the one on matrix (6). The right hand side (RHS) of the system should has an unity value in the line containing the derivative that is being estimated and zero for the other one. For example, when estimating the gradient on the west face the RHS should be [0 1 0 0]'.

$$\begin{bmatrix} 1 & 1 & 1 & 1 \\ (x_{i+1}-x_w) & (x_i-x_w) & (x_{i-1}-x_w) & (x_{i-2}-x_w) \\ (x_{i+1}-x_w)^2 & (x_i-x_w)^2 & (x_{i-1}-x_w)^2 & (x_{i-2}-x_w)^2 \\ (x_{i+1}-x_w)^3 & (x_i-x_w)^3 & (x_{i-1}-x_w)^3 & (x_{i-2}-x_w)^3 \end{bmatrix} \begin{bmatrix} A \\ B \\ C \\ D \end{bmatrix} = \begin{bmatrix} 0 \\ 1 \\ 0 \\ 0 \end{bmatrix} \tag{6}$$

The west gradient will then be given by Eq. (7), with the values of A, B, C and D taken from inverting the local matrix.

$$\nabla\phi_w = A\phi_{i+1} + B\phi_i + C\phi_{i-1} + D\phi_{i-2} \tag{7}$$

By combining the Taylor series from four different locations (all centered around face f), it is possible to impose the first, third, and fourth terms of their combination to be zero. So, the lowest order term that will contribute to the error will be the fifth

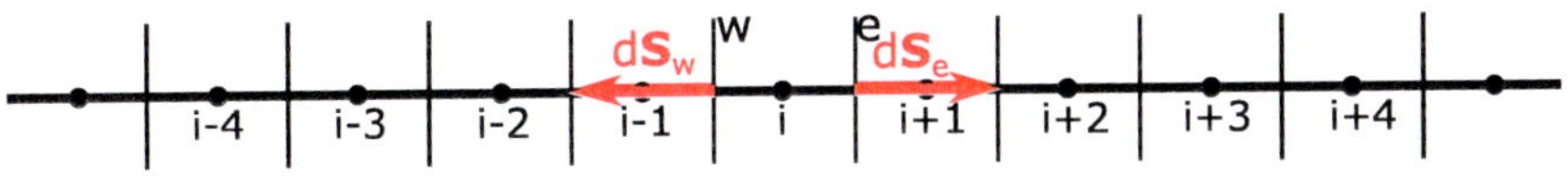

Fig. 1 Finite volume discretization over cell i, on a one-dimensional grid (• representing the cell centers)

term of the Taylor series combination, that is directly proportional to the reference length to the fourth (h^4). If instead, the objective was to determine the property value and not its gradient, the only difference would be on the right hand side, by swapping the first and second terms ([1 0 0 0]'). In this particular 1D method, the number of required points should be equal to the convergence order o, as $O(h^o)$.

Once the local coefficients have been obtained, the results will be passed to the global matrix $\boldsymbol{A}$ of the system $\boldsymbol{A\Phi} = \boldsymbol{b}$, following Eq. (4).

2.2 Contributions from the Boundaries

To fully close the problem, it is necessary to impose boundary conditions. They will turn the partial differential equations problem into a well posed one. Two types of boundary conditions can be considered, the first is Dirichlet and it is used when the value at the boundary is known $\phi = \phi_b$. The second boundary condition is Neumann and it is used when the flux at the border is imposed or known $\nabla\phi = \nabla\phi_b$. The border faces can also be included into the local matrix to obtain the coefficients, however their analytical value is not added to the global matrix, but instead goes to the right hand side of the equation $\boldsymbol{A\Phi} = \boldsymbol{b}$.

2.3 The Peclet Number

Throughout this work, the Peclet number (Pe) will be used. This number is defined by the convective velocity times a local reference length (δx) divided by the diffusion coefficient, see Eq. (8). Here δx will be the local grid size of the mesh. This non-dimension number provides information about the phenomenon (convection or diffusion) that most influences the problem.

$$P_e = \frac{|\mathbf{V}|\delta x}{\Gamma} \tag{8}$$

Another relevant number is the Pe_L, which is similar to the Peclet number from Eq. (8) but considers the length of the domain L instead of the grid size δx. This is a useful parameter, since it gives the convection-diffusion ratio for a given solution, regardless of the grid size used.

2.4 Tested Grids in 1D Results

In this sub-section, the grids used to analyze the algorithm performance are presented. One sample per grid type is shown, with approximately similar cell sizes. For convergence order testing, geometrical identical grids with smaller cell sizes were used.

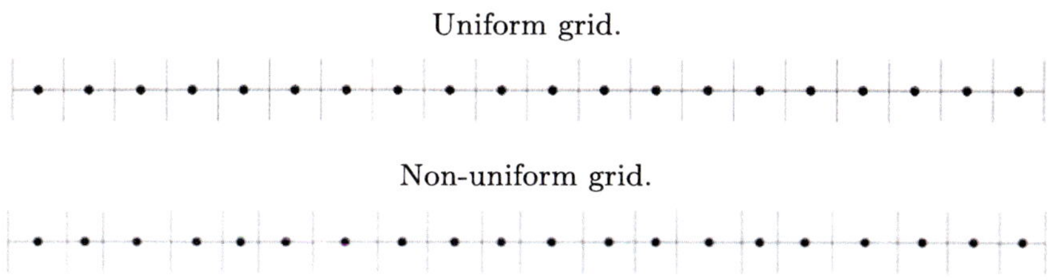

Fig. 2 Samples of 1D grids used in this work

The uniform grid (Fig. 2) has its cell size constant through all the domain, while the non-uniform grid (Fig. 2) has a variable cell length. The presented grids were created in Matlab®, with the non-uniformity generated by adding random displacements on the uniform cells faces and posteriorly recalculating the cells centroids and volumes.

3 Implementation Verification with 1D Grids

Several numerical tests were performed to check the 1D implementation on non-uniform grids. The used analytical solution is given by Eq. (9).

$$\phi(x) = \exp\left(-\frac{(x-0.5)^2}{0.0175}\right) \tag{9}$$

Figure 3 shows the obtained results with a convective coefficient equal to one ($V = 1$), a unity diffusive coefficient ($\Gamma = 1$) and Dirichlet boundary condition on all domain's faces. The solid line corresponds to the acquired results, while the dashed line represents the theoretical convergence order. The error behaviour of all orders suggests a correct implementation and that the implicit solving of a one-dimensional convection-diffusion equation can be used for this problem.

To test the effect of convection-diffusion ratio, two additional simulations were made with larger ratios. Figure 4 shows the results for the previous analytical function, but with a convection-diffusion ratio of 10^3, while Fig. 5 shows the results for a ratio of 10^6.

From these results, it is possible to conclude that high convection-diffusion ratios ($\geq 10^6$) cause some oscillations, mainly on second and fourth orders. Moreover, the fourth order scheme only achieves its theoretical value for the finer grids, for the highest convection-diffusion ratio. The sixth and eighth orders are not as much affected. Only the eighth order round off error starts earlier, due to the double precision limit.

The previous results were obtained for non-uniform grids, since these are the ones that are subjected to a more unstable behaviour. Nonetheless, the uniform

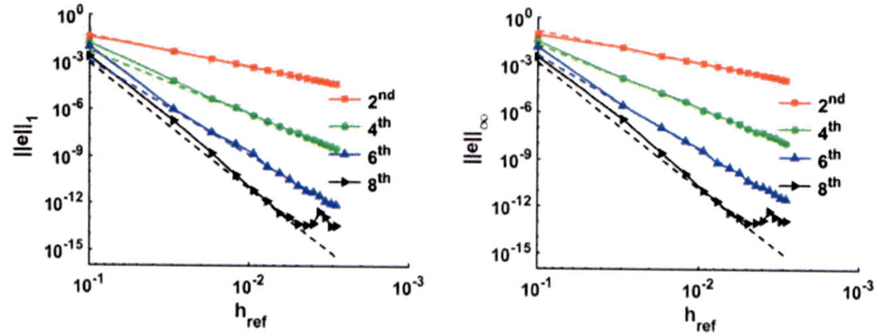

Fig. 3 Convergence curves of $\|e\|_1$ and $\|e\|_\infty$, for all schemes ($V = 1$ and $\Gamma = 1$), where the dashed lines represent the theoretical order

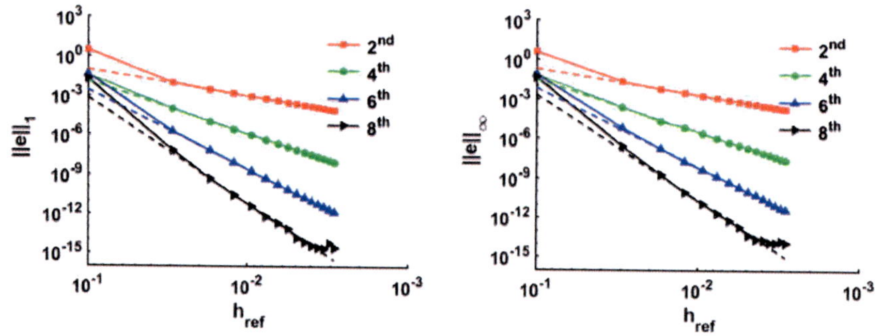

Fig. 4 Convergence curves of $\|e\|_1$ and $\|e\|_\infty$, for all schemes ($V = 100$ and $\Gamma = 0.1$), where the dashed lines represent the theoretical order

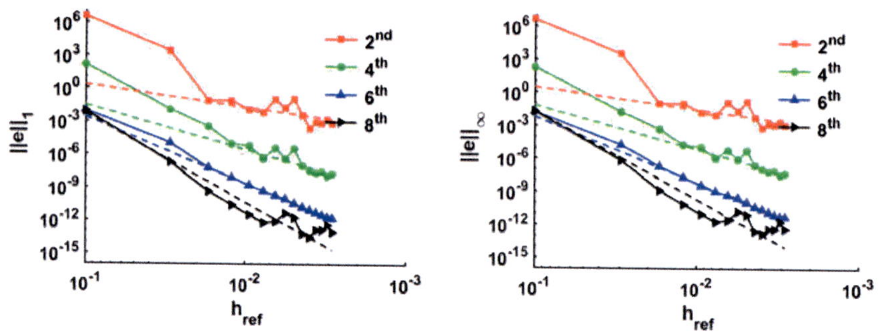

Fig. 5 Convergence curves of $\|e\|_1$ and $\|e\|_\infty$, for all schemes ($V = 1000$ and $\Gamma = 0.001$), where the dashed lines represent the theoretical order

grids were also tested, presenting a smaller number of oscillations for the higher convection-diffusion ratios and with smaller amplitudes.

4 Results with Polyhedral Unstructured Grids

To understand the Peclet number effect in the 2D numerical convection scheme, an analytical solution with a null source term was selected from the Jasak Ph.D. Thesis [14]. Contrary to other analytical solutions from the literature, this one does not have sharp gradients near the boundaries, uses a null Neumann boundary condition at the outlet and the distribution does not vary with the Peclet number.

The analytical solution is presented in Eq. (10), where $S = 16.67$ (source strength), K_0 is the modified Bessel function of second kind and zero order, V_x is the velocity magnitude and Γ is the diffusive coefficient. It was chosen to move the singularity further away from the domain, 0.15 units more than Jasak's original function, to improve the solution distribution relatively to the computational domain. It was also opted to divide the solution by its maximum value inside the domain, located at $x = 0$ and $y = 0.5$.

$$\phi(x, y) = \frac{S}{2\pi\Gamma} K_0 \left(V_x \frac{\sqrt{(x + 0.15)^2 + (y - 0.5)^2}}{2\Gamma} \right) e^{\frac{V_x x}{2\Gamma}} \tag{10}$$

The presented 2D results were obtained using the SOL code and the available AZTEC libraries, see [15] for more details. The least-squares reconstruction technique required for polyhedral unstructured grids is explained in the previous works [8] and [16]. The same reconstructed polynomial is used to compute the diffusive and convective face fluxes.

The results for a regular triangular grid, using $V_x = 10$ and diffusive coefficient of $\Gamma = 0.1$ are shown in Table 1 and Fig. 6. The second order scheme is slightly above its expected value, mainly on the norm-∞ of the error. Both fourth and sixth orders present the expected theoretical values for the norm-1 of the error, but they are under-convergent for the norm-∞. The eighth order scheme never reaches the value of eight, obtaining at best the 7.56 value. Figure 7 presents the obtained numerical result using a triangular grid of 3726 cells and the sixth order scheme, to better visualize the considered analytical solution.

Other grid types, such as Cartesian and regular polyhedrons, were also tested for the same convection-diffusion ratio. The results are shown in Table 2 and Fig. 8. For the Cartesian grid, the second and fourth order schemes results are within the expected values. On the other hand, the sixth order is close to the desired value of six, but never achieves it and the eighth order scheme is more than one order under the theoretical value, with a norm-1 of the error very close to the ones from the sixth order scheme. In the case of the polyhedral grid, the results are close to the theoretical ones, even showing some over-convergence, mainly on second and

Table 1 Obtained $\|e\|_1$, $\|e\|_\infty$, O_1 and O_∞ for solution (10) with $V_x = 10$, $V_y = 0.0$ and $\Gamma = 0.1$, using a regular triangular grid type

Triangular		FLS2		FLS4		FLS6		FLS8	
n_{cells}	h_{ref}	$\|e\|_1$	O_1	$\|e\|_1$	O_1	$\|e\|_1$	O_1	$\|e\|_1$	O_1
3726	1.43E−02	1.06E−04	2.47	5.87E−06	4.30	7.47E−07	5.86	1.55E−07	7.30
14,812	7.20E−03	2.42E−05	2.15	3.51E−07	4.09	1.16E−08	6.05	8.47E−10	7.56
59,385	3.60E−03	5.91E−06	2.03	2.11E−08	4.06	1.68E−10	6.11	5.22E−12	7.34
n_{cells}	h_{ref}	$\|e\|_\infty$	O_∞	$\|e\|_\infty$	O_∞	$\|e\|_\infty$	O_∞	$\|e\|_\infty$	O_∞
3726	1.43E−02	2.66E−03	2.41	3.52E−04	3.91	8.00E−05	5.38	1.68E−05	6.94
14,812	7.20E−03	5.50E−04	2.29	3.71E−05	3.27	2.85E−06	4.84	2.14E−07	6.34
59,385	3.60E−03	1.25E−04	2.14	2.56E−06	3.85	5.40E−08	5.72	1.29E−09	7.37

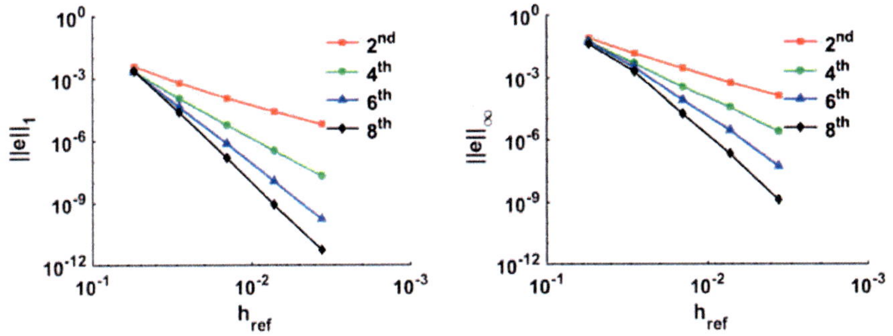

Fig. 6 The $\|e\|_1$ and $\|e\|_\infty$ for all schemes on a regular triangular grid, with the Jasak analytical solution (10) ($V_x = 10$, $V_y = 0.0$ and $\Gamma = 0.1$)

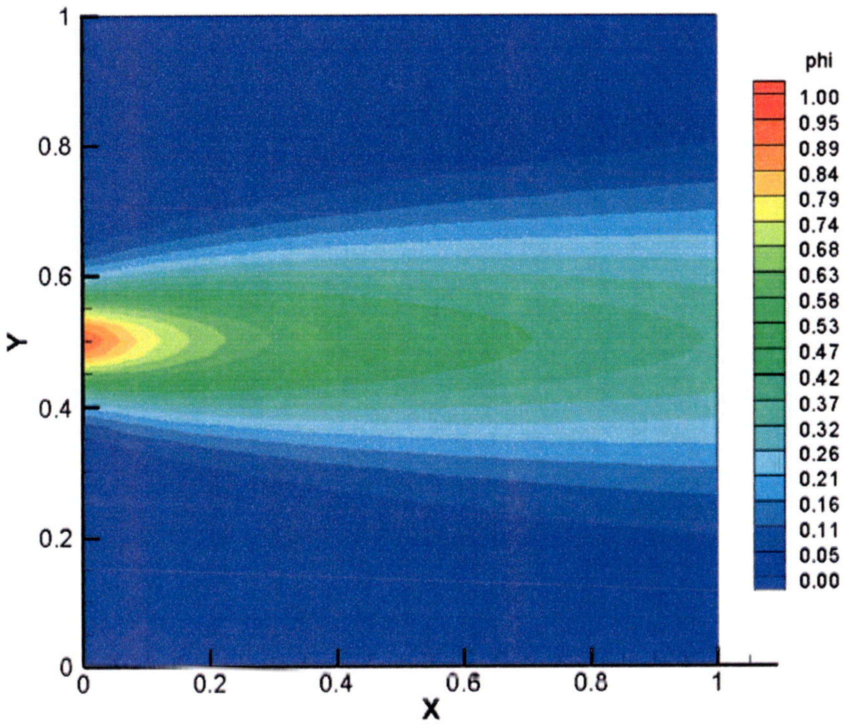

Fig. 7 Obtained sixth order scheme result of Eq. (10), using a regular triangular mesh with 3726 cells.The horizontal velocity is set to 10 and the diffusive coefficient is set to 0.1

Table 2 Obtained $\|e\|_1$ and O_1 for solution (10) with $V_x = 10$, $V_y = 0.0$ and $\Gamma = 0.1$, using Cartesian and Polyhedral grid types

Cartesian		FLS2		FLS4		FLS6		FLS8	
n_{cells}	h_{ref}	$\|e\|_1$	O_1	$\|e\|_1$	O_1	$\|e\|_1$	O_1	$\|e\|_1$	O_1
1600	2.50E−02	4.63E−04	2.08	2.16E−05	3.96	1.60E−06	5.61	1.93E−06	4.52
6400	1.25E−02	1.15E−04	2.02	1.39E−06	3.97	4.03E−08	5.31	2.63E−08	6.20
25,600	6.25E−03	2.86E−05	2.00	8.84E−08	3.97	8.71E−10	5.53	2.14E−10	6.94
Polyhedral		**FLS2**		**FLS4**		**FLS6**		**FLS8**	
n_{cells}	h_{ref}	$\|e\|_1$	O_1	$\|e\|_1$	O_1	$\|e\|_1$	O_1	$\|e\|_1$	O_1
1908	2.43E−02	3.35E−04	2.20	2.12E−05	3.90	1.00E−06	6.29	1.91E−07	7.70
7491	1.23E−02	8.15E−05	2.08	1.45E−06	3.95	1.60E−08	6.09	1.53E−09	7.11
29,695	6.22E−03	1.88E−05	2.14	8.44E−08	4.15	1.85E−10	6.51	6.38E−12	7.98

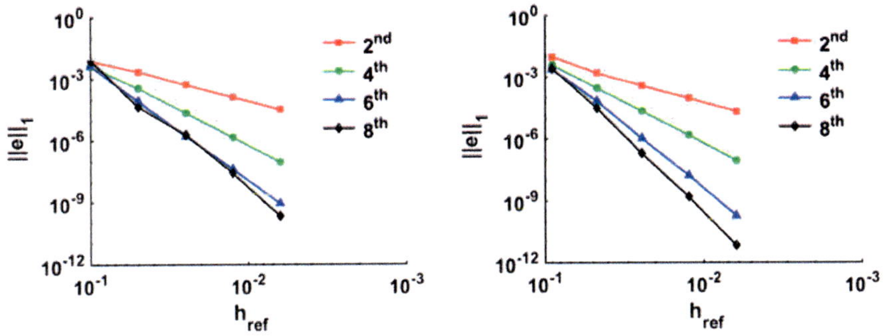

Fig. 8 The $\|e\|_1$ for all schemes in the Cartesian and polyhedral grids, with the Jasak analytical solution (10) ($V_x = 10$, $V_y = 0.0$ and $\Gamma = 0.1$)

sixth order schemes. For the grids tested, the triangular is the one with best results, followed by the polyhedral grid. The Cartesian grid type, presents the poorest results and does not allowed the scheme to achieve its eighth order.

Finally, to understand if there is a limit to the Peclet number (Pe) for these numerical schemes. Other tests were performed with convection-diffusion ratios (or Pe_L) of 500, 750, 1000 and 1200. Since the Cartesian grid did not achieved eighth order, it was opted to use the regular triangular grid to study this possible limit.

Figure 9 shows the obtained results and all plots have an unstable behaviour for coarser grids, that becomes asymptotic and stable during the grid refinement. In Fig. 9, vertical lines were drawn where this asymptotic behaviour starts for the second order schemes, using as criterion that the order at one point should not vary more than 20% relative to the previous one. The obtained grid size values were grouped and plotted together in Fig. 10 as a function of the Pe_L. Here the fitting of the function $h_{ref} = a/Pe_L$ is also included, where the a is the Pe_{Limit}. The estimated value for the second order scheme is $\mathrm{Pe}_{Limit} = 11.23$. For higher orders, it is harder to quantify this value and so the Pe_{Limit} will not be estimated. Nonetheless, from Fig. 9 it is clear that the higher-order schemes begin their asymptotic behaviour earlier than the second order counterpart. This demonstrating that high-order schemes are more robust and allow for higher Pe_{Limit} values.

5 Conclusions

A new convection scheme is introduced for the finite volume method and polyhedral unstructured grids in the point-wise framework. The scheme is verified for both 1D and 2D cases. The theoretical order of convergence was reached for different grid topologies.

The Peclet number analysis suggests that high-order schemes are less susceptible to convection instabilities and present higher Peclet limits than second-order one.

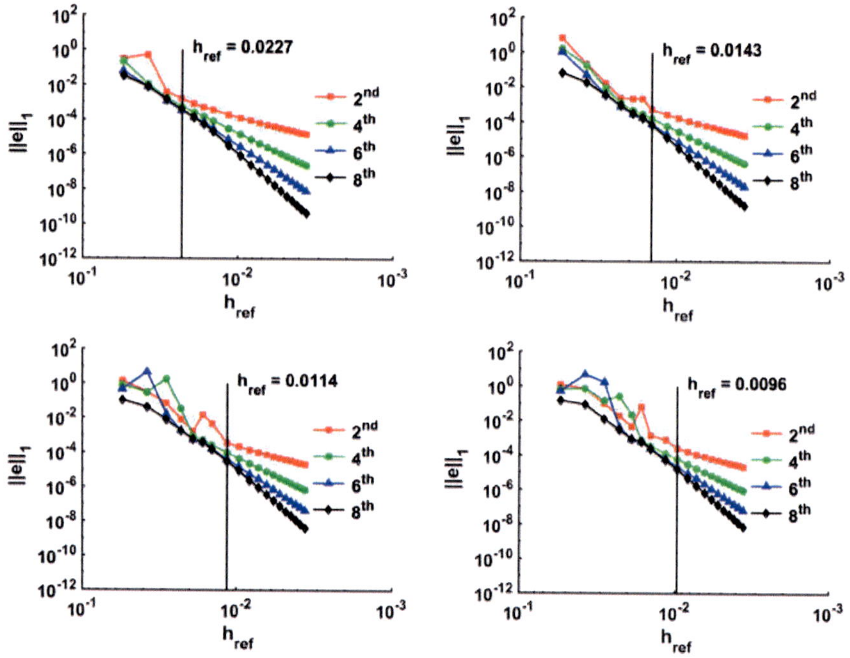

Fig. 9 Convergence curves of $\|e\|_1$ for different convection-diffusion ratios, on a regular triangular grid type using the Jasak analytical solution (10)

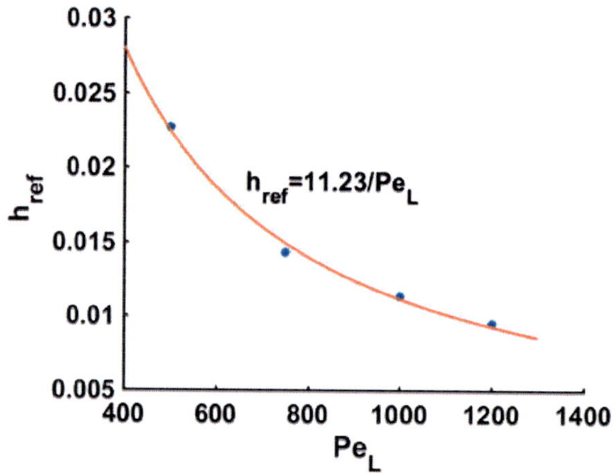

Fig. 10 Fitting of the function $h_{ref} = a/Pe_L$ to the data points acquired from the second order schemes in Fig. 9, with $a = 11.23$

The used analytical solution presents characteristic typically not shown in the literature: its distribution does not vary with the Peclet, it avoids sharp gradients near the domain's boundaries and it requires a null Neumann condition at the outlet, where the imposed velocity leaves the domain.

Acknowledgments Both authors acknowledge the support received by FCT under the portuguese research project: High-order immersed boundary for moving body problems—HIBforMBP—with the following reference PTDC/EME-EME/32315/2017. Institutional acknowledgement: This work was supported by FCT, through IDMEC, under LAETA, project UIDB/50022/2020.

References

1. Barth, T., Frederickson, P.: Higher order solution of the Euler equations on unstructured grids using quadratic reconstruction. In: 28th Aerospace Sciences Meeting from AIAA (1990)
2. Ollivier-Gooch, C., Van Altena, M.: A High-order-accurate unstructured mesh finite-volume scheme for the advaction-diffusion equation. J. Comput. Phys. **181** (2002)
3. Sejekan, C.B., Ollivier-Gooch, C.F.: Improving finite-volume diffusive fluxes through better reconstruction. Comput. Fluids. **139**, 216–232 (2016)
4. Jalali, A., Ollivier-Gooch, C.: Higher-order unstructured finite volume RANS solution of turbulent compressible flows. Comput. Fluids. **147** (2017)
5. Nogueira, X., Colominas, I., Cueto-Felgueroso, L., Khelladi, S.: On the simulation of wave propagation with a higher-order finite volume scheme based on Reproducing Kernel Methods. Comput. Methods Appl. Mech. Eng. **199**, 1471–1490 (2010)
6. Chassaing, J., Khelladi, S., Nogueira, X.: Accuracy assessment of a high-order moving least squares finite volume method for compressible flows. Comput. Fluids. **71**, 41–53 (2013)
7. Clain, S., Machado, G.J., Nobrega, J.M., Pereira, R.M.S.: A sixth-order finite volume method for multidomain convection?diffusion problem with discontinuous coefficients, Comput. Methods Appl. Mech. Eng. **267** (2013)
8. Vasconcelos, A.G.R., Albuquerque, D.M.S., Pereira, J.C.F.: A very high-order finite volume method based on weighted least squares for elliptic on polyhedral unstructured grids. Comput. Fluids **181**, 383–402 (2019)
9. Costa, R., Nobrega, J., Clain, S., Machado, G.J.: Efficient very high-order accurate polyhedral mesh finite volume scheme for 3D conjugate heat transfer problems in curved domains. J. Comput. Phys. **445** (2021)
10. Costa, P.M.P., Albuquerque, D.M.S.: A novel approach for temporal simulations with very high-order finite volume schemes on polyhedral unstructured grids. J. Comput. Phys. **453** (2022)
11. Albuquerque, D.M.S., Pereira, J.M.C., Pereira, J.C.F.: Residual least-squares error estimate for unstructured h adaptive meshes. Numer. Heat Transfer Part B: Fundam. **67**, 187–210 (2015)
12. Martins, D.M.C., Albuquerque, D.M.S., Pereira, J.C.F.: On the use of polyhedral unstructured grids with moving immersed boundary method. Comput. Fluids **174**, 78–88 (2018)
13. Moukalled, F., Mangani, L., Darwish, M.: The Finite Volume Method in Computational Fluid Dynamics: An Advanced Introduction with OpenFOAM and Matlab. Springer, Berlin (2015)
14. Jasak, H.: Error analysis and estimation for the finite volume method with applications to fluid flows. Ph.D. Thesis in Imperial College of Science, Technology and Medicine (1996)
15. Pereira, J.C.F., Pereira, J.M.C., Leite, A.L.A., Albuquerque, D.M.S.: Calculation of spotting particles maximum distance in idealised forest fire scenarios. J. Combustion **2015**, Article ID 513576 (2015)

16. Albuquerque, D.M.S., Vasconcelos, A.G.R., Pereira, J.C.F.: A Novel Eighth-Order Diffusive Scheme for Unstructured Polyhedral Grids Using the Weighted Least-Squares Method. Lecture Notes in Computational Science and Engineering, vol. 134 (2020). http://dx.doi.org/10.1007/978-3-030-39647-3_27

Flexible Weights for High Order Face Based Finite Element Interpolation

Ana Alonso Rodríguez, Ludovico Bruni Bruno, and Francesca Rapetti

1 Introduction

The *weights* of a differential k-form are its integrals over some k-simplices, usually called *small k-simplices*, defined in the tetrahedra of a simplicial mesh in the considered domain $\Omega \subset \mathbb{R}^3$. The idea of using such quantities as degrees of freedom in $P_r^- \Lambda^k(\Omega)$, the space of trimmed polynomial differential forms (see [4]) that are a generalization of Whitney forms to arbitrary polynomial degrees $r \geq 1$, is first presented in [8]. There, small simplices are built over a principal lattice of fixed order in each tetrahedron of the mesh; however, such a construction perfectly accomodates different choices of other nodal lattices. This remark has been discussed for $k = 1$ in [3] where new degrees of freedom for the first family of Nédélec edge based finite elements have been proposed. Several sets of unisolvent 1-simplices have been built starting from a principal lattice and then modifying the small edge extremities to insist in distributions of nodes that are optimal for nodal interpolation, such as symmetrised Lobatto and warp and blend nodes (see [9]). The quality of the resulting new sets of small edges has been discussed in terms of the *generalised Lebesgue constant* introduced in [1].

In this work we extend the above analysis to Nédélec first family face based finite elements, that are proxies for trimmed polynomial differential 2-forms. The starting steps consist in selecting a subset of the weights for $k = 2$ defined in [8] and in proving that this subset is unisolvent and minimal for $\mathcal{P}_r^- \Lambda^2(T)$, with T a

A. Alonso Rodríguez · L. Bruni Bruno (✉)
Dipartimento di Matematica, Università degli Studi di Trento, Povo, Italy
e-mail: ana.alonso@unitn.it; ludovico.brunibruno@unitn.it

F. Rapetti
Département de Mathématiques, Université Côte d'Azur, Nice, France
e-mail: francesca.rapetti@univ-cotedazur.fr

J. M. Melenk et al. (eds.), *Spectral and High Order Methods for Partial Differential Equations ICOSAHOM 2020+1*, Lecture Notes in Computational Science and Engineering 137, https://doi.org/10.1007/978-3-031-20432-6_5

tetrahedron. Then, following the same procedure as in [3], we exploit this choice to build new unisolvent and minimal weights that involve 2-simplices whose vertices coincide with three-dimensional symmetrised Lobatto or warp and blend nodes. We estimate the generalised Lebesgue constants for the associated interpolation operators and comment on the results.

2 Distributions of Unisolvent and Minimal 2-Simplices

Let T denote an n-simplex of $\mathbb{R}^n$, with $n = 2$ or $n = 3$. Let $\{\lambda_i(\mathbf{x})\}_{i=0}^n$ be the barycentric coordinates of $\mathbf{x} \in T$. The principal lattice of order $r \geq 1$ is the set of points

$$L_r(T) := \{\mathbf{x} \in T : \lambda_i(\mathbf{x}) \in \{0, \frac{1}{r}, \dots, \frac{j}{r}, \dots, \frac{r-1}{r}, 1\},\ i = 0, \dots, n\}$$

and we introduce $Z_r(T) := \{\mathbf{x} \in L_r(T) : \lambda_0(\mathbf{x}) \neq 0\}$. The set $Z_r(T)$ consists of $\binom{n+r-1}{r-1}$ points of T and the only polynomial of degree lower or equal than $r-1$ in n variables that is equal to zero at each point of $Z_r(T)$ is the zero polynomial.

2.1 *With Overlapping*

For each $\xi \in \mathbb{R}^n$ we consider the map obtained as the composition of the homothety $\mathbf{x} \mapsto \mathbf{x}_0 + \lambda_0(\xi)(\mathbf{x} - \mathbf{x}_0)$ and the translation $\mathbf{x} \mapsto \mathbf{x} + (\xi - \mathbf{x}_0)$, namely

$$\tau_\xi : \mathbf{x} \mapsto \tau_\xi(\mathbf{x}) = \mathbf{x}_0 + \lambda_0(\xi)(\mathbf{x} - \mathbf{x}_0) + (\xi - \mathbf{x}_0).$$

We thus have

$$\tau_\xi(\mathbf{x}) = \xi + \lambda_0(\xi)(\mathbf{x} - \mathbf{x}_0) = \sum_{i=0}^{n} \lambda_i(\xi)\mathbf{x}_i + \lambda_0(\xi)(\mathbf{x} - \mathbf{x}_0) = \lambda_0(\xi)\mathbf{x} + \sum_{i=1}^{n} \lambda_i(\xi)\mathbf{x}_i .$$

We denote $T_\xi := \tau_\xi(T)$. If $\lambda_0(\xi) \neq 0$ then T_ξ is a n-simplex, similar to T with n vertices in the $(n-1)$-dimensional affine subspace π_T containing $\{\mathbf{x}_1, \dots, \mathbf{x}_n\}$. Otherwise $T_\xi = \xi$. (See Fig. 1, left, for a depiction in the case of $n = 2$.)

Let $\Delta_k(T)$ be the set of k-subsimplices of T. For each $0 \leq k \leq n$ we denote

$$\Sigma_r^k(T) = \{\tau_\xi(F) : F \in \Delta_k(T),\ \xi \in Z_r(T)\}.$$

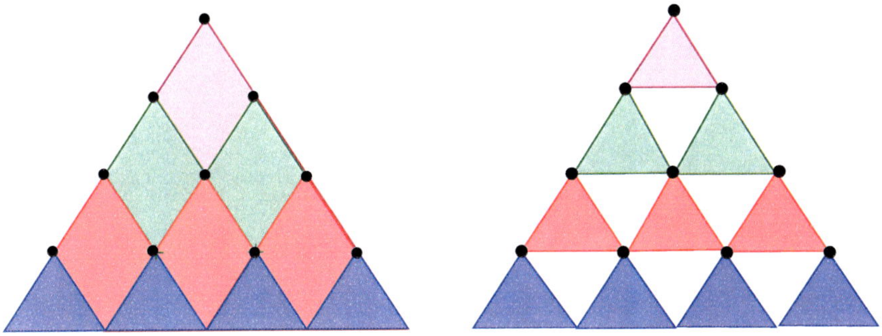

Fig. 1 Black dots represent the set $Z_r(T)$ for a 2-simplex T in $\mathbb{R}^2$ when $r = 4$. The coloured 2-simplices of the figure on the left are $\tau_{\boldsymbol{\xi}}(T)$ for $\boldsymbol{\xi} \in Z_r(T)$. Figure on the right represents the set of small 2-simplices $\beta_{\boldsymbol{\xi}}(T)$. Their formal definition will be introduced in the next subsection

For each $k_0 \in \mathbb{N}$ with $0 \leq k_0 \leq n$ and each $F \in \Delta_{k_0}(T)$ we can consider the principal lattice of order r of F,

$$L_r(F) := \{\mathbf{x} \in F : \lambda_i^F(\mathbf{x}) \in \{0, \frac{1}{r}, \dots, \frac{j}{r}, \dots, \frac{r-1}{r}, 1\},\ i = 0, \dots, k_0\}$$

and define analogously

$$Z_r(F) := \left\{\mathbf{x} \in F \ : \ \lambda_0^F(\mathbf{x}) \neq 0 \quad \text{and} \quad \lambda_i^F(\mathbf{x}) \in L_r(F)\right\},$$

being $\lambda_i^F(\mathbf{x})$ the barycentric coordinate of $\mathbf{x}$ on F for $i \in \{0, 1, \dots, k_0\}$, and

$$\Sigma_r^k(F) = \{\tau_{\boldsymbol{\xi}}(f) \ : \ f \in \Delta_k(F),\ \boldsymbol{\xi} \in Z_r(F)\}, \qquad 0 \leq k \leq k_0.$$

It is worth noting that for $n = 3$ the set of elements of $\Sigma_r^2(T)$ supported on $F \in \Delta_2(T)$ contains $\Sigma_r^2(F)$. These two sets coincide if and only if $\mathbf{x}_0 \in F$.

We now recall the definition of unisolvence and gather results that show the unisolvence of the set of k-simplices $\Sigma_r^k(T)$ for the space of trimmed polynomial differential k-forms $\mathcal{P}_r^- \Lambda^k(T)$ (see [4] for a formal construction).

Definition 1 A set of k-simplices $S^k(T)$ is said to be *unisolvent* for $\mathcal{P}_r^- \Lambda^k(T)$ if, for any $\omega \in \mathcal{P}_r^- \Lambda^k(T)$, $\int_{\sigma_i} \omega = 0$ for each $\sigma_i \in S^k(T)$ implies that $\omega = 0$. Such a set is said to be *minimal* if its cardinality coincides with the dimension of $\mathcal{P}_r^- \Lambda^k(T)$, namely if $\# S^k(T) = \dim \mathcal{P}_r^- \Lambda^k(T)$.

The following result is proved in [2].

Propostion 2 *If $\omega \in \mathcal{P}_r^- \Lambda^k(T)$ is such that $\int_\sigma \omega = 0$ for all $\sigma \in \Sigma_r^k(T)$ then $\omega = 0$.*

This result states that the set of k-simplices $\Sigma_r^k(T)$ is unisolvent for $\mathcal{P}_r^- \Lambda^k(T)$, however it is easy to check that it is not minimal if $1 \leq k \leq n-1$.

Proposition 2 can be refined, for what concerns $k = 2$ and $\mathbb{R}^3$, as one has that the set

$$\Sigma^2_{r,min}(T) := \{f \in \Sigma^2_r(T) \ : \ f \not\subset T - [\mathbf{x}_0]\} \bigcup \Sigma^2_r(T - [\mathbf{x}_0])$$

is unisolvent for $\mathcal{P}_r^- \Lambda^2(T)$, namely it is such that

$$\left(\omega \in \mathcal{P}_r^- \Lambda^2(T) \quad \texttt{and} \quad \int_\sigma \omega = 0, \ \forall \sigma \in \Sigma^2_{r,min}(T) \right) \quad \Longrightarrow \quad \omega = 0.$$

In fact, assume $\omega \in \mathcal{P}_r^- \Lambda^2(T)$. If $\int_\sigma \omega = 0$ for all $\sigma \in \Sigma^2_{r,min}(T)$, then in particular $\int_\sigma \omega = 0$ for all σ in $\Sigma^2_r(T - [\mathbf{x}_0])$ and hence the resctriction of ω to $T - [\mathbf{x}_0]$, that belongs to $\mathcal{P}_r^- \Lambda^2(T - [\mathbf{x}_0])$, is zero and $\int_\sigma \omega = 0$ for all σ in $\Sigma^2_r(T)$. Thus Proposition 2 yields $\omega = 0$.

To show the minimality of such a set, we compute

$$\#\Sigma^2_{r,min}(T) = \binom{r+2}{r-1}\binom{3}{2} + \binom{r+1}{r-1} = \binom{r+1}{2}\binom{r+3}{1} = \dim \mathcal{P}_r^- \Lambda^2(T).$$

We hence have proved the following.

Propostion 3 *Let $T \subset \mathbb{R}^3$. The set $\Sigma^2_{r,min}(T)$ is unisolvent and minimal for $\mathcal{P}_r^- \Lambda^2(T)$.*

Remark 4 The set $\Sigma^2_{r,min}(T)$ contains all the boundaries of the small tetrahedra homothetic to T constructed using the map $\tau_{\boldsymbol{\xi}}(T)$ except those contained in F_0, the face opposite to the vertex $\mathbf{x}_0$. On that face only small 2-simplices constructed as homotheties of F_0 by using $\tau_{\boldsymbol{\xi}}(F_0)$ are considered.

2.2 With Holes

This minimal set helps us to select a minimal subset of the small simplices $X_r^2(T)$ introduced in [8], which we recall in the sequel.

For each $\boldsymbol{\xi} \in \mathbb{R}^n$ we consider the map $\beta_{\boldsymbol{\xi}}$ obtained as the composition of the homothety of factor $\frac{1}{r}$ that reads $\mathbf{x} \mapsto \mathbf{x}_0 + \frac{1}{r}(\mathbf{x} - \mathbf{x}_0)$ and the translation defined by $\mathbf{x} \mapsto \mathbf{x} + (\boldsymbol{\xi} - \mathbf{x}_0)$, namely

$$\beta_{\boldsymbol{\xi}} : \mathbf{x} \mapsto \beta_{\boldsymbol{\xi}}(\mathbf{x}) = \mathbf{x}_0 + \frac{1}{r}(\mathbf{x} - \mathbf{x}_0) + (\boldsymbol{\xi} - \mathbf{x}_0) = \frac{1}{r}(\mathbf{x} - \mathbf{x}_0) + \boldsymbol{\xi}.$$

The small k-simplices for each $0 \le k \le n$ are defined as

$$X_r^k(T) = \left\{\beta_{\xi}(F) \; : \; F \in \Delta_k(T), \; \xi \in Z_r(T)\right\}$$

and $X_r^k(F)$ is defined similarly by replacing T with $F \in \Delta_{k_0}(T)$, for any $0 \le k \le k_0$. (See Fig. 1, right.)

The following result proved in [6] is the analogous of Proposition 2.

Propostion 5 *If* $\omega \in \mathcal{P}_r^- \Lambda^k(T)$ *is such that* $\int_\sigma \omega = 0$ *for all* $\sigma \in X_r^k(T)$ *then* $\omega = 0$.

Each element of both $\Sigma_r^k(T)$ and $X_r^k(T)$ is characterized by a pair $(F, \xi) \in \Delta_k(T) \times Z_r(T)$. This fact induces a natural bijection between the elements of $\Sigma_r^k(T)$ and those of $X_r^k(T)$.

We define also in this case the analogous of $\Sigma^2_{r,min}(T)$, namely

$$X^2_{r,min}(T) := \left\{f \in X_r^2(T) \; : \; f \not\subset T - [\mathbf{x}_0]\right\} \bigcup X_r^2(T - [\mathbf{x}_0]).$$

Remark 6 The set $X^2_{r,min}(T)$ contains all the boundaries of the small tetrahedra homothetic to T constructed using the map $\beta_\xi(T)$ except those in the interior of T that are parallel to F_0, the face opposite to the vertex $\mathbf{x}_0$.

We claim that $X^2_{min}(T)$ is unisolvent for $\mathcal{P}_r^- \Lambda^2(T)$. One clearly has $\#X^2_{r,min}(T) = \#\Sigma^2_{r,min}(T)$, hence minimality is granted.

Propostion 7 *If* $\omega \in \mathcal{P}_r^- \Lambda^2(T)$ *is such that* $\int_\sigma \omega = 0$ *for all* $\sigma \in X^2_{r,min}(T)$ *then* $\omega = 0$.

In fact, we show that if $\omega \in \mathcal{P}_r^- \Lambda^2(T)$ is such that $\int_\sigma \omega = 0$ for all $\sigma \in X^2_{r,min}(T)$ then $\int_{\sigma'} \omega = 0$ for all $\sigma' \in \Sigma^2_{r,min}(T)$. The result will thus follow from Proposition 3.

Assume $\omega \in \mathcal{P}_r^- \Lambda^2(T)$ is such that $\int_\sigma \omega = 0$ for each $\sigma \in X_r^2(F)$ and $F \in \Delta_2(T)$. By replacing T with F in Proposition 5, it follows that $\int_{\sigma'} \omega = 0$ for any 2-simplex σ' supported in F. In particular

$$\int_{\sigma'} \omega = 0 \qquad \text{for each } \sigma' \in \Sigma^2_{r,min}(T) \cap F, \; F \in \Delta_2(T). \tag{1}$$

For each face $F_i = T - [x_i]$ with $i \in \{1, 2, 3\}$ and $j = \{1, \dots, r-1\}$ consider $\pi_{i,j} := \{\mathbf{x} \in T \; : \; \lambda_i(\mathbf{x}) = \frac{j}{r}\}$. Since $i \neq 0$, $X^2_{r,min}(T) \cap \pi_{i,j} = X_r^2(T) \cap \pi_{i,j} = X^2_{r-j}(\pi_{i,j})$. Thus the hypothesis yields

$$\int_\sigma \omega = 0 \qquad \text{for each } \sigma \in X^2_{r-j}(\pi_{i,j}). \tag{2}$$

Since $\omega|_{F_i} = 0$ for $i \in \{1, 2, 3\}$, then $\omega|_{\pi_{i,1}} \in \mathcal{P}_{r-1}^{-}\Lambda^2(\pi_{i,1})$, hence by Proposition 5 we conclude that $\omega|_{\pi_{i,1}} = 0$. As a consequence, $\int_{\sigma'} \omega = 0$ for each $\sigma' \in \Sigma^2_{r-1,min}(T) \cap \pi_{i,1}$. One can continue in this way for $j = 2, \ldots, r-1$ to deduce with the same argument that $\omega|_{\pi_{i,j}} = 0$, hence $\int_{\sigma'} \omega = 0$ for each $\sigma' \in \Sigma^2_{r,min}(T) \cap \pi_{i,j}$, for $i \in \{1, 2, 3\}$ and $j \in \{1, \ldots, r-1\}$. Together with (1) we have $\int_{\sigma'} \omega = 0$ for each $\sigma' \in \Sigma^2_{r,min}(T)$, hence one has $\omega = 0$ by Proposition 3.

3 Towards Non-uniform Distributions of Small Faces

We analyse the quality of the interpolation on the uniform distributions $\Sigma^2_{r,min}(T)$ and $X^2_{r,min}(T)$ relying on the generalised Lebesgue constant defined in [1], which we recall. Given an unisolvent set of k-simplices, $S^k = \{\sigma_1, \ldots, \sigma_{N_k}\}$, the associated Lebesgue function $\mathcal{L}_{S^k} : C_k(T) \to \mathbb{R}^+$ is

$$c \mapsto \mathcal{L}_{S^k}(c) = \sum_{j=1}^{N_k} |\sigma_j|_0 \left| \int_c \mathrm{w}_j^{S^k} \right|, \tag{3}$$

where $C_k(T)$ is the set of k-chains supported in T, $|\cdot|_0$ denotes the measure of a k-simplex, and $\{\mathrm{w}_j^{S^k}\}_{j=1}^{N_k} \subset \mathcal{P}_r^{-}\Lambda^k(T)$ is the dual basis associated with S^k, namely

$$\int_{\sigma_i} \mathrm{w}_j^{S^k} = \delta_{i,j}.$$

The Lebesgue constant is then computed by taking the supremum over the elements of $C_k(T)$, namely $\Lambda_{S^k} = \sup_{c \in C_k(T)} \frac{\mathcal{L}_{S^k}(c)}{|c|_0}$. When $k = 0$ this coincides with the classical definition of Lebesgue constant.

To estimate the Lebesgue constant associated with an unisolvent set of 2-simplices S^2 we proceed as follows. We consider a fine test mesh τ; hence we replace the supremum in (3) by the maximum reached by Λ_{S^2} on the 2-skeleton of τ, namely the set of faces of τ. We call this the *estimated* Lebesgue constant. The results in Table 1 and subsequent Tables are computed with respect to a mesh that has 3876 faces with area ranging between $a_{\min} = 0.0017$ and $a_{\max} = 0.0030$.

In the first two columns of Table 1 the estimated Lebesgue constants for $X^2_{r,min}(T)$ and $\Sigma^2_{r,min}(T)$ are reported. It is evident that overlapping simplices do not offer good interpolation properties.

In order to avoid overlappings, which yield unsatisfactory results, we consider *tiles*, roughly speaking the parallelepipeds appearing in Fig. 1, whose existence in this context is motivated by the linearity of the integral. The related results are

Table 1 Lebesgue constants for $k = 2$ in a *tetrahedron* T, associated with simplices and tiles built upon uniform distributions

r	Uniform, with holes	Uniform, with overlapping	Uniform tiles
2	7.51	28.17	12.67
3	13.04	114.11	40.56
4	22.24	329.78	100.62
5	35.79	781.88	214.92
6	55.07	1635.72	433.61

included in Table 1 and show an improvement of the associated Lebesgue constants, which makes them in any case far from $X^2_{r,min}(T)$.

The set of simplices $\Sigma^2_{r,min}(T)$ has been useful to identify $X^2_{r,min}(T)$ but seems to be not suitable for interpolation. In the following we focus on the original definition with holes given in [8], which is based on $X^2_{r,min}(T)$. Then, we study how to improve the choice of the small 2-simplices relying on the idea developed in [2] for small 1-simplices.

It is well known that Lagrangian interpolation at uniformly distributed points in the mesh elements can yield poor approximation. This is due to a rapid increase of the Lebesgue constant. For this reason there have been several attempts in the literature to produce families on nodal sets in tetrahedra that reduce the growth of the Lebesgue constant (see [5], for instance). Concerning polynomial interpolation of differential 2-forms we propose to consider families of 2-simplices with those as vertices. A possibility is to introduce a vertex bijection φ from $L_r(T)$ to one of these nodal sets in tetrahedra. Under appropriate hypotheses this vertex bijection induces an isomorphism of simplicial complexes Φ between $\mathbb{X} := X^0_r \cup X^1_r \cup X^2_{r,min}$ and its image $\mathbb{Y} := \Phi(\mathbb{X})$. The new set of 2-simplices is the 2-skeleton of $\mathbb{Y}$, which we denote for short by $Y^2_{r,min}(T)$. See Fig. 2 for an example.

The arguments used to prove the unisolvence of $X^2_{r,min}(T)$ are not suitable in this context, since elements of $Y^2_{r,min}(T)$ do not necessarily lie on planes parallel to the faces $F_i \in \Delta_2(T)$ anymore (see Fig. 3, right). It is thus worth noting the following. Let $\{\omega_j\}_{j=1}^{N_k}$ be a basis for $\mathcal{P}_r^- \Lambda^k(T)$ and $S^k = \{\sigma_1, \ldots, \sigma_{N_k}\}$ be a set of k-simplices with $N_k = \dim \mathcal{P}_r^- \Lambda^k(T)$. The corresponding *generalised Vandermonde matrix* $V \in \mathbb{R}^{N_k \times N_k}$ has entries

$$V_{i,j} = \int_{\sigma_i} \omega_j. \tag{4}$$

It is easy to prove that S^k is unisolvent if and only if V is non-singular.

A suitable vertex map φ is that defined in [7, Chapter 10], which is a bijection that move points of $L_r(T)$ to a set of nodes that depends on a parameter α. This maps has a warping and blending effect which partially restores a rotational symmetry in the small simplices. When $\alpha = 0$ such a choice mimics *symmetrised Lobatto* points

Fig. 2 Different configurations of small 2-simplices: on the left, the elements of $X^2_{r,min}(T)$ of degree $r = 4$ in a 2-simplex T. On the right 2-simplices associated with warp and blend nodes of the same degree

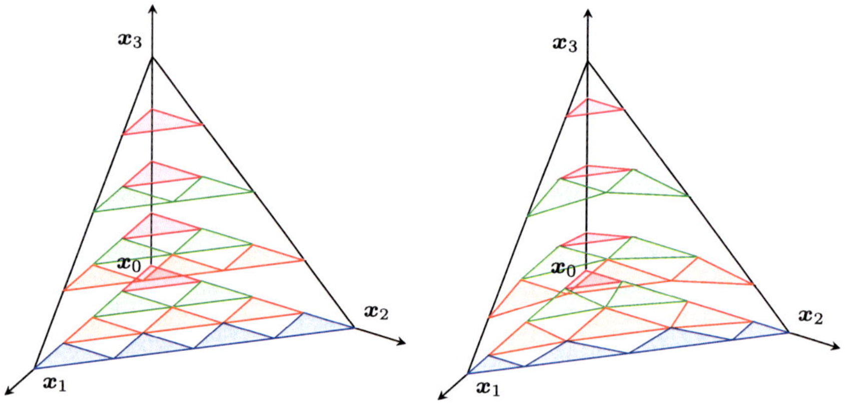

Fig. 3 On the left, in a generic tetrahedron, we show some small 2-simplices $X^2_{r,min}(T)$ associated with the uniform distribution of points $L_r(T)$. They are those parallel to a face F_i with $i \neq 0$. Note that they lie on parallel planes. On the right the corresponding small 2-simplices obtained by moving points of $L_r(T)$ by means of the vertex bijection φ to the warp and blend nodes. Such small simplices cannot be collected anymore on planes parallel to faces F_i. Note that the number of vertices, small 1-simplices and small 2-simplices lying on each F_i is preserved. Here $r = 4$

and when α is optimised (with respect to the degree r) it yields the *warp and blend* points.

We now consider the cases $\alpha = 0$ and its optimised values $\alpha = \alpha_{opt}$ of the above mentioned vertex map φ. The latter case, $\alpha = \alpha_{opt}$, is depicted in Fig. 3 and there compared with the set $X^2_{r,min}(T)$ of the same degree. Conditioning numbers of the generalised Vandermonde matrix are reported in Table 2 and show that these choices are unisolvent.

Table 2 Conditioning numbers of the generalised Vandermonde matrix associated with the basis $\boldsymbol{\lambda}^{\alpha}\omega_F$, as defined in [2]

r	Uniform	Symmetrised Lobatto	Warp and blend
3	5.2716×10^1	5.2716×10^1	5.2716×10^1
4	3.1195×10^2	2.4998×10^2	2.4998×10^2
5	2.4726×10^3	1.8703×10^3	1.8704×10^3
6	1.7750×10^4	1.1114×10^4	1.1404×10^4
7	1.2803×10^5	8.1111×10^4	8.8343×10^4
8	1.0204×10^6	6.2049×10^5	6.7417×10^5
9	9.4665×10^6	5.1412×10^6	5.6041×10^6

4 Numerical Results

The approach just discussed is viable as the computed Lebesgue constant turns out to be very sensible to the choice of the supports of the degrees of freedom, which is in accordance with the case $k = 0$. Table 3 shows the estimated Lebesgue constant for three different choices of nodes, including the uniform ones. The test mesh τ considered is the same as the one adopted in Sect. 2, with 3876 faces whose areas range between $a_{\min} = 0.0017$ and $a_{\max} = 0.0030$.

In Fig. 4 we compare the results obtained for the case $k = 2$ with those known for $k = 0$ (see [7]) and $k = 1$ (see [3]) for uniform nodes (hence 1-simplices and 2-simplices whose vertices belong to L_r) and warp and blend nodes (hence 1-simplices and 2-simplices whose vertices belong to $\varphi(L_r)$, φ being the vertex map discussed in Sect. 3 for $\alpha = \alpha_{opt}$). In a semilogarithmic plot, lines corresponding with the same set of nodes (that for $k > 0$ corresponds with the set of vertices of the small k-simplices) are parallel.

The generalised Lebesgue constant also measures the stability of the interpolation operator even when $k > 0$, as shown in [3] for the case of $k = 1$. In fact, let $\Pi\omega$ be the unique trimmed polynomial differential form such that

$$\int_\sigma \Pi\omega = \int_\sigma \omega, \qquad \forall \sigma \in S^k.$$

Table 3 Lebesgue constants for $k = 2$ in a *tetrahedron* T, associated with simplices built upon uniform, symmetrised Lobatto and warp and blend distributions

r	Uniform	Symmetrised Lobatto	Warp and blend
3	13.04	13.04	13.04
4	22.24	19.78	19.78
5	35.79	27.01	27.01
6	55.07	33.66	32.87
7	83.40	41.86	37.70
8	120.84	48.00	42.00
9	190.02	58.90	51.28

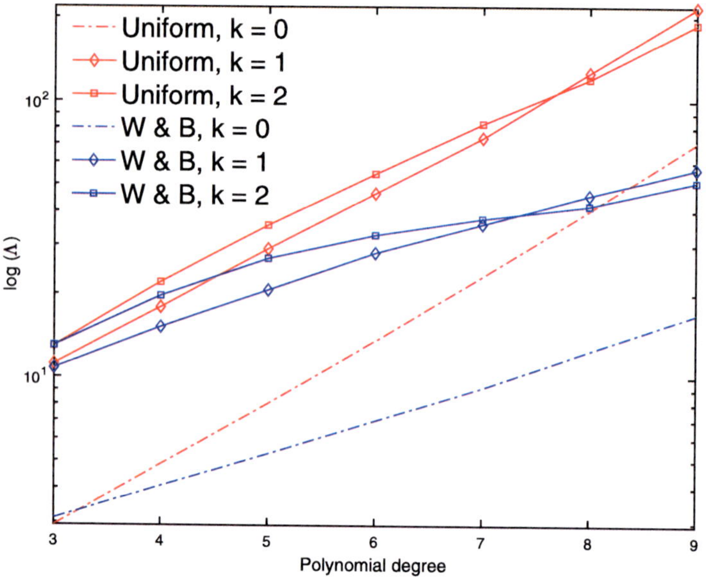

Fig. 4 Red lines represent the estimated Lebesgue constants for $k = 0$, $k = 1$ and $k = 2$ on uniformly distributed nodes, blue lines interpolate the estimated Lebesgue constants for $k = 0$, $k = 1$ and $k = 2$ for the warp and blend nodes, with respect to the total degree r. This plot is in semilogarithmic scale

One has

$$||\Pi\omega - \Pi\widetilde{\omega}||_0 \leq \varepsilon\Lambda, \tag{5}$$

where ε is a perturbation on the data ω such that $||\omega - \widetilde{\omega}||_0 \leq \varepsilon$. All the quantities have to be understood in terms of the 0-norm, which for k-forms reads as

$$||\omega||_0 = \sup_{c\in C_K(T)} \frac{1}{|c|_0}\int_c \omega.$$

On the numerical side, $||\omega||_0$ is estimated by using the mesh τ introduced before. We hence exploit (5) to test the results of Table 3 for the case $k = 2$. In what follows, we let ε vary between 10^{-1} and 10^{-8}, adding then a random perturbation $\theta \in [0, 1]$ such that

$$\frac{1}{|\sigma|_0}\int_\sigma (\omega - \widetilde{\omega}) = (2\theta - 1)\varepsilon.$$

The right hand side of the above equation in `Matlab` reads $(2 * \texttt{rand}(1) - 1) * \varepsilon$. In Table 4 we report and compare the quantities $\frac{1}{\varepsilon}||\Pi\omega - \Pi\widetilde{\omega}||_0$ for

Table 4 The quantity $\frac{1}{\varepsilon}||\Pi\omega - \Pi\widetilde{\omega}||_0$ for weights associated with the uniform distribution (left) and the warp and blend distribution (right)

	Uniform			Warp and blend		
r	$\varepsilon = 10^{-2}$	$\varepsilon = 10^{-5}$	$\varepsilon = 10^{-8}$	$\varepsilon = 10^{-2}$	$\varepsilon = 10^{-5}$	$\varepsilon = 10^{-8}$
2	1.8062	2.5604	3.1250	1.1672	2.3161	2.1164
3	3.6174	3.6340	2.8400	5.2595	3.5963	2.6999
4	4.7001	6.8761	2.9441	3.6823	4.4570	3.2353
5	4.8769	5.3975	5.1765	2.9470	3.4403	2.8560
6	7.2511	5.6625	11.8428	5.8220	5.2760	4.1077
7	8.8414	7.4621	7.2553	4.6160	4.4468	3.3946
8	10.7360	17.1331	9.9696	5.4941	4.0359	4.1365
9	25.9319	20.5579	20.8289	5.2935	5.8337	3.6155

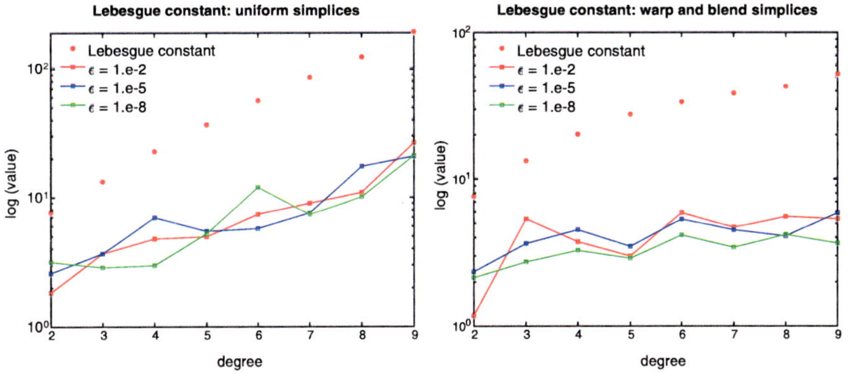

Fig. 5 Stability comparison between estimated Lebesgue constants (marked with a star) and the quantities reported in above tables. On the left we offer the comparison for uniform faces, on the right that for warp and blend ones

$\varepsilon = 10^{-2}, 10^{-5}, 10^{-8}$ and in Fig. 5 we match them with the Lebesgue constants offered in Table 3. For the sake of brevity we just compare results relative to uniform and warp and blend weights. It is worth noting that in semilogarithmic scale the growth of $\frac{1}{\varepsilon}||\Pi\omega - \Pi\widetilde{\omega}||_0$ as a function of the polynomial degree r is parallel to that of the Lebesgue constant. This indicates that estimate (5) is sharp. This is in accordance with the cases $k = 0$ and $k = 1$, see [3].

5 Conclusions

We have considered the weights on 2-simplices as degrees of freedom for polynomial interpolation of differential 2-forms. This offers a great flexibility in the choice of the supports, hence in that of the associated degrees of freedom. We compared different choices of weights (with overlappings, with holes, uniform and non uniform) in terms of the generalised Lebesgue constant, which measures in fact the stability of the interpolation. The best results are obtained for the classical small simplices with holes whose vertices are non uniformly distributed in T. This non uniformity corresponds with distributions of nodes that have been optimised for nodal interpolation (such as symmetrised Lobatto and warp and blend, since they are provided with an explicit formula). Numerical results showed that these choices are also suitable for polynomial interpolation of differential 2-forms, as it has been already observed in the case of 1-forms. Although the nodes here presented optimise the Lebesgue constant for functions (i.e. 0-forms), there is no reason for concluding that they do the same for 2-forms. The configuration of 2-simplices that minimises the generalised Lebesgue constant is thus still unknown and shall be determined. Tests on the stability of the interpolation confirm the solid behaviour of the nonuniform distributions of supports.

Acknowledgments This research was supported by the Italian project PRIN-201752HKH8 of the Università degli Studi di Trento (Italy), and by the French program MathIT, through the ANR-15-IDEX-01 of the Université Côte d'Azur in Nice (France).

Reference

1. Alonso Rodríguez, A., Rapetti, F.: On a generalization of the Lebesgue's constant. J. Comput. Phys. **428**, 109964 (2021). https://doi.org/10.1016/j.jcp.2020.109964
2. Alonso Rodríguez, A., Bruni Bruno, L., Rapetti, F.: Minimal Sets of Unisolvent Weights for High Order Whitney Forms on Simplices. Lecture Notes in Computational Science and Engineering, vol. 139. Springer, Berlin (2020). https://doi.org/10.1007/978-3-030-55874-1_18
3. Alonso Rodríguez, A., Bruni Bruno, L., Rapetti, F.: Towards nonuniform distributions of unisolvent weights for high-order Whitney edge elements. (2020, submitted). https://hal.inria.fr/hal-03114568/
4. Arnold, D.N., Falk, R.S., Winther, R.: Finite element exterior calculus, homological techniques, and applications. Acta Numer. **15**, 1–155 (2006). https://doi.org/10.1017/S0962492906210018
5. Blyth, M.G., Luo, H., Pozrikidis, C.: A comparison of interpolation grids over the triangle or the tetrahedron. J. Eng, Math. **56**, 263–272 (2006). https://doi.org/10.1007/s10665-006-9063-0
6. Christiansen, S.H., Rapetti, F.: On high order finite element spaces of differential forms. Math. Comput. **85**(298), 517–548 (2016). https://doi.org/10.1090/mcom/2995
7. Hesthaven, J.S., Warburton, T.: Nodal Discontinuous Galerkin Methods: Algorithms, Analysis, and Applications. Texts in Applied Mathematics, vol. 54. Springer, New York (2008). https://doi.org/10.1007/978-0-387-72067-8
8. Rapetti, F., Bossavit, A.: Whitney forms of higher degree. SIAM J. Numer. Anal. **47**(3), 2369–2386 (2009). https://doi.org/10.1137/070705489
9. Warburton, T.: An explicit construction of interpolation nodes on the simplex. J. Eng. Math. **56**(3), 247–262 (2006). https://doi.org/10.1007/s10665-006-9086-6

Taming the CFL Number for Discontinuous Galerkin Methods by Local Exponentiation

Daniel Appelö, Mingyu Hu, and Maxim Zinchenko

1 Introduction

It was recognized by Kreiss and Oliger [11] almost half a century ago that high-order methods are superior in terms of accuracy for propagating waves over long distances. Since then, much research has been devoted to spectral and high-order methods and as a result, many highly accurate finite-difference, finite-element, spectral element and discontinuous Galerkin (DG) methods have been developed. Moreover, advances in computational hardware with its ever increasing level of parallelism have favored methods that are robustly stable, geometrically flexible and suitable to implement on parallel computers. The DG method posses all these qualities and has become popular among practitioners, for example in computational electromagnetics where it is gradually replacing the Yee—FDTD scheme. Although DG methods are spectrally convergent in the order q of the approximation, very high orders are rarely used in practice. A reason for this is that polynomials on bounded domains, used as bases in DG methods, have boundary layers at the element edges, [14]. These boundary layers cause the norms of matrices corresponding to differentiation to grow fast with the order, q resulting in numerical stiffness. In turn, this numerically induced stiffness forces the use of excessively small

D. Appelö (✉)
Michigan State University, East Lansing, MI, USA
e-mail: appeloda@msu.edu

M. Hu
University of Colorado, Boulder, CO, USA
e-mail: mingyu.hu@colorado.edu

M. Zinchenko
University of New Mexico, Albuquerque, NM, USA
e-mail: maxim@math.unm.edu

J. M. Melenk et al. (eds.), *Spectral and High Order Methods for Partial Differential Equations ICOSAHOM 2020+1*, Lecture Notes in Computational Science and Engineering 137, https://doi.org/10.1007/978-3-031-20432-6_6

time steps, discouraging the use of very high order methods. Since polynomials vary slowly near the middle of the reference interval, methods such as co-volume filtering has been suggested to tame the CFL condition [14]. Similar ideas were also used in [12], where Li et al. formulated central DG methods for solving ideal magnetohydrodynamic (MHD) equations on overlapping meshes. Yet another way to allow for larger time step sizes is to modify the numerical flux as proposed in [1].

Another source of stiffness, not specific to DG methods, arises from small cells needed to resolve geometric features. For geometric stiffness, local time-stepping and implicit time stepping can be advantageous. Diaz and Grote [4] and Grote and Mitkova [7] proposed energy conserving explicit local time steppers for second order wave equation and Maxwell's equation, respectively. Local time-stepping using linear multistep methods has also been considered by Goedel et al. [6]. Locally implicit energy conserving methods have been proposed in [3, 5, 13], and locally implicit upwind DG methods were considered by Hockbruck and Sturm [10].

Yet another class of time stepping methods, exponential time integrators, are capable of handling both numerically and geometrically induced stiffness. For a spatially discretized linear autonomous system of differential equations, the solution at each time step can be calculated exactly (up to machine precision) and the time stepping is stable for any choice of time step size. This is particularly beneficial for stiff problems. The literature on exponential integrators is extensive but summarized in the review paper by Hochbruck and Ostermann in [9]. In this paper we focus on the time evolution of a system of linear differential equations,

$$\mathbf{u}_t = A\mathbf{u},$$

arising from a spatial DG discretization of a linear, variable coefficient, hyperbolic PDE. We refer to entries in $\mathbf{u}$ as *degrees of freedom* or shorter, DOF. An exponential time-stepping scheme would then advance the DOF at time t_n by the use of the identity

$$\mathbf{u}(t_{n+1}) = e^{A\Delta t}\mathbf{u}(t_n) \equiv Q\mathbf{u}(t_n),$$

where thus Q is the matrix exponential solution operator corresponding to a timestep Δt. The main drawback with this approach is that the exponentiation of a matrix is an expensive and memory demanding operation. In addition as the matrix Q is in general dense, efficient parallelization can be difficult.

In this paper, to reduce the computational cost and memory requirements for finding and storing Q, we propose a local approximation to the exponential time integrator. This approximation still tames the CFL condition of DG schemes but is amenable to parallelization by domain decomposition techniques and is inexpensive to apply. The method is motivated by the finite-speed-of-propagation intrinsic to wave propagation. That is, the value of a solution at some location x and time t will only influence the solution nearby during a time interval Δt. This locality can be used to find an approximation, $\tilde{Q}$, to, Q, one row at a time by considering matrix exponentiation of low rank versions of $\Delta t A$. These low rank matrices will be chosen so that the time update of a DOF will only use nearby DOF.

The rest of the paper is organized as follows. In Sect. 2, we present the construction of our local exponential time integrator. We also present an error bound and discuss computational cost of the local approximation. In Sect. 3, we test the local exponential integrator for the one-dimensional Maxwell's equation discretized in space using DG. In Sect. 4, we conclude this work by pointing out potential applications of our local exponential time integrator and an outlook to future work.

2 Time-Stepping by Local Approximation to the Matrix Exponential

Consider a DG semi-discretization of a linear first order hyperbolic system on a mesh in one, two or three dimensions. This semi-discretization can be expressed as linear system of ordinary differential equations

$$\mathbf{u}_t = A\mathbf{u}, \quad \mathbf{u} \in \mathbb{R}^M, \quad A \in \mathbb{R}^{M\times M},$$

where M is the number of degrees of freedom in $\mathbf{u}$. To identify nearby elements and their degrees of freedom we will let $\rho(j,k)$ be the Euclidian distance between the element centers of elements j and k. The set

$$D(k,r) = \{j \in [1,\ldots,M] \,:\, \rho(k,j) \leq r\},$$

thus represents a ball of radius r in the physical space. We note that the matrix A acting on $\mathbf{u}$ can be expressed as

$$(A\mathbf{u})_k = \sum_{j=1}^{M} a_{kj} u_j, \quad k = 1,\ldots,M.$$

However, as we consider DG methods for hyperbolic problems most of the elements in A are zeros. In fact, for a degree of freedom k on a (finite) element E, only matrix elements $a_{k,j}$ with degree of freedom j belonging to the element E or its nearest neighbor elements would be nonzero. Let R_k be the radius of a circle with a center coinciding with the element center of E and let R_k be large enough for the circle to enclose the nearest neighbor elements (note that we may take R_k larger to improve the timestep constraints, see below). This allows us write the formula above as

$$(A\mathbf{u})_k = \sum_{j\in D(k,R_k)} a_{kj} u_j, \quad k = 1,\ldots,M.$$

2.1 Local Exponentiation

We now define the local exponentiation of the matrix A. For a given k, we fix R_k and form the $M \times M$ matrix $\tilde{A}^{(k)}$ by copying all the rows whose row-index belong to $D(k, R_k)$ into the $M \times M$ zero matrix one row at a time. We write this as a matrix operation $\tilde{A}^{(k)} = P^{(k)} A$ so that $P^{(k)}$ is the matrix that extracts the rows in A that corresponds to the indices in the set $D(k, R_k)$ and sets all the other rows to zero. In other words $P^{(k)}$ is a diagonal matrix with ones in the diagonal elements corresponding to the integers contained in the set $D(k, R_k)$.

With $\tilde{A}^{(k)}$ computed we can now build one part of the approximation, $\tilde{Q} \approx Q \equiv e^{\Delta t A}$. Precisely, the kth row in $\tilde{Q}$ is defined to be the kth row of the matrix

$$e^{\Delta t \tilde{A}^{(k)}}.$$

Expressing this construction in a formula we have

$$\tilde{Q}^T \delta_k = (e^{\Delta t \tilde{A}^{(k)}})^T \delta_k = e^{\Delta t \tilde{A}^{(k)T}} \delta_k = e^{\Delta t A^T P^{(k)}} \delta_k, \quad k \in \Omega, \tag{1}$$

where δ_k is the kth unit vector.

Remark 1 Equation (1) is defined one row / DOF at a time and the set $D(k, R_k)$ above is defined so that it may contain a fraction of the number of DOF in an element as this makes the formulas short. However, in practice we work on one element at a time and process all rows of that element, one at a time, before moving to the next set of degrees of freedoms (rows).

Remark 2 In terms of matrix operations on the vector $\mathbf{u}$, the degrees of the freedom in A that are to be multiplied with the corresponding entries in $\mathbf{u}$ should be columns of A and one may consider the above approach but with the construction done column-by-column instead. However, our numerical experiments show that $\tilde{Q}$ constructed using rows of A yields better approximation to the full exponential time integrator Q.

Remark 3 Since the construction of the rows in $\tilde{Q}$ is local, depending only on degrees of freedom in A of the neighboring elements inside the circle of radius R_k of the center of the element that hold the degree of freedom k, the computation for each k can trivially be carried out in parallel.

2.2 Exponential Locality and Decay of Discretizations of Hyperbolic Operators

The following theorem shows that for R_k large the locally exponentiated $\tilde{Q}$ is close to $Q = e^{\Delta t A}$ in the operator norm.

Theorem 4 *Let* $\Omega = \{1, \ldots, M\}$ *and let* $W = \max_{k\in\Omega} |D(k, R_k)|$ *be the largest number of degrees of freedom included for the construction of any row. Further let* $m = \lfloor R_k \rfloor$ *and* $T_m(z) = \sum_{k=m+1}^{\infty} \frac{z^k}{k!}$, *then*

$$\|Q - \tilde{Q}\| \leq (W + 1)T_m(\|\Delta t A\|) \leq (W + 1)e^{\|\Delta t A\|} \frac{\|\Delta t A\|^{m+1}}{(m + 1)!}.$$

Proof It suffices to show the above norm estimate for the transposed matrices. First, note that A^T has the same nearest neighbor structure as A and hence $A^T\delta_k$ is supported on $D(k, R_k)$, that is, $(A^T\delta_k)_j = 0$ for all $j \in \Omega\backslash D(k, R_k)$. Using the triangle inequality for the metric ρ we obtain that $(A^T)^2\delta_k$ is supported on $D(k, 2R_k)$ and by induction $(A^T)^n\delta_k$ is supported on $D(k, nR_k)$. Then for $n \leq m$ we have $n \leq R_k$ and hence $P_k(A^T)^n\delta_k = (A^T)^n\delta_k$. This implies

$$(\tilde{A}^{(k)T})^n\delta_k = (A^T P^{(k)})^n\delta_k = (A^T)^n\delta_k, \quad n \leq m, \; k \in \Omega.$$

Next, consider an arbitrary vector $\mathbf{u} \in \mathbb{R}^M$ such that $\mathbf{u} = \sum_{k\in\Omega} u_k\delta_k$. Then

$$(Q^T - \tilde{Q}^T)\mathbf{u} = \sum_{k\in\Omega} u_k(Q^T - \tilde{Q}^T)\delta_k = \sum_{k\in\Omega} u_k(e^{\Delta t A^T} - e^{\Delta t\tilde{A}^{(k)T}})\delta_k.$$

Using the Taylor series expansion for $e^{\Delta t A^T}$ and $e^{\Delta t \tilde{A}^{(k)T}}$, we get

$$(Q^T - \tilde{Q}^T)\mathbf{u} = \sum_{k\in\Omega} u_k \sum_{n=0}^{m} \frac{(\Delta t A^T)^n - (\Delta t \tilde{A}^{(k)T})^n}{n!}\delta_k + \sum_{k\in\Omega} u_k\big(T_m(\Delta t A^T) - T_m(\Delta t \tilde{A}^{(k)T})\big)\delta_k.$$

Since $(\Delta t\tilde{A}^{(k)T})^n\delta_k = (\Delta t A^T)^n\delta_k$ for all $n \leq m$, the first double sum in the above equality is zero, so we have

$$(Q^T - \tilde{Q}^T)\mathbf{u} = T_m(\Delta t A^T)\mathbf{u} - \sum_{k\in\Omega} u_k T_m(\Delta t \tilde{A}^{(k)T})\delta_k.$$

Using the triangle inequality, the norm of the first term is easily estimated by $\|T_m(\Delta t A^T)u\| \leq T_m(\|\Delta t A\|)\|u\|$. To estimate the norm of the second term, we first estimate the magnitude of its individual components. Noting that $(\Delta t\tilde{A}^{(k)T})^n\delta_k$ is supported on $D(k, R_k)$ for every $n \geq 0$ and that $\|\Delta t\tilde{A}^{(k)T}\| \leq \|\Delta t A\|$ we have

$$\begin{aligned}\left|\Big(\sum_{k\in\Omega} u_k T_m(\Delta t\tilde{A}^{(k)T})\delta_k\Big)_j\right| &= \left|\sum_{k\in D(j,R_k)} u_k T_m(\Delta t\tilde{A}^{(k)T})\delta_k\right| \\ &\leq T_m(\|\Delta t A\|) \sum_{k\in D(j,R_k)} |u_k|, \quad j \in \Omega.\end{aligned}$$

Then, using the Cauchy-Schwarz inequality and the fact that each disk $D(j, R_k)$ contains at most W points and hence each point $k \in \Omega$ lies in at most W distinct disks $D(j, R_k)$ we get

$$\Big\| \sum_{k\in\Omega} u_k T_m(\Delta t \tilde{A}^{(k)T})\delta_k \Big\|^2 \le T_m(\|\Delta t A\|)^2 \sum_{j\in\Omega}\Big(\sum_{k\in D(j,R_k)} |u_k|\Big)^2$$
$$\le T_m(\|\Delta t A\|)^2 \sum_{j\in\Omega} W \sum_{k\in D(j,R_k)} |u_k|^2 \le T_m(\|\Delta t A\|)^2 W^2 \|\mathbf{u}\|^2.$$

Combining the two estimates then yields

$$\big\|(Q^T - \tilde{Q}^T)\mathbf{u}\big\| \le T_m(\|\Delta t A\|)(1+W)\|\mathbf{u}\|.$$

Since u is an arbitrary vector we get

$$\big\|Q - \tilde{Q}\big\| = \big\|Q^T - \tilde{Q}^T\big\| \le T_m(\|\Delta t A\|)(1+W).$$

2.3 Cost and Complexity for Local Exponentiation

The spatial discretization matrix A of size $M \times M$ is a sparse matrix thus the local discretization matrix for the k-th DOF, $A^{(k)}$, will be even more sparse as it is composed of only a few of the entries in A. Moreover, as the vast majority of the eigenvalues of $A^{(k)}$ will be zero the degree of the minimal polynomial $\mathcal{M}_{A^{(k)}}(z)$ will be *very small* compared to M. It is well known, [2], that if the degree of $\mathcal{M}_{A^{(k)}}(z)$ is $m_{\min}$ then the matrix exponent can be explicitly computed by the formula

$$e^{\Delta t A^{(k)}} = \sum_{s=0}^{m_{\min}-1} f_s(\Delta t)\left(A^{(k)}\right)^s,$$

where $f_0(\Delta t), f_1(\Delta t), \ldots, f_{m_{\min}-1}(\Delta t)$ are analytical functions whose values can be computed by solving a linear system of equations.

Suppose $\tilde{A}^{(k)}$ has m non-zero columns with at most n non-zero entries in each column, where $m, n \ll M$. Then the cost of computing all of $e^{\Delta t A^{(k)}}$ would scale as $O((\min\{m,n\})^3 + \min\{m,n\}m_{\min})$, where the first term represents the cost of directly solving for $f_0, \ldots$ and the second term corresponds to the applications of the matrix powers of the sum to the kth unit vector. The cost of assembling all of the rows of $\tilde{Q}$ would increase this by a factor of M but this is still much less than computing $e^{\Delta t A}$ and, again, can be trivially parallelized.

3 Numerical Experiments in One Dimension: Maxwell's Equations

We consider example 2.6 from [8], Maxwell's equations in one-dimension domain $x \in [-\pi, \pi]$

$$\epsilon(x)\frac{\partial E(x,t)}{\partial t} = -\frac{\partial H(x,t)}{\partial x}, \tag{2a}$$

$$\mu(x)\frac{\partial H(x,t)}{\partial t} = -\frac{\partial E(x,t)}{\partial x}, \tag{2b}$$

with homogenous boundary conditions on the electric field and initial conditions

$$\begin{aligned} \text{BC}: \quad & E(0,t) = E(2\pi,t) = 0, \\ \text{IC}: \quad & E(x,0) = \sin(x), \ H(x,0) = 0. \end{aligned} \tag{3}$$

Here E and H are the electric and magnetic fields, and ϵ and μ are permittivity and permeability. To this end, we take $\epsilon(x)$ and $\mu(x)$ to be 1. We note that this system assumes solutions to the three dimensional Maxwell system where the electric and magnetic fields are of the form $\mathbf{E}(x, y, z, t) = (0, E(x,t), 0)$, $\mathbf{H}(x, y, z, t) = (0, 0, H(x,t))$. The exact solution to this particular problem is the standing wave $E(x,t) = \cos(t)\sin(x)$, $H(x,t) = -\sin(t)\cos(x)$.

We numerically approximate the solution with a modal DG method using Legendre polynomial bases on a uniform spatial grid with n elements. The standard upwind discontinuous Galerkin method takes the form

$$M_E\mathbf{E}_t = S\mathbf{H} + (H_R^- - H^*)\mathbf{L}_R - (H_L^- - H^*)\mathbf{L}_L, \tag{4a}$$

$$M_H\mathbf{H}_t = S\mathbf{E} + (E_R^- - E^*)\mathbf{L}_R - (E_L^- - E^*)\mathbf{L}_L, \tag{4b}$$

where M_E and M_H are the mass matrices for Eqs. (2a) and (2b) receptively. The numerical fluxes are $E^* = \{\{E\}\} - [H]$ and $H^* = \{\{H\}\} - [E]$, where the $\{\{\cdot\}\}$ and $[\cdot]$ denotes the usual average and jump. The $-$ superscript indicates that the value is taken from the solution on the current element, and the subscripts indicate evaluation at the left and right end points of the element. The left and right lift vectors, $\boldsymbol{L}_L$ and $\boldsymbol{L}_R$, are composed of Legendre polynomials up to the q-th order evaluated at the left and right boundaries.

We refer to the modal coefficients as degrees of freedom of the problem. Then Eqs. (4a) and (4b) is rewritten as

$$\mathbf{u}_t = A\mathbf{u}, \tag{5}$$

where $\hat{\mathbf{u}}$ contains all the degrees of freedom from both $\mathbf{E}$ and $\mathbf{H}$. The matrix A is the differentiation matrix for the entire system. It is observed that A is block-tridiagonal

Table 1 Base testing parameter values for 1D Maxwell's equations

Parameter	Value	Parameter (cont'd)	Value
n	20	# of d.e.	3
q	3	# of quadrature points	$q+4$
t_{final}	20	n_{Taylor}	4
h_t, uniform	$0.05hn_{\text{Taylor}}/q$		

if the degrees of freedom in $\mathbf{u}$ are ordered by variables. If the degrees of freedom in $\mathbf{u}$ are ordered by elements, A is block-diagonal. For the rest of the paper, the former ordering of degrees of freedom is used. Since the system is autonomous, the exact numerical time-stepping solution is given by

$$\mathbf{u}^{(n+1)} = \exp(\Delta t A)\mathbf{u}^{(n)} = Q\mathbf{u}^{(n)}, \tag{6}$$

where Δt is the time step size, $\mathbf{u}^{(n)}$ is the numerical solution at time t_n, and Q is the matrix representing the exponential time integrator. All matrix exponentiation in our numerical examples are computed using the MATLAB built-in function `expm`. With the proposed local exponentiation construction in Sect. 2, stability of time-stepping with the local exponentiation matrix $\tilde{Q}$ is investigated and the convergence of error is compared with n_{Taylor}-stage Taylor time-stepping method. We also compare the performance of $\tilde{Q}_{col}$, which is constructed using column degrees of freedom of A, to that of $\tilde{Q}_{row}$, which is constructed using rows degrees of freedom of A. While $\tilde{Q}_{col}$ is more physically intuitive in terms of matrix-vector multiplication, however, we will show that $\tilde{Q}_{row}$ demonstrates better approximation.

The parameters in the tests are the number of elements, denoted by n, time step size Δt, the highest degree of Legendre polynomial q, and the number of neighbor elements (dependent element(s)) denoted by d.e. taken on each side of the k-th element. For the base case, the parameters take the values shown in Table 1.

3.1 Comparison of Spectra

In this section, we investigate the stability of the local exponentiation time-stepping based on the eigenvalue of the time-stepping operator $\tilde{Q}$. Eigenvalues with modulus greater than 1 will lead to instability during time-stepping. Figure 1 shows $\xi(\text{.d.e}) = (|\lambda(\tilde{Q})|_{max} - 1)$ as a function of d.e. in the construction of $\tilde{Q}$ using row degree of freedom.

The eigenvalues with imaginary part falling in $(-0.01, 0.01)$ are excluded from the plot as those eigenvalues are not computed accurately numerically. Except for the varying parameters, all other parameters take the base values listed in Table 1. By varying different parameters, we also justify that our choices of base parameter values construct a stable local exponential time integrator $\tilde{Q}$ with minimum cost in

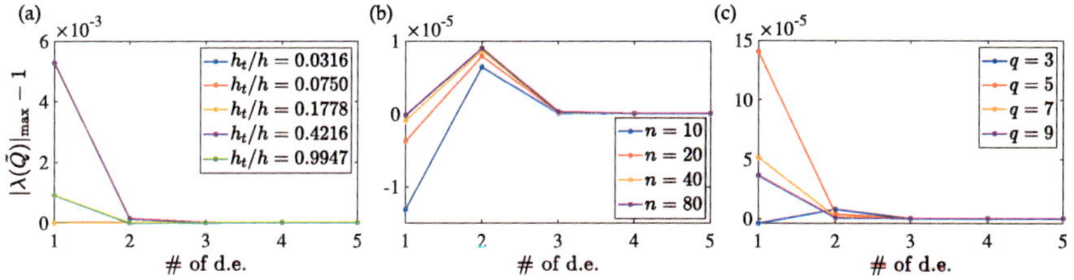

Fig. 1 Eigenvalue discrepancy of $\tilde{Q}_{row}$ v.s. # of d.e. for different (**a**) $\Delta t/h$, (**b**) n, (**c**) q. Note that $\Delta t = h_t$ in the figure

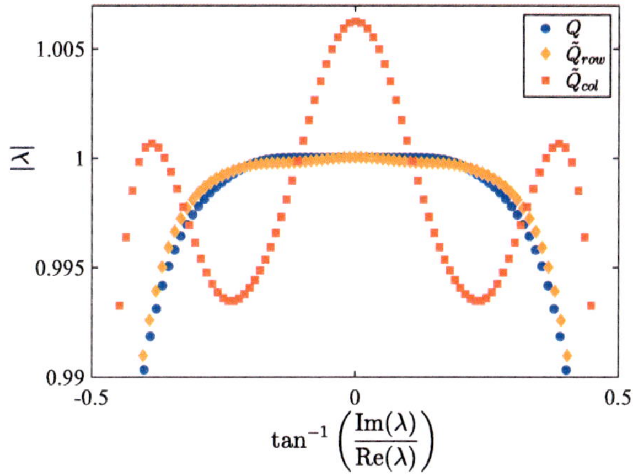

Fig. 2 Distribution of eigenvalues whose modulus is close to 1

its computation. In particular, Fig. 1b shows that the number of d.e., independent of the refinement of the spatial discretization, is the key variable for the stability of $\tilde{Q}$. Overall, the function ξ is small for high-order polynomial basis and relaxed CFL condition. Thus the local exponentiation time stepper is expected to be stable with appropriate combinations of d.e., Δt, and q.

When only one d.e. is taken on each side of each element in the construction of $\tilde{Q}$, Fig. 2 shows that $\tilde{Q}_{row}$, in general, is a better approximation to Q than $\tilde{Q}_{col}$. The latter clearly would lead to an unstable scheme in this case but that also $\tilde{Q}_{col}$ will become stable if d.e. is large enough.

3.2 *Convergence Properties of the Method*

The convergence of error for both row neighbors and column neighbors is investigated in this section and is compared to fourth-order Taylor time-stepping. Figure 3 shows that, compared to column degree of freedom, row degree of freedom gives

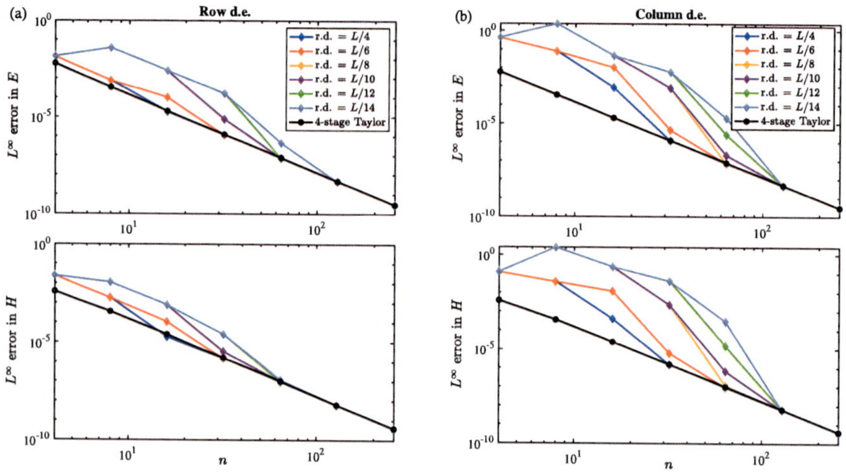

Fig. 3 Error convergence with respect to n on the uniform grid: (**a**) L^∞ errors, row degree of freedom; (**b**) L^∞ errors, column degree of freedom

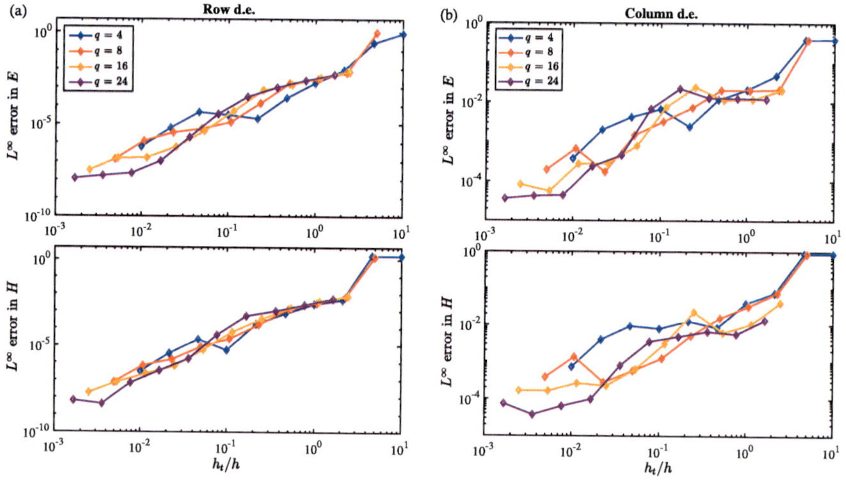

Fig. 4 Error convergence with respect to h_t/h on the uniform grid: (**a**) L^∞ errors, row degree of freedom; (**b**) L^∞ errors, column degree of freedom

smaller error and the error converges to the 4−stage Taylor time-stepping error on a coarser grid. Figure 4 shows that while both row-neighbor- and column-neighbor-constructed local exponential time integrators show advantage in stability, the errors are smaller with row degrees of freedom. In general, the local exponential time integrator demonstrates the possibility to take much larger time step size than that is allowed by other numerical time-stepping methods and the method is robust so high-order spatial approximation can be adopted.

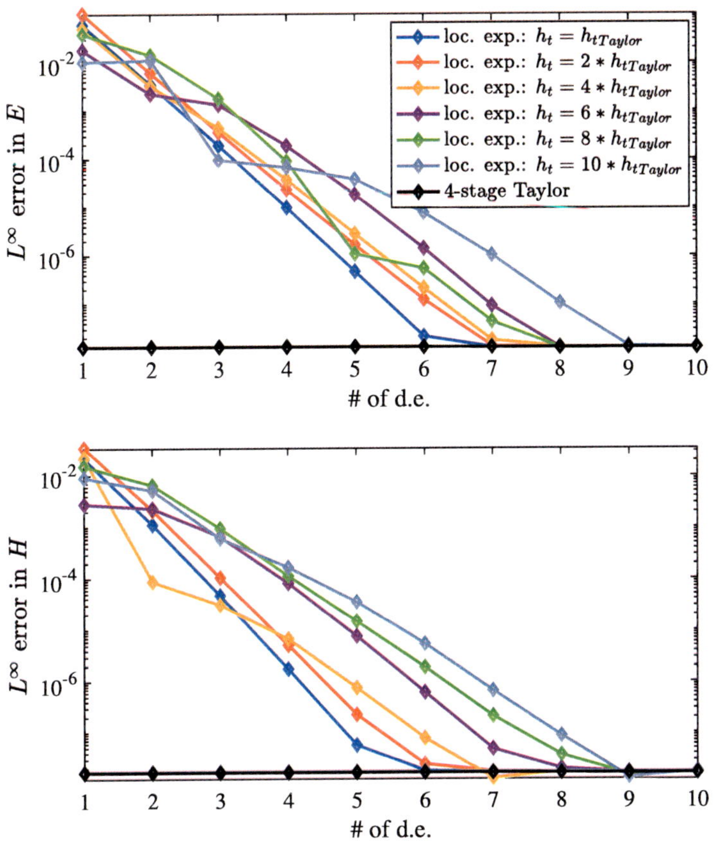

Fig. 5 Error convergence with respect to the number of d.e. with row degree of freedom on a uniform grid

Base on the above observations, we conclude that in the construction of $\tilde{A}^{(k)}$, rows should be taken as dependent degrees of freedom in the assembly of $\tilde{Q}$. In Fig. 5, we display the error convergence versus the number of d.e. for different CFL numbers to demonstrate the efficiency of the local exponentiation time integrator in taming the CFL restriction which is inherited from the DG discretization. The number of elements used is $n = 100$ for more prominent effect. With less than 20% of total degrees of freedom, the local exponentiation time stepper is able to achieve the same or smaller error compared to the explicit 4-stage Taylor time-stepping but with 10 times larger CFL number than permitted by the Taylor time-stepping.

4 Conclusion

An approximate exponential time integrator is proposed and tested for Maxwell's equation in 1D. The construction of this approximate is local, leading to cheaper computation compared to the exponential integrator and potential parallel implementations. The size and structure of the domain of dependency in constructing the local exponential time integrator will be determined specifically based on different physical settings. Our numerical experiments have shown that this local exponentiation time-stepper is robust to high-order spatial discretization and is stable with at least 10 times larger the CFL number required by explicit time-stepping methods.

References

1. Chalmers, N., Krivodonova, L., Qin, R.: Relaxing the CFL number of the discontinuous Galerkin method. SIAM J. Sci. Comput. **36**(4), A2047–A2075 (2014)
2. Cheng, H.-W., Yau, S. S.-T.: More explicit formulas for the matrix exponential. Linear Algebra Appl. **262**, 131–163 (1997)
3. Descombes, S., Lanteri, S., Moya, L.: Locally implicit time integration strategies in a discontinuous Galerkin method for Maxwell's equations. J. Sci. Comput. **56**(1), 190–218 (2013)
4. Diaz, J., Grote, M.J.: Energy conserving explicit local time stepping for second-order wave equations. SIAM J. Sci. Comput. **31**(3), 1985–2014 (2009)
5. Dolean, V., Fahs, H. , Fezoui, L., Lanteri, S.: Locally implicit discontinuous Galerkin method for time domain electromagnetics. J. Comput. Phys. **229**(2), 512–526 (2010)
6. Gödel, N., Schomann, S., Warburton, T., Clemens, M.: Local timestepping discontinuous Galerkin methods for electromagnetic RF field problems. In: 3rd European Conference on Antennas and Propagation, 2009 (EuCAP 2009), pp. 2149–2153. IEEE, Piscataway (2009)
7. Grote, M.J., Mitkova, T.: Explicit local time-stepping methods for Maxwell's equations. J. Comput. Appl. Math. **234**(12), 3283–3302 (2010)
8. Hesthaven, J.S., Warburton, T.: Nodal Discontinuous Galerkin Methods: Algorithms, Analysis, and Applications. Springer, New York (2007)
9. Hochbruck, M., Ostermann, A.: Exponential integrators. Acta Numer. **19**, 209–286 (2010)
10. Hochbruck, M., Sturm, A.: Upwind discontinuous Galerkin space discretization and locally implicit time integration for linear Maxwell's equations. Math. Comput. **88**, 1 (2018)
11. Kreiss, H.-O., Oliger, J.: Comparison of accurate methods for the integration of hyperbolic equations. Tellus **24**(3), 199–215 (1972)
12. Li, F., Xu, L., Yakovlev, S.: Central discontinuous Galerkin methods for ideal MHD equations with the exactly divergence-free magnetic field. J. Comput. Phys. **230**(12), 4828–4847 (2011)
13. Piperno, S.: Symplectic local time-stepping in non-dissipative DGTD methods applied to wave propagation problems. ESAIM: Math. Model. Numer. Anal. **40**(5), 815–841 (2006)
14. Warburton, T., Hagstrom, T.: Taming the CFL number for discontinuous Galerkin methods on structured meshes. SIAM J. Numer. Anal. **46**(6), 3151–3180 (2008)

hp-Finite Elements with Decoupled Constraints for Elastoplasticity

Patrick Bammer, Lothar Banz, and Andreas Schröder

1 Introduction

Elastoplasticity with hardening appears in many problems of mechanical engineering [12]. An important case of elastoplasticity with *linear kinematic hardening* is the holonomic constitutive law which is used to model elastoplastic deformation in an incremental sense [10–12]. In this paper we present an *hp*-finite element discretization for a model problem of elastoplasticity with linear kinematic hardening. The model problem can be seen as one pseudo time-step of the elastoplastic model and can equivalently be formulated either as a variational inequality of second kind or as a mixed variational formulation. Due to the non-differentiability of the plasticity functional involved in the model, the computation of a discrete solution of the variational inequality is not straightforward and may require special solution schemes like Bundle-Newton methods [13] or special discretization techniques. As proposed in [9] we apply an appropriate interpolation for the Frobenius norm appearing in the definition of the plasticity functional. Thereby, we obtain discretizations of the variational inequality and the mixed variational formulation whose solutions coinside. The discrete mixed formulation can be solved for instance by an Uzawa method, see [16], or a semi-smooth Newton method, see [3]. The novelty in the presented approach consists in the use of biorthogonal basis functions in the discrete mixed variational formulation, see e.g. [4]. This allows the decoupling of the constraints associated to the involved Lagrange multiplier. Herewith, the inequality constraints can be reformulated as a system of decoupled nonlinear equations, which in fact, simplifies the use of solution schemes like semi-smooth Newton-solvers. We demonstrate the applicability of a semi-smooth Newton solver

P. Bammer (✉) · L. Banz · A. Schröder
University of Salzburg, Salzburg, Austria
e-mail: patrick.bammer@plus.ac.at; lothar.banz@plus.ac.at; andreas.schroeder@plus.ac.at

J. M. Melenk et al. (eds.), *Spectral and High Order Methods for Partial Differential Equations ICOSAHOM 2020+1*, Lecture Notes in Computational Science and Engineering 137, https://doi.org/10.1007/978-3-031-20432-6_7

to these equations. Furthermore, its robustness to mesh size, polynomial degree and projection parameters are shown with some numerical examples. A priori and a posteriori error results for the primal and mixed variational discretization schemes as well as a proof of the local superlinearly convergence of the semi-smooth Newton solver will be presented in some forthcoming papers [2, 3].

h- and *hp*-finite elements for problems of elastopasticity are discussed, for instance, in [1, 5] and [8, 15], respectively and for adaptive approaches we refer to [6]. For the application of semi-smooth Newton solvers to elastoplastic (contact) problems, see [7, 14]. The beneficial use of biorthogonal basis functions in the context of *hp*-discretizations of elastoplastic problems has not been discussed so far.

2 The Model Problem and Its *hp*-FE Discretization

Let $\Omega \subset \mathbb{R}^d$, $d \in \{2, 3\}$, be a bounded, polygonal domain with Lipschitz-boundary $\Gamma := \partial\Omega$ which has the outer unit normal $\mathfrak{n}$. The model problem of elastoplasticity with linear kinematic hardening is to find a displacement field $\mathfrak{u} \in H^1(\Omega, \mathbb{R}^d)$ and a plastic strain $\boldsymbol{p} \in Q := L^2(\Omega, \mathbb{S}_{d,0})$ where $\mathbb{S}_{d,0}$ denotes the set of symmetric $d \times d$ matrices over $\mathbb{R}$ with vanishing trace $\mathrm{tr}(\boldsymbol{q}) := \sum_{j=1}^d q_{jj}$, $\boldsymbol{q} = (q_{ij}) \in \mathbb{R}^{d\times d}$, such that

$$-\operatorname{div} \boldsymbol{\sigma}(\mathfrak{u}, \boldsymbol{p}) = \mathfrak{f} \qquad \text{in } \Omega \tag{1a}$$

$$\mathfrak{u} = \mathfrak{o} \qquad \text{on } \Gamma_D \tag{1b}$$

$$\boldsymbol{\sigma}(\mathfrak{u}, \boldsymbol{p})\mathfrak{n} = \mathfrak{g} \qquad \text{on } \Gamma_N \tag{1c}$$

$$\boldsymbol{\sigma}(\mathfrak{u}, \boldsymbol{p}) - \mathbb{H}\boldsymbol{p} \in \partial j(\boldsymbol{p}) \qquad \text{in } \Omega. \tag{1d}$$

The volume forces $\mathfrak{f} \in L^2(\Omega, \mathbb{R}^d)$ and the surface force $\mathfrak{g} \in L^2(\Gamma_N, \mathbb{R}^d)$ are given where $\Gamma_N := \Gamma \setminus \Gamma_D$ is the complement of the Dirichlet boundary part $\Gamma_D \subset \Gamma$ of positive surface measure. The stress tensor is defined as $\boldsymbol{\sigma}(\mathfrak{u}, \boldsymbol{p}) := \mathbb{C}(\boldsymbol{\varepsilon}(\mathfrak{u}) - \boldsymbol{p})$ with the linearized strain tensor $\boldsymbol{\varepsilon}(\mathfrak{u}) := \frac{1}{2}(D\mathfrak{u} + (D\mathfrak{u})^\top)$ and the elasticity tensor $\mathbb{C}$. The latter is assumed to be symmetric, i.e. $\mathbb{C}_{ijkl} = \mathbb{C}_{jilk} = \mathbb{C}_{klij}$, and uniformly elliptic with entries $\mathbb{C}_{ijkl} \in L^\infty(\Omega)$. The hardening tensor $\mathbb{H}$ with entries $\mathbb{H}_{ijkl} \in L^\infty(\Omega)$ is also assumed to be symmetric and uniformly elliptic. Furthermore, $\partial j(\cdot)$ denotes the subdifferential of the non-differentiable part of the plastic dissipation function $j(\cdot)$ given by $j(\boldsymbol{q}) := \sigma_y \|\boldsymbol{q}\|_F$. Here, $\sigma_y \in L^\infty(\Omega)$ is the yield stress in uniaxial tension with $\sigma_y \geq \sigma_0 > 0$ and $\|\cdot\|_F$ denotes the Frobenius norm induced by the inner product $\boldsymbol{p} : \boldsymbol{q} := \sum_{i,j=1}^d p_{ij} q_{ij}$ for $\boldsymbol{p} = (p_{ij}), \boldsymbol{q} = (q_{ij}) \in \mathbb{R}^{d\times d}$. Throughout this paper we set $V := \{\mathfrak{v} \in H^1(\Omega, \mathbb{R}^d) : \mathfrak{v} = \mathfrak{o} \text{ on } \Gamma_D\}$ and denote the usual L^2-scalar product for scalar-, vector- and matrix-valued functions over Ω by $(\cdot, \cdot)$.

It is well known, see e.g. [12], that a weak formulation to (1) is to find a pair $(\mathfrak{u}, \boldsymbol{p}) \in V \times Q$ for which the variational inequality of second kind

$$a\big((\mathfrak{u}, \boldsymbol{p}), (\mathfrak{v} - \mathfrak{u}, \boldsymbol{q} - \boldsymbol{p})\big) + \psi(\boldsymbol{q}) - \psi(\boldsymbol{p}) \geq \ell(\mathfrak{v} - \mathfrak{u}) \tag{2}$$

holds for all $(\mathfrak{v}, \boldsymbol{q}) \in V \times Q$. Here, the bilinear form $a(\cdot, \cdot)$, the plasticity functional $\psi(\cdot)$ and the linear form $\ell(\cdot)$ are given by

$$\begin{aligned} a\big((\mathfrak{u}, \boldsymbol{p}), (\mathfrak{v}, \boldsymbol{q})\big) &:= \big(\boldsymbol{\sigma}(\mathfrak{u}, \boldsymbol{p}), \boldsymbol{\varepsilon}(\mathfrak{v}) - \boldsymbol{q}\big) + (\mathbb{H}\boldsymbol{p}, \boldsymbol{q}), \\ \psi(\boldsymbol{q}) &:= (\sigma_y, \|\boldsymbol{q}\|_F), \\ \ell(\mathfrak{v}) &:= (\mathfrak{f}, \mathfrak{v}) + \int_{\Gamma_N} \mathfrak{g}^\top \mathfrak{v} \, ds, \end{aligned}$$

respectively. One possibility of handling the non-differentiability of the plasticity functional $\psi(\cdot)$ is to introduce a Lagrange multiplier $\boldsymbol{\lambda}$ which leads to the mixed variational formulation of finding a triple $(\mathfrak{u}, \boldsymbol{p}, \boldsymbol{\lambda}) \in V \times Q \times \Lambda$ such that

$$a\big((\mathfrak{u}, \boldsymbol{p}), (\mathfrak{v}, \boldsymbol{q})\big) + (\boldsymbol{\lambda}, \boldsymbol{q}) = \ell(\mathfrak{v}) \tag{3a}$$

$$(\boldsymbol{\mu} - \boldsymbol{\lambda}, \boldsymbol{p}) \leq 0 \tag{3b}$$

for all $(\mathfrak{v}, \boldsymbol{q}, \boldsymbol{\mu}) \in V \times Q \times \Lambda$. Here, we choose the non-empty, closed and convex set $\Lambda := \{\boldsymbol{\mu} \in Q : (\boldsymbol{\mu}, \boldsymbol{q}) \leq \psi(\boldsymbol{q}) \text{ for all } \boldsymbol{q} \in Q\}$ as set of admissible Lagrange multipliers.

It can be shown that the primal formulation (2) and the mixed variational formulation (3) are equivalent, i.e. if $(\mathfrak{u}, \boldsymbol{p})$ solves (2), then $(\mathfrak{u}, \boldsymbol{p}, \boldsymbol{\lambda})$ with $\boldsymbol{\lambda} := \operatorname{dev}(\boldsymbol{\sigma}(\mathfrak{u}, \boldsymbol{p}) - \mathbb{H}\boldsymbol{p})$ is a solution to (3), where $\operatorname{dev}(\boldsymbol{q}) := \frac{1}{d}(\boldsymbol{q} - \operatorname{tr}(\boldsymbol{q})\boldsymbol{I})$ denotes the deviatoric part of $\boldsymbol{q} \in L^2(\Omega, \mathbb{R}^{d \times d})$ and $\boldsymbol{I}$ represents the identity matrix. On the other hand, if $(\mathfrak{u}, \boldsymbol{p}, \boldsymbol{\lambda})$ solves (3), then $(\mathfrak{u}, \boldsymbol{p})$ is a solution to (2) and there holds $\boldsymbol{\lambda} = \operatorname{dev}(\boldsymbol{\sigma}(\mathfrak{u}, \boldsymbol{p}) - \mathbb{H}\boldsymbol{p})$, see e.g. [2, 16].

For the discretization with *hp*-finite elements let $\mathcal{T}_h$ be a locally quasi-uniform finite element mesh consisting of quadrilaterals or hexahedrons. We set $h := (h_T)_{T \in \mathcal{T}_h}$ and $p := (p_T)_{T \in \mathcal{T}_h}$ where h_T and p_T denote the local element size and the local polynomial degree, respectively. Furthermore, let $\mathcal{F}_T : \widehat{T} \to T$ be the bi/tri-linear bijective mapping from the reference element $\widehat{T} = [-1, 1]^d$ onto the physical element $T \in \mathcal{T}_h$. For the discretization of the displacement field and the plastic strain we use the *hp*-finite element spaces

$$\begin{aligned} V_{hp} &:= \left\{\mathfrak{v}_{hp} \in V : \mathfrak{v}_{hp}|_T \circ \mathcal{F}_T \in \big(\mathbb{P}_{p_T}(\widehat{T})\big)^d \text{ for all } T \in \mathcal{T}_h\right\}, \\ Q_{hp} &:= \left\{\boldsymbol{q}_{hp} \in Q : \boldsymbol{q}_{hp}|_T \circ \mathcal{F}_T \in \big(\mathbb{P}_{p_T - 1}(\widehat{T})\big)^{d \times d} \text{ for all } T \in \mathcal{T}_h\right\}. \end{aligned}$$

For simplicity, we assume that $\sigma_y|_T \circ \mathcal{F}_T \in \mathbb{P}_{\max(p_T+1-d,0)}(\widehat{T})$ for the rest of this paper, which covers the important case of (piecewise) constant yield stress σ_y. We approximate the Frobenius norm in the definition of $\psi(\cdot)$ by using an appropriate interpolation operator. For this purpose, let $\{\hat{\mathfrak{x}}_{k,T}\}_{k=1,\dots,n_T} \subset \widehat{T}$ be the tensor product Gauss quadrature points with positive weights $\{\omega_{k,T}\}_{k=1,\dots,n_T} \subset \mathbb{R}$ where $n_T := p_T^d$ and let $\{\widehat{\phi}_{k,T}\}_{k=1,\dots,n_T}$ be the Lagrange basis functions on $\widehat{T}$ defined in these Gauss points, i.e. $\widehat{\phi}_{k,T} \in \mathbb{P}_{p_T-1}(\widehat{T})$ and $\widehat{\phi}_{k,T}(\hat{\mathfrak{x}}_{l,T}) = \delta_{kl}$ for $1 \le k, l \le n_T$ (where δ is the usual Kronecker delta and $T \in \mathcal{T}_h$). For $q \in L^2(\Omega)$ with $q|_T \in C^0(T)$ for all $T \in \mathcal{T}_h$, we define the interpolation operator $\mathcal{J}_{hp}(\cdot)$ piecewisely by

$$\mathcal{J}_{hp}(q)|_T := \sum_{k=1}^{n_T} q\big(\mathcal{F}_T(\hat{\mathfrak{x}}_{k,T})\big)\, \widehat{\phi}_{k,T} \circ \mathcal{F}_T^{-1}, \quad T \in \mathcal{T}_h.$$

This yields an approximation $\psi_{hp} : Q_{hp} \to \mathbb{R}$ of $\psi(\cdot)$ given by $\psi_{hp}(\boldsymbol{q}_{hp}) := (\sigma_y, \mathcal{J}_{hp}(\|\boldsymbol{q}_{hp}\|_F))$. Note that $|\det D\mathcal{F}_T| \in \mathbb{P}_{d-1}(\widehat{T})$ for a convex element $T \in \mathcal{T}_h$. Due to the polynomial assumption on σ_y, we can evaluate $\psi_{hp}(\boldsymbol{q}_{hp})$ exactly by a Gauss quadrature formula. If $\mathcal{T}_{h,1}$ denotes the set of those elements $T \in \mathcal{T}_h$ for which $|\det D\mathcal{F}_T| \in \mathbb{P}_1(\widehat{T})$ or $p_T \ge 2$, and $\mathcal{T}_{h,2}$ denotes the set of the remaining elements for which $|\det D\mathcal{F}_T| \in \mathbb{P}_2(\widehat{T}) \setminus \mathbb{P}_1(\widehat{T})$ and $p_T = 1$ there holds

$$\begin{aligned}\psi_{hp}(\boldsymbol{q}_{hp}) = &\sum_{T \in \mathcal{T}_{h,2}} |T|\, \sigma_y\big(\mathcal{F}_T(\mathfrak{o})\big) \big\|\boldsymbol{q}_{hp}\big(\mathcal{F}_T(\mathfrak{o})\big)\big\|_F \\ &+ \sum_{T \in \mathcal{T}_{h,1}} \sum_{k=1}^{n_T} \omega_{k,T}\, |\det D\mathcal{F}_T(\hat{\mathfrak{x}}_{k,T})|\, \sigma_y\big(\mathcal{F}_T(\hat{\mathfrak{x}}_{k,T})\big) \big\|\boldsymbol{q}_{hp}\big(\mathcal{F}_T(\hat{\mathfrak{x}}_{k,T})\big)\big\|_F.\end{aligned}$$

Hence, the approximation $\psi_{hp}(\cdot)$ can in fact be interpreted as a quadrature formula for the plasticity functional $\psi(\cdot)$. These preparations yield the following discretization of the variational inequality (2): Find a pair $(\mathfrak{u}_{hp}, \boldsymbol{p}_{hp}) \in V_{hp} \times Q_{hp}$ such that

$$a\big((\mathfrak{u}_{hp}, \boldsymbol{p}_{hp}), (\mathfrak{v}_{hp} - \mathfrak{u}_{hp}, \boldsymbol{q}_{hp} - \boldsymbol{p}_{hp})\big) + \psi_{hp}(\boldsymbol{q}_{hp}) - \psi_{hp}(\boldsymbol{p}_{hp}) \ge \ell(\mathfrak{v}_{hp} - \mathfrak{u}_{hp}) \tag{4}$$

for all $(\mathfrak{v}_{hp}, \boldsymbol{q}_{hp}) \in V_{hp} \times Q_{hp}$. Since $\psi_{hp}(\cdot)$ is convex, continuous and subdifferentiable, the discrete variational inequality (4) has a unique solution [2]. As $\psi_{hp}(\cdot)$ is evaluable for $\boldsymbol{q}_{hp}$ but still non-differentiable, the solution procedure for (4) remains challenging. We therefore consider a discretization of the mixed variational formulation (3). For this purpose, let

$$\Lambda_{hp} := \big\{\boldsymbol{\mu}_{hp} \in Q_{hp} : (\boldsymbol{\mu}_{hp}, \boldsymbol{q}_{hp}) \le \psi_{hp}(\boldsymbol{q}_{hp}) \text{ for all } \boldsymbol{q}_{hp} \in Q_{hp}\big\} \tag{5}$$

be the closed and convex set of admissible discrete Lagrange multipliers. Thereby, the discretization of (3) is to find a triple $(\mathfrak{u}_{hp}, \boldsymbol{p}_{hp}, \boldsymbol{\lambda}_{hp}) \in V_{hp} \times Q_{hp} \times \Lambda_{hp}$ such that

$$a\big((\mathfrak{u}_{hp}, \boldsymbol{p}_{hp}), (\mathfrak{v}_{hp}, \boldsymbol{q}_{hp})\big) + (\boldsymbol{\lambda}_{hp}, \boldsymbol{q}_{hp}) = \ell(\mathfrak{v}_{hp}) \tag{6a}$$

$$(\boldsymbol{\mu}_{hp} - \boldsymbol{\lambda}_{hp}, \boldsymbol{p}_{hp}) \leq 0 \tag{6b}$$

for all $(\mathfrak{v}_{hp}, \boldsymbol{q}_{hp}, \boldsymbol{\mu}_{hp}) \in V_{hp} \times Q_{hp} \times \Lambda_{hp}$. We emphasize that (4) and (6) are equivalent, i.e. if $(\mathfrak{u}_{hp}, \boldsymbol{p}_{hp}) \in V_{hp} \times Q_{hp}$ solves (4) then $(\mathfrak{u}_{hp}, \boldsymbol{p}_{hp}, \boldsymbol{\lambda}_{hp})$ with $\boldsymbol{\lambda}_{hp} := \operatorname{dev}(\boldsymbol{\sigma}(\mathfrak{u}_{hp}, \boldsymbol{p}_{hp}) - \mathbb{H}\boldsymbol{p}_{hp})$ is a solution to (6). Moreover, if $(\mathfrak{u}_{hp}, \boldsymbol{p}_{hp}, \boldsymbol{\lambda}_{hp}) \in V_{hp} \times Q_{hp} \times \Lambda_{hp}$ is a solution to (6) then $(\mathfrak{u}_{hp}, \boldsymbol{p}_{hp})$ solves (4) and there holds $\boldsymbol{\lambda}_{hp} = \operatorname{dev}(\boldsymbol{\sigma}(\mathfrak{u}_{hp}, \boldsymbol{p}_{hp}) - \mathbb{H}\boldsymbol{p}_{hp})$, see [2].

The constraints in Λ_{hp} and (6b) can be decoupled, which in fact is the foundation of the nonlinear equation and the semi-smooth Newton solver proposed in Sect. 3. To this end, let $\phi_1, \ldots, \phi_N$ be piecewisely defined as

$$\phi_{\zeta(k,T')}|_T := \begin{cases} \widehat{\phi}_{k,T'} \circ \mathcal{F}_{T'}^{-1}, & \text{if } T = T', \\ 0, & \text{if } T \neq T' \end{cases} \qquad T, T' \in \mathcal{T}_h,\ 1 \leq k \leq n_T$$

where $\zeta : \{(k, T) : T \in \mathcal{T}_h, 1 \leq k \leq n_T\} \to \{1, \ldots, N\}$ with $N := \sum_{T \in \mathcal{T}_h} n_T$ is a one to one numbering. Obviously, $\{\phi_1, \ldots, \phi_N\}$ forms a basis of the space $W_{hp} := \{q \in L^2(\Omega) : q|_T \circ \mathcal{F}_T \in \mathbb{P}_{p_T - 1}(\widehat{T}) \text{ for all } T \in \mathcal{T}_h\}$. Furthermore, we introduce biorthogonal functions $\varphi_1, \ldots, \varphi_N$ that are uniquely determined by $\varphi_{\zeta(k,T)}|_T \circ \mathcal{F}_T \in \mathbb{P}_{p_T - 1}(\widehat{T})$ for $T \in \mathcal{T}_h$, $1 \leq k \leq n_T$ and $(\phi_i, \varphi_j) = \delta_{ij}(\phi_i, 1) =: \delta_{ij} D_i$ for $1 \leq i, j \leq N$. Clearly, the functions $\varphi_1, \ldots, \varphi_N$ are linearly independent and, thus, form a basis of W_{hp} as well. Moreover, $\operatorname{supp}(\varphi_i) = \operatorname{supp}(\phi_i)$ for $1 \leq i \leq N$ and in the cases that $|\det D\mathcal{F}_T| \in \mathbb{P}_1(\widehat{T})$ or $p_T = 1$ we have $\varphi_{\zeta(k,T)} = \phi_{\zeta(k,T)}$ for $1 \leq k \leq n_T$. Note that $D_i > 0$ for all $1 \leq i \leq N$.

Lemma 1 *For $\boldsymbol{q}_{hp} = \sum_{i=1}^N \boldsymbol{q}_i \phi_i \in Q_{hp}$ there holds*

$$\mathcal{J}_{hp}(\|\boldsymbol{q}_{hp}\|_F) = \sum_{i=1}^N \|\boldsymbol{q}_i\|_F \phi_i.$$

Proof For $T \in \mathcal{T}_h$ and $1 \leq k \leq p_T$ we have

$$\begin{aligned} \|\boldsymbol{q}_{hp}(\mathfrak{x}_{k,T})\|_F^2 &= \sum_{i,j=1}^N (\boldsymbol{q}_i : \boldsymbol{q}_j) \phi_i(\mathfrak{x}_{k,T}) \phi_j(\mathfrak{x}_{k,T}) \\ &= \sum_{i,j=1}^N (\boldsymbol{q}_i : \boldsymbol{q}_j) \delta_{i,\zeta(k,T)} \delta_{j,\zeta(k,T)} = \boldsymbol{q}_{\zeta(k,T)} : \boldsymbol{q}_{\zeta(k,T)} = \|\boldsymbol{q}_{\zeta(k,T)}\|_F^2 \end{aligned}$$

where $\mathfrak{x}_{k,T} := \mathcal{F}_T(\hat{\mathfrak{x}}_{k,T})$. Therefore,

$$\mathcal{J}_{hp}(\|\boldsymbol{q}_{hp}\|_F) = \sum_{T\in\mathcal{T}_h}\sum_{k=1}^{n_T} \|\boldsymbol{q}_{hp}(\hat{\mathfrak{x}}_{k,T})\|_F \phi_{\zeta(k,T)} = \sum_{T\in\mathcal{T}_h}\sum_{k=1}^{n_T} \|\boldsymbol{q}_{\zeta(k,T)}\|_F \phi_{\zeta(k,T)}$$

which completes the proof since $\zeta(\cdot,\cdot)$ is one to one. □

Defining the quantities $\sigma_i := D_i^{-1}(\sigma_y, \phi_i)$ and observing that $Q_{hp} = \{\sum_{i=1}^N \boldsymbol{q}_i\phi_i : \boldsymbol{q}_i \in \mathbb{S}_{d,0}\} = \{\sum_{i=1}^N \boldsymbol{\mu}_i\varphi_i : \boldsymbol{\mu}_i \in \mathbb{S}_{d,0}\}$ we are able to decouple the constraints in Λ_{hp} and (6b).

Theorem 2 *It holds*

$$\Lambda_{hp} = \left\{\sum\nolimits_{i=1}^N \boldsymbol{\mu}_i\varphi_i : \boldsymbol{\mu}_i \in \mathbb{S}_{d,0} \textit{ and } \|\boldsymbol{\mu}_i\|_F \le \sigma_i\right\}. \tag{7}$$

Furthermore, $\boldsymbol{\lambda}_{hp} = \sum_{i=1}^N \boldsymbol{\lambda}_i\varphi_i \in \Lambda_{hp}$ *satisfies the inequality* (6b) *if and only if* $\boldsymbol{\lambda}_i : \boldsymbol{p}_i = \sigma_i\|\boldsymbol{p}_i\|_F$ *for all* $1 \le i \le N$ *where* $\boldsymbol{p}_{hp} = \sum_{i=1}^N \boldsymbol{p}_i\phi_i \in Q_{hp}$.

Proof Writing $\boldsymbol{\mu}_{hp}, \boldsymbol{q}_{hp} \in Q_{hp}$ as $\boldsymbol{\mu}_{hp} = \sum_{i=1}^N \boldsymbol{\mu}_i\varphi_i$ and $\boldsymbol{q}_{hp} = \sum_{i=1}^N \boldsymbol{q}_i\phi_i$ the biorthogonality of the basis functions yields

$$(\boldsymbol{\mu}_{hp}, \boldsymbol{q}_{hp}) = \sum_{i,j=1}^N \boldsymbol{\mu}_i : \boldsymbol{q}_j \int_\Omega \varphi_i\phi_j \,\mathrm{d}\mathfrak{x} = \sum_{i=1}^N \boldsymbol{\mu}_i : \boldsymbol{q}_i D_i \tag{8}$$

and from Lemma 1 we obtain

$$\psi_{hp}(\boldsymbol{q}_{hp}) = \big(\sigma_y, \mathcal{J}_{hp}(\|\boldsymbol{q}_{hp}\|_F)\big) = \sum_{i=1}^N \|\boldsymbol{q}_i\|_F\,(\sigma_y, \phi_i) = \sum_{i=1}^N \|\boldsymbol{q}_i\|_F D_i\sigma_i. \tag{9}$$

Hence, if $\boldsymbol{\mu}_{hp} \in \Lambda_{hp}$ we have

$$\sum_{i=1}^N \boldsymbol{\mu}_i : \boldsymbol{q}_i D_i \le \sum_{i=1}^N \|\boldsymbol{q}_i\|_F D_i\sigma_i.$$

Thus, choosing $\boldsymbol{q}_{hp} := \boldsymbol{\mu}_j\phi_j$ yields $\|\boldsymbol{\mu}_j\|_F \le \sigma_j$ for all $1 \le j \le N$ as $D_j > 0$. Conversely, if $\|\boldsymbol{\mu}_i\|_F \le \sigma_i$ for all $1 \le i \le N$ the Cauchy-Schwarz-inequality gives

$$(\boldsymbol{\mu}_{hp}, \boldsymbol{q}_{hp}) = \sum_{i=1}^N \boldsymbol{\mu}_i : \boldsymbol{q}_i D_i \le \sum_{i=1}^N \|\boldsymbol{\mu}_i\|_F\|\boldsymbol{q}_i\|_F D_i \le \sum_{i=1}^N \|\boldsymbol{q}_i\|_F D_i\sigma_i = \psi_{hp}(\boldsymbol{q}_{hp})$$

by using (8) and (9). To verify the second assertion let $\boldsymbol{\lambda}_{hp} \in \Lambda_{hp}$ and $\boldsymbol{p}_{hp} \in Q_{hp}$ satisfy (6b). Therefore, (8) yields

$$0 \geq (\boldsymbol{\mu}_{hp} - \boldsymbol{\lambda}_{hp}, \boldsymbol{p}_{hp}) = \sum_{i=1}^{N} (\boldsymbol{\mu}_i - \boldsymbol{\lambda}_i) : \boldsymbol{p}_i \, D_i . \tag{10}$$

Choosing $\boldsymbol{\mu}_{hp} := \sum_{j=1, j\neq i}^{N} \boldsymbol{\lambda}_j \varphi_j + \boldsymbol{\mu}_i \varphi_i$ with $\boldsymbol{\mu}_i := \pm\sigma_i \boldsymbol{p}_i \|\boldsymbol{p}_i\|_F^{-1}$ if $\boldsymbol{p}_i \neq \boldsymbol{0}$ and $\boldsymbol{\mu}_i := \boldsymbol{0}$ if $\boldsymbol{p}_i = \boldsymbol{0}$ for any $1 \leq i \leq N$ gives $\boldsymbol{\mu}_{hp} \in \Lambda_{hp}$ and (10) immediately yields $\sigma_i \|\boldsymbol{p}_i\|_F = \boldsymbol{\lambda}_i : \boldsymbol{p}_i$ as $D_i > 0$. Conversely, if $\boldsymbol{\lambda}_{hp} \in \Lambda_{hp}$ satisfies $\sigma_i \|\boldsymbol{p}_i\|_F = \boldsymbol{\lambda}_i : \boldsymbol{p}_i$ for all $1 \leq i \leq N$ the Cauchy-Schwarz-inequality together with (8) shows

$$\begin{aligned}(\boldsymbol{\mu}_{hp} - \boldsymbol{\lambda}_{hp}, \boldsymbol{p}_{hp}) &= \sum_{i=1}^{N} (\boldsymbol{\mu}_i - \boldsymbol{\lambda}_i) : \boldsymbol{p}_i \, D_i \\ &\leq \sum_{i=1}^{N} (\|\boldsymbol{\mu}_i\|_F \|\boldsymbol{p}_i\|_F - \sigma_i \|\boldsymbol{p}_i\|_F) D_i \leq 0\end{aligned}$$

for $\boldsymbol{\mu}_{hp} \in \Lambda_{hp}$ as $D_i > 0$ and (due to the first assertion) $\|\boldsymbol{\mu}_i\|_F \leq \sigma_i$ for all $1 \leq i \leq N$. □

Note that $\sigma_{\zeta(k,T)} = \sigma_y(\mathcal{F}_T(\hat{\mathfrak{x}}_{k,T})) > 0$ for all $T \in \mathcal{T}_h$ and $1 \leq k \leq n_T$ due to the polynomial assumption on σ_y and, thus, $\Lambda_{hp} \neq \emptyset$. Moreover, if $\|\boldsymbol{\lambda}_i\|_F \leq \sigma_i$ the Cauchy-Schwarz-inequality implies the equivalence of the condition $\boldsymbol{\lambda}_i : \boldsymbol{p}_i = \sigma_i \|\boldsymbol{p}_i\|_F$ to the two implications

$$\|\boldsymbol{\lambda}_i\|_F < \sigma_i \quad \Rightarrow \quad \boldsymbol{p}_i = \boldsymbol{0} \tag{11a}$$

$$\|\boldsymbol{\lambda}_i\|_F = \sigma_i \quad \Rightarrow \quad \exists c \geq 0 \text{ with } \boldsymbol{p}_i = c\boldsymbol{\lambda}_i . \tag{11b}$$

3 Representation as Nonlinear Equation

The decoupling of the constraints stated in Theorem 2 allows the introduction of a nonlinear, semi-smooth NCP-function $\chi(\cdot)$ to write the inequality constraints as a nonlinear equation. For this purpose we specify an adequate basis for the space of symmetric and trace-free matrices of $\mathbb{S}_{d,0}$. For $d = 2$ we choose $\boldsymbol{\Phi}_1 := \begin{pmatrix} 1 & 0 \\ 0 & -1 \end{pmatrix}$, $\boldsymbol{\Phi}_2 := \begin{pmatrix} 0 & 1 \\ 1 & 0 \end{pmatrix}$ and for $d = 3$ we take

$$\boldsymbol{\Phi}_1 := \tfrac{1}{\sqrt{2}} \begin{pmatrix} 1 & 0 & 0 \\ 0 & -1 & 0 \\ 0 & 0 & 0 \end{pmatrix}, \quad \boldsymbol{\Phi}_2 := \tfrac{1}{\sqrt{6}} \begin{pmatrix} 1 & 0 & 0 \\ 0 & 1 & 0 \\ 0 & 0 & -2 \end{pmatrix}, \quad \boldsymbol{\Phi}_3 := \tfrac{1}{\sqrt{2}} \begin{pmatrix} 0 & 1 & 0 \\ 1 & 0 & 0 \\ 0 & 0 & 0 \end{pmatrix},$$

$$\boldsymbol{\Phi}_4 := \tfrac{1}{\sqrt{2}} \begin{pmatrix} 0 & 0 & 1 \\ 0 & 0 & 0 \\ 1 & 0 & 0 \end{pmatrix}, \quad \boldsymbol{\Phi}_5 := \tfrac{1}{\sqrt{2}} \begin{pmatrix} 0 & 0 & 0 \\ 0 & 0 & 1 \\ 0 & 1 & 0 \end{pmatrix}.$$

Note that these basis functions are orthonormal with respect to the Frobenius inner product. Thereby, we write $\boldsymbol{p}_{hp} = \sum_{i=1}^{N} \boldsymbol{p}_i \phi_i$ and $\boldsymbol{\lambda}_{hp} = \sum_{i=1}^{N} \boldsymbol{\lambda}_i \varphi_i$ as

$$\boldsymbol{p}_{hp} = \sum_{i=1}^{N} \sum_{k=1}^{L} p_{L(i-1)+k} \boldsymbol{\Phi}_k \phi_i, \qquad \boldsymbol{\lambda}_{hp} = \sum_{i=1}^{N} \sum_{k=1}^{L} \lambda_{L(i-1)+k} \boldsymbol{\Phi}_k \varphi_i,$$

where $L := 2^{-1}(d-1)(d+2)$. In view of (11) it seems natural to define the nonlinear function $\boldsymbol{\chi}_i : \mathbb{S}_{d,0} \times \mathbb{S}_{d,0} \to \mathbb{S}_{d,0}$ as

$$\boldsymbol{\chi}_i(\boldsymbol{p}_i, \boldsymbol{\lambda}_i) := \max\left\{\sigma_i, \|\boldsymbol{\lambda}_i + \rho \boldsymbol{p}_i\|_F\right\} \boldsymbol{\lambda}_i - \sigma_i(\boldsymbol{\lambda}_i + \rho \boldsymbol{p}_i) \tag{12}$$

for $1 \leq i \leq N$ and some $\rho > 0$. With $\boldsymbol{\chi}_i(\cdot)$ at hand the following theorem enables to reformulate the discrete mixed problem (6) as a system of decoupled nonlinear equations.

Theorem 3 *$\boldsymbol{\lambda}_{hp} \in \Lambda_{hp}$ satisfies (6b) if and only if $\boldsymbol{\chi}_i(\boldsymbol{p}_i, \boldsymbol{\lambda}_i) = \mathbf{0}$ for all $1 \leq i \leq N$.*

Proof Let $\boldsymbol{\chi}_i(\boldsymbol{p}_i, \boldsymbol{\lambda}_i) = \mathbf{0}$. If $\|\boldsymbol{\lambda}_i + \rho \boldsymbol{p}_i\|_F < \sigma_i$, then

$$\mathbf{0} = \boldsymbol{\chi}_i(\boldsymbol{p}_i, \boldsymbol{\lambda}_i) = \sigma_i \boldsymbol{\lambda}_i - \sigma_i(\boldsymbol{\lambda}_i + \rho \boldsymbol{p}_i) = -\sigma_i \boldsymbol{p}_i$$

yields $\boldsymbol{p}_i = \mathbf{0}$ so that $\|\boldsymbol{\lambda}_i\|_F < \sigma_i$. If $\|\boldsymbol{\lambda}_i + \rho \boldsymbol{p}_i\|_F \geq \sigma_i$ we have

$$\mathbf{0} = \boldsymbol{\chi}_i(\boldsymbol{p}_i, \boldsymbol{\lambda}_i) = \|\boldsymbol{\lambda}_i + \rho \boldsymbol{p}_i\|_F \boldsymbol{\lambda}_i - \sigma_i(\boldsymbol{\lambda}_i + \rho \boldsymbol{p}_i) \tag{13}$$

which is equivalent to $\boldsymbol{\lambda}_i = \sigma_i \|\boldsymbol{\lambda}_i + \rho \boldsymbol{p}_i\|_F^{-1}(\boldsymbol{\lambda}_i + \rho \boldsymbol{p}_i)$. Hence, $\|\boldsymbol{\lambda}_i\|_F = \sigma_i$ and (13) simplifies to $\boldsymbol{\lambda}_i(\|\boldsymbol{\lambda}_i + \rho \boldsymbol{p}_i\|_F - \sigma_i) = \sigma_i \rho \boldsymbol{p}_i$ which implies

$$\boldsymbol{p}_i = \frac{1}{\rho \sigma_i}\left(\|\boldsymbol{\lambda}_i + \rho \boldsymbol{p}_i\|_F - \sigma_i\right)\boldsymbol{\lambda}_i =: c\boldsymbol{\lambda}_i, \qquad \text{where } c \geq 0.$$

Therefore, $\boldsymbol{\lambda}_{hp}$ satisfies (6b) by Theorem 2 and using the equivalence stated in (11). Conversely, let $\|\boldsymbol{\lambda}_i\|_F \leq \sigma_i$. If $\|\boldsymbol{\lambda}_i\|_F < \sigma_i$ (11a) implies $\boldsymbol{p}_i = \mathbf{0}$ and, thus,

$$\boldsymbol{\chi}_i(\boldsymbol{p}_i, \boldsymbol{\lambda}_i) = \max\{\sigma_i, \|\boldsymbol{\lambda}_i\|_F\}\boldsymbol{\lambda}_i - \sigma_i \boldsymbol{\lambda}_i = \mathbf{0}.$$

If $\|\boldsymbol{\lambda}_i\|_F = \sigma_i$ the implication (11b) gives $\boldsymbol{p}_i = c\boldsymbol{\lambda}_i$ and therefore

$$\boldsymbol{\chi}_i(\boldsymbol{p}_i, \boldsymbol{\lambda}_i) = \max\{\sigma_i, (1 + c\rho)\sigma_i\}\boldsymbol{\lambda}_i - \sigma_i(1 + c\rho)\boldsymbol{\lambda}_i = \mathbf{0}.$$

Finally, (11) together with Theorem 2 yields the assertion. □

Let $\mathfrak{e}_k$ denote the k-th Euclidean unit vector and let $\vartheta_1, \ldots, \vartheta_M$ be scalar valued functions such that $\{\mathfrak{e}_k \vartheta_i\}_{k=1,\ldots,d,\, i=1,\ldots,M}$ forms a basis of V_{hp}. Then,

$$\mathfrak{u}_{hp} = \sum_{i=1}^{M} \sum_{k=1}^{d} u_{d(i-1)+k} \mathfrak{e}_k \vartheta_i$$

and $(\mathfrak{u}_{hp}, \boldsymbol{p}_{hp}, \boldsymbol{\lambda}_{hp})$ can completely be represented by the vectors $\mathfrak{u} := (u_1, \ldots, u_{dM})^\top$, $\mathfrak{p} := (p_1, \ldots, p_{LN})^\top$ and $\underline{\lambda} := (\lambda_1, \ldots, \lambda_{LN})^\top$. Therewith, the discrete mixed variational problem (6) can equivalently be written as

$$\mathfrak{F}(\mathfrak{u}, \mathfrak{p}, \underline{\lambda}) := \begin{pmatrix} \boldsymbol{K}\mathfrak{u} - \boldsymbol{B}\mathfrak{p} - \mathfrak{l} \\ -\boldsymbol{B}^\top \mathfrak{u} + \boldsymbol{C}\mathfrak{p} + \boldsymbol{D}\underline{\lambda} \\ \underline{\chi}_1 \left(\sum_{k=1}^{L} p_k \boldsymbol{\Phi}_k, \sum_{k=1}^{L} \lambda_k \boldsymbol{\Phi}_k \right) \\ \vdots \\ \underline{\chi}_N \left(\sum_{k=1}^{L} p_{L(N-1)+k} \boldsymbol{\Phi}_k, \sum_{k=1}^{L} \lambda_{L(N-1)+k} \boldsymbol{\Phi}_k \right) \end{pmatrix} = \mathfrak{o}, \qquad (14)$$

where $\boldsymbol{K} \in \mathbb{R}^{dM \times dM}$ and $\boldsymbol{C} \in \mathbb{R}^{LN \times LN}$ with components

$$K_{d(i-1)+k, d(j-1)+l} = a\big((\mathfrak{e}_l \vartheta_j, \mathbf{0}), (\mathfrak{e}_k \vartheta_i, \mathbf{0})\big),$$
$$C_{L(i-1)+k, L(j-1)+l} = a\big((\mathfrak{o}, \boldsymbol{\Phi}_l \phi_i), (\mathfrak{o}, \boldsymbol{\Phi}_k \phi_j)\big),$$

are both symmetric, positive definite matrices, $\boldsymbol{D} \in \mathbb{R}^{LN \times LN}$ with components

$$D_{L(i-1)+k, L(j-1)+l} = \delta_{lk} \delta_{ij} D_i,$$

is a positive definite diagonal matrix, the coupling matrix $\boldsymbol{B} \in \mathbb{R}^{dM \times LN}$ is given component-wise by

$$B_{d(i-1)+k, L(j-1)+l} = a\big((\mathfrak{o}, \boldsymbol{\Phi}_l \phi_j), (\mathfrak{e}_k \vartheta_i, \mathbf{0})\big),$$

and $\mathfrak{l} \in \mathbb{R}^{dM}$ is the loadvector with components $l_{d(i-1)+k} = \ell(\mathfrak{e}_k \vartheta_i)$. Note that $\boldsymbol{\chi}_i(\boldsymbol{p}_i, \boldsymbol{\lambda}_i)$ is a linear combination of the orthonormal matrices $\boldsymbol{\Phi}_1, \ldots, \boldsymbol{\Phi}_L$. Thus, writing $\boldsymbol{\chi}_i(\boldsymbol{p}_i, \boldsymbol{\lambda}_i) = \sum_{k=1}^{L} \chi_{i,k} \boldsymbol{\Phi}_k$ and taking the Frobenius inner product of $\boldsymbol{\chi}_i(\boldsymbol{p}_i, \boldsymbol{\lambda}_i)$ with each $\boldsymbol{\Phi}_l$ yields the l-th component of $\underline{\chi}_i := (\chi_{i,1}, \ldots, \chi_{i,L})^\top$:

$$\begin{aligned} \chi_{i,l} &= \left(\sum_{k=1}^{L} \max \left\{ \sigma_i, \left(\sum_{k=1}^{L} \left(\lambda_{L(i-1)+k} + \rho p_{L(i-1)+k} \right)^2 \right)^{1/2} \right\} \lambda_{L(i-1)+k} \boldsymbol{\Phi}_k \right. \\ &\qquad \left. - \sigma_i \left(\lambda_{L(i-1)+k} + \rho p_{L(i-1)+k} \right) \boldsymbol{\Phi}_k \right) : \boldsymbol{\Phi}_l \\ &= \max\{\sigma_i, |\underline{\lambda}_i + \rho \mathfrak{p}_i|\} \lambda_{L(i-1)+l} - \sigma_i \left(\lambda_{L(i-1)+l} + \rho p_{L(i-1)+l} \right), \end{aligned}$$

where $|\cdot|$ denotes the Euclidean norm in $\mathbb{R}^L$, $\underline{\lambda}_i := (\lambda_{L(i-1)+1}, \ldots, \lambda_{Li})^\top$ and $\mathfrak{p}_i := (p_{L(i-1)+1}, \ldots, p_{Li})^\top$. Direct calculations lead to the partial derivatives

$$\frac{\partial}{\partial p_j}\chi_{i,l} = -\rho\sigma_i\delta_{j,L(i-1)+l}, \qquad \frac{\partial}{\partial \lambda_j}\chi_{i,l} = 0,$$

if $|\underline{\lambda}_i + \rho\mathfrak{p}_i| < \sigma_i$ and

$$\frac{\partial}{\partial p_j}\chi_{i,l} = \rho\,\frac{\lambda_{L(i-1)+l}(\lambda_j + \rho p_j)}{|\underline{\lambda}_i + \rho\mathfrak{p}_i|} - \rho\sigma_i\delta_{j,L(i-1)+l},$$

$$\frac{\partial}{\partial \lambda_j}\chi_{i,l} = \frac{\lambda_{L(i-1)+l}(\lambda_j + \rho p_j)}{|\underline{\lambda}_i + \rho\mathfrak{p}_i|} + (|\underline{\lambda}_i + \rho\mathfrak{p}_i| - \sigma_i)\delta_{j,L(i-1)+l}$$

if $|\underline{\lambda}_i + \rho\mathfrak{p}_i| > \sigma_i$ for $j = L(i-1)+1, \ldots, Li$ and $i = 1, \ldots, N$. The partial derivatives for the remaining indices vanish. In the case that $|\underline{\lambda}_i + \rho\mathfrak{p}_i| = \sigma_i$ the function $\mathfrak{F}(\cdot)$ is not differentiable in the classical sense. Therefore, we resort to a semi-smooth Newton method (SSN) to solve (14) which is given by

$$\left(\mathfrak{u}^{(k+1)}, \mathfrak{p}^{(k+1)}, \underline{\lambda}^{(k+1)}\right)^\top := \left(\mathfrak{u}^{(k)}, \mathfrak{p}^{(k)}, \underline{\lambda}^{(k)}\right)^\top - \boldsymbol{H}_k^{-1}\mathfrak{F}\left(\mathfrak{u}^{(k)}, \mathfrak{p}^{(k)}, \underline{\lambda}^{(k)}\right) \tag{15}$$

with $\boldsymbol{H}_k \in \partial\mathfrak{F}(\mathfrak{u}^{(k)}, \mathfrak{p}^{(k)}, \underline{\lambda}^{(k)})$, where $\partial\mathfrak{F}(\cdot)$ denotes the Clarke subdifferential of $\mathfrak{F}(\cdot)$. We show in [3] that this method is locally well defined and converges superlinearly for any matrix $\partial\mathfrak{F}(\mathfrak{u}^{(k)}, \mathfrak{p}^{(k)}, \underline{\lambda}^{(k)})$. In the following numerical examples $\boldsymbol{H}_k$ is chosen to have the largest possible number of zero entries and fewest possible distinctions of cases.

4 Numerical Examples

The following setting is used for all numerical experiments: $\Omega := (0,1)^2 \setminus [0.5, 1]^2$, $\Gamma_D := [0, 0.5] \times \{1\}$, $\mathfrak{f} := \mathfrak{o}$, $\mathfrak{g} := (0, -1.25)^\top$ on $[0,1] \times \{0\}$ and zero elsewhere, $\mathbb{C}\boldsymbol{\tau} := \lambda\,\mathrm{tr}(\boldsymbol{\tau})\boldsymbol{I} + 2\mu\boldsymbol{\tau}$ with Lamé constants $\lambda := \mu := 1000$, $\mathbb{H}\boldsymbol{\tau} := 100\boldsymbol{\tau}$ and $\sigma_y := 1.25$, see [16]. For the application of the SSN (15) we set $\rho := 100$ and take $(\mathfrak{u}^{(0)}, \mathfrak{p}^{(0)}) := \begin{pmatrix} \boldsymbol{K} & -\boldsymbol{B} \\ -\boldsymbol{B}^\top & \boldsymbol{C} \end{pmatrix}^{-1} (\mathfrak{l}, \mathfrak{o})^\top$ and $\underline{\lambda}^{(0)} := \mathfrak{o}$ as initial solutions. The stopping criteria for the SSN method is chosen as $|\mathfrak{F}(\mathfrak{u}^{(k)}, \mathfrak{p}^{(k)}, \underline{\lambda}^{(k)})| < 2^{1/2} \cdot 10^{-10}$.

Figure 1 shows the deformation of Ω resulting from the displacement field $\mathfrak{u}$. Furthermore, $\|\boldsymbol{p}_{hp}\|_F$ and $\|\boldsymbol{\lambda}_{hp}\|_F$ are depicted by colors in Fig. 1a, b, respectively. In Fig. 1c the region with purely elastic deformation is highlighted in blue and its complement in yellow. In this region there holds $\|\boldsymbol{\lambda}_{hp}\|_F < \sigma_y - 2.22 \cdot 10^{-15}$.

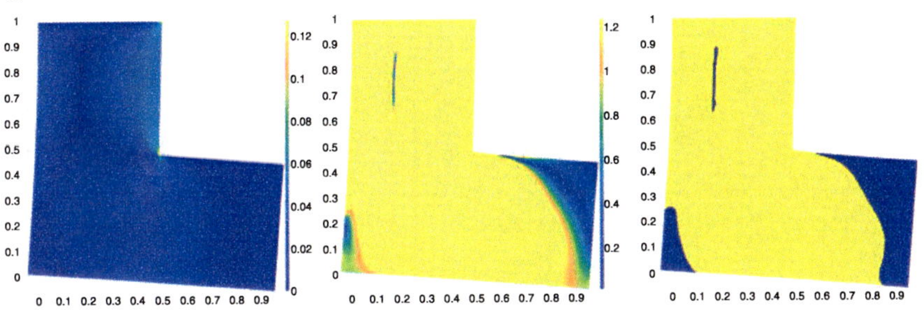

Fig. 1 Deformation of Ω for uniform mesh with $h = 2^{-8}$ and $p = 1$. **(a)** $\|\boldsymbol{p}_{hp}\|_F$. **(b)** $\|\boldsymbol{\lambda}_{hp}\|_F$. **(c)** Elastic region (blue)

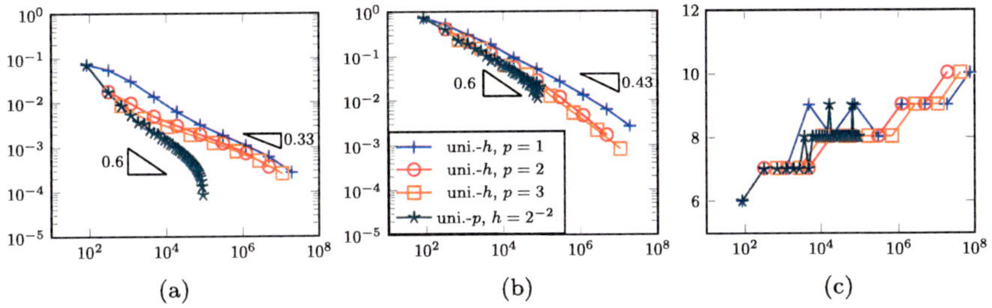

Fig. 2 Error reduction and number of semi-smooth Newton iterations. **(a)** $(e_{\mathrm{u}}^2 + e_{\boldsymbol{p}}^2)^{1/2}$. **(b)** $e_{\boldsymbol{\lambda}}$. **(c)** SSN iterations

Figure 2a, b display the reduction of $e_{\mathrm{u}} := \|\mathrm{u}_{\mathrm{fine}} - \mathrm{u}_{hp}\|_{H^1(\Omega)}$, $e_{\boldsymbol{p}} := \|\boldsymbol{p}_{\mathrm{fine}} - \boldsymbol{p}_{hp}\|_{L^2(\Omega)}$ and $e_{\boldsymbol{\lambda}} := \|\boldsymbol{\lambda}_{\mathrm{fine}} - \boldsymbol{\lambda}_{hp}\|_{L^2(\Omega)}$ respectively, where *fine* indicates the finest discrete solution which is used instead of the unknown exact solution. Due to the singular behavior of the solution at the reentrant corner we expect a convergence rate (EOC) of at most $1/3$ w.r.t. the degrees of freedom (DOF) for uniform h-refinements with $p = 1, 2, 3$. For uniform increasing p the EOC is 0.56 w.r.t. DOF. It is well known that the EOC for such a reentrant corner singularity would be $2/3$ for uniform p-versions in the purely elastic case. Hence, the reduced EOC results from the singularity associated to elasticity-plasticity transmission, which is not appropriately resolved by the uniformly refined mesh, see Fig. 1c. The number of SSN iterations increases with the DOF, but the increases is only 4 (from 6 to 10 iterations) while the DOF increase from 10^2 to 10^8, see Fig. 2c. Therefore, the number of iterations seems to be robust to the (uniform) discretization scheme.

In Fig. 3 we plot the quotients

$$q_{\mathrm{SSN}}(\alpha) := \frac{\left|\left(\mathrm{u}^{(k+1)} - \mathrm{u}^{(\infty)}, \mathrm{p}^{(k+1)} - \mathrm{p}^{(\infty)}, \underline{\lambda}^{(k+1)} - \underline{\lambda}^{(\infty)}\right)\right|}{\left|\left(\mathrm{u}^{(k)} - \mathrm{u}^{(\infty)}, \mathrm{p}^{(k)} - \mathrm{p}^{(\infty)}, \underline{\lambda}^{(k)} - \underline{\lambda}^{(\infty)}\right)\right|^{\alpha}}$$

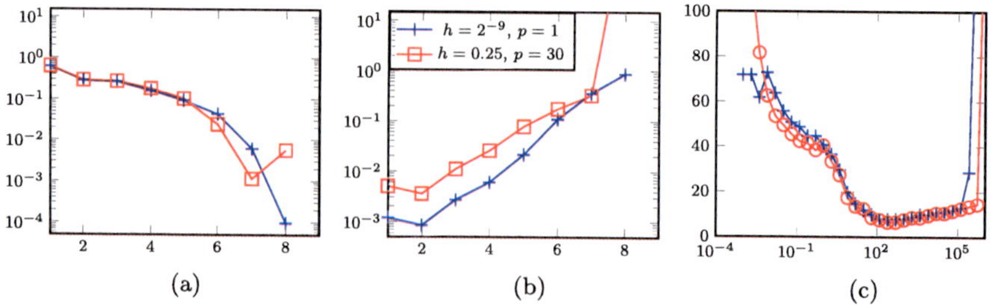

Fig. 3 Convergence properties of semi-smooth Newton method. **(a)** $q_{\mathrm{SSN}}(1)$. **(b)** $q_{\mathrm{SSN}}(2)$. **(c)** SSN iterations vs. ρ

for two uniform discretizations. One with small $h = 2^{-9}$ and small $p = 1$ (295.680 DOF), and the other with large $h = 2^{-2}$ and large $p = 30$ (29.040 DOF). We clearly see that the SSN method converges locally superlinearly as $q_{\mathrm{SSN}}(1) \to 0$ but not quadratic as $q_{\mathrm{SSN}}(2) \to \infty$. The remaining Fig. 3c shows the dependency of the number of SSN iterations on ρ. For very small ρ we observe that more than the first 50% of iterations is characterized by the need of damping the SSN. For very large ρ the conditioning of the algorithm seems to worsen which does not play well with a small stopping criterion. We point out that if ρ is sufficiently large varying ρ does not effect the number of iterations significantly. Moreover, the optimal ρ seems to be independent of h, p and, thus, of DOF.

References

1. Alberty, J., Carstensen, C.: Numerical analysis of time-depending primal elastoplasticity with hardening. SIAM J. Numer. Anal. **37**, 1271–1294 (2000)
2. Bammer, P., Banz, L., Schröder, A.: *hp*-FEM for a mixed variational approach in elastoplasticity (in preperation)
3. Bammer, P., Banz, L., Schröder, A.: A posteriori error estimates and a semi-smooth Newton solver for *hp*-FEM in elastoplasticity (in preperation)
4. Banz, L., Schröder, A.: Biorthogonal basis functions in *hp*-adaptive FEM for elliptic obstacle problems. Comput. Math. Appl. **70**, 1721–1742 (2015)
5. Carstensen, C.: Numerical analysis of the primal problem of elastoplasticity with hardening. Numer. Math. **82**, 577–597 (1999)
6. Carstensen, C., Schröder, A., Wiedemann, S.: An optimal adaptive finite element method for elastoplasticity. Numer. Math. **132**, 131–154 (2016)
7. Christensen, P.W.: A nonsmooth Newton method for elastoplastic problems. Comput. Methods Appl. Mech. Eng. **191**, 1189–1219 (2002)
8. Gruber, P., Kienesberger, J., Langer, U., Schöberl, J., Valdman, J.: Fast solvers and a posteriori error estimates in elastoplasticity. In: Numerical and Symbolic Scientific Computing, pp. 45–63. Springer, New York (2012)
9. Gwinner, J.: *hp*-FEM convergence for unilateral contact problems with Tresca friction in plane linear elastostatics. J. Comput. Appl. Math. **254**, 175–184 (2013)

10. W. Han.: Finite element analysis of a holonomic elastic-plastic problem. Numer. Math. **60**, 493–508 (1991)
11. Han, W., Reddy, B.D.: On the finite element method for mixed variational inequalities arising in elastoplasticity. SIAM J. Numer. Anal. **32**, 1778–1807 (1995)
12. Han, W., Reddy, B.D.: Plasticity: Mathematical Theory and Numerical Analysis, 2nd edn. Springer, New York (2013)
13. Haslinger, J., Panagiotopoulos, P.D.: Finite element method for hemivariational inequalities. In: Theory, Methods and Applications. Springer, Berlin (1999)
14. Hintermüller, M., Rösel, S.: A duality-based path-following semismooth Newton method for elasto-plastic contact problems. J. Comput. Appl. Math. **292**, 150–173 (2016)
15. Kienesberger, J., Valdman, J.: An efficient solution algorithm for elastoplasticity and its first implementation towards uniform h- and p- mesh refinements. In: Numerical Mathematics and Advanced Applications, pp. 1117–1125. Springer, Berlin (2006)
16. A. Schröder, S. Wiedemann.: Error estimates in elastoplasticity using a mixed method. Appl. Numer. Math. **61**, 1031–1045 (2011)

Convergence of Finite Difference Schemes: Matrix Versus Kernel Analysis

M. Ben-Artzi, J.-P. Croisille, and D. Fishelov

1 Biharmonic Time Dependent Problem, Continous and Semidiscrete

The convergence analysis of finite difference schemes attracts the interest of the numerical analysis community for several decades (see for example [10]). A renew of interest invoking various analytical and discrete frameworks is currently observed, (see [7, 12, 14, 15]). Here we wish to draw attention to the concurrent "languages" (analysis versus algebra oriented) tools, which may be used to establish the same convergence rate result. We would like to inspect the advantages versus disadvantages for each approach.

As a model problem we consider the evolution equation (T is a fixed final time),

$$\frac{\partial}{\partial t}u = -\left(\frac{\partial}{\partial x}\right)^4 u, \quad x \in \Omega = [0, 1], \; t \in [0, T], \tag{1}$$

M. Ben-Artzi
Institute of Mathematics, The Hebrew University, Jerusalem, Israel
e-mail: mbartzi@math.huji.ac.il

J.-P. Croisille (✉)
University of Lorraine, CNRS, IECL, Metz, France
e-mail: jean-pierre.croisille@univ-lorraine.fr

D. Fishelov
Afeka Tel Aviv Academic College for Engineering, Tel Aviv, Israel
e-mail: daliaf@afeka.ac.il

J. M. Melenk et al. (eds.), *Spectral and High Order Methods for Partial Differential Equations ICOSAHOM 2020+1*, Lecture Notes in Computational Science and Engineering 137, https://doi.org/10.1007/978-3-031-20432-6_8

subject to initial condition and boundary conditions

$$\begin{cases} u(x,0) = u_0(x), \ \ x \in \Omega = [0,1], \\ u(0,t) = u_x(0,t) = u(1,t) = u_x(1,t) = 0, \ \ t \in [0,T]. \end{cases} \tag{2}$$

When there is no risk of confusion we shall write $u(t)$ for $u(x,t)$. Note that (1) is well-posed in the space $L^2(\Omega)$. In other words, the semigroup $e^{-t(\frac{\partial}{\partial x})^4}$ is a continuous contraction semigroup in this space. The domain of its generator is $H^4(\Omega) \cap H_0^2(\Omega)$. Moreover, $H^s(\Omega)$ is a persistence space for every $s \geq 0$, where H^s is the Sobolev space of order s. This is readily seen by casting the semigroup $e^{-t(\frac{\partial}{\partial x})^4}$ in terms of Fourier series. The following finite difference operators are involved[1] (see [3, 5]).

- The standard three point Laplacian is δ_x^2 and the centered difference δ_x are defined by

$$(\delta_x^2 \mathfrak{v})_j = \frac{\mathfrak{v}_{j+1} + \mathfrak{v}_{j-1} - 2\mathfrak{v}_j}{h^2}, \quad (\delta_x \mathfrak{v})_j = \frac{\mathfrak{v}_{j+1} - \mathfrak{v}_{j-1}}{2h}, \quad 1 \leq j \leq N-1. \tag{3}$$

- The *Hermitian derivative* is $\tilde{\delta}_x \mathfrak{v}$, defined as a function of $\mathfrak{v}$ by

$$(\sigma_x \tilde{\delta}_x \mathfrak{v})_j = (\delta_x \mathfrak{v})_j, \quad j = 1, \ldots, N-1, \tag{4}$$

where

$$(\sigma_x \mathfrak{w})_j = \frac{1}{6}(\mathfrak{w}_{j-1} + 4\mathfrak{w}_j + \mathfrak{w}_{j+1}), \quad j = 1, \ldots, N-1. \tag{5}$$

Equivalently, $\tilde{\delta}_x = \sigma_x^{-1}\delta_x$. Here we assume that $\mathfrak{w}_0 = \mathfrak{w}_N = (\tilde{\delta}_x \mathfrak{v})_0 = (\tilde{\delta}_x \mathfrak{v})_N = 0$.

- The Discrete Biharmonic Operator (DBO) δ_x^4 is defined by Ben-Artzi and Katriel [1] and Stephenson [13]

$$(\delta_x^4 \mathfrak{v})_j = \frac{12}{h^2}(\delta_x \tilde{\delta}_x \mathfrak{v} - \delta_x^2 \mathfrak{v})_j, \quad j = 1, \ldots, N-1. \tag{6}$$

Equation (1) is approximated in space on a uniform grid $x_j = jh, h = 1/N, \ \ j = 0, 1, \ldots, N$ by $t \mapsto \mathfrak{v}(t) = [\mathfrak{v}_1(t), \ldots, \mathfrak{v}_{N-1}(t)]^T$. The gridfunction $\mathfrak{v}(t)$ is solution of the discrete analog to Eq. (1) is

$$\mathfrak{v}_t = -\delta_x^4 \mathfrak{v}, \quad t \in [0,T], \tag{7}$$

[1] Gridfunctions are denoted by the fraktur font.

The function $\mathfrak{v}(t)$ depends smoothly on $t \in [0, T]$ and is subject to the initial condition $\mathfrak{v}(0) = u_0^*$, and boundary condition $\mathfrak{v}_0(t) = (\mathfrak{v}_x)_0(t) = \mathfrak{v}_N(t) = (\mathfrak{v}_x)_N(t) = 0, \ t \in [0, T]$.[2]

The error $\mathfrak{e}(t) = \mathfrak{v}(t) - u^*(t)$ satisfies

$$\partial_t \mathfrak{e} + \delta_x^4 \mathfrak{e} = \tau, \tag{8}$$

where $\tau(t)$ is the truncation error. We adopt the notation $\mu \hookrightarrow M$ to express that the gridfunction (resp. finite difference operator) μ corresponds to the vector (resp. matrix) M. Here $\mathfrak{e}(t) \hookrightarrow E(t) \in \mathbf{R}^{N-1}$, $\delta_x^4 \hookrightarrow B \in \mathbf{R}^{(N-1)\times(N-1)}$ and $\tau(t) \hookrightarrow R(t) \in \mathbf{R}^{N-1}$. The vector form of (8) is

$$\partial_t E(t) + BE(t) = R(t). \tag{9}$$

We now consider the convergence of $\mathfrak{v}(t)$ to $u(t)$ as $N \to \infty$. In the rest of the paper, we establish the convergence to 0 of the error $\mathfrak{e}(t)$, or equivalently of $E(t)$, for $t \in [0, T]$ when $h \to 0$. We will need the following

Claim 1 *Let*

$$\left(\frac{\partial}{\partial x}\right)^4 u = f, \quad x \in \Omega = [0, 1],$$

and assume that $f \in C^4(\Omega)$. Then operating on the grid function u^* by the discrete operator δ_x^4 leads to the following truncation errors.

$$(\delta_x^4 u^*)_j = f_j^* + O(h^4), \quad \forall j \in \{2, \ldots, N-2\}, \tag{10}$$

and on near-boundary points

$$(\delta_x^4 u^*)_1 = f_1^* + O(h), \qquad (\delta_x^4 u^*)_{N-1} = f_{N-1}^* + O(h). \tag{11}$$

2 Error Analysis for the Discrete Biharmonic Operator (DBO)

We consider the equation

$$\left(\frac{\partial}{\partial x}\right)^4 u = f, \quad x \in \Omega = [0, 1]. \tag{12}$$

It is well-known that the kernel of $\left(\frac{\partial}{\partial x}\right)^{-4}$ is given by the following claim [1, Claim 5.1].

[2] For $u(x)$ a given function, u^* is the gridfunction defined by $u^* = [u(x_1), \ldots u(x_{N-1})]$.

Claim 2 *The solution of* (12) *is given by*

$$u(x) = \int_0^1 K(x, y) f(y) dy, \tag{13}$$

where

$$K(x, y) = \begin{cases} \frac{1}{6}(1-x)^2 y^2 [2x(1-y) + x - y], & y < x \\ \frac{1}{6}x^2 (1-y)^2 [2y(1-x) + y - x], & x < y \end{cases}. \tag{14}$$

Now consider the discrete analog

$$\delta_x^4 \mathfrak{v} = f^*. \tag{15}$$

We wish to estimate the difference of grid functions $\mathfrak{v} - u^*$. Corollary 5.2 in [1] gives the explicit form for the kernel of the inverse discrete operator $(\delta_x^4)^{-1} : l_{h,0}^2 \to l_{h,0}^2$:

Claim 3 *The discrete operator* $(\delta_x^4)^{-1} : l_{h,0}^2 \to l_{h,0}^2$ *is represented by a matrix* $\left\{K_{i,j}^h\right\}_{1 \le i, j \le N-1}$, *explicitly given by*

$$K_{i,j}^h = hK(x_i, x_j), \quad 1 \le i, j \le N-1, \tag{16}$$

where $K(x, y)$ *is the resolvent kernel of* $\left(\frac{\partial}{\partial x}\right)^{-4}$, *as in Eq.* (14).

Note that the matrix K^h can be written as

$$K_{i,j}^h = K_{j,i}^h = \frac{h}{6} \psi(x_i) \cdot \theta(x_j), \quad 1 \le i \le j \le N-1, \tag{17}$$

where the product in the right-hand side is the scalar product in $\mathbf{R}^2$ of the two vector functions:

$$\psi(x) = x^2 \Big(-x, 3 \Big),$$
$$\theta(y) = (1-y)^2 \Big(2y + 1, y \Big).$$

Thus, the matrix K^h is quasi-separable of rank 2 [6, Section 4.2]. It simplifies (computationally) the evaluation of the matrix. Based on these facts, we can prove a convergence theorem for a general continuous function $f(x)$.

Theorem 4 *Let $f \in C[0,1]$. Let $u(x)$ be the solution to* (12) *and let $\mathfrak{v}$ be the solution to the discrete equation* (15). *Then*

$$\lim_{h\to 0} |\mathfrak{v} - u^*|_\infty = 0. \tag{18}$$

Proof In view of Claims 2 and 3 we have

$$\mathfrak{v}_i - u_i^* = \sum_{j=1}^{N-1} hK(x_i, x_j) f_j^* - \int_0^1 K(x_i, y) f(y) dy, \quad i = 1, 2, \ldots, N-1. \tag{19}$$

The first term in the right hand side is the Riemann sum of the second term, so that

$$\max\left\{|\mathfrak{v}_i - u_i^*|, \ i = 1, 2, \ldots, N-1\right\} \le C\omega(h), \tag{20}$$

where $\omega(h)$ is the supremum (over $x \in [0,1]$) of the oscillation functions $K(x,y)f(y)$ as functions of $y \in [0,1]$.

We can now state our optimal convergence estimate, in the case where the function f is sufficiently smooth. This estimate was first obtained in [9] by using a detailed analysis of the matrix associated with the inverse of the DBO. The proof presented here is much shorter and gives a better insight into the analytical character of the DBO.

Propostion 5 *Suppose that $f \in C^4[0,1]$ and let $u(x)$ be the solution to* (12). *Let $\mathfrak{v}$ be the solution to the discrete equation* (15). *Then there exists a constant $C > 0$ depending on f but not on h, j such that*

$$|\mathfrak{v}_j - u_j^*| \le Ch^4, \quad j = 1, 2, \ldots, N-1. \tag{21}$$

Proof Let $\mathfrak{e} = \mathfrak{v} - u^*$ and $\mathfrak{r} = \delta_x^4 \mathfrak{e}$. In view of Claim 1

$$\mathfrak{r}_j = O(h^4), \quad \forall j \in \{2, \ldots, N-2\}, \quad \mathfrak{r}_1 = O(h), \qquad \mathfrak{r}_{N-1} = O(h). \tag{22}$$

In view of Claim 3 we have

$$\mathfrak{e}_i = \sum_{j=1}^{N-1} hK(x_i, x_j)\mathfrak{r}_j, \quad i = 1, 2, \ldots, N-1 \tag{23}$$

By Eq. (14) we have (uniformly in $i \in \{1, 2, \ldots, N-1\}$)

$$K(x_i, x_j) = \begin{cases} O(h^2), & j = 1, N-1, \\ O(1), & j \in \{2, 3, \ldots, N-2\}. \end{cases} \tag{24}$$

Combining (22) and (24) we conclude

$$|\mathfrak{e}_i| \le Ch^4 + C\sum_{j=2}^{N-2} h\cdot h^4 \le Ch^4, \quad i = 1, 2, \ldots, N-1. \tag{25}$$

which gives (30).

3 Convergence of the Discrete Time-Evolution Solution by Purely Matrix Method

Here we prove optimal convergence in the error $\mathfrak{e}(t)$ (Eq. (8)). For this we invoke a slightly different version of Proposition 5, (see [3, 9]).

Propostion 6 (Optimal Convergence Theorem for the DBO) *Assume that the vector $\mathfrak{r}$, which contains the truncation errors, satisfy the following estimates:*

$$\begin{aligned} &|(\sigma_x\mathfrak{r})_j| \le Ch^4, \quad j = 2, \ldots, N-2, \\ &|(\sigma_x\mathfrak{r})_1| \le Ch, \quad |(\sigma_x\mathfrak{r})_{N-1}| \le Ch. \end{aligned} \tag{26}$$

The DBO operator δ_x^4 is invertible and its inverse is denoted by $\delta_x^{-4} = (\delta_x^4)^{-1}$. Then, the vector $\delta_x^{-4}\mathfrak{r} = [\delta_x^{-4}\mathfrak{r}_1, \ldots, \delta_x^{-4}\mathfrak{r}_{N-1}]^T$ satisfies

$$|(\delta_x^{-4}\mathfrak{r})_j| \le Ch^4, \qquad j = 1, \ldots, N-1. \tag{27}$$

The following Lemma results from the optimal theorem above (see also [3, 7, 8]).

Claim 7 (Vector Form of the Optimal Convergence Theorem for the DBO) *Assume that $R \in \mathbf{R}^{N-1}$ satisfies*

$$PR = [O(h), O(h^4), \ldots, O(h^4), O(h)]^T, \tag{28}$$

then the vector $B^{-1}R$ satisfies

$$|(B^{-1}R)_j| \le Ch^4, \quad j = 1, \ldots, N-1. \tag{29}$$

We state and prove now the following theorem.

Theorem 8 *Suppose that u is a solution to the problem (1) so that $u(t) \in C^8(\mathbf{R})$ continuously in t, then the error $\mathfrak{e}(t)$, satisfying Eq. (7) is bounded by*

$$\max_{0<t<T} |\mathfrak{e}(t)|_h \le C(T)h^4, \tag{30}$$

where $|\mathfrak{g}(t)|_h = \sqrt{\sum_{j=1}^{N-1} h|\mathfrak{g}_j(t)|^2}$ and C is depends only on u_0 and T.

Proof Here we consider a proof by a matrix approach. Consider (9). Since $E(0) = 0$, the Duhamel formula gives

$$E(t) = \int_0^t e^{-(t-\rho)B} R(\rho) d\rho = \int_0^t e^{-\rho B} R(t-\rho) d\rho. \tag{31}$$

The goal is to obtain an error estimate of the form

$$\|E(t)\|_2 \leq C(T) h^{3.5}. \tag{32}$$

This is implied by the point-wise estimate $E_j(t) = O(h^4)$, $j = 1, \ldots, N-1$. Rewriting the integrand in (31) as $e^{-\rho B} R(t-\rho) = e^{-\rho B} B B^{-1} R(t-\rho)$, the error $E(t)$ may be expressed as

$$E(t) = \int_0^t \big(e^{-\rho B} B\big)\big(B^{-1} R(t-\rho)\big) d\rho. \tag{33}$$

We denote the eigenvalues of B by

$$0 < \lambda_1 < \cdots < \lambda_{N-1}, \tag{34}$$

and denote $\tilde{\Lambda}(\rho) = \operatorname{diag}\{e^{-\rho\lambda_1}\lambda_1, \ldots, e^{-\rho\lambda_{N-1}}\lambda_{N-1}\}$. Let $Q \in \mathbf{R}^{(N-1)\times(N-1)}$ be orthogonal such that $B = Q \operatorname{diag}\{\lambda_1, \ldots, \lambda_{N-1}\} Q^T$. Then,

$$e^{-\rho B} B = Q\, \tilde{\Lambda}(\rho)\, Q^T. \tag{35}$$

Inserting Eq. (35) in (33), we have

$$E(t) = \int_0^t Q\, \tilde{\Lambda}(\rho)\, Q^T\, B^{-1} R(t-\rho) d\rho. \tag{36}$$

Denoting $\Lambda(\rho) = \operatorname{diag}\{e^{-\rho\lambda_1}, \ldots, e^{-\rho\lambda_{N-1}}\}$, we have

$$\tilde{\Lambda}(\rho) = -\frac{d}{d\rho} \Lambda(\rho). \tag{37}$$

Inserting (37) into (36), and integrating by parts, yields

$$\begin{aligned} E(t) &= -\int_0^t \frac{d}{d\rho}(Q\Lambda(\rho)\, Q^T) B^{-1} R(t-\rho) d\rho \\ &= \underbrace{\big[-(Q\Lambda(\rho)\, Q^T) B^{-1} R(t-\rho)\big]_0^t}_{E^{(1)}(t)} + \underbrace{\int_0^t (Q\Lambda(\rho)\, Q^T) B^{-1} \Big(\frac{d}{d\rho} R(t-\rho)\Big) d\rho}_{E^{(2)}(t)}. \end{aligned} \tag{38}$$

Since $R(0) = 0$, we have

$$E^{(1)}(t) = (Q\ \Lambda(0)\ Q^T)B^{-1}R(t). \tag{39}$$

Using Claim 7 and [7, 8], we have that

$$\|B^{-1}\ R(t)\|_2 \le C_1\|B^{-1}\ R(t)\|_2 \le C(t)h^{3.5}.$$

In addition, $\|Q\|_2 = \|Q^T\|_2 = 1$ and $\|\Lambda(\rho)\|_2 \le 1$, for $\rho \ge 0$. Thus, we will have

$$\|E^{(1)}(t)\|_2 \le C(t)h^{3.5}. \tag{40}$$

We turn now to $E^{(2)}(t)$. The components of $P\frac{d}{d\rho}R(t-\rho)$ are of the same order (as powers of h) as the components of $PR(\rho)$. This will yield (using Claim 7) that

$$\|B^{-1}\ \frac{d}{d\rho}R(t-\rho)\|_2 \le C(t)h^{3.5}. \tag{41}$$

Using again $\|Q\|_2 = \|Q^T\|_2 = 1$ and $\|\Lambda(\rho)\|_2 \le 1$, for $0 \le \rho \le t$, we have for some $\bar{\rho} \in [0, t]$,

$$\begin{aligned}\|E^{(2)}(t)\|_2 &= C(t)\max_{0\le\rho\le t}\|\ B^{-1}\ \frac{dR}{d\rho}(\rho)\|_2 = C(t)\|\ B^{-1}\ \frac{dR}{d\rho}(\bar{\rho})\|_2 \\ &\le C_1(t)\|\ B^{-1}\ \frac{dR}{d\rho}(\bar{\rho})\|_2\end{aligned} \tag{42}$$

Therefore,

$$\|E^{(2)}(t)\|_2 \le C(t)h^{3.5}. \tag{43}$$

Combining Eqs. (38), (40), and (43), we conclude that

$$\|E(t)\|_2 \le C(t)h^{3.5} \text{ or equivalently } |\mathfrak{e}(t)|_h \le C(t)h^4. \tag{44}$$

We comment now how (30) can be extended to the case of a nonlinear perturbation. Consider the equation

$$\frac{\partial}{\partial t}u = -\Big(\frac{\partial}{\partial x}\Big)^4 u - \alpha(u) \quad x \in \Omega = [0,1],\ t \in [0,T], \tag{45}$$

together with zero boundary conditions on $u, \partial_x u$. Here we assume that the perturbation $\alpha(u)$ is such that (45) has a regular solution $u(x,t)$ on $\Omega \times [0,T]$. The numerical scheme is

$$\mathfrak{v}_t = -\delta_x^4\mathfrak{v} - \alpha(\mathfrak{v}), \quad t \in [0,T],\ t \ge 0. \tag{46}$$

It is assumed that $t \mapsto \mathfrak{v}(t)$ is a regular function of time, existing on $[0, T]$ for all value of h. The error $\mathfrak{e}(t) = \mathfrak{v}(t) - u^*(t)$ evolves for $t \in [0, T]$ as

$$\partial_t \mathfrak{e} + \delta_x^4 \mathfrak{e} = \tau_1 + \tau_2, \tag{47}$$

where (the time dependence is dropped) $\tau_1 = (\partial_x^4 u)^* - \delta_x^4 u^*$ and $\tau_2 = \alpha(u^*) - \alpha(\mathfrak{v})$. In matrix/vector form, (9) becomes

$$\partial_t E(t) + BE(t) = R_1(t) + R_2(t), \tag{48}$$

where $\tau_{1,2} \hookrightarrow R_{1,2}$. The term R_1 represents the truncation error of δ_x^4. It depends only on u^*. The term R_2 involves u^* and $\mathfrak{v}$. The Duhamel formula is expressed as

$$E(t) = \underbrace{\int_0^t e^{-\rho B} R_1(t-\rho)d\rho}_{E_a(t)} + \underbrace{\int_0^t e^{-\rho B} R_2(t-\rho)d\rho}_{E_b(t)}. \tag{49}$$

so that

$$\|E(t)\|_2 \le \|E_a(t)\|_2 + \int_0^t \|e^{-(t-\rho)B}\|_2 \|R_2(\rho)\|_2 d\rho \tag{50}$$

We next proceed along the lines of [2, 11]. Fix $h > 0$. Using $E(0) = 0$ and the continuity of $t \mapsto E(t)$, define $t_0(h)$ by

$$t_0(h) = \sup\{t > 0 \ s.t. \ \|E(t)\|_2 < 1\}. \tag{51}$$

This implies $|u_j^*(t) - \mathfrak{v}_j(t)| < 1$ for $1 \le j \le N-1$, $0 \le t < t_0(h)$ and

$$\max_{\xi \in [u_j^*(t), \mathfrak{v}_j(t)]} |\alpha'(\xi)| \le \max_{|\xi - u_j^*(t)| < 1} |\alpha'(\xi)| \le C'(T). \tag{52}$$

where $C'(T)$ depends only on $u(x,t)$. Therefore, for $0 \le t < t_0(h)$,

$$|R_{2,j}(t)| = |\alpha(u_j^*(t)) - \alpha(\mathfrak{v}_j(t))| \le C'(T)|E_j(t)|, \quad 1 \le j \le N-1. \tag{53}$$

The spectrum of B satisfies $\text{spec}(B) \subset [\lambda_{\min}, \lambda_{\max}]$. We have with $\lambda_{\min} = O(1)$ and $\lambda_{\max} = O(1/h^4)$. It results from [1, Theorem 7.7] that $\lambda_{\min} > b$, with some constant $b > 0$ uniform in h . By (44), $\|E_a(t)\|_2 \le C(T)h^{3.5}$. This gives in (50),

$$\|E(t)\|_2 \le C(T)h^{3.5} + C'(T) \int_0^t e^{-b(t-\rho)} \|E(\rho)\|_2 d\rho. \tag{54}$$

Using Gronwall's inequality, we obtain that for $0 \le t < t_0(h)$,

$$\|E(t)\|_2 \le C(T)h^{3.5} \exp\Big(\int_0^t e^{-b(t-\rho)}d\rho\Big) \le C''(T)h^{3.5}. \quad (55)$$

where $C''(T)$ depends only on $u(x,t)$. It results from (55) that there exists h_0 small enough such that $C''(T)h_0^{3.5} < 1/2$. Thus, for all $h < h_0$, we have that $t_0(h) > T$ (proceed by contradiction). Therefore for h small enough, $\max_{0\le t\le T} \|E(t)\|_2 \le C''(T)h^{3.5}$ and (30) holds for (45).

4 Convergence of the Discrete Time-Evolution Solution by Kernel Analysis

As in the elliptic case discussed in the previous sections we can study this issue either under "minimal regularity" assumptions or "high regularity" leading to "optimal" fourth-order convergence. We begin with the latter case, analogous to Proposition 5. The "minimal regularity" case is postponed to Theorem 10. The following proposition improves the "almost optimal" estimate obtained in [4]. The same result has been obtained in [9] using matrix techniques. We recall that $H^9(\Omega)$ is the Sobolev space of order 9.

Propostion 9 *Assume that $u_0(x) \in C^9(\Omega)$ and*

$$u_0(0) = u_0'(0) = u_0(1) = u_0'(1) = 0.$$

Let $u(t)$ be the solution to (1), and $\mathfrak{v}(t)$ the solution to (7). There exists a constant $C > 0$ depending on u_0, T but not on h, j such that

$$|\mathfrak{v}_j(t) - u_j^*(t)| \le Ch^4, \quad j = 1, 2, \dots, N-1, \quad t \in [0, T]. \quad (56)$$

Proof As observed in Sect. 1, the Sobolev space H^9 is a persistence space for the solution. Thus for $u_0 \in C^9(\Omega) \subseteq H^9(\Omega)$ the function $u(t)$ is continuous on $[0, T]$ into H^9. The Sobolev embedding theorem implies that it is also continuous into $C^8(\Omega)$. Thus, let $\mathfrak{e}(t) = \mathfrak{v}(t) - u^*(t)$ be the error function and

$$\mathfrak{r}(t) = \Big[(\frac{\partial}{\partial x})^4 u\Big]^* - \delta_x^4 u^*(t).$$

The error function satisfies the equation

$$\mathfrak{e}_t(t) + \delta_x^4 \mathfrak{e}(t) = \mathfrak{r}(t). \quad (57)$$

Since the function $t \rightarrow u(x,t) \in C^8(\Omega)$ is continuous the family $\{u(t),\ t \in [0,T]\} \subseteq C^8(\Omega)$ is compact and Claim 1 can be applied uniformly to this family. It follows that, uniformly in $t \in [0,T]$,

$$\mathfrak{r}_j(t) = O(h^4), \quad \forall j \in \{2,\dots,N-2\}, \quad \mathfrak{r}_1(t) = O(h), \qquad \mathfrak{r}_{N-1}(t) = O(h). \tag{58}$$

Equation (57) can be rewritten as

$$\frac{d}{dt}\Big[(\delta_x^4)^{-1}\mathfrak{e}(t)\Big] + \mathfrak{e}(t) = (\delta_x^4)^{-1}\mathfrak{r}(t). \tag{59}$$

In light of (25) we have

$$\mathfrak{w}(t) := (\delta_x^4)^{-1}\mathfrak{r}(t) = O(h^4) \tag{60}$$

uniformly in $t \in [0,T]$.

Let

$$\Lambda_h = \left\{\lambda_{1,h} < \lambda_{2,h} < \dots < \lambda_{N-1,h}\right\} \tag{61}$$

be the eigenvalues[3] of δ_x^4 with corresponding normalized eigenvectors $\left\{\mathfrak{c}_h^1,\dots,\mathfrak{c}_h^{N-1}\right\}$. We can expand

$$\mathfrak{e}(t) = \sum_{k=1}^{N-1} a_{k,h}(t)\mathfrak{c}_h^k, \quad \mathfrak{w}(t) = \sum_{k=1}^{N-1} b_{k,h}(t)\mathfrak{c}_h^k \tag{62}$$

and projecting Eq. (59) on the k-th eigenvector yields

$$\lambda_{k,h}^{-1}\frac{d}{dt}a_{k,h}(t) + a_{k,h}(t) = b_{k,h}(t), \quad k = 1,2,\dots,N-1. \tag{63}$$

From (60) we get, for a constant $C > 0$ depending only on T (and not on N)

$$|b_{k,h}(t)| \le Ch^4, \quad k = 1,2,\dots,N-1,\ t \in [0,T].$$

We obtain (since $\mathfrak{e}(0) = 0$)

$$a_{k,h}(t)e^{\lambda_{k,h}t} = \int_0^t \lambda_{k,h}e^{\lambda_{k,h}s}b_{k,h}(s)ds, \quad t \in [0,T].$$

[3] The eigenvalues $\lambda_{k,h}$ are identical to the λ_k in (34).

In view of the estimate for $b_{k,h}(s)$ the right-hand side can be estimated by

$$\Big|\int_0^t \lambda_k e^{\lambda_{k,h}s} b_{k,h}(s)ds\Big| \le Ch^4 e^{\lambda_{k,h}t},$$

hence also

$$|a_{k,h}(t)| \le Ch^4, \quad k = 1, 2, \ldots, N-1, \ t \in [0, T].$$

We now proceed to study the convergence of the discrete solution to the continuous one, when the initial data is just continuous. Of course, such convergence is not expected to be "optimal" as in Proposition 9.

As pointed out already in the beginning of this section, Eq. (1) is not well-posed in $C(\Omega)$, due to the lack of a maximum principle. On the other hand the space $H^1(\Omega)$ is a persistence space for the solution. Recall that $H^1(\Omega)$ is the Sobolev space of L^2 functions whose distributional derivatives are also in L^2. The domain of the generator is $H^4(\Omega) \cap H_0^2(\Omega)$. In particular, this space contains all continuous piecewise linear ("zigzag") functions. By the Sobolev embedding theorem $H^1(\Omega) \subseteq C(\Omega)$.

For the discrete semigroup we use the operator notation $e^{-t\delta_x^4}$. We first state the following coercivity property.

$$(\delta_x^4 \mathfrak{z}, \mathfrak{z})_h \ge C\Big(|\mathfrak{z}|_h^2 + |\delta_x^2 \mathfrak{z}|_h^2 + |\delta_x \tilde{\delta}_x \mathfrak{z}|_h^2\Big), \tag{64}$$

valid for any grid function $\mathfrak{z} \in l_{h,0}^2$ such that also $\tilde{\delta}_x \mathfrak{z} \in l_{h,0}^2$.

Theorem 10 *Let $u_0(x) \in C^1(\Omega) \subseteq H^1(\Omega)$ and $u(x,t)$ the solution to (1). Let $\mathfrak{v}(t) = e^{-t\delta_x^4} u_0^*$ be the corresponding discrete solution. Then, uniformly in $t \in [0, T]$,*

$$\lim_{h\to 0} |\mathfrak{v}(t) - u^*(t)|_h = 0, \quad t \in [0, T]. \tag{65}$$

Proof Pick $\varepsilon > 0$. Let $\widetilde{u}(x,t)$ be solution to (1) with initial data $\widetilde{u_0} \in C^9(\Omega)$. For notational simplicity we occasionally designate $\widetilde{u}(t)$ for $\widetilde{u}(x,t)$ and $u(t)$ for $u(x,t)$. Due to the continuity of the solution of (1) in H^1 we can assume that $\widetilde{u}(t) \in H^9(\Omega)$ satisfies

$$\|\widetilde{u}(t) - u(t)\|_{H^1} < \varepsilon, \quad t \in [0, T]. \tag{66}$$

The Sobolev embedding theorem implies

$$\sup_{t\in[0,T]} \|\widetilde{u}(t) - u(t)\|_{C(\Omega)} < \varepsilon. \tag{67}$$

Since $\widetilde{u}(x,t)$ is sufficiently regular, Proposition 9 can be invoked. Let $\widetilde{\mathfrak{v}}(t) = e^{-t\delta_x^4}\widetilde{u_0}^*$ be the corresponding discrete solution. There exists $h_0 > 0$ such that

$$\sup_{t\in[0,T]} |\widetilde{\mathfrak{v}}(t) - \widetilde{u}(t)^*|_\infty < \varepsilon, \quad 0 < h < h_0. \tag{68}$$

Finally, the positivity of δ_x^4 (see (64)) implies that the semigroup $e^{-t\delta_x^4}$ is contractive on $l^2_{h,0}$ hence

$$|\widetilde{\mathfrak{v}}(t) - \mathfrak{v}(t)|_h \leq |\widetilde{u_0}^* - u_0^*|_h < \varepsilon. \tag{69}$$

Combining (67), (68) and (69) we obtain (65).

5 Comments and Perspectives

In summary, the matrix approach benefits from tools consisting of matrix theory of linear algebra and from known results of the solution of a set of ordinary differential equations. It may be extended to nonlinear (see Sect. 3) and multidimensional problems [7]. In particular, there is no need to derive the appropriate kernel of the discrete problem, which is in general a difficult task. On the contrary, the functional approach utilizes the connection between the discrete and the continuous problem. It may require the knowledge of the discrete kernel. An important aspect of the functional approach is its ability to deal with low regularity data (see Theorem 4). Note finally that an important topic for convergence analysis is the notion of consistency (see also [14]). In our context, this notion requires further studies.

References

1. Ben-Artzi, M., Katriel, G.: Spline functions, the biharmonic operator and approximate eigenvalues. Numer. Math. **141**(4), 839–879 (2019)
2. Ben-Artzi, M., Croisille, J.P., Fishelov, D.: Convergence of a compact scheme for the pure streamfunction formulation of the unsteady Navier–Stokes system. SIAM J. Numer. Anal. **44**, 1997–2024 (2006)
3. Ben-Artzi, M., Croisille, J.P., Fishelov, D.: Navier–Stokes Equations in Planar Domains. Imperial College Press, London (2013)
4. Ben-Artzi, M., Croisille, J.P., Fishelov, D.: Time evolution of discrete fourth-order elliptic operators. Numer. Methods PDEs **35**(4), 1429–1457 (2019)
5. Davies, E.: Spectral Theory and Differential Operators. Blackie and Son, Glasgow (1961)
6. Eidelman, Y., Gohberg, I., Haimovici, I.: Separable Type Representations of Matrices and Fast Algorithms, vol. I. Birkhäuser, Basel (2014)
7. Fishelov, D., Croisille, J.P.: Optimal convergence for time dependent Stokes equation: a new approach. J. Sci. Comput. **89**, 66 (2021)

8. Fishelov, D., Croisille, J.P.: Time dependent two dimensional fourth-order problems: optimal convergence. In: Vermolen, F.J., Vuik, C. (eds.) Numerical Mathematics and Advanced Application, ENUMATH 2019. Lecture Notes in Computational Science and Engineering, vol. 139, pp. 429–438. Springer, Berlin (2021)
9. Fishelov, D., Ben-Artzi, M., Croisille, J.P.: Recent advances in the study of a fourth-order compact scheme for the one-dimensional biharmonic equation. J. Sci. Comput. **53**(1), 55–79 (2012)
10. Gustafsson, B.: The convergence rate for difference approximations to mixed initial boundary value prolems. Math. Comput. **29**, 396–406 (1975)
11. Hou, T., Wetton, B.: Convergence of a finite difference scheme for the Navier–Stokes equations using vorticity boundary conditions stream function-only formulation. SIAM J. Numer. Anal. **29**, 615–639 (1992)
12. Müller, S., Schweiger, F., Süli, E.: Optimal-order finite difference approximation of generalized solutions to the biharmonic equation in a cube. SIAM J. Numer. Anal. **58**, 298–329 (2020)
13. Stephenson, J.W.: Single cell discretizations of order two and four for biharmonic problems. J. Comput. Phys. **55**, 65–80 (1984)
14. Svärd, M., Nordström, J.: On the convergence rates of energy-stable finite-difference schemes. J. Comput. Phys. **397**, 108819 (2019)
15. Wang, S., Kreiss, G.: Convergence of summation-by-parts finite difference methods for the wave equation. J. Sci. Comput. **71**(1), 219–245 (2017)

Shape Optimization with Nonlinear Conjugate Gradient Methods

Sebastian Blauth

1 Introduction

Shape optimization problems arise in many industrial applications, such as the design optimization of airplanes [27], automobiles [23], electric motors [11], microchannel systems [5, 6], polymer spin packs [16, 20], and melting furnaces [19]. For the solution of such problems, shape optimization based on shape calculus (see, e.g., [8]) has attracted lots of research interest in recent years, with particular regards to the development of efficient solution algorithms. This can be seen, e.g., in [28] and [29], where Newton and limited memory BFGS (L-BFGS) methods for shape optimization are proposed, respectively, and in [9, 22], where special mesh deformation procedures for increasing the mesh quality and the avoidance of remeshing are investigated.

In this chapter, we consider the recently proposed nonlinear conjugate gradient (NCG) methods for shape optimization from [4], which are numerical solution algorithms for such problems, and investigate their performance. One of the benefits of these NCG methods is that they require only slightly more memory than the popular gradient descent method, while being significantly more efficient. Here, we compare the numerical performance of the NCG methods to the already established gradient descent and L-BFGS methods for shape optimization. For this numerical comparison we utilize our software package cashocs [3] which implements these methods and allows for a detailed comparison.

This chapter is structured as follows. In Sect. 2, we briefly present some theoretical background on shape optimization and shape calculus. This is required for the presentation of the nonlinear conjugate gradient methods in Sect. 3. Finally,

S. Blauth (✉)
Fraunhofer ITWM, Kaiserslautern, Germany
e-mail: sebastian.blauth@itwm.fraunhofer.de

J. M. Melenk et al. (eds.), *Spectral and High Order Methods for Partial Differential Equations ICOSAHOM 2020+1*, Lecture Notes in Computational Science and Engineering 137, https://doi.org/10.1007/978-3-031-20432-6_9

the numerical examples, which showcase the capabilities of the NCG methods, can be found in Sect. 4.

2 Theoretical Background

In this section, we present some theoretical background on shape optimization and shape calculus which we require for our presentation of the NCG methods in Sect. 3.

2.1 Fundamentals of Shape Optimization

A general shape optimization problem with a partial differential equation (PDE) as constraint is given by

$$\begin{aligned} &\min_{\Omega, u} \mathcal{J}(\Omega, u) \\ &\text{subject to} \qquad e(\Omega, u) = 0, \\ &\qquad\qquad\qquad\quad \Omega \in \mathcal{A}. \end{aligned} \tag{1}$$

Here, $\mathcal{J}$ is a cost functional which we want to minimize over a set of admissible geometries $\mathcal{A} \subset \{ \Omega \subset D \}$ for some given bounded hold-all domain $D \subset \mathbb{R}^d$ and u is the so-called state variable which lies in the state space $U(\Omega)$. Additionally, $e(\Omega, \cdot) \colon U(\Omega) \to V(\Omega)^*$, where $V(\Omega)^*$ is the dual space of $V(\Omega)$, is an operator that models the PDE constraint, which we interpret in the weak form

$$\text{Find } u \in U(\Omega) \text{ such that} \quad \langle e(\Omega, u), v \rangle_{V(\Omega)^*, V(\Omega)} = 0 \quad \text{for all } v \in V(\Omega). \tag{2}$$

We assume that the state equation (2) admits a unique solution $u = u(\Omega)$ for all $\Omega \in \mathcal{A}$ so that we have $e(\Omega, u(\Omega)) = 0$. With this, we can introduce the reduced cost functional $J(\Omega) = \mathcal{J}(\Omega, u(\Omega))$ and rephrase (1) equivalently as

$$\min_{\Omega} J(\Omega) \quad \text{subject to} \quad \Omega \in \mathcal{A}, \tag{3}$$

where we have formally eliminated the PDE constraint.

2.2 Shape Calculus

To obtain efficient solution algorithms for problem (1), one can use techniques from shape calculus to derive sensitivities of the cost functional $J(\Omega)$ w.r.t. variations of the domain Ω. In the following, we briefly recall these techniques and we refer to, e.g., [8] for an exhaustive treatment of this topic.

We define a family of deformed domains Ω_t for $t \geq 0$ as

$$\Omega_t = F_t(\Omega) = \{ F_t(x) \mid x \in \Omega \}.$$

In this chapter, we use the so-called perturbation of identity (cf. [8]), in which the transformation F_t is given by

$$F_t(x) = (I + t\mathcal{V})(x),$$

where I denotes the identity matrix in $\mathbb{R}^d$ and $\mathcal{V}$ is a vector field in $C_0^k(D; \mathbb{R}^d)$ for some $k \geq 1$, i.e., the space of k-times continuously differentiable functions from D to $\mathbb{R}^d$ with compact support. Note, that there exist other, equivalent approaches for calculating first order shape derivatives, such as the speed method (cf. [8]), but for the sake of brevity we only consider the perturbation of identity in this chapter. Now, we can define the shape derivative as follows (cf. [8, Chapter 9, Definition 3.4]).

Definition 1 Let $\tau > 0$ be sufficiently small, $\mathcal{A} \subset \{ \Omega \subset D \}$, $J \colon \mathcal{A} \to \mathbb{R}$, and $\Omega \in \mathcal{A}$. Additionally, let $\mathcal{V} \in C_0^k(D; \mathbb{R}^d)$ with $k \geq 1$, let $F_t = I + t\mathcal{V}$ be the perturbation of identity with $\mathcal{V}$, and assume that $\Omega_t = F_t(\Omega) \in \mathcal{A}$ for all $t \in [0, \tau]$.

We say that J has a Eulerian semi-derivative at Ω in direction $\mathcal{V}$ if the limit

$$dJ(\Omega)[\mathcal{V}] := \lim_{t \searrow 0} \frac{J(F_t(\Omega)) - J(\Omega)}{t} = \frac{d}{dt} J(F_t(\Omega)) \bigg|_{t=0^+}$$

exists. Moreover, let Ξ be a topological vector subspace of $C_0^\infty(D; \mathbb{R}^d)$. We say that J is shape differentiable at Ω w.r.t. Ξ if it has a Eulerian semi-derivative at Ω in all directions $\mathcal{V} \in \Xi$ and, additionally, the mapping

$$dJ(\Omega) \colon \Xi \to \mathbb{R}; \quad \mathcal{V} \mapsto dJ(\Omega)[\mathcal{V}]$$

is linear and continuous. In this case, we call $dJ(\Omega)[\mathcal{V}]$ the shape derivative of J at Ω w.r.t. Ξ in direction $\mathcal{V} \in \Xi$.

Note, that there exist several possibilities for calculating shape derivatives in the context of PDE constrained shape optimization, an overview of which can, e.g., be found in [30]. These methods involve solving a so-called adjoint equation to calculate the shape derivative, which is usual in PDE constrained optimization (cf. [15]).

An important result from shape calculus is Hadamard's structure theorem, which we briefly recall here.

Theorem 2 (Structure Theorem) *Let J be a shape functional which is shape differentiable at some $\Omega \subset D$ and let $\Gamma = \partial\Omega$ be compact. Further, let $k \geq 0$*

be the smallest integer such that $dJ(\Omega)\colon C_0^\infty(D;\mathbb{R}^d) \to \mathbb{R};\ \mathcal{V} \mapsto dJ(\Omega)[\mathcal{V}]$ *is continuous w.r.t. the* $C_0^k(D;\mathbb{R}^d)$ *topology, and assume that* Γ *is of class* C^{k+1}*. Then, there exists a continuous, linear functional* $g : C^k(\Gamma) \to \mathbb{R}$ *such that*

$$dJ(\Omega)[\mathcal{V}] = g[\mathcal{V} \cdot n],$$

where n *is the outer unit normal vector on* Γ*. In particular, if* $g \in L^1(\Gamma)$*, it holds that*

$$dJ(\Omega)[\mathcal{V}] = \int_\Gamma g\, \mathcal{V} \cdot n \, ds.$$

Proof The proof can be found in [8, Theorem 3.6 and Corollary 1, pp. 479–481].

2.3 A Riemannian View on Shape Optimization and Steklov-Poincaré-Type Metrics

In order to formulate the nonlinear CG methods for shape optimization, we now briefly recall the Riemannian view on shape optimization from [28] as well as the corresponding Steklov-Poincaré-type metrics from [29].

We consider compact and connected subsets $\Omega \subset D \subset \mathbb{R}^2$ with C^∞ boundary, where D is, again, some bounded hold-all domain. As in [28], we define

$$B_e(S^1;\mathbb{R}^2) := \mathrm{Emb}(S^1;\mathbb{R}^2)/\mathrm{Diff}(S^1),$$

i.e., the set of all equivalence classes of C^∞ embeddings of the unit circle $S^1 \subset \mathbb{R}^2$ into $\mathbb{R}^2$, given by $\mathrm{Emb}(S^1;\mathbb{R}^2)$, where the equivalence relation is defined via the set of all C^∞ diffeomorphisms of S^1 into itself, given by $\mathrm{Diff}(S^1)$. Note, that this equivalence relation factors out reparametrizations as these do not change the underlying shape. It is shown in [18] that B_e, which is the set of all smooth two-dimensional shapes, is a smooth manifold. An element of $B_e(S^1;\mathbb{R}^2)$ is represented by a smooth curve $\Gamma\colon S^1 \to \mathbb{R}^2;\ \theta \mapsto \Gamma(\theta)$. Due to the equivalence relation, the tangent space at $\Gamma \in B_e$ is isomorphic to the set of all C^∞ normal vector fields along Γ, i.e.,

$$T_\Gamma B_e \cong \left\{ h \;\middle|\; h = \alpha n, \alpha \in C^\infty(\Gamma;\mathbb{R}) \right\} \cong \left\{ \alpha \;\middle|\; \alpha \in C^\infty(\Gamma;\mathbb{R}) \right\}.$$

As in [29], we consider the following Steklov-Poincaré-type metric g_Γ^S at some $\Gamma \in B_e$, which is defined as

$$g_\Gamma^S\colon H^{1/2}(\Gamma) \times H^{1/2}(\Gamma) \to \mathbb{R};\ (\alpha, \beta) \mapsto \int_\Gamma \alpha \left(S_\Gamma^p\right)^{-1} \beta \, ds. \tag{4}$$

Here, S_Γ^p is a symmetric and coercive operator defined by

$$S_\Gamma^p \colon H^{-1/2}(\Gamma) \to H^{1/2}(\Gamma); \ \alpha \mapsto U \cdot n,$$

where $U \in H^1(\Omega)^d$ solves the problem

$$\text{Find } U \in H^1(\Omega)^d \text{ such that} \quad a_\Omega(U, V) = \int_\Gamma \alpha \, (V \cdot n) \, \mathrm{d}s \quad \text{for all } V \in H^1(\Omega)^d, \tag{5}$$

for a symmetric, continuous, and coercive bilinear form $a_\Omega \colon H^1(\Omega)^d \times H^1(\Omega)^d \to \mathbb{R}$. To define a Riemannian metric on B_e, we restrict g^S to the tangent space $T_\Gamma B_e$.

Let us now briefly discuss the relation between the metric g_Γ^S and shape calculus. To do so, we assume that the shape functional J is shape differentiable and has a shape derivative of the form

$$dJ(\Omega)[\mathcal{V}] = \int_\Gamma g \, \mathcal{V} \cdot n \, \mathrm{d}s,$$

with $g \in L^2(\Gamma)$ (cf. Theorem 2). Then, the Riemannian shape gradient w.r.t. g_Γ^S is given by $\gamma \in T_\Gamma B_e$, which is the solution of

$$\text{Find } \gamma \in T_\Gamma B_e \text{ such that} \quad g_\Gamma^S(\gamma, \phi) = \int_\Gamma g\phi \, \mathrm{d}s \quad \text{for all } \phi \in T_\Gamma B_e. \tag{6}$$

Due to the definition of g_Γ^S, the solution of (6) is given by $\gamma = S_\Gamma^p g$, in particular, we have that $\gamma = \mathcal{G} \cdot n$, where $\mathcal{G}$ solves

$$\text{Find } \mathcal{G} \in H^1(\Omega)^d \text{ such that} \quad a_\Omega(\mathcal{G}, \mathcal{V}) = dJ(\Omega)[\mathcal{V}] \quad \text{for all } \mathcal{V} \in H^1(\Omega)^d. \tag{7}$$

Due to the Lax-Milgram lemma, this problem has a unique solution $\mathcal{G}$ which we call the gradient deformation of J at Ω. The gradient deformation $\mathcal{G}$ can be interpreted as an extension of the shape gradient γ to the entire domain Ω. Since a_Ω is coercive, there exists a constant $C > 0$ so that

$$dJ(\Omega)[-\mathcal{G}] = a_\Omega(-\mathcal{G}, \mathcal{G}) \leq -C \, ||\mathcal{G}||^2_{H^1(\Omega)} \leq 0, \tag{8}$$

i.e., an infinitesimal perturbation of identity with the negative gradient deformation yields a descent in the shape functional J. This fact is often used for the solution of shape optimization problems with a gradient descent method (see, e.g., [9, 16]).

3 Nonlinear CG Methods for Shape Optimization

We now use the theoretical framework introduced in Sect. 2 to formulate our NCG methods for shape optimization from [4]. To do so, we consider only reduced shape optimization problems of the form (3). For the sake of brevity, we only focus on a theoretical description of the methods. However, a detailed description of our numerical implementation of the methods can be found in our previous work [4].

Starting from an initial guess Ω_0, the k-th iteration of the NCG methods consists of the following steps: First, we compute the shape derivative $dJ(\Omega)[\cdot]$ of our cost functional J which involves solving the state and adjoint equation on the domain Ω_k (cf. [4]). Next, we compute the Riemannian shape gradient γ_k via Eq. (6). Note, that this involves the computation of the gradient deformation (cf. Eq. (7)), which is beneficial for our numerical implementation of the method as described detailedly in [4]. The next step of the methods involves the computation of the search direction δ_k, which is defined as

$$\delta_k = -\gamma_k + \beta_k \mathcal{T}_{\eta_{k-1}} \delta_{k-1},$$

where we set $\delta_0 = -\gamma_0$. Here, the parameter η_{k-1} is defined as $\eta_{k-1} = t_{k-1}\delta_{k-1}$ with the previous step size t_{k-1} (see below) and $\mathcal{T}$ denotes a vector transport (see, e.g., [1, 26] for more details). Moreover, the parameter β_k is an update parameter for the NCG method, which is detailed below. After we have obtained the search direction, we have to compute a feasible step size for updating our domain. This can be accomplished with, e.g., an Armijo line search (cf. [4] for more details) and yields the step size t_k for iteration k. The geometry is then updated with the help of an retraction R (see, e.g., [1, 26] for more details), i.e.,

$$\Gamma_{k+1} = R_{\Gamma_k} \eta_k,$$

where η_k is the scaled search direction for iteration k, i.e., $\eta_k = t_k \delta_k$. Finally, we increment k and proceed as before, until an appropriate stopping criterion is reached (cf. [4]). Note, that an algorithmic description of this procedure is given in Algorithm 1.

Algorithm 1: Nonlinear CG methods for shape optimization

Input: Initial geometry Ω_0, represented by its boundary Γ_0
for *k=0,1,2,…,* k_{max} **do**
  Compute the shape derivative $dJ(\Omega_k)[\cdot]$
  Compute the Riemannian shape gradient γ_k by solving (6)
  if *Stopping criterion is satisfied* **then**
    Stop with approximate solution Γ_k
  Compute the search direction $\delta_k = -\gamma_k + \beta_k \mathcal{T}_{\eta_{k-1}} \delta_{k-1}$
  Compute a feasible step size t_k
  Set $\eta_k = t_k \delta_k$ and update the geometry $\Gamma_{k+1} = R_{\Gamma_k} \eta_k$

We consider five different NCG variants, namely the NCG methods of Fletcher and Reeves (FR) [10], Polak and Ribiere (PR) [25], Hestenes and Stiefel (HS) [14], Dai and Yuan (DY) [7], and Hager and Zhang (HZ) [12]. Note, that an overview over these methods for finite-dimensional problems can be found, e.g., in [13]. The update parameters β in the context of shape optimization are given by

$$\beta_k^{\mathrm{FR}} = \frac{g^S_{\Gamma_k}(\gamma_k, \gamma_k)}{g^S_{\Gamma_k}\left(\mathcal{T}_{\eta_{k-1}}\gamma_{k-1}, \mathcal{T}_{\eta_{k-1}}\gamma_{k-1}\right)},$$

$$\beta_k^{\mathrm{PR}} = \frac{g^S_{\Gamma_k}(\gamma_k, y_{k-1})}{g^S_{\Gamma_k}\left(\mathcal{T}_{\eta_{k-1}}\gamma_{k-1}, \mathcal{T}_{\eta_{k-1}}\gamma_{k-1}\right)},$$

$$\beta_k^{\mathrm{HS}} = \frac{g^S_{\Gamma_k}(\gamma_k, y_{k-1})}{g^S_{\Gamma_k}\left(\mathcal{T}_{\eta_{k-1}}d_{k-1}, y_{k-1}\right)},$$

$$\beta_k^{\mathrm{DY}} = \frac{g^S_{\Gamma_k}(\gamma_k, \gamma_k)}{g^S_{\Gamma_k}\left(\mathcal{T}_{\eta_{k-1}}d_{k-1}, y_{k-1}\right)},$$

$$\beta_k^{\mathrm{HZ}} = g^S_{\Gamma_k}\left(y_{k-1} - 2\mathcal{T}_{\eta_{k-1}}d_{k-1}\frac{g^S_{\Gamma_k}(y_{k-1}, y_{k-1})}{g^S_{\Gamma_k}\left(\mathcal{T}_{\eta_{k-1}}d_{k-1}, y_{k-1}\right)}, \frac{\gamma_k}{g^S_{\Gamma_k}\left(\mathcal{T}_{\eta_{k-1}}d_{k-1}, y_{k-1}\right)}\right),$$

where we use

$$y_{k-1} = \gamma_k - \mathcal{T}_{\eta_{k-1}}\gamma_{k-1}.$$

A particular advantage of the NCG methods is the following: The NCG methods only require one or two additional vectors of storage compared to the popular gradient descent method, while being substantially more efficient, as is shown in Sect. 4. The L-BFGS methods with memory size m, on the other hand, require $2m$ additional vectors of storage, which can be prohibitive for very large scale problems, such as the ones arising from industrial applications (see [17]). Hence, the NCG methods are particularly interesting for these kinds of problems, where memory requirements are of great importance.

4 Numerical Examples

In this section, we investigate the previously introduced NCG methods numerically on two benchmark problems. In Sect. 4.1, we consider a two-dimensional shape optimization problem with a Poisson equation as PDE constraint and in Sect. 4.2, we consider the drag minimization in a three-dimensional pipe. For both test cases, we compare the five NCG variants from Sect. 3 to the gradient descent and L-BFGS methods. The numerical implementation is done in our software cashocs [3], which is based on the finite element software FEniCS [2, 21], and we refer to our previous work [4] for a comprehensive description of our implementation of the NCG methods.

4.1 Shape Optimization with a Poisson Equation

The first test case is taken from [4, 9] and is given by

$$\begin{aligned} \min_{\Omega} J(\Omega, u) &= \int_{\Omega} u \, \mathrm{d}x \\ \text{subject to} \quad -\Delta u &= f \quad \text{in } \Omega, \\ u &= 0 \quad \text{on } \Gamma, \end{aligned} \tag{9}$$

where we consider the problem in two dimensions and use

$$f(x) = 2.5 \left(x_1 + 0.4 - x_2^2\right)^2 + x_1^2 + x_2^2 - 1.$$

Our initial guess Ω_0 is given by the unit disc in $\mathbb{R}^2$. We discretize the PDE constraint with piecewise linear Lagrange elements, for which we use a uniform mesh consisting of 7651 nodes and 15,000 triangles.

We solve this problem with the gradient descent (GD) method, a L-BFGS method with memory size $m = 5$ (L-BFGS 5), and the five NCG methods presented in Sect. 3. The history of the optimization can be seen in Fig. 1, where the evolution of the cost functional (Fig. 1a) and relative gradient norm (Fig. 1b) are shown. Here, we have highlighted the graphs of the gradient descent, L-BFGS 5, and the NCG variant of Dai and Yuan (NCG DY), as the latter performed best of all NCG methods. For the sake of better readability, the remaining NCG methods are shown in transparent colors. Here, we observe that all NCG methods perform significantly

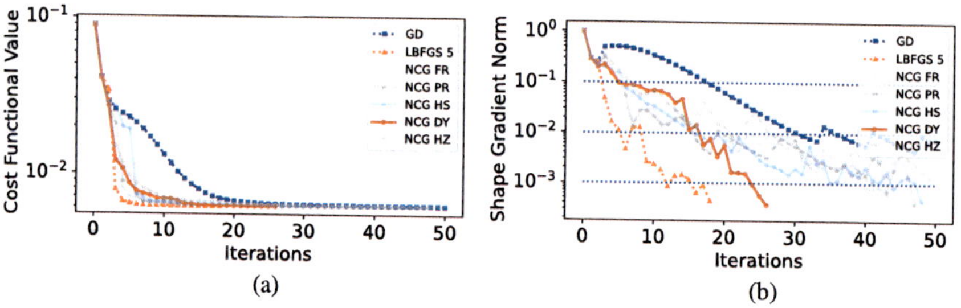

Fig. 1 History of the optimization methods for problem (9). (**a**) Cost functional (shifted by +0.1). (**b**) Relative gradient norm

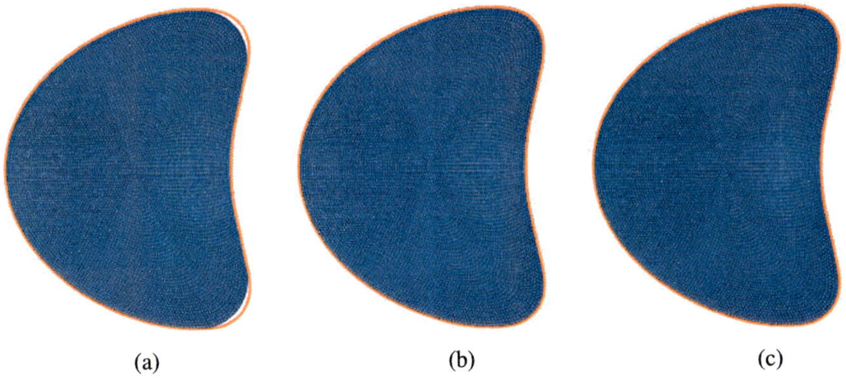

Fig. 2 Optimized Shapes (blue) compared to the reference solution (orange) for the Poisson problem (9). (**a**) Gradient descent method. (**b**) Fletcher-Reeves NCG method. (**c**) Polak-Ribière NCG method

better than the gradient descent method, as they reach the optimal function value faster and also have lower gradient norms throughout the optimization. On average, the NCG methods achieve a gradient norm that is one order of magnitude smaller than the gradient norm obtained with the gradient descent method. However, the performance of the L-BFGS 5 method is still slightly better than that of the NCG methods, but this comes at the cost of a higher memory usage, as discussed previously.

A visual inspection of the optimized geometries, which can be found in Fig. 2, shows that the NCG methods perform very well. Here, the optimized geometries for the gradient descent, Fletcher-Reeves NCG and Polak-Ribiere NCG methods are shown and compared to the reference solution (computed with the L-BFGS 5 method). We observe, that there are still visible differences between the reference solution and the one obtained with the gradient descent method, particularly in the right corners of the geometry. The two NCG methods, however, show no visible differences compared to the reference domain, which underlines their improved convergence behavior compared to the gradient descent method.

4.2 Shape Optimization of Energy Dissipation in a Pipe

For our second test case, we consider the problem of reducing the energy dissipation of a fluid in a three-dimensional pipe, which is taken from [24]. Here, the flow of the fluid is governed by the Navier-Stokes equations. The corresponding shape optimization problem is given by

$$\begin{aligned}
\min_{\Omega}\ J(\Omega, u) &= \frac{1}{\mathrm{Re}} \int_{\Omega} \varepsilon(u) : \varepsilon(u) \,\mathrm{d}x + \frac{\gamma}{2}\left(\int_{\Omega} 1 \,\mathrm{d}x - \int_{\Omega_0} 1 \,\mathrm{d}x \right)^2 \\
\text{subject to} \quad -\frac{2}{\mathrm{Re}} \nabla \cdot \varepsilon(u) + (u \cdot \nabla)\, u + \nabla p &= 0 \quad \text{in } \Omega, \\
\nabla \cdot u &= 0 \quad \text{in } \Omega, \\
u &= u^{\text{in}} \quad \text{on } \Gamma^{\text{in}}, \\
u &= 0 \quad \text{on } \Gamma^{\text{wall}}, \\
\frac{2}{\mathrm{Re}} \varepsilon(u) n - pn &= 0 \quad \text{on } \Gamma^{\text{out}}.
\end{aligned} \tag{10}$$

Here, u denotes the flow velocity, p the pressure and $\varepsilon(u) = 1/2(\nabla u + \nabla u^T)$ is the symmetric gradient of u. Note, that the cost functional consists of two terms, where the first one measures the energy dissipation in the pipe and the second one is a regularization of a volume equality constraint. The latter is used to constrain the volume of the pipe to its initial volume. Moreover, the pipe's boundary Γ is partitioned into the inlet Γ^{in} and the wall boundary Γ^{wall}, where we use Dirichlet boundary conditions, as well as the outlet Γ^{out}, where we use a do-nothing boundary condition. We discretize the geometry with 17,873 nodes and 82,422 tetrahedrons. Additionally, we discretize the Navier-Stokes equations with the inf-sup-stable Taylor-Hood elements, i.e., piecewise quadratic Lagrange elements for the velocity and piecewise linear Lagrange elements for the pressure. Note, that a plot of the initial and optimized geometries is shown in Fig. 3, and that the velocity magnitude on these domains is visualized in Fig. 4.

As before, we solve this shape optimization problem with the gradient descent, L-BFGS 5, and the NCG methods, where we consider the case of laminar flow and use a Reynolds number of $\mathrm{Re} = 1$ as well as a penalty parameter of $\gamma = 100$. Note, that the choice of γ is sufficiently large to ensure a relative volume difference below 0.5% between the initial and optimized geometries for all methods, so that the equality constraint is satisfied numerically. The corresponding results of the optimization are shown in Fig. 5, where, again, the history of the cost functional

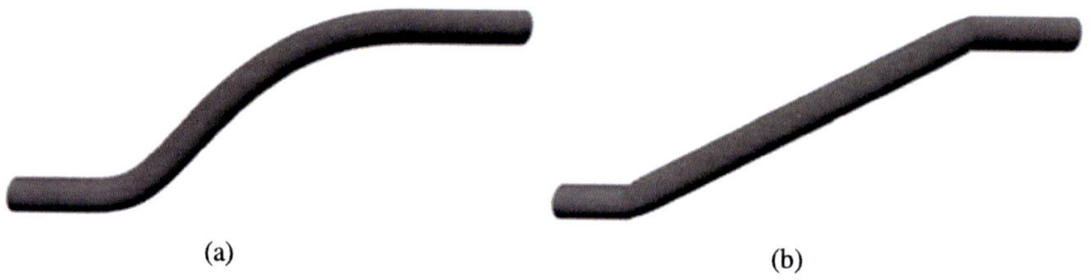

Fig. 3 Initial and optimized geometries for problem (10). (**a**) Initial geometry of the pipe. (**b**) Optimized geometry of the pipe (obtained with the NCG HS method)

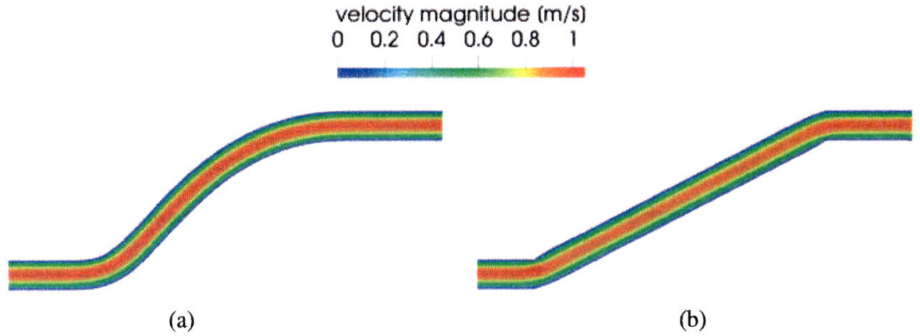

Fig. 4 Velocity magnitude on the initial and optimized geometries, shown as slice through the middle of the geometry. (**a**) Velocity magnitude on the initial geometry. (**b**) Velocity magnitude on the optimized geometry (obtained with the NCG HS method)

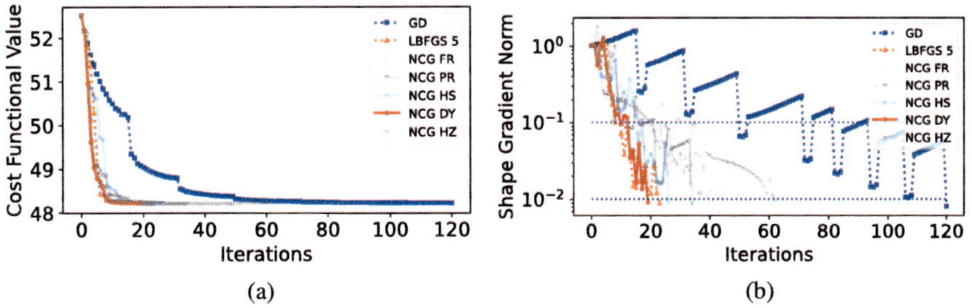

Fig. 5 History of the optimization methods for problem (10). (**a**) Evolution of the cost functional. (**b**) Evolution of the relative gradient norm

(cf. Fig. 5a) and the relative gradient norm (cf. Fig. 5b) are shown. Here, we again observe that the NCG methods are very efficient. The best performing NCG method is, again, the one of Dai and Yuan (DY), which even slightly outperformed the L-BFGS 5 method. Additionally, the method of Hestenes and Stiefel (HS) also performed very well and was only slightly worse than the L-BFGS method. The remaining NCG variants performed a bit worse, but all of them were substantially better than the gradient descent method as they required less than half the amount of iterations to reach the prescribed relative tolerance for this problem.

5 Conclusions

In this chapter, we have presented and investigated the nonlinear conjugate gradient (NCG) methods for shape optimization from [4]. After recalling recent results from shape optimization and shape calculus, we formulated the NCG methods in the Riemannian setting for shape optimization introduced in [28, 29]. Afterwards, we investigated these methods numerically and compared them to the already established gradient descent and L-BFGS methods for shape optimization. The results show that the NCG methods perform substantially better than the popular gradient descent method and that their performance is comparable to the one of the L-BFGS methods from [29]. Hence, the NCG methods could be particularly interesting for large-scale industrial problems due to their efficiency and low memory requirements.

References

1. Absil, P.A., Mahony, R., Sepulchre, R.: Optimization Algorithms on Matrix Manifolds. Princeton University Press, Princeton (2008)
2. Alnæs, M.S., Blechta, J., Hake, J., Johansson, A., Kehlet, B., Logg, A., Richardson, C., Ring, J., Rognes, M.E., Wells, G.N.: The FEniCS project version 1.5. Archive Numer. Softw. **3**(100), 9–23 (2015)
3. Blauth, S.: Cashocs: a computational, adjoint-based shape optimization and optimal control software. SoftwareX **13**, 100646 (2021)
4. Blauth, S.: Nonlinear conjugate gradient methods for PDE constrained shape optimization based on Steklov-Poincaré-type metrics. SIAM J. Optim. **31**(2), 1658–1689 (2021)
5. Blauth, S., Leithäuser, C., Pinnau, R.: Model hierarchy for the shape optimization of a microchannel cooling system. ZAMM Z. Angew. Math. Mech. **101**, e202000166 (2020)
6. Blauth, S., Leithäuser, C., Pinnau, R.: Shape sensitivity analysis for a microchannel cooling system. J. Math. Anal. Appl. **492**(2), 124476 (2020)
7. Dai, Y.H., Yuan, Y.: A nonlinear conjugate gradient method with a strong global convergence property. SIAM J. Optim. **10**(1), 177–182 (1999)
8. Delfour, M.C., Zolésio, J.P.: Shapes and geometries. In: Advances in Design and Control, vol. 22, 2nd edn. Society for Industrial and Applied Mathematics (SIAM), Philadelphia (2011)
9. Etling, T., Herzog, R., Loayza, E., Wachsmuth, G.: First and second order shape optimization based on restricted mesh deformations. SIAM J. Sci. Comput. **42**(2), A1200–A1225 (2020)
10. Fletcher, R., Reeves, C.M.: Function minimization by conjugate gradients. Comput. J. **7**, 149–154 (1964)
11. Gangl, P., Langer, U., Laurain, A., Meftahi, H., Sturm, K.: Shape optimization of an electric motor subject to nonlinear magnetostatics. SIAM J. Sci. Comput. **37**(6), B1002–B1025 (2015)
12. Hager, W.W., Zhang, H.: A new conjugate gradient method with guaranteed descent and an efficient line search. SIAM J. Optim. **16**(1), 170–192 (2005)
13. Hager, W.W., Zhang, H.: A survey of nonlinear conjugate gradient methods. Pac. J. Optim. **2**(1), 35–58 (2006)
14. Hestenes, M.R., Stiefel, E.: Methods of conjugate gradients for solving linear systems. J. Res. Nat. Bur. Standards **49**, 409–436 (1952)
15. Hinze, M., Pinnau, R., Ulbrich, M., Ulbrich, S.: Optimization with PDE constraints. In: Mathematical Modelling: Theory and Applications, vol. 23. Springer, New York (2009)
16. Hohmann, R., Leithäuser, C.: Shape optimization of a polymer distributor using an Eulerian residence time model. SIAM J. Sci. Comput. **41**(4), B625–B648 (2019)
17. Kelley, C.T.: Iterative methods for optimization. In: Frontiers in Applied Mathematics, vol. 18. Society for Industrial and Applied Mathematics (SIAM), Philadelphia (1999)
18. Kriegl, A., Michor, P.W.: The convenient setting of global analysis. In: Mathematical Surveys and Monographs, vol. 53. American Mathematical Society, Providence (1997)
19. Leithäuser, C., Pinnau, R.: Energy-Efficient High Temperature Processes via Shape Optimization, pp. 123–143. Springer, Cham (2021)
20. Leithäuser, C., Pinnau, R., Feßler, R.: Designing polymer spin packs by tailored shape optimization techniques. Optim. Eng. **19**(3), 733–764 (2018)

21. Logg, A., Mardal, K.A., Wells, G.N., et al.: Automated Solution of Differential Equations by the Finite Element Method. Springer, Berlin (2012)
22. Müller, P.M., Kühl, N., Siebenborn, M., Deckelnick, K., Hinze, M., Rung, P.: A novel p-harmonic descent approach applied to fluid dynamic shape optimization. Struct. Multidisciplinary Optim. **64**, 3489–3503 (2021)
23. Othmer, C.: Adjoint methods for car aerodynamics. J. Math. Ind. **4**, 6, 23 (2014)
24. Paganini, A., Wechsung, F.: Fireshape: a shape optimization toolbox for Firedrake. Struct. Multidiscip. Optim. **63**, 2553–2569 (2021)
25. Polak, E., Ribière, G.: Note sur la convergence de méthodes de directions conjuguées. Rev. Française Informat. Recherche Opérationnelle **3**(16), 35–43 (1969)
26. Ring, W., Wirth, B.: Optimization methods on Riemannian manifolds and their application to shape space. SIAM J. Optim. **22**(2), 596–627 (2012)
27. Schmidt, S., Ilic, C., Schulz, V., Gauger, N.R.: Three-dimensional large-scale aerodynamic shape optimization based on shape calculus. AIAA J. **51**(11), 2615–2627 (2013)
28. Schulz, V.H.: A Riemannian view on shape optimization. Found. Comput. Math. **14**(3), 483–501 (2014)
29. Schulz, V.H., Siebenborn, M., Welker, K.: Efficient PDE constrained shape optimization based on Steklov-Poincaré-type metrics. SIAM J. Optim. **26**(4), 2800–2819 (2016)
30. Sturm, K.: Shape differentiability under non-linear PDE constraints. In: New Trends in Shape Optimization. International Series of Numerical Mathematics, vol. 166, pp. 271–300. Birkhäuser, Cham (2015)

SpecSolve: Spectral Methods for Spectral Measures

Matthew J. Colbrook and Andrew Horning

2010 Mathematics Subject Classification: 47A10, 46N40, 47N50, 65N35, 81Q10

1 Introduction

Any finite and self-adjoint matrix $A \in \mathbb{C}^{n\times n}$ has an orthonormal basis of eigenfunctions. This basis diagonalizes A by decomposing the space $\mathbb{C}^n$ into a sum of orthogonal eigenspaces. However, many applications require us to study a self-adjoint operator $\mathcal{L}$ with domain $\mathcal{D}(\mathcal{L}) \subset \mathcal{H}$ on an *infinite-dimensional* Hilbert space $\mathcal{H}$ with inner product $\langle\cdot,\cdot\rangle$. Even when we are given a finite matrix, it is often an approximation or discretization of an underlying infinite-dimensional operator. In infinite dimensions, there may not exist a basis of eigenfunctions since $\mathcal{L}$ can have a continuous spectral component. This situation arises in, for example, stochastic processes and signal-processing [1, 2] and [3,Ch. 7], scattering in particle physics [4, 5], density-of-states in materials [6, 7], and many other areas [8–11].

Instead of eigenfunctions, $\mathcal{L}$ can be diagonalized through spectral measures supported on its spectrum $\Lambda(\mathcal{L}) \subset \mathbb{R}$ (see Sect. 2 and Eq. (3)). While efficient methods for computing spectral measures of (even very large) finite matrices exist [7], the infinite-dimensional case is more subtle. Most existing methods focus on specific operators where analytical formulas are available or perturbations of such

M. J. Colbrook (✉)
Department of Applied Mathematics and Theoretical Physics, University of Cambridge, Cambridge, UK

Centre Sciences des Données, Ecole Normale Supérieure, Paris, France
e-mail: m.colbrook@damtp.cam.ac.uk

A. Horning
Department of Mathematics, Massachusetts Institute of Technology, Cambridge, MA, USA
e-mail: horninga@mit.edu

J. M. Melenk et al. (eds.), *Spectral and High Order Methods for Partial Differential Equations ICOSAHOM 2020+1*, Lecture Notes in Computational Science and Engineering 137, https://doi.org/10.1007/978-3-031-20432-6_10

cases [12, Section 3]. Recently, [12] developed methods for computing spectral measures of general ODEs and integral operators using just two ingredients:[1]

1. A numerical solver for shifted linear equations $(\mathcal{L} - z)u = f$ for $z \in \mathbb{C}\backslash\Lambda(\mathcal{L})$.
2. Numerical approximations to inner products of the form $\langle u, f\rangle$.

The software `SpecSolve` [18] implements these ingredients using spectral methods.

This article extends `SpecSolve` to two important classes of operators with continuous spectra: singular integro-differential operators and operator pencils. Leveraging sparse spectral methods for the Hilbert transform on the real line, we compute spectral measures of singular integral operators such as

$$[\mathcal{L}u](x) = a(x)u(x) + \frac{1}{\pi i}\int_{\mathbb{R}} \frac{G(x, y)}{y - x} u(y)\, dy, \tag{1}$$

where $G(x, y) = \overline{G(y, x)}$ and real-valued $a(x)$ satisfy appropriate regularity constraints on $\mathbb{R}$. Additional differential terms are straightforward to incorporate to tackle a broad class of singular integro-differential operators. We also extend the two-step framework to compute spectral measures associated with the generalized spectral problem $\mathcal{A}v = \lambda\mathcal{B}v$, for operators $\mathcal{A}$ and $\mathcal{B}$. The two essential computational steps are performed with off-the-shelf spectral methods, illustrating the power and flexibility of `SpecSolve`'s "discretization-oblivious" paradigm.

2 Spectral Measures

The spectral theorem for a finite self-adjoint matrix $A \in \mathbb{C}^{n\times n}$ states that there exists an orthonormal basis of eigenvectors $v_1, \ldots, v_n$ for $\mathbb{C}^n$ such that

$$v = \left(\sum_{k=1}^{n} v_k v_k^*\right) v, \quad v \in \mathbb{C}^n \quad \text{and} \quad Av = \left(\sum_{k=1}^{n} \lambda_k v_k v_k^*\right) v, \quad v \in \mathbb{C}^n, \tag{2}$$

where $\lambda_1, \ldots, \lambda_n$ are eigenvalues of A, i.e., $Av_k = \lambda_k v_k$ for $1 \le k \le n$. In other words, the projections $v_k v_k^*$ decompose $\mathbb{C}^n$ and diagonalize A.

Switching to infinite dimensions, associated with the operator $\mathcal{L}$ is a projection-valued measure, $\mathcal{E}$ [19, Theorem VIII.6], whose support is the spectrum $\Lambda(\mathcal{L})$. The measure $\mathcal{E}$ assigns an orthogonal projector to each Borel subset of $\mathbb{R}$ such that

$$f = \left(\int_{\mathbb{R}} d\mathcal{E}(y)\right) f, \quad f \in \mathcal{H} \quad \text{and} \quad \mathcal{L}f = \left(\int_{\mathbb{R}} y\, d\mathcal{E}(y)\right) f, \quad f \in \mathcal{D}(\mathcal{L}). \tag{3}$$

[1] See also [13] which also includes computing spectral type in the context of the SCI hierarchy [14–16], and [17] for applications in physics.

Here, $\mathcal{D}(\mathcal{L})$ denotes the domain of the operator $\mathcal{L}$. Analogous to (2), the relations in (3) show how $\mathcal{E}$ decomposes $\mathcal{H}$ and diagonalizes the operator $\mathcal{L}$.

Of particular interest are the (scalar-valued) spectral measures of $\mathcal{L}$ with respect to $f \in \mathcal{H}$, given by $\mu_f(\Omega) := \langle \mathcal{E}(\Omega) f, f \rangle$, for Borel-measurable sets $\Omega \subset \mathbb{R}$. Lebesgue's decomposition of μ_f is

$$d\mu_f(y) = \underbrace{\sum_{\lambda \in \Lambda^{\mathrm{p}}(\mathcal{L})} \langle \mathcal{P}_\lambda f, f \rangle \, \delta(y-\lambda) dy}_{\text{discrete part}} + \underbrace{\rho_f(y)\, dy + d\mu_f^{(\mathrm{sc})}(y)}_{\text{continuous part}} .$$

The discrete part of μ_f is a sum of Dirac delta distributions, supported on the set of eigenvalues of $\mathcal{L}$, which we denote by $\Lambda^{\mathrm{p}}(\mathcal{L})$. The coefficient of each δ in the sum is $\langle \mathcal{P}_\lambda f, f \rangle = \|\mathcal{P}_\lambda f\|^2$, where $\mathcal{P}_\lambda$ is the orthogonal spectral projector associated with the eigenvalue λ, and $\|\cdot\| = \sqrt{\langle \cdot, \cdot \rangle}$ is the norm on $\mathcal{H}$. The continuous part of μ_f consists of an absolutely continuous[2] part with Radon–Nikodym derivative $\rho_f \in L^1(\mathbb{R})$ and a singular continuous component $\mu_f^{(\mathrm{sc})}$. Without loss of generality, we assume throughout that $\|f\| = 1$, which ensures that μ_f is a probability measure.

Computing μ_f is important in many applications, and can be considered an infinite-dimensional analogue of computing eigenvectors. We aim to evaluate smoothed approximations of μ_f. We compute a smooth function μ_f^ϵ, with smoothing parameter $\epsilon > 0$, that converges weakly to μ_f [20, Ch. 1]. That is,

$$\int_{\mathbb{R}} \phi(y) \mu_f^\epsilon(y)\, dy \to \int_{\mathbb{R}} \phi(y)\, d\mu_f(y), \qquad \text{as} \qquad \epsilon \downarrow 0, \tag{4}$$

for any bounded, continuous function ϕ.

3 Algorithmic Framework for SpecSolve

Our key ingredient is the resolvent $(\mathcal{L}-z)^{-1} = \int_{\Lambda(\mathcal{L})} (\lambda - z)^{-1} d\mathcal{E}(\lambda)$ for $z \notin \Lambda(\mathcal{L})$. Stone's formula [21] links the resolvent to convolution with the Poisson kernel:

$$\mu_f^\epsilon(x) = \frac{-1}{\pi} \mathrm{Im}\left(\langle (\mathcal{L} - (x - \epsilon i))^{-1} f, f \rangle \right) = \int_{\mathbb{R}} \frac{\epsilon \pi^{-1}}{(x-\lambda)^2 + \epsilon^2}\, d\mu_f(\lambda). \tag{5}$$

As $\epsilon \downarrow 0$, this approximation converges weakly to μ_f. To compute $(\mathcal{L} - (x - \epsilon i))^{-1} f$ we must somehow discretize the operator. However, for a given discretization size, if ϵ is too small, the approximation via (5) becomes unstable [12, Section 4.3] due to the discrete spectrum of the discretization. We must adaptively increase the discretization/truncation size as $\epsilon \downarrow 0$ and there is an

[2] We take "absolutely continuous" to be with respect to the Lebesgue measure.

Algorithm 1 A computational framework for evaluating an approximate spectral measure of an operator $\mathcal{L}$ at $x_0 \in \mathbb{R}$ with respect to a vector $f \in \mathcal{H}$

Input: $\mathcal{L}$, $f \in \mathcal{H}$, $x_0 \in \mathbb{R}$, $a_1, \ldots, a_m \in \{z \in \mathbb{C} : \mathrm{Im}(z) > 0\}$, and $\epsilon > 0$.
1: Solve the Vandermonde system (6) for the residues $\alpha_1, \ldots, \alpha_m \in \mathbb{C}$.
2: Solve $(\mathcal{L} - (x_0 - \epsilon a_j))u_j^\epsilon = f$ for $1 \le j \le m$.
3: Compute $\mu_f^\epsilon(x_0) = \frac{-1}{\pi}\mathrm{Im}\left(\sum_{j=1}^m \alpha_j \langle u_j^\epsilon, f\rangle\right)$.
Output: The approximate spectral measure $\mu_f^\epsilon(x_0)$.

increased computational cost for smaller ϵ. Therefore, replacing the Poisson kernel with higher-order rational kernels is advantageous. These kernels have better convergence rates as $\epsilon \downarrow 0$, allowing a larger ϵ to be used for a given accuracy, and thus a lower computational burden.

Let $\{a_j\}_{j=1}^m$ be distinct points in the upper half plane and suppose that the constants $\{\alpha_j\}_{j=1}^m$ satisfy the following (transposed) Vandermonde system:

$$\begin{pmatrix} 1 & \ldots & 1 \\ a_1 & \ldots & a_m \\ \vdots & \ddots & \vdots \\ a_1^{m-1} & \ldots & a_m^{m-1} \end{pmatrix} \begin{pmatrix} \alpha_1 \\ \alpha_2 \\ \vdots \\ \alpha_m \end{pmatrix} = \begin{pmatrix} 1 \\ 0 \\ \vdots \\ 0 \end{pmatrix}. \tag{6}$$

Then the kernel

$$K(x) = \frac{1}{2\pi i}\sum_{j=1}^m \frac{\alpha_j}{x - a_j} - \frac{1}{2\pi i}\sum_{j=1}^m \frac{\overline{\alpha_j}}{x - \overline{a_j}} \quad \text{with } K_\epsilon(x) = \epsilon^{-1}K(x\epsilon^{-1})$$

is an mth order kernel, and we have the following generalization of Stone's formula

$$\mu_f^\epsilon(x) = [K_\epsilon * \mu_f](x) = \frac{-1}{\pi}\sum_{j=1}^m \mathrm{Im}\left(\alpha_j \langle (\mathcal{L} - (x - \epsilon a_j))^{-1} f, f\rangle\right). \tag{7}$$

This provides $O(\epsilon^m \log(\epsilon^{-1}))$ convergence in (4) if ϕ is sufficiently regular, and similar rates for $\mu_f^\epsilon(x) \to \rho_f(x)$ if μ_f is sufficiently regular near x [12].

We consider the choice $a_j = 2j/(m+1) - 1 + i$ and the framework for evaluating μ_f^ϵ is summarized in Algorithm 1. This algorithm forms the foundation of `SpecSolve` [18] and can be performed in parallel for several x_0. We compute an accurate value of μ_f^ϵ provided that the resolvent is applied with sufficient accuracy. For an efficient adaptive implementation, `SpecSolve` constructs a fixed discretization, solves linear systems at each required complex shift, and checks the approximation error at each shift. If further accuracy is needed at a subset of the shifts, then the discretization size is doubled, applied at these shifts, and the error is recomputed. This process is repeated until the resolvent is computed accurately at all shifts.

4 Singular Integro-Differential Operators

Singular integral operators of Cauchy type play a pivotal role in the classical theory of PDEs and their spectral properties [22]. They appear in a wide range of physical models, along with their integro-differential and nonlinear counterparts [23].

Consider the self-adjoint singular integral operator $\mathcal{L}$ in (1) with $G(x, y) = \overline{G(y, x)}$, and $a(x)$ real, continuously differentiable, and bounded. To compute spectral measures of $\mathcal{L}$ in the SpecSolve framework, we must compute inner products between functions in $L^2(\mathbb{R})$ and solve linear equations with a complex shift z, e.g.,

$$(a(x) - z)u(x) + \frac{1}{\pi i} \int_{\mathbb{R}} \frac{G(x, y)}{y - x} u(y)\, dy = f(x). \tag{8}$$

We discretize $L^2(\mathbb{R})$ with the orthogonal rational basis functions $\rho_n(x) = \frac{1}{\sqrt{\pi}} \frac{(1+ix)^n}{(1-ix)^{n+1}}$, for $n \in \mathbb{Z}$. These functions have excellent approximation properties, are associated with banded differentiation and multiplication matrices, and expansion coefficients can be computed from function samples in quasi-linear time with the FFT [24]. Moreover, they diagonalize the Hilbert transform and lead to banded discretizations of (8) when $G(x, y)$ is sufficiently smooth and of low numerical rank [25].

Both the multiplicative and integral components of $\mathcal{L}$ can contribute continuous spectrum. When $G(x, y) = k(x)k(y)$ is a rank one kernel with $k(x) > 0$, the spectrum fills the interval $[\min |a(x) - k(x)|, \max |a(x) + k(x)|]$ [26]. Figure 1 (left) shows the spectral measures μ_f of $\mathcal{L}$, with $f(x) = \sqrt{2/\pi}(1 - x^2)^{-1}$,

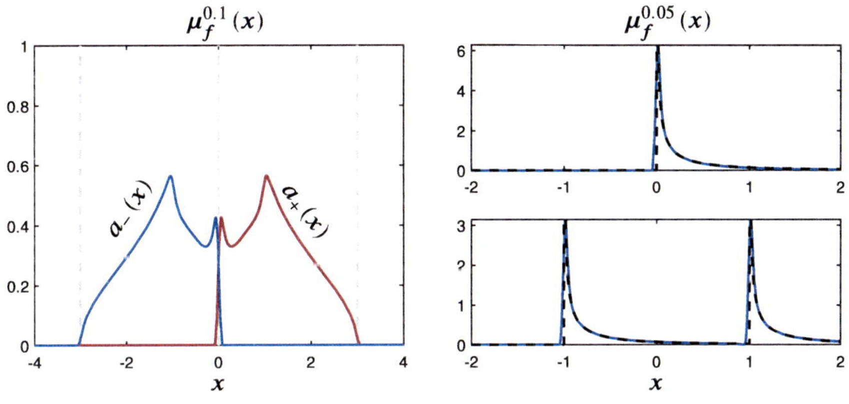

Fig. 1 Left: the smoothed spectral measures, $\mu_f^{0.1}$, computed with a 4th order kernel are supported on the intervals $[\min |a(x) - k(x)|, \max |a(x) + k(x)|]$. Right: the smoothed spectral measures, $\mu_f^{0.05}$, computed with a 4th order kernel for $-d^2/dx^2$ (top) and $-d^2/dx^2 + (1/\pi i) \int_{\mathbb{R}} (y - x)^{-1} dy$ (bottom) are compared with analytical solutions (dashed lines)

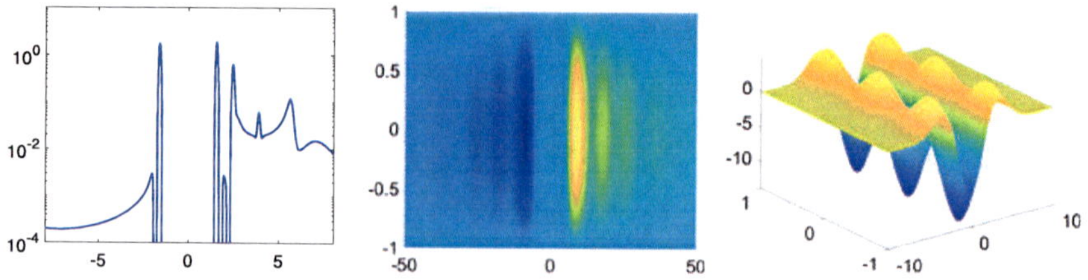

Fig. 2 Left: the smoothed spectral measure, $\mu_f^{0.2}$, of the partial integro-differential operator in (9) computed with a 6th order kernel. Middle: spectral projection $\mathcal{E}([2.1, 3])f$ of $f(x, y) = (1 + x)(1 + x^2)^{-1}\cos(\pi y/2)/\sqrt{\pi}$ associated with the third resonance peak from the left in the plot of $\mu_f^{0.2}$. Right: the potential energy landscape $v(x, y)$ for the operator in (9)

$k(x) = e^{-x^2}$, and $a_{\pm}(x) = \pm 2/(1 + x^2)^2$. The dashed grey lines highlight the support of the measures in the expected interval. We can also tackle singular integro-differential operators. Figure 1 (right) compares the spectral measures of $-d^2/dx^2$ and $-d^2/dx^2 + (1/\pi i)\int_{\mathbb{R}}(y-x)^{-1}dy$ with respect to f. Both the second derivative and the singular integral are diagonalized by the Fourier transform, and the spectral measures can be computed analytically (dashed lines). The integral perturbation breaks the symmetry between positive and negative Fourier modes, which effectively splits the spectral measure of $-d^2/dx^2$ into two duplicate peaks of half height at ± 1.

The `SpecSolve` framework can also compute spectral projections $\mathcal{E}([a, b])$ associated with the projection-valued measure by omitting the inner product step in Algorithm 1 and applying endpoint corrections [27]. Figure 2 displays a scalar spectral measure and spectral projection for the partial integro-differential operator

$$-\Delta u + v(x, y)u + \frac{4}{\pi i}\int_{\mathbb{R}} \frac{\exp(-x^2 - s^2)}{s - x} u(s, y)ds, \quad \mathcal{H} = L^2(\mathbb{R} \times [-1, 1]), \tag{9}$$

and the function $f(x, y) = (1+x)(1+x^2)^{-1}\cos(\pi y/2)/\sqrt{\pi}$. The potential function $v(x, y)$ is also plotted in Fig. 2. The operator is discretized with a tensor product basis of the rational orthogonal functions $\{\rho_n\}$ and ultraspherical polynomials [28], resulting in a sparse and banded discretization (we use basis reordering to reduce the bandwidth). In Fig. 2, narrow peaks in the scalar measure reveal scattering resonances of the partial integro-differential operator and the associated spectral projections uncover wave-packet modes that are highly concentrated within the potential well.

5 Linear Operator Pencils

For matrices $A, B \in \mathbb{C}^n$, the classical generalized eigenvalue problem is the problem of finding $v \in \mathbb{C}^n \backslash \{0\}$ and $\lambda \in \mathbb{C}$ such that $Av = \lambda Bv$. For example, this problem arises in finite element discretizations of eigenproblems for elliptic partial differential operators, where A corresponds to the "stiffness matrix" and B corresponds to the "mass matrix" [29]. Another example is linearization methods for non-linear eigenvalue problems [30]. For many applications, A and B are finite approximations of (possibly unbounded) operators $\mathcal{A}$ and $\mathcal{B}$ acting on a separable Hilbert space. We consider the case that $\mathcal{A}$ and $\mathcal{B}$ are both self-adjoint and that $\mathcal{B}$ is positive and invertible. We study the generalized spectral problem through the operator formally defined as $\mathcal{L} = \mathcal{B}^{-1}\mathcal{A}$.

5.1 Recovering a Self-Adjoint Operator

It is well-known that $\mathcal{D}(\mathcal{B}^{1/2})$ is complete with respect to the norm $\|f\|_{\mathcal{B}} := \langle \mathcal{B}^{1/2} f, \mathcal{B}^{1/2} f\rangle$ [31, Theorem 4.4.2]. We denote the induced Hilbert space by $\mathcal{H}_{\mathcal{B}}$. The operator $\mathcal{B}^{-1}\mathcal{A}$ with domain $\mathcal{D}(\mathcal{A}) \cap \mathcal{D}(\mathcal{B}^{1/2})$ is symmetric in $\mathcal{H}_{\mathcal{B}}$.[3] However, to apply the spectral theorem, we need a self-adjoint operator. We assume that $\mathcal{D}(\mathcal{A}) \cap \mathcal{D}(\mathcal{B}^{1/2})$ is a dense subspace of the Hilbert space $\mathcal{H}_{\mathcal{B}}$. Since $\mathcal{B}^{-1}\mathcal{A}$ is symmetric in $\mathcal{H}_{\mathcal{B}}$, it is closable. We define the symmetric closed operator

$$\mathcal{L} = \overline{\mathcal{B}^{-1}\mathcal{A}|_{\mathcal{D}(\mathcal{A})\cap\mathcal{D}(\mathcal{B}^{1/2})}}, \tag{10}$$

where the closure is performed with respect to $\mathcal{H}_{\mathcal{B}}$. This allows us to perform *numerical computations* with $\mathcal{L}$ by restricting to the subspace $\mathcal{D}(\mathcal{A}) \cap \mathcal{D}(\mathcal{B}^{1/2})$. To do this, we consider the inner product space $\{f : f \in \mathcal{H}\}$ with inner product $\langle f, g\rangle_{\mathcal{B}^{-1}} = \langle \mathcal{B}^{-1/2} f, \mathcal{B}^{-1/2} g\rangle$. We take the completion of this space, $\mathcal{H}_{\mathcal{B}^{-1}}$. $\mathcal{D}(\mathcal{B})$ is dense in $\mathcal{H}_{\mathcal{B}}$ and hence $\mathcal{B}$ can be extended to an invertible isometry from $\mathcal{H}_{\mathcal{B}}$ to $\mathcal{H}_{\mathcal{B}^{-1}}$, and $\mathcal{H}_{\mathcal{B}^{-1}}$ can be identified with the dual of $\mathcal{H}_{\mathcal{B}}$. We assume that $\mathcal{A}|_{\mathcal{D}(\mathcal{A})\cap\mathcal{D}(\mathcal{B}^{1/2})} : \mathcal{H}_{\mathcal{B}} \to \mathcal{H}_{\mathcal{B}^{-1}}$ is closable, with closure denoted by $\mathcal{A}_{\mathcal{B}}$. We can now define

$$\begin{aligned} &\mathcal{T}(z) : \mathcal{D}(\mathcal{A}_{\mathcal{B}}) \to \mathcal{H}_{\mathcal{B}^{-1}}, \quad f \hookrightarrow (\mathcal{A}_{\mathcal{B}} - z\mathcal{B})f, \\ &\Lambda(\mathcal{A}, \mathcal{B}) = \{z \in \mathbb{C} : \mathcal{T}(z) \text{ does not have bounded inverse}\}. \end{aligned} \tag{11}$$

---

[3] Suppose that $f, g \in \mathcal{D}(\mathcal{A}) \cap \mathcal{D}(\mathcal{B}^{1/2})$. Then $\langle \mathcal{B}^{1/2}(\mathcal{B}^{-1}\mathcal{A})g, \mathcal{B}^{1/2} f\rangle = \langle \mathcal{B}^{-1/2}\mathcal{A}g, \mathcal{B}^{1/2} f\rangle = \langle \mathcal{A}g, f\rangle$. The first equality follows since $\mathcal{B}^{-1}\mathcal{A}g \in \mathcal{D}(\mathcal{B}^{1/2})$, whereas the second follows since $\mathcal{B}^{-1/2}$ is a bounded self-adjoint operator on $\mathcal{H}$. Similarly, we have that $\langle \mathcal{B}^{1/2} g, \mathcal{B}^{1/2}(\mathcal{B}^{-1}\mathcal{A}) f\rangle = \langle g, \mathcal{A}f\rangle$.

Propostion 1 *For any* $z \in \mathbb{C}$, $\mathcal{D}(\mathcal{T}(z)) = \mathcal{D}(\mathcal{L})$ *and* $\mathcal{T}(z) = \mathcal{B}(\mathcal{L} - z)$. *Moreover,* $\Lambda(\mathcal{A}, \mathcal{B}) = \Lambda(\mathcal{L})$ *and if* $z \in \mathbb{C} \setminus \Lambda(\mathcal{L})$, *then* $(\mathcal{L} - z)^{-1} = \mathcal{T}(z)^{-1}\mathcal{B}$.

Proof Let $z \in \mathbb{C}$ and $f \in \mathcal{D}(\mathcal{L})$. Then there exists $f_n \in \mathcal{D}(\mathcal{A}) \cap \mathcal{D}(\mathcal{B}^{1/2})$ such that $\lim_{n\to\infty} f_n = f$ (in $\mathcal{H}_{\mathcal{B}}$) and $\lim_{n\to\infty}(\mathcal{L} - z) f_n = (\mathcal{L} - z) f$ (in $\mathcal{H}_{\mathcal{B}}$). Since $\|\mathcal{T}(z) f_n - \mathcal{T}(z) f_m\|_{\mathcal{B}^{-1}} = \|(\mathcal{L} - z) f_n - (\mathcal{L} - z) f_m\|_{\mathcal{B}}$, it follows that $\{\mathcal{T}(z) f_n\}$ is Cauchy in $\mathcal{H}_{\mathcal{B}^{-1}}$ and hence converges to some $g \in \mathcal{H}_{\mathcal{B}^{-1}}$. Since $\mathcal{T}(z)$ is closed, $f \in \mathcal{D}(\mathcal{T}(z))$ and $\mathcal{T}(z) f = g$. Moreover, $\|(\mathcal{L}-z) f_n - \mathcal{B}^{-1} g\|_{\mathcal{B}} = \|\mathcal{T}(z) f_n - g\|_{\mathcal{B}^{-1}}$ converges to zero. Since $\mathcal{L}$ is closed, $(\mathcal{L}-z) f = \mathcal{B}^{-1} g$ and hence that $\mathcal{B}(\mathcal{L}-z) f = \mathcal{T}(z) f$. A similar argument shows that $\mathcal{D}(\mathcal{T}(z)) \subset \mathcal{D}(\mathcal{L})$. Hence, $\mathcal{D}(\mathcal{T}(z)) = \mathcal{D}(\mathcal{L})$ and $\mathcal{T}(z) = \mathcal{B}(\mathcal{L} - z)$. The proposition follows since $\mathcal{B} : \mathcal{H}_{\mathcal{B}} \to \mathcal{H}_{\mathcal{B}^{-1}}$ is an isometry. □

The following theorem gives sufficient conditions for $\mathcal{L}$ to be self-adjoint. Common examples of these conditions include when $\mathcal{A}$ and $\mathcal{B}$ are suitable elliptic PDEs of the same differentiation order (see condition (C1)), $\mathcal{A}$ is bounded (see condition (C2)), and $\mathcal{B}$ is a suitable weight function (see condition (C3))

Theorem 2 *Consider the operators* $\mathcal{A}$, $\mathcal{B}$ *and* $\mathcal{L}$ *above. Suppose that any of the following conditions hold:*

(C1) *There exist constants* $a, b > 0$ *such that for any* $f \in \mathcal{D}(\mathcal{A}) \cap \mathcal{D}(\mathcal{B}^{1/2})$

$$\|\mathcal{B}^{-1/2}\mathcal{A} f\| \leq a\|f\| + b\|\mathcal{B}^{1/2} f\|. \tag{12}$$

(C2) $\mathcal{A}$ *is a relatively bounded perturbation of* $\mathcal{B}$, *meaning that* $\mathcal{D}(\mathcal{B}) \subset \mathcal{D}(\mathcal{A})$ *and there exist constants* $a, b > 0$ *such that for any* $f \in \mathcal{D}(\mathcal{B})$

$$\|\mathcal{A} f\| \leq a\|f\| + b\|\mathcal{B} f\|. \tag{13}$$

(C3) $\mathrm{Sp}(\mathcal{A}) \neq \mathbb{R}$ *and* $\mathcal{B}$ *is a relatively bounded perturbation of* $\mathcal{A}$, *meaning that* $\mathcal{D}(\mathcal{A}) \subset \mathcal{D}(\mathcal{B})$ *and there exist constants* $a, b > 0$ *such that for any* $f \in \mathcal{D}(\mathcal{A})$

$$\|\mathcal{B} f\| \leq a\|f\| + b\|\mathcal{A} f\|. \tag{14}$$

Then $\mathcal{L}$ *is self-adjoint on* $\mathcal{H}_{\mathcal{B}}$. *Moreover, when* (C1) *holds,* $\mathcal{L}$ *is bounded.*

Proof Suppose first that $(C1)$ holds. Since $\mathcal{B}^{1/2}$ is strictly positive, (12) implies that there exists a positive constant c such that $\|\mathcal{B}^{-1/2}\mathcal{A} f\| \leq c\|\mathcal{B}^{1/2} f\|$ for any $f \in \mathcal{D}(\mathcal{A}) \cap \mathcal{D}(\mathcal{B}^{1/2})$. This is equivalent to boundedness of $\mathcal{B}^{-1}\mathcal{A}|_{\mathcal{D}(\mathcal{A})\cap\mathcal{D}(\mathcal{B}^{1/2})}$ in the Hilbert space $\mathcal{H}_{\mathcal{B}}$, and hence $\mathcal{L}$ is bounded and self-adjoint on $\mathcal{H}_{\mathcal{B}}$.

For $(C2)$ or $(C3)$, we claim that it is enough to show that there exists some $\gamma > 0$ and $\kappa \in \mathbb{R}$ such that the operators

$$T_{\pm} = \mathcal{A} + \kappa I \pm i\gamma\mathcal{B}, \quad \mathcal{D}(T_{\pm}) = \mathcal{D}(\mathcal{A}) \cap \mathcal{D}(\mathcal{B}) \tag{15}$$

are closable (in $\mathcal{H}$), and that their closures, denoted $\mathcal{T}_\pm$, are invertible (in $\mathcal{H}$). To see this, suppose that these conditions hold. Let $g \in \mathcal{D}(\mathcal{B})$ and set $f^\pm = T_\pm^{-1}\mathcal{B}g$. Then, by definition of the closure, there exists $f_n^\pm \in \mathcal{D}(\mathcal{A}) \cap \mathcal{D}(\mathcal{B}) \subset \mathcal{D}(\mathcal{A}) \cap \mathcal{D}(\mathcal{B}^{1/2})$ such that $f_n^\pm \to f^\pm$ and $T_\pm f_n^\pm \to \mathcal{B}g$ as $n \to \infty$ (with convergence in $\mathcal{H}$). Thus, $\|\mathcal{B}^{-1}(\mathcal{A}+\kappa I)f_n^\pm \pm i\gamma f_n^\pm - g\|_{\mathcal{B}} = \|\mathcal{B}^{-1/2}\left(T_\pm f_n^\pm - \mathcal{B}g\right)\| \to 0$ as $n \to \infty$. Since $\mathcal{D}(\mathcal{B})$ is dense in $\mathcal{H}_{\mathcal{B}}$, it follows that the ranges of $\gamma^{-1}\mathcal{B}^{-1}(\mathcal{A}+\kappa I) \pm iI$ are also dense in $\mathcal{H}_{\mathcal{B}}$. It follows that $\gamma^{-1}\mathcal{B}^{-1}(\mathcal{A}+\kappa I)$ is essentially self-adjoint in $\mathcal{H}_{\mathcal{B}}$ [19, p. 257], and hence so is $\mathcal{B}^{-1}\mathcal{A}$. This proves the claim.

Now suppose that $(C2)$ holds. Since $\mathcal{B}$ is strictly positive, (13) implies that there exists a positive constant $c < 1$ and $\gamma > 0$ such that $\|\mathcal{A}f\| \le c\|\gamma\mathcal{B}f\|$ for any $f \in \mathcal{D}(\mathcal{B}) \subset \mathcal{D}(\mathcal{A})$. Hence $\mathcal{A}$ is a relatively bounded perturbation of $i\gamma\mathcal{B}$, with $i\gamma\mathcal{B}$-bound less than 1. Stability of bounded invertibility [[32],Theorem IV.4.1.16] implies that $T_\pm$ in (15) (with $\kappa = 0$) are closed and invertible (in $\mathcal{H}$).

Finally, suppose that $(C3)$ holds. Choose $\kappa \in \mathbb{R}$ with $-\kappa \notin \mathrm{Sp}(\mathcal{A})$ so that $\mathcal{A}+\kappa I$ is invertible, and set $C = \mathcal{A}+\kappa I$. For any $f \in \mathcal{D}(\mathcal{A})$ and $\gamma > 0$, (14) implies that

$$\|\gamma\mathcal{B}f\| \le \gamma(a+|\kappa|)\|f\| + \gamma b\|Cf\|. \tag{16}$$

Choose $\gamma > 0$ so that $\gamma(a+|\kappa|)\|C^{-1}\| + \gamma b < 1$. The stability of bounded invertibility [32, Theorem IV.4.1.16] and (16) imply that $T_\pm$ are closed and invertible. □

5.2 *Framework for Generalized Spectral Measures*

To extend SpecSolve to the above pencil problem, we simply apply (7) with the operator $\mathcal{L}$ defined in (10) and the Hilbert space $\mathcal{H}_{\mathcal{B}}$. We suppose for simplicity that $f \in \mathcal{D}(\mathcal{B})$. Using Theorem 1 and (7) and the self-adjointness of $\mathcal{B}^{1/2}$, we have

$$\begin{aligned}\mu_f^\epsilon(x) = [K_\epsilon * \mu_f](x) &= \frac{-1}{\pi}\sum_{j=1}^m \mathrm{Im}\left(\alpha_j\left\langle(\mathcal{B}^{1/2}\mathcal{T}(x-\epsilon a_j)^{-1}\mathcal{B}f, \mathcal{B}^{1/2}f\right\rangle\right)\\ &= \frac{-1}{\pi}\sum_{j=1}^m \mathrm{Im}\left(\alpha_j\left\langle(\mathcal{T}(x-\epsilon a_j)^{-1}\mathcal{B}f, \mathcal{B}f\right\rangle\right),\end{aligned} \tag{17}$$

where we use that $\mathcal{B}^{1/2}$ is self-adjoint in the second line and $\langle\cdot,\cdot\rangle$ denotes the inner product on $\mathcal{H}$. This leads to Algorithm 2, which generalizes Algorithm 1. To apply Algorithm 2, we only need to compute approximations of $g = \mathcal{B}f$, solve the systems $(\mathcal{A} - (x_0 - \epsilon a_j)\mathcal{B})u_j^\epsilon = g$, and then compute inner products. We approximate u_j^ϵ using spectral methods and compute inner products using quadrature.

Algorithm 2 A computational framework for evaluating an approximate spectral measure of an operator $\mathcal{L}$ in (10) corresponding to the pencil $\mathcal{A} - \lambda\mathcal{B}$ at $x_0 \in \mathbb{R}$ with respect to a vector $f \in \mathcal{D}(\mathcal{B})$

Input: $\mathcal{A}$, $\mathcal{B}$, $f \in \mathcal{D}(\mathcal{B})$, $x_0 \in \mathbb{R}$, $a_1, \ldots, a_m \in \{z \in \mathbb{C} : \mathrm{Im}(z) > 0\}$, and $\epsilon > 0$.

Compute $g = \mathcal{B}f$.

Solve the Vandermonde system (6) for the residues $\alpha_1, \ldots, \alpha_m \in \mathbb{C}$.

Solve $(\mathcal{A} - (x_0 - \epsilon a_j)\mathcal{B})u_j^\epsilon = g$ for $1 \le j \le m$.

Compute $\mu_f^\epsilon(x_0) = \frac{-1}{\pi}\mathrm{Im}\left(\sum_{j=1}^m \alpha_j \langle u_j^\epsilon, g\rangle\right)$.

Output: The approximate spectral measure $\mu_f^\epsilon(x_0)$.

5.3 Examples

We now present two examples, using Fourier spectral methods and a spectral element method, respectively. Both examples fall into the setup of Theorem 2.

Pseudo-Differential Operators and Internal Waves Spectral properties of 0th order pseudo-differential operators arise naturally in fluid mechanics [33] and pseudoparabolic equations [34]. See [35] for the study of internal waves and [36] for connections with scattering resonances. As a simple example, we consider

$$\mathcal{A} = -i(1 + \cos(x)/2)\partial_y, \quad \mathcal{B} = (1 - \partial_y^2)^{1/2}, \quad x, y \in [-\pi, \pi]_{\mathrm{per}},$$

where the initial Hilbert space is $\mathcal{H} = L^2([-\pi, \pi]_{\mathrm{per}}^2)$. To solve the linear systems in Algorithm 2, we use the standard tensor product Fourier basis.

Figure 3 (left) shows the smoothed spectral measures computed using $\epsilon = 0.01$, and the first and sixth-order kernels for $f(x, y) = C\exp(\sin(x + y))/(2 + \cos(y))$,

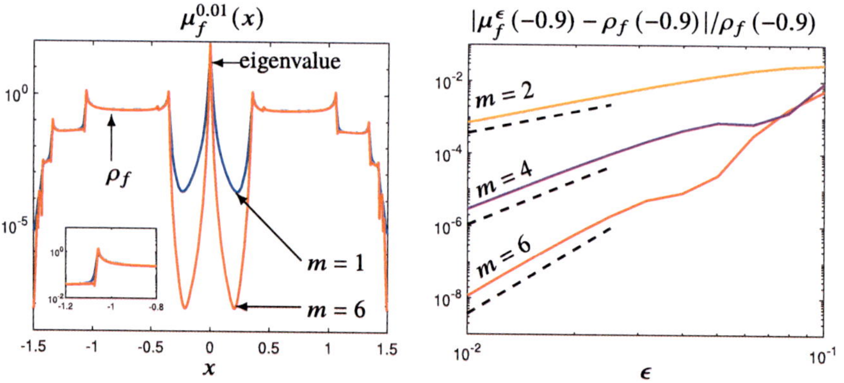

Fig. 3 Left: smoothed spectral measures, $\mu_f^{0.01}$, computed using the 1st and 6th order kernels. The zoomed-in section shows better resolution of jump discontinuities in ρ_f for larger m. Right: relative pointwise convergence to ρ_f and expected rates shown as dashed black lines

where C is a normalization constant so that μ_f is a probability measure. The spectral measure has an absolutely continuous component (with piecewise continuous Radon–Nikodym derivative), and an eigenvalue at 0. The higher order kernel ($m = 6$) provides a better localization of the singular part of the spectral measure at 0, and also a better resolution of jumps in the Radon–Nikodym derivative (see zoomed-in section). Figure 3 (right) shows the pointwise convergence to the Radon–Nikodym derivative and the expected rates of convergence for $m = 2, 4$ and 6.

Elliptic Differential Operator Preconditioners A common use of $\mathcal{L}$ in (10) is preconditioning, where $\mathcal{B}$ is a preconditioner of $\mathcal{A}$ [37]. For example, sometimes one can prove mesh-independent bounds on condition numbers for methods such as finite elements [38], which are useful for applying Krylov space methods. The papers [39, 40] discuss the spectrum of $\mathcal{L}$ in this context. The spectral measure of $\mathcal{L}$ and its discretizations determine the behavior of Krylov subspace methods. See [39, Section 2] for a instructive example for which the spectrum is not enough.

We follow [41] and consider a bounded Lipschitz domain $\Omega \subset \mathbb{R}^2$. We take

$$\mathcal{A}u = -\nabla \cdot [(1 + \exp(-x^2 - y^2))\nabla u], \quad \mathcal{B}u = -\nabla^2 u,$$

both with zero Dirichlet boundary conditions. The spectrum of $\mathcal{L}$ is the interval [41] $\Lambda(\mathcal{L}) = [\inf_{(x,y)\in\Omega} 1 + \exp(-x^2 - y^2), \sup_{(x,y)\in\Omega} 1 + \exp(-x^2 - y^2)]$, but the spectral measure is unknown. To solve the linear systems in Algorithm 2, we use the (hp-adaptive and sparse) ultraspherical spectral element method [42].

We take Ω to be a regular n-gon and set $f = C(\Omega)\mathcal{B}^{-1}g$, where $g(x, y) = x^2 + y^2$ and $C(\Omega)$ are normalization constants so that each μ_f is a probability measure. Figure 4 (left) shows these f and Fig. 4 (right) shows the smoothed spectral measures. The endpoints of the spectrum are shown as vertical dashed lines.

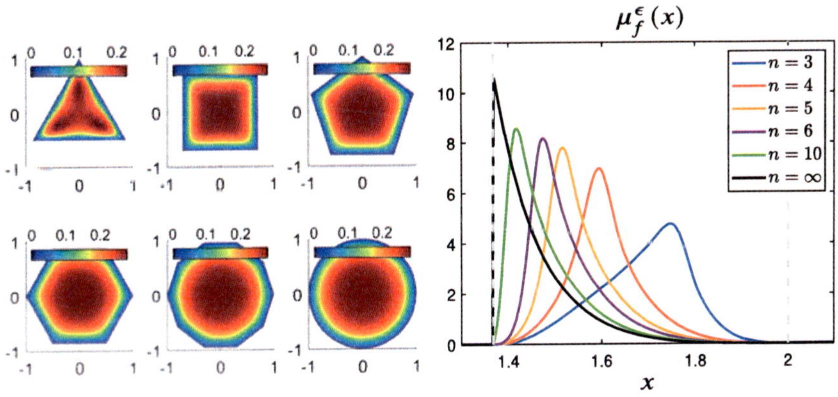

Fig. 4 Left: functions f for different n. Right: smoothed spectral measures, μ_f^ϵ, for different n-gons computed using the 6th order kernel. For $n < \infty$ we use $\epsilon = 0.05$ and for the circle we use $\epsilon = 0.001$. The dashed vertical lines are the endpoints of the spectrum

The measures appear to be absolutely continuous and converge to the corresponding measure for the disk ($n = \infty$) as n gets larger. To deal with the disk, we use separation of variables and solve the resulting radial ODEs using the ultraspherical spectral method [28].

Acknowledgments MJC is supported by a Research Fellowship at Trinity College, Cambridge, and a Fondation Sciences Mathématiques de Paris Postdoctoral Fellowship at École Normale Supérieure. We thank Alex Townsend for pointing out that separation of variables efficiently deals with the $n = \infty$ case in Fig. 4 and for reading a draft version of the article. We thank Zdenek Strakos for discussions on the preconditioner example and for reading a draft version of the article.

References

1. Kallianpur, G., Mandrekar, V.: Spectral theory of stationary H-valued processes. J. Multivar. Anal. **1**, 1–16 (1971)
2. Girardin, V., Senoussi, R.: Semigroup stationary processes and spectral representation. Bernoulli **9**, 857–876 (2003)
3. Rosenblatt, M.: Stochastic curve estimation. In: NSF-CBMS Regional Conference Series in Probability and Statistics, vol. 3 (1991)
4. Efros, V.D., Leidemann, W., Orlandini, G., Barnea, N.: The Lorentz integral transform (LIT) method and its applications to perturbation-induced reactions. J. Phys. G **34**, R459 (2007)
5. Efros, V.D., Leidemann, W., Orlandini, G.: Response functions from integral transforms with a Lorentz kernel. Phys. Lett. B **338**, 130–133 (1994)
6. Haydock, R., Heine, V., Kelly, M. J.: Electronic structure based on the local atomic environment for tight-binding bands. J. Phys. C Solid State Phys. **5**, 2845 (1972)
7. Lin, L., Saad, Y., Yang, C.: Approximating spectral densities of large matrices. SIAM Rev. **58**, 34–65 (2016)
8. Wilkening, J., Cerfon, A.: A spectral transform method for singular Sturm–Liouville problems with applications to energy diffusion in plasma physics. SIAM J. Appl. Math. **75**, 350–392 (2015)
9. Killip, R., Simon, B.: Sum rules for Jacobi matrices and their applications to spectral theory. Ann. Math. **158**, 253–321 (2003)
10. Dombrowski, J., Nevai, P.: Orthogonal polynomials, measures and recurrence relations. SIAM J. Math. Anal. **17**, 752–759 (1986)
11. Trogdon, T., Olver, S., Deconinck, B.: Numerical inverse scattering for the Korteweg–de Vries and modified Korteweg–de Vries equations. Phys. D Nonlinear Phenom. **241**, 1003–1025 (2012)
12. Colbrook, M.J., Horning, A., Townsend, A.: Computing spectral measures of self-adjoint operators. SIAM Rev. **63**, 489–524 (2021)
13. Colbrook, M.J.: Computing spectral measures and spectral types. Commun. Math. Phys. **384**(1), 433–501 (2021)
14. Colbrook, M.J., Hansen, A.C.: The foundations of spectral computations via the Solvability Complexity Index hierarchy. J. Eur. Math. Soc. (2022). https://ems.press/journals/jems/articles/8215362
15. Colbrook, M.J.: On the computation of geometric features of spectra of linear operators on Hilbert spaces. Found. Comput. Math., Springer, 1–82 (2022). https://link.springer.com/article/10.1007/s10208-022-09598-0
16. Ben-Artzi, J., Colbrook, M.J., Hansen, A., Nevanlinna, O., Seidel, M.: Computing Spectra – on the solvability complexity index hierarchy and towers of algorithms (2020). arXiv:1508.03280

17. Johnstone, D., Colbrook, M.J., Nielsen, A., Öhberg, P., Duncan, C.: Bulk localised transport states in infinite and finite quasicrystals via magnetic aperiodicity. Phys. Rev. B. **106**(4), 045149 (2022)
18. Colbrook, M. J., Horning, A.: https://github.com/SpecSolve/SpecSolve (2020)
19. Reed, M., Simon, B.: Methods of Modern Mathematical Physics I, 2nd edn. Academic Press, New York (1980)
20. Billingsley, P.: Convergence of Probability Measures, 2nd edn. Wiley, Hoboken (1999)
21. Stone, M.: Linear Transformations in Hilbert Space, vol. 15. American Mathematical Society Colloquium Publications (1990)
22. Muskhelishvili, N.I., Radok, J.R.M.: Singular Integral Equations: Boundary Problems of Function Theory and Their Application to Mathematical Physics. Noordhoff, Groningen (1953)
23. Cuminato, J.A., Fitt, A.D., McKee, S.: A review of linear and nonlinear Cauchy singular integral and integro-differential equations arising in mechanics. J. Integral Equ. Appl. **19**, 163–207 (2007)
24. Iserles, A., Webb, M.: A family of orthogonal rational functions and other orthogonal systems with a skew-Hermitian differentiation matrix. J. Fourier Anal. Appl. **26**, 1–28 (2020)
25. Slevinsky, R.M., Olver, S.: A fast and well-conditioned spectral method for singular integral equations. J. Comput. Phys. **332**, 290–315 (2017)
26. Koppelman, W.: On the spectral theory of singular integral operators. Trans. Am. Math. Soc. **97**, 35–63 (1960)
27. Colbrook, M.J., Horning, A., Thicke, K., Watson, A.B.: Computing spectral properties of topological insulators without artificial truncation or supercell approximation (2021). arXiv:2112.03942
28. Olver, S., Townsend, A.: A fast and well-conditioned spectral method. SIAM Rev. **55**, 462–489 (2013)
29. Boffi, D.: Finite element approximation of eigenvalue problems. Acta Numer. **19**, 1–120 (2010)
30. Güttel, S., Tisseur, F.: The nonlinear eigenvalue problem. Acta Numer. **26**, 1–94 (2017)
31. Davies, E.B.: Spectral Theory and Differential Operators. Cambridge University Press, Cambridge (1996)
32. Kato, T.: Perturbation Theory for Linear Operators. Springer, New York (1976)
33. Ralston, J.: On stationary modes in inviscid rotating fluid. J. Math. Anal. Appl. **44**, 366–383 (1973)
34. Showalter, R., Ting, T.: Pseudoparabolic partial differential equations. SIAM J. Math. Anal. **1**, 1–26 (1970)
35. Colin De Verdiere, Y., Saint-Raymond, L.: Attractors for two-dimensional waves with homogeneous hamiltonians of degree 0. Commun. Pure Appl. Math. **73**, 421–462 (2020)
36. Galkowski, J., Zworski, M.: Viscosity limits for 0th order pseudodifferential operators. Commun. Pure Appl. Math. **75**(8), 1798–1869 (2022). https://doi.org/10.1002/cpa.22072
37. Málek, J., Strakoš, Z.: Preconditioning and the Conjugate Gradient Method in the Context of Solving PDEs. SIAM, Philadelphia (2014)
38. Mardal, K.-A., Winther, R.: Preconditioning discretizations of systems of partial differential equations. Numer. Linear Algebra Appl. **18**, 1–40 (2011)
39. Gergelits, T., Mardal, K.-A., Nielsen, B., Strakos, Z.: Laplacian preconditioning of elliptic PDEs: localization of the eigenvalues of the discretized operator. SIAM J. Numer. Anal. **57**, 1369–1394 (2019)
40. Gergelits, T., Nielsen, B., Strakos, Z.: Generalized spectrum of second order differential operators. SIAM J. Numer. Anal. **58**, 2193–2211 (2020)
41. Gergelits, T., Nielsen, B., Strakoš, Z.: Numerical approximation of the spectrum of self-adjoint operators and operator preconditioning (2021). arXiv:2103.00849
42. Fortunato, D., Hale, N., Townsend, A.: The ultraspherical spectral element method. J. Comput. Phys. **436**, 110087 (2021)

Assessment of a Wall Distance Free Transition Model Based on the Laminar Kinetic Energy in a Discontinuous Galerkin Solver

Alessandro Colombo, Antonio Ghidoni, and Gianmaria Noventa

1 Introduction

In the near future, the predicting capabilities of standard industrial CFD solvers will still rely on Reynolds-Averaged Navier-Stokes (RANS) equations [5, 12], for the excessive computational resources required by high-fidelity simulations, e.g., Direct Numerical simulations (DNS) and Large Eddy Simulations (LES) [2–4]. Where, the greater accuracy and geometrical flexibility guaranteed by discontinuous Galerkin (dG) methods in solving RANS equations could represent an appealing solution to enhance the predicting capabilities of standard industrial CFD codes, without resorting to intensive computational approaches, such as DNS and LES.

In this context, numerical models able to accurately predict transitional flows are mandatory to overcome the limits of turbulence models. In fact, many reliable and robust turbulence models are nowadays available for RANS equations to accurately simulate a wide range of engineering flows. However, turbulence models are not able to correctly predict flow phenomena with low to moderate Reynolds numbers, which are characterized by strong transitions. Laminar-turbulent transition is common in aerospace, turbomachinery, maritime, automotive, and cooling applications. As a consequence, numerical models able to accurately predict transitional flows are

A. Colombo
Department of Engineering and Applied Sciences, Università degli Studi di Bergamo, Dalmine, Italy
e-mail: alessandro.colombo@unibg.it

A. Ghidoni · G. Noventa (✉)
Department of Mechanical and Industrial Engineering, Università degli Studi di Brescia, Brescia, Italy
e-mail: antonio.ghidoni@unibs.it; gianmaria.noventa@unibs.it

J. M. Melenk et al. (eds.), *Spectral and High Order Methods for Partial Differential Equations ICOSAHOM 2020+1*, Lecture Notes in Computational Science and Engineering 137, https://doi.org/10.1007/978-3-031-20432-6_11

mandatory to overcome the limits of turbulence models for the efficient design of many industrial applications.

Literature classifies transition models into low-Reynolds, non-local and local models. The non-local models are based on correlations, which relate the momentum thickness Reynolds number to local free-stream conditions, such as turbulence intensity and pressure gradient, where several correlations have been developed for different transition mechanisms. The main drawback is due to the fact that the non-local formulation needs the information on the integral thickness of the boundary layer and the state of the flow at the edge of the boundary layer. While the local models are based on transport equations for turbulent or transitional variable, similarly to the classical turbulence model, and require only local variables. The local formulation guarantees better robustness, accuracy and easiness of implementation in modern solvers. These models can be divided into empirical correlation models, proposed by Menter and Langtry [8, 11], and phenomenological transition models, proposed by Walters and Leylek [13, 14].

The phenomenological models aim at incorporating the physics of boundary layer transition, and for this reasons is an attractive but challenging research field. Many authors in the last decade proposed different versions of the three-equation phenomenological transition model based on the concept of the laminar kinetic energy (LKE) proposed by Mayle [10]. These models have been proposed in the finite volume context to predict the laminar-turbulent transition, but the increasing required level of resolution naturally leads to consider discretization methods with a higher order of accuracy, such as dG methods. In particular, the use of a dG solver for the implementation of transition models can guarantee greater accuracy also on stretched, and, eventually, bad quality meshes, and relax the constraint on the non-dimensional height $y^+ < 1$ of the first element near the wall, resorting to a higher polynomial representation of the solution.

In dG methods the solution of the weak or variational form of a partial differential problem is approximated by polynomial functions over the elements, similarly to the classical continuous finite element method (FEM). However, unlike continuous finite element methods, dG methods use an approximation that is in general discontinuous at the element interfaces. The coupling of the approximate solutions between neighbouring elements is (weakly) enforced by interface, or numerical, fluxes. An appropriate definition of numerical flux guarantees the consistency and stability of the dG numerical approximation. The main drawback of this higher accuracy is the increased computational cost compared to standard finite volume methods (FVM), but the compact stencil of dG spatial approximation, involving only one element and its neighbours, makes the methods very well suited for massively parallel computer platforms. Furthermore, the computational efficiency of dG methods can be substantially improved by resorting to multilevel solution approaches, such as the p-multigrid algorithms [6].

Almost all the transition models rely on the computation of the wall distance to define some critical terms of the model that allow to correctly predict transition, i.e., the production, destruction and transfer terms of the local turbulent transport variables. The estimation of the distance can be critical in the dG context for the high-order representation of the boundaries, which can become very expensive and high-memory consuming.

To overcome these issues, a wall distance free version of the k_L-k_T-$\tilde{\omega}$ model [1, 9] is proposed and implemented in a high-order accurate dG solver. Lorini et al. in [1, 9] proposed the local and phenomenological k_L-k_T-$\tilde{\omega}$ transition model, where k_L and k_T are the laminar and turbulent kinetic energy. Numerical results on different benchmark cases, from flat plates to turbine cascade with different inflow conditions and transition modes, demonstrate the higher accuracy of the model. This model is strongly based on the wall distance, and the anisotropic part of the dissipation terms of the original formulation, proposed by Walters and Leylek [13, 14], is difficult to implement in a high-order solver. Moreover, the transition model is implemented by limiting the values of k_T and k_L, and by using $\tilde{\omega} = \log \omega$, as suggested in [1], to guarantee the positiveness and smoothness of the turbulent working variables.

This work can be considered as the first attempt to write a wall distance-free transition model, as no other examples are available in literature. All terms are defined in terms of local turbulent variables, e.g, the laminar k_L and turbulent k_T kinetic energy, the specific dissipation rate $\tilde{\omega}$ and some combinations of the velocity derivatives. Different corrections are introduced in the model to adjust the wall distance free terms both in a laminar, transitional and turbulent boundary layer. The prediction capabilities of this model are assessed by computing the flow over flat plates of the ERCOFTAC T3 series with zero pressure gradients and different values of the freestream turbulence intensity [7].

2 Model

The transitional model presented in this section is based on the work of Lorini et al. [1, 9], where the RANS system is closed with three additional equations for the turbulent kinetic energy k_T, the laminar kinetic energy k_L, and the specific dissipation rate $\omega = \epsilon / k$, where ϵ is the dissipation. The complete set of the wall distance free RANS and k_L-k_T-$\tilde{\omega}$ model equations for a compressible flow can be written as

$$
\begin{aligned}
\frac{\partial \rho}{\partial t} + \frac{\partial}{\partial x_j}(\rho u_j) &= 0 \\
\frac{\partial}{\partial t}(\rho u_i) + \frac{\partial}{\partial x_j}(\rho u_j u_i) &= -\frac{\partial p}{\partial x_i} + \frac{\partial \widehat{\tau}_{ji}}{\partial x_j}, \\
\frac{\partial}{\partial t}(\rho u_i) + \frac{\partial}{\partial x_j}(\rho u_j u_i) &= -\frac{\partial p}{\partial x_i} + \frac{\partial \widehat{\tau}_{ji}}{\partial x_j}, \\
\frac{\partial}{\partial t}(\rho E) + \frac{\partial}{\partial x_j}(\rho u_j H) &= \frac{\partial}{\partial x_j}\left[u_i \widehat{\tau}_{ij} - \widehat{q}_j\right] \\
&\quad - \rho\left(P_{k_T} - D_{k_T}\right) - \rho\left(P_{k_L} - D_{k_L}\right),
\end{aligned}
$$

$$\frac{\partial}{\partial t}(\rho k_T) + \frac{\partial}{\partial x_j}(\rho u_j k_T) = \frac{\partial}{\partial x_j}\left[\left(\mu + \frac{\rho \alpha_T}{\sigma_k}\right)\frac{\partial k_T}{\partial x_j}\right] + \rho\left(P_{k_T} + R_{BP} + R_{NAT} - D_{k_T}\right),$$

$$\frac{\partial}{\partial t}(\rho k_L) + \frac{\partial}{\partial x_j}(\rho u_j k_L) = \frac{\partial}{\partial x_j}\left[\mu \frac{\partial k_L}{\partial x_j}\right] + \rho\left(P_{k_L} - R_{BP} - R_{NAT} - D_{k_L}\right),$$

$$\frac{\partial}{\partial t}(\rho \tilde{\omega}) + \frac{\partial}{\partial x_j}(\rho u_j \tilde{\omega}) = \frac{\partial}{\partial x_j}\left[\left(\mu + \frac{\rho \alpha_T}{\sigma_\omega}\right)\frac{\partial \tilde{\omega}}{\partial x_j}\right] + \rho\left(P_{\tilde{\omega}} - D_{\tilde{\omega}}\right) + \left(\mu + \frac{\rho \alpha_T}{\sigma_\omega}\right)\frac{\partial \tilde{\omega}}{\partial x_k}\frac{\partial \tilde{\omega}}{\partial x_k},$$

where u_i is the absolute velocity, μ the molecular viscosity, α_T an effective diffusivity, $\hat{\tau}_{ij}$ the total stress tensor, and σ_k and σ_ω are model constants.

The total energy, E, the total enthalpy, H, the pressure, p, and the heat flux vector, $\hat{q}$ are given by

$$E = \hat{e} + u_k u_k/2, \qquad H = h + u_k u_k/2,$$

$$p = (\gamma - 1)\rho\left(E - u_k u_k/2\right), \qquad \widehat{q}_j = -\left(\frac{\mu}{\mathrm{Pr}} + \rho \alpha_\theta\right)\frac{\partial h}{\partial x_j},$$

where $\hat{e}$ is the internal energy, h the enthalpy, γ the ratio of gas specific heats, α_θ the turbulent thermal diffusivity, and Pr the molecular Prandtl number. The transition model is implemented by limiting the values of k_T and k_L, i.e., $\overline{k_T} = \max(0, k_T)$ and $\overline{k_L} = \max(0, k_L)$ and by using the logarithm of ω instead of ω itself, i.e., $\tilde{\omega} = \log(\omega)$ [1]. These choices guarantee the positiveness and smoothness of the turbulent working variables increasing the robustness of the convergence solution, especially for high-order solvers [1].

P_{k_T}, P_{k_L} and $P_{\tilde{\omega}}$ represent the production terms of the model, while D_{k_T}, D_{k_L} and $D_{\tilde{\omega}}$ are the destruction terms. The transition process itself is represented by a transfer of energy from the laminar kinetic energy k_L to the turbulent kinetic energy k_T, through the terms R_{BP} and R_{NAT}, respectively due to the bypass and natural transition mode. The model constants are tabulated in Table 1, where in comparison with the original version [1, 9]only the constant c_λ is different.

This model takes into account two different contributions to the total kinetic energy (TKE or k_{TOT}), i.e. the turbulent, k_T, and the laminar, k_L, kinetic energy, and two different contributions to the turbulent kinetic energy, the small-scale energy, k_s, and large-scale energy, k_l. The first contribution interacts with the mean-flow as a typical turbulence energy, whereas the second only contributes to the production of k_L. The production of the turbulent kinetic energy due to the small-scale turbulent fluctuations and of the laminar kinetic energy due to the large-scale turbulent

Table 1 Model constants

$A_0 = 2.04$	$A_S = 2.12$	$A_\nu = 5.20$
$A_{BP} = 0.2$	$A_{NAT} = 1\,000$	$A_{TS} = 200$
$C_{BP,crit} = 1.75$	$C_{NC} = 0.1$	$C_{NAT,crit} = 9\,969.04$
$C_{INT} = 0.95$	$C_{TS,crit} = 6\,225.025$	$C_{R,NAT} = 0.03$
$C_{l1} = 3.4 \times 10^{-6}$	$C_{l2} = 2.58 \times 10^{-12}$	$C_R = 0.32$
$C_{\alpha,\theta} = 0.030$	$C_{SS} = 1.7$	$C_{\tau,l} = 4\,360$
$C_{\omega 1} = 0.44$	$C_{\omega 2} = 0.92$	$C_{\omega 3} = 0.3$
$C_{\omega R} = 1.5$	$C_\lambda = 2.145$	$C_{\mu,std} = 0.09$
$Pr_T = 0.85$	$\sigma_k = 1$	$\sigma_\omega = 1.17$

fluctuations are modelled as

$$P_{k_T} = \tau_{ij,s}\frac{\partial u_i}{\partial x_j}, \qquad P_{k_L} = \tau_{ij,l}\frac{\partial u_i}{\partial x_j} \tag{1}$$

where $\tau_{ij,s}$ and $\tau_{ij,l}$ are the small- and large-scale components of the turbulent stress tensor, $\tau_{ij} = \tau_{ij,s} + \tau_{ij,l}$, and are defined as

$$\tau_{ij,s} = 2\nu_{T,s}\left(S_{ij} - \frac{1}{3}\frac{\partial u_k}{\partial x_k}\delta_{ij}\right) - \frac{2}{3}\rho k_s \delta_{ij}, \tag{2}$$

$$\tau_{ij,l} = 2\nu_{T,l}\left(S_{ij} - \frac{1}{3}\frac{\partial u_k}{\partial x_k}\delta_{ij}\right) - \frac{2}{3}\rho k_l \delta_{ij}, \tag{3}$$

and S_{ij} is the mean strain-rate tensor, $\nu_{T,s}$ and $\nu_{T,l}$ are the small-scale and large-scale components of the eddy-viscosity, respectively, and δ_{ij} the Kronecker symbol. The small-scale eddy-viscosity is defined as

$$\nu_{T,s} = f_\nu f_{INT} C_\mu \sqrt{k_s}\lambda_{eff}, \tag{4}$$

where $k_s = f_{SS} f_W k_T$. The kinematic wall effect is included through an effective wall-limited turbulence length scale λ_{eff} and a damping function f_W defined as

$$\lambda_{eff} - \frac{\sqrt{k_L + k_T}}{e^{\omega}} \min\left(C_\lambda \sqrt{\frac{e^{\tilde{\omega}}}{S}}, 1\right), \qquad f_W = \left(\frac{\lambda_{eff}}{\lambda_{TOT}}\right)^{\frac{2}{3}}, \tag{5}$$

where the lengthscale $\lambda_T = \sqrt{k_T}/e^{\tilde{\omega}}$ due only to turbulent kinetic energy is here substitute with the total lengthscale $\lambda_{TOT} = \sqrt{k_L + k_T}/e^{\tilde{\omega}} = \sqrt{k_{TOT}}/e^{\tilde{\omega}}$. The viscous wall effect and the shear-sheltering effect are incorporated through the damping functions f_ν and f_{SS}, respectively, i.e.

$$f_\nu = 1 - \exp\left(-\frac{\sqrt{Re_T}}{A_\nu}\right), \quad f_{SS} = \exp\left[-\left(\frac{C_{SS}\nu\Omega}{k_T}\right)^2\right] = \exp\left[-\left(\frac{C_{SS}}{Re_\Omega}\right)^2\right],$$

where $Re_T = f_W^2 k_T/\nu e^{\tilde{\omega}}$ is the effective turbulence Reynolds number, $\Omega = \sqrt{2\Omega_{ij}\Omega_{ij}}$ is the vorticity magnitude, and Re_Ω the vorticity Reynolds number. In the original model, alternatively instead of Re_Ω can be used the wall-distance Reynolds number $Re_d = \sqrt{k_T}d/\nu$ adjusting the model constant C_{SS}. Here the vorticity Reynolds number is considered in order to reduce the influence of the wall distance on the model.

To satisfy the realizability constraint the turbulent viscosity coefficient, $C_\mu = 1/(A_0 + A_S S/\omega)$, is added where A_0 and A_S are model constants. Intermittency effects on the production of turbulence are included through the intermittency damping function $f_{INT} = \min(k/(C_{INT}k_{TOT}), 1)$.

The production of the laminar kinetic energy, k_L, is assumed to be given by the correlation of the pre-transitional fluctuation growth with free-stream low-frequency wall-normal turbulent fluctuations. The large-scale turbulence contribution is defined as $k_l = k - k_s$, and the large-scale eddy-viscosity is defined as

$$\nu_{T,l} = \min\left\{f_{\tau l}C_{l1}Re_{\Omega,eff}\sqrt{k_l}\lambda_{eff} + \beta_{TS}C_{l2}Re_{\Omega,eff}\lambda_{eff}^2\Omega, \frac{k_L + k_l}{2S}\right\}, \tag{6}$$

where

$$\beta_{TS} = 1 - \exp\left(\frac{-\max\left(Re_{\Omega,eff} - C_{TS,crit}, 0\right)^2}{A_{TS}}\right), \qquad f_{\tau,l} = 1 - \exp\left(-C_{\tau,l}\frac{k_l}{\lambda_{eff}^2\Omega^2}\right),$$

and $Re_{\Omega,eff} = \Omega\lambda_{eff}^2/\nu$ is the Reynolds number based on the vorticity and the effective wall-limited turbulence length scale. The large-scale component of the eddy-viscosity, $\nu_{T,l}$, is made of two contributions, respectively due to the development of the Klebanoff modes and the self-excited, or natural, modes.

The dissipation terms for the laminar and turbulent kinetic energy equation are

$$D_{k_T} = \overline{k_T} e^{\tilde{\omega}} + 2\nu\frac{k}{d_{eff}^2}, \qquad D_{k_L} = 2\nu\frac{k_L}{d_{eff}^2} \tag{7}$$

where $d_{eff} = \lambda_{eff}/C_\lambda$ is an effective distance from the wall used to correctly predict fully turbulent free shear flows [13]. The original terms included by Walters et al. [13], i.e., $\nu\partial\sqrt{k_T}/\partial x_j\partial\sqrt{k_T}/\partial x_j$ and $\nu\partial\sqrt{k_L}/\partial x_j\partial\sqrt{k_L}/\partial x_j$, are here defined according to the low-Reynolds Chien $k-\epsilon$ model [5], with the effective wall distance d_{eff} instead the wall distance, to increase the robustness of the convergence solution, especially for a high-order solver. The anisotropic terms are used to model the increased viscous dissipation in the sub-layer.

The transport equations for the turbulent and laminar kinetic energy are coupled by the transfer terms R_{BP} and R_{NAT}, which represent the break-down process of the pre-transitional fluctuations.These terms, where active ($\beta_{BP} > 0$ and/or $\beta_{NAT} > 0$), become a sink for the k_L transport equation and a source for the k equation. R_{BP} and R_{NAT} account for the bypass and natural transition mechanism, respectively, and are defined as

$$R_{BP} = C_R \beta_{BP} k_L \Omega / f_W, \qquad R_{NAT} = C_{R,NAT} \beta_{NAT} k_L \Omega, \tag{8}$$

where

$$\beta_{BP} = 1 - \exp\left(-\frac{\phi_{BP}}{A_{BP}}\right), \qquad \phi_{BP} = \max\left[\left(Re_\Omega - C_{BP,crit}\right), 0\right], \tag{9}$$

$$\beta_{NAT} = 1 - \exp\left(-\frac{\phi_{NAT}}{A_{NAT}}\right), \qquad \phi_{NAT} = \max\left(Re_\Omega - \frac{C_{NAT,crit}}{f_{NAT,crit}}, 0\right), \tag{10}$$

$$f_{NAT,crit} = 1 - \exp\left(-C_{NC}\frac{\sqrt{k_L} d_{eff}}{\nu}\right). \tag{11}$$

Similarly to the shear-sheltering damping function the wall distance Reynolds number Re_d can be used instead the vorticy Reynolds number Re_Ω. Where, as good practice rule, the same Reynolds number should be use in the shear-sheltering damping function and in the threshold functions. The transition process in both the cases is assumed to start when the characteristic time-scale for turbulence production is smaller than the viscous diffusion time-scale of the pre-transitional fluctuations.

The production and dissipation terms for $\tilde{\omega}$ is defined as

$$P_{\tilde{\omega}} = \frac{e^{\tilde{\omega}}}{k_T} P_k + \left(\frac{C_{\tilde{\omega}R}}{f_W} - 1\right)\frac{e^{\tilde{\omega}}}{k_T} R_{BP}, \qquad D_{e^{\tilde{\omega}}} = f_2 e^{\tilde{\omega}} e^{\tilde{\omega}}, \qquad f_2 = C_{\omega 2} f_W^2, \tag{12}$$

where the first term represents an increment of the dissipation rate due to turbulence production, while the second term a reduction in the turbulence length scale during the transition breakdown process. In particular this term is active only for the bypass transition mode differently from the model of Walters et al. [13]

The turbulent transport terms in the k and $\tilde{\omega}$ equations include a turbulent effective diffusivity, $\alpha_T = f_\nu C_{\mu,std}\sqrt{k_s}\lambda_{eff}$. The influence of turbulent and laminar fluctuations on the mean-flow and energy equations is accounted for by defining the turbulent stress tensor as

$$\tau_{ij} = \tau_{ij,l} + \tau_{ij,s} = 2\nu_T\left(S_{ij} - \frac{1}{3}\frac{\partial u_k}{\partial x_k}\delta_{ij}\right) - \frac{2}{3}\rho k_T \delta_{ij}, \tag{13}$$

where $\nu_T = \nu_{T,s} + \nu_{T,l}$ is the turbulent viscosity. When including heat transfer effects, the turbulent thermal diffusivity, α_θ, is modelled as

$$\alpha_\theta = \frac{\nu_{T,s}}{Pr_T} + C_{\alpha,\theta}\sqrt{k_L}\lambda_{eff}, \tag{14}$$

where Pr_T is the turbulent Prandtl number. The thermal diffusivity equation was rewritten in the original version of this model [1, 9], in order to simplify the formulation and to improve the prediction capabilities of the model for the thermal energy transport in a pre-transitional and fully turbulent boundary layer.. The homogeneous Neumann condition is imposed at the solid walls for the specific dissipation rate $\partial\omega/\partial n = 0$, as suggested in [13]. While both the laminar and the turbulent kinetic energy are imposed zero at the walls due to the no-slip condition.

At the inflow or freestream boundary the laminar kinetic energy is set to zero, while the turbulent kinetic energy and the specific dissipation rate values are computed according to the definition of turbulence intensity and eddy viscosity ratio. A deeply explanation of the boundary conditions can be found in [1, 9].

3 Numerical Framework

The governing equations for m variables in $d = 2, 3$ dimensions can be written in compact form as

$$\mathbf{P}(\mathbf{w})\frac{\partial \mathbf{w}}{\partial t} + \nabla \cdot \mathbf{F}_c(\mathbf{w}) + \nabla \cdot \mathbf{F}_v(\mathbf{w}, \nabla\mathbf{w}) + \mathbf{s}(\mathbf{w}, \nabla\mathbf{w}) = \mathbf{0}, \tag{15}$$

where $\mathbf{w} = [p, u_i, T, k_T, k_L, \tilde{\omega}]^T \in \mathbb{R}^m$ is the unknown solution vector, $\mathbf{F}_c, \mathbf{F}_v \in \mathbb{R}^m \otimes \mathbb{R}^d$ are the convective and viscous flux functions, $\mathbf{s} \in \mathbb{R}^m$ is the vector of source terms, and $\mathbf{P}(\mathbf{w}) \in \mathbb{R}^m \otimes \mathbb{R}^m$ is the transformation matrix arising from the use of the primitive variables.

By assembling together all the elemental contributions of the discretized equations, a system of ordinary differential equations governing the evolution in time of the discrete solution is obtained

$$\mathbf{M_P}(\mathbf{W})\frac{\mathrm{d}\mathbf{W}}{\mathrm{d}t} + \mathbf{R}(\mathbf{W}) = \mathbf{0}, \tag{16}$$

where $\mathbf{R}(\mathbf{W})$ is the vector of residuals and $\mathbf{M_P}(\mathbf{W})$ is the global block diagonal matrix. The Eq. (16) is instead discretized in time by means of a linearized backward Euler (LBE) scheme with a pseudo-transient continuation strategy [1]. The linear system arising from the spatial and temporal discretization is solved at each time step by means of GMRES algorithm, where linear algebra and parallelization are handled through PETSc library.

4 Results

The proposed wall distance free version of the k_L-k_T-$\tilde{\omega}$ model is here validated by computing the transitional flow over the flat plates with zero pressure gradients of the ERCOFTAC T3 series [7]. These test cases are mainly characterized by bypass transition, due to the levels of freestream turbulence intensity, i.e., $Tu \geq 1\%$. All the computations are carried out in parallel by initializing the piecewise constant $\mathbb{P}^0$ solution from the uniform flow at the inflow conditions and the higher-order solutions from the lower-order ones. The maximum order of the polynomial solution approximation is $\mathbb{P}^4$.

Three different cases are considered, namely the T3A, T3B and T3A- flat plates, which are characterized by different values of the velocity and turbulence intensity at the leading edge. The inlet turbulent quantities are set in order to match the turbulence intensity at the plate leading edge and the correct decay of the turbulent kinetic energy along the plate. The computational mesh is made of 8 800 quadrilateral elements with linear edges, and is used for all the testcases. The results of the wall distance free model here proposed are compared with the available experimental data of Coupland [7], the numerical solution of Langtry et al. [8, 11] and the numerical solution of the original version of the model [1, 9].

Figure 1 shows the skin friction coefficient, c_f, distributions on the plate for each case and with different solution approximation. All these distribution are in good agreement with the experimental data [7] starting from $\mathbb{P}^2$ or $\mathbb{P}^3$ solution approximation, showing a slight under-prediction of the c_f in the turbulent region. The results show a lower accuracy in the prediction of the transition flow region compared to the original model [1, 9], but however an higher accuracy compared to the γ-Re_θ SST turbulence model [8, 11]. Figures 2 and 3 shows the laminar and turbulent kinetic energy contours, the production term of laminar and turbulent kinetic energy contours and the by-pass source term contour near the transition region of the flow $0.32 < x/L < 0.47$ in order to understand the behaviour of the model. In order to compare the two versions of the k_L-k_T-$\tilde{\omega}$ model Fig. 4 shows also convergence history on the T3A case of the L^2 norm of the density residuals, in terms of the iterations, with different order of the solution, from $\mathbb{P}^0$ to $\mathbb{P}^4$ where the higher-order solutions are initialized from the lower-order one, and the same parameters of the convergence algorithm. The wall distance free model needs a lower number of iterations to convergence, however the original version guarantees an higher robustness with the capability to convergence also with heavy parameters of the convergence algorithm.

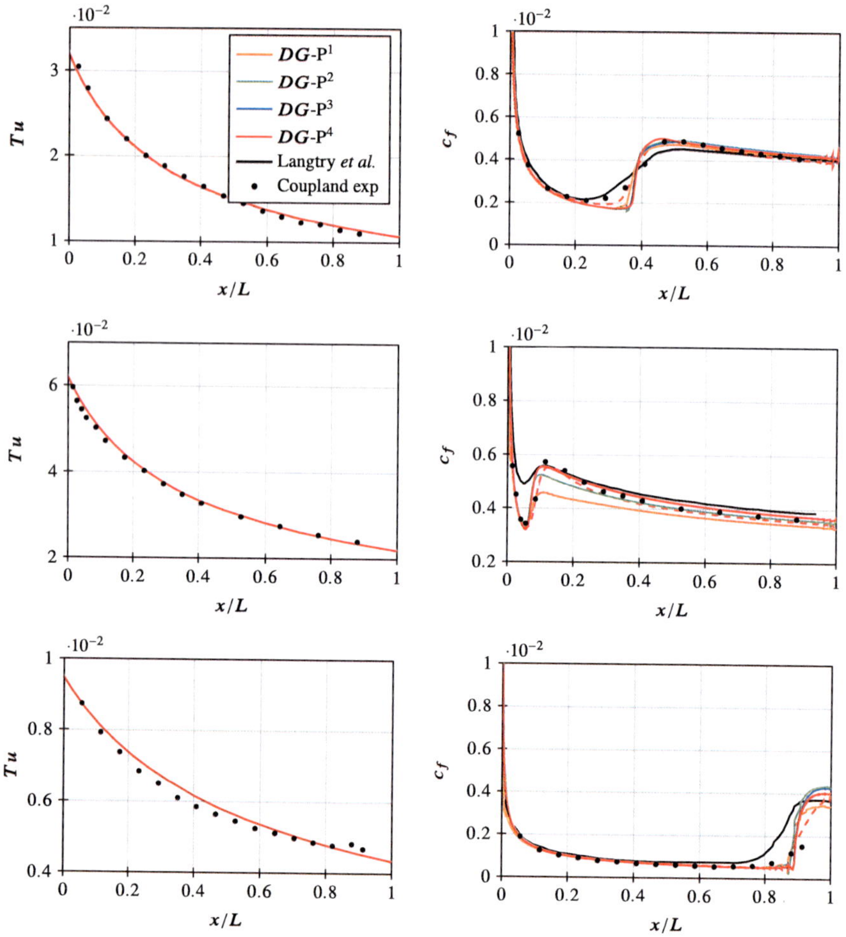

Fig. 1 Zero pressure gradient flat plates. Turbulent intensity Tu along the domain and skin friction coefficient c_f along the plate for the zero pressure gradient flat plate, T3A (top), T3B and T3A- (bottom) respectively, from $\mathbb{P}^1$ to $\mathbb{P}^4$. The dashed line represents the solution with the original version, i.e., the wall distance based version, of the model with $\mathbb{P}^4$ solution

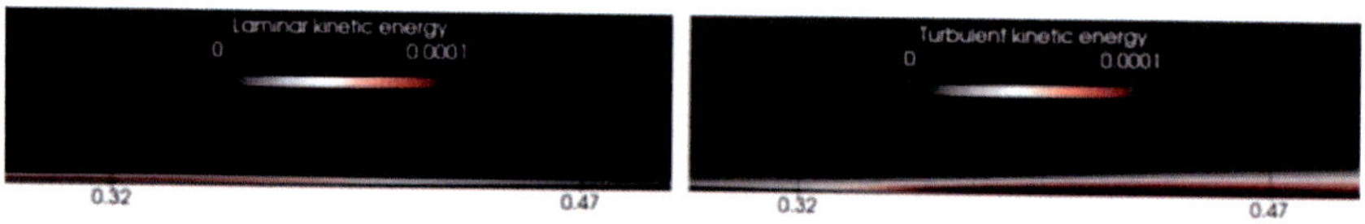

Fig. 2 T3A. Laminar k_L (left) and turbulent k_T (right) kinetic energy contours near the transition region of the flow $0.32 < x/L < 0.47$

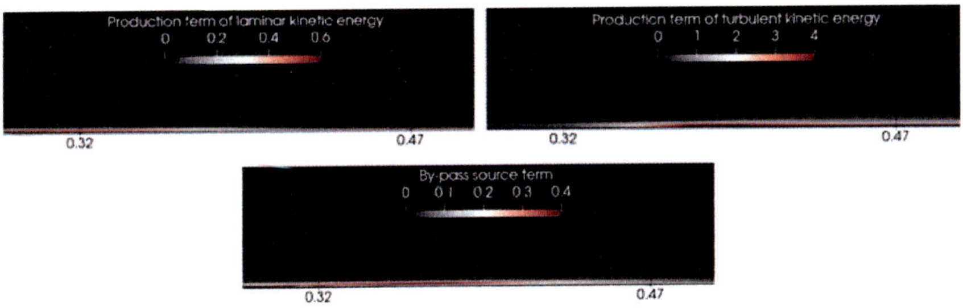

Fig. 3 T3A. Production term of laminar k_L (top, left) and turbulent k_T (top, right) kinetic energy contours and By-pass source term R_{BP} contour near the transition region of the flow $0.32 < x/L < 0.47$

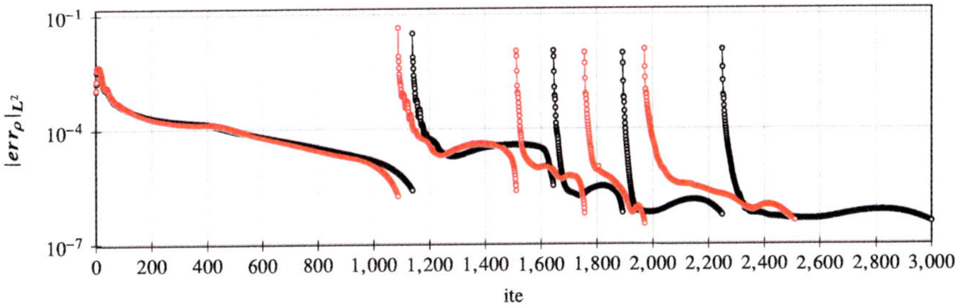

Fig. 4 T3A. The convergence history of the L^2 norm of the density residuals, in terms of the iterations, with different order of the solution, from $\mathbb{P}^0$ to $\mathbb{P}^4$ where the higher-order solutions are initialized from the lower-order one, and the same parameters of the convergence algorithm. The gray lines represent the original version, i.e. the wall distance based version, of the model

5 Conclusions

The implementation of a wall distance free version of the k_L-k_T-$\tilde{\omega}$ transition model [1, 9] based on the LKE concept into a high-order accurate dG code is here presented. The work represents one of the first attempts to write a wall distance free version of a transition model, furthermore in a high-order numerical context.

The validation of the model is carried out by computing the transitional flow over flat plates with zero pressure gradient and different boundary conditions. The higher accuracy of this transition model is confirmed also here with the wall distance free version, in fact the numerical results are compared with the reference data from the literature, numerical and experimental, and the numerical results of the original version of the model. Despite this the wall distance free version is not so robustness and is more dependent to the solution approximation, or in general spatial accuracy, in comparison with the original version [1, 9].

The work is in progress to solve the equations with a p-multigrid approach to substantially improve the computational efficiency by resorting to multilevel solutions [6]. p-multigrid algorithms represent an additional motivation to implement numerical models in high-order solvers, where their use is mandatory to overcome the main drawback of the high-order solvers if compared to standard FVM, i.e., the higher computational cost.

References

1. Bassi, F., Colombo, A., Ghidoni, A., Lorini, M., Noventa, G.: Discontinuous Galerkin solution of the Reynolds-Averaged Navier-Stokes and $k_L - k_T - \tilde{\omega}$ transition model equations. In: ECCOMAS Congress 2016 - Proceedings of the 7th European Congress on Computational Methods in Applied Sciences and Engineering, vol. 2, pp. 2632–2647 (2016). https://doi.org/10.7712/100016.1986.6971
2. Bassi, F., Colombo, A., Crivellini, A., Fidkowski, K.J., Franciolini, M., Ghidoni, A., Noventa, G.: An entropy-adjoint p-adaptive discontinuous Galerkin method for the under-resolved simulation of turbulent flows. AIAA Aviation 2019 Forum, pp. 1–17 (2019). https://doi.org/10.2514/6.2019-3418
3. Bassi, F., Botti, L., Colombo, A., Crivellini, A., Franciolini, M., Ghidoni, A., Noventa, G.: A p-adaptive matrix-free discontinuous Galerkin method for the implicit LES of incompressible transitional flows. Flow Turbulence Combust. **105**(2), 437–470 (2020). https://doi.org/10.1007/s10494-020-00178-2
4. Bassi, F., Colombo, A., Crivellini, A., Fidkowski, K.J., Franciolini, M., Ghidoni, A., Noventa, G.: Entropy-adjoint p-adaptive discontinuous Galerkin method for the under-resolved simulation of turbulent flows. AIAA J. **58**(9), 3963 – 3977 (2020). https://doi.org/10.2514/1.J058847
5. Chien, K.Y.: Predictions of channel and boundary-layer flows with a low-Reynolds-number turbulence model. AIAA J. **20**(1), 33–38 (1982). https://doi.org/10.2514/3.51043
6. Colombo, A., Ghidoni, A., Noventa, G., Rebay, S.: p-multigrid high-order discontinuous Galerkin solution of compressible flows. In: Efficient High-Order Discretizations for Computational Fluid Dynamics, vol. 602 (2021). https://doi.org/10.1007/978-3-030-60610-7_4
7. Coupland, J.: ERCOFTAC special interest group on laminar to turbulent transition and retransition: T3A and T3B test cases. Technical Report (1990)
8. Langtry, R.B., Menter, F.R., Likki, S.R., Suzen, Y.B., Huang, P.G., Völker, S.: A correlation-based transition model using local variables - part II: test cases and industrial applications. J. Turbomach. **128**(3), 423–434 (2004). https://doi.org/10.1115/1.2184353
9. Lorini, M., Bassi, F., Colombo, A., Ghidoni, A., Noventa, G.: Discontinuous Galerkin solution of the RANS and $k_L - k_T - \tilde{\omega}$ equations for natural and bypass transition. Comput. Fluids **214**, 104767 (2021). https://doi.org/10.1016/j.compfluid.2020.104767
10. Mayle, R.E.: The 1991 IGTI scholar lecture: the role of laminar-turbulent transition in gas turbine engines. J. Turbomach. **113**(4), 509–536 (1991)
11. Menter, F.R., Langtry, R.B., Likki, S.R., Suzen, Y.B., Huang, P.G., Völker, S.: A correlation-based transition model using local variables - part I: model formulation. J. Turbomach. **128**(3), 413–422 (2004). https://doi.org/10.1115/1.2184352
12. Noventa, G., Massa, F., Rebay, S., Bassi, F., Ghidoni, A.: Robustness and efficiency of an implicit time-adaptive discontinuous Galerkin solver for unsteady flows. Comput. Fluids **204**, 104529 (2020). https://doi.org/10.1016/j.compfluid.2020.104529
13. Walters, D.K., Cokljat, D.: A three-equation eddy-viscosity model for Reynolds-Averaged Navier-Stokes simulations of transitional flow. J. Fluids Eng. **130**(12), 1–14 (2008). https://doi.org/10.1115/1.2979230
14. Walters, D.K., Leylek, J.H.: A new model for boundary layer transition using a single-point RANS approach. J. Turbomach. **126**(1), 193–202 (2004). https://doi.org/10.1115/1.1622709

A Posteriori Error Estimate and Mesh Adaptation for the Numerical Solution of the Richards Equation

Vít Dolejší and Hyun-Geun Shin

1 Introduction

We deal with the numerical solution of the Richards equation that describes the motion of unsaturated flows in porous media, which have many applications in hydrology, soluble and nonsoluble contaminant transport studies, design of dangerous waste storage facilities, and others. The Richards equation is a quasilinear degenerate parabolic equation whose accurate and efficient numerical computation is a challenging task namely due to the strong nonlinearities of the constitute relations and the possible degeneracy.

In the last decades, various numerical methods have been proposed for the solution of degenerate parabolic problems, e.g., the conforming finite element methods developed [24], the mixed finite element methods [3, 27, 30], the finite volume based technique [16, 17] and the discontinuous Galerkin method [4].

In [11, 12], we developed and tested the discretization of the Richards equation by the *space-time discontinuous Galerkin* (STDG) method which is based on a discontinuous piecewise polynomial approximation with respect to the space and time coordinates. The advantage of the STDG method is a natural treatment of different meshes on different time levels, which is a favourable property for the mesh adaptive solution of time-dependent PDEs. The high-order polynomial approximations in combination with the anisotropic mesh adaptation showed a potential of the proposed approach. However, the presented mesh adaptation technique was based on a heuristic approach, namely the residual based error estimators from [9] which do not provide a guaranteed bound of the error.

V. Dolejší (✉) · H.-G. Shin
Charles University, Faculty of Mathematics and Physics, Prague, Czech Republic
e-mail: dolejsi@karlin.mff.cuni.cz

J. M. Melenk et al. (eds.), *Spectral and High Order Methods for Partial Differential Equations ICOSAHOM 2020+1*, Lecture Notes in Computational Science and Engineering 137, https://doi.org/10.1007/978-3-031-20432-6_12

There are number of results devoted to a posteriori error estimates of nonlinear (and possible degenerated) parabolic problems, we cite only few of them, e.g., [8, 19, 21, 25, 29] and quite recently [23]. However, the majority of the papers deals with the low (first or second) order time discretization only. In [13], we derived guaranteed a posteriori error estimates for the nonlinear convection-diffusion equation discretized by STDG method of arbitrary orders with respect to the space and time. This technique employs a special construction of the dual norm of residuals from [8] and the flux reconstruction with respect to space from [5, 15, 22]. Additionally, we employed the reconstruction with respect to time, which can be made with the aid of interpolation on right Gauss-Radau quadrature nodes. This type of reconstruction is known from the analysis, e.g., in [1, 14, 20].

In this paper we present a posteriori error analysis of STDGM applied to the Richards equation by generalizing the results from [8]. In Sect. 2, we introduce the continuous problem, whose space-time Galerkin discretization is treated in Sect. 3. In Sect. 4, we describe the error measure as the dual norm of residual with respect to the reformulated problem. In Sect. 5, we derive an a posteriori upper bound using the reconstructions of the solution with respect to time and space. The efficiency of the error estimates is justified by numerical experiments in Sect. 6. Finally, Sect. 7 contains an application of this technique to a real numerical simulation.

2 Continuous Problem

Let $\Omega \subset \mathbb{R}^d$ ($d = 2, 3$) be a bounded polyhedral domain with Lipschitz continuous boundary $\partial\Omega$ and $T > 0$. The Richards equation, describing variably-saturated porous media flows, can be written in the form

$$\tfrac{\partial}{\partial t}\vartheta(u) - \nabla \cdot (\mathbf{K}(u)\nabla u) = g \quad \text{in } \Omega \times (0, T), \tag{1}$$

where $u = u(x, t) : \Omega \times (0, T) \to \mathbb{R}$ is the sought function representing hydraulic head (=normalized pressure) with the physical unit L, $g : \Omega \times (0, T) \to \mathbb{R}$ is the source term, $\vartheta : \mathbb{R} \to \mathbb{R}$ is the dimensionless active pore volume and $\mathbf{K} : \mathbb{R} \to \mathbb{R}^{d\times d}$ is the hydraulic conductivity with the physical unit $\mathrm{L} \cdot \mathrm{T}^{-1}$. We note that in order to be precise, we should write $\vartheta = \vartheta(u - z)$ and $\mathbf{K} = \mathbf{K}(u - z)$ where z is the vertical coordinates. The nonlinear functions ϑ and $\mathbf{K}$ are given by constitutive relations. We assume that

- ϑ is Lipschitz continuous, non-negative and non-decreasing function,
- $\mathbf{K}(u) = \mathbf{K}_s K_r(u)$, where $\mathbf{K}_s$ is the saturated conductivity tensor and K_r is continuous, non-negative and non-decreasing continuous function bounded from above.

The Richards equation (1) has to be accompanied by initial and boundary conditions. For simplicity, we consider the homogeneous Dirichlet boundary condition over all $\Gamma := \partial\Omega$, i.e.,

$$u(x,t) = 0 \quad \text{on } \Gamma \times (0,T), \qquad u(x,0) = u^0(x) \quad \text{in } \Omega, \tag{2}$$

where the initial condition $u^0 \in H_0^1(\Omega)$.

We denote the weak time derivative $\vartheta'(u) = \frac{\partial\vartheta(u)}{\partial t} = \frac{\mathrm{d}\vartheta}{\mathrm{d}u}\frac{\partial u}{\partial t}$ and define spaces

$$\begin{aligned} X &= L^2(0,T,H_0^1(\Omega)), \qquad & H(\mathrm{div},\Omega) &= \{v \in L^2(\Omega)^d : \mathrm{div}\, v \in L^2(\Omega)\}, \\ Y &= \{v \in X : \vartheta'(v) \in L^2(0,T,L^2(\Omega))\}, \qquad & Y^0 &= \{v \in Y : v(0) = u^0\}. \end{aligned} \tag{3}$$

We note that if $v \in Y$ then $\vartheta(v) \in C([0,T], L^2(\Omega))$. For shorter notation, we define the physical flux

$$\sigma(u,\nabla u) := \mathbf{K}(u)\nabla u. \tag{4}$$

Definition 1 We say that the function $u \in Y^0$ is the weak solution of (1), if

$$\int_0^T (\vartheta'(u), v) + (\sigma(u,\nabla u), \nabla v) - (g, v)\,\mathrm{d}t = 0 \quad \forall v \in X. \tag{5}$$

The existence, uniqueness, and regularity of the weak solution of (5) were studied in [2, 26]. We assume that u is sufficiently regular such that $\sigma(u,\nabla u) \in L^2(0,T,H(\mathrm{div},\Omega))$.

3 STDGM Discretizations

In order to discretize problem (5), we consider a time partition of the interval $(0,T)$ as $0 = t_0 < t_1 < \ldots < t_r = T$ and put $I_m = (t_{m-1}, t_m)$, $\tau_m = |I_m| = t_m - t_{m-1}$ and $\tau = \max_{m=1,\ldots,r} \tau_m$. For each $m = 0,\ldots,r$, we consider a mesh $\mathscr{T}_h^m$ covering $\overline{\Omega}$ and consisting of a finite number of closed, d-dimensional simplices with mutually disjoint interiors. We assume that $\mathscr{T}_h^m$, $m = 0,\ldots,r$ are conforming, i.e., neighbouring elements share an entire edge or face. We denote the edges of the mesh by e and set $h_e = \mathrm{diam}(e)$, $h_K = \mathrm{diam}(K)$ and $h = \max_{K\in\mathscr{T}_h^m,\, m=0,\ldots,r} h_K$.

We assume that $\mathscr{T}_h^m$, $m = 0,\ldots,r$ are *shape regular* and *locally quasi-uniform*, i.e., $h_K \le C h_{K'}$ for any neigbouring elements K and K', where constant C does not depend on $h \in (0,h_0)$, $m = 0,\ldots,r$. For each edge/face e, let $n = n_e$ denote an unit normal vector to e with arbitrary but fixed directions for the inner edges and with outer direction on $\partial\Omega$. Moreover, for each $K \in \mathscr{T}_h^m$, n_K is the unit outer normal vector to K.

We define the space of discontinuous piecewise polynomial functions on $\mathscr{T}_h^m$ by

$$S_{hp,m} = \{v \in L^2(\Omega) : v|_K \in P_p(K),\ K \in \mathscr{T}_h^m\}, \qquad m = 0, \ldots, r \tag{6}$$

where p is the approximation polynomial degree with respect to spatial coordinates and $P_p(K)$ denotes the space of polynomials on K up to the degree $p \geq 1$. Furthermore, we consider the space of *space-time* discontinuous piecewise polynomial functions

$$S_{hp}^{\tau q} = \{v \in L^2(\Omega \times (0, T)) : \ v|_{I_m} \in P_q(I_m, S_{hp,m}),\ m = 1, \ldots, r\}. \tag{7}$$

For $v \in S_{hp}^{\tau q}$, we define the one-sided limits and jumps with respect to time by

$$\begin{aligned} v_+^m &= \lim_{t \to t_m^+} v(t), \ \ m = 0, \ldots, r-1, \qquad v_-^m = \lim_{t \to t_m^-} v(t), \ \ m = 1, \ldots, r, \\ \{v\} &= v_+^m - v_-^m, \quad m = 1, \ldots, r-1, \qquad v_-^0 = u^0, \qquad \{v\}_0 = v_+^0 - u^0, \end{aligned} \tag{8}$$

where u^0 is the initial condition. In order to simplify the notation, we set the local L^2-scalar products

$$(u, v)_M = \int_M uv \,\mathrm{d}x, \qquad (u, v)_{M,m} = \int_{M \times I_m} uv \,\mathrm{d}x\,\mathrm{d}t, \quad m = 1, \ldots, r, \tag{9}$$

where $M \subset \overline{\Omega}$ and the corresponding norms are denoted by $\|.\|_M$, $\|.\|_{M,m}$. The symbol $\sum_{K,m}$ means a sum over all space-time elements $K \times I_m$, where $K \in \mathscr{T}_h^m$ and $m = 1, \ldots, r$.

We define one-sided values, jumps and mean values with respect to space for $v \in S_{hp,m}$ on the inner edges with the unit normal n

$$\begin{aligned} v_L(x) &= \lim_{s \to 0+} v(x - ns), \qquad v_R(x) = \lim_{s \to 0+} v(x + ns), \\ [v] &= (v_L - v_R)n, \qquad \langle v \rangle = (v_L + v_R)/2. \end{aligned} \tag{10}$$

The value $[v]$ is independent on the orientation of n. For the boundary edges, we put $v_R = 0$, $\langle v \rangle = v_L$ and $[v] = v_L n$, where n is the unit outer normal to Ω.

For each space-time element $K \times I_m$, $K \in \mathscr{T}_h^m$, $m = 1, \ldots, r$, we define the forms

$$\begin{aligned} a_{K,m}(u, v) &:= (\mathbf{K}(u)\nabla u, \nabla v)_{K,m} = (\sigma(u, \nabla u), \nabla v)_{K,m} - (g, v)_{K,m}, \\ A_{K,m}(u, v) &:= (\mathbf{K}(u)\nabla u, \nabla v)_{K,m} - (\langle \mathbf{K}(u)\nabla u \rangle \cdot n - \alpha[u] \cdot n, v)_{\partial K,m} \\ &\quad + (\beta - \tfrac{1}{2})\,(\mathbf{K}(u)[u], \nabla v)_{\partial K,m} - (g, v)_{K,m}, \end{aligned} \tag{11}$$

where $\alpha > 0$ is a penalization parameter large enough to ensure the ellipticity of the discretization of the elliptic term $\mathbf{K}(u)\nabla u$ and $\beta \in \{0, \frac{1}{2}, 1\}$ corresponds to the choice of the variant of the interior penalty discretization (SIPG with $\beta = 0$, IIPG with $\beta = 1/2$ and NIPG with $\beta = 1$), see, e.g., [7, Chapter 2].

Now, we introduce the space-time discontinuous Galerkin discretization of (1).

Definition 2 We say that the function $u_h^\tau \in S_{hp}^{\tau q}$ is the approximate solution of (5) obtained by the space-time discontinuous Galerkin method (STDGM), if

$$\sum_{K,m} B_{K,m}(u_h^\tau, v) = 0 \qquad \forall v \in S_{hp}^{\tau q}, \tag{12}$$

where

$$B_{K,m}(u, v) := \left(\vartheta'(u), v\right)_{K,m} + A_{K,m}(u, v) + \left(\{\vartheta(u)\}_{m-1}, v_+^{m-1}\right)_K \tag{13}$$

with $A_{K,m}$ given by (11) and $\{\cdot\}$ defined by (8).

4 Error Measure

The weak solution u defined by (5) is continuous in time, more precisely, $\vartheta(u) \in C([0, T], L^2(\Omega))$. For the purpose of a posteriori error analysis, we reformulate the definition of the weak solution. We define the space

$$Y^\tau = \{v \in X : \vartheta'(u)|_{I_m} \in L^2(I_m, L^2(\Omega))\} \tag{14}$$

of piecewise continuous functions with respect to time. From the definition of the space Y^τ we can see that

$$Y^0 \subset Y \subset Y^\tau \subset X, \qquad S_{hp}^{\tau q} \subset Y^\tau. \tag{15}$$

This space is suitable for redefinition of the weak formulation given by Definition 1, which is replaced by the following one.

Definition 3 Function $u \in Y^\tau$ is the weak solution of (1) if

$$\sum_{K,m} b_{K,m}(u, v) = 0 \quad \forall v \in Y^\tau, \tag{16}$$

where

$$b_{K,m}(u, v) := \left(\vartheta'(u), v\right)_{K,m} + a_{K,m}(u, v) + \left(\{\vartheta(u)\}_{m-1}, v_+^{m-1}\right)_K \tag{17}$$

with $a_{K,m}$ given by (11) and $\{\cdot\}$ defined by (8).

It is possible to prove that if $u \in Y^0$ is the unique solution of problem (5), then u is the unique solution of problem (16).

In virtue of [8, § 2.3.1], we define a positive parameter associated to space-time element $K \times I_m$ denoted hereafter by $d_{K,m}$. The forthcoming analysis will be independent of the choice of this parameter. The parameter $d_{K,m}$ exhibits a user-dependent weight, typically with units $(T\,L)^{1/2}$ so that the error measure has the same units as the energy norm. The possible choice is

$$d_{K,m} := \left(h_K^{-2}\|\mathbf{K}\|_{K,m,\infty} + \tau_m^{-2} T \|\mathrm{d}\vartheta/\mathrm{d}u\|_{K,m,\infty}\right)^{-1/2}, \tag{18}$$

where $\|\cdot\|_{K,m,\infty} := \|\cdot\|_{L^\infty(K\times I_m)}$. This choice covers a possible degeneracy when $\|\mathbf{K}\|_{K,m,\infty} \to 0$ or $\|\mathrm{d}\vartheta/\mathrm{d}u\|_{K,m,\infty} \to 0$ for some $K \times I_m$. We equip the space Y^τ with the norm

$$\|v\|_{Y^\tau}^2 = \sum_{K,m} \|v\|_{Y^\tau,K,m}^2, \quad \|v\|_{Y^\tau,K,m}^2 = d_{K,m}^{-2}\left(h_K^2\|\nabla v\|_{K,m}^2 + \tau_m^2\|v'\|_{K,m}^2\right). \tag{19}$$

Following formulation (16) and using norm $\|\cdot\|_{Y^\tau}$, we introduce the first ingredient of the error measure as a dual norm of the residual

$$\mathcal{R}(u_h^\tau) = \sup_{0\neq v\in Y^\tau} \frac{\sum_{K,m} b_{K,m}(u_h^\tau, v)}{\|v\|_{Y^\tau}}, \tag{20}$$

where $b_{K,m}$ is given by (17). The form $\mathcal{R}(u_h^\tau)$ represents a natural error measure for $u - u_h^\tau$, see comments in [8, Remark 2.3] for example. In Sect. 5, we estimate $\mathcal{R}(u_h^\tau)$ for u_h^τ being the solution of (12).

Since the approximate solution u_h^τ lives in the space of discontinuous function $S_{hp}^{\tau q}$, it is necessary to introduce the second building block measuring the nonconformity of the solution, i.e., the distance of u_h^τ from Y^τ. Hence we define

$$\mathcal{J}(v) = \sum_{K,m} J_{K,m}(v), \quad J_{K,m}(v) = d_{K,m}^2\, \tau_m^{-1}\, h_K^{-2}\, C_{K,m,\mathbf{K},\alpha} \|[v]\|_{\partial K,m}^2, \tag{21}$$

where $C_{K,m,\mathbf{K},\alpha} = \alpha^2 + \|\mathbf{K}(u_h^\tau)\|_{L^\infty(K\times I_m)}^2$. The scaling factors are chosen such that $\mathcal{J}(v)^{1/2}$ has the same physical units as $\mathcal{R}(u_h^\tau)$. We note that $\mathcal{J}(v)$ has to be modified for nonhomogeneous Dirichlet boundary condition.

The final error measure is then defined by

$$\mathcal{E}(u_h^\tau) := \left(\mathcal{R}(u_h^\tau)^2 + \mathcal{J}(u_h^\tau)\right)^{1/2}, \tag{22}$$

where $\mathcal{R}(u_h^\tau)$ is given by (20) and $\mathcal{J}(u_h^\tau)$ by (21). There holds $\mathcal{E}(u_h^\tau) = 0$ if and only is $u_h^\tau = u$ is the weak solution u given by (5). Indeed, if $u_h^\tau = u$ then $\mathcal{J}(u_h^\tau) = 0$ and

$\mathcal{R}(u_h^\tau) = 0$ due to (16). On the other hand, if $\mathcal{J}(u_h^\tau) = 0$ then $u_h^\tau \in Y^\tau$. Moreover, $\mathcal{R}(u_h^\tau) = 0$ and the uniqueness of (16) implies that u_h^τ is the weak solution (5).

5 Error Estimates

Whereas the part of the error (22) represented by $\mathcal{J}(u_h^\tau)$ can be directly evaluated, we have to estimate its residual part given by $\mathcal{R}(u_h^\tau)$. We adopt the reconstruction technique from [13] where the space and time equilibrated flux reconstructions are developed for a nonlinear convection-diffusion equation.

5.1 *Temporal Flux Reconstruction of the Solution*

The goal of the time reconstruction is to define a function $R_h^\tau = R_h^\tau(x,t)$ that mimics function $\vartheta(u_h^\tau)$, $u_h^\tau \in S_{hp}^{\tau q}$ and is continuous with respect to the time. Thus, we employ the right Radau polynomial on I_m denoted as $r_m \in P_{q+1}(I_m)$ such that $r_m(t_{m-1}) = 1$, $r_m(t_m) = 0$ and r_m is orthogonal to $P_{q-1}(I_m)$ with respect to the $L^2(I_m)$-inner product. Then the polynomial reconstruction R_h^τ is set on each interval I_m by

$$R_h^\tau(x,t) = \vartheta(u_h^\tau(x,t)) - \{\vartheta(u_h^\tau)\}_{m-1}(x) r_m(t), \quad x \in \Omega,\ t \in I_m, \tag{23}$$

where $\{\cdot\}$ is given by (8). The resulting function R_h^τ is continuous in time and satisfies the initial condition in the sense (cf. 8)

$$\begin{aligned} R_h^\tau(x,0) &= \vartheta(u_h^\tau(x,0)) - \{\vartheta(u_h^\tau)\}_0(x) r_1(0) \\ &= \vartheta(u_h^\tau(x,0)) - (\vartheta(u_h^\tau(x,0)) - \vartheta(u^0(x))) = \vartheta(u^0(x)). \end{aligned} \tag{24}$$

Moreover, using the integration per-partes and employing $r_m(t_{m-1}) = 1$, $r_m(t_m) = 0$, we get from (23) the identity

$$\begin{aligned} \left((R_h^\tau - \vartheta(u_h^\tau))', v\right)_{m,K} &= -\left(r_m'\{\vartheta(u_h^\tau)\}_{m-1}, v\right)_{m,K} \\ &= \left(r_m\{\vartheta(u_h^\tau)\}_{m-1}, v'\right)_{m,K} - r_m(t_m)\left(\{\vartheta(u_h^\tau)\}_{m-1}, v_-^m\right)_K \\ &\quad + r_m(t_{m-1})\left(\{\vartheta(u_h^\tau)\}_{m-1}, v_+^{m-1}\right)_K \\ &= \left(r_m\{\vartheta(u_h^\tau)\}_{m-1}, v'\right)_{m,K} + \left(\{\vartheta(u_h^\tau)\}_{m-1}, v_+^{m-1}\right)_K, \quad v \in Y^\tau \end{aligned} \tag{25}$$

which together with (23) leads to

$$\left((R_h^\tau - \vartheta(u_h^\tau))', v\right)_{m,K} - \left(\{\vartheta(u_h^\tau)\}_{m-1}, v_+^{m-1}\right)_K = -\left(R_h^\tau - \vartheta(u_h^\tau), v'\right)_{m,K} \quad \forall v \in Y^\tau. \tag{26}$$

Furthermore, employing the fact that r_m is orthogonal to $P_{q-1}(I_m)$, we obtain from (25), the expression

$$\left((R_h^\tau - \vartheta(u_h^\tau))', v\right)_{m,K} = \left(\{\vartheta(u_h^\tau)\}_{m-1}, v_+^{m-1}\right)_K \quad \forall v \in P_q(I_m, L^2(K)), \tag{27}$$

which implies that u_h^τ being the approximate solution defined by (12) satisfies

$$((R_h^\tau)', v)_{K,m} = (\vartheta'(u_h^\tau), v)_{K,m} + (\{\vartheta(u_h^\tau)\}_{m-1}, v_+^{m-1})_K = -A_{K,m}(u_h^\tau, v) \tag{28}$$
$$\forall v \in P_q(I_m, P_p(K)).$$

Finally, we note that the reconstruction R_h^τ is locally defined and explicit so its computation is cheap and easy to implement.

5.2 *Spatial Flux Reconstruction of the Solution*

Let $u_h^\tau \in S_{hp}^{\tau q}$ be the approximate solution then the corresponding flux $\sigma(u_h^\tau, \nabla u_h^\tau) = \mathbf{K}(u_h^\tau)\nabla u_h^\tau \notin L^2(0, T, H(\mathrm{div}, \Omega))$ in general. Therefore, in the same spirit as the temporal reconstruction, we reconstruct the spatial fluxes of the solution in such a way that $\sigma_h^\tau \in L^2(0, T, H(\mathrm{div}, \Omega))$ and $\sigma(u_h^\tau, \nabla u_h^\tau) \approx \sigma_h^\tau$.

We employ the Raviart-Thomas-Nedelec finite elements, cf. [6] for example. Let $\mathrm{RTN}_p(K) = P_p(K)^d + xP_p(K)$ be space of order p for element $K \in \mathscr{T}_h^m$, we define the spatial reconstruction element-wise: for each $K \in \mathscr{T}_h^m$, find $\sigma_h^\tau|_{K\times I_m} \in P_q(I_m, \mathrm{RTN}_p(K))$ such that

$$(\sigma_h^\tau \cdot n, v)_{e,m} = \left(\langle \mathbf{K}(u_h^\tau)\nabla u_h^\tau\rangle \cdot n - \alpha[u_h^\tau] \cdot n, v\right)_{e,m} \quad \forall v \in P_q(I_m, P_p(e)),\ e \subset K$$
$$(\sigma_h^\tau, v)_{K,m} = \left(\mathbf{K}(u_h^\tau)\nabla u_h^\tau, \nabla v\right)_{K,m} + (\beta - \tfrac{1}{2})\left(\mathbf{K}(u_h^\tau)[u_h^\tau], \nabla v\right)_{\partial K,m} \tag{29}$$
$$\forall v \in P_q(I_m, P_{p-1}(K)^d).$$

Here, the numerical flux $\langle \mathbf{K}(u_h^\tau)\nabla u_h^\tau\rangle \cdot n - \alpha[u_h^\tau] \cdot n$ is conservative on interior edges, which implies that the resulting reconstruction $\sigma_h^\tau \in L^2(0, T, H(\mathrm{div}, \Omega))$ globally. From the Green theorem, (29), (11) and (28), we obtain

$$\begin{aligned}\left(\nabla \cdot \sigma_h^\tau, v\right)_{K,m} &= -\left(\sigma_h^\tau, \nabla v\right)_{K,m} + \left(\sigma_h^\tau \cdot n_K, v\right)_{\partial K,m} \\ &= -\left(\mathbf{K}(u_h^\tau)\nabla u_h^\tau, \nabla v\right)_{K,m} - (\beta - \tfrac{1}{2})\left(\mathbf{K}(u_h^\tau)[u_h^\tau], \nabla v\right)_{\partial K,m}\end{aligned} \tag{30}$$

$$+ \left(\langle \mathbf{K}(u_h^\tau)\nabla u_h^\tau \rangle \cdot n - \alpha [u_h^\tau] \cdot n, v\right)_{\partial K,m}$$
$$= -A_{K,m}(u_h^\tau, v) - (g, v) = \left((R_h^\tau)' - g, v\right)_{K,m}$$
$$\times \quad \forall v \in P_q(I_m, P_p(K)), \;\; K \in \mathscr{T}_h^m.$$

We note that this reconstruction is not robust with respect to the polynomial approximation degree and can not be used for the hp-mesh adaptation. It is possible to use the another variant of reconstruction proposed in [10] for linear elliptic problems and in [13] for nonlinear convection-diffusion equations. This will be the subject of further research. Moreover, (30) is valid only for "exactly" computed approximate solution which does not suffer from algebraic errors.

5.3 Upper Bound

We are ready to preset the main result of this paper, namely a posteriori upper bound for $\mathcal{R}(u_h^\tau)$, cf. (20).

Theorem 4 *Let $u \in Y$ be the solution of (5) and $u_h^\tau \in S_{hp}^{\tau q}$ be the approximate solution given by (12). Let R_h^τ and σ_h^τ be the reconstructions computed from u_h^τ by (23) and (29), respectively. Then*

$$\mathcal{R}(u_h^\tau)^2 \le \eta^2 := \sum_{K,m} \eta_{K,m}^2, \qquad \eta_{K,m} := C_P\, \eta_{R,K,m} + \left(\eta_{S,K,m}^2 + \eta_{T,K,m}^2\right)^{1/2}, \tag{31}$$

where $C_P \le 1/\pi$ is the constant from the Poincaré inequality, and

$$\eta_{R,K,m} = d_{K,m} \|(R_h^\tau)' - \nabla \cdot \sigma_h^\tau - g\|_{K,m}, \tag{32}$$
$$\eta_{S,K,m} = \frac{d_{K,m}}{h_K} \|\sigma_h^\tau - \sigma(u_h^\tau, \nabla u_h^\tau)\|_{K,m},$$
$$\eta_{T,K,m} = \frac{d_{K,m}}{\tau_m} \|R_h^\tau - \vartheta(u_h^\tau)\|_{K,m}, \qquad K \in \mathscr{T}_h^m,\; m = 1, \ldots, r.$$

Proof Let $v \in Y^\tau$, using Green's theorem and the fact that $\sigma_h^\tau \in L^2(0, T, H(\mathrm{div}, \Omega))$, we split the numerator of $\mathcal{R}(u_h^\tau)$ (cf. (20)) as

$$\sum_{K,m} b_{K,m}(u_h^\tau, v) \tag{33}$$
$$= \sum_{K,m} \left(\left(\vartheta'(u_h^\tau) - g, v\right)_{K,m} + \left(\sigma(u_h^\tau, \nabla u_h^\tau), \nabla v\right)_{K,m} + \left(\{\vartheta(u_h^\tau)\}_{m-1}, v_+^{m-1}\right)_K \right)$$

$$
\begin{aligned}
&= \sum_{K,m} \left((R_h^\tau)' - \nabla\cdot\sigma_h^\tau - g, v\right)_{K,m} - \sum_{K,m} \left(\sigma_h^\tau - \sigma(u_h^\tau, \nabla u_h^\tau, \nabla v\right)_{K,m} \\
&\quad - \sum_{K,m} \left(\left((R_h^\tau - \vartheta(u_h^\tau))', v\right)_{K,m} - \left(\{\vartheta(u_h^\tau)\}_{m-1}, v_+^{m-1}\right)_K\right) =: \xi_1 + \xi_2 + \xi_3.
\end{aligned}
$$

We estimate individual terms ξ_1, ξ_2 and ξ_3. Let $v_{K,m}$ be the L^2-orthogonal projection of $v \in Y^\tau$ on $P_0(K \times I_m)$. Using (30), the space-time scaled Poincaré inequality from [8, Lemma 2.2], (18)–(19), the Cauchy inequality and (32), we obtain

$$
\begin{aligned}
|\xi_1| &\leq \sum_{K,m} \left| \left((R_h^\tau)' - \nabla\cdot\sigma_h^\tau - g, v\right)_{K,m} \right| \\
&= \sum_{K,m} \left| \left((R_h^\tau)' - \nabla\cdot\sigma_h^\tau - g, v - v_{K,m}\right)_{K,m} \right| \qquad (34) \\
&\leq \sum_{K,m} C_P \, \|(R_h^\tau)' - \nabla\cdot\sigma_h^\tau - g\|_{K,m} \, (h_K^2 \|\nabla v\|_{K,m}^2 + \tau_m^2 \|v'\|_{K,m}^2)^{1/2} \\
&= \sum_{K,m} C_P \, d_{K,m} \, \|(R_h^\tau)' - \nabla\cdot\sigma_h^\tau - g\|_{K,m} \, \|v\|_{Y^\tau,K,m} = \sum_{K,m} C_P \, \eta_{R,K,m}.
\end{aligned}
$$

Furthermore, the Cauchy inequality and (32) give

$$
\begin{aligned}
|\xi_2| &\leq \sum_{K,m} \left| \left(\sigma_h^\tau - \sigma(u_h^\tau, \nabla u_h^\tau), \nabla v\right)_{K,m} \right| \qquad (35) \\
&\leq \sum_{K,m} \frac{d_{K,m}}{h_K} \|\sigma_h^\tau - \sigma(u_h^\tau, \nabla u_h^\tau)\|_{K,m} \frac{h_K}{d_{K,m}} \|\nabla v\|_{K,m} \\
&= \sum_{K,m} \eta_{S,K,m} \frac{h_K}{d_{K,m}} \|\nabla v\|_{K,m}.
\end{aligned}
$$

Finally, the integration per-partes, (26), the Cauchy inequality and (32) imply

$$
\begin{aligned}
|\xi_3| &\leq \sum_{K,m} \left(\left| \left((R_h^\tau - \vartheta(u_h^\tau))', v\right)_{K,m} - \left(\{\vartheta(u_h^\tau)\}_{m-1}, v_+^{m-1}\right)_K \right| \right. \qquad (36) \\
&= \sum_{K,m} \left| \left(R_h^\tau - \vartheta(u_h^\tau), v'\right)_{K,m} \right| \leq \sum_{K,m} \frac{d_{K,m}}{\tau_m} \|R_h^\tau - \vartheta(u_h^\tau)\|_{K,m} \frac{\tau_m}{d_{K,m}} \|v'\|_{K,m} \\
&= \sum_{K,m} \eta_{T,K,m} \frac{\tau_m}{d_{K,m}} \|v'\|_{K,m}.
\end{aligned}
$$

Using estimates (35) and (36), we obtain by the Cauchy inequalities

$$|\xi_2| + |\xi_3| \leq \sum_{K,m} \left(\eta_{S,K,m} \frac{h_K}{d_{K,m}} \|\nabla v\|_{K,m} + \eta_{T,K,m} \frac{\tau_m}{d_{K,m}} \|v'\|_{K,m} \right) \tag{37}$$

$$\leq \sum_{K,m} \left(\eta_{S,K,m}^2 + \eta_{T,K,m}^2 \right)^{1/2} \|v\|_{Y^\tau,K,m},$$

which together with (34) implies (31).

Obviously, (22) and (31) give the estimate of the total error measure

$$\mathcal{E} \leq (\eta^2 + \mathcal{J}(u_h^\tau))^{1/2}. \tag{38}$$

5.4 *Lower Bound*

The proof of the local efficiency (lower bound) is much more technical. It is possible to adopt the approach from [13] developed for nonlinear convection-diffusion equation. Defining the local variant of the error measure (20) by

$$\mathcal{R}_{M,m}(u_h^\tau) = \sup_{0 \neq v \in Y^\tau} \frac{\sum_{K \in M} b_{K,m}(u_h^\tau, v)}{\|v\|_{Y^\tau}}, \qquad m = 1, \ldots, r, \tag{39}$$

where M is a union of some elements of mesh $\mathcal{T}_h^m$, we expect the estimate

$$\eta_{R,K,m} \leq c \mathcal{R}_{\mathcal{T}_K^2,m}(u_h^\tau), \quad \eta_{S,K,m} \leq c \mathcal{R}_{\mathcal{T}_K^2,m}(u_h^\tau), \quad \eta_{T,K,m} \leq c \mathcal{R}_{\mathcal{T}_K,m}(u_h^\tau),$$

for $K \in \mathcal{T}_h^m$, $m = 1, \ldots, r$, where $c > 0$ is a constant, $\mathcal{T}_K$ is a patch consisting of elements sharing at least a vertex with K and $\mathcal{T}_K^2$ is a union of patches $\mathcal{T}_{K'}$ for $K' \subset \mathcal{T}_K$. The local efficiency is supported by numerical experiments in the following Sect. 6, which show the rate of the convergence of the error estimator.

6 Rates of Convergence of the Error Estimator

We consider the Tracy problem [28] which is a classical benchmark problem that uses Gardner's constitutive relations [18] for ϑ and $\mathbf{K}$. The computational domain is $\Omega := (0, 1) \times (0, 1)$ m. The initial condition is a constant hydraulic head $u_0 = -10.0$ m and the boundary condition $u = u_0$ on the left, bottom and right parts of $\partial\Omega$ and a parabolic profile on the top part of $\partial\Omega$. Due to the inconsistency of the initial and boundary conditions, the most challenging is the computations close to

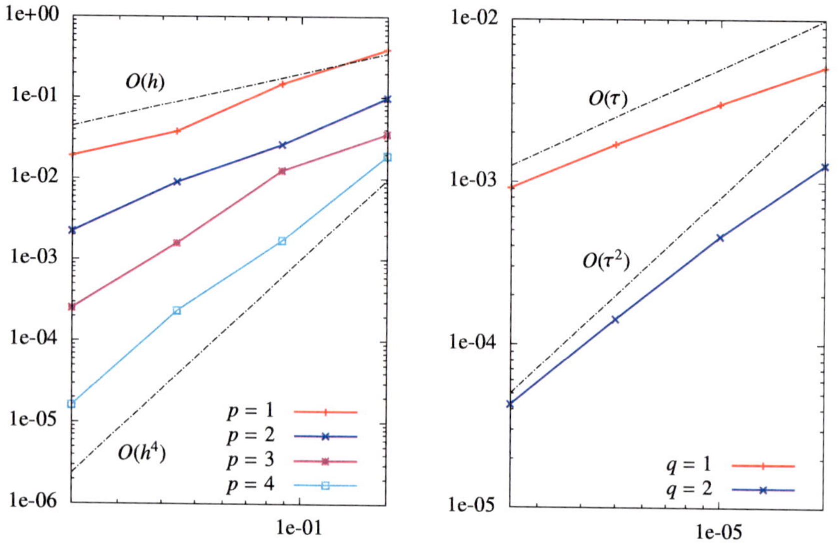

Fig. 1 Convergence of estimator η with respect to h (case (C1), left) and τ (case (C2), right)

$t = 0$. On the other hand, in order to demonstrate the efficiency of the higher order discretization, we investigate the error on the interval $T \in [10^{-5}, 1.1 \cdot 10^{-4}]$.

We carried out two type of computations:

(C1) sequence of uniformly refined grids with P_p, $p = 1, \dots, 4$ approximation with respect to the space and an overkill with respect to the time,

(C2) sequence of constant time steps with P_q, $q = 1, 2$ approximation with respect to the time and an overkill with respect to the space.

The convergence of the error estimator η with respect to h and τ is shown in Fig. 1. We observe that error estimator behaves as $O(h^p + \tau^q)$ which is in agreement with theoretical expectations.

7 Simulation of the Single Ring Infiltration Process

In order to demonstrate the usability of the proposed error estimate for practical problems, we present a numerical simulation of the single ring infiltration through an anisotropic porous medium which was considered in [12]. We combine the presented a posteriori estimator with the anisotropic hp-mesh adaptation technique used in [12]. The local error estimators $\eta_{K,m}^2 + J_{K,m}$, $K \in \mathscr{T}_h^m$, $m = 1, \dots, r$ (cf. (31) and (21)) are used for the setting of the sizes of mesh elements. The shapes of elements and the polynomial approximation degrees are controlled by the interpolation error estimates. We note that here we make an inaccuracy since the

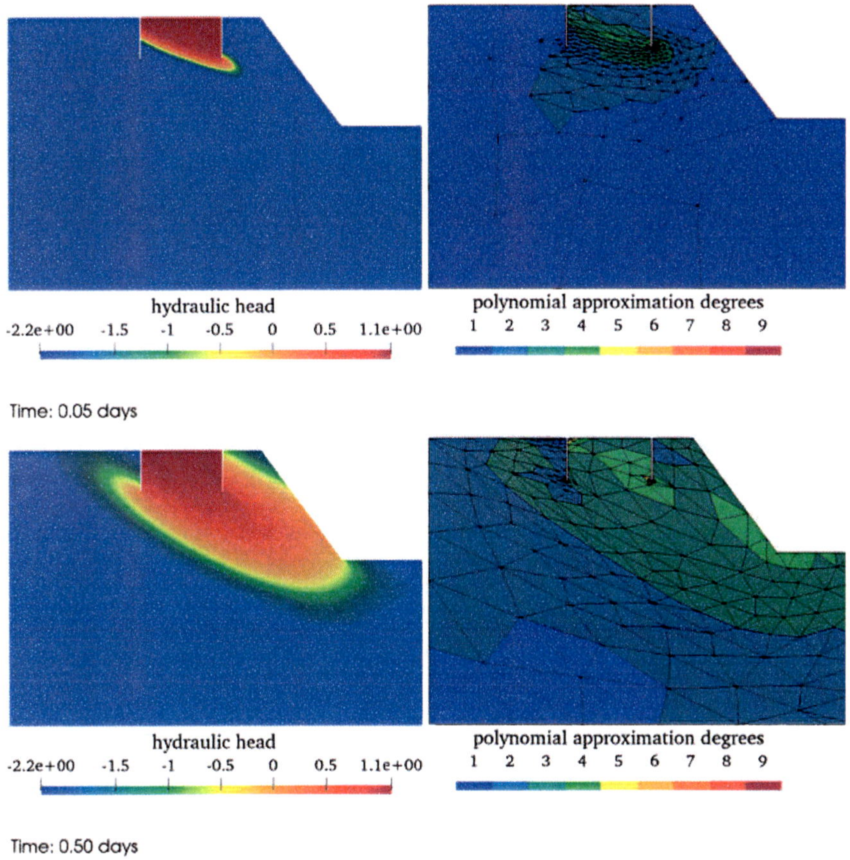

Fig. 2 Ring infiltration process, hydraulic head (left) and the *hp*-mesh (right) at $t = 0.05$ days (top) and $t = 0.5$ days (bottom)

reconstruction (29) does not guarantee that $\sigma_h^\tau \in L^2(0, T, H(\mathrm{div}, \Omega))$ for varying polynomial degree.

Figure 2 shows the distribution of the hydraulic head in the computational domain at two different time levels and the corresponding *hp*-mesh generated by the algorithm. We observe the propagation of the fluid from the inlet to the interior of the domain and the corresponding mesh refinement in the vicinity of the interface.

Acknowledgments This research was supported by the grants of the Czech Science Foundation No. 20-01074S.

References

1. Akrivis, G., Makridakis, C., Nochetto, R.H.: Galerkin and Runge-Kutta methods: unified formulation, a posteriori error estimates and nodal superconvergence. Numer. Math. **118**(3), 429–456 (2011)
2. Alt, H.W., Luckhaus, S.: Quasilinear elliptic-parabolic differential equations. Math. Z. **183**(3), 311–341 (1983)
3. Arbogast, T., Wheeler, M., Zhang, N.Y.: A nonlinear mixed finite element method for a degenerate parabolic equation arising in flow in porous media. SIAM J. Numer. Anal. **33**(4), 1669–1687 (1996)
4. Bastian, P.: A fully-coupled discontinuous Galerkin method for two-phase flow in porous media with discontinuous capillary pressure. Comput. Geosci. **18**(5), 779–796 (2014)
5. Braess, D., Pillwein, V., Schöberl, J.: Equilibrated residual error estimates are p-robust. Comput. Methods Appl. Mech. Eng. **198**(13–14), 1189–1197 (2009)
6. Brezzi, F., Fortin, M.: Mixed and hybrid finite element methods. Springer Series in Computational Mathematics, vol. 15. Springer, New York (1991)
7. Dolejší, V., Feistauer, M.: Discontinuous Galerkin Method – Analysis and Applications to Compressible Flow. Springer Series in Computational Mathematics, vol. 48. Springer, Cham (2015)
8. Dolejší, V., Ern, A., Vohralík, M.: A framework for robust a posteriori error control in unsteady nonlinear advection-diffusion problems. SIAM J. Numer. Anal. **51**(2), 773–793 (2013)
9. Dolejší, V., Roskovec, F., Vlasák, M.: Residual based error estimates for the space-time discontinuous Galerkin method applied to the compressible flows. Comput. Fluids **117**, 304–324 (2015)
10. Dolejší, V., Ern, A., Vohralík, M.: hp-adaptation driven by polynomial-degree-robust a posteriori error estimates for elliptic problems. SIAM J. Sci. Comput. **38**(5), A3220–A3246 (2016)
11. Dolejší, V., Kuráž, M., Solin, P.: Adaptive higher-order space-time discontinuous Galerkin method for the computer simulation of variably-saturated porous media flows. Appl. Math. Model. **72**, 276–305 (2019)
12. Dolejší, V., Kuráž, M., Solin, P.: Numerical simulation of a single ring infiltration experiment with hp-adaptive space-time discontinuous galerkin method. Acta Polytech. **61**, 59–67 (2021)
13. Dolejší, V., Roskovec, F., Vlasák, M.: A posteriori error estimates for higher order space-time Galerkin discretizations of nonlinear parabolic problems. SIAM J. Numer. Anal. **59**(3), 1486–1509 (2021)
14. Eriksson, K., Johnson, C., Thomée, V.: Time discretization of parabolic problems by the discontinuous Galerkin method. RAIRO, Modélisation Math. Anal. Numér. **19**, 611–643 (1985)
15. Ern, A., Vohralík, M.: Polynomial-degree-robust a posteriori estimates in a unified setting for conforming, nonconforming, discontinuous Galerkin, and mixed discretizations. SIAM J. Numer. Anal. **53**(2), 1058–1081 (2015)
16. Eymard, R., Gutnic, M., Hilhorst, D.: The finite volume method for richards equation. Comput. Geosci. **3**(3–4), 259–294 (1999)
17. Eymard, R., Hilhorst, D., Vohralík, M.: A combined finite volume-nonconforming/mixed-hybrid finite element scheme for degenerate parabolic problems. Numer. Math. **105**(1), 73–131 (2006)
18. Gardner, W.R.: Some steady state solutions of the unsaturated moisture flow equation with application to evaporation from a water table. Soil Sci. **85**, 228–232 (1958)
19. Georgoulis, E.H., Lakkis, O.: A posteriori error bounds for discontinuous Galerkin methods for quasilinear parabolic problems. In: Numerical Mathematics and Advanced Applications 2009. Proceedings of ENUMATH 2009, Uppsala, Sweden, June 29–July 3, 2009, pp. 351–358. Springer, Berlin (2010)

20. Hulme, B.L.: One-step piecewise polynomial Galerkin methods for initial value problems. Math. Comput. **26**, 415–426 (1972)
21. Kreuzer, C.: Reliable and efficient a posteriori error estimates for finite element approximations of the parabolic p-Laplacian. Calcolo **50**(2), 79–110 (2013)
22. Luce, R., Wohlmuth, B.: A local a posteriori error estimator based on equilibrated fluxes. SIAM J. Numer. Anal. **42**(4), 1394–1414 (2004)
23. Mitra, K., Vohralík, M.: A posteriori error estimates for the Richards equation. Tech. Rep. hal-03328944, INRIA (2021)
24. Nochetto, R.H., Verdi, C.: Approximation of degenerate parabolic problems using a numerical integration. SIAM J. Numer. Anal. **25**(4), 784–814 (1988)
25. Nochetto, R., Schmidt, A., Verdi, C.: A posteriori error estimation and adaptivity for degenerate parabolic problems. Math. Comput. **69**(229), 1–24 (2000)
26. Otto, F.: L^1-contraction and uniqueness for quasilinear elliptic-parabolic equations. J. Differ. Equ. **131**(1), 20–38 (1996)
27. Radu, F., Pop, I., Knabner, P.: Error estimates for a mixed finite element discretization of some degenerate parabolic equations. Numer. Math. **109**(2), 285–311 (2008)
28. Tracy, F.T.: Clean two- and three-dimensional analytical solutions of Richards equation for testing numerical solvers. Water Resour. Res. **42**(8), 8503 (2006)
29. Verfürth, R.: A posteriori error estimates for nonlinear problems: $L^r(0, T; W^{1,\rho}(\Omega))$-error estimates for finite element discretizations of parabolic equations. Numer. Methods Part. Differ. Equ. **14**, 487–518 (1998)
30. Woodward, C., Dawson, C.: Analysis of expanded mixed finite element methods for a nonlinear parabolic equation modeling flow into variably saturated porous media. SIAM J. Numer. Anal. **37**(3), 701–724 (2000)

A Data-Driven Partitioned Approach for the Resolution of Time-Dependent Optimal Control Problems with Dynamic Mode Decomposition

Eleonora Donadini, Maria Strazzullo, Marco Tezzele, and Gianluigi Rozza

1 Introduction

Scientific and industrial contexts need to represent natural phenomena through mathematical models. Historically, this role has been played by partial differential equations (PDEs) and by their numerical simulations. Yet, the model equations, in some frameworks, are not enough to well describe the complexity of the natural phenomenon one is dealing with. To increase the reliability of a proposed PDEs-based model, a very classical and elegant mathematical tool has been exploited: optimal control. This technique responds to the need for reducing the gap between PDEs and collected data. Indeed, the data information, given by some previous knowledge on the system, is exploited in order to achieve a *desired configuration*: a convenient profile similar to the behaviour observed or expected in nature. Namely, through a PDE-constrained minimization strategy, optimal control problems (OCPs) are able to steer the solution of the problem at hand towards a target profile. OCPs governed by PDEs are widespread in many scientific applications: some examples in shape optimization may be found in [10, 16, 25] or, in fluid dynamics, in [8, 9, 28] where a parametrized setting is discussed. In the same framework, we can cite biomedical applications [2, 7, 13, 22, 39, 47], or environmental ones [3, 30, 31, 36, 40]. From this references, it is clear that OCPs are of indisputable usefulness in many applications. Yet, they still are very complex to analyse and to simulate, most of all if time-dependency is taken into consideration, since it requires larger computational costs. Despite these difficulties, time-dependent OCPs have been treated in many works: here a far from exhaustive list [17, 20, 23, 34, 35, 37, 38, 40]. In contrast with the aforementioned works, we tackle the study of the evolution of OCPs trough the

E. Donadini · M. Strazzullo · M. Tezzele · G. Rozza (✉)
Mathematics Area, mathLab, SISSA, Trieste, Italy
e-mail: edonadini@sissa.it, mstrazzu@sissa.it; mtezzele@sissa.it; grozza@sissa.it

J. M. Melenk et al. (eds.), *Spectral and High Order Methods for Partial Differential Equations ICOSAHOM 2020+1*, Lecture Notes in Computational Science and Engineering 137, https://doi.org/10.1007/978-3-031-20432-6_13

employment of dynamic mode decomposition with control (DMDc) [29], which is a data-driven technique for system identification in the context of feedback and control. The idea is to use the desired state from the OCPs framework as forcing term within the DMDc. Then, with a partitioned approach, we use the solution snapshots to construct different DMD operators. With such operators, given new actuation snapshots, we perform future predictions for all the variables of interest, in a data-driven fashion. This results in great advantages in terms of computational time, keeping the accuracy within a certain threshold, as usual in the reduced order modeling framework [32]. The main novelty of this work, thus, is the recasting of an OCP in a DMDc framework: to the best of our knowledge, indeed, this is the first time that it is highlighted and investigated. We propose two test cases: a boundary control governed by a Graetz flow and a distributed control with Stokes constraints.

2 Time-Dependent Optimal Control Problems

We provide the continuous formulation for general time dependent OCP. Let us consider the spatial domain $\Omega \subset \mathbb{R}^d$, with $d = 2, 3$: here, the analysed physical phenomenon, described by a linear time-dependent PDE, is taking place in the time interval $[0, T]$. Furthermore, we denote with Γ_D and Γ_N two non-overlapping boundary portions where homogeneous Dirichlet and Neumann boundary conditions apply. In the context of constrained optimization, we consider an Hilbert space Y such that $Y \hookrightarrow H \hookrightarrow Y^*$ for some suitable H. Moreover, we define

$$\mathcal{Y}_t = \{y \in L^2(0, T; Y) \text{ such that } y_t \in L^2(0, T; Y^*) \text{ with } y(t) = 0\} \subset Q,$$

with $Q = L^2(0, T; Y)$. The problem *state variable* y is sought in $\mathcal{Y}_0$. In addition, we need to define $\mathcal{U} = L^2(0, T; U)$ as the space for the *control variable* u, with U another suitable Hilbert space.

The control acts on $\Omega_u \subseteq \Omega$ of a portion of its boundary, say $\Gamma_C \subset \partial\Omega$ with $\Gamma_C \cap \Gamma_N \cap \Gamma_D = \emptyset$ while $\Gamma_D \cup \Gamma_N \cup \Gamma_C = \partial\Omega$. In the first case, we say that the control problem is *distributed*, while in the second case, we say that the problem is a *boundary control*. With no distinctions, from now on, we will call Ω_u or Γ_C the *control domain*.

In order to change the classical solution behaviour, we need to define a *controlled system* of the following form, for all q in Q, considering once again y_0 as initial condition

$$\int_0^T \langle y_t, q\rangle_{Y^*,Y} + \int_0^T a(y, q)dt = \int_0^T c(u, q)dt + \int_0^T \langle G, q\rangle_{Y^*,Y}dt, \tag{1}$$

with initial condition y_0 where, $a : Y \times Y \to \mathbb{R}$ is a coercive and continuous bilinear form and $G \in Y^*$ represents the forcing and the boundary terms of the problem at hand, and $c : U \times Y \to \mathbb{R}$ is the L^2 product on the control domain. For the sake of notation, we define the controlled equation as $\mathcal{E} : (\mathcal{Y} \times \mathcal{U}) \times Q \to \mathbb{R}$. Namely, (1) is verified when $\mathcal{E}((y, u), q) = 0$ for all q in Q.

The goal of an OCP is to steer the solution towards a desired profile $y_d \in \mathcal{Z} \supseteq \mathcal{Y}$ by minimizing a convex *cost functional* $J : \mathcal{Y}_0 \times \mathcal{U} \to \mathbb{R}$ over $\mathcal{Y}_0 \times \mathcal{U}$ such that (1) is verified. The functional $J(y, u)$ must be defined case by case. For the well-posedness of the problem, the interested reader may refer to [18]. The minimization problem can be recast in an unconstrained fashion exploiting a Lagrangian approach [18, 43]. Thus, let us consider an arbitrary *adjoint variable* $z \in \mathcal{Y}_T \subset Q$ and define the following Lagrangian functional

$$\mathscr{L}(y, u, z) = J(y, u) + \mathcal{E}((y, u), z).$$

It is well known in literature [18], that the aforementioned minimization of problem is equivalent to the following system: find $(y, u, z) \in \mathcal{Y}_0 \times \mathcal{U} \times \mathcal{Y}_T$ such that

$$\begin{cases} D_y\mathscr{L}(y, u, z)[\omega] = 0 & \forall \omega \in Q \quad \text{(adjoint equation)}, \\ D_u\mathscr{L}(y, u, z)[\kappa] = 0 & \forall \kappa \in \mathcal{U} \quad \text{(optimality equation)}, \\ D_z\mathscr{L}(y, u, z)[\zeta] = 0 & \forall \zeta \in Q \quad \text{(state equation)}. \end{cases} \tag{2}$$

In this context, the terms D_y, D_u and D_z will denote the differentiation with respect to the state, the control and the adjoint variables.

We underline that, over the control domain, the optimality equation in strong form reads:

$$\alpha u - z = 0, \tag{3}$$

where the role of α will be clarified in the next sections. We decided to solve (2) using a *space-time discretization*. The space-time techniques have been successfully applied in many contexts, from parabolic equations [14, 44–46] to time-dependent PDE-constrained optimization [17, 19, 37, 38, 40]. For the sake of brevity, we are not presenting the details of such an approach. The main idea is to treat the three-equations system (2) as a steady one, where all the time instances are sought all-at-once by means of a direct solver. The reader may refer to the previous cited bibliography for an insight to the method

For our purposes, it is enough to define a time discretization over $[0, T]$ divided in N_t sub-intervals. Namely, we consider $\Delta t > 0$ and the time instance $t_k = k\Delta t$ for $k = 0, \dots, N_t$. Let us focus on the state, adjoint and control variables evaluated at t_k. At each t_k, the variables can be expressed through the corresponding spatial bases, in our case Finite Element (FE) bases. From now on, for the sake of notation, y_k will denote the column vector of FE coefficients of the FE expansion. The same argument applies to control and adjoint variables at t_k, denoted by u_k and p_k. We remark that state and adjoint have been discretized with the same discretized function space [37] to guarantee the well-posedness of the problem (2). To simulate the solution of the OCP we solve a system of dimension $N_t(2N_y + N_u) \times N_t(2N_y +$

$\mathcal{N}_u$) through a direct solver. This may lead to time consuming simulations to understand the field dynamic. To lighten these computational costs, one may use multigrid approaches combined with proper preconditioners, see e.g. [4, 33, 35] and the references therein. However, these iterative methods might only partially solve this issue. DMD-based techniques respond to the need of a fast prediction tool to avoid a complete simulation over the time interval $[0, T]$.

3 Dynamic Mode Decomposition with Control

Dynamic mode decomposition (DMD) is a powerful method to identify and approximate dynamical systems using only few spatiotemporal coherent structures [5, 21]. It is ideally suited for time-dependent problems, and its data-driven nature makes it very versatile, also in industrial contexts [11, 41, 42].

When dealing with actuated systems, unfortunately, DMD fails to properly reconstruct the underlying dynamics since it is incapable of incorporating the contribution of the forcing term. To this end, dynamic mode decomposition with control (DMDc) [29] has been developed to overcome this issue. By including the actuation snapshots in the analysis it is able to provide reduced order representations for input-output systems. The DMDc method quantifies the effect of control inputs on the state of the system and computes the underlying dynamics without being confounded by the effect of external control. Recently, an extension for quantum control problems, called bilinear DMD, has been proposed in [15]. A local version of DMDc for predictive control of hydraulic fracturing can be found in [27], while for compressive system identification see [1].

Let us denote the snapshot representing the state of a system at the i-th time instant with $x_i \in \mathbb{R}^{\mathcal{N}}$, where $\mathcal{N}$ represents the number of degrees of freedom of our system. We collect a set of N_t equispaced snapshots $\{x_i\}_{i=1}^{N_t}$ and we arrange them by column in two matrices, X and X', where X' is the time-shifted version of X. We also collect the input control snapshots $\{\eta_i\}_{i=1}^{N_t-1}$, with $\eta_i \in \mathbb{R}^{\mathcal{L}}$, in the matrix Υ, obtaining the following matrices:

$$X = \begin{bmatrix} | & | & & | \\ x_1 & x_2 & \dots & x_{N_t-1} \\ | & | & & | \end{bmatrix}, \quad X' = \begin{bmatrix} | & | & & | \\ x_2 & x_3 & \dots & x_{N_t} \\ | & | & & | \end{bmatrix}, \quad \Upsilon = \begin{bmatrix} | & | & & | \\ \eta_1 & \eta_2 & \dots & \eta_{N_t-1} \\ | & | & & | \end{bmatrix}.$$

We remark that the snapshots can represent data coming from experiments, from simulations, or even sensors and acquired in real-time.

DMDc is a regression-based approach to system identification that is able to disambiguate the intrinsic dynamics, described by the matrix A, and the effects of control, described by the matrix B, as in the following

$$X' \approx AX + B\Upsilon = \begin{bmatrix} A & B \end{bmatrix} \begin{bmatrix} X \\ \Upsilon \end{bmatrix} := G\Xi. \tag{4}$$

We seek an approximation of the linear mappings A and B using only the three data matrices. We start by computing the SVD of the matrix Ξ in (4), which contains both the state and control snapshot information, as $\Xi \approx U\Sigma V^*$. With the symbol * we denote the conjugate transpose, and we truncate the SVD keeping only the first r_Ξ modes. We can thus express the best-fit matrix $G \in \mathbb{R}^{\mathcal{N}\times(\mathcal{N}+\mathcal{L})}$ as

$$G \approx X'\Xi^{-1} = X'V\Sigma^{-1}U^*. \tag{5}$$

By rewriting $U^* = \begin{bmatrix} U_1^* & U_2^* \end{bmatrix}$, and by computing the SVD of the output matrix $X' \approx U_{X'}\Sigma_{X'}V_{X'}^*$, we can compute the reduced order approximation of A and B as

$$\tilde{A} = U_{X'}^* X'V\Sigma^{-1}U_1^*U_{X'}, \tag{6}$$

$$\tilde{B} = U_{X'}^* X'V\Sigma^{-1}U_2^*. \tag{7}$$

We remark that the truncation rank $r_{X'}$ of the SVD of the output matrix has to be less or equal to r_Ξ. With this reduced order operators we can write $\tilde{x}_{k+1} = \tilde{A}\tilde{x}_k + \tilde{B}\eta_k$, where $\tilde{x}_k = U_{X'}x_k$. The dynamic modes of A can be computed from the eigenvectors w of $\tilde{A}$ as

$$\phi = X'V\Sigma^{-1}U_1^*U_{X'}w. \tag{8}$$

4 A Partitioned Approach for Time-Dependent OCPs

In this section we are going to explain how to handle OCPs with DMDc. Our goal is to characterize the relationship between the current measurements of the time-dependent OCP x_k, the future one x_{k+1}, and the current input η_k, given by the desired state y_d at each time instant t_k for a finite number of steps $k = 1, \dots, N_t$:

$$x_{k+1} = Ax_k + B\eta_k, \quad \forall k = 1, \dots, N_t - 1.$$

We remark that the current input for our OCP system is the desired state y_d. In fact, it represents the target state that we want to achieve and hence all the OCP variables (state, control, and adjoint) depend on it. Due to such a dependency, a modification in the desired state will cause a change in the other variables. We present a *partitioned approach* in order to identify two separate dynamical systems: one for the state, and one for the adjoint variables. The control can then be reconstructed through α and the adjoint variable by using the linear relation in (3). For the state variable, we use as measurement matrices $X_y \in \mathbb{R}^{N_y\times(N_t-1)}$ and $X_y' \in \mathbb{R}^{N_y\times(N_t-1)}$ given by:

$$X_y = \begin{bmatrix} | & | & & | \\ y_1 & y_2 & \dots & y_{N_t-1} \\ | & | & & | \end{bmatrix}, \qquad X_y' = \begin{bmatrix} | & | & & | \\ y_2 & y_3 & \dots & y_{N_t} \\ | & | & & | \end{bmatrix}, \tag{9}$$

where each row represents the time series of the state measurement for a particular spatial point given by the space-time discretization. On the other hand, each column contains the coefficients of the FE expansion of the state space-time variable. The final model we obtain is:

$$X'_y = A_y X_y + B_y \Upsilon_d. \tag{10}$$

For the adjoint variable we have that $X_z \in \mathbb{R}^{N_y \times (N_t - 1)}$, $X'_z \in \mathbb{R}^{N_y \times (N_t - 1)}$ where each row represents the time series of the adjoint measurement and each column collects the coefficients of the FE expansion of the adjoint space-time variable. We obtain the DMDc model as:

$$X'_z = A_z X_z + B_z \Upsilon_d. \tag{11}$$

We emphasize that the input matrix Υ_d is the same for all the models.

5 Numerical Results

In this section, we present the numerical results to validate the DMDc approach applied to OCPs. The first example takes into consideration an unsteady Graetz Flow boundary OCP. A second test case deals with a time-dependent OCP governed by Stokes equations. The experiments validate the performances of DMDc strategy in term of reconstruction and prediction. Indeed, we define the time-pointwise relative error E_k as:

$$E_k = \frac{\|x_k - \tilde{x}_k\|_2}{\|x_k\|_2}, \tag{12}$$

where x_k represents the generic true snapshot at time t_k, and $\tilde{x}_k$ the approximated variable. This error is shown for the reconstruction analysis. For the prediction, we average (12) over all the time instances in the test data set. We vary the size of the training data set and we keep fixed 20 as dimension for the test data set, to understand the sensitivity of the method with respect to the data needed for an accurate prediction. For the computations we used the following libraries: multiphenics [26], which is an implementation in FEniCS [24] for block-based systems, and PyDMD [12].

5.1 *Boundary OCP Governed by a Graetz Flow*

Let us consider a boundary control governed by a Graetz Flow in the time interval $[0, T] = [0, 1]$ and in the rectangular space domain $\Omega = (0, 3) \times (0, 1) \in \mathbb{R}^2$ as depicted in Fig. 1.

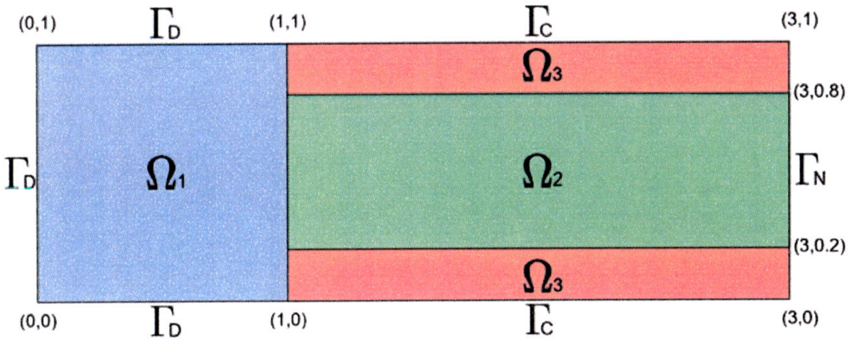

Fig. 1 *Graetz Flow*: space domain

The control domain is

$$\Gamma_C = ([1, 3] \times \{0\}) \cup ([1, 3] \times \{1\}),$$

while the observation domain is

$$\Omega_3 = ((1, 3) \times (0, 0.2)) \cup ((1, 3) \times (0.8, 1)).$$

The Neumann and the Dirichlet conditions are applied, respectively, to

$$\Gamma_N = \{3\} \times [0, 1] \text{ and } \Gamma_D = \partial\Omega \setminus (\Gamma_C \cup \Gamma_N).$$

We recall that $Y = H^1_{\Gamma_D^0}(\Omega)$, i.e. the set of functions that are H^1 in the domain but null on Γ_D, and the control space $U = L^2(\Gamma_C)$. We recall that we can recover an homogeneous problem of the form presented in Sect. 2 after a lifting procedure [30]. Thus, the Lagrangian formulation of this OCP reads: find $(y, u) \in \mathcal{Y}_0 \times \mathcal{U}$ which solves:

$$\min_{(y,u)\in\mathcal{Y}_0\times\mathcal{U}} J(y, u) = \frac{1}{2}\int_0^T \int_{\Omega_3} (y - y_d)^2 dxdt + \frac{\alpha}{2}\int_0^T \int_{\Gamma_C} u^2 dsdt \tag{13}$$

constrained to

$$\begin{cases} y_t - \epsilon\Delta y + \beta \cdot \nabla y = 0 & \text{in } \Omega \times (0, T), \\ y = 1 & \text{on } \Gamma_D \times (0, T), \\ \epsilon\dfrac{\partial y}{\partial n} = u & \text{on } \Gamma_C \times (0, T), \\ \epsilon\dfrac{\partial y}{\partial n} = 0 & \text{on } \Gamma_N \times (0, T), \\ y = y_0 & \text{in } \Omega \times \{0\}, \end{cases} \tag{14}$$

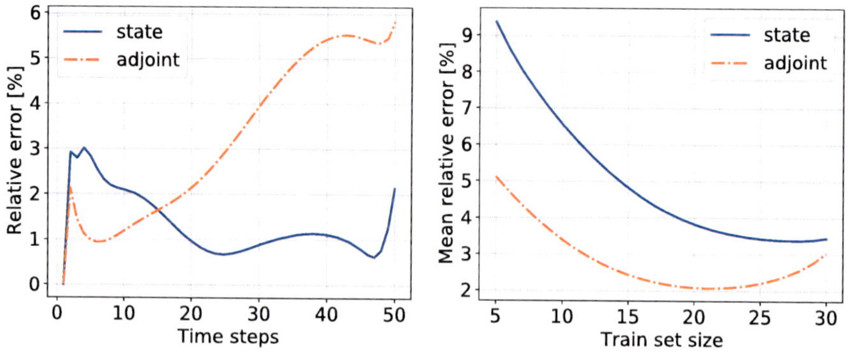

Fig. 2 On the left the L^2 relative reconstruction error for the Graetz case for state and adjoint variables. On the right the L^2 mean relative prediction error. The average is done over 20 snapshots in the future

where $\epsilon = \frac{1}{12}$, the vector filed β is defined as $[x_2(1 - x_2), 0]$, and x_2 as vertical spatial coordinate. The desired state profile is $y_d = 1+t$, $\alpha = 10^{-2}$ is a penalization parameter on the control, and y_0 is a null function verifying the boundary conditions. For the discretization, we employed $\mathbb{P}^1$ elements for all the variables with $\mathcal{N}_y = \mathcal{N}_u = 2304$. Moreover, in terms of time discretization, we consider $\Delta t = 0.02$, leading to $N_t = 50$ time instances for the time interval [0, 1]. Thus, the global space-time dimension is 345,600. In the left panel of Fig. 2 we plotted the relative reconstruction error E_k for all the variables of interest and for every $k = 1, \dots, 50$. We used 4 DMD modes for the state and 3 for the adjoint. The mean relative error for the state is 1.3%, while for the adjoint variable is 3.2%.

In Fig. 2 (right panel) we plotted the L^2 mean relative prediction error over 20 future states for the state, and adjoint variable, varying the dimension of the train dataset. We see that for the state variable we need at least 20 snapshots in order to have a satisfactory prediction accuracy below 4%, while for the adjoint variable 10 snapshots are sufficient.

5.2 *OCP Governed by Time-Dependent Stokes Equations*

We now deal with a distributed control problem governed by time-dependent Stokes equation. Let us consider the spatial domain Ω as the unit square in $\mathbb{R}^2$ an the time interval [0, 1]. For this specific case $\Gamma_D = \partial\Omega$, and we consider

$$\mathcal{V}_0 \doteq \left\{ v \in L^2(0, T; V) \text{ such that } v_t \in L^2(0, T; V^*) \text{ such that } v(0) = 0 \right\}$$

with

$$V \doteq [H^1_{\Gamma_D}(\Omega)]^2 \text{ and } \mathcal{P} = L^2(0, T; L^2_0(\Omega)),$$

where $L^2_0(\Omega)$ is the space of functions that have null mean over Ω with L^2 regularity.[1] We seek the state-control variable in $\mathcal{Y}_0 \times \mathcal{U}$ where $\mathcal{Y}_0 = \mathcal{V}_0 \times \mathcal{P}$ and $\mathcal{U} = [L^2(\Omega)]^2$. The specific problem we are dealing with is to find the minimum of

$$\min_{((v,p),u)\in\mathcal{Y}_0\times\mathcal{U}} J((v,p),u) := \frac{1}{2}\int_0^T \int_\Omega (v - v_d)^2 dx dt + \frac{\alpha}{2}\int_0^T \int_\Omega u^2 dx dt, \tag{16}$$

constrained to the equations:

$$\begin{cases} v_t - \Delta v + \nabla p = u & \text{in } \Omega \times (0,T), \\ -\nabla \cdot v = 0 & \text{in } \Omega \times (0,T), \\ v = 0 & \text{on } \partial\Omega \times (0,T), \\ v(0) = 0 & \text{in } \Omega \times \{0\}, \end{cases} \tag{17}$$

where

$$v_d = \left[10(1+t)\left(1 + \frac{1}{2}\cos(4\pi t - \pi)\right), 0\right] \in L^2(0,T;[L^2(\Omega)]^2) \tag{18}$$

is considered on the whole space-time domain, and $\alpha = 10^{-5}$. For the space discretization we employed $\mathbb{P}^2 - \mathbb{P}^1$ polynomials for the velocity and pressure fields both for the state and the adjoint variables, while for the control we used $\mathbb{P}^2$ polynomials. Thus, the FE dimension is 674 for state/adjoint velocity and control variables, while it is 337 for the state and adjoint pressure. In terms of time discretization, in the interval [0, 1] we considered $N_t = 50$ time instances, with $\Delta t = 0.02$. The final dimension of the system is 134,800.

In the left panel of Fig. 3 we plotted the relative reconstruction error E_k for all the variables of interest and for every $k = 1, \ldots, 50$. The error at the first time step is exactly 0, while the spike at the last time instant is due to the nature of DMD which divides the snapshots data into two shifted matrices and we have the accumulation of all the residuals. The maximum error is registered for the adjoint velocity and it is below 1.5%, while on average all the variables present an error below 0.6%. We remark that the adjoint pressure snapshots have been normalized removing the mean of all the states before applying the DMDc. Even if such preprocessing can be done only for reconstruction, we emphasize that the adjoint pressure in this case is not a

[1] Numerically, to enforce the condition of null mean for the state pressure variable, we employed Lagrange multipliers. Namely, the condition is weakly imposed in integral form

$$\int_\Omega p\lambda \, dx = 0, \quad \forall \lambda \in \mathbb{R}. \tag{15}$$

This constraint reflects in the derivation of the adjoint equation and an extra term appears in the divergence free constraint in weak form.

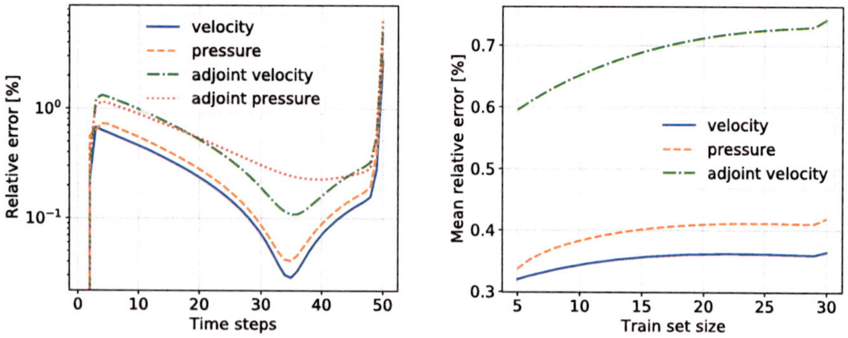

Fig. 3 On the left the L^2 relative reconstruction error for the Stokes case for velocity, pressure, and the adjoint variables. On the right the L^2 mean relative prediction error. The average is done over 20 snapshots in the future

variable of interest for prediction purposes since only the adjoint velocity affects the control in the optimality equation. Nonetheless we present its relative reconstruction error in order to show that the partitioned approach is able to deal also with such variable.

In Fig. 3 (right panel) we plotted the L^2 mean relative prediction error over 20 future states for velocity, pressure, and adjoint velocity, varying the dimension of the train dataset. We notice an almost stationary behaviour of the mean error for all the variables of interest, with all the values below 0.8%. This suggests that the future state prediction is robust with respect to the amount of available data.

5.3 *Speedup Considerations*

The computational gain provided by the data-driven approach is considerable for both the test problems. The one-shot approach takes between 13 and 18 minutes, depending on the problem at hand, in order to compute all the variables. The partitioned approach exploiting DMDc, instead, requires less than 1 second for the construction of the reduced operators, and approximately 2 seconds to predict all the quantities on interest for the entire time spans considered. This results in a speedup $\sim O(10^2)$.

6 Conclusions and Perspectives

In this work we showed the potential of data-driven methods for the resolution of time-dependent optimal control problems with PDE constraints. A further improvement of the work might concern a deeper analysis of the sensitivity of the problems with respect to the penalization parameter α which drastically changes the magnitude of the adjoint variable. Another possible approach could be to use the data-driven sparse identification of nonlinear dynamics with control method [6], instead of DMDc. Lastly, we stress that even if the test cases are quite academic, we believe in the potential of such an approach in more complex setting based on data collection. Indeed, this represents a first attempt for a deeper study of this formulation, which might be of interest in many interdisciplinary research fields.

Acknowledgments This work was partially funded by European Union Funding for Research and Innovation—Horizon 2020 Program—in the framework of European Research Council Executive Agency: H2020 ERC CoG 2015 AROMA-CFD project 681447 "Advanced Reduced Order Methods with Applications in Computational Fluid Dynamics" P.I. Professor Gianluigi Rozza.

References

1. Bai, Z., Kaiser, E., Proctor, J.L., Kutz, J.N., Brunton, S.L.: Dynamic mode decomposition for compressive system identification. AIAA J. **58**(2), 561–574 (2020). https://doi.org/10.2514/1.J057870
2. Ballarin, F., Faggiano, E., Manzoni, A., Quarteroni, A., Rozza, G., Ippolito, S., Antona, C., Scrofani, R.: Numerical modeling of hemodynamics scenarios of patient-specific coronary artery bypass grafts. Biomech. Model. Mechanobiol. **16**(4), 1373–1399 (2017). https://doi.org/10.1007/s10237-017-0893-7
3. Ballarin, F., Rozza, G., Strazzullo, M.: Chapter 9 - Space-time POD-Galerkin approach for parametric flow control. In: Trélat, E., Zuazua, E. (eds.) Numerical Control: Part A, Handbook of Numerical Analysis, vol. 23, pp. 307–338. Elsevier (2022). https://doi.org/10.1016/bs.hna.2021.12.009
4. Benzi, M., Golub, G.H., Liesen, J.: Numerical solution of saddle point problems. Acta Numer. **14**, 1–137 (2005). https://doi.org/10.1017/S0962492904000212
5. Brunton, S.L., Kutz, J.N.: Data-driven Science and Engineering: Machine Learning, Dynamical Systems, and Control. Cambridge University Press (2019). https://doi.org/10.1017/9781108380690
6. Brunton, S.L., Proctor, J.L., Kutz, J.N.: Sparse identification of nonlinear dynamics with control (SINDYc). IFAC-PapersOnLine **49**(18), 710–715 (2016). https://doi.org/10.1016/j.ifacol.2016.10.249
7. Carere, G., Strazzullo, M., Ballarin, F., Rozza, G., Stevenson, R.: A weighted pod-reduction approach for parametrized PDE-constrained optimal control problems with random inputs and applications to environmental sciences. Comput. Math. Appl. **102**, 261–276 (2021). https://doi.org/10.1016/j.camwa.2021.10.020
8. Dedè, L.: Optimal flow control for Navier-Stokes equations: drag minimization. Int. J. Numer. Methods Fluids **55**(4), 347–366 (2007). https://doi.org/10.1002/fld.1464
9. Dedè, L.: Reduced basis method and a posteriori error estimation for parametrized linear-quadratic optimal control problems. SIAM J. Sci. Comput. **32**(2), 997–1019 (2010). https://doi.org/10.1137/090760453
10. Delfour, M.C., Zolésio, J.: Shapes and Geometries: Metrics, Analysis, Differential Calculus, and Optimization, vol. 22. SIAM, Philadelphia (2011). https://doi.org/10.1137/1.9780898719826
11. Demo, N., Tezzele, M., Gustin, G., Lavini, G., Rozza, G.: Shape optimization by means of proper orthogonal decomposition and dynamic mode decomposition. In: Technology and

Science for the Ships of the Future: Proceedings of NAV 2018: 19th International Conference on Ship & Maritime Research, pp. 212–219. IOS Press (2018). https://doi.org/10.3233/978-1-61499-870-9-212
12. Demo, N., Tezzele, M., Rozza, G.: PyDMD: python dynamic mode decomposition. J. Open Source Softw. **3**(22), 530 (2018). https://doi.org/10.21105/joss.00530
13. Fevola, E., Ballarin, F., Jiménez-Juan, L., Fremes, S., Grivet-Talocia, S., Rozza, G., Triverio, P.: An optimal control approach to determine resistance-type boundary conditions from in-vivo data for cardiovascular simulations. Int. J. Numer. Methods Biomed. Eng. **37**(10) (2021). https://doi.org/10.1002/cnm.3516
14. Glas, S., Mayerhofer, A., Urban, K.: Two Ways to Treat Time in Reduced Basis Methods, pp. 1–16. Springer International Publishing, Cham (2017). https://doi.org/10.1007/978-3-319-58786-8_1
15. Goldschmidt, A., Kaiser, E., Dubois, J.L., Brunton, S.L., Kutz, J.N.: Bilinear dynamic mode decomposition for quantum control. New J. Phys. **23**(3), 033035 (2021). https://doi.org/10.1088/1367-2630/abe972
16. Haslinger, J., Mäkinen, R.A.E.: Introduction to Shape Optimization: Theory, Approximation, and Computation. Advances in Design and Control. SIAM, Philadelphia (2003). https://doi.org/10.1137/1.9780898718690
17. Hinze, M., Köster, M., Turek, S.: A hierarchical space-time solver for distributed control of the Stokes equation. Technical Report, SPP1253-16-01 (2008)
18. Hinze, M., Pinnau, R., Ulbrich, M., Ulbrich, S.: Optimization with PDE Constraints. Mathematical Modelling: Theory and Applications, vol. 23. Springer Netherlands (2009). https://doi.org/10.1007/978-1-4020-8839-1
19. Hinze, M., Köster, M., Turek, S.: A space-time multigrid method for optimal flow control. In: Constrained Optimization and Optimal Control for Partial Differential Equations, p. 147. Springer (2012). https://doi.org/10.1007/978-3-0348-0133-1_8
20. Iapichino, L., Trenz, S., Volkwein, S.: Reduced-order multiobjective optimal control of semilinear parabolic problems. In: Karasözen, B., Manguoğlu, M., Tezer-Sezgin, M., Göktepe, S., Uğur, Ö. (eds.) Numerical Mathematics and Advanced Applications ENUMATH 2015, pp. 389–397. Springer International Publishing, Cham (2016). https://doi.org/10.1007/978-3-319-39929-4_37
21. Kutz, J.N., Brunton, S.L., Brunton, B.W., Proctor, J.L.: Dynamic Mode Decomposition: Data-driven Modeling of Complex Systems. SIAM (2016). https://doi.org/10.1137/1.9781611974508
22. Lassila, T., Manzoni, A., Quarteroni, A., Rozza, G.: A reduced computational and geometrical framework for inverse problems in hemodynamics. Int. J. Numer. Methods Biomed. Eng. **29**(7), 741–776 (2013). https://doi.org/10.1002/cnm.2559
23. Leugering, G., Benner, P., Engell, S., Griewank, A., Harbrecht, H., Hinze, M., Rannacher, R., Ulbrich, S.: Trends in PDE Constrained Optimization. International Series of Numerical Mathematics, vol. 165. Springer, New York (2014). https://doi.org/10.1007/978-3-319-05083-6
24. Logg, A., Mardal, K., Wells, G.: Automated Solution of Differential Equations by the Finite Element Method. Springer, Berlin (2012)
25. Mohammadi, B., Pironneau, O.: Applied Shape Optimization for Fluids. Oxford University Press, New York (2010). https://doi.org/10.1093/acprof:oso/9780199546909.001.0001
26. multiphenics - easy prototyping of multiphysics problems in fenics, https://mathlab.sissa.it/multiphenics
27. Narasingam, A., Kwon, J.S.I.: Development of local dynamic mode decomposition with control: Application to model predictive control of hydraulic fracturing. Comput. Chem. Eng. **106**, 501–511 (2017). https://doi.org/10.1016/j.compchemeng.2017.07.002
28. Negri, F., Manzoni, A., Rozza, G.: Reduced basis approximation of parametrized optimal flow control problems for the Stokes equations. Comput. Math. Appl. **69**(4), 319–336 (2015). https://doi.org/10.1016/j.camwa.2014.12.010

29. Proctor, J.L., Brunton, S.L., Kutz, J.N.: Dynamic mode decomposition with control. SIAM J. Appl. Dynam. Syst. **15**(1), 142–161 (2016). https://doi.org/10.1137/15M1013857
30. Quarteroni, A., Rozza, G., Dedè, L., Quaini, A.: Numerical approximation of a control problem for advection-diffusion processes. In: Ceragioli, F., Dontchev, A., Futura, H., Marti, K., Pandolfi, L. (eds.) System Modeling and Optimization. International Federation for Information Processing, CSMO Conference on System Modeling and Optimization, vol. 199, pp. 261–273. Springer, Boston (2005)
31. Quarteroni, A., Rozza, G., Quaini, A.: Reduced basis methods for optimal control of advection-diffusion problems. In: Advances in Numerical Mathematics, no. 2006-003 in CMCS-CONF, pp. 193–216. RAS and University of Houston (2007)
32. Rozza, G., Hess, M., Stabile, G., Tezzele, M., Ballarin, F.: Basic ideas and tools for projection-based model reduction of parametric partial differential equations. In: Benner, P., Grivet-Talocia, S., Quarteroni, A., Rozza, G., Schilders, W.H.A., Silveira, L.M. (eds.) Model Order Reduction, vol. 2, chap. 1, pp. 1–47. De Gruyter, Berlin, Boston (2020). https://doi.org/10.1515/9783110671490-001
33. Schöberl, J., Zulehner, W.: Symmetric indefinite preconditioners for saddle point problems with applications to PDE-constrained optimisation problems. SIAM J. Matrix Anal. Appl. **29**(3), 752–773 (2007). https://doi.org/10.1137/060660977
34. Seymen, Z.K., Yücel, H., Karasözen, B.: Distributed optimal control of time-dependent diffusion–convection–reaction equations using space–time discretization. J. Comput. Appl. Math. **261**, 146–157 (2014). https://doi.org/10.1016/j.cam.2013.11.006
35. Stoll, M., Wathen, A.: All-at-once solution of time-dependent Stokes control. J. Comput. Phys. **232**(1), 498–515 (2013). https://doi.org/10.1016/j.jcp.2012.08.039
36. Strazzullo, M., Ballarin, F., Mosetti, R., Rozza, G.: Model reduction for parametrized optimal control problems in environmental marine sciences and engineering. SIAM J. Sci. Comput. **40**(4), B1055–B1079 (2018). https://doi.org/10.1137/17M1150591
37. Strazzullo, M., Ballarin, F., Rozza, G.: POD-Galerkin model order reduction for parametrized time dependent linear quadratic optimal control problems in saddle point formulation. J. Sci. Comput. **83**(55) (2020). https://doi.org/10.1007/s10915-020-01232-x
38. Strazzullo, M., Ballarin, F., Rozza, G.: A certified reduced basis method for linear parametrized parabolic optimal control problems in space-time formulation (2021). Submitted. https://arxiv.org/abs/2103.00460
39. Strazzullo, M., Zainib, Z., Ballarin, F., Rozza, G.: Reduced order methods for parametrized nonlinear and time dependent optimal flow control problems: towards applications in biomedical and environmental sciences. In: Numerical Mathematics and Advanced Applications ENUMATH 2019 (2021). https://doi.org/10.1007/978-3-030-55874-1_83
40. Strazzullo, M., Ballarin, F., Rozza, G.: POD-Galerkin model order reduction for parametrized nonlinear time-dependent optimal flow control: an application to shallow water equations. J. Numer. Math. **30**(1), 63–84 (2022). https://doi.org/10.1515/jnma-2020-0098
41. Tezzele, M., Demo, N., Stabile, G., Mola, A., Rozza, G.: Enhancing CFD predictions in shape design problems by model and parameter space reduction. Advanced Model. Simul. Eng. Sci. **7**(40) (2020). https://doi.org/10.1186/s40323-020-00177-y
42. Tezzele, M., Demo, N., Mola, A., Rozza, G.: An integrated data-driven computational pipeline with model order reduction for industrial and applied mathematics. In: Günther, M., Schilders, W. (eds.) Novel Mathematics Inspired by Industrial Challenges, no. 38 in Mathematics in Industry. Springer International Publishing (2022). https://doi.org/10.1007/978-3-030-96173-2_7
43. Tröltzsch, F.: Optimal Control of Partial Differential Equations: Theory, Methods and Applications. Graduate Studies in Mathematics, vol. 112 (2010). https://doi.org/10.1090/gsm/112
44. Urban, K., Patera, A.T.: A new error bound for reduced basis approximation of parabolic partial differential equations. C. R. Math. **350**(3–4), 203–207 (2012). https://doi.org/10.1016/j.crma.2012.01.026

45. Yano, M.: A space-time Petrov–Galerkin certified reduced basis method: Application to the Boussinesq equations. SIAM J. Sci. Comput. **36**(1), A232–A266 (2014). https://doi.org/10.1137/120903300
46. Yano, M., Patera, A.T., Urban, K.: A space-time hp-interpolation-based certified reduced basis method for Burgers' equation. Math. Models Methods Appl. Sci. **24**(09), 1903–1935 (2014). https://doi.org/10.1142/S0218202514500110
47. Zainib, Z., Ballarin, F., Fremes, S., Triverio, P., Jiménez-Juan, L., Rozza, G.: Reduced order methods for parametric optimal flow control in coronary bypass grafts, toward patient-specific data assimilation. Int. J. Numer. Methods Biomed. Eng. e3367 (2020). https://doi.org/10.1002/cnm.3367

A New Multiscale Discontinuous Galerkin Method for a Class of Second-Order Equations with Oscillatory Solutions in Two-Dimensional Space

Bo Dong and Wei Wang

1 Introduction

In this paper, we develop and numerically test a new multiscale discontinuous Galerkin (DG) methods for a class of stationary second-order equations in two dimensions,

$$\begin{cases} -\varepsilon^2 \Delta u - f u = 0, & \text{in } \Omega, \\ u = u_D & \text{on } \Gamma_D, \\ \frac{\partial u}{\partial n} - i\omega u = 2i\omega g & \text{on } \Gamma_N = \partial\Omega \setminus \Gamma_D, \end{cases} \tag{1}$$

where Ω is a rectangular domain in $\mathbb{R}^2$, $\varepsilon > 0$ is a small parameter, and $f > 0$ is a real-valued smooth function. One example of such equations is the stationary Schrödinger equation with open boundary conditions in the modeling of quantum transport in nanoscale semiconductors [2, 9, 13, 14].

It is easy to see that the problem (1) has oscillatory solutions with frequency $1/\varepsilon$. When ε is small, it is challenging to capture the high frequency wave solutions using traditional numerical methods because they require extremely refined meshes. Moreover, in applications such as the simulation of nanoscale semiconductor

Research was supported by the National Science Foundation (grant DMS-1818998).

B. Dong (✉)
University of Massachusetts, Dartmouth, MA, USA
e-mail: bdong@umassd.edu

W. Wang
Florida International University, Miami, FL, USA
e-mail: weiwang1@fiu.edu

J. M. Melenk et al. (eds.), *Spectral and High Order Methods for Partial Differential Equations ICOSAHOM 2020+1*, Lecture Notes in Computational Science and Engineering 137, https://doi.org/10.1007/978-3-031-20432-6_14

devices, this type of equations need to be numerically solved for many times. Thus, it is important to devise numerical methods that can efficiently solve the problem on coarse meshes.

One popular way to design such efficient numerical schemes is to incorporate the multiscale information of the problem into the finite element spaces and choose proper non-polynomial basis functions. There have been continuous efforts on developing multiscale methods along this line, including [1–3, 5, 6, 11, 12, 14–21]. Since DG methods [4] have advantages such as high-order accuracy, hp-adaptivity, and easy construction of basis in multidimensional spaces, third-order multiscale methods have been developed in the DG framework for one-dimensional stationary Schrödinger equations in [10, 17] and for the two-dimensional problem (1) in [13]. In [8], we generalized the low-order multiscale DG methods [10, 17] in one dimension to higher orders by defining two different families of multiscale basis functions, namely E^k and T^k, where the E^k basis consists of two exponential functions and polynomials and the T^k basis contains exponential functions for any odd number k. In [9], we developed a high-order multiscale DG method for the two-dimensional problem (1) by using the tensor product of the E^k basis and the polynomial basis. Numerical results showed that the method had better accuracy than traditional polynomial DG methods on both coarse and fine meshes.

In this work, we extend the definition of the one-dimensional T^k basis to any integer $k \geq 1$ and use it to design another new high-order multiscale DG for the two-dimensional problem (1). Briefly speaking, in the direction that the solution has stronger oscillations, we use the exponential basis functions in T^k and in the other direction we use polynomial basis for our finite element space. Our numerical experiments show that the method converges on coarse meshes where traditional DG methods do not. On fine meshes where traditional DG methods converge, our method has errors that are several magnitudes smaller than those of traditional DG methods with the same number of basis functions. We also apply our method to solve two-dimensional stationary Schrödinger equation in quantum transport, and we observe that our method produce better results than traditional DG methods on coarse meshes. The results of this new method are comparable to that in [9], both better than traditional DG using polynomials. It provides us another new competitive tool for solving equations with oscillatory solutions.

We organize the rest of the paper as follows. In Sect. 2, we define our multiscale DG methods. In Sect. 3, we display numerical results. Finally, we conclude in Sect. 4.

2 Multiscale DG Method

We partition the rectangular domain Ω using Cartesian meshes, letting $\Omega_h = \{K_{ij} = I_i \times J_j \mid I_i = (x_{i-\frac{1}{2}}, x_{i+\frac{1}{2}}), J_j = (y_{j-\frac{1}{2}}, y_{j+\frac{1}{2}}), i = 1, \cdots, N_x, j = 1, \ldots, N_y\}$, $\partial\Omega_h := \{\partial K_{ij} | K_{i,j} \in \Omega_h\}$ be the set of the boundaries of all elements, $\mathcal{E}_h$ be the set of all faces, and $\mathcal{E}_h^i := \mathcal{E}_h \setminus \partial\Omega$ be the set of all interior faces. We use the notation $x_i = \frac{1}{2}(x_{i-\frac{1}{2}} + x_{i+\frac{1}{2}})$, $y_j = \frac{1}{2}(y_{j-\frac{1}{2}} + y_{j+\frac{1}{2}})$, $\Delta x_i = x_{i+\frac{1}{2}} - x_{i-\frac{1}{2}}$, and $\Delta y_j = y_{j+\frac{1}{2}} - y_{j-\frac{1}{2}}$.

Next, we introduce the finite element spaces and the weak formulation of the approximate solutions, as well as the numerical traces that appear in the weak formulation.

For the multiscale finite element spaces in two-dimensional space, since the solution of the problem (1) oscillates mainly in x-direction due to the boundary condition, we use non-polynomial oscillatory basis functions in x-direction and the polynomial basis in y-direction. We define

$$\mathcal{T}^k := \{v \in L^2(\Omega_h) \mid v|_{K_{ij}} \in T^k|_{I_i} \otimes P^k|_{I_j}, \forall K_{ij} \in \Omega_h\},$$

where

$$P^k|_{J_j} = \text{span}\{1, \frac{y - y_j}{\Delta y_j}, \cdots, (\frac{y - y_j}{\Delta y_j})^k\}$$

and

$$T^k|_{I_i} = \begin{cases} \text{span}\{e^{\pm i\omega(x-x_i)}, \cdots, e^{\pm i\ell\omega(x-x_i)}\} & \text{if } k \text{ is odd} \\ \text{span}\{e^{\pm i\omega(x-x_i)}, \cdots, e^{\pm i\ell\omega(x-x_i)}, e^{i(\ell+1)\omega(x-x_i)}\} & \text{if } k \text{ is even} \end{cases}$$

for $k \geq 1$, where $\ell = \left\lfloor \frac{k+1}{2} \right\rfloor$. Here, the notation $\lfloor x \rfloor$ is the floor function that gives the largest integer that is less than or equal to x. For vector-valued functions, we define the finite element space

$$\mathbf{T}^k := \{\boldsymbol{w} = (w_1, w_2) \mid w_i \in \mathcal{T}^k, i = 1, 2\}.$$

Note that the T^k basis above is the same as in [8] when k is odd. Now we extend its definition to include even k. Also note that $T^1 = E^1$, and thus the multiscale finite element space $\mathcal{T}^1 = M^1$ and $\mathbf{T}^1 = \mathbf{M}^1$ in [9]. We would like to remark that if the solution mainly oscillates in y-direction, then the space is defined in a similar way with T^k in y-direction and P^k in x-direction.

To define the weak formulation of our multiscale DG method, we rewrite the second-order equation in (1) into the mixed form

$$\boldsymbol{q} - \varepsilon \nabla u = 0, \tag{2a}$$

$$-\varepsilon \nabla \cdot \boldsymbol{q} - fu = 0, \tag{2b}$$

with the boundary conditions

$$u = u_D \quad \text{on } \Gamma_D, \tag{2c}$$

$$\boldsymbol{q} \cdot \boldsymbol{n} - \boldsymbol{i}\varepsilon\omega u = 2\boldsymbol{i}\varepsilon\omega g \quad \text{on } \Gamma_N. \tag{2d}$$

The weak formulation of our DG methods for (2) is to find approximate solutions $(u_h, \boldsymbol{q}_h) \in \mathcal{T}^k \times \mathbf{T}^k$ such that

$$(\boldsymbol{q}_h, \boldsymbol{w})_{\Omega_h} + (\varepsilon u_h, \nabla \cdot \boldsymbol{w})_{\Omega_h} - \langle \varepsilon \widehat{u}_h, \boldsymbol{w} \cdot \boldsymbol{n} \rangle_{\partial \Omega_h} = 0 \tag{3a}$$

$$(\varepsilon \boldsymbol{q}_h, \nabla v)_{\Omega_h} - \langle \varepsilon \widehat{\boldsymbol{q}}_h, v\,\boldsymbol{n} \rangle_{\partial \Omega_h} - (f u_h, v)_{\Omega_h} = 0 \tag{3b}$$

for any test functions $v \in \mathcal{T}^k$ and $\boldsymbol{w} \in \mathbf{T}^k$. Here, we have used the notation

$$(\varphi, v)_{\Omega_h} = \sum_{i,j} \int_{K_{i,j}} \varphi \bar{v}\, dxdy, \qquad (\boldsymbol{\psi}, \boldsymbol{w})_{\Omega_h} = \sum_{i,j} \int_{K_{i,j}} \boldsymbol{\psi} \cdot \overline{\boldsymbol{w}}\, dxdy,$$

$$\langle \varphi, v \rangle_{\partial \Omega_h} = \sum_{i,j} \int_{\partial K_{i,j}} \varphi \bar{v} ds, \qquad \langle \boldsymbol{\psi}, \boldsymbol{w} \rangle_{\partial \Omega_h} = \sum_{i,j} \int_{\partial K_{i,j}} \boldsymbol{\psi} \cdot \overline{\boldsymbol{w}} ds,$$

where $\overline{v}$ is the complex conjugate of any function v and $\boldsymbol{n}$ is the unit outward normal vector.

As to the definition of the numerical traces in (3), we need introduce some notation. At any interior face $e \in \mathcal{E}_h^i$ shared by two elements K_1 and K_2, the average and jump of a scalar function v are given by

$$\{v\} = \frac{1}{2}(v_1 + v_2), \quad [\![v\boldsymbol{n}]\!] = v_1 \boldsymbol{n}_1 + v_2 \boldsymbol{n}_2,$$

where $v_i = v|_{\partial K_i}$ and $\boldsymbol{n}_i$ is the unit normal vector on e pointing exterior to K_i for $i = 1, 2$. Similarly, for a vector-valued function $\boldsymbol{w}$, we let

$$\{\boldsymbol{w}\} = \frac{1}{2}(\boldsymbol{w}_1 + \boldsymbol{w}_2), \quad [\![\boldsymbol{w} \cdot \boldsymbol{n}]\!] = \boldsymbol{w}_1 \cdot \boldsymbol{n}_1 + \boldsymbol{w}_2 \cdot \boldsymbol{n}_2.$$

At any boundary edge $e \in \partial\Omega$, we define

$$[\![v\boldsymbol{n}]\!] = v\boldsymbol{n}, \quad [\![\boldsymbol{w} \cdot \boldsymbol{n}]\!] = \boldsymbol{w} \cdot \boldsymbol{n}.$$

In our multiscale DG schemes, we define the numerical traces as follows. At any interior element interface $e \in \mathcal{E}_h^i$, we let

$$\widehat{u}_h = u_h^- - \boldsymbol{i}\,\beta\,[\![\boldsymbol{q}_h \cdot \boldsymbol{n}]\!], \tag{4a}$$

$$\widehat{\boldsymbol{q}}_h = \boldsymbol{q}_h^+ + \boldsymbol{i}\,\alpha\,[\![u_h \boldsymbol{n}]\!], \tag{4b}$$

where we use the notation

$$v^- = \{v\} + \boldsymbol{\rho}_0 \cdot [\![v\boldsymbol{n}]\!], \qquad \boldsymbol{w}^+ = \{\boldsymbol{w}\} - \boldsymbol{\rho}_0[\![\boldsymbol{w} \cdot \boldsymbol{n}]\!].$$

The $\boldsymbol{\rho}_0$ term above is any function on $\mathscr{E}_h$ such that for any point $e \in \partial K \cap \mathscr{E}_h$,

$$\boldsymbol{\rho}_0 \cdot \boldsymbol{n}_K(e) = \frac{1}{2}\text{sign}(\boldsymbol{w}_0 \cdot \boldsymbol{n}_K(e)),$$

where $\boldsymbol{w}_0$ is any nonzero piecewise constant vector in $\mathbf{H}(\text{div}, \Omega)$. In (4), we include the penalty terms so that the numerical traces are defined in a general way. Numerical tests in [10] showed that nonzero penalty terms in the numerical traces helped reduce resonance errors when the mesh size is comparable to the wave length. When the penalty parameters α and β are taken to be zero, the numerical traces are in the same form as the alternating fluxes of the standard Minimal Dissipation Local DG in [7].

At the domain boundary Γ_D, we define the numerical traces as

$$\langle \widehat{u}_h, v\rangle_{\Gamma_D} = \langle u_D, v\rangle_{\Gamma_D}, \tag{5a}$$

$$\widehat{\boldsymbol{q}}_h = \boldsymbol{q}_h + \boldsymbol{i}\theta(u_h - \widehat{u}_h)\boldsymbol{n} \;\; \text{on } \Gamma_D. \tag{5b}$$

On the boundary Γ_N, we let

$$\langle \widehat{u}_h, v\rangle_{\Gamma_N} = \langle u_h + \boldsymbol{i}\frac{\gamma}{\varepsilon\omega}\mathbf{R}_h \cdot \boldsymbol{n}, v\rangle_{\Gamma_N}, \tag{5c}$$

$$\langle \widehat{\boldsymbol{q}}_h, \boldsymbol{w}\rangle_{\Gamma_N} = \langle \boldsymbol{q}_h - (1-\gamma)\mathbf{R}_h, \boldsymbol{w}\rangle_{\Gamma_N}, \tag{5d}$$

where γ can be any real constant in $(0, 1)$ and $\mathbf{R}_h := \boldsymbol{q}_h - \boldsymbol{i}\varepsilon\omega(u_h + 2g)\boldsymbol{n}$. The numerical traces in (5) are defined so that they satisfy the following equation,

$$\langle \widehat{\boldsymbol{q}}_h \cdot \boldsymbol{n} - \boldsymbol{i}\varepsilon\omega\, \widehat{u}_h, v\rangle_{\Gamma_N} = \langle 2\boldsymbol{i}\varepsilon\omega g, v\rangle_{\Gamma_N} \quad \text{for any } v \in V_h,$$

which is the discrete version of the boundary condition (2d) for the exact solutions. Again, in implementation the penalty parameter γ can be taken as zero for simplicity.

3 Numerical Results

In this section, we perform several numerical experiments to test the performance of our multiscale DG methods for two dimensional problems on different meshes. The test examples are the same as those in [9]. The solutions of the test problems have oscillations mainly in x-direction, so our multiscale DG method uses the multiscale finite element spaces $\mathcal{T}^1 = T^1 \otimes P^1$, $\mathcal{T}^2 = T^2 \otimes P^2$ and $\mathcal{T}^3 = T^3 \otimes P^3$. For simplicity, we take the penalty parameters in the numerical traces to be zero in the numerical tests.

In the first example, we show that our multiscale DG method can capture the oscillatory solution exactly (up to round-off errors) if the solution is in the multiscale finite element spaces. In the second example, we test the convergence orders of our multiscale DG method for two levels of ε. In the third example, we apply the method in the simulation of a two-dimensional Schrödinger problem. In the second and third examples, we also compare our multiscale DG method using $\mathcal{T}^k$ with the traditional minimal-dissipation local DG method [7] using Q^k, where Q^k is the space of piecewise polynomials up to degree k in each variable and therefore has the same number of basis functions as $\mathcal{T}^k$. Numerical results show that our multiscale DG method provides more accurate approximations than traditional DG methods for these oscillatory solutions on both cases meshes and fine meshes. In all three examples, we can see that the performance of our new method is comparable to that in [9], which is better than traditional DG using polynomials.

3.1 Constant f and $\omega = \frac{\sqrt{f}}{\varepsilon}$

Example In the first example, we consider the simple case of Eq. (1) with constant function $f(x) = 10$. The boundary conditions are given as

$$\begin{cases} u_x(0, y) + i\omega u(0, y) = 2i\omega \\ u_x(1, y) - i\omega u(1, y) = 0 \\ u(x, 0) = e^{i\frac{\sqrt{10}}{\varepsilon}x} \\ u(x, 1) = e^{i\frac{\sqrt{10}}{\varepsilon}x} \end{cases},$$

where $\omega = \sqrt{10}/\varepsilon = \sqrt{f}/\varepsilon$. It has the exact solution $u = e^{i\frac{\sqrt{10}}{\varepsilon}x}$. In this example, the exact solution lies in the multiscale finite element spaces. Thus the proposed multiscale DG methods with these spaces are able to compute the solution exactly with round-off errors. The L^2-errors of the multiscale DG with multiscale spaces $\mathcal{T}^1, \mathcal{T}^2$ and $\mathcal{T}^3$ are shown in Table 1 for two different levels of ε, i.e. $\varepsilon = 0.1$ and a smaller $\varepsilon = 0.03$. We see nearly round-off errors in double precision for all results.

Table 1 Example 3.1: L^2-errors by multiscale DG methods for $f(x) = 10$

	$\varepsilon = 0.1$			$\varepsilon = 0.03$		
$N_x \times N_y$	$\mathcal{T}^1$	$\mathcal{T}^2$	$\mathcal{T}^3$	$\mathcal{T}^1$	$\mathcal{T}^2$	$\mathcal{T}^3$
4×16	1.80E-11	4.26E-12	6.34E-13	1.64E-11	2.70E-11	4.43E-11
8×32	1.24E-13	4.66E-13	3.65E-13	3.43E-12	8.36E-12	3.68E-11

We remark that in our computations, the integrals involving exponential functions are integrated numerically by quadrature rules which produces small errors.

3.2 Constant f, $\omega \neq \frac{\sqrt{f}}{\varepsilon}$

Example In this example, we consider a constant $f(x) = 10+\varepsilon^2\pi^2$. The boundary conditions are given as

$$\begin{cases} u_x(0, y) + i\omega u(0, y) = 2i\omega \sin(\pi y) \\ u_x(1, y) - i\omega u(1, y) = 0 \\ u(x, 0) = 0 \\ u(x, 1) = 0 \end{cases},$$

where $\omega = \sqrt{10}/\varepsilon$. It has the exact solution $u = e^{i\frac{\sqrt{10}}{\varepsilon}x}\sin(\pi y)$ which contains a highly oscillatory wave function in the x-direction and a sine function in the y-direction. The multiscale finite element spaces are able to capture the oscillatory wave function in the x-direction. In the numerical results, we compare multiscale DG methods with traditional polynomial DG methods in the same order of finite elements spaces. For multiscale DG, we use coarse meshes in the x-direction since the multiscale finite element spaces are designed to capture the oscillatory solution in the x-direction.

Tables 2, 3, 4 and Tables 5, 6, 7 show the comparisons between multiscale DG methods with multiscale spaces $\mathcal{T}^1$, $\mathcal{T}^2$ and $\mathcal{T}^3$ and traditional polynomial DG methods with Q^1, Q^2 and Q^3 for $\varepsilon = 0.1$ and a smaller $\varepsilon = 0.03$, respectively. From Tables 2, 3, and 4, we can see that when $\varepsilon = 0.1$, the mesh size is smaller than ε, both multiscale DG methods with multiscale spaces $\mathcal{T}^1$, $\mathcal{T}^2$ and $\mathcal{T}^3$ and traditional polynomial DG methods with Q^1, Q^2 and Q^3 achieve the optimal $(k+1)$-th order of convergence. However, the errors of multiscale DG methods are several magnitudes smaller than those of traditional polynomial DG methods even though the mesh size of the multiscale DG in x-direction is four times larger. For a smaller ε, $\varepsilon = 0.03$, in

Table 2 Example 3.2. Comparison between multiscale DG $\mathcal{T}^1$ and polynomial DG Q^1 for $\varepsilon = 0.1$

Multiscale $\mathcal{T}^1$			Polynomial Q^1		
$N_x \times N_y$	Error	Order	$N_x \times N_y$	Error	Order
			8×8	7.55E-01	
4×16	1.83E-03	–	16×16	2.55E-01	1.56
8×32	4.42E-04	2.05	32×32	4.53E-02	2.49
16×64	1.05E-04	2.08	64×64	1.09E-02	2.06

Table 3 Example 3.2: Comparison between multiscale DG $\mathcal{T}^2$ and polynomial DG Q^2 for $\varepsilon = 0.1$

Multiscale $\mathcal{T}^2$			Polynomial Q^2		
$N_x \times N_y$	Error	Order	$N_x \times N_y$	Error	Order
			8×8	2.49E-01	
4×16	3.50E-05	–	16×16	2.93E-02	3.09
8×32	3.29E-06	3.41	32×32	3.38E-03	3.12
16×64	4.08E-07	3.01	64×64	4.18E-04	3.02

Table 4 Example 3.2: Comparison between multiscale DG $\mathcal{T}^3$ and polynomial DG Q^3 for $\varepsilon = 0.1$

Multiscale $\mathcal{T}^3$			Polynomial Q^3		
$N_x \times N_y$	Error	Order	$N_x \times N_y$	Error	Order
			8×8	5.50E-02	
4×16	5.63E-07	–	16×16	3.23E-03	4.09
8×32	2.48E-08	4.50	32×32	2.02E-04	4.00
16×64	1.32E-09	4.23	64×64	–	–

Table 5 Example 3.2: Comparison between multiscale DG $\mathcal{T}^1$ and polynomial DG Q^1 for $\varepsilon = 0.03$

Multiscale $\mathcal{T}^1$			Polynomial Q^1		
$N_x \times N_y$	Error	Order	$N_x \times N_y$	Error	Order
			8×8	7.06E-01	
4×16	2.21E-03		16×16	7.15E-01	-0.02
8×32	4.29E-04	2.37	32×32	6.98E-01	0.03
16×64	1.05E-04	2.03	64×64	3.16E-01	1.14

Table 6 Example 3.2: Comparison between multiscale DG $\mathcal{T}^2$ and polynomial DG Q^2 for $\varepsilon = 0.03$

Multiscale $\mathcal{T}^2$			Polynomial Q^2		
$N_x \times N_y$	Error	Order	$N_x \times N_y$	Error	Order
			8×8	7.22E-01	
4×16	3.64E-05	–	16×16	7.03E-01	0.04
8×32	7.71E-06	2.24	32×32	4.86E-01	0.53
16×64	4.37E-07	4.14	64×64	1.59E-02	4.93

Tables 5, 6, and 7 we can still see the $(k+1)$-th order of convergence for multiscale DG methods at all mesh levels. However, standard polynomial DG methods can not approximate the solutions well until the mesh is refined to $h < \varepsilon$. For example, polynomial DG methods with Q^1 and Q^2 do not have any order of convergence until the mesh is refined to 64×64. We remark that due to the limitation of the single processor, we are not able to compute polynomial DG methods with Q^3 on the 64×64 mesh in Tables 4 and 7.

Table 7 Example 3.2: Comparison between multiscale DG $\mathcal{T}^3$ and polynomial DG Q^3 for $\varepsilon = 0.03$

Multiscale $\mathcal{T}^3$			Polynomial Q^3		
$N_x \times N_y$	Error	Order	$N_x \times N_y$	Error	Order
			8×8	7.16E-01	
4×16	6.45E-07	–	16×16	7.32E-01	-0.03
8×32	2.22E-08	4.86	32×32	3.16E-02	4.54
16×64	1.28E-09	4.11	64×64	–	–

3.3 *Applications to 2D Schrödinger Equation*

Example In this example, we apply the multiscale method to a two-dimensional stationary Schrödinger problem on the domain $[0, 1] \times [0, 1]$,

$$\begin{aligned}
&-\varepsilon^2 \Delta\phi + V(x, y)\phi = E\phi,\\
&\phi_x(0, y) + i\omega\phi(0, y) = 2i\omega,\\
&\phi_x(1, y) - i\omega\phi(1, y) = 0,\\
&\phi(x, 0) = 0,\\
&\phi(x, 1) = 0.
\end{aligned}$$

We let $\varepsilon = 0.03$, the energy $E = 1$, the external potential $V(x, y) = \frac{1}{2}\sin x \cos y - 1$, and $\omega = \sqrt{2}/\varepsilon$.

We do not have the formula of the exact solution. The reference solutions are computed by the local DG method with the polynomial space P^2 on the mesh with 80×80 elements; see Fig. 1. On Fig. 2, we compare the numerical results by the multiscale DG using the $\mathcal{T}^1$ space and the standard local DG using the polynomial Q^1 space on the same 32×32 coarse mesh. We can see that using the same degree of freedom, multiscale DG with $\mathcal{T}^1$ is able to capture the shape and height of the waves pretty well. However, the polynomial DG produces very spurious waves. This is because standard polynomial DG methods need to have the mesh size smaller than the wave length in order to resolve the oscillations in the solution. When the mesh is refined, standard polynomial DG will be able to approximate the solution well. Next, we compare the numerical results by the multiscale DG with $\mathcal{T}^2$ and the polynomial DG with Q^2 on the 16×32 coarse mesh; see Fig. 3. Again, we observe that multiscale DG with $\mathcal{T}^2$ is able to capture the oscillatory waves very well, but the polynomial DG with Q^2 produces spurious waves. Similar results are observed in Fig. 4 when we compare the multiscale DG with $\mathcal{T}^3$ and the polynomial DG with Q^3 on the 16×16 mesh. Therefore, the multiscale DG is able to approximate the solution on coarse meshes and uses less degree of freedom than standard polynomial DG methods. Thus it is more efficient and accurate than the standard DG methods for solving problems involving small scales.

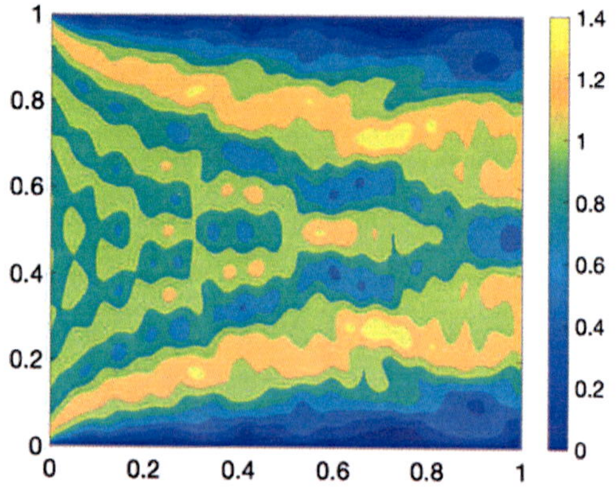

Fig. 1 Example 3.3: reference solution

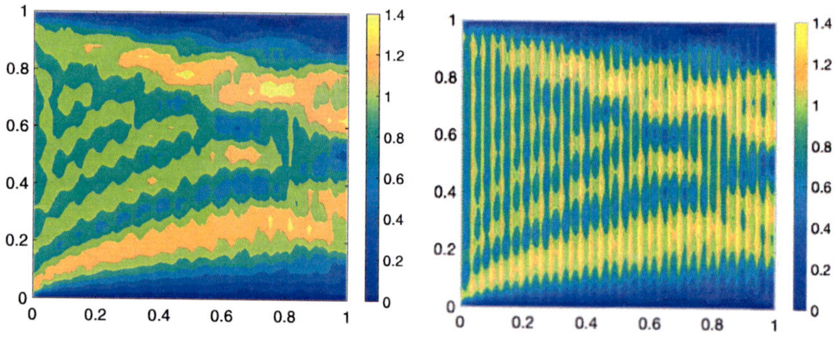

Fig. 2 Example 3.3: Numerical solutions on the 32 × 32 mesh. Left: multiscale DG using $\mathcal{T}^1$. Right: polynomial DG using Q^1

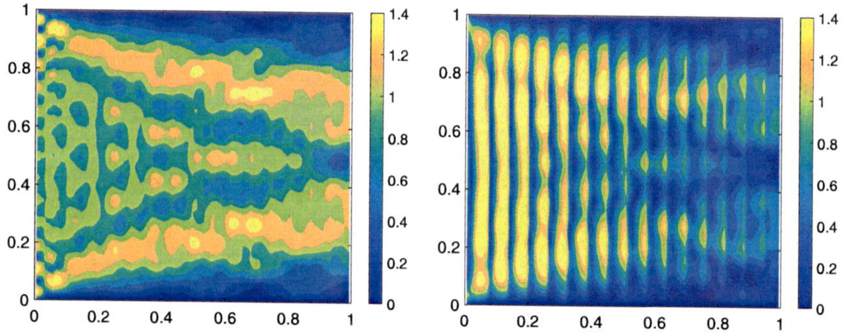

Fig. 3 Example 3.3: Numerical solutions on the 16 × 32 mesh. Left: multiscale DG using $\mathcal{T}^2$. Right: polynomial DG using Q^2

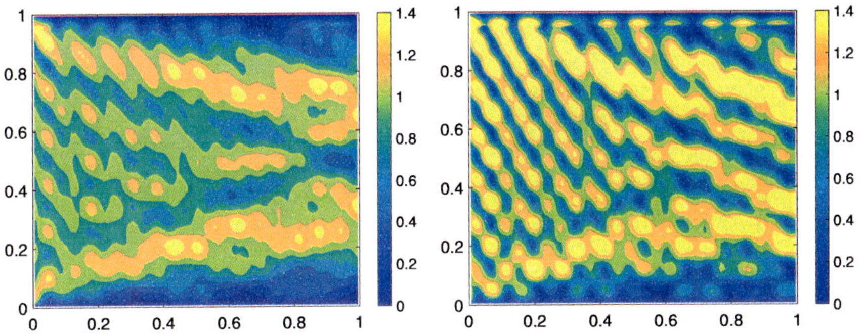

Fig. 4 Example 3.3: Numerical solutions on the 16×16 mesh. Left: multiscale DG using $\mathcal{T}^3$. Right: polynomial DG using Q^3

4 Concluding Remarks

In this paper, we extend our previous work of high order multiscale DG methods for one-dimensional stationary Schrödinger equations in [8] to two-dimensional space. Since the solution under consideration has frequency change mainly in one direction, we use oscillatory non-polynomial basis functions T^k in that direction and the polynomial basis in the other direction. Numerically we observe that the method converges on coarse meshes and has an optimal convergence order when the mesh size is refined to the scale of the wave length. We compare them with traditional polynomial DG methods and see that our multiscale method has better performance in approximating oscillatory solutions. We also demonstrate their ability to capture high frequency waves in the application of the stationary Schrödinger equations. In the future work, we would like to develop the multiscale discontinuous Galerkin method for more general two-dimensional problems with oscillatory solutions in both directions.

References

1. Aarnes, J., Heimsund, B.-O.: Multiscale discontinuous Galerkin methods for elliptic problems with multiple scales. In: Multiscale Methods in Science and Engineering 1–20, Lect. Notes Comput. Sci. Eng., vol. 44. Springer, Berlin (2005)
2. Abdallah, N.B., Pinaud, O.: Multiscale simulation of transport in an open quantum system: Resonances and WKB interpolation. J. Comput. Phys. **213**, 288–310 (2006)
3. Abdallah, N.B., Mouis, M., Negulescu, C.: An accelerated algorithm for 2D simulations of the quantum ballistic transport in nanoscale MOSFETs. J. Comput. Phys. **225**, 74–99 (2007)
4. Arnold, D.N., Brezzi, F., Cockburn, B., Marini, L.D.: Unified analysis of discontinuous Galerkin methods for elliptic problems. SIAM J. Numer. Anal. **39**, 1749–1779 (2002)
5. Arnold, A., Abdallah, N.B., Negulescu, C.: WKB-based schemes for the oscillatory 1D Schrödinger equation in the semiclassical limit. SIAM J. Numer. Anal., **49**, 1436–1460 (2011)

6. Buffa, A., Monk, P.: Error estimates for the Ultra Weak Variational Formulation of the Helmholtz equation. ESAIM: M2AN Math. Model. Numer. Anal. **42**, 925–940 (2008)
7. Cockburn, B., Dong, B.: An analysis of the minimal dissipation local discontinuous Galerkin method for convection-diffusion problems. J. Sci. Comput. **32**, 233–262 (2007)
8. Dong, B., Wang, W.: High-order multiscale discontinuous Galerkin methods for the one-dimensional stationary Schrodinger equation. J. Comput. Appl. Math. **380**, 1–11 (2020)
9. Dong, B., Wang, W.: A high-order multiscale discontinuous Galerkin method for two-dimensional Schrödinger equation in quantum transport. J. Comput. Appl. Math. **418**, 114701 (2023). ISSN 0377–0427. https://doi.org/10.1016/j.cam.2022.114701
10. Dong, B., Shu, C.-W., Wang, W.: A new multiscale discontinuous Galerkin method for the one-dimensional stationary Schrodinger equation. J. Sci. Comput. **66**, 321–345 (2016)
11. Gabard, G.: Discontinuous Galerkin methods with plane waves for time-harmonic problems. J. Comput. Phys. **225**, 1961–1984 (2007)
12. Gittelson, C., Hiptmair, R., Perugia, I.: Plane wave discontinuous Galerkin methods: Analysis of the h-version. ESAIM: M2AN Math. Model. Numer. Anal. **43**, 297–331 (2009)
13. Guo, L., Xu, Y.: Local discontinuous galerkin methods for the 2D simulation of quantum transport phenomena on quantum directional coupler. Commun. Comput. Phys. **15**, 1012–1028 (2014)
14. Negulescu, C.: Numerical analysis of a multiscale finite element scheme for the resolution of the stationary Schrödinger equation. Numer. Math. **108**, 625–652 (2008)
15. Negulescu, C., Abdallah, N.B., Polizzi, E., Mouis, M.: Simulation schemes in 2D nanoscale MOSFETs: A WKB based method. J. Comput. Electron. **3**, 397–400 (2004)
16. Polizzi, E., Abdallah, N.B.: Subband decomposition approach for the simulation of quantum electron transport in nanostructures. J. Comput. Phys. **202**, 150–180 (2005)
17. Wang, W., Shu, C.-W.: The WKB local discontinuous Galerkin method for the simulation of Schrödinger equation in a resonant tunneling diode. J. Sci. Comput. **40**, 360–374 (2009)
18. Wang, W., Guzmán, J., Shu, C.-W.: The multiscale discontinuous Galerkin method for solving a class of second order elliptic problems with rough coefficients. Int. J. Numer. Anal. Model **8**, 28–47 (2011)
19. Yuan, L., Shu, C.-W.: Discontinuous Galerkin method based on non-polynomial approximation spaces. J. Comput. Phys. **218**, 295–323 (2006)
20. Yuan, L., Shu, C.-W.: Discontinuous Galerkin method for a class of elliptic multi-scale problems. Int. J. Numer. Methods Fluids **56**, 1017–1032 (2008)
21. Zhang, Y., Wang, W., Guzmán, J., Shu, C.-W.: Multi-scale discontinuous Galerkin method for solving elliptic problems with curvilinear unidirectional rough coefficients. J. Sci. Comput. **61**, 42–60 (2014)

Hybrid High-Order Methods for Elliptic PDEs on Curved and Complicated Domains

Zhaonan Dong and Zuodong Wang

1 Introduction

Hybrid high-order (HHO) methods were introduced for solving the linear diffusion problems in [10] and for locking-free linear elasticity problems in [9]. In the HHO method, the discrete unknowns are attached to the mesh cells, and the mesh faces. The two key ingredients for HHO methods are local reconstruction operators and local stabilization operators in each mesh cell. HHO methods offer various attractive features, such as the support of polytopal meshes, optimal error estimates, local conservation properties, and computational efficiency due to compact stencils and local elimination of the cell unknowns by static condensation. We refer to two monographs [6, 7] for a comprehensive review.

The present work aims to analyze and test a variant of the HHO method employing Nitsche's techniques for solving the Poisson problem with mixed boundary conditions on curved and complicated domains. The complicated domain in this work is denoted as the Lipschitz domain with a possibly arbitrary number of $(d-1)$-dimensional tiny faces. Using Nitsche's techniques for weakly enforcing the boundary conditions or the interface conditions for HHO methods has been applied for second-order, and fourth-order elliptic PDEs on polygonal and polyhedral meshes cells [2, 3, 5, 11, 12]. The HHO methods mentioned above can deal with

Z. Dong (✉)
Inria, Paris, France

CERMICS, École des Ponts, Marne-la-Vallée, France
e-mail: zhaonan.dong@inria.fr

Z. Wang
École des Ponts, Marne-la-Vallée, France
e-mail: zuodong.wang@eleves.enpc.fr

J. M. Melenk et al. (eds.), *Spectral and High Order Methods for Partial Differential Equations ICOSAHOM 2020+1*, Lecture Notes in Computational Science and Engineering 137, https://doi.org/10.1007/978-3-031-20432-6_15

the curved interface or domain boundary without computing the parameterized mappings. However, to the best of the authors' knowledge, there is no HHO methods in the literature can solve the Poisson problem on a complicated Lipschitz domain containing a lot of small faces on the Neumann boundary without being influenced by the number of faces. In this work, we focus on the complicated domain with arbitrary number of planar faces only.

There are two main novelties of the HHO methods in this work: Firstly, the proposed HHO methods do not contain any face unknowns on the Dirichlet and Neumann boundary faces. There is no need to compute the parameterized mapping to satisfy the boundary condition on the curved domain. Secondly, the size of the linear system is independent of the number of faces on the domain boundary, which is essential for solving PDEs on a complicated domain with a lot of tiny faces. Although, the computational cost for computing the face integrals still depends on the number of faces on the domain boundary. Finally, we present one numerical example to test the proposed method.

2 Model and Discrete Setting

We follow the standard notations in Sobolev space, $\|\cdot\|$ means the standard L^2 norm and $|\cdot|_{H^{t+1}}$ means the standard H^{t+1} seminorm, and we simplify the notation of L^2 inner product as $(\cdot,\cdot)$, if there is no special statement.

2.1 Model Problem

Let Ω be a bounded Lipschitz domain in $\mathbb{R}^d$, $d \in \{2, 3\}$, $\Gamma = \Gamma_{\mathrm{D}} \cup \Gamma_{\mathrm{N}}$, where Γ_{D} is the closed Dirichlet boundary satisfying $|\Gamma_{\mathrm{D}}| > 0$ and Γ_{N} is the Neumann boundary. We consider the solution $u \in H^1(\Omega)$ of the following elliptic boundary problem:

$$
\begin{gathered}
-\Delta u = f \quad \text{in} \quad \Omega, \qquad (1)\\
u = g_{\mathrm{D}} \quad \text{on} \quad \Gamma_{\mathrm{D}}, \qquad \nabla u \cdot \boldsymbol{n}_{\Omega} = g_{\mathrm{N}} \quad \text{on} \quad \Gamma_{\mathrm{N}}
\end{gathered}
$$

with $f \in L^2(\Omega)$, $g_{\mathrm{D}} \in H^{\frac{1}{2}}(\Gamma_{\mathrm{D}})$, $g_{\mathrm{N}} \in L^2(\Gamma_{\mathrm{N}})$ and $\boldsymbol{n}_{\Omega}$ to be the unit outward normal vector of domain Ω.

The weak form of the above problem is defined as: Find $u \in H^1(\Omega)$, satisfying $u|_{\Gamma_{\mathrm{D}}} = g_{\mathrm{D}}$,

$$
a(u, v) = \ell(v), \qquad \forall v \in H^1_{\Gamma_{\mathrm{D}}}(\Omega),
$$

with $a(u, v) = (\nabla u, \nabla v)_\Omega$ and $\ell(v) = (f, v)_\Omega + (g_\mathrm{N}, v)_{\Gamma_\mathrm{N}}$, where $H^1_{\Gamma_\mathrm{D}}(\Omega) = \{v \in H^1(\Omega) | v = 0 \text{ on } \Gamma_\mathrm{D}\}$.

2.2 *Polytopal and Curved Meshes*

Let $\{\mathcal{T}_h\}_{h>0}$ be a mesh family such that each mesh $\mathcal{T}_h$ covers Ω exactly. A generic mesh cell is denoted by $T \in \mathcal{T}_h$, its diameter by h_T, and its unit outward normal vector by $\boldsymbol{n}_T$. We partition the boundary ∂T of any mesh cell $T \in \mathcal{T}_h$ by means of the two subsets $\partial T^\mathrm{i} = \overline{\partial T \cap \Omega}$, $\partial T^\mathrm{b} = \partial T \cap \partial\Omega$. Similarly, we partition the mesh as $\mathcal{T}_h = \mathcal{T}_h^\mathrm{i} \cup \mathcal{T}_h^\mathrm{b}$, where $\mathcal{T}_h^\mathrm{b}$ is the collection of all the mesh cells T such that ∂T^b has positive measure. Moreover, we further split ∂T^b into two nonoverlapping subsets $\partial T^\mathrm{D} = \partial T^\mathrm{b} \cap \Gamma_\mathrm{D}$ and $\partial T^\mathrm{N} = \partial T^\mathrm{b} \cap \Gamma_\mathrm{N}$. The mesh faces are collected in the set $\mathcal{F}_h$, which is split as $\mathcal{F}_h = \mathcal{F}_h^\mathrm{i} \cup \mathcal{F}_h^\mathrm{b}$, where $\mathcal{F}_h^\mathrm{i}$ is the collection of the interior faces (shared by two distinct mesh cells) and $\mathcal{F}_h^\mathrm{b}$ the collection of the boundary faces. For all $F \in \mathcal{F}_h$, we orient F by means of the fixed unit normal vector $\boldsymbol{n}_F$ whose direction is arbitrary for all $F \in \mathcal{F}_h^\mathrm{i}$ and $\boldsymbol{n}_F = \boldsymbol{n}_\Omega$ for all $F \in \mathcal{F}_h^\mathrm{b}$. For any mesh cell $T \in \mathcal{T}_h$, the mesh faces composing its boundary ∂T are collected in the set $\mathcal{F}_{\partial T}$, which is partitioned as $\mathcal{F}_{\partial T} = \mathcal{F}_{\partial T^\mathrm{i}} \cup \mathcal{F}_{\partial T^\mathrm{D}} \cup \mathcal{F}_{\partial T^\mathrm{N}}$ with obvious notation.

Assumption 1

(1) Any interior element $T \in \mathcal{T}_h^\mathrm{i}$ is a polytope with planar faces, and the sequence of interior meshes $\mathcal{T}_h^\mathrm{i}$ is shape-regular in the sense of [9].
(2) For any boundary mesh cell $T \in \mathcal{T}_h^\mathrm{b}$, all the faces in $\mathcal{F}_{\partial T^\mathrm{i}}$ are planar with diameter uniformly equivalent to h_T. Moreover, for each ∂T^b, T can be decomposed into a finite union of nonoverlapping subsets, $\{T_{\partial T^\mathrm{b},m}\}_{m \in \{1,\ldots,n_{T,\partial T^\mathrm{b}}\}}$, such that $T_{\partial T^\mathrm{b},m}$ is star-shaped with respect to an interior ball of radius $r_{T,m}$ satisfying $\rho h_T \le r_{T,m}$, with the mesh-regularity parameter $\rho > 0$. We refer to Fig. 1 for an illustration with $n_{T,\partial T^\mathrm{i}} = 1$ (left) and $n_{T,\partial T^\mathrm{i}} = 3$ (right).

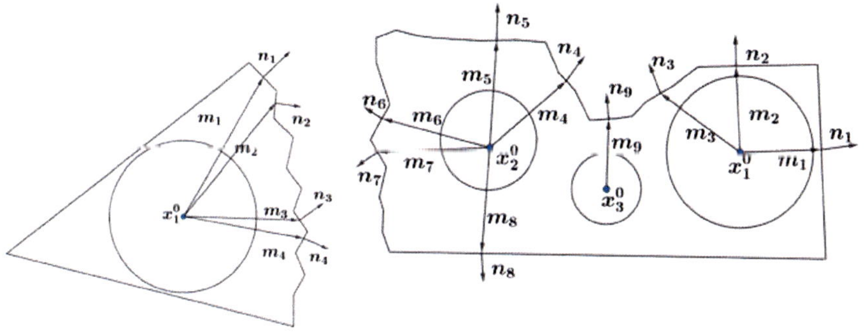

Fig. 1 Examples for the mesh assumptions

Remark 2 (Mesh Assumptions) The above mesh assumptions on the mesh sequence are pretty general. In this work, we assume the interior faces of the mesh are planar and shape-regular. The reason is that those face unknowns on interior faces will be coupled globally after the static condensation, and we try to avoid using the meshes with an unbounded number of tiny faces in the interior of the domain. However, HHO methods have been proven to be stable and converge optimally on the meshes with an arbitrary number of tiny faces under certain star-shapedness assumptions. We refer to the recent work [13] for a detailed discussion.

On the other hand, using a Nitsche-like penalty method avoids introducing discrete face unknowns on the boundary faces. Therefore, the face on the domain boundary can be curved or complicated. The star-shapedness assumption (2) on the mesh boundary cells is introduced to invoke a Poincaré-type inequality, discrete trace inequality and multiplicative trace inequality to hold in such cells.

Remark 3 (Limitation of the Domain) In this work, the complicated domain is denoted as the Lipschitz domain with a possibly arbitrary number of $(d-1)$-dimensional tiny planar faces. Following the technical result in [1, Theorem 2.1], a Lipschitz domain can be covered by a finite number of subdomains that are star-shaped with respect to an open ball. This result implies that the above mesh assumption is not very restrictive for the complicated Lipschitz domains. However, if the domain is non-Lipschitz, the above meshes assumption may fail to hold.

2.3 Analysis Tools

Let us briefly review the main analysis tools used in this section. We only state the results. In what follows, we always consider a shape-regular mesh sequence satisfying the above assumptions. Moreover, in various bounds, we use $a \lesssim b$ to denote $a \leq Cb$ with C to be any generic constant (its value can change at each occurrence) that is independent of mesh size $h > 0$, but may depend on the shape-regularity of the mesh sequence and the polynomial degree for all $T \in \mathcal{T}_h$. Let $k \in \mathbb{N}$ be the polynomial degree and $\mathbb{P}_{k,d}(T)$ to be the polynomial space of order at most k in any cell $T \in \mathcal{T}_h$.

Lemma 4 (Discrete Trace Inequalities) *For any $T \in \mathcal{T}_h$ satisfying the mesh assumptions, the following relation holds for all $v \in \mathbb{P}_{k,d}(T)$:*

$$\|v\|_{\partial T} \lesssim h_T^{-\frac{1}{2}} \|v\|_T. \tag{2}$$

Lemma 5 (Multiplicative Trace Inequality) *Let element $T \in \mathcal{T}_h$. Then, the following statement holds true for all $v \in H^1(\Omega)$:*

$$\|v\|_{\partial T} \lesssim (h_T^{-\frac{1}{2}} \|v\|_T + \|v\|_T^{\frac{1}{2}} \|\nabla v\|_T^{\frac{1}{2}}). \tag{3}$$

Lemma 6 (Polynomial Approximation) *Let $k \geq 0$ be the polynomial degree, and there is a real number $t \in [0, k+1]$, for all $m \in \{0, \ldots, \lfloor t \rfloor\}$, and $v \in H^t(T)$,*

$$|v - \Pi_T^k(v)|_{H^m(T)} \lesssim h_T^{t-m} |v|_{H^t(T)}, \tag{4}$$

where Π_T^k denoted the L^2-orthogonal projection onto $\mathbb{P}_{k,d}(T)$.

The following approximation result holds by using (4) combined with the multiplicative trace inequality (3): Let $k \geq 0$ be the polynomial degree, the following estimate holds for all $v \in H^t(T)$,

$$\begin{aligned} \|v - \Pi_T^k(v)\|_T + h_T^{\frac{1}{2}} \|v - \Pi_T^k(v)\|_{\partial T} + h_T \|\nabla(v - \Pi_T^k(v))\|_T \\ + h_T^{\frac{3}{2}} \|\nabla(v - \Pi_T^k(v))\|_{\partial T} \lesssim h_T^t |v|_{H^t(T)} \end{aligned} \tag{5}$$

Remark 7 We briefly comment on the proofs of the above lemmas. For Lemmas 4 and 5, the proof on mesh cells having flat faces can be found in [8, Sec. 1.4.3]. On the mesh cells having a curved face, these results can be found in [2, 17] assuming that the curved face is a C^2 manifold. More recently, these results were extended in [4] with fully explicit constants to Lipschitz manifolds satisfying the mild additional geometric assumptions.

Concerning Lemma 6, the key step is to establish the Poincaré inequality since (4) can then be derived by using recursively the Poincaré inequality. On the interior mesh cells, which can be decomposed as a finite union of subsimplices, this latter inequality is established by proceeding as in [14, 16]. On the boundary mesh cells, which can have a curved face, one invokes the star-shapedness assumption with respect to a ball. We refer the reader to [18] for the derivation of this inequality with an explicitly determined constant under such an assumption.

3 HHO Discretization

In this section, we will introduce the mixed-order HHO method employing Nitsche's boundary penalty techniques. We first start to introduce the HHO space. Let $k \geq 0$ be the polynomial degree. We consider a pair $\hat{v}_T^k = (v_T^{k+1}, v_{\partial T^{\mathrm{i}}}^k)$, where v_T^{k+1} is defined on T and $v_{\partial T^{\mathrm{i}}}^k$ is defined on the inner faces (facewise) $F \in \mathcal{F}_{\partial T^{\mathrm{i}}}$ composing the boundary ∂T^{i} of T. We define the local HHO space

$$\hat{V}_T^k = \mathbb{P}_{k+1,d}(T) \times \mathbb{P}_{k,d-1}(\mathcal{F}_{\partial T^{\mathrm{i}}}),$$

noting the above local HHO space does not contain face unknowns on the domain boundary.

Next, we denote by ∂_n the (scalar-valued) outward normal derivative on ∂T for all $T \in \mathcal{T}_h$. The construction of the HHO methods are as follows: We start with the reconstruction of the gradient operator and then add the stabilization operator. The reconstruction operator is defined as: $R^i_T : \hat{V}^k_T \to \mathbb{P}_{k+1,d}(T)$ s.t. for every pair $\hat{v}^k_T = (v^{k+1}_T, v^k_{\partial T^{\rm i}}) \in \hat{V}^k_T$, the function $R^i_T(\hat{v}^k_T)$ is s.t. for all $q \in \mathbb{P}_{k+1,d}(T)$,

$$\begin{aligned}(\nabla R^i_T(\hat{v}^k_T), \nabla q)_T &:= (\nabla v^{k+1}_T, \nabla q)_T - (v^{k+1}_T - v^k_{\partial T^{\rm i}}, \partial_n q)_{\partial T^{\rm i}} - (v^{k+1}_T, \partial_n q)_{\partial T^{\rm D}} \\ &= -(v^{k+1}_T, \Delta q)_T + (v^k_{\partial T^{\rm i}}, \partial_n q)_{\partial T^{\rm i}} + (v^{k+1}_T, \partial_n q)_{\partial T^{\rm N}} \end{aligned} \tag{6}$$

together with $(R^i_T(\hat{v}^k_T) - v^{k+1}_T, 1)_T = 0$. In addition, we define the lifting operator $L^{k+1}_T : L^2(\partial T^{\rm D}) \to \mathbb{P}_{k+1,d}(T)$ for all $T \in \mathcal{T}_h$ such that, for all $g_{\rm D} \in L^2(\partial T^{\rm D})$ and all $q \in \mathbb{P}_{k+1,d}(T)$,

$$(\nabla L^{k+1}_T(g_{\rm D}), \nabla q)_T = (g_{\rm D}, \partial_n q)_{\partial T^{\rm D}}, \tag{7}$$

together with the condition $(L^{k+1}_T(g_{\rm D}), 1)_T = 0$. Noticing that $L^{k+1}_T(g_{\rm D}) = 0$ for all interior cell $T \in \mathcal{T}^{\rm i}_h$.

Next, we consider the stabilization operators. For the interior faces $F \in \mathcal{F}_{\partial T^{\rm i}}$, the LS stabilization operators are defined as

$$S^i_T(\hat{v}^k_T) = \Pi^k_{\partial T^{\rm i}}(v^{k+1}_T - v^k_{\partial T^{\rm i}}).$$

where $\Pi^k_{\partial T^{\rm i}}$ denotes the L^2-orthogonal projection onto the broken polynomial space $\mathbb{P}_{k,d-1}(\mathcal{F}_{\partial T^{\rm i}})$.

For the Dirichlet face $F \in \mathcal{F}_{\partial T^{\rm D}}$, we employ Nitsche's boundary penalty techniques, i.e., $S^D_T(\hat{v}^k_T) = v^{k+1}_T|_{\partial T^{\rm D}}$.

Then, the local bilinear form $\hat{a}_T$ and local linear form ℓ_T in each cell T is defined as follows:

$$\begin{aligned}\hat{a}_T(\hat{u}^k_T, \hat{v}^k_T) &= (\nabla R^i_T(\hat{u}^k_T), \nabla R^i_T(\hat{v}^k_T))_T \\ &\quad + h_T^{-1}(S^i_T(\hat{u}^k_T), S^i_T(\hat{v}^k_T))_{\partial T^{\rm i}} + h_T^{-1}(u^{k+1}_T, v^{k+1}_T)_{\partial T^{\rm D}}, \\ \ell_T(\hat{v}^k_T) &= (f, v^{k+1}_T)_T + (g_{\rm N}, v^{k+1}_T)_{\partial T^{\rm N}} - (g_{\rm D}, \partial_n R^i_T(\hat{v}^k_T) - h_T^{-1} v^{k+1}_T)_{\partial T^{\rm D}},\end{aligned}$$

for all $\hat{u}^k_T, \hat{v}^k_T$ in $\hat{V}^k_T$.

Next, let us define the global HHO space as

$$\hat{V}^k_h = \mathbb{P}_{k+1,d}(\mathcal{T}_h) \times \mathbb{P}_{k,d-1}(\mathcal{F}^{\rm i}_h),$$

then the global bilinear forms $\hat{a}_h$, and linear form ℓ_h as follows:

$$\begin{aligned}\hat{a}_h(\hat{u}^k_h, \hat{v}^k_h) &= \sum_{T \in \mathcal{T}_h} \hat{a}_T(\hat{u}^k_T, \hat{v}^k_T), \\ \ell_h(\hat{v}^k_h) &= \sum_{T \in \mathcal{T}_h} \ell_T(\hat{v}^k_T),\end{aligned} \tag{8}$$

for all $\hat{u}_h^k$, $\hat{v}_h^k$ in $\hat{V}_h^k$.

Finally, the discrete problem can be defined as follows:

$$\begin{cases} \text{Find } \hat{u}_h^k \in \hat{V}_h^k \text{ such that} \\ \hat{a}_h(\hat{u}_h^k, \hat{v}_h^k) = \ell_h(\hat{v}_h^k), \ \forall \hat{v}_h^k \in \hat{V}_h^k. \end{cases} \tag{9}$$

Remark 8 We mention that there is another version for HHO method with Nitsche's boundary penalty techniques which does not contain the lifting operator, see section 2 in [5]. Without the lifting operator, the stability of the method in [5] is achieved by choosing the penalty parameter large enough. On the contrary, the proposed HHO methods do not suffer from it.

4 Main Results

4.1 Stability and Well-Posedness

The stability and approximation results presented in this section follows the same technique as in [15, Chapter 39]. For simplification of the presentation, the details of some proofs are not presented in this work.

For all $T \in \mathcal{T}_h$ and all $\hat{v}_T^k \in \hat{V}_T^k$, we define the local seminorm as:

$$|\hat{v}_T^k|^2_{\hat{V}_T^k} = \|\nabla v_T^{k+1}\|^2_T + h_T^{-1}\|v^k_{\partial T^{\mathrm{i}}} - v_T^{k+1}\|^2_{\partial T^{\mathrm{i}}} + h_T^{-1}\|v_T^{k+1}\|^2_{\partial T^{\mathrm{D}}}.$$

In addition, we define the norm for the space $\hat{V}_h^k$ as

$$\|\hat{v}_h^k\|^2_{\hat{V}_h^k} = \sum_{T\in\mathcal{T}_h} |\hat{v}_T^k|^2_{\hat{V}_T^k}.$$

Lemma 9 (Local Stability and Boundedness) *Let all $T \in \mathcal{T}_h$ satisfy the mesh assumptions. Then, the local stability holds*

$$|\hat{v}_T^k|^2_{V_T^k} \lesssim \hat{a}_T(\hat{v}_T^k, \hat{v}_T^k) \lesssim |\hat{v}_T^k|^2_{V_T^h}. \tag{10}$$

Proof The proof follows the same techniques as in [6, Sec. 2.2]. □

An immediate consequence of Lemma 9 is the following bound establishing that the discrete bilinear form $\hat{a}_h$ is coercive on $\hat{V}_h^k$:

$$\|\hat{v}_h^k\|^2_{\hat{V}_h^k} \lesssim \hat{a}_h(\hat{v}_h^k, \hat{v}_h^k), \qquad \forall \hat{v}_h^k \in \hat{V}_h^k.$$

Invoking the Lax–Milgram lemma the (9) is well-posed.

4.2 Approximation and H^1 Error Estimate

For any $T \in \mathcal{T}_h$, we define the local reduction operator $\hat{I}_T^k : H^1(T) \to \mathbb{P}_{k+1,d}(T)$, for all $v \in H^1(T)$,

$$\hat{I}_T^k(v) = (\Pi_T^{k+1}(v), \Pi_{\partial T^{\mathrm{i}}}^k(v)) \in \hat{V}_T^k.$$

We then define the energy projection operator $\mathcal{E}_T : H^1(T) \to \mathbb{P}_{k+1,d}(T)$ such that

$$\mathcal{E}_T(v) = R_T^i \circ \hat{I}_T^k(v) + L_T^{k+1}(v), \qquad \forall v \in H^1(T).$$

Similarly, the global energy projection is defined by $\mathcal{E}(v)|_T = \mathcal{E}_T(v)$, for all $T \in \mathcal{T}_h$.

Lemma 10 (Energy Projection) *Let $\mathcal{E}_T$ be the energy projection. Then, the following relation holds,*

$$(\nabla(\mathcal{E}_T(v) - v), \nabla q)_T = (\Pi_T^{k+1}(v) - v, \partial_n q)_{\partial T^{\mathrm{N}}}, \tag{11}$$

and $(\mathcal{E}_T(v), 1)_T = (v, 1)_T$ for all $q \in \mathbb{P}_{k+1,d}(T)$ and $v \in H^1(T)$.

Proof Let $v \in H^1(T)$ and $\phi = R_T^i \circ \hat{I}_T^k(v)$. Using the definition of the reconstruction operator (6), we infer that for all $q \in V_T^{k+1}$,

$$\begin{aligned}(\nabla\phi, \nabla q)_T &= -(\Pi_T^{k+1}(v), \Delta q)_T + (\Pi_{\partial T^{\mathrm{i}}}^k(v), \partial_n q)_{\partial T^{\mathrm{i}}} + (\Pi_T^{k+1}(v), \partial_n q)_{\partial T^{\mathrm{N}}} \\ &= -(v, \Delta q)_T + (v, \partial_n q)_{\partial T^{\mathrm{i}}} + (v, \partial_n q)_{\partial T^{\mathrm{N}}} + (\Pi_T^{k+1}(v) - v, \partial_n q)_{\partial T^{\mathrm{N}}} \\ &= (\nabla v, \nabla q)_T - (v, \partial_n q)_{\partial T^{\mathrm{D}}} + (\Pi_T^{k+1}(v) - v, \partial_n q)_{\partial T^{\mathrm{N}}},\end{aligned}$$

since $\partial_n q_{|F}$ is a polynomial of order k on $F \in \partial T^{\mathrm{i}}$. Then, using the definition of lifting operator, we have $(v, \partial_n q)_{\partial T^{\mathrm{D}}} = (\nabla L_T^{k+1}(v), \nabla q)_T$, this gives $(\nabla(\mathcal{E}_T(v) - v), \nabla q)_T = (\Pi_T^{k+1}(v) - v, \partial_n q)_{\partial T^{\mathrm{N}}}$. Moreover, $(R_T^i \circ \hat{I}_T^i(v) + L_T^{k+1}(v), 1)_T = (\Pi_T^{k+1}(v), 1)_T = (v, 1)_T$ by the definition of the reconstruction operator and the lifting operator. □

Remark 11 By noticing the term $(\Pi_T^{k+1}(v) - v, \partial_n q)_{\partial T^{\mathrm{N}}}$ depends on the Neumann boundary, $\mathcal{E}_T(v)$ is the elliptic projection for the function v on the cells which does not contain any face on the Neumann boundary.

We start with the approximation properties of operators.

Lemma 12 (Approximation Results) *For all $v \in H^{t+1}(\Omega), t > \frac{1}{2}$ and $v|_T \in H^{k+2}(T)$ for all $T \in \mathcal{T}_h$, we have*

$$\|\nabla(\mathcal{E}_T(v) - v)\|_T + h_T^{\frac{1}{2}}\|\nabla(\mathcal{E}_T(v) - v)\|_{\partial T^{\mathrm{i}} \cup \partial T^{\mathrm{D}}} \lesssim h_T^{k+1}|v|_{H^{k+2}(T)}, \tag{12}$$

$$h_T^{-\frac{1}{2}}\|S_T^i(\hat{I}_T^k(v))\|_{\partial T^{\mathrm{i}}} + h_T^{-\frac{1}{2}}\|S_T^D(\hat{I}_T^k(v)) - v\|_{\partial T^{\mathrm{D}}} \lesssim h_T^{k+1}|v|_{H^{k+2}(T)}. \tag{13}$$

Proof Invoking the discrete trace inequality (2), multiplicative trace inequality (3) and approximation result (4) The above results follow the same steps as in [6, Sec. 2.3]. □

We define the consistency error $\delta_h \in (\hat{V}_h^k)'$ which is the dual space of $\hat{V}_h^k$ such that, for all $\hat{v}_h^k \in \hat{V}_h^k$,

$$\langle \delta_h, \hat{v}_h^k \rangle = \sum_{T \in \mathcal{T}_h} \left(\hat{a}_T^k(\hat{I}_T^k(u), \hat{v}_T^k) - \ell_T(\hat{v}_T^k) \right), \tag{14}$$

where the brackets refer to the duality pairing between $(\hat{V}_h^k)'$ and $\hat{V}_h^k$.

Lemma 13 (Consistency and Boundedness) *Assume that $u \in H^{t+1}(\Omega)$ with $t > \frac{1}{2}$ and $\hat{v}_h^k \in \hat{V}_h^k$. The following holds true*

$$|\langle \delta_h, \hat{v}_h^k \rangle| \leq \Big(\sum_{T \in \mathcal{T}_h} \|g_T\|_{*T}^2 + \|\xi_T\|_{\#T}^2 \Big)^{\frac{1}{2}} \times \|\hat{v}_h^k\|_{\hat{V}_h^k},$$

where

$$\|g_T\|_{*T}^2 = \|\nabla(u - \mathcal{E}(u))\|_T^2 + h_T\|\nabla(u - \mathcal{E}(u))\|_{\partial T^{\mathrm{i}} \cup \partial T^{\mathrm{D}}}^2,$$
$$\|\xi_T\|_{\#T}^2 = h_T^{-1}\|u - \Pi_T^{k+1}(u)\|_{\partial T^{\mathrm{i}}}^2 + h_T^{-1}\|u - \Pi_T^{k+1}(u)\|_{\partial T^{\mathrm{D}}}^2.$$

Proof Let $\hat{v}_h^k \in \hat{V}_h^k$. Recalling that $-\Delta u = f$, we obtain $\langle \delta_h, \hat{v}_h^k \rangle = \Psi_1 + \Psi_2$, where

$$\Psi_1 = \sum_{T \in \mathcal{T}_h} \Big((\nabla R_T^i \circ \hat{I}_T^i(u), \nabla R_T^i(\hat{v}_T^k))_T + (\Delta u, v_T^{k+1})_T + (\nabla L_T^{k+1}(g_D), \nabla R_T^i(\hat{v}_T^k))_T - (g_{\mathrm{N}}, v_T^{k+1})_{\partial T^{\mathrm{N}}} \Big), \tag{15}$$

$$\Psi_2 = \sum_{T \in \mathcal{T}_h} \Big(h_T^{-1}(S_T^i \circ \hat{I}_T^k(u), S_T^i(\hat{v}^k))_{\partial T^{\mathrm{i}}} + h_T^{-1}(\Pi_T^{k+1}(u) - g_{\mathrm{D}}, v_T^{k+1})_{\partial T^{\mathrm{D}}} \Big). \tag{16}$$

Integrating by parts in every mesh cell $T \in \mathcal{T}_h$, using the exact solution of PDE, $\sum_{T \in \mathcal{T}_h} (\nabla u \cdot \boldsymbol{n}_T, v_{\partial T^{\mathrm{i}}}^k)_{\partial T^{\mathrm{i}}} = 0$, since $\partial_n u$ is meaningful and single valued on all interior faces $F \in \partial T^{\mathrm{i}}$, and the definition of the reconstruction operator, we have

$$\Psi_1 = \sum_{T\in\mathcal{T}_h} \Big(-(\nabla(\mathcal{E}(u)-u)\cdot \boldsymbol{n}_T, v_T^{k+1})_{\partial T^{\mathrm{D}}} + (\nabla(\mathcal{E}(u)-u), \nabla v_T^{k+1})_T + (\nabla(\mathcal{E}(u)-u)\cdot \boldsymbol{n}_T, v_{\partial T^{\mathrm{i}}}^k - v_T^{k+1})_{\partial T^{\mathrm{i}}}\Big).$$

Noticing $u = g_D$ on Γ_D, we have

$$\Psi_2 = \sum_{T\in\mathcal{T}_h} \Big(h_T^{-1}(\Pi_{\partial T^{\mathrm{i}}}^k(u-\Pi_T^{k+1}(u)), v_{\partial T^{\mathrm{i}}}^k - v_T^{k+1})_{\partial T^{\mathrm{i}}} + h_T^{-1}(\Pi_T^{k+1}(u)-u, v_T^{k+1})_{\partial T^{\mathrm{D}}}\Big).$$

Then, the Cauchy-Schwarz inequality and discrete trace inequality (2) finish this proof. □

Theorem 14 (Broken H^1 Error Estimate) *In the framework of Lemma 13, with $\hat{u}_T^k$ solution of the discrete problem (9) on each cell $T \in \mathcal{T}_h$, we have*

$$\sum_{T\in\mathcal{T}_h} \|\nabla(u_{|T} - R_T^i(\hat{u}_T^k) - L_T^{k+1}(g_{\mathrm{D}}))\|_T^2 \lesssim \sum_{T\in\mathcal{T}_h} \Big(\|g_T\|_{*T}^2 + \|\xi_T\|_{\#T}^2\Big). \tag{17}$$

Moreover, if $u|_T \in H^{k+2}(T)$ for all $T \in \mathcal{T}_h$, with $k \geq 0$, we have

$$\sum_{T\in\mathcal{T}_h} \|\nabla(u_{|T} - R_T^i(\hat{u}_T^k) - L_T^{k+1}(g_{\mathrm{D}}))\|_T^2 \lesssim \sum_{T\in\mathcal{T}_h} \Big(h_T^{2(k+1)}|u|_{H^{k+2}(T)}^2\Big). \tag{18}$$

Proof First, let us set $\hat{e}_h^k = \hat{I}_h^k(u) - \hat{u}_h^k \in \hat{V}_h^k$, so that $\hat{e}_T^k = \hat{I}_T^k(u|_T) - \hat{u}_T^k \in \hat{V}_T^k$ for all $T \in \mathcal{T}_h$. The property (10) and the identity $a_h(\hat{e}_h^k, \hat{e}_h^k) = -\langle \delta_h, \hat{e}_h^k\rangle$ imply that

$$\sum_{T\in\mathcal{T}_h} \|\nabla(R_T^i(\hat{e}_T^k))\|_T^2 \lesssim \|\hat{e}_h^k\|_{\hat{V}_h^k}^2 \lesssim a_h(\hat{e}_h^k, \hat{e}_h^k) = -\langle \delta_h, \hat{e}_h^k\rangle.$$

Owing to the Lemma 13, we infer that

$$\sum_{T\in\mathcal{T}_h} \|\nabla(R_T^i(\hat{e}_T^k))\|_T^2 \lesssim \sum_{T\in\mathcal{T}_h} (\|g_T\|_{*T}^2 + \|\xi_T\|_{\#T}^2).$$

By adding and subtracting $R_T^i(\hat{I}_T^k(u|_T))$, we have

$$u_{|T} - R_T^i(\hat{u}_T^k) - L_T^{k+1}(u|_T) = (u_{|T} - \mathcal{E}_T(u|_T)) + R_T^i(\hat{e}_T^k).$$

Using the triangle inequality, the error estimate (17) is derived. Furthermore, the error bound (18) results from (17) and the approximation results in Lemma 12. □

5 Numerical Experiment

We will test the proposed HHO-Nitsche methods on the domain containing a lot of tiny faces. The computational domain is constructed as follow: (1) Construct a unit disc centred at origin, with a circular hole centred at origin and radius 0.3 removed. (2) Remove the domain on the right-hand side of the curve $x = 0.05\sin(4\pi y - 4\pi) + 0.7$. Finally, randomly perturb the 4010 boundary vertices with the amplitude less or equal to 0.001. This results a polygon centred at (0,0) with saw-like boundary consisting of 4010 small edges which does not embedded on any smooth manifold, see Fig. 2. The HHO-Nitsche method is tested with the exact solution $u = \sin(\pi x)\sin(\pi y)$, with Dirichlet boundary condition imposed on 753 boundary edges allocated inside the domain $-1 \leq x \leq -0.8$, and Neumann boundary condition on the remaining boundary edges.

We consider a quasi-uniform sequence of meshes composed of 217, 515, 2058, 5455, and 8992 triangular cells. All meshes fit the domain Ω exactly, and for every mesh in the sequence, each interior cell has only straight edges, whereas each boundary cell has a lot of small edges that exactly fit the boundary of Ω. The convergence result is given in Fig. 3. The key observation is that the convergence rates are $O(h^{k+2})$ and $O(h^{k+1})$ for the L^2 error and the energy error, respectively. We emphasize that the total number of global coupled unknowns for the proposed HHO methods is independent of the number of edges on the domain boundary.

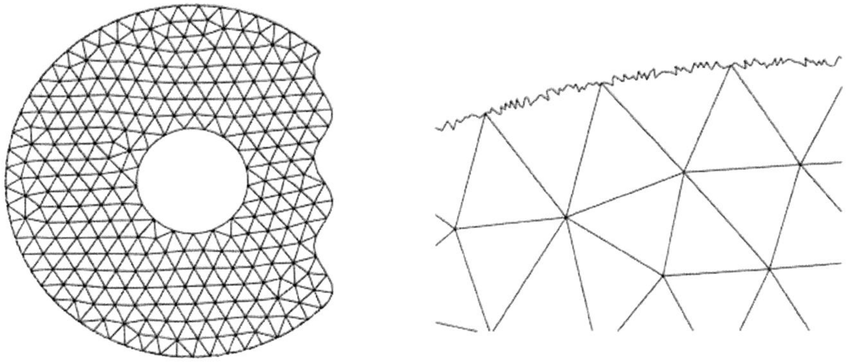

Fig. 2 Complicated domain

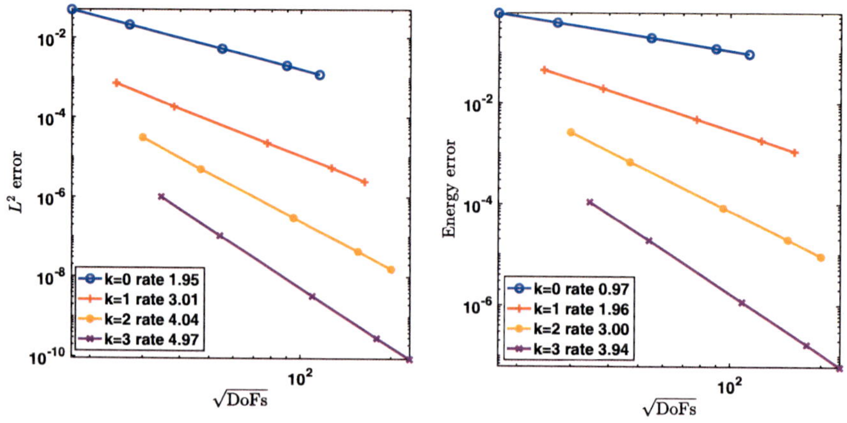

Fig. 3 Convergence rates in the L^2-norm and energy seminorm for $k = 0, 1, 2, 3$

Acknowledgments The authors would like to thank Professor Alexandre Ern (École des Ponts & Inria, France) for his valuable advice and helpful discussion.

References

1. Amrouche, C., Ciarlet, P.G., Mardare, C.: On a lemma of Jacques-Louis Lions and its relation to other fundamental results. J. Math. Pures Appl. (9) **104**(2), 207–226 (2015)
2. Burman, E., Ern, A.: An unfitted hybrid high-order method for elliptic interface problems. SIAM J. Numer. Anal. **56**(3), 1525–1546 (2018)
3. Burman, E., Cicuttin, M., Delay, G., Ern, A.: An unfitted hybrid high-order method with cell agglomeration for elliptic interface problems. SIAM J. Sci. Comput. **43**(2), A859–A882 (2021)
4. Cangiani, A., Dong, Z., Georgoulis, E.H.: *hp*-version discontinuous Galerkin methods on essentially arbitrarily-shaped elements. Math. Comput. **91**(333), 1–35 (2022)
5. Cascavita, K.L., Chouly, F., Ern, A.: Hybrid high-order discretizations combined with Nitsche's method for Dirichlet and Signorini boundary conditions. IMA J. Numer. Anal. **40**(4), 2189–2226 (2020)
6. Cicuttin, M., Ern, A., Pignet, N.: Hybrid high-order methods. a primer with application to solid mechanics. SpringerBriefs in Mathematics (2021)
7. Di Pietro, D.A., Droniou, J.: The Hybrid High-order Method for Polytopal Meshes, volume 19 of MS&A. Modeling, Simulation and Applications. Springer, Cham (2020) ©2020. Design, analysis, and applications
8. Di Pietro, D.A., Ern, A.: Mathematical Aspects of Discontinuous Galerkin Methods, vol. 69. Springer Berlin Heidelberg, Berlin, Heidelberg (2012)
9. Di Pietro, D.A., Ern, A.: A hybrid high-order locking-free method for linear elasticity on general meshes. Comput. Methods Appl. Mech. Eng. **283**, 1–21 (2015)
10. Di Pietro, D.A., Ern, A., Lemaire, S.: An arbitrary-order and compact-stencil discretization of diffusion on general meshes based on local reconstruction operators. Comput. Methods Appl. Math. **14**(4), 461–472 (2014)
11. Dong, Z., Ern, A.: Hybrid high-order and weak Galerkin methods for the biharmonic problem. SIAM J. Numer. Anal. **60**(5), 2626–2656 (2022). https://doi.org/10.1137/21M1408555

12. Dong, Z., Ern, A.: Hybrid high-order method for singularly perturbed fourth-order problems on curved domains. ESAIM Math. Model. Numer. Anal. **55**(6), 3091–3114 (2021)
13. Droniou, L., Yemm, J.: Robust hybrid high-order method on polytopal meshes with small faces. Comput. Methods Appl. Math. **22**(1), 47–71 (2022)
14. Ern, A., Guermond, J.-L.: Finite element quasi-interpolation and best approximation. ESAIM Math. Model. Numer. Anal. (M2AN) **51**(4), 1367–1385 (2017)
15. Ern, A., Guermond, J.-L.: Finite Elements II: Galerkin Approximation, Elliptic and Mixed PDEs, volume 73 of Texts in Applied Mathematics. Springer Nature, Cham (2021)
16. Veeser, A., Verfürth, R.: Poincaré constants for finite element stars. IMA J. Numer. Anal. **32**(1), 30–47 (2012)
17. Wu, H., Xiao, Y.: An unfitted hp-interface penalty finite element method for elliptic interface problems. J. Comput. Math. **37**(3), 316–339 (2019)
18. Zheng, W., Qi, H.: On Friedrichs-Poincaré-type inequalities. J. Math. Anal. Appl. **304**(2), 542–551 (2005)

On Higher Order Passivity Preserving Schemes for Nonlinear Maxwell's Equations

Herbert Egger and Vsevolod Shashkov

1 Introduction

We consider the propagation of high-intensity electromagnetic waves through dielectric media, which is of relevance in typical applications of nonlinear optics [7, 21]. The underlying physics are modelled by Maxwell's equations

$$\partial_t \mathbf{d} = \operatorname{curl} \mathbf{h}, \qquad \partial_t \mathbf{b} = -\operatorname{curl} \mathbf{e}, \tag{1}$$

with $\mathbf{e}$ and $\mathbf{h}$ denoting the electric and magnetic field intensities, and $\mathbf{d}$ and $\mathbf{b}$ the corresponding fluxes. For the following discussion, we assume that the relation between fields and fluxes is given by

$$\mathbf{d} = \mathbf{d}(\mathbf{e}) := \epsilon_0(\chi^{(1)} + \chi^{(3)}|\mathbf{e}|^2)\,\mathbf{e}, \qquad \mathbf{b} = \mathbf{b}(\mathbf{h}) := \mu_0 \mathbf{h}, \tag{2}$$

which describes the instantaneous electric and magnetic response of a nonlinear Kerr-type medium. Here ϵ_0, μ_0 represent the permittivity and permeability of vacuum and the constants $\chi^{(1)}$ and $\chi^{(3)}$ describe the linear and nonlinear dielectric effects. Let us note that more general nonlinear constitutive equations as well as lossy materials can be considered with slight modifications of our arguments.

H. Egger (✉)
Johann Radon Institute for Computational and Applied Mathematics, Linz, Austria

H. Egger (✉)
Johannes Kepler Universität Linz, Linz, Austria
e-mail: herbert.egger@jku.at

V. Shashkov
TU Darmstadt, Darmstadt, Germany
e-mail: shashkov@mathematik.tu-darmstadt.de

J. M. Melenk et al. (eds.), *Spectral and High Order Methods for Partial Differential Equations ICOSAHOM 2020+1*, Lecture Notes in Computational Science and Engineering 137, https://doi.org/10.1007/978-3-031-20432-6_16

Energy Balance The constitutive equations (2) allow to eliminate the fluxes $\mathbf{d}$, $\mathbf{b}$ in Eq. (1) and to rewrite the system solely in terms of field intensities $\mathbf{e}$, $\mathbf{h}$ as

$$\mathbf{d}'(\mathbf{e})\partial_t \mathbf{e} = \operatorname{curl} \mathbf{h}, \tag{3}$$

$$\mu_0 \partial_t \mathbf{h} = -\operatorname{curl} \mathbf{e}, \tag{4}$$

where $\mathbf{d}'(\mathbf{e}) = \epsilon_0(\chi^{(1)} + 3\chi^{(3)}|\mathbf{e}|^2)$ denotes the incremental permittivity. For the further discussion, we restrict our considerations to a bounded domain Ω and we assume homogeneous magnetic boundary conditions

$$\mathbf{n} \times \mathbf{h} = \mathbf{0} \qquad \text{on } \partial\Omega. \tag{5}$$

For prescribed fields $\mathbf{e}$, $\mathbf{h}$, the electromagnetic energy of the system is then given by

$$\mathcal{E}(\mathbf{e}, \mathbf{h}) = \int_\Omega w_E(\mathbf{e}) + w_M(\mathbf{h})\, d\mathbf{x} \tag{6}$$

with energy densities $w_E(\mathbf{e}) = \frac{\epsilon_0}{2}\left(\chi^{(1)}|\mathbf{e}|^2 + \frac{3\chi^{(3)}}{2}|\mathbf{e}|^4\right)$ and $w_M(\mathbf{h}) = \frac{\mu_0}{2}|\mathbf{h}|^2$. Let us note that these are convex functions of their arguments and further observe that $w_E'(\mathbf{e}) = \mathbf{d}'(\mathbf{e})\mathbf{e}$ and $w_M'(\mathbf{h}) = \mu_0 \mathbf{h} = \mathbf{b}'(\mathbf{h})\mathbf{h}$. This allows to show that

$$\mathcal{E}(\mathbf{e}(t), \mathbf{h}(t)) = \mathcal{E}(\mathbf{e}(s), \mathbf{h}(s)) \tag{7}$$

for any solution $(\mathbf{e}, \mathbf{h})$ of (3)–(5) and all $t \geq s$; see Sect. 2. Hence the energy of the system is conserved for all time and, in particular, the system is passive.

Passivity Preserving Discretization The finite-difference time-domain method is certainly today's industry standard for the numerical solution of Maxwell's equations in time domain. Various extensions to nonlinear media have been proposed, e.g., in [13, 17, 22, 23]. Let us refer to [19] for comparison of different approaches and to [6, 16] for recent developments. More flexible finite volume and related discontinuous Galerkin approximations have been discussed in [3] and [15]. A combination of mixed finite element or discontinuous Galerkin approximations with Runge-Kutta time-stepping schemes has been studied in [1, 12] and [4, 5]. While these methods are capable of providing higher order approximations, their strict passivity for higher order approximations in time seems not completely settled.

Contributions In this paper, we propose two strategies that allow to construct provably passivity preserving discretization schemes of arbitrary high order. A key ingredient here is the use of a variational time integration methodology [2, 18]. As illustrated in [10, 11], this allows to obtain conservative or dissipative numerical approximations for a large class of nonlinear evolution problems. Our first approach is based on a mixed finite element approximation of the **e**–**h** formulation (3)–(5) in space and a discontinuous Galerkin method in time. The resulting scheme is slightly dissipative and leads to a discrete analogue of (7) with an inequality instead of equality. The second scheme is based on a reformulation of the problem in terms of the electric field $\mathbf{e}$ and the magnetic vector potential $\mathbf{a}$. We consider a discretization of this **e**–**a** formulation in space with the same finite element spaces for both variables and a Petrov-Galerkin time integration. This leads to a scheme with exact energy conservation.

Outline In Sect. 2, we derive the variational form of (3)–(5) and prove the energy balance (7). Section 3 then discusses the discretization of the **e**–**h** formulation and presents the corresponding discrete energy inequality. In Sect. 4, we derive the equivalent **e**–**a** formulation, its variational formulation, and the corresponding energy balance. Section 5 is devoted to the discretization of this problem for which we prove exact energy conservation on the discrete level. In Sect. 6, we illustrate our theoretical findings by numerical tests and comment on the convergence properties and performance of the proposed methods. We then close with a short summary.

2 The e–h Formulation

Let us briefly introduce the most relevant notation used in the rest of the paper. We write $L^2(\Omega)$ for the space of square integrable scalar or vector valued functions and denote by $H(\text{curl};\Omega) = \{\mathbf{w} \in L^2(\Omega) : \text{curl}\,\mathbf{w} \in L^2(\Omega)\}$ the set of functions with square integrable curl. Furthermore, we use $\langle \mathbf{f}, \mathbf{g}\rangle = \int_\Omega \mathbf{f}\cdot\mathbf{g}\,d\mathbf{x}$ to abbreviate the L^2-scalar product. The following lemma summarizes the main properties of solutions to the **e**-**h** formulation that we will use in the following.

Lemma 1 *Let* $(\mathbf{e}, \mathbf{h})$ *be a smooth solution of* (3)–(5). *Then*

$$\langle \mathbf{d}'(\mathbf{e}(t))\partial_t \mathbf{e}(t), \mathbf{w}\rangle = \langle \mathbf{h}(t), \text{curl}\,\mathbf{w}\rangle \tag{8}$$

$$\langle \mu_0 \partial_t \mathbf{h}(t), \mathbf{q}\rangle = -\langle \text{curl}\,\mathbf{e}(t), \mathbf{q}\rangle \tag{9}$$

for all $\mathbf{w} \in H(\text{curl};\Omega)$, $\mathbf{q} \in L^2(\Omega)$, *and all* $t \geq 0$. *Moreover,*

$$\mathcal{E}(\mathbf{e}(t), \mathbf{h}(t)) = \mathcal{E}(\mathbf{e}(s), \mathbf{h}(s)) \qquad \forall t \geq s. \tag{10}$$

Proof The second equation follows immediately by multiplying (4) with the test function $\mathbf{q}$ and integration over the domain Ω. To verify the first identity, we multiply Eq. (3) by $\mathbf{w}$, integrate over Ω, and observe that

$$\langle \mathbf{d}'(\mathbf{e})\partial_t \mathbf{e}, \mathbf{w}\rangle = \langle \text{curl}\,\mathbf{h}, \mathbf{w}\rangle = \langle \mathbf{h}, \text{curl}\,\mathbf{w}\rangle + \int_{\partial\Omega} \mathbf{n} \times \mathbf{h} \cdot \mathbf{w}\, ds(\mathbf{x}).$$

Here we used integration-by-parts in the second step. Due to (5), the boundary term vanishes and we obtain (8). By formal differentiation of the energy, we further get

$$\frac{d}{dt}\mathcal{E}(\mathbf{e},\mathbf{h}) = \langle w_E'(\mathbf{e}), \partial_t \mathbf{e}\rangle + \langle w_M'(\mathbf{h}), \partial_t \mathbf{h}\rangle = \langle \mathbf{d}'(\mathbf{e})\mathbf{e}, \partial_t \mathbf{e}\rangle + \langle \mu_0 \mathbf{h}, \partial_t h\rangle$$
$$= \langle \mathbf{d}'(\mathbf{e})\partial_t \mathbf{e}, \mathbf{e}\rangle + \langle \mu_0 \partial_t \mathbf{h}, \mathbf{h}\rangle = \langle \mathbf{h}, \operatorname{curl} \mathbf{e}\rangle - \langle \operatorname{curl} \mathbf{e}, \mathbf{h}\rangle = 0.$$

In the first line, we used the definition of the energy functional and the relations between the energy densities and the material parameters stated in the introduction. In the second line, we first exchanged the order of the functions in the scalar product and then used the variational identities (8)–(9) with test functions $\mathbf{w} = \mathbf{e}$ and $\mathbf{q} = \mathbf{h}$. The energy identity then follows immediately by integration over time. □

Let us note that in the essential steps of the above proof, we only made use of the variational form (8)–(9) of the problem with test functions $\mathbf{w} = \mathbf{e}(t)$ and $\mathbf{q} = \mathbf{h}(t)$. This motivates to consider a variational discretization scheme in the following.

3 Discretization of the e–h Formulation

Let $W_h \subset H(\text{curl}; \Omega)$ and $Q_h \subset L^2(\Omega)$ denote some finite dimensional subspaces and let $I_\tau = \{t^n : 0 \le n \le N\}$ be a sequence of discrete time steps $t^n = n\tau$ with $\tau = T/N$. We write $I^n = [t^{n-1}, t^n]$ for the nth time interval and denote by $P_k(I^n; X)$ the space of polynomial functions $v : I_n \to X$ with values in some vector space X. By $(*)|_{t^n}$ we mean the evaluation of the time dependent expression $(*)$ at $t = t^n$. For discretization of problem (3)–(5), we then consider the following method.

Problem 2 Let the discrete initial values $\mathbf{e}_h^0 \in W_h$ and $\mathbf{h}_h^0 \in Q_h$ be given. Then for $1 \le n \le N$ find $\mathbf{e}_h^n \in P_k(I^n; W_h)$ and $\mathbf{h}_h^n \in P_k(I^n; Q_h)$ such that

$$\int_{I_n} \langle \mathbf{d}'(\mathbf{e}_h^n)\partial_t \mathbf{e}_h^n, \mathbf{w}_h\rangle - \langle \mathbf{h}^n, \operatorname{curl} \mathbf{w}_h\rangle dt = \langle \mathbf{d}'(\mathbf{e}_h^n)(\mathbf{e}_h^{n-1} - \mathbf{e}_h^n), \mathbf{w}_h\rangle|_{t^{n-1}} \tag{11}$$

$$\int_{I_n} \langle \mu_0 \partial_t \mathbf{h}_h^n, \mathbf{q}_h\rangle dt + \langle \operatorname{curl} e_h^n, \mathbf{q}_h\rangle = \langle \mu_0(\mathbf{h}_h^{n-1} - \mathbf{h}_h^n), \mathbf{q}_h\rangle|_{t^{n-1}} \tag{12}$$

holds for all test functions $\mathbf{w}_h \in P_k(I^n; W_h)$ and $\mathbf{q}_h \in P_k(I^n; Q_h)$.

This scheme is based on a Galerkin approximation of (8)–(9) in space and a discontinuous Galerkin method in time [2]. It emerges as a particular example of an abstract discretization framework for dissipative evolution problems; see [10, Sec. 8]. From the theoretical results derived in this reference, we conclude the following.

Lemma 3 *Let* $(\mathbf{e}_h^n, \mathbf{h}_h^n)_n$ *denote a solution of Problem* 2. *Then*

$$\mathcal{E}(\mathbf{e}_h^n(t^n), \mathbf{h}_h^n(t^n)) \leq \mathcal{E}(\mathbf{e}_h^m(t^m), \mathbf{h}_h^m(t^m)) \qquad \forall m \leq n. \tag{13}$$

Proof For $m = n - 1$, the inequality above can be verified by testing the variational principle (11)–(12) with $\mathbf{w}_h = \mathbf{e}_h^n$ and $\mathbf{q}_h = \mathbf{h}_h^n$ and using convexity of the energy densities $w_E(\mathbf{e})$ and $w_M(\mathbf{h})$. The case $m < n - 1$ can then be treated by induction. For details of the arguments used in the proof, we refer to [10, Sec. 3 and 4]. □

Let us emphasize that the variational form of the time integration scheme is important here to prove passivity in the case of nonlinear constitutive equations.

4 The e–a Formulation

We now present an alternative approach towards the passivity preserving discretization of the problem, which is based on a standard reformulation in terms of the magnetic vector potential. Let us start with an auxiliary observation.

Lemma 4 *Let* $(\mathbf{e}, \mathbf{h})$ *be smooth functions satisfying identity* (4). *Furthermore, define* $\mathbf{a}(t) = \mathbf{a}_0 - \int_0^t \mathbf{e}(s)ds$ *with* $\mathbf{a}_0$ *chosen such that* $\operatorname{curl} \mathbf{a}_0 = \mu_0 \mathbf{h}(0)$. *Then*

$$\mathbf{e}(t) = -\partial_t \mathbf{a}(t) \qquad \textit{and} \qquad \mu_0 \mathbf{h}(t) = \operatorname{curl} \mathbf{a}(t) \qquad \forall t \geq 0.$$

Proof The first identity is clear. Using that μ_0 is constant and employing (4), we deduce that $\mu_0 \partial_t \mathbf{h} = -\operatorname{curl} \mathbf{e} = \operatorname{curl} \partial_t \mathbf{a}$. By integration in time, we then get

$$\mu_0 \mathbf{h}(t) = \mu_0 \mathbf{h}(0) + \int_0^t \mu_0 \partial_t \mathbf{h}\, dt = \operatorname{curl} \mathbf{a}_0 + \int_0^t \operatorname{curl} \partial_t \mathbf{a}\, dt = \operatorname{curl} \mathbf{a}(t),$$

which already proves the second identity of the lemma. □

Let us note that $\operatorname{curl} \mathbf{a} = \mu_0 \mathbf{h} = \mathbf{b}$, i.e., $\mathbf{a}$ is just the usual magnetic vector potential frequently used in the magneto-quasi static setting. Using the above observations, we can now reformulate the system (3)–(4) equivalently as

$$-\mathbf{d}'(\mathbf{e})\partial_t \mathbf{a} = \mathbf{d}'(\mathbf{e})\mathbf{e}, \tag{14}$$

$$\mathbf{d}'(\mathbf{e})\partial_t \mathbf{e} = \operatorname{curl}(\nu_0 \operatorname{curl} \mathbf{a}), \tag{15}$$

where we introduced $\nu_0 = \mu_0^{-1}$ for convenience. The particular choice of the multiplying factors in the first equation will become clear from the proof of

Lemma 5 below. The boundary condition (5) further translates to

$$\mathbf{n} \times \operatorname{curl} \mathbf{a} = 0 \qquad \text{on } \partial\Omega. \tag{16}$$

For obvious reasons, we call (14)–(16) the **e**–**a** formulation of our problem. As a final step, we also rewrite the energy functional in terms of the fields $(\mathbf{e}, \mathbf{a})$, i.e.,

$$\mathcal{H}(\mathbf{e}, \mathbf{a}) = \mathcal{E}(\mathbf{e}, \nu_0 \operatorname{curl}(a)) = \int_\Omega w_E(\mathbf{e}) + \tfrac{\nu_0}{2} |\operatorname{curl} \mathbf{a}|^2 d\mathbf{x}. \tag{17}$$

We again briefly summarize the basic properties of this alternative formulation.

Lemma 5 *Let* $(\mathbf{e}, \mathbf{a})$ *denote a sufficiently smooth solution of* (14)–(16). *Then*

$$-\langle \mathbf{d}'(\mathbf{e}(t))\partial_t \mathbf{a}(t), \mathbf{w}\rangle = \langle \mathbf{d}'(\mathbf{e}(t))\mathbf{e}(t), \mathbf{w}\rangle \tag{18}$$

$$\langle \mathbf{d}'(\mathbf{e}(t))\partial_t \mathbf{e}(t), \mathbf{z}\rangle = \langle \nu_0 \operatorname{curl} \mathbf{a}(t), \operatorname{curl} \mathbf{z}\rangle \tag{19}$$

for all test functions $\mathbf{w}, \mathbf{z} \in H(\text{curl}; \Omega)$ *and all* $t \geq 0$. *Furthermore*

$$\mathcal{H}(\mathbf{e}(t), \mathbf{a}(t)) = \mathcal{H}(\mathbf{e}(s), \mathbf{a}(s)) \qquad \forall s \leq t. \tag{20}$$

Proof The first equation follows immediately by multiplying (14) with $\mathbf{w}$ and integration over the domain. In the same manner, we deduce from Eq. (15) that

$$\begin{aligned}\langle \mathbf{d}'(\mathbf{e})\partial_t \mathbf{e}, \mathbf{z}\rangle &= \langle \operatorname{curl}(\nu_0 \operatorname{curl} \mathbf{a}), \mathbf{z}\rangle \\ &= \langle \nu_0 \operatorname{curl} \mathbf{a}, \operatorname{curl} \mathbf{z}\rangle + \int_{\partial\Omega} \mathbf{n} \times (\nu_0 \operatorname{curl} \mathbf{a}) \cdot \mathbf{z} \, ds(\mathbf{x}).\end{aligned}$$

In the second step, we used integration-by-parts. The boundary term vanishes due to the boundary condition (16), which already leads to (19). By formal differentiation of the energy functional and the relation between the electric energy functional $w_E(\mathbf{e})$ and the constitutive law $\mathbf{d}(\mathbf{e})$, we can further see that

$$\begin{aligned}\frac{d}{dt}\mathcal{H}(\mathbf{e}, \mathbf{a}) &= \langle \mathbf{w}_E'(\mathbf{e}), \partial_t \mathbf{e}\rangle + \langle \nu_0 \operatorname{curl}(\mathbf{a}), \operatorname{curl} \partial_t \mathbf{a}\rangle \\ &= \langle \mathbf{d}'(\mathbf{e})\mathbf{e}, \partial_t \mathbf{e}\rangle + \langle \nu_0 \operatorname{curl}(\mathbf{a}), \operatorname{curl} \partial_t \mathbf{a}\rangle.\end{aligned}$$

This corresponds to the sum of the two terms on the right hand side of (18)–(19) with test functions $\mathbf{w} = \partial_t \mathbf{e}$ and $\mathbf{z} = \partial_t \mathbf{a}$. As a consequence, we thus obtain

$$\frac{d}{dt}\mathcal{H}(\mathbf{e}, \mathbf{a}) = -\langle \mathbf{d}'(\mathbf{e})\partial_t \mathbf{a}, \partial_t \mathbf{e}\rangle + \langle \mathbf{d}'(\mathbf{e})\partial_t \mathbf{e}, \partial_t \mathbf{a}\rangle = 0.$$

The energy identity (20) now follows immediately by integration over time. □

Note that in the basic step of the proof, we again simply utilized the variational identities (18)–(19), here with the particular test functions $\mathbf{w} = \partial_t \mathbf{e}(t)$ and $\mathbf{z} = \partial_t \mathbf{a}(t)$. This motivates the variational discretization scheme of the following section.

5 Discretization of e − a Formulation

As before, let $W_h \subset H(\text{curl}; \Omega)$ denote some finite dimensional subspace and further recall the notation about the time grid from Sect. 3. For the numerical approximation of problem (14)–(16), we then consider the following method.

Problem 6 Let $\mathbf{e}_h^0, \mathbf{a}_h^0 \in W_h$ be given and for $1 \leq n \leq N$, find $\mathbf{e}_h^n, \mathbf{a}_h^n \in P_{k+1}(I_n; W_h)$ such that $\mathbf{e}_h^n(t^{n-1}) = \mathbf{e}_h^{n-1}(t^{n-1})$ and $\mathbf{a}_h^n(t^{n-1}) = \mathbf{a}_h^{n-1}(t^{n-1})$ as well as

$$-\int_{I^n} \langle \mathbf{d}'(\mathbf{e}_h^n)\partial_t \mathbf{a}_h^n, \tilde{\mathbf{w}}_h\rangle dt = \int_{I^n} \langle \mathbf{d}'(\mathbf{e}_h^n)\mathbf{e}_h^n, \tilde{\mathbf{w}}_h\rangle dt \qquad \forall \tilde{\mathbf{w}} \in P_k(I^n; W_h) \tag{21}$$

$$\int_{I^n} \langle \mathbf{d}'(\mathbf{e}_h^n)\partial_t \mathbf{e}_h^n, \tilde{\mathbf{z}}_h\rangle = \int_{I^n} \langle \nu_0 \,\text{curl}\, \mathbf{a}_h^n, \text{curl}\, \tilde{\mathbf{z}}_h\rangle dt \qquad \forall \tilde{\mathbf{z}} \in P_k(I^n; W_h). \tag{22}$$

This scheme is based on a Galerkin approximation of (18)–(19) in space together with a Petrov-Galerkin time discretization [2]. Similar methods can be applied for the numerical solution of a wide class of evolution problems; see [11] for examples. By construction, the discrete solution is continuous in time and can be computed by an implicit time stepping algorithm. The most important property of the method can be summarized as follows.

Lemma 7 *Let $(\mathbf{e}_h^n, \mathbf{a}_h^n)_{n\geq 0}$ denote a solution of Problem 6. Then*

$$\mathcal{H}(\mathbf{e}_h^n(t^n), \mathbf{a}_h^n(t^n)) = \mathcal{H}(\mathbf{e}_h^m(t^m), \mathbf{a}_h^m(t^m)) \qquad \forall n \geq m. \tag{23}$$

Proof By the fundamental theorem of calculus, we obtain

$$\mathcal{H}(\mathbf{e}_h^n(t^n), \mathbf{a}_h^n(t^n)) - \mathcal{H}(\mathbf{e}_h^n(t^{n-1}), \mathbf{a}_h^n(t^{n-1})) = \int_{I^n} \frac{d}{dt}\mathcal{H}(\mathbf{e}_h^n, \mathbf{a}_h^n)\, dt$$

$$= \int_{I^n} \langle w_E'(\mathbf{e}_h^n), \partial_t e_h^n\rangle + \langle \nu_0 \,\text{curl}\, \mathbf{a}_h^n, \partial_t \mathbf{a}_h^n\rangle dt =: (*).$$

Let us recall that $w'_E(\mathbf{e}) = \mathbf{d}'(\mathbf{e})\mathbf{e}$, which allows us to conclude that

$$(*) = \int_{I^n} \langle d'(\mathbf{e}_h^n)\mathbf{e}_h^n, \partial_t e_h^n\rangle + \langle \nu_0 \operatorname{curl} \mathbf{a}_h^n, \partial_t \mathbf{a}_h^n\rangle dt$$

$$= -\int_{I^n} \langle \mathbf{d}'(\mathbf{e}_h^n)\partial_t \mathbf{a}_h^n, \partial_t \mathbf{e}_h^n\rangle - \langle \mathbf{d}'(\mathbf{e}_h^n)\partial_t \mathbf{e}_h^n, \partial_t \mathbf{a}_h^n\rangle dt = 0.$$

In the second step, we here used the variational identities (21)–(22) with $\tilde{\mathbf{w}}_h = \partial_t \mathbf{e}_h^n$ and $\tilde{\mathbf{z}}_h = \partial_t \mathbf{a}_h^n$, which is admissible by the choice of approximation and test spaces. From the conditions for $\mathbf{e}_h^n(t^{n-1})$ and $\mathbf{a}_h^n(t^{n-1})$, we can then deduce that

$$\mathcal{H}(\mathbf{e}_h^n(t^n), \mathbf{a}_h^n(t^n)) = \mathcal{H}(\mathbf{e}_h^n(t^{n-1}), \mathbf{a}_h^n(t^{n-1})) = \mathcal{H}(\mathbf{e}_h^{n-1}(t^{n-1}), \mathbf{a}_h^{n-1}(t^{n-1})).$$

This already verifies the energy identity for $m = n - 1$. The general case $m < n - 1$ can finally again be obtained by induction. □

6 Numerical Validation

We now illustrate our theoretical results by some numerical tests and comment on the implementation of the proposed methods and their convergence behavior.

Test Problem For simplicity, we consider in the sequel a one-dimensional version of problem (1) over the domain $\Omega = (0, 1)$. In that case $H(\text{curl}; \Omega) = H^1(\Omega)$ and the two curl operators in (14)–(15) reduce to ∂_x and $-\partial_x$, respectively. The parameters in the material laws (2) are chosen as $\epsilon_0 = \mu_0 = \chi^{(1)} = 1$ and in order to illustrate the effect of the nonlinear material response, we will consider the two choices $\chi^{(3)} = 0$ and $\chi^{(3)} = 0.1$ below. Note that the problem is linear in the first case. The initial values are finally set to $\mathbf{e}(0, x) = \exp(-100x^2)$ and $\mathbf{a}(0) = \mathbf{h}(0) = \mathbf{0}$.

Spatial Approximation Let $\mathcal{T}_h$ be a uniform mesh with grid points $x_i = ih$ and uniform mesh size $h = 1/M$. We use piecewise polynomial spaces

$$W_h = P_p(\mathcal{T}_h) \cap H(\text{curl}; \Omega), \qquad Q_h = P_{p-1}(\mathcal{T}_h)$$

over the grid $\mathcal{T}_h$ for the space discretization with polynomial degree $p \geq 1$.

Remark 8 In order to facilitate the implementation of the proposed methods, the scalar products $\langle \cdot, \cdot \rangle$ in the discrete variational problems (11)–(12) and (21)–(22) are approximated by inexact versions $\langle \cdot, \cdot \rangle_h$, which are realized by numerical quadrature. In our computations, we use the Gauss-Lobatto formula with $p + 1$ nodes on every element, which is a standard choice; see [8, 9, 14]. Let us note that the discrete energy inequality (13) and the identity (23) remain valid, if the same quadrature rule is used for defining the discrete energy functionals.

Time Integration Since the material law $\mathbf{d} = \mathbf{d}(\mathbf{e})$ only involves polynomial nonlinearities, all time integrals in our discretization methods can be computed exactly by numerical quadrature. For the solution of the nonlinear systems in Problems 2 and 6, we utilize a simple fixed point iteration with tolerance set to 10^{-12}.

Comparison of Linear and Nonlinear Material Behavior In Fig. 1 we display some snapshots $\mathbf{e}_h^n$ of the numerical approximations for the electric field in case of a medium with linear and a nonlinear material behaviour, respectively.

In the absence of nonlinearities, the signal simply propagates to the right-hand side without changing its shape. The nonlinear medium response, on the other hand, leads to a slowing down of the propagation at high intensities, which results in a precursor at the beginning and a ramp-up at the end of the pulse.

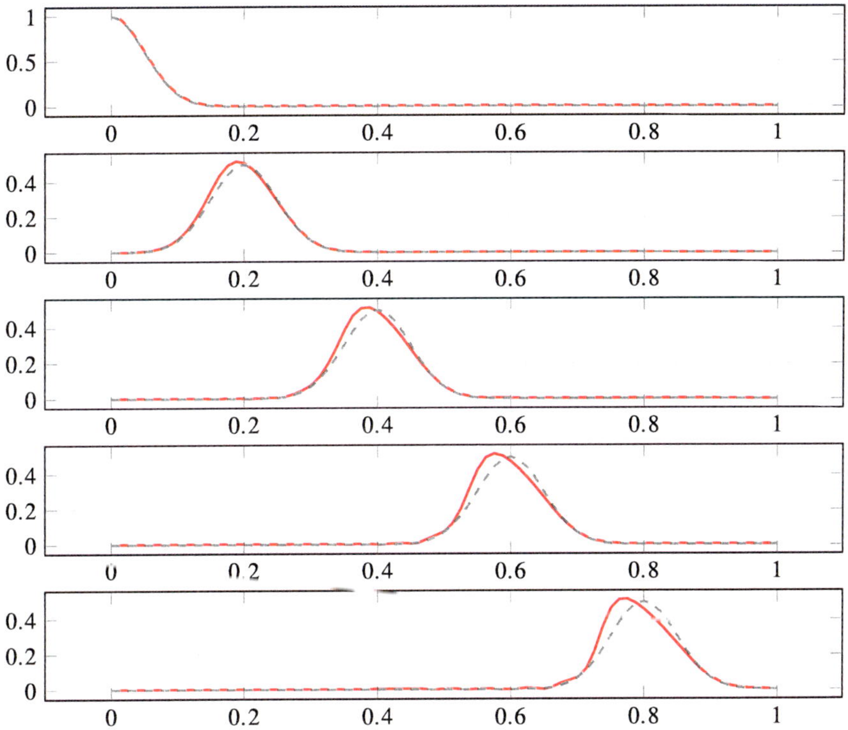

Fig. 1 Snapshot $\mathbf{e}_h^n(t^n)$ of the numerical solution at time steps $t^n = 0.0, 0.2, 0.4, 0.6, 0.8$ obtained with the method of Sect. 5 for two scenarios: linear case ($\chi^{(3)} = 0$; black dashed) and nonlinear case ($\chi^{(3)} = 0.1$; red solid). In both cases, the discrete energy is preserved exactly

6.1 Results for the e–h Formulation

Let us note that in the linear case ($\chi^{(3)} = 0$), the method (11)–(12) corresponds to a mixed finite element approximation in space with time stepping by the Radau-IIA method with $s = k + 1$ stages [2]. From the error analysis in [8, 14, 20], we expect that the error behaves like

$$err_{h,\tau}^{p,k} := \max_{0 \le t^n \le T} \|\mathbf{e}_h^n(t^n) - \mathbf{e}_{h/2}^{2n}(t^n)\|_{h/2} = O(h^{p+1} + \tau^{2k+1}). \tag{24}$$

Here p and k are the polynomial degrees of the spatial and temporal approximation, respectively. Further $\mathbf{e}_h^n$ is the numerical solution on the interval $I_n = [t^{n-1}, t^n]$ with $t^n = n\tau$, while $\mathbf{e}_{h/2}^{2n}$ is the corresponding solution obtained on uniformly refined mesh and with time step $\tilde{\tau} = \tau/2$. The approximation $\|\cdot\|_{h/2}$ for the L^2-norm is computed by numerical quadrature on the finer grid $\mathcal{T}_{h/2}$. In Tables 1 and 2, we display the errors obtained in our numerical tests for the nonlinear case $\chi^{(3)} = 0.1$ for different approximation orders p and k in space and time.

The convergence rates that are expected for the linear case are also observed for the nonlinear case. In all computations, the discrete energy decays monotonically. Since the solutions are uniformly bounded, one can see that, similar to the linear case, the energy error behaves like

$$|\mathcal{E}(\mathbf{e}, \mathbf{h}) - \mathcal{E}(\tilde{\mathbf{e}}, \tilde{\mathbf{h}}))| \approx \|\mathbf{e} - \tilde{\mathbf{e}}\|^2 + \|\mathbf{h} - \tilde{\mathbf{h}}\|^2.$$

Table 1 Convergence of the method of Sect. 3 with respect to the mesh size h for different polynomial orders p used for the spatial approximation

	$p = 1$		$p = 2$		$p = 3$	
h	err $\times 10^{-1}$	e.o.c.	err $\times 10^{-2}$	e.o.c.	err $\times 10^{-3}$	e.o.c.
0.05	0.248186	–	0.387722	–	0.417564	–
0.025	0.071272	1.80	0.003319	3.54	0.018346	4.50
0.0125	0.018438	1.95	0.000299	3.47	0.000950	4.27
0.00625	0.004641	1.99	0.000034	3.11	0.000058	4.02

Table 2 Convergence of the method of Sect. 3 with respect to the time step size τ for different polynomial orders k used for the temporal approximation

	$k = 0$		$k = 1$		$k = 2$	
τ	err $\times 10^{-1}$	e.o.c.	err $\times 10^{-2}$	e.o.c.	err $\times 10^{-3}$	e.o.c.
0.025	0.257057	–	0.280420	–	0.550798	–
0.0125	0.171025	0.61	0.038358	2.87	0.019199	4.84
0.00625	0.100673	0.76	0.004879	2.98	0.000610	4.97
0.003125	0.054697	0.88	0.000612	3.00	0.000019	5.00

As a consequence, we expect and observe very small energy errors. In particular for higher order methods, the numerical dissipation seems therefore negligible.

6.2 *Numerical Results for the* e–a *Formulation*

As a next step, we investigate the convergence of our second discretization scheme. The spatial approximation here is a standard finite element method and, in the linear case ($\chi^{(3)} = 0$), the time discretization amounts to the Lobatto-IIIA method with $s = k + 1$ stages. We thus expect that the error behaves like

$$err_{h,\tau}^{p,k} := \max_{0 \le t^n \le T} \|\mathbf{e}_h^n(t^n) - \mathbf{e}_{h/2}^{2n}(t^n)\|_{h/2} = O(h^{p+1} + \tau^{2k+2}). \tag{25}$$

In Tables 3 and 4, we display the numerical errors obtained with the method of Sect. 5 for different approximation orders p and k in space and time.

As can be easily be deduced from the tables, the convergence rates are again exactly as expected. Let us further mention that the discrete energy was preserved up to round-off errors in all our computational tests with this method.

Table 3 Convergence of the method of Sect. 5 with respect to the mesh size h for different polynomial orders p used for the spatial approximation

	$p = 1$		$p = 2$		$p = 3$	
h	err $\times 10^{-3}$	e.o.c.	err $\times 10^{-3}$	e.o.c.	err $\times 10^{-3}$	e.o.c.
0.05	0.412735	–	0.297889	–	0.277589	–
0.025	0.127333	1.70	0.022976	3.69	0.011844	4.55
0.0125	0.033235	1.94	0.002874	2.99	0.000747	3.99
0.00625	0.008372	1.99	0.000359	3.00	0.000046	3.99

Table 4 Convergence of the method of Sect. 5 with respect to the time step size τ for different polynomial orders k used for the temporal approximation

	$k = 0$		$k = 1$		$k = 2$	
τ	err $\times 10^{-1}$	e.o.c.	err $\times 10^{-3}$	e.o.c.	err $\times 10^{-4}$	e.o.c.
0.05	0.801343	–	0.611080	–	0.368882	–
0.025	0.226615	1.82	0.040060	3.93	0.006549	5.81
0.0125	0.057709	1.97	0.002538	3.98	0.000108	5.93
0.00625	0.014537	1.99	0.000160	3.98	0.000002	5.96

7 Discussion

In this paper, we presented two possible approaches towards the construction of higher order provably passivity preserving numerical schemes for Maxwell's equations in nonlinear media. A key ingredient was the use of appropriate variational space and time discretization schemes which allowed us to rigorously prove fully discrete energy identities, respectively, inequalities on the discrete level. Both approaches investigated in the paper lead to implicit time-stepping schemes, which for linear media coincide with certain Runge-Kutta methods. The proposed schemes further show the expected convergence behavior for linear as well as for nonlinear problems. A full error analysis should be possible but is left for future research.

Acknowledgments Support by the German Science Foundation (DFG) and the Austrian Science Fund (FWF) via grants TRR 361 and SFB F90, project C02 is gratefully acknowledged.

References

1. Aanes, A., Angermann, L.: Energy-stable time-domain finite element methods for the 3D nonlinear Maxwell's equations. IEEE Photonics J. **12**, 6500415 (2020)
2. Akrivis, G., Makridakis, C., Nochetto, R.H.: Galerkin and Runge–Kutta methods: unified formulation, a posteriori error estimates and nodal superconvergence. Numer. Math. **118**, 429–456 (2011)
3. Aregba-Driollet, D.: Godunov scheme for Maxwell's equations with Kerr nonlinearity. Commun. Math. Sci **13**, 2195–2222 (2015)
4. Blank, E.: The Discontinuous Galerkin method for Maxwell's equations: Application to bodies of revolution and Kerr-nonlinearities. Ph.D. thesis, KIT (2013)
5. Bokil, V.A., Cheng, Y., Jiang, Y., Li, F.: Energy stable discontinuous Galerkin methods for Maxwell's equations in nonlinear optical media. J. Comput. Phys. **350**, 420–452 (2017)
6. Bokil, V.A., Cheng, Y., Jiang, Y., Li, F., Sakkaplangkul, P.: High spatial order energy stable FDTD methods for Maxwell's equations in nonlinear optical media in one dimension. J. Sci. Comput. **77**, 330–371 (2018)
7. Boyd, R.W.: Nonlinear Optics, 3rd edn. Academic Press (2008)
8. Cohen, G.: Higher-Order Numerical Methods for Transient Wave Equations. Springer, Heidelberg (2002)
9. Cohen, G., Monk, P.: Gauss point mass lumping schemes for Maxwell's equations. Numer. Methods Partial Differential Equations **14**, 63–88 (1998)
10. Egger, H.: Structure preserving approximation of dissipative evolution problems. Numer. Math. **143**, 85–106 (2019)
11. Egger, H., Habrich, O., Shashkov, V.: On the energy stable approximation of Hamiltonian and gradient systems. J. Comput. Methods Appl. Math. **21**, 335–349 (2021)
12. Fisher, A., White, D., Rodrigue, G.: An efficient vector finite element method for nonlinear electromagnetic modeling. J. Comput. Phys. **225**, 1331–1346 (2007)
13. Fujii, M., Tahara, M., Sakagami, I., Freude, W., Russer, P.: High-order FDTD and auxiliary differential equation formulation of optical pulse propagation in 2-D Kerr and Ramann nonlinear dispersive media. J. Quantum Electron. **40**, 175–182 (2004)
14. Geevers, S., Mulder, W., van der Vegt, J.: New higher-order mass-lumped tetrahedral elements for wave propagation modelling. SIAM J. Sci. Comput. **40**, A2830–A2857 (2018)

15. Huang, J., Shu, C.W.: A second-order asymptotic-preserving and positivity-preserving discontinuous Galerkin scheme for the Kerr–Debye model. M3AS **27**, 549–579 (2017)
16. Jia, H., Li, J., Fang, Z., Li, M.: A new FDTD scheme for Maxwell's equations in Kerr-type nonlinear media. Numer. Algorithms **81**, 223–243 (2019)
17. Joseph, R.M., Taflove, A.: FDTD Maxwell's equations models for nonlinear electrodynamics and optics. IEEE Trans. Antennas Prop. **45**, 364–374 (1997)
18. Lew, A., Marsden, J.E., Ortiz, M., West, M.: Variational time integrators. Int. J. Numer. Methods Eng. **60**, 153–212 (2004)
19. Maksymov, I.S., Sukhorokov, A.A., Lavrinenko, A.V., Kivshar, Y.S.: Comparative study of FDTD-adopted numerical algorithms for Kerr nonlinearities. IEEE Antennas Wirel. Propag. Lett. **10**, 143–146 (2011)
20. Monk, P.: Analysis of a finite element method for Maxwell's equations. SIAM J. Numer. Anal. **29**, 714–729 (1992)
21. Shen, Y.R.: The Principles of Nonlinear Optics. Wiley Interscience (1994)
22. Spachmann, H., Gutschling, S., Krüger, H., Weiland, T.: FIT-formulation for non-linear dispersive media. Int. J. Numer. Model. **12**, 81–92 (1999)
23. Ziolkowski, R.W., Judkins, J.B.: Full-wave vector Maxwell equation modeling of the self-focusing of ultrashort optical pulses in a nonlinear Kerr medium exhibiting a finite response time. J. Opt. Soc. Am. B **10**, 186–198 (1993)

Electromagnetic Displacements Rotating Inside an Annular Region

Lorella Fatone and Daniele Funaro

1 Introduction

The aim of this paper is to further refine and specialize the results discussed in [4–8]. In [9], some exact solutions in the whole 3D space for the classical set of equations ruling electrodynamics have been introduced. These belong to spherical eigenfunctions and describe the evolution of electromagnetic fields inside a rotating sphere around the polar axis. The goal is to extend these fields outside such a sphere, in order to avoid peripheral signals on the equatorial plane travelling at velocities arbitrarily exceeding the speed of light in vacuum. The trick is to embed the main sphere into another domain, inside which the information rotates at lower angular velocity. The problem is how to match the signals in order to avoid discontinuities. By a suitable construction, we propose a solution where the dynamics of the electromagnetic fields between the two boundaries recalls that of a ball bearing set. In this way, it is possible to transfer the information outward in continuous fashion, roughly maintaining the limits imposed by the speed of light. To fulfill this achievement it is necessary to combine eigenfunctions having the same eigenvalue (namely: $-\omega^2$). A combination of a rotatory movement and a radial oscillating behavior between the internal and the external boundaries actually requires four degrees of freedom, so that the multiplicity of $-\omega^2$ will be also required to be equal to four. Thus, our aim here is to find external domains that allow for the existence

L. Fatone (✉)
Dipartimento di Matematica, Università degli Studi di Camerino, Camerino, Italy
e-mail: lorella.fatone@unicam.it

D. Funaro
Dipartimento di Scienze Chimiche e Geologiche, Università degli Studi di Modena e Reggio Emilia, Modena, Italy
e-mail: daniele.funaro@unimore.it

J. M. Melenk et al. (eds.), *Spectral and High Order Methods for Partial Differential Equations ICOSAHOM 2020+1*, Lecture Notes in Computational Science and Engineering 137, https://doi.org/10.1007/978-3-031-20432-6_17

of such types of eigenfunctions. For simplicity, we carry out our analysis on annular domains lying on the equatorial plane, although the study can be generalized within the full 3D context.

Although the present paper does not contain explicit applications or the discussion of numerical techniques, the possibility of having suitable sets of orthogonal eigenfunctions clearly opens the way to numerous implementations in the framework of spectral methods. As pointed out in [9], there is a strict link between the solutions defined within the sphere and the equations of Magneto-HydroDynamics (MHD), particularly in relation to the induction equation. In solar MHD, a comprehensive text is [14], where for instance the linearised equations are proposed and examined in Sect. 7.2. Other possible applications are found in the context of plasma cosmology (see, e.g., [13]) or in the study of ball lightning phenomena as in [10]. Moreover, our approach may be helpful in devising high-order approximation schemes aimed to study plasma configurations outside the body of a star. To this regard, Sect. 5 of this paper provides hints about a hypothetical characterization of the plasma displacement surrounding the Sun. Only a simple linear approximation is discussed. The result can be however improved upon by solving the appropriate nonlinear model equations, and representing the solution as a suitable combination of the basis functions here introduced.

2 Preliminaries

The classical electromagnetic fields in vacuum evolve according to the equations:

$$\frac{\partial \mathbf{E}}{\partial t} = c^2 \mathrm{curl}\mathbf{B}, \qquad \mathrm{div}\mathbf{E} = 0, \tag{1}$$

$$\frac{\partial \mathbf{B}}{\partial t} = -\,\mathrm{curl}\mathbf{E}, \qquad \mathrm{div}\mathbf{B} = 0, \tag{2}$$

where $\mathbf{E} = (E_1, E_2, E_3)$ denotes the electric field and $\mathbf{B} = (B_1, B_2, B_3)$ the magnetic one. Here c is the speed of light. It is well know that combining the above equations one gets:

$$\frac{\partial^2 \mathbf{E}}{\partial t^2} = c^2\, \Delta \mathbf{E}, \qquad \frac{\partial^2 \mathbf{B}}{\partial t^2} = c^2\, \Delta \mathbf{B}, \tag{3}$$

where the symbol Δ denotes the vector Laplacian in spherical coordinates.

By looking for solution in the framework of vector spherical eigenfunctions, and by using spherical coordinates (r, θ, ϕ), an interesting displacement is provided by the following analytic solution:

$$\mathbf{E} = (E_1, E_2, E_3) = \frac{1}{\omega}\left(\frac{H}{r}\sqrt{1-x^2}\left[S_3' - \frac{x}{1-x^2}S_3 + \frac{m}{1-x^2}S_2\right]\cos\zeta,\right.$$

$$\left.\left(H' + \frac{H}{r}\right)S_3\cos\zeta,\ \left(H' + \frac{H}{r}\right)S_2\sin\zeta\right), \tag{4}$$

$$\mathbf{B} = (B_1, B_2, B_3) = \frac{1}{c}\Big(0,\ H(r)S_2(\cos\theta)\cos\zeta,\ H(r)S_3(\cos\theta)\sin\zeta\Big), \tag{5}$$

where $H = H(r)$ is a function of the variable r, and S_2, S_3 are functions of the variable $x = \cos\theta$. In (4), S_3' denotes the derivative of S_3 with respect to x, and H' the derivative of H with respect to r. These functions will be specified here below. Finally, we have $\zeta = c\omega t - m\phi$, where $\omega > 0$ is a given parameter and m is integer.

It is worthwhile to observe that $\mathbf{E} \cdot \mathbf{B} \neq 0$. Hence, we are proposing an example where electric and magnetic fields are not orthogonal. Let us also point out that, at the radial points where $H = 0$, we obtain $\mathbf{B} = 0$, whereas $\mathbf{E}$ is tangential to the corresponding sphere (see Fig. 1).

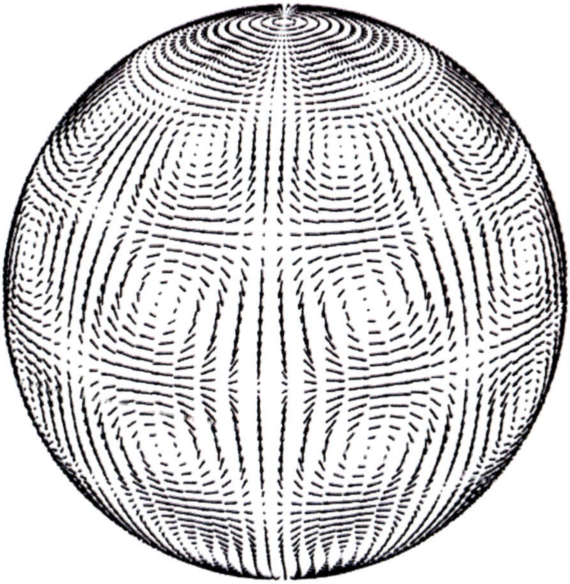

Fig. 1 Surface distribution of the electric field when the radius of the sphere is a zero of the function H. The magnetic field is uniformly zero. The number of vortexes is related to the parameters m and ℓ. In this example we have $m = 4$ and $\ell = 6$. Viewed from a framework at rest, during the rotation about the vertical axis, the electric field oscillates at the equatorial plane tangentially to the circumference

The functions H, S_2, S_3 can be deduced from the magnetic field $\mathbf{B}$ in (5) by imposing that:

$$\frac{\partial^2 \mathbf{B}}{\partial t^2} = -c^2\omega^2\,\mathbf{B}, \qquad \Delta\mathbf{B} = -\omega^2\mathbf{B}. \tag{6}$$

After calculations, it follows that, for a fixed integer $\ell \geq 0$, H is solution of the following eigenvalue problem:

$$H'' + \frac{2H'}{r} - \ell(\ell+1)\frac{H}{r^2} = -\omega^2 H. \tag{7}$$

Up to multiplicative constants, the above equation is satisfied by the functions $J_{\ell+\frac{1}{2}}(\omega r)/\sqrt{\omega r}$ and $Y_{\ell+\frac{1}{2}}(\omega r)/\sqrt{\omega r}$, known as spherical Bessel's functions of the first and the second type, respectively.

Regarding the variable θ, we define S to be an eigenfunction of the following eigenvalue problem:

$$(1-x^2)S'' - 2xS' - \frac{m^2}{1-x^2}S = -\lambda S, \tag{8}$$

with $m \geq 0$ integer. In this fashion we get the so called Associated Legendre polynomials (see, e.g., [1]). The eigenvalue λ must be of the form $\ell(\ell+1)$, where ℓ is an integer with $\ell \geq m$. From S we compute the two functions S_2 and S_3 in (5), that are given by:

$$S_2 = \frac{S}{\sqrt{1-x^2}}, \qquad S_3 = -\frac{S'}{m}\sqrt{1-x^2}. \tag{9}$$

The vector displacement so far introduced describes the time evolution of electromagnetic fields rotating about the axis joining the poles of a sphere. We remark that Bessel's functions of type Y (second kind) are singular at the origin, so that a bounded solution exists only if we assume that H is of type J (first kind). In this way the fields are defined on the whole space $\mathbf{R}^3$. This is however not realistic in practice, since the information would travel along circumferences of arbitrary size, as we move away from the origin. We suggest cutting the sphere at the first zero z_ℓ of the function $J_{\ell+\frac{1}{2}}(\omega r)$, although other possibilities may be investigated. In this case, as specified above, on the surface of the sphere the magnetic field $\mathbf{B}$ is identically zero and the electric field $\mathbf{E}$ is tangential. According to Fig. 1, the number of vortexes along the azimuth direction is related to the parameter m, and the number of vortexes associated with the altitude angle is related to ℓ. For $\ell \geq 1$, the modulus of the peripheral velocity at the equator turns out to be equal to $v_\ell = cz_\ell/\ell$. We have for instance: $v_2 \approx 2.88c$, $v_3 \approx 2.32c$, $v_4 \approx 2.04c$, $v_5 \approx 1.87c$. These values are of the same order of magnitude of the speed of light c and approach c as ℓ grows.

3 Extension on the Equatorial Plane Outside the Sphere

We will consider (4) and (5) in the special case where $m = \ell \geq 1$. With this particular choice we have:

$$S_2(\cos\theta) = (\sin\theta)^{m-1}, \qquad S_3(\cos\theta) = \cos\theta(\sin\theta)^{m-1}. \tag{10}$$

For large m, the function S_2 assumes the value 1 for $\theta = \pi/2$ and decays fast to zero when approaching $\theta = 0$ or $\theta = \pi$. The fields are then concentrated within a flat annular region around the equator. Thus, if $H = 0$ and $\theta = \pi/2$, the electric field is lined up with the equator and oscillates according to the rule: $\cos\zeta = \cos(c\omega t - m\phi)$.

We would like to guess a possible development of the fields outside the sphere. For simplicity, this analysis is reduced to the equatorial plane. When r is a zero of the function H we expect a change of regime, in order to avoid peripheral velocities arbitrarily large. Note that boundary conditions are tangential and, point-wise, the electric field oscillates. Along the equator, this behavior recalls that of a series of contiguous dipoles activated at different phases. The situation is rather different from that of the *lighthouse paradox* in special relativity (see, e.g., [11]), where instead boundary conditions are supposed to have a non vanishing radial component. The aim is to construct solutions defined inside a circular crown, in such a way that the signals turn around at velocities comparable in magnitude to that of light, both at the internal and external boundaries. As we will see, this construction relies on the possibility of finding eigenfunctions of the Laplace operator, corresponding to eigenvalues of multiplicity four (at least). A problem of this type was preliminary studied in [6]. By applying separation of variables, in the special case $m = \ell$, with some approximations we can deal with a scalar equation in two dimensions. Instead, if $m < \ell$, the discussion involves the full 3D environment, bringing to a problem which is too complex to be handled in this short report.

We suppose that the domain is an annular region between the radii $r_{\min}$ and $r_{\max}$, with $r_{\max} > r_{\min}$. Since the equations are linear, it is not restrictive to assume $r_{\min} = 1$. Tangential boundary constraints are imposed that correspond to homogeneous Dirichlet boundary conditions for H. Indeed, when $H = 0$, the first component in (4) vanishes. In proximity of the inner boundary, we want the shift of the signal to be governed by the rule $\cos(c\omega t - m_1\phi)$, where $m_1 \geq 1$ is an integer. At the outside boundary we require instead a behavior of the type $\cos(c\omega t \pm m_2\phi)$, where m_2, with $m_2 > m_1$, is another integer. Depending on the sign $\pm$ of the last expression, the outer boundary co-rotates or counter-rotates with respect to the inner one. For reasons that will be detailed later on, the parameters m_1, m_2, ω and $r_{\max}$, must be detected in order to have a basis of at least four orthogonal eigenfunctions associated with the same eigenvalue. Suitable dynamical behaviors will be then recovered from linear combinations of these eigenfunctions.

Since we are now far from the origin, it is possible to use Bessel's functions J_α and Y_α (for a given α), both of the first and the second kind. We then start by defining:

$$F_m(r) = \frac{1}{\sqrt{\omega r}}\Big(Y_{m+\frac{1}{2}}(\omega r_{\min})J_{m+\frac{1}{2}}(\omega r) - J_{m+\frac{1}{2}}(\omega r_{\min})Y_{m+\frac{1}{2}}(\omega r)\Big). \tag{11}$$

It is straightforward to check that: $F_m(r_{\min}) = 0$. We want to determine our parameters in order to have:

$$F_{m_1}(r_{\max}) = 0, \qquad F_{m_2}(r_{\max}) = 0. \tag{12}$$

In this way, we are requiring homogeneous Dirichlet conditions both at the internal and external boundaries of the annular domain. This has to be true for different modes m_1 and m_2 at the same time. We observe that such a nonlinear problem does not always admit a solution.

Afterwards, we define:

$$\Phi_m(r,\phi) = \alpha_m F_m(r)\cos(m\phi), \qquad \Psi_m(r,\phi) = \beta_m F_m(r)\sin(m\phi), \tag{13}$$

for some arbitrary multiplicative constants α_m and β_m. These are orthogonal eigenfunctions with eigenvalue $-\omega^2$. Successively, solutions to the wave equation are deduced by introducing a dependence on time. This is done by writing:

$$\Phi_{m_1}\sin(c\omega t)+\Phi_{m_2}\cos(c\omega t)+\Psi_{m_1}\sin(c\omega(t+t_0))+\Psi_{m_2}\cos(c\omega(t+t_0)), \tag{14}$$

where t_0 is an arbitrary time shift. In the examples that follow, we will take $t_0 = 0$. By suitably adjusting the magnitude of α_m and β_m, we actually get interesting evolution patterns. Some results are discussed in the next section.

4 Determination of the Parameters

Using numerical tools, we looked for sets of parameters satisfying both the equations in (12). We fixed m_1 and m_2 and investigated on the possibility of finding the corresponding values ω and $r_{\max}$. Once again we remark that a solution to this problem does not always exists. We provide in Fig. 2 some peculiar situations where we have $F_{m_1}(r_{\max}) = F_{m_2}(r_{\max}) = 0$. In these plots the two functions F_{m_1}, F_{m_2} have been normalized in the maximum norm.

The results of a more systematic analysis are given in Table 1, which reports some of the values for which the search was successful. Some cases are also visualized in the plots of Figs. 3, 4, 5, 6, 7, and 8, where the corresponding eigenfunctions (suitably normalized) are shown. These exactly correspond to those already considered in Fig. 2. We recall that the four orthogonal functions have

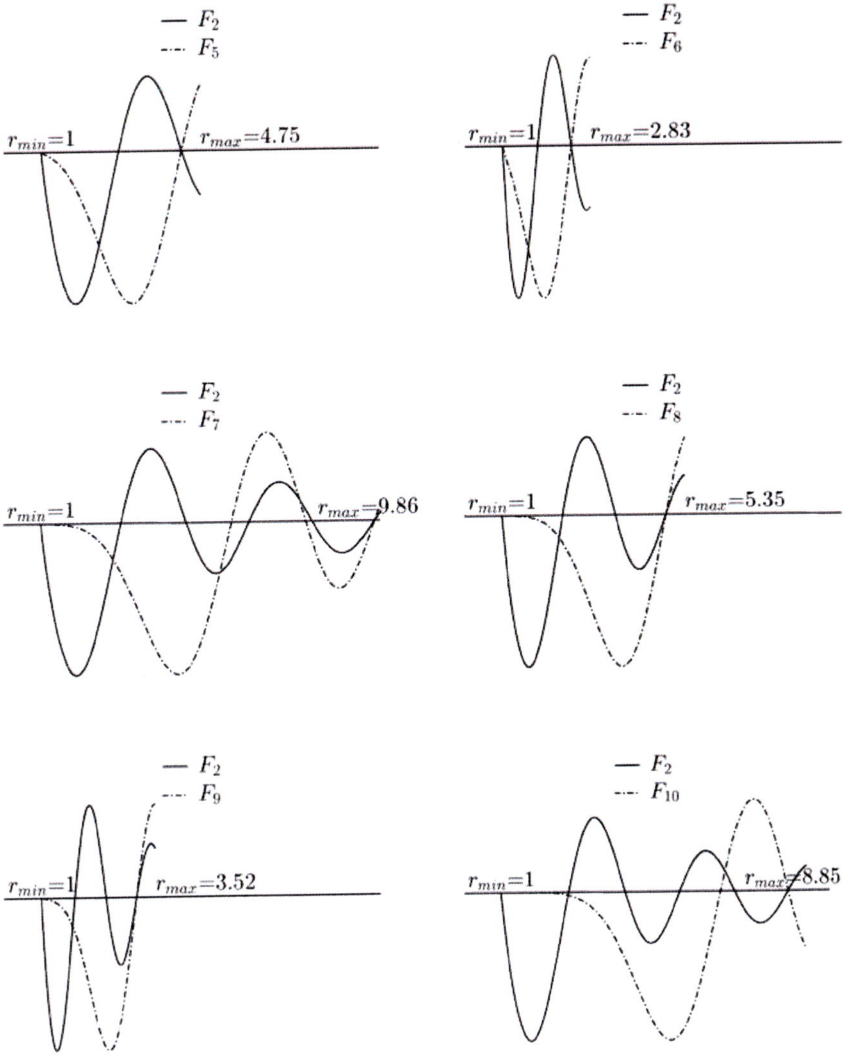

Fig. 2 A survey on radial functions satisfying $F_{m_1}(r_{\max}) = F_{m_2}(r_{\max}) = 0$

a common eigenvalue equal to $-\omega^2$. According to the notation of (13), up to multiplicative constants, the eigenfunctions in Figs. 3, 4, 5, 6, 7, and 8 are (from left to right): Φ_{m_1}, Ψ_{m_1}, Φ_{m_2}, Ψ_{m_2}. When the frequency is 'high' in the variable ϕ, there is usually a single bump in the radial direction. Conversely, for 'low' frequencies in the variable ϕ, the functions exhibit more oscillations in the variable r. Of course, if we switch m_1 and m_2, we get identical values of ω and $r_{\max}$.

Table 1 Some choices of the parameters allowing to get $F_{m_1}(r_{\max}) = F_{m_2}(r_{\max}) = 0$

m_1	m_2	ω	$r_{\max}$	m_1	m_2	ω	$r_{\max}$	m_1	m_2	ω	$r_{\max}$
1	4	1.88	6.23	5	1	3.35	3.88	9	1	2.20	9.74
	5	3.35	3.88		2	1.97	4.75		2	3.95	3.52
	6	4.71	3.03		10	3.16	8.27		3	4.14	4.31
	7	3.36	7.62		19	3.86	7.62		4	3.44	7.23
2	5	1.97	4.75	6	1	4.71	3.03	10	1	3.94	6.63
	6	3.72	2.83		2	3.72	2.83		2	2.15	8.85
	7	1.92	9.86		3	2.32	4.53		3	4.72	3.18
	8	2.39	5.35		9	5.44	3.28		4	4.85	3.92
3	6	2.32	4.53	7	1	3.36	7.62	11	1	3.35	4.82
	7	4.56	2.56		2	1.92	9.86		2	2.67	6.05
	8	2.14	9.43		3	4.56	2.56		3	2.26	8.94
	9	4.14	4.31		4	4.17	3.70		4	5.44	2.97
4	1	1.88	6.23	8	1	3.88	4.29	12	1	2.81	8.93
	7	4.17	3.70		2	2.39	5.35		2	4.14	4.17
	8	5.38	2.38		3	2.14	9.43		3	3.07	5.62
	9	3.44	7.23		4	5.38	2.38		4	4.15	6.04

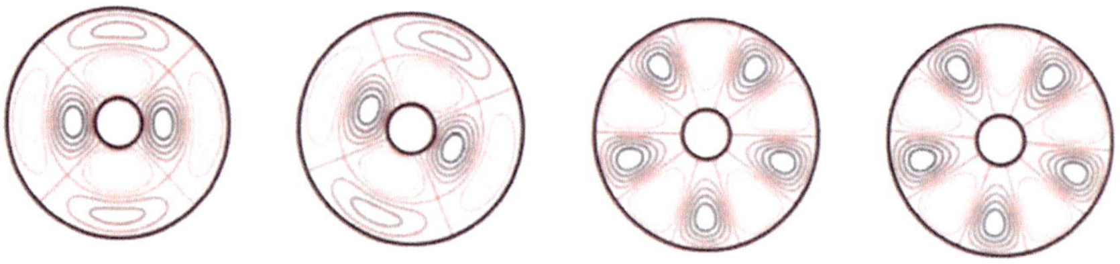

Fig. 3 Contour lines of the eigenfunctions Φ_{m_1}, Ψ_{m_1}, Φ_{m_2}, Ψ_{m_2} for $m_1 = 2, m_2 = 5, \omega = 1.97, r_{\max} = 4.75$

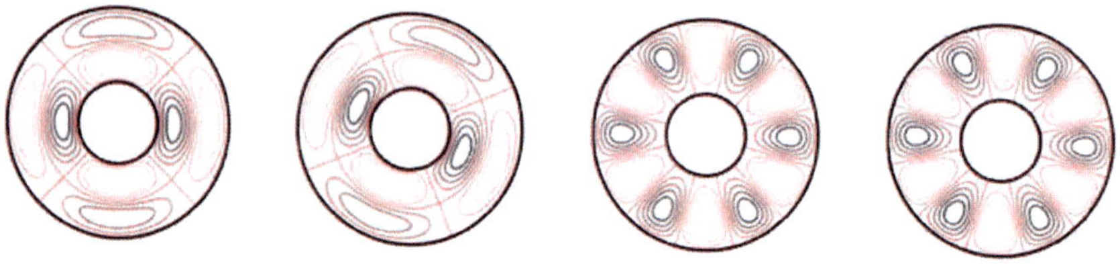

Fig. 4 Contour lines of the eigenfunctions Φ_{m_1}, Ψ_{m_1}, Φ_{m_2}, Ψ_{m_2} for $m_1 = 2, m_2 = 6, \omega = 3.72, r_{\max} = 2.83$

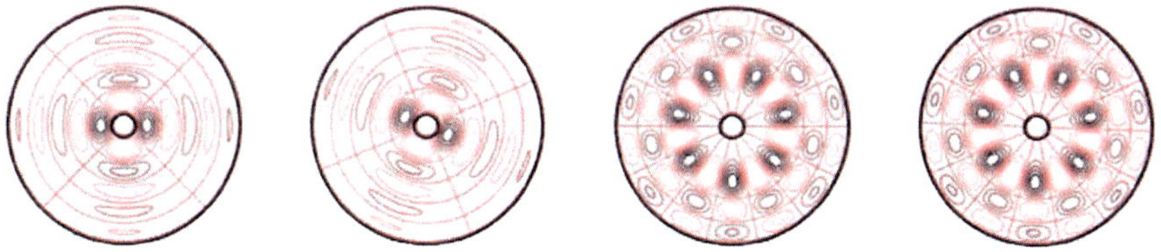

Fig. 5 Contour lines of the eigenfunctions Φ_{m_1}, Ψ_{m_1}, Φ_{m_2}, Ψ_{m_2} for $m_1 = 2, m_2 = 7, \omega = 1.92, r_{\max} = 9.86$

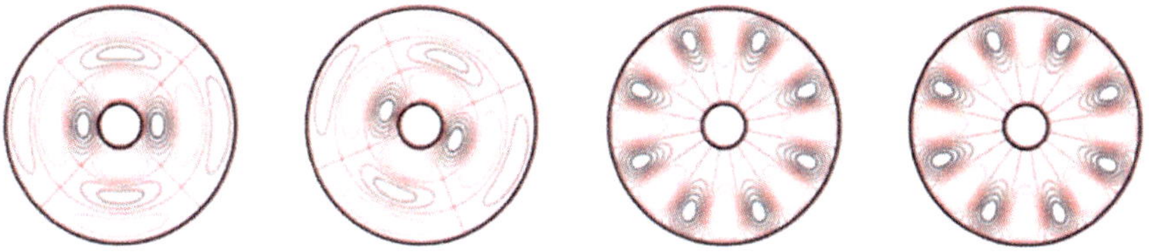

Fig. 6 Contour lines of the eigenfunctions Φ_{m_1}, Ψ_{m_1}, Φ_{m_2}, Ψ_{m_2} for $m_1 = 2, m_2 = 8, \omega = 2.39, r_{\max} = 5.35$

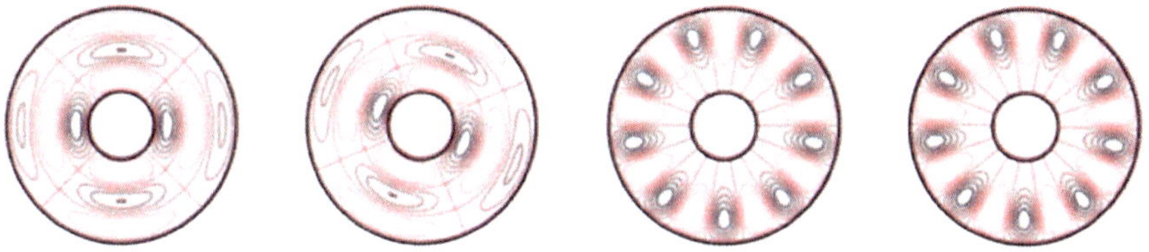

Fig. 7 Contour lines of the eigenfunctions Φ_{m_1}, Ψ_{m_1}, Φ_{m_2}, Ψ_{m_2} for $m_1 = 2, m_2 = 9, \omega = 3.95, r_{\max} = 3.52$

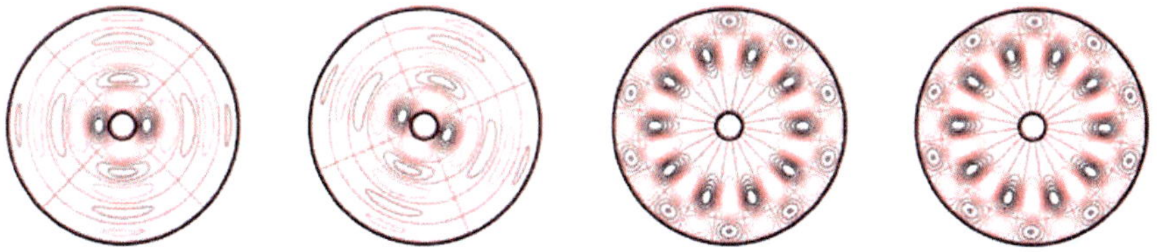

Fig. 8 Contour lines of the eigenfunctions Φ_{m_1}, Ψ_{m_1}, Φ_{m_2}, Ψ_{m_2} for $m_1 = 2, m_2 = 10, \omega = 2.15, r_{\max} = 8.85$

Finally, in Fig. 9, we show the electric field distribution, obtained through (4). We fixed $m_1 = 2$ and $m_2 = 8$. In the first two displacements are displayed the configurations corresponding to Φ_{m_1} and Φ_{m_2}. By combining these eigenfunctions according to (14) we get the successive displacements. We switched the sign of the angle ϕ regarding the eigenfunctions Φ_{m_2} and Ψ_{m_2}, so that the plot in the middle of Fig. 9 rotates anticlockwise. A snapshot of the vector configuration at a given

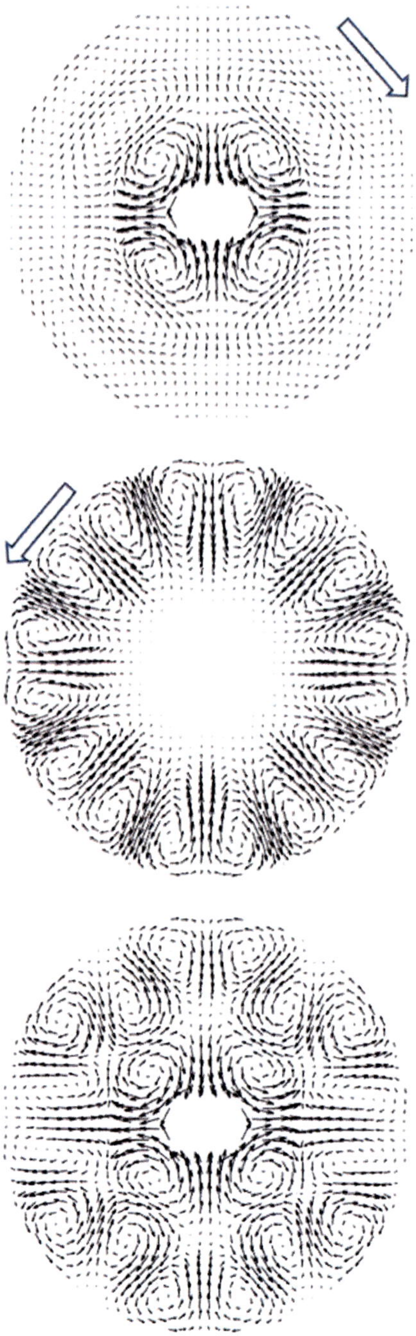

Fig. 9 Electric field distributions corresponding to (4), when $\theta = \pi/2$, $m_1 = 2$, $m_2 = 8$. We have $r_{\min} = 1$ and $r_{\max} = 5.35$. Eigenfunctions of type Φ_{m_1} and Φ_{m_2} are shown on top and centre, respectively. A suitable combination of Φ_{m_1} and Φ_{m_2} is shown on the bottom. As time runs, the two components rotate in different directions, inducing the vortexes to move around like in a ball bearing set. The information travels approximately at the same speed near the inner and the outer boundaries, without exceeding the speed of light contrary to what would happen in the rotation of a rigid body

time is shown in Fig. 9, bottom. The multiplicative constants α_{m_1}, α_{m_2} β_{m_1} β_{m_2}, are in principle arbitrary. Here, they have been chosen in order to allow a clear graphical view. The time evolution of these patterns is rather involved and can be fully appreciated only with the help of a movie.

5 Final Comments and Speculations

In a very rough approximation, the electromagnetic environment here introduced represents a possible background distribution of the cosmic neutral plasma outside a rotating star (our Sun for example; see, e.g., [15]). Due to the small concentration of charged particles, we are basically in vacuum and the information evolves at speeds comparable to that of light. We observe that, first of all, a star emits photons, which are energy packets produced as a consequence of chemical or subatomic reactions. Photons actually constitute the main detectable part. *Solar wind* (see, e.g., [5]) is also a typical electromagnetic manifestation, being related to the called *Parker's spiral* (see [12]).

In addition to the above mentioned instances, the complex electrodynamical evolution at the surface of a star may induce the formation of complicated whirls in the surrounding plasma. These reasonably follow a behavior of the type described in the previous sections, so generating a kind of spherical shell around the main body of the star. Nothing prevents us to think that such a process can be repeated at the new external boundary, where the electric field is again tangential and oscillating. We find ourselves with a sequence of encapsulated shells, whose size grows geometrically. The transition between a shell and the contiguous one is then realized in a continuous way, without outrageously exceeding the speed of light. Indeed, these systems evolve at an averaged speed comparable to that of light, but when they become larger, their angular velocity diminishes.

Perhaps, we can appreciate the existence of the set of shells in an indirect way. In fact, this construction may contribute to explain the formation of planets, initially in a state of massive fragments, successively aggregated by gravitational forces and the action of an organized plasma (see, e.g., [2, 3]). The analysis of the first mode in the variable θ (see the case $m = \ell$) tells us that a privileged framework is that of the equatorial plane. In the case of our heliosphere this could in part clarify the coplanar distribution of the planets. Moreover, due to the geometrical growth of the shells, we may also contribute to the understanding of the *Titius-Bode law*, that rules the averaged distance of the planets from the Sun according to the geometric growth rate: $0.4 + 0.3 \times 2^k$, where the unity of measure is expressed in Astronomical Units. The planet Neptune is the only one that does not conform to this law. However, it displays anomalous magnetic surroundings that may agree with an intermediate location within the shell situated between Uranus and Pluto.

The above observations are only qualitative, since, for a better understanding of the phenomenon, one has to resort to the nonlinear models proposed for instance in MHD. Our suggestion is to face such more advanced problems in the framework of spectral methods, by using as basis functions a set of eigenfunctions similar to those here proposed.

References

1. Abramowitz, M., Stegun, I.A. (eds.): Handbook of Mathematical Functions with Formulas, Graphs and Mathematical Tables, Applied Mathematics Series, vol. 55. Dover Publications (1965)
2. Alfvén, H.: On the Origin of the Solar System, 1st edn. Clarendon Press (1954)
3. Alfvén, H.: Cosmology in the plasma universe: An introductory exposition. IEEE Trans. Plasma Sci. **18**, 5–10 (1990)
4. Chinosi, C., Della Croce, L., Funaro, D.: Rotating electromagnetic waves in toroid-shaped regions. Int. J. Mod. Phys. C **21**(1), 11–32 (2010)
5. Ferraro, V.C.A., Bhatia, V.B.: Corotation and solar wind in the solar corona and interplanetary medium. Astrophys. J. **147**, 220–229 (1967)
6. Funaro, D.: Trapping electromagnetic solitons in cylinders. Math. Model. Anal. **19**(1), 44–51 (2014)
7. Funaro, D.: From Photons to Atoms, The Electromagnetic Nature of Matter. World Scientific, Singapore (2019)
8. Funaro, D.: Electromagnetic waves in annular regions. Appl. Sci. MDPI **10**(5), 1780 (2020)
9. Fatone, L., Funaro, D.: Electromagnetic fields simulating a rotating sphere and its exterior with implications to the modeling of the heliosphere. Math. Methods Appl. Sci. **46**(2), 1952–1963 (2022)
10. Funaro, D.: Ball lightning as plasma vortexes: A reinforcement of the conjecture. Appl. Sci. **12**(7), 3451 (2022)
11. Maudlin, T.: Quantum Non-locality and Relativity, Metaphysical Intimations of Modern Physics, 3rd edn. Basil Blackwell, Oxford (2011)
12. Parker, E.N.: Dynamics of the interplanetary gas and magnetic fields. Astrophys. J. **128**, 664 (1958)
13. Peratt A.L.: Physics of the Plasma Universe, 2nd edn. Springer, New York (2015)
14. Priest E.R.: Magnetohydrodynamics of the Sun. Cambridge University Press (2014)
15. Stix, M.: The Sun, An Introduction. Springer, Berlin Heidelberg (1989)

Exponential Convergence of *hp*-FEM for the Integral Fractional Laplacian in 1D

Björn Bahr, Markus Faustmann, Carlo Marcati, Jens Markus Melenk, and Christoph Schwab

1 Introduction

The (numerical) analysis of non-integer powers of elliptic differential operators has garnered a lot of interest recently. The prototype of such an operator is the so called fractional Laplacian $(-\Delta)^s$ for $s \in (0, 1)$, which can be defined in several different ways. On the full space, a classical way is to define it as an operator with Fourier symbol $|\xi|^{2s}$, but (equivalent) alternatives via semi-group theory, operator theory, or via singular integrals exist as well, [19]. Here, we consider the so called integral fractional Laplacian, using the singular integral definition.

Solutions to equations involving the integral fractional Laplacian on bounded domains usually are non-smooth at the boundary and feature only finite regularity even if the data is smooth, [17]; we mention finite regularity results in Hölder spaces [17, 21] (smooth domains), in isotropic weighted Sobolev spaces [1] or unweighted Besov spaces [6] (Lipschitz domains). In our recent work, [12], we prove weighted analytic regularity estimates in two space dimensions that reflect both the interior analyticity and the anisotropic nature of the solution near the boundary. In the present article, we provide similar weighted analytic estimates for the one dimensional case.

The numerical approximation of fractional PDEs by means of finite element techniques is in principle well understood; we mention the survey articles [5, 7, 9, 20] and references therein. An important active field is the design of suitable meshes

B. Bahr · M. Faustmann · J. M. Melenk (✉)
Institute of Analysis and Scientific Computing, TU Wien, Vienna, Austria
e-mail: bjoern.bahr@tuwien.ac.at; markus.faustmann@tuwien.ac.at; melenk@tuwien.ac.at

C. Marcati · C. Schwab
Seminar for Applied Mathematics, ETH Zürich, Zürich, Switzerland
e-mail: carlo.marcati@sam.math.ethz.ch; schwab@sam.math.ethz.ch

J. M. Melenk et al. (eds.), *Spectral and High Order Methods for Partial Differential Equations ICOSAHOM 2020+1*, Lecture Notes in Computational Science and Engineering 137, https://doi.org/10.1007/978-3-031-20432-6_18

to counteract the singular behavior at the boundary, [1, 5]. We mention [15], where a sharp analysis of vertex and edge singularities via Mellin techniques is used to derive the correct mesh grading for methods converging at the optimal algebraic rate. In this article, we leverage our weighted analytic regularity estimates to design an exponentially convergent method by means of *hp*-finite element approximation in one dimension. The generalization to two dimensions is the topic of our follow-up work [13].

The techniques employed in the present work are closely related to the higher dimensional case done in [12], but the one dimensional case allows for simplifications and a clearer presentation of the main concepts. A crucial step is a reformulation of the fractional PDE as a Dirichlet to Neumann operator of a degenerate *local* elliptic PDE, the so-called Caffarelli-Silvestre extension, [8]. For this extension, the second crucial step is that a (global) regularity shift of essentially $1/2$ can be obtained using difference quotient techniques, [22]. We derive interior regularity estimates of Caccioppoli type and bootstrap these regularity results to obtain weighted analytic regularity.

2 Model Problem and Main Results

2.1 *The Fractional Laplacian*

We consider the integral fractional Laplacian defined pointwise as the principal value integral

$$(-\Delta)^s u(x) := C(s)\ \text{P.V.} \int_{\mathbb{R}} \frac{u(x)-u(z)}{|x-z|^{1+2s}}\, dz \quad \text{with} \quad C(s) := -2^{2s}\frac{\Gamma(s+1/2)}{\pi^{1/2}\Gamma(-s)},$$

where $\Gamma(\cdot)$ denotes the Gamma function.

Fractional differential equations are conveniently described using fractional Sobolev spaces. Let $\Omega \subset \mathbb{R}$ be an open, bounded interval. Denoting by $H^t(\Omega)$ the classical integer order Sobolev spaces with $t \in \mathbb{N}_0$, fractional order Sobolev spaces for $t \in (0, 1)$ are defined in terms of the Aronstein-Slobodeckij seminorm $|\cdot|_{H^t(\Omega)}$ and the corresponding full norm $\|\cdot\|_{H^t(\Omega)}$, which are given by

$$|v|^2_{H^t(\Omega)} = \int_\Omega \int_\Omega \frac{|v(x)-v(z)|^2}{|x-z|^{1+2t}}\, dz\, dx, \qquad \|v\|^2_{H^t(\Omega)} = \|v\|^2_{L^2(\Omega)} + |v|^2_{H^t(\Omega)}.$$

For $t \in (0, 1)$, we let $H^t(\mathbb{R})$ be the usual Sobolev space (defined, e.g., in terms of the Fourier transform) and we employ the spaces $\widetilde{H}^t(\Omega) := \{u \in H^t(\mathbb{R}) : u \equiv 0 \text{ on } \mathbb{R}\backslash\overline{\Omega}\}$ with norm given by

$$\|v\|^2_{\widetilde{H}^t(\Omega)} := \|v\|^2_{H^t(\Omega)} + \|v/r^t\|^2_{L^2(\Omega)},$$

where $r(x) := \mathrm{dist}(x, \partial\Omega)$ denotes the Euclidean distance of a point $x \in \Omega$ from the boundary $\partial\Omega$. For $t > 0$, the space $H^{-t}(\Omega)$ denotes the dual space of $\widetilde{H}^t(\Omega)$, and we write $\langle\cdot,\cdot\rangle_{L^2(\Omega)}$ for the duality pairing that extends the $L^2(\Omega)$-inner product.

On $\Omega \subset \mathbb{R}$, we consider the fractional differential equation

$$(-\Delta)^s u = f \qquad \text{in } \Omega, \tag{1}$$

$$u = 0 \qquad \text{in } \Omega^c := \mathbb{R}\backslash\overline{\Omega}, \tag{2}$$

where $s \in (0, 1)$ and $f \in H^{-s}(\Omega)$ is a given right-hand side. The weak form of (1), (2) is to find $u \in \widetilde{H}^s(\Omega)$ such that

$$a(u, v) := \frac{C(s)}{2}\int_{\mathbb{R}}\int_{\mathbb{R}} \frac{(u(x) - u(z))(v(x) - v(z))}{|x - z|^{1+2s}}\,dz\,dx = \langle f, v\rangle_{L^2(\Omega)} \tag{3}$$

for all $v \in \widetilde{H}^s(\Omega)$. Existence and uniqueness of $u \in \widetilde{H}^s(\Omega)$ follow from the Lax–Milgram Lemma for any $f \in H^{-s}(\Omega)$, upon the observation that the bilinear form $a(\cdot,\cdot) : \widetilde{H}^s(\Omega) \times \widetilde{H}^s(\Omega) \to \mathbb{R}$ is continuous and coercive, see, e.g., [1, Sec. 2.1].

2.2 *Weighted Analytic Regularity*

Our first main result, Theorem 1, asserts analytic regularity of the solution $u \in \widetilde{H}^s(\Omega)$ to our model problem (3) in scales of weighted Sobolev spaces, provided f in (3) is analytic in $\overline{\Omega}$.

Theorem 1 *Let the data $f \in C^\infty(\overline{\Omega})$ satisfy, for constants $C_f, \gamma_f > 0$,*

$$\forall p \in \mathbb{N}_0\colon \quad \|D^p f\|_{L^2(\Omega)} \le C_f \gamma_f^p p!. \tag{4}$$

Let u solve (3). Then, there is γ (depending only on γ_f, s, Ω) such that for any $\varepsilon > 0$ there exists a constant $C_\varepsilon > 0$ (depending on C_f, s, Ω, ε) such that

$$\forall p \in \mathbb{N}\colon \quad \|r^{p-1/2-s+\varepsilon} D^p u\|_{L^2(\Omega)} \le C_\varepsilon \gamma^p p!.$$

With the space $\mathcal{B}^1_\beta := \{u \in L^2(\Omega) : \|r^{n+\beta} D^{n+1} u\|_{L^2(\Omega)} \le C\gamma^n n! \quad \forall n \in \mathbb{N}_0\}$, we have $u \in \mathcal{B}^1_\beta$ for $\beta = 1/2 - s + \varepsilon$. In particular, $u \in C(\overline{\Omega})$.

2.3 *Exponential Convergence of hp-FEM*

Once weighted regularity results are available, numerical approximation by means of the hp-FEM, [23], can be analyzed. In fact, employing geometric meshes and piecewise polynomials of higher degree, our second main result, Theorem 3, states exponential convergence of the hp-FEM for the integral fractional Laplacian.

Definition 2 [Geometric Mesh on $(-1, 1)$] On $(-1, 1)$, for a *grading factor* $\sigma \in (0, 1)$ and a number L *of layers of geometric refinement*, the *geometric mesh* $\mathcal{T}^L_{geo,\sigma} = \{T_i : i = 1, \ldots, 2L + 2\}$ with $2L + 2$ elements $T_i = (x_{i-1}, x_i)$ is given by the nodes

$$x_0 := -1, \quad x_i = -1 + \sigma^{L-i+1}, \quad i = 1, \ldots, L, \\ x_{i+1} = 1 - \sigma^{i-L}, \quad i = L, \ldots, 2L, \quad x_{2L+2} := 1. \tag{5}$$

Geometric partitions $\mathcal{T}^L_{geo,\sigma}$ on bounded intervals $\Omega \subset \mathbb{R}$ are obtained from (5) by translation and dilation. Key features of a geometric partition $\mathcal{T}^L_{geo,\sigma}$ are a) elements $T_i \in \mathcal{T}^L_{geo,\sigma}$ with $\overline{T}_i \cap \partial\Omega = \emptyset$ satisfy $\operatorname{diam}(T_i) \sim \operatorname{dist}(T_i, \partial\Omega)$ and b) elements $T_i \in \mathcal{T}^L_{geo,\sigma}$ with $\overline{T}_i \cap \partial\Omega \neq \emptyset$ satisfy $\operatorname{diam}(T_i) = O(\sigma^L)$.

On a geometric mesh $\mathcal{T}^L_{geo,\sigma}$ and for a polynomial degree $p \in \mathbb{N}$, we introduce the spline space $S^{p,1}(\mathcal{T}^L_{geo,\sigma}) := \{v \in H^1(\Omega) : v|_{T_i} \in \mathcal{P}_p(T_i)\ \forall T_i \in \mathcal{T}^L_{geo,\sigma}\}$. Here, $\mathcal{P}_p(T_i)$ is the space of all polynomials of degree (at most) p on T_i. The subspace with zero boundary conditions is $S^{p,1}_0(\mathcal{T}^L_{geo,\sigma}) := \{v \in S^{p,1}(\mathcal{T}^L_{geo,\sigma}) \ : \ v|_{\partial\Omega} = 0\}$. We note $N := \dim S^{p,1}_0(\mathcal{T}^L_{geo,\sigma}) \sim pL$.

The hp-FEM approximation u_N is the Galerkin discretization of (3):

$$u_N \in S^{p,1}_0(\mathcal{T}^L_{geo,\sigma}) : \quad a(u_N, v_N) = \langle f, v_N \rangle_{L^2(\Omega)} \ \ \forall v_N \in S^{p,1}_0(\mathcal{T}^L_{geo,\sigma}). \tag{6}$$

Theorem 3 *Let f satisfy (4). Let $\mathcal{T}^L_{geo,\sigma}$ be a geometric mesh on the interval Ω with grading factor $\sigma \in (0, 1)$ and L layers of refinement towards the boundary points. Let $u_N \in S^{p,1}_0(\mathcal{T}^L_{geo,\sigma})$ solve (6) and u solve (3). Then, there are constants C_{apx}, $b > 0$ independent of p and L such that, with $\beta = 1/2 - s + \varepsilon$, $\varepsilon > 0$, given by Theorem 1,*

$$\|u - u_N\|_{\widetilde{H}^s(\Omega)} \leq C_{\mathrm{apx}}(e^{-bp} + \sigma^{(1-\beta-s)L}).$$

The choice $L \sim p$ leads to convergence $\|u - u_N\|_{\widetilde{H}^s(\Omega)} \leq C \exp(-b'\sqrt{N})$, where $N = \dim S^{p,1}_0(\mathcal{T}^L_{geo,\sigma})$ and C, b' are constants independent of N.

The rest of this note will provide short proofs of Theorems 1 and 3.

3 Regularity Results

3.1 The Caffarelli-Silvestre Extension

The main tool in our regularity analysis is the very influential reformulation of the nonlocal fractional Laplacian as the Dirichlet-to-Neumann operator of a local, degenerate elliptic PDE posed on a half space in one additional space dimension, the so-called *Caffarelli-Silvestre extension*, [8].

For its formulation, let $\alpha := 1 - 2s$ and write $\omega^+ := \omega \times \mathbb{R}_+$ for any measurable subset $\omega \subset \mathbb{R}$, where $\mathbb{R}_+ = (0, \infty)$. We define $L^2_\alpha(\omega^+)$ as the space of square-integrable functions with respect to the weight y^α with the norm

$$\|U\|^2_{L^2_\alpha(\omega^+)} := \int_{y\in\mathbb{R}_+} y^\alpha \int_{x\in\omega} |U(x, y)|^2 dx\, dy.$$

We introduce the Beppo-Levi space $\mathrm{BL}^1_\alpha := \{U \in L^2_{loc}(\mathbb{R} \times \mathbb{R}_+) \,:\, \nabla U \in L^2_\alpha(\mathbb{R} \times \mathbb{R}_+)\}$. For elements of BL^1_α, one can give meaning to their trace at $y = 0$, which is denoted $\operatorname{tr} U$. In fact, $\operatorname{tr} U \in H^s_{loc}(\mathbb{R})$ (see, e.g., [18, Lem. 3.8]) with $|\operatorname{tr} U|_{H^s(\mathbb{R})} \lesssim \|\nabla U\|_{L^2_\alpha(\mathbb{R}\times\mathbb{R}_+)}$. Finally, we require the space $\mathrm{BL}^1_{\alpha,0,\Omega} := \{V \in \mathrm{BL}^1_\alpha \,:\, \operatorname{tr} V = 0 \text{ on } \Omega^c\}$.

Let $H > 0$ be fixed. (We refer to [12, Rem. 3.2] for a discussion of this parameter.) Let data $F \in C_0^\infty(\mathbb{R}^2)$ with $\operatorname{supp} F \subset \mathbb{R} \times [0, H]$ and $f \in C^\infty(\overline{\Omega})$ be given. The Caffarelli-Silvestre extension problem reads: Find $U = U(x, y) \in \mathrm{BL}^1_\alpha$ such that

$$\begin{aligned} -\operatorname{div}(y^\alpha \nabla U) &= F && \text{in } \mathbb{R} \times (0, \infty), \\ \partial_{n_\alpha} U(\cdot, 0) &= f && \text{in } \Omega, \\ \operatorname{tr} U &= 0 && \text{on } \Omega^c, \end{aligned} \tag{7}$$

where $\partial_{n_\alpha} U(x, 0) = -d_s \lim_{y\to 0} y^\alpha \partial_y U(x, y)$ for $d_s = 2^{2s-1}\Gamma(s)/\Gamma(1-s)$. The weak form of (7) is: Find $U \in \mathrm{BL}^1_{\alpha,0,\Omega}$ such that for all $V \in \mathrm{BL}^1_{\alpha,0,\Omega}$

$$b(U, V) := \int_{\mathbb{R}\times\mathbb{R}_+} y^\alpha \nabla U \cdot \nabla V dx dy = \int_{\mathbb{R}\times(0,H)} F V dx dy + \int_\Omega f \operatorname{tr} V dx. \tag{8}$$

We refer to [12, Sec. 3] for the fact that $b(\cdot, \cdot)$ is an inner product on the Hilbert space $\mathrm{BL}^1_{\alpha,0,\Omega}$, and that there exists a constant $C > 0$ depending on H such that

$$\forall V \in \mathrm{BL}^1_{\alpha,0,\Omega}: \quad \|V\|_{L^2_\alpha(\mathbb{R}\times(0,H))} \le C\|\nabla V\|_{L^2_\alpha(\mathbb{R}\times\mathbb{R}_+)}. \tag{9}$$

Therefore, by Lax-Milgram, (8) is uniquely solvable with

$$\|\nabla U\|_{L^2_\alpha(\mathbb{R}\times\mathbb{R}_+)} \le C\left[\|f\|_{H^{-s}(\Omega)} + \|F\|_{L^2_{-\alpha}(\mathbb{R}\times(0,H))}\right]. \tag{10}$$

Finally, the solution u to (3) is given by $u = \operatorname{tr} U$ with U being the unique solution of (7) with $F = 0$.

3.2 Global Regularity

The following Lemma 4 provides additional global regularity in the x-variable. Although it is a special case of [12, Lem. 3.2], which is valid for any space dimension $d \ge 1$, we sketch the proof as it is crucial for this article.

Lemma 4 *Let $f \in C^\infty(\overline{\Omega})$, $F \in C_0^\infty(\mathbb{R}^2)$ with* $\operatorname{supp} F \subset \mathbb{R} \times [0, H]$. *Let U solve (7). Then, for $t \in [0, 1/2)$, there is $C_t > 0$ depending only on s, t, H, and Ω, such that*

$$\int_{\mathbb{R}_+} y^\alpha \|\nabla U(\cdot, y)\|^2_{H^t(\Omega)} dy < C_t N^2(F, f)$$

$$\text{with} \qquad N^2(F, f) := \|f\|^2_{H^1(\Omega)} + \|F\|^2_{L^2_{-\alpha}(\mathbb{R}\times(0,H))}. \tag{11}$$

Proof (Sketch) The idea is to apply the difference quotient argument from [22] only in the x-direction. For $h \in \mathbb{R}$, denote $T_h U := \eta U_h + (1 - \eta)U$, where $U_h(x, y) := U(x + h, y)$ and η is a cut-off function that localizes to a fixed interval $B_{2\rho}(x_0)$ in the first variable and that is constant in the second variable.

The main result of [22] is that estimates for the modulus $\omega(U)$ defined by

$$\omega(U) := \sup_{h\in D\setminus\{0\}} \frac{b(T_hU, T_hU) - b(U, U) + \int_{\mathbb{R}\times(0,H)} F(T_hU - U) + \int_\Omega f \operatorname{tr}(T_hU - U)}{|h|}$$

can be used to derive regularity results in Besov spaces. Here, $D \subset \mathbb{R}$ denotes a set of admissible directions h. These directions are chosen such that the function T_hU is an admissible test function, i.e., $T_hU \in \mathrm{BL}^1_{\alpha,0,\Omega}$. In the present case this set can easily be characterized as $h \in [-\rho, \rho]$, if $B_{4\rho}(x_0) \subset \Omega$ or $\operatorname{dist}(B_{3\rho}(x_0), \Omega) \ge \rho$. In the other cases, we can take $h \in [0, \rho]$, if the right endpoint of the interval Ω is in the intersection of $B_{4\rho}(x_0) \cap \Omega$ or $h \in [-\rho, 0]$, if the left endpoint of the interval Ω is in the intersection of $B_{4\rho}(x_0) \cap \Omega$.

Step 1. (Estimate of $b(\cdot,\cdot)$-Terms) Using support properties of η as well as the estimate $\|U(\cdot,y) - U_h(\cdot,y)\|_{L^2(B_{2\rho})} \lesssim |h| \|\nabla U(\cdot,y)\|_{L^2(B_{3\rho})}$, one can deduce

$$|b(T_h U, T_h U) - b(U,U)| \lesssim |h| \int_{B_{3\rho}^+} y^\alpha |\nabla U|^2 \, dx\, dy.$$

Step 2. (Estimate of F-Integral) Similarly to the first step, one can estimate

$$\begin{aligned}\left| \int_{\mathbb{R}\times(0,H)} F(U - T_h U)\, dx\, dy \right| &\leq \|F\|_{L^2_{-\alpha}(B_{2\rho}\times(0,H))} \|U - U_h\|_{L^2_\alpha(B_{2\rho}^+)} \\ &\lesssim |h| \|F\|_{L^2_{-\alpha}(B_{2\rho}\times(0,H))} \|\nabla U\|_{L^2_\alpha(B_{3\rho}^+)}.\end{aligned}$$

Step 3. (Estimate of f-Integral) With the trace inequality from [18, Lem. 3.3] and (9), we obtain by exploiting the regularity of f (cf. [Step 3 of Lem. 3.3, 12] for details)

$$\left| \int_\Omega f \operatorname{tr}(U - T_h U)\, dx \right| \lesssim |h| \|f\|_{H^1(B_{4\rho})} \|\nabla U\|_{L^2_\alpha(\mathbb{R}\times\mathbb{R}_+)}. \tag{12}$$

Step 4. (Application of the Abstract Framework of [22]) We introduce the seminorms $[U]^2 := \int_{\mathbb{R}\times\mathbb{R}_+} y^\alpha |\nabla U|^2 \, dxdy$. By the coercivity of $b(\cdot,\cdot)$ on $\mathrm{BL}^1_{\alpha,0,\Omega}$ with respect to $[\cdot]^2$ and the abstract estimates in [22, Sec. 2], we have

$$\begin{aligned}&[U - T_h U]^2 \lesssim \omega(U)|h| \\ &\quad\lesssim |h| \|\nabla U\|_{L^2_\alpha(\mathbb{R}\times\mathbb{R}_+)} \left(\|\nabla U\|_{L^2_\alpha(B_{3\rho}^+)} + \|F\|_{L^2_{-\alpha}(B_{2\rho}\times(0,H))} + \|f\|_{H^1(B_{4\rho})} \right).\end{aligned}$$

Employing the a priori estimate (10) and using $\eta \equiv 1$ on $B_\rho^+(x_0)$ leads to

$$\int_{B_\rho^+} y^\alpha |\nabla U - \nabla U_h|^2 \, dx\, dy \leq [U - T_h U]^2 \leq |h| \, N^2(F,f). \tag{13}$$

Step 5a: ($H^t(\Omega)$-Estimate) Thus far, we only considered one sided difference quotients, i.e., $h \in D$ in (13). The restriction $h \in D$ in (13) can be lifted as shown in [12]. In the present 1D situation, the key observation is that when computing the Aronstein-Slobodeckij norm, one can write for functions v defined on $\mathbb{R}$

$$\begin{aligned}\int_{x\in\mathbb{R}} \int_{|h|\leq h_0} \frac{|v(x+h) - v(x)|^2}{|h|^{1+2\sigma}}\, dh\, dx &= \int_{x\in\mathbb{R}} \int_{h=0}^{h_0} \frac{|v(x+h) - v(x)|^2}{h^{1+2\sigma}}\, dh\, dx \\ &+ \int_{x\in\mathbb{R}} \int_{h=0}^{h_0} \frac{|v(x-h) - v(x)|^2}{h^{1+2\sigma}}\, dh\, dx,\end{aligned}$$

and a change of variables $x - h = x'$ in the second integral leads again to a one-sided difference quotient. For simplicity of presentation, we will therefore assume in the following Step 5b that (13) holds for all $|h| \le h_0$, and we will assume that B_ρ in (13) can be replaced by Ω; this is possible by covering with suitable localizations and using (13).

Step 5b: ($H^t(\Omega)$-Estimate) For $t < 1/2$ and $\widetilde{R}$ large enough (s.t. $\Omega \subset (-\widetilde{R}, \widetilde{R})$), we estimate with the Aronstein-Slobodeckij seminorm

$$\int_{\mathbb{R}_+} |\nabla U(\cdot, y)|^2_{H^t(\Omega)}\, dy \le \int_{\mathbb{R}_+} \int_\Omega \int_{|h| \le \widetilde{R}} \frac{|\nabla U(x+h, y) - \nabla U(x, y)|^2}{|h|^{1+2t}}\, dh\, dx\, dy.$$

The integral in h is split into the range $|h| \le \varepsilon$ for some fixed $\varepsilon > 0$, for which (13) can be brought to bear, and $\varepsilon < |h| < \widetilde{R}$, for which a triangle inequality can be used. This gives the sought estimate.

□

3.3 *Interior Regularity*

We require Caccioppoli type estimates that allow us to control higher order derivatives by lower order derivatives on slightly enlarged intervals.

Lemma 5 *Let $B_R \subset \Omega$ be a ball of radius R and B_{cR} be the concentrically scaled ball of radius cR. Let U satisfy (7) with given data f and F. There is a constant C_{Cac} depending only on Ω, s such that for every $c \in (0, 1)$*

$$\|\nabla \partial_x U\|_{L^2_\alpha(B^+_{cR})} \le C_{\mathrm{Cac}}\Big(((1-c)R)^{-1} \|\nabla U\|_{L^2_\alpha(B^+_R)} + \|\partial_x f\|_{L^2(B_R)} + \|F\|_{L^2_{-\alpha}(B_R \times (0,H))}\Big). \tag{14}$$

With a constant $\gamma > 0$ depending only on s, Ω, and c, it holds for all $p \in \mathbb{N}$

$$\|\nabla \partial_x^p U\|_{L^2_\alpha(B^+_{cR})} \le (\gamma p)^p R^{-p} \|\nabla U\|_{L^2_\alpha(B^+_R)} + \sum_{i=1}^{p} (\gamma p)^{p-i} R^{i-p} \big(\|\partial_x^i f\|_{L^2(B_R)} + \|\partial_x^{i-1} F\|_{L^2_{-\alpha}(B_R \times (0,H))}\big). \tag{15}$$

Proof Estimate (14) follows from [12, Lem. 3.5], which holds for any space dimension d, using difference quotient techniques as previously employed in [14]. Note that the test function V in the proof of [Lem. 3.5, 12] is an admissible test function by [(3.14), 12].

We now show (15): As the x-derivatives commute with the differential operator in (7), we have that $\partial_x^i U$ solves (7) on B_R^+ instead of $\mathbb{R} \times \mathbb{R}_+$ with data $\partial_x^i F$ and $\partial_x^i f$ for any i.

For given $c > 0$, we choose sets $B_{c_i R}$ with $c_i = c + (i-1)\frac{(1-c)}{p}$, which implies $c_i < c_{i+1} < 1$ for all $i < p$. Then, we have $c_{i+1}R - c_i R = \frac{(1-c)R}{p}$. Applying (14) to $\partial_x^{p-1} U$ with the sets $B_{c_1 R}$, $B_{c_2 R}$ leads to the estimate

$$\|\nabla \partial_x^p U\|_{L^2_\alpha(B^+_{c_1 R})} \le C_{\mathrm{Cac}} \Big(\frac{p}{(1-c)} R^{-1} \|\nabla \partial_x^{p-1} U\|_{L^2_\alpha(B^+_{c_2 R})} + \|\partial_x^p f\|_{L^2(B_R)} + \|\partial_x^{p-1} F\|_{L^2_{-\alpha}(B_R \times (0,H))} \Big).$$

Inductively applying (14) to control $\|\nabla \partial_x^{p-j+1} U\|_{L^2_\alpha(B^+_{c_j R})}$ for $2 \le j \le p$ with sets $B_{c_j R}$, $B_{c_{j+1} R}$ provides the claimed estimate with $\gamma = C_{\mathrm{Cac}}/(1-c)$. □

The right-hand side of the Caccioppoli estimates (14), (15) suggests that we need to control $R^{-1}\|\nabla U\|_{L^2_\alpha(B_R^+)}$. This term is actually small for $R \to 0$ in the presence of regularity of U, which was asserted in Lemma 4.

Lemma 6 *Let $S_R := \{x \in \Omega : r(x) < R\}$. Let U solve (7). Then, for $t \in [0, 1/2)$, there exists $C_{\mathrm{reg}} > 0$ depending only on s, t, and Ω such that, with the constant $C_t > 0$ from Lemma 4 and $N^2(F, f)$ given by (11), we have*

$$R^{-2t} \|\nabla U\|^2_{L^2_\alpha(S_R^+)} \le \|r^{-t} \nabla U\|^2_{L^2_\alpha(\Omega^+)} \le C_{\mathrm{reg}} C_t N^2(F, f). \tag{16}$$

Proof The first estimate in (16) is trivial. For the second bound, we start by noting that Lemma 4 provides the global regularity $\int_{\mathbb{R}_+} y^\alpha \|\nabla U(\cdot, y)\|^2_{H^t(\Omega)} dy \le C_t N^2(F, f)$. For $t \in [0, 1/2)$ and any $v \in H^t(\Omega)$, we have by, e.g., [16, Thm. 1.4.4.3] the embedding result $\|r^{-t} v\|_{L^2(\Omega)} \le C_{\mathrm{reg}} \|v\|_{H^t(\Omega)}$. Applying this embedding to $\nabla U(\cdot, y)$, multiplying by y^α, and integrating in y yields the claimed estimate. □

With the Caccioppoli estimate, we obtain estimates for the derivatives.

Lemma 7 *Let U solve (7) with data f, F satisfying for some $C_f, C_F, \gamma > 0$*

$$\forall p \in \mathbb{N}_0: \quad \|\partial_x^p f\|_{L^2(\Omega)} \le C_f \gamma^p p^p, \qquad \|\partial_x^p F\|_{L^2_{-\alpha}(\Omega \times (0,H))} \le C_F \gamma^p p^p.$$

Then, there is $\tilde{\gamma} > 0$ (depending on γ, s, Ω) and, for every $\varepsilon \in (0, 1)$, $t \in [0, 1/2)$ a constant C_ε (depending only on ε, s, t, Ω) such that

$$\forall p \in \mathbb{N}_0: \quad \|r^{p-t+\varepsilon} \nabla \partial_x^p U\|_{L^2_\alpha(\Omega^+)} \le C_\varepsilon \tilde{\gamma}^p p! (C_f + C_F + N(F, f)).$$

Proof The case $p = 0$ follows immediately from Lemma 6. For $p \geq 1$ and for any ball $B_R \subset \Omega$ and $c > 0$, Lemma 5 combined with Lemma 6 gives

$$\|\nabla \partial_x^p U\|_{L^2_\alpha(B^+_{cR})} \lesssim \gamma^p p^p R^{t-p} N(F, f) + \sum_{i=1}^{p} \gamma^{p-i} p^{p-i} R^{i-p} \Big(\|\partial_x^i f\|_{L^2(B_R)} + \|\partial_x^{i-1} F\|_{L^2_{-\alpha}(B_R \times (0,H))} \Big). \tag{17}$$

In order to obtain an estimate for the L^2-norm over $\Omega = (a, b)$, we dyadically cover $\Omega \subset \bigcup_{i \in \mathbb{N}} B_{cr_i}(x_i)$ with intervals $B_{cr_i}(x_i) \subset \Omega$, where the points x_i run through the set $\{a+(b-a)2^{-j} : j \in \mathbb{N}\} \cup \{b-(b-a)2^{-j} : j \in \mathbb{N}\}$, $c \in (1/2, 1)$ is fixed and $r_i = \operatorname{dist}(x_i, \partial\Omega)$. A geometric series argument gives $\sum_i r_i^\varepsilon < \infty$ for any chosen $\varepsilon > 0$. Using the assumption on the data f, F and (17), we obtain using $t < 1/2$

$$\begin{aligned} \|r^p \nabla \partial_x^p U\|_{L^2_\alpha(B_{cr_i}(x_i)^+)} &\lesssim (1 + c)^p r_i^p \|\nabla \partial_x^p U\|_{L^2_\alpha(B_{cr_i}(x_i)^+)} \\ &\lesssim \tilde{\gamma}^p p^p r_i^t (C_f + C_F + N(F, f)) \end{aligned} \tag{18}$$

for suitable $\tilde{\gamma}$. Therefore, we have for fixed $\varepsilon > 0$

$$\begin{aligned} \|r^{p-t+\varepsilon}(\nabla \partial_x^p U)\|^2_{L^2(\Omega^+)} &\lesssim \sum_{i \in \mathbb{N}} r_i^{2p-2t+2\varepsilon} \|\nabla \partial_x^p U\|^2_{L^2(B_{cr_i}(x_i)^+)} \\ &\overset{(18)}{\lesssim} \tilde{\gamma}^{2p} p^{2p} \sum_{i \in \mathbb{N}} r_i^{2\varepsilon} (C_f + C_F + N(F, f))^2 = C_\varepsilon^2 \tilde{\gamma}^{2p} p^{2p} (C_f + C_F + N(F, f))^2, \end{aligned}$$

which finishes the proof after noting $p^p \leq p! e^p$ and adjusting $\tilde{\gamma}$. □

Taking traces, we obtain weighted analytic estimates for the fractional Laplacian:

Proof *(of Theorem 1)* [18, Lem. 3.7] gives

$$|V(x, 0)|^2 \lesssim \|V(x, \cdot)\|^{1-\alpha}_{L^2_\alpha(\mathbb{R}_+)} \|\partial_y V(x, \cdot)\|^{1+\alpha}_{L^2_\alpha(\mathbb{R}_+)} + \|V(x, \cdot)\|^2_{L^2_\alpha(\mathbb{R}_+)}.$$

Using this trace estimate with $V = \partial_x^p U$, additionally multiplying with $r^{2p-1-2s+2\varepsilon}$, and using $\alpha = 1 - 2s$ provides

$$\begin{aligned} r^{2p-1-2s+2\varepsilon} |\partial_x^p U(x, 0)|^2 &\lesssim \|r^{p-1/2-s+\varepsilon} \partial_x^p U(x, \cdot)\|^2_{L^2_\alpha(\mathbb{R}_+)} \\ &\quad + \|r^{p-3/2+\varepsilon} \partial_x^p U(x, \cdot)\|^{1-\alpha}_{L^2_\alpha(\mathbb{R}_+)} \|r^{p-1/2+\varepsilon} \partial_y \partial_x^p U(x, \cdot)\|^{1+\alpha}_{L^2_\alpha(\mathbb{R}_+)}. \end{aligned}$$

Integration over Ω gives in view of $u(x) = U(x, 0)$

$$\begin{aligned} \|r^{p-1/2-s+\varepsilon} \partial_x^p u\|^2_{L^2(\Omega)} &\lesssim \|r^{p-1/2-s+\varepsilon} \partial_x^p U(x, \cdot)\|^2_{L^2_\alpha(\Omega^+)} \\ &\quad + \|r^{p-3/2+\varepsilon} \partial_x^p U\|^{1-\alpha}_{L^2_\alpha(\Omega^+)} \|r^{p-1/2+\varepsilon} \partial_y \partial_x^p U\|^{1+\alpha}_{L^2_\alpha(\Omega^+)}. \end{aligned}$$

Note that for $p \geq 1$, we have $|\partial_x^p U| \leq |\nabla \partial_x^{p-1} U|$ and $|\partial_y \partial_x^p U| \leq |\nabla \partial_x^p U|$. Applying Lemma 7 with $t = 1/2 - \varepsilon/2$ and $\varepsilon/2$ instead of ε therein for the two terms with weights $r^{p-3/2+\varepsilon}$ and $r^{p-1/2+\varepsilon}$ and $t = \max(0, s - 1/2)$ for the term with the weight $r^{p-1/2-s+\varepsilon}$ provides the desired estimate.

The statement $u \in \mathcal{B}^1_\beta$ follows by definition. The assertion $u \in C(\overline{\Omega})$ is implied by the observation $u \in C^\infty(\Omega)$ together with $u \in L^2(\Omega)$ and $r^{1/2-s+\varepsilon} u' \in L^2(\Omega)$. □

3.4 Exponential Convergence of hp-FEM

For $\beta' \in [0, 1)$, it is convenient to introduce the norm $\|\cdot\|_{H^1_{\beta'}(\Omega)}$ by

$$\|v\|^2_{H^1_{\beta'}(\Omega)} := \|r^{\beta'} v'\|^2_{L^2(\Omega)} + \|r^{\beta'-1} v\|^2_{L^2(\Omega)}.$$

Lemma 8 *Let $\beta' \in [0, 1)$, $\sigma \in (0, 1 - \beta')$. Then, there is $C_{\beta',\sigma} > 0$ such that*

$$\|v\|_{\widetilde{H}^\sigma(\Omega)} \leq C_{\beta',\sigma} \left[\|r^{\beta'} v'\|_{L^2(\Omega)} + \|r^{\beta'-1} v\|_{L^2(\Omega)} \right] \tag{19}$$

for all v such that right-hand side is finite.

Proof For two continuously embedded Banach spaces $X_1 \subset X_0$ and for $v \in X_0$, $t > 0$, we define the K-functional by $K(v, t) := \inf_{w \in X_1} \|v - w\|_{X_0} + t\|w\|_{X_1}$. For $\theta \in (0, 1)$ and fine index $q \in [1, \infty]$, the interpolation spaces (e.g. [24, Lecture 22]) $X_{\theta,q} := (X_0, X_1)_{\theta,q}$ are given by the norm

$$\|v\|^q_{X_{\theta,q}} := \int_{t=0}^{\infty} \left(t^{-\theta} K(v, t)\right)^q \frac{dt}{t}, \quad q \in [1, \infty), \qquad \|v\|_{X_{\theta,\infty}} := \sup_{t \in (0,\infty)} t^{-\theta} K(v, t).$$

We use the fact that, since $X_1 \subset X_0$, replacing the integration and the supremum limit ∞ by a finite number T leads to an equivalent norm, [11, Chap. 6, Sec. 7]. Let $S_t := \{x \in \Omega \mid r(x) < t\}$ denote a t-neighborhood of $\partial\Omega$. For each $t > 0$ sufficiently small, we may choose $\chi_t \in C_0^\infty(\mathbb{R})$ such that $\chi_t \equiv 0$ on $S_{t/2}$ and $\chi_t \equiv 1$ on $\Omega \setminus S_t$ as well as $\|\chi_t^{(j)}\|_{L^\infty(\mathbb{R})} \leq C t^{-j}$, $j \in \{0, 1\}$. Decomposing $v = \chi_t v + (1 - \chi_t) v$, we have $\chi_t v \in H_0^1(\Omega)$ and $(1 - \chi_t) v \in L^2(\Omega)$. A calculation reveals

$$\|(\chi_t v)'\|_{L^2(\Omega)} \leq C t^{-\beta'} \|v\|_{H^1_{\beta'}(\Omega)},$$

$$\|(1 - \chi_t) v\|_{L^2(\Omega)} \leq C \|v\|_{L^2(S_t)} \leq t^{1-\beta'} \|r^{\beta'-1} v\|_{L^2(\Omega)}.$$

This implies for $X_0 = L^2(\Omega)$, $X_1 = H_0^1(\Omega)$ that $K(v,t) \leq Ct^{1-\beta'}\|v\|_{H^1_{\beta'}(\Omega)}$. For the Besov space $\widetilde{B}^{1-\beta'}_{2,\infty}(\Omega) := (L^2(\Omega), H_0^1(\Omega))_{1-\beta',\infty}$, we obtain $\|v\|_{\widetilde{B}^{1-\beta'}_{2,\infty}(\Omega)} \leq C\|v\|_{H^1_{\beta'}(\Omega)}$. We conclude the proof by noting

$$\widetilde{H}^\sigma(\Omega) \overset{[24]}{=} (L^2(\Omega), H_0^1(\Omega))_{\sigma,2} \overset{\sigma<1-\beta'}{\subset} (L^2(\Omega), H_0^1(\Omega))_{1-\beta',\infty} = \widetilde{B}^{1-\beta'}_{2,\infty}(\Omega).$$

□

Lemma 9 *Let $\beta' \in [0,1)$, $\varepsilon > 0$. Then, there is $C_{\beta',\varepsilon} > 0$ such that the following holds: For $\widehat{v} \in C([0,1])$ let $I\widehat{v}$ be the linear interpolant in the endpoints 0, 1. Then, provided the right-hand side is finite, there holds for $e := \widehat{v} - I\widehat{v}$*

$$\|x^{\beta'-1}e\|_{L^2(0,1)} + \|x^{\beta'}e'\|_{L^2(0,1)} \leq C_{\beta',\varepsilon}\|x^{\min\{\beta'+1,3/2-\varepsilon\}}\widehat{v}''\|_{L^2(0,1)}.$$

Proof *Step 1:* Let $\widetilde{\pi}_1\widehat{v} \in \mathcal{P}_1$ be the linear interpolant of $\widehat{v}$ in the points 1/2 and 1. From [4, Lemma 15, (A.4)], we get for any $\alpha > -1$

$$\|x^{\alpha/2}(\widehat{v} - \widetilde{\pi}_1\widehat{v})'\|_{L^2(0,1)} \leq C_\alpha\|x^{\alpha/2+1}\widehat{v}''\|_{L^2(0,1)}. \tag{20}$$

A maximum norm estimate is obtained from

$$|(\widehat{v} - \widetilde{\pi}_1\widehat{v})(x)| \leq \int_x^1 |(\widehat{v} - \widetilde{\pi}_1\widehat{v})'(t)|\,dt$$

$$\leq \sqrt{\int_0^1 t^{-1+2\varepsilon}\,dt \int_0^1 t^{1-2\varepsilon}|(\widehat{v} - \widetilde{\pi}_1\widehat{v})'(t)|^2\,dt} \overset{(20)}{\leq} C_\varepsilon\|x^{3/2-\varepsilon}\widehat{v}''\|_{L^2(0,1)}.$$

Step 2 ($\beta' < 1/2$): Abbreviate $e := \widehat{v} - I\widehat{v}$ and note $e(x) = \int_0^x e'(t)\,dt$. For $\beta' \in [0, 1/2)$, Hardy's inequality [11, Chap. 2, Thm. 3.1] is applicable and yields

$$\|x^{\beta'-1}e\|_{L^2(0,1)} \leq C\|x^{\beta'}e'\|_{L^2(0,1)}.$$

We estimate with $w = \widetilde{\pi}_1\widehat{v} \in \mathcal{P}_1$

$$\begin{aligned}\|x^{\beta'}e'\|_{L^2(0,1)} &\leq \|x^{\beta'}(\widehat{v} - w)'\|_{L^2(0,1)} + \|x^{\beta'}(I(\widehat{v} - w))'\|_{L^2(0,1)} \\ &\overset{\mathcal{P}_1 \text{ finite dimensional}}{\lesssim} \|x^{\beta'}(\widehat{v} - w)'\|_{L^2(0,1)} + \|\widehat{v} - w\|_{L^\infty(0,1)} \\ &\overset{\text{Step 1, } w=\widetilde{\pi}_1\widehat{v}}{\lesssim} \|x^{\min\{\beta'+1,3/2-\varepsilon\}}\widehat{v}''\|_{L^2(0,1)}.\end{aligned}$$

Step 3 ($\beta' \geq 1/2$): From the representation $e(x) = \int_0^x e'(t)\,dt$ we get for any $\alpha \in [0, 1/2)$ by the Cauchy-Schwarz inequality $|e(x)| \leq C_\alpha x^{1/2-\alpha}\|x^\alpha e'\|_{L^2(0,1)}$.

Hence,

$$\|x^{\beta'-1}e\|_{L^2(0,1)} \lesssim \sqrt{\int_0^1 x^{2\beta'-2+1-2\alpha}\,dx}\;\|x^{\alpha}e'\|_{L^2(0,1)} \lesssim \|x^{\alpha}e'\|_{L^2(0,1)}.$$

We conclude, since $\alpha < 1/2 \leq \beta'$, that $\|x^{\beta'-1}e\|_{L^2(0,1)} + \|x^{\beta'}e'\|_{L^2(0,1)} \lesssim \|x^{\alpha}e'\|_{L^2(0,1)}$. Applying Step 2 with α taking the role of β' there, we get, by selecting α sufficiently close to $1/2$

$$\|x^{\alpha}e'\|_{L^2(0,1)} \lesssim \|x^{3/2-\varepsilon}u''\|_{L^2(0,1)} \lesssim \|x^{\min\{\beta'+1,3/2-\varepsilon\}}u''\|_{L^2(0,1)},$$

which concludes the proof. □

Proof (of Theorem 3) Since the bilinear form $a(\cdot,\cdot)$ is elliptic on $\widetilde{H}^s(\Omega)$, by Céa's lemma it suffices to construct a function $v \in S_0^{p,1}(\mathcal{T}^L_{geo,\sigma})$ that satisfies the stated error bound. By Theorem 1, we have $u \in \mathcal{B}^1_\beta(\Omega)$ for $\beta = 1/2-s+\varepsilon$, $\varepsilon > 0$ arbitrary and $u \in C(\overline{\Omega})$. We approximate $u \in C(\overline{\Omega})$ by a $v \in S^{p,1}(\mathcal{T}^L_{geo,\sigma})$ defined as the linear interpolant in the elements abutting $\partial\Omega$ and the Gauss-Lobatto interpolant of degree p on all other elements. With $\beta' := 1 - s - \varepsilon'$ for ε' sufficiently small, Lemma 8 shows $\|u - v\|_{\widetilde{H}^s(\Omega)} \leq C\|u - v\|_{H^1_{\beta'}(\Omega)}$, which can be estimated by summing elementwise error contributions. For the approximation on the elements abutting $\partial\Omega$, we observe that, for sufficiently small ε, we can select ε' to ensure

$$3/2 - s + \varepsilon = \beta + 1 < \min\{\beta' + 1, 3/2 - \varepsilon\} = \min\{2 - s - \varepsilon', 3/2 - \varepsilon\}. \tag{21}$$

In Lemma 9, we may therefore replace $x^{\min\{\beta'+1,3/2-\varepsilon\}}$ by $x^{\beta+1}$. Combining then Lemma 9 with a scaling argument, we get for T_i with $\overline{T}_i \cap \partial\Omega \neq \emptyset$

$$\|r^{\beta'-1}(u-v)\|_{L^2(T_i)} + \|r^{\beta'}(u-v)'\|_{L^2(T_i)} \leq Ch_i^{1-\beta-(1-\beta')}\|r^{\beta+1}u''\|_{L^2(T_i)}.$$

For the remaining elements T_i with $\overline{T}_i \cap \partial\Omega = \emptyset$, we use $u \in \mathcal{B}^1_\beta(\Omega)$ and get following [2, Thm. 3.13] with a scaling argument

$$\|r^{\beta'-1}(u-v)\|_{L^2(T_i)} + \|r^{\beta'}(u-v)'\|_{L^2(T_i)} \leq Ch_i^{1-\beta-(1-\beta')}e^{-bp}$$

for some $b > 0$ independent of p. Noting that $\beta' - \beta = 1/2 - \varepsilon - \varepsilon' > 0$, we obtain from a geometric series argument by summation over all elements and using $h_i = O(\sigma^L)$ for the elements abutting on $\partial\Omega$ that

$$\|u - v\|_{H^1_{\beta'}(\Omega)} \lesssim \sigma^{L(\beta'-\beta)} + e^{-bp} = \sigma^{L(1/2-\varepsilon-\varepsilon')} + e^{-bp}.$$

The proof of Theorem 3 is completed by suitably adjusting ε. □

4 Numerical Example

On the interval $(-1, 1)$ with $f = 1$, the exact solution to (3) is $u(x) = 2^{-2s}\sqrt{\pi}(\Gamma(s+1/2)\Gamma(1+s))^{-1}(1-x^2)^s$. The singularity of $u(x) \sim \mathrm{dist}(x, \partial\Omega)^s$ is generic for the Dirichlet problem of the integral fractional Laplacian, also in higher dimensions, [21]. We employ the geometric mesh $\mathcal{T}^L_{geo,\sigma}$ that is graded towards ± 1 with grading factor σ and L levels of refinement.

Example 10 We utilize two *hp*-FEM spaces: (a) the space $S_0^{L,1}(\mathcal{T}^L_{geo,\sigma})$ of piecewise polynomials of degree $p = L$ and (b) the subspace $\widetilde{S}^L \subset S_0^{L,1}(\mathcal{T}^L_{geo,\sigma})$ obtained by lowering the polynomial degree from $p = L$ to 1 in the two elements touching $\partial\Omega$. The grading factor is $\sigma = 0.6$. The *hp*-FEM approximation is computed in MATLAB by using an implementation from [10].

In Fig. 1, the energy norm error $\sqrt{a(u - u_N, u - u_N)} = \sqrt{a(u, u) - a(u_N, u_N)}$ between the exact solution $u \in \widetilde{H}^s(\Omega)$ and the *hp*-FEM approximation u_N is plotted versus L. Note that $a(v, v) \sim \|v\|^2_{\widetilde{H}^s(\Omega)}$. The left panel shows the performance of *hp*-FEM based on $S_0^{L,1}(\mathcal{T}^L_{geo,\sigma})$ whereas the right panel that of *hp*-FEM based on $\widetilde{S}^L$. As predicted by Theorem 3, we observe exponential convergence with respect to the number of layers L. In fact, the convergence is close to $O(\sigma^{L/2}L^{-1})$. The additional factor L^{-1} is due to the fact that we approximate by polynomials of degree $p = L$ on the boundary elements, whereas the proof of Theorem 3 employed the linear interpolant on these elements, i.e., an approximation

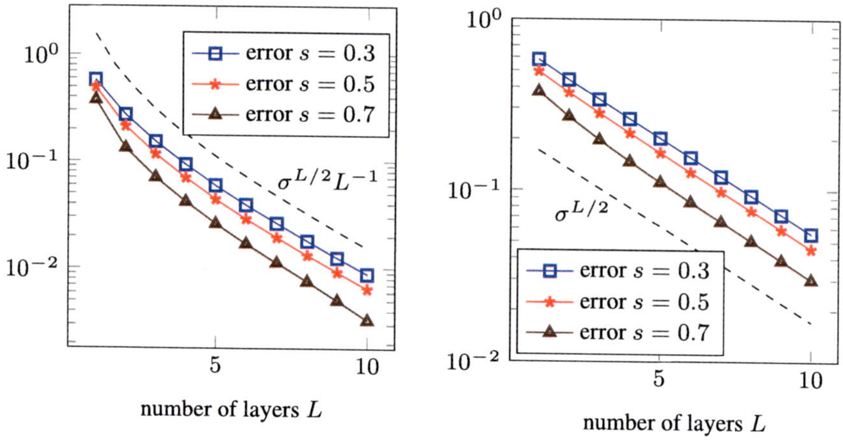

Fig. 1 Exponential energy norm error convergence of *hp*-FEM on geometric mesh with grading factor $\sigma = 0.6$ for $s \in \{0.3, 0.5, 0.7\}$, $\Omega = (-1, 1)$, $f = 1$. Left: *hp*-FEM based on $S_0^{p,1}(\mathcal{T}^L_{geo,\sigma})$ with $p = L$. Right: *hp*-FEM based on subspace $\widetilde{S}^L \subset S_0^{L,1}(\mathcal{T}^L_{geo,\sigma})$

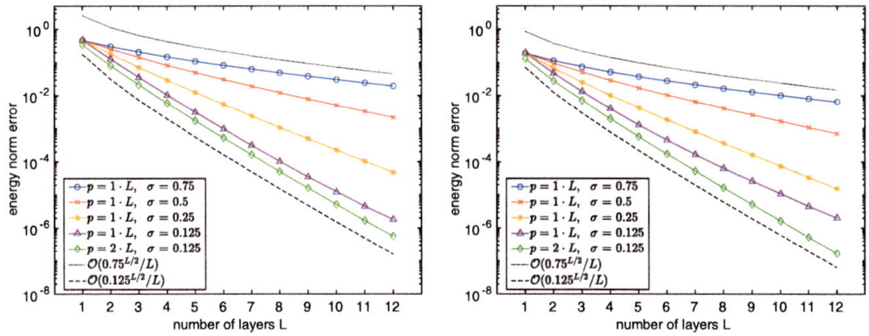

Fig. 2 Exponential energy norm error convergence of *hp*-FEM based on $S_0^{p,1}(\mathcal{T}^L_{geo,\sigma})$ for grading factor $\sigma \in \{0.125, 0.25, 0.5, 0.75\}$ and $p = L$ or $p = 2L$ as indicated; $\Omega = (-1.1)$, $f = 1$. Left: $s = 0.25$. Right: $s = 0.75$

in $\widetilde{S}^L$. The right panel of Fig. 1 shows the convergence behavior $O(\sigma^{L/2})$ predicted by Theorem 3.

Example 11 For $s = 0.25$ and $s = 0.75$ Fig. 2 shows the *hp*-FEM energy norm error vs. L based on the spaces $S_0^{p,1}(\mathcal{T}^L_{geo,\sigma})$ for different values of $\sigma \in \{0.125, 0.25, 0.5, 0.75\}$. We choose $p = L$ except for $\sigma = 0.125$, where additionally $p = 2L$ is shown. An explanation for this last choice is as follows: An inspection of the proof of Theorem 3 shows that the constant b deteriorates as $\sigma \to 0$ so that for small σ a convergence behavior $O(\sigma^L/L)$ can only be achieved by increasing the proportionality constant c in the relation $p = cL$. The numerical realization of this example is based on [3].

Acknowledgments JMM gladly acknowledges support by the Austrian Science Fund (FWF) through project F 65.

References

1. Acosta, G., Borthagaray, J.: A fractional Laplace equation: regularity of solutions and finite element approximations. SIAM J. Numer. Anal. **55**(2), 472–495 (2017)
2. Apel, T., Melenk, J.: Interpolation and quasi-interpolation in h- and hp-version finite element spaces (extended version). Tech. Rep. 39, Institute of Analysis and Scientific Computing, TU Wien (2015). http://www.asc.tuwien.ac.at/preprint/2015/asc39x2015.pdf
3. Bahr, B., Faustmann, M., Melenk, J.M.: An implementation of *hp*-FEM for the fractional Laplacian. In preparation
4. Banjai, L., Melenk, J.M., Nochetto, R.H., Otárola, E., Salgado, A.J., Schwab, C.: Tensor FEM for spectral fractional diffusion. Found. Comput. Math. **19**(4), 901–962 (2019)
5. Bonito, A., Borthagaray, J., Nochetto, R., Otárola, E., Salgado, A.: Numerical methods for fractional diffusion. Comput. Vis. Sci. **19**(5–6), 19–46 (2018)

6. Borthagaray, J., Nochetto, R.: Besov regularity for the Dirichlet integral fractional Laplacian in Lipschitz domains. arXiv e-prints p. arXiv:2110.02801 (2021)
7. Borthagaray, J., Li, W., Nochetto, R.: Linear and nonlinear fractional elliptic problems. In: 75 years of mathematics of computation, Contemp. Math., vol. 754, pp. 69–92. Amer. Math. Soc., Providence, RI (2020)
8. Caffarelli, L., Silvestre, L.: An extension problem related to the fractional Laplacian. Commun. Partial Differential Equations **32**(7–9), 1245–1260 (2007)
9. D'Elia, M., Du, Q., Glusa, C., Gunzburger, M., Tian, X., Zhou, Z.: Numerical methods for nonlocal and fractional models. Acta Numer. **29**, 1–124 (2020)
10. Dadic, A.: Aspekte einer hp-FEM Implementierung für den fraktionalen Laplace-Operator. Bachelor thesis, Department of Mathematics, TU Wien (2021)
11. DeVore, R., Lorentz, G.: Constructive Approximation. Springer Verlag (1993)
12. Faustmann, M., Marcati, C., Melenk, J., Schwab, C.: Weighted analytic regularity for the integral fractional Laplacian in polygons. SIAM J. Math. Anal. **54**(6), 6323–6357 (2022)
13. Faustmann, M., Marcati, C., Melenk, J., Schwab, C.: Exponential convergence of hp-FEM for the integral fractional Laplacian in polygons (2022). arXiv e-prints p. arXiv:2209.11468
14. Faustmann, M., Melenk, J., Praetorius, D.: Quasi-optimal convergence rate for an adaptive method for the integral fractional Laplacian. Math. Comput. **90**, 1557–1587 (2021)
15. Gimperlein, H., Stephan, E., Štoček, J.: Corner singularities for the fractional Laplacian and finite element approximation. Preprint (2020). https://mat1.uibk.ac.at/heiko/corners.pdf
16. Grisvard, P.: Elliptic Problems in Nonsmooth Domains. Classics in Applied Mathematics, vol. 69. Society for Industrial and Applied Mathematics (SIAM), Philadelphia, PA (2011)
17. Grubb, G.: Fractional Laplacians on domains, a development of Hörmander's theory of μ-transmission pseudodifferential operators. Adv. Math. **268**, 478–528 (2015)
18. Karkulik, M., Melenk, J.: $\mathcal{H}$-matrix approximability of inverses of discretizations of the fractional Laplacian. Adv. Comput. Math. **45**(5–6), 2893–2919 (2019)
19. Kwaśnicki, M.: Ten equivalent definitions of the fractional Laplace operator. Fract. Calc. Appl. Anal. **20**(1), 7–51 (2017)
20. Lischke, A., Pang, G., Gulian, M., Song, F., Glusa, C., Zheng, X., Mao, Z., Cai, W., Meerschaert, M.M., Ainsworth, M., Karniadakis, G.: What is the fractional Laplacian? A comparative review with new results. J. Comput. Phys. **404**, 109009, 62 (2020)
21. Ros-Oton, X., Serra, J.: The Dirichlet problem for the fractional Laplacian: regularity up to the boundary. J. Math. Pures Appl. (9) **101**(3), 275–302 (2014)
22. Savaré, G.: Regularity results for elliptic equations in Lipschitz domains. J. Funct. Anal. **152**(1), 176–201 (1998)
23. Schwab, C.: p- and hp-Finite Element Methods. Numerical Mathematics and Scientific Computation. The Clarendon Press/Oxford University Press, New York (1998). Theory and applications in solid and fluid mechanics
24. Tartar, L.: An Introduction to Sobolev Spaces and Interpolation Spaces. Lecture Notes of the Unione Matematica Italiana, vol. 3. Springer, Berlin (2007)

Discontinuous Galerkin Method for Linear Wave Equations Involving Derivatives of the Dirac Delta Distribution

Scott E. Field, Sigal Gottlieb, Gaurav Khanna, and Ed McClain

1 Introduction

In this article we describe a discontinuous Galerkin (DG) method [5–8, 15, 20] for solving the wave equation

$$-\partial_t^2\psi + \partial_x^2\psi + V(x)\psi = \sum_{n=0}^{N} a_n(t,x)\delta^{(n)}(x)\,, \tag{1}$$

where $x \in [a, b]$, V is a potential, $\delta^{(n)}(x) = \partial_x^n \delta(x)$ is the nth distributional derivative [14] of a dirac delta distribution $\delta(x)$, and $a_n(t, x)$ are arbitrary (classical) functions. We let the functions $\psi_0(x) = \psi(0, x)$ and $\dot{\psi}_0(x) = \partial_t\psi(0, x)$ specify the initial data. Differential equations of the form (1) arise when modeling phenomena driven by well-localized sources and have found applications as diverse as neuroscience [1, 4], seismology [18, 21], and gravitational wave physics [19, 24]. As one example, when a rotating blackhole is perturbed by a small, compact object

S. E. Field (✉) · S. Gottlieb · E. McClain
Department of Mathematics, Center for Scientific Computing & Visualization Research, University of Massachusetts, Dartmouth, MA, USA
e-mail: sfield@umassd.edu; sgottlieb@umassd.edu; emcclain@umassd.edu

G. Khanna
Department of Mathematics, Center for Scientific Computing & Visualization Research, University of Massachusetts, Dartmouth, MA, USA

Department of Physics, University of Rhode Island, Kingston, RI, USA
e-mail: gkhanna@uri.edu

J. M. Melenk et al. (eds.), *Spectral and High Order Methods for Partial Differential Equations ICOSAHOM 2020+1*, Lecture Notes in Computational Science and Engineering 137, https://doi.org/10.1007/978-3-031-20432-6_19

the relevant (Teukolsky) equation features terms proportional to $\delta^{(2)}(x)$ on the right-hand-side [24]. To solve Eq. (1), various "regularized" numerical approaches [9, 26] and schemes [2, 3, 13, 16, 17, 22, 24, 25] have been proposed. Most of these methods only treat source terms proportional to $\delta(x)$ and $\delta^{(1)}(x)$ and do not achieve spectral accuracy at $x = 0$.

Discontinuous Galerkin methods are especially well suited for solving Eq. (1) and, more broadly, problems with delta distributions. Indeed, the solution's non-smoothness can be "hidden" at an interface between subdomains. Furthermore, the DG method solves the weak form of the problem, a natural setting for the delta distribution.

To the best of our knowledge, two distinct DG-based strategies have appeared in the literature for solving hyperbolic equations with δ-singularities. Yang and Shu [27] show that when the source term features a dirac delta distribution (but no distributional derivatives of them), by using kth degree polynomials the error will converge in a negative-order norm. Post-processing techniques are then used to recover high-order accuracy in the L_2 norm so long as the solution is not required near the singularity. A different approach, and the one we follow here, is based on the observation that (1) the solution is smooth to the left and the right of the singularity and (2) if the singularity is collocated with a subdomain interface then the effect of the dirac delta distribution is to modify the numerical flux. This framework was originally proposed by Fan et al. [10] for the Schrödinger equation sourced by a delta distribution $\delta(x)$ and later extended by Field et al. [11] to solve Eq. (1) with source terms proportional to $\delta(x)$ and $\delta^{(1)}(x)$. Building on previous work [10, 11], the main contributions of our paper are (1) to show how the DG method proposed in Refs. [10, 11] can be readily extended to solve Eq. (1) as well as (2) to clarify the importance of satisfying a distributional constraint that arises when performing a first-reduction of this equation. We also derive equations that directly relate the coefficients $a_n(t, x)$ to the numerical flux modification rule.

To motivate the main idea, consider the advection equation

$$\partial_t \psi + \partial_x \psi = \cos(t)\delta^{(1)}(x) , \tag{2}$$

whose inhomogeneous solution is

$$\psi(t, x) = \cos(t)\delta(x) + H(x)\sin(t - x) , \tag{3}$$

where the Heaviside step function obeys $H(x) = 0$ for $x < 0$, and $H(x) = 1$ for $x \geq 0$. Away from $x = 0$ the solution is smooth, suggesting a spectrally-convergent basis should be used. At $x = 0$ the solution is both discontinuous and contains a term proportional to $\delta(x)$.

Our proposed DG scheme proceeds as follows. First, we use a change of variable $\bar{\psi} = \psi - \cos(t)\delta(x)$ to remove the singular term and solve the modified equation,

$$\partial_t \bar{\psi} + \partial_x \bar{\psi} = \sin(t)\delta(x)\,. \tag{4}$$

We will show this is always possible, works for singular source terms $\delta^{(n)}(x)$, and is easy to enact. Next, we use a non-overlapping, multi-domain setup such that $x = 0$ is one of interface locations; for example, using two subdomains we would have $\mathsf{D}^1 = [a, 0]$ and $\mathsf{D}^2 = [0, b]$. In each element we expand $\bar{u}$ in a polynomial basis of degree k denoting this numerical approximation $\bar{\psi}_h \in V_h$, where

$$V_h = \{v : v|_{\mathsf{D}^j} \in \mathcal{P}^k(\mathsf{D}^j),\, j = 1, \ldots, N\}\,, \tag{5}$$

and $\mathcal{P}^k(\mathsf{D}^j)$ denotes the space of polynomials of degree at most k defined on subdomain D^j. In each element, we then follow the standard DG procedure by integrating each elementwise residual against all test functions $v(x) \in V_h$. A key ingredient of any DG method is the choice of numerical flux, denoted here by ψ^*, which couples neighboring subdomains. As we will show later on, the δ distribution's effect on the scheme is to modify the numerical flux according to $\psi^* \to \psi^* + \sin(t)$ at the left side of D^2. This extra $\sin(t)$ term arises from the evaluation of the integral,

$$\begin{aligned}\int_a^b \sin(t)\delta(x)v(x)dx &= \int_{\mathsf{D}^1} \sin(t)\delta(x)v_1(x)dx + \int_{\mathsf{D}^2} \sin(t)\delta(x)v_2(x)dx \\ &= \sin(t)v_2(0)\,, \end{aligned} \tag{6}$$

in a manner that is consistent with the defining property of a delta distribution, $\int f(x)\delta(x)dx = f(0)$, and the expectation that the delta distribution should only affect the solution to the right of $x = 0$ (i.e. upwinding). Here $v_{1,2} = v|_{\mathsf{D}^{1,2}}$ denote the global test function restricted to a subdomain. Figure 1 provides a schematic of this procedure. The numerical solution to Eq. (2) is then $\psi_h = \bar{\psi}_h + \cos(t)\delta(x)$.

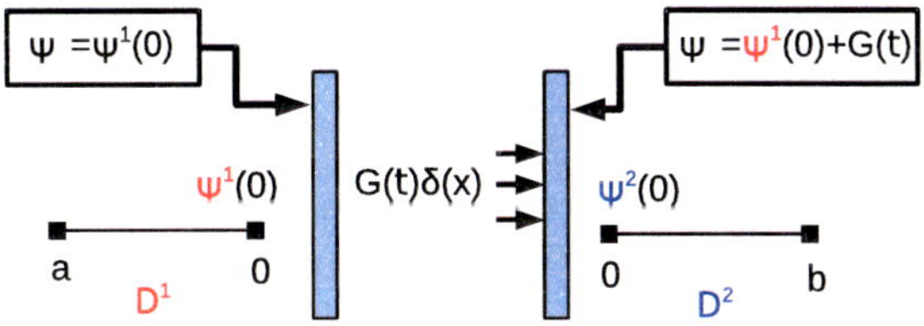

Fig. 1 Cartoon of the numerical flux modification enacted at $x = 0$ to solve Eq. (4). Here we show a generic time-dependent source term $G(t)\delta(x)$ "located between" subdomain D^1 and D^2. Interfaces are represented schematically as light blue rectangles. The advection equation (2) moves information from left-to-right, and accordingly the upwinded numerical flux is modified to the right of the delta distribution. Section 1 provides further discussion of the setup

! Remark

When the coefficients $a_n(t, x)$ appearing in Eq. (1) are functions of both independent variables, they can be put into the form assumed by Theorem 1 by the selection property of delta distributions. For example, $a_0(t, x)\delta(x) = a_0(t, 0)\delta(x)$ or $a_1(t, x)\delta^{(1)}(x) = a_1(t, 0)\delta^{(1)}(x) - \partial_x a_1(t, x)|_{x=0}\delta(x)$. Consequently, to streamline the discussion, we will exclusively focus on source terms of the form $\sum_{n=0}^{N} a_n(t)\delta^{(n)}(x)$ for the remainder of this paper.

2 Reduction to a First-Order System

Throughout, we use both an over–dot and superscript to denote $\partial/\partial t$ differentiation, for example $\partial a/\partial t = \dot{a}(t) = a^{(1)}(t)$, and both a prime and superscript to denote $\partial/\partial x$, for example $\partial\delta/\partial x = \delta'(x) = \delta^{(1)}(x)$.

2.1 Removing Singular Behavior

The simple advection example introduced in Sect. 1 demonstrates how to remove a certain amount of singular behavior in the source term such that the new dependent variable, $\bar{\psi}$, contains no terms proportional to $\delta(x)$. The following theorem shows how to apply the same procedure to Eq. (1).

Theorem 1 *Consider Eq. (1) with $V = 0$. Assume $a_n(t)$ has at least n derivatives. If ψ is the exact solution to Eq. (1), then*

$$\bar{\psi} = \psi - \sum_{i=0}^{\frac{N-1}{2}-1} \left[\delta^{(2i)}(x) \sum_{n=0}^{\frac{N-1}{2}-1-i} a_{2n+2i+2}^{(2n)}(t) + \delta^{(2i+1)}(x) \sum_{n=0}^{\frac{N-1}{2}-1-i} a_{2n+2i+3}^{(2n)}(t) \right], \tag{7}$$

solves

$$-\partial_t^2\bar{\psi} + \partial_x^2\bar{\psi} = G(t)\delta(x) + F(t)\delta'(x), \tag{8}$$

where

$$G(t) = \sum_{n=0}^{(N-1)/2} a_{2n}^{(2n)}(t), \qquad F(t) = \sum_{n=0}^{(N-1)/2} a_{2n+1}^{(2n)}(t). \tag{9}$$

Proof Assume N is odd, which is always possible by taking $a_N(t, x) = 0$ if necessary. Equation (1) can be transformed into Eq. (8) by a sequence of $(N-1)/2$ substitutions of the form

$$\bar{\psi}_{i+1} = \bar{\psi}_i - \sum_{n=0}^{N-2i-2} a_{n+2i+2}^{(2i)}(t)\delta^{(n)}(x),$$

where we define $\bar{\psi}_0 = \psi$. For example, we first substitute $\bar{\psi}_1$ into Eq. (1), generating a new wave equation for $\bar{\psi}_1$ with source terms proportional to $\delta^{N(x)}$ and $\delta^{N-1}(x)$ removed. Next, we substitute $\bar{\psi}_2$ into the PDE for $\bar{\psi}_1$, generating a new wave equation for ${}^{(2)}\bar{\psi}$ with the source terms proportional to $\delta^{N-2}(x)$ and $\delta^{N-3}(x)$ removed. This process continues until we arrive at an equation for $\bar{\psi} = \bar{\psi}_{(N-1)/2}$ whose source terms are $G(t)\delta(x)$ and $F(t)\delta^{(1)}(x)$. The final result can also be checked by direct computation.

! Remark

When $V \neq 0$ we can still remove singular behavior from the source term, although we do not provide a general expression as in Theorem 1. As a direct example, and assuming V is differentiable, the differential equation

$$-\partial_t^2\psi+\partial_x^2\psi + V(x)\psi = a_0(t)\delta(x) + a_1(t)\delta^{(1)}(x) + a_2(t)\delta^{(2)}(x) + a_3(t)\delta^{(3)}(x), \tag{10}$$

can be transformed into

$$\begin{aligned} -\partial_t^2\bar{\psi} + \partial_x^2\bar{\psi} + V(x)\bar{\psi} &= \left[a_0 + a_2^{(2)} - a_2 V(0) + a_3 V^{(1)}(0)\right]\delta \\ &+ \left[a_1 + a_3^{(2)} - a_3 V(0)\right]\delta^{(1)}, \end{aligned} \tag{11}$$

where $\bar{\psi} = \psi - a_2(t)\delta(x) - a_3(t)\delta^{(1)}(x)$.

2.2 *First-Order Reduction*

Thanks to Theorem 1, we can develop our DG method for the modified problem (8) and numerically solve for $\bar{\psi}$. At this point, we follow the approach of Ref. [11] and introduce the auxiliary variables,

$$\hat{\phi} = \partial_x\bar{\psi}, \qquad \pi = -\partial_t\bar{\psi}, \tag{12}$$

from which the original second-order wave equation (1) can be rewritten as the following first-order system,

$$\begin{bmatrix} \bar{\psi} \\ \pi \\ \hat{\phi} \end{bmatrix}_t + \begin{bmatrix} 0 & 0 & 0 \\ 0 & 0 & 1 \\ 0 & 1 & 0 \end{bmatrix} \begin{bmatrix} \bar{\psi} \\ \pi \\ \hat{\phi} \end{bmatrix}_x + \begin{bmatrix} \pi \\ V\bar{\psi} \\ 0 \end{bmatrix} = \begin{bmatrix} 0 \\ G(t)\delta(x) + F(t)\delta'(x) \\ 0 \end{bmatrix} . \tag{13}$$

We find it convenient (this will be helpful in when deriving the numerical flux in Sect. 3.2) to define a new distributional auxiliary variable

$$\phi = \hat{\phi} - F(t)\delta(x) , \tag{14}$$

which removes all terms proportional to $\delta'(x)$,

$$\begin{bmatrix} \bar{\psi} \\ \pi \\ \phi \end{bmatrix}_t + \begin{bmatrix} 0 & 0 & 0 \\ 0 & 0 & 1 \\ 0 & 1 & 0 \end{bmatrix} \begin{bmatrix} \bar{\psi} \\ \pi \\ \phi \end{bmatrix}_x + \begin{bmatrix} \pi \\ V\bar{\psi} \\ 0 \end{bmatrix} = \begin{bmatrix} 0 \\ G(t)\delta(x) \\ -\dot{F}(t)\delta(x) \end{bmatrix} . \tag{15}$$

System (15) can be written more compactly as,

$$\partial_t U + \partial_x F + \hat{V} = S(t)\delta(x) , \tag{16}$$

for the system vector U, flux vector F, potential $\hat{V}$, and source vector S:

$$\begin{aligned} U &= \left[\bar{\psi}, \pi, \phi\right]^T , \qquad & F(U) &= [0, \phi, \pi]^T , \\ \hat{V} &= \left[\pi, V\bar{\psi}, 0\right]^T , \qquad & S &= \left[0, G(t), -\dot{F}(t)\right]^T . \end{aligned} \tag{17}$$

2.3 Distributional Constraint

A solution to the first-order system (15) is also a solution to the original PDE (1) provided the distributional constraint

$$C(t, x) = \phi - \partial_x \bar{\psi} + F(t)\delta(x) , \tag{18}$$

vanishes. One can show that $C(t, x)$ obeys (upon setting $G = V = 0$ for simplicity):

$$-\partial_t^2 C + \partial_x^2 C = 0 . \tag{19}$$

Thus we conclude that if the initial data implies $C(0, x) = \dot{C}(0, x) = 0$, and our physical boundary condition is compatible with Eq. (18), then $C = 0$ for all future times.

For certain applications, the exact initial data is unknown. If one is interested in the late-time behavior of the problem due to the forcing term, trivial initial data ($\bar{\psi} = \phi = \pi = 0$) is often supplied instead. Trivial data results in an impulsive (i.e. discontinuous in time) start-up, and a key question is if a physical solution eventually emerges from such trivial initial data? The answer is unfortunately no. Under the assumption of trivial initial data we have $C(0, x) = F(0)\delta(x)$ and $\dot{C}(0, x) = 0$.[1] In numerical simulations, this manifests as an extremely localized feature that advects off the computational grid. It also implies, a time-independent constraint-violating spurious (or "junk") solution will develop in it's wake. In Sect. 5 we provide a numerical experiment that confirms this spurious solution as well as two ways to remove it.

3 Discontinuous Galerkin Method

To solve the wave equation (1) we first transform it into a simpler form using Theorem 1 then carry out a fully first-order reduction. This section describes the nodal DG method we have implemented to numerically solve the resulting system (16) subject to the constraint (18). The method is exactly the one first proposed in Ref. [11], and so we only briefly summarize the key ideas. Indeed, a key contribution of our paper is to show the methods of Ref. [11] continue to be applicable for more challenging problems such as Eq. (1).

3.1 The Source-Free Method

We divide the spatial domain into N non-overlapping subdomains $a = x_0 < x_1 < \cdots < x_N = b$ and denote $\mathsf{D}^j = [x_{j-1}, x_j]$ as the jth subdomain. In this one-dimensional setup, the points $\{x_i\}_{i=1}^{N-1}$ locate the internal subdomain interfaces, and we require one of them to be $x = 0$. In each subdomain, each component of the vectors U, F, and V are expanded in a polynomial basis, which are taken to be degree-k Lagrange interpolating polynomials $\{\ell_i(x)\}_{i=0}^{k}$ defined from Legendre-Gauss-Lobatto nodes. The time-dependent coefficients of this expansion (e.g. on subdomain j: $\pi_h^j = \sum_{i=0}^{k} \pi_i(t)\ell_i(x)$) are the unknowns we solve for. We directly approximate $\bar{\psi}$, π, ϕ, and V and other terms arising in Eq. (16), such as $\bar{\psi}V$, are achieved through pointwise products, for example $\bar{\psi}_h V_h$.

[1] The term $\dot{\phi}(0, x)$ is found from evaluation of the evolution equation (15), $\dot{\phi} = -\pi' - \dot{F}(t)\delta(x)$, at $t = 0$.

On each subdomain, we follow the standard DG procedure by requiring the residual to satisfy

$$\int_{\mathsf{D}^j}\left[\frac{\partial U_h^j}{\partial_t}+\frac{\partial F_h^j}{\partial_x}+V_h^j\right]\ell_i^j dx=\left[\left(F_h^j-F^*\right)\ell_i^j\right]_{x_{j-1}}^{x^j}, \tag{20}$$

for all basis functions. Here we use an upwind numerical flux, $F^*(U_h)$, that depends on the values of the numerical solution from both sides of the interface. These integrals can be pre-computed on a reference interval, leading to a coupled system of ordinary differential equations (see Eq. 47 of Ref. [11]) that can be integrated in time.

3.2 Modifications to the Numerical Flux

With non-zero source terms, the numerical flux evaluated at the interface $x = 0$ will be modified through additional terms. The form of these new terms were derived in Eq. 58 of Ref. [11]. Instead of reproducing those results here, we provide an alternative viewpoint leading to the same result.

We begin by writing Eq. (16) in terms of characteristic variables,

$$\begin{aligned}\partial_t w^+ + \partial_x w^+ + \frac{1}{2}V\bar{\psi} &= \frac{1}{2}G(x)\delta(x) - \frac{1}{2}\dot{F}(t)\delta(x) = R(t)\delta(x)\\ \partial_t w^- - \partial_x w^- + \frac{1}{2}V\bar{\psi} &= \frac{1}{2}G(x)\delta(x) + \frac{1}{2}\dot{F}(t)\delta(x) = L(t)\delta(x)\end{aligned} \tag{21}$$

where $w^{\pm} = (\pi \pm \phi)/2$ and note that $\pi = w^+ + w^-$ and $\phi = w^+ - w^-$. We now have two copies of an advection equation with source terms proportional to $\delta(x)$. The equation for w^+ (w^-) describes a wave moving from left-to-right (right-to-left). This allows us to apply the DG method described in the introduction: viewing the source term "between" the two subdomains, we associate the entire contribution of $R(t)$ with the subdomain to the right and the entire contribution of $L(t)$ with the subdomain to the left. Figure 2 provides a schematic of this procedure for a simple 2 domain setup as well as the corresponding modification to the upwind numerical flux $(w^{\pm})^*$. Returning the original system, the numerical flux is modified to the left (right boundary point of D^1) and right (left boundary point of D^2) of $\delta(x)$ by

$$F^*_{\text{left}} \to F^*_{\text{left}} + [0, -L(t), L(t)]^T\,, \qquad F^*_{\text{right}} \to F^*_{\text{right}} + [0, R(t), R(t)]^T\,. \tag{22}$$

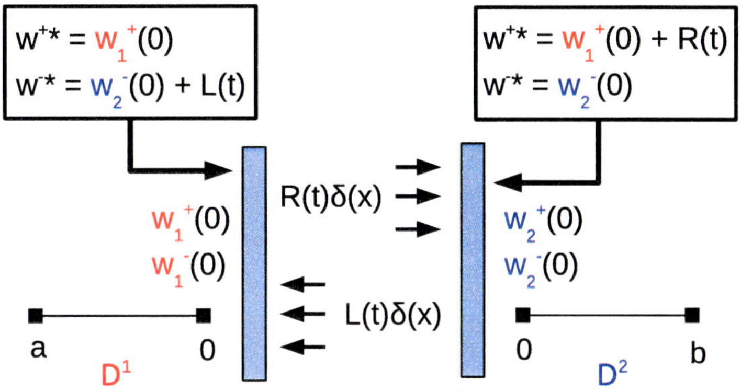

Fig. 2 Cartoon of the numerical flux modification enacted at $x = 0$ to solve Eq. (1) after its been rewritten in terms of characteristic variables (21). Here we show a generic time-dependent source term "located between" subdomain D^1 and D^2. Interfaces are represented schematically as light blue rectangles. The system (21) includes two characteristic variables. The term $R(t)\delta(x)$ sources the right moving wave and $L(t)\delta(x)$ sources the left moving wave. The upwinded numerical flux is modified according to this directionality. Section 3.2 provides further discussion of the setup

4 Distributional Solutions to the 1+1 Wave Equation

This section presents exact solutions to the distributionally-forced $1 + 1$ wave equation. These solutions will be used in Sect. 5 for testing our numerical scheme.

Our recipe for solving

$$-\partial_t^2\Psi(t,x) + \partial_x^2\Psi(t,x) = F(t)\delta^{(s)}(x)\,, \tag{23}$$

amounts to first solving

$$-\partial_t^2\Psi(t,x;c,s) + \partial_x^2\Psi(t,x;c,s) = F(t)\delta^{(s)}(x+c)\,, \tag{24}$$

for $s = 0$ followed by an application of Eq. (26). We shall view c and s as parameters, and the solution $\Psi(t,x;c,s)$ as parameterized by them.

A solution to (24) can be found by the method of Green's function. Recall the fundamental solution, $G(t,x;\tilde{t},\tilde{x})$, for

$$-\partial_t^2 G + \partial_x^2 G = \delta(t-\tilde{t})\delta(x-\tilde{x})\,,$$

can be written in terms of the Heaviside [23]

$$G(t,x;\tilde{t},\tilde{x}) = -\frac{1}{2}H(t-\tilde{t}-|x-\tilde{x}|)\,.$$

Thus the solution to Eq. (24) with $s = 0$ can be written as

$$\Psi(t, x; c, s = 0) = -\frac{1}{2}\int_0^{t-|x+c|} F(t - |x + c| - y)dy\,, \tag{25}$$

where $y = t - \tilde{t} - |x + c|$, and we have restricted to times $t \geq 0$ for which the Heaviside is zero whenever $t - |x + c| < 0$. This $s = 0$ solution generates an entire family of solutions corresponding to $s > 0$. Clearly $\partial_c^s \Psi(t, x; c, s = 0)$ solves Eq. (24), and so the particular solution of

$$-\partial_t^2 \Psi(t, x; 0, s) + \partial_x^2 \Psi(t, x; 0, s) = F(t)\delta^s(x)\,,$$

is given by

$$\Psi(t, x; 0, s) = \partial_c^s \Psi(t, x; c, 0)\Big|_{c=0}\,. \tag{26}$$

We now provide explicit constructions for the cases considered in the numerical experiment section.

Let $F(t) = \cos(t)$ and $s = 0$, then the generating function is

$$\Psi_c(t, x; c, 0) = -\frac{1}{2}\sin(t - |x + c|)\,,$$

and setting $c = 0$ gives

$$\Psi(t, x) = -\frac{1}{2}\sin(t - |x|)\,.$$

Let $F(t) = \cos(t)$ and $s = 1$, then the generating function is

$$\partial_c \Psi_c(t, x; c, 0) = \frac{1}{2}\text{sgn}(x + c)\cos(t - |x + c|)\,,$$

and setting $c = 0$ gives

$$\Psi(t, x) = \frac{1}{2}\text{sgn}(x)\cos(t - |x|)\,. \tag{27}$$

Let $F(t) = \cos(t)$ and $s = 2$, then the generating function is

$$\begin{aligned}\partial_c^2 \Psi_c(t, x; c, 0) &= \delta(x + c)\cos(t - |x + c|) \\ &\quad + \frac{1}{2}\mathrm{sgn}(x + c)\sin(t - |x + c|)\mathrm{sgn}(x + c) \\ &= \delta(x + c)\cos(t) + \frac{1}{2}\sin(t - |x + c|)\,,\end{aligned}$$

and setting $c = 0$ gives

$$\Psi(t, x) = \delta(x)\cos(t) + \frac{1}{2}\sin(t - |x|)\,. \tag{28}$$

5 Numerical Results

5.1 *Wave Equation with a $\delta^{(2)}(x)$ Source Term*

We consider

$$-\partial_t^2 \psi + \partial_x^2 \psi = \cos(t)\delta^{(2)}(x)\,, \tag{29}$$

whose solution is given by Eq. (28). We will check the convergence of our numerically generated solution against this exact solution. Before discretization, we remove some of the equation's singular structure by Theorem 1: let $\bar{\psi} = \psi - \cos(t)\delta(x)$ and solve Eq. (11) after setting $V = a_0 = a_1 = a_3 = 0$. At the physical boundary points we choose fluxes that enforce simple Sommerfeld boundary conditions and take the initial data from the exact solution. We solve our problem on the domain $[a, b] = [-10, 10]$, set the final time T = 10, and choose a Δt sufficiently small such that the Runge–Kutta's timestepping error is below the spatial discretization error. Despite the exact solution being both non-smooth and containing a term proportional to $\delta(x)$, Fig. 3 shows the spectral (exponential) convergence in the approximation. For a fixed polynomial degree k, the scheme's rate of convergence is observed to be $k+1$. Please see Ref. [11] for numerical experiments with a non-zero potential.

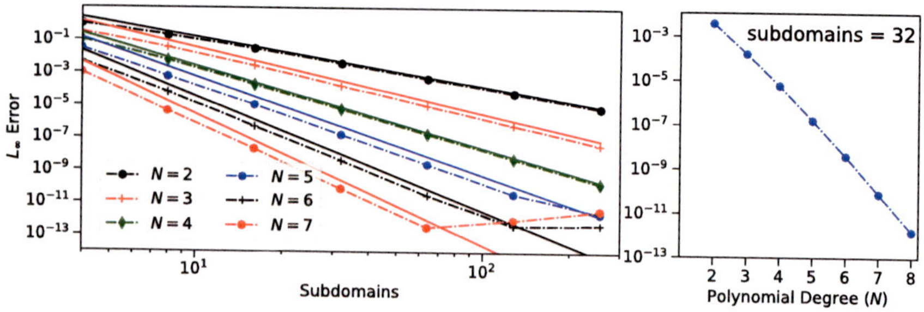

Fig. 3 Convergence of the numerical solution for a sequence of grids for the problem setup described in Sect. 5.1. We consider convergence by increasing the number of elements (left) and polynomial degree of the numerical approximation (right). These error profiles, computed as $\max_{x\in[-10,10]}\left|\bar{\psi}_{\mathrm{exact}}(10,x)-\bar{\psi}_{\mathrm{numerical}}(10,x)\right|$, are typical of our DG scheme when the solution is smooth. Notably, the error is also monitored at $x = 0$. **Left panel**: For a fixed value of k, the approximation error decreases with a power law (dashed line) at a rate which closely matches the expected rate of $-(k+1)$ (solid line). For the cases $k = \{6, 7\}$, round-off error effects become noticeable around 10^{-13}. **Right panel**: The DG scheme achieves exponential convergence in the approximation error as the polynomial degree k is increased

5.2 *Persistent Spurious Solutions from Distributional Constraint Violations*

We consider

$$-\partial_t^2\psi + \partial_x^2\psi = a_1(t)\delta^{(1)}(x)\,, \tag{30}$$

for two different cases: $a_1(t) = \sin(t)$ and $a_1(t) = \cos(t)$, the latter's exact solution is given by Eq. (27). We solve our problem on the domain $[a, b] = [-10, 10]$ and we choose fluxes that enforce simple Sommerfeld boundary conditions. When the initial data is taken from the exact solution the scheme's convergence properties are identical to those shown in Fig. 3. In light of the Sect. 2.3's discussion on distributional constraint violations, we will check the for the appearance of persistent spurious solutions when supplying trivial initial data.

When $a_1(t) = \cos(t)$, the constraint violation is $C(t,x) = 0.5\,[\delta(x+t) + \delta(x-t)]$ and the spurious solution is the offset that develops inside the future domain influenced by $(t, x) = 0, 0$. Figure 4 (left) shows the numerical solution (black circles) offset from the exact solution (dashed black line) at $T = 8$, and by $T = 40$ the spurious solution has contaminated the entire computational domain (black circles). When $a_1(t) = \sin(t)$, the constraint violation and spurious solution vanish; this too is confirmed by Fig. 4 (red data on both left and right panels). And so we see that the problematic spurious solution can be made to vanish if it is possible to arrange the problem such that $F(0) = 0$.

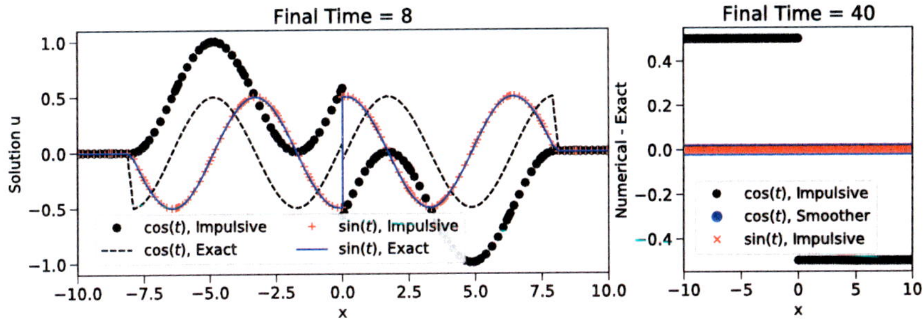

Fig. 4 For certain applications, the exact initial data is unknown. Here we consider possible effects on the numerical solution when the initial data (here taken to be trivial data) leads to an impulsively started problem. The left panel shows a snapshot of the numerical solution at $T = 8$ where we see that for a $\cos(t)\delta^{(1)}(x)$ source term the numerical and exact solution do not agree. In the right panel we plot the difference between the numerical and exact solution and find that the "time-independent" spurious solution that has developed is $a_1(0)\,[H(-x) - H(x)]/2$. This is not a generic feature of the problem: no spurious solution appears when the source term is smoothly turned on (right figure; blue circles) or when the when the source term is instead taken to be $\sin(t)\delta^{(1)}(x)$

With neither the correct initial data nor the ability to arrange $F(0) = 0$, a more general solution to this problem is to modify the source term

$$F(t) \to \begin{cases} f(t;\tau,\delta)F(t) & \text{for } 0 \le t \le \tau \\ F(t) & \text{for } t > \tau, \end{cases} \tag{31}$$

where $f(t;\tau,\delta) = \frac{1}{2}[\mathrm{erf}(\sqrt{\delta}(t - \tau/2) + 1]$ turns on the source term [12] over the timescale τ. We select $\tau = 30$ and $\delta = 0.15$, which yields $f(0) \approx 10^{-16}$ and $f(t) = 1$ for $t > 30$. Both the constraint violation and spurious solution now vanish, as shown in the right panel of Fig. 4 (Blue circles).

6 Final Remarks

We have shown that the high–order accurate discontinuous Galerkin method developed in Ref. [11] is applicable to the wave equation (1) when written in fully first-order form (16). In particular, Ref. [11] considered a wave equation with source terms of the form $a(t, x)\delta(x) + a_1(t, x)\delta'(x)$. In Theorem 1 we show that one can always write Eq. (1) in this form, allowing for immediate application of their method to this generalized problem. The method maintains pointwise spectral convergence even at the source's location where the solution may be discontinuous, singular, or both. The numerical error has been quantified by comparing against exact distributional solutions, and we have presented a procedure for finding particular solutions for any $\delta^{(n)}(x)$.

Our choice for writing the second-order scalar equation (1) in first-order form (16) relies on an auxiliary variable that must satisfy a distributional constraint (18). While this constraint vanishes for all times if it does at the initial time, for many realistic problems the initial data is not known and satisfying the distributional constraint may be challenging. For trivial initial data, we show the constraint violation advects off the computational grid, leaving behind a time-independent constraint-violating spurious (or "junk") solution in it's wake. We discuss two remedies that can be used to prevent the problematic spurious solution from appearing.

Acknowledgments We thank Manas Vishal for providing an independent check of Theorem 1 using Mathematica. The authors acknowledge support of NSF Grants No. PHY-2010685 (G.K) and No. DMS-1912716 (S.F, S.G, and G.K), AFOSR Grant No. FA9550-18-1-0383 (S.G) and Office of Naval Research/Defense University Research Instrumentation Program (ONR/DURIP) Grant No. N00014181255. This material is based upon work supported by the National Science Foundation under Grant No. DMS-1439786 while a subset of the authors were in residence at the Institute for Computational and Experimental Research in Mathematics in Providence, RI, during the Advances in Computational Relativity program.

References

1. Cáceres, M.J., Carrillo, J.A., Perthame, B.: Analysis of nonlinear noisy integrate & fire neuron models: blow-up and steady states. J. Math. Neurosci. **1**(1), 1–33 (2011)
2. Canizares, P., Sopuerta, C.F.: Simulations of extreme-mass-ratio inspirals using pseudospectral methods. In: Journal of Physics: Conference Series, vol. 154, p. 012053. IOP Publishing (2009)
3. Canizares, P., Sopuerta, C.F., Jaramillo, J.L.: Pseudospectral collocation methods for the computation of the self-force on a charged particle: Generic orbits around a schwarzschild black hole. Phys. Rev. D **82**(4), 044023 (2010)
4. Casti, A.R.R., Omurtag, A., Sornborger, A., Kaplan, E., Knight, B., Victor, J., Sirovich, L.: A population study of integrate-and-fire-or-burst neurons. Neural Comput. **14**(5), 957–986 (2002)
5. Cockburn, B.: An introduction to the discontinuous Galerkin method for convection-dominated problems. In: Advanced Numerical Approximation of Nonlinear Hyperbolic Equations: Lectures given at the 2nd Session of the Centro Internazionale Matematico Estivo (C.I.M.E.) held in Cetraro, Italy, June 23–28, 1997. Lecture Notes in Physics, pp. 150–268. Springer, Berlin (1998). https://doi.org/10.1007/BFb0096353
6. Cockburn, B.: Devising discontinuous Galerkin methods for non-linear hyperbolic conservation laws. J. Comput. Appl. Math. **128**, 187 (2001). https://doi.org/10.1016/S0377-0427(00)00512-4
7. Cockburn, B., Shu, C.W.: The Runge-Kutta discontinuous Galerkin method for conservation laws V. Multidimensional systems. J. Comput. Phys. **141**, 199 (1998). https://doi.org/10.1006/jcph.1998.5892
8. Cockburn, B., Karniadakis, G.E., Shu, C.W.: The development of discontinuous Galerkin methods. In: Cockburn, B., Karniadakis, G.E., Shu, C.W. (eds.) Discontinuous Galerkin Methods: Theory, Computation and Applications, Lecture Notes in Computational Science and Engineering, pp. 3–50. Springer, Berlin (2000). https://doi.org/10.1007/978-3-642-59721-3_1
9. Engquist, B., Tornberg, A.K., Tsai, R.: Discretization of dirac delta functions in level set methods. J. Comput. Phys. **207**(1), 28–51 (2005)

10. Fan, K., Cai, W., Ji, X.: A generalized discontinuous Galerkin (GDG) method for Schrödinger equations with nonsmooth solutions. J. Comput. Phys. **227**(4), 2387–2410 (2008)
11. Field, S.E., Hesthaven, J.S., Lau, S.R.: Discontinuous Galerkin method for computing gravitational waveforms from extreme mass ratio binaries. Class. Quantum Grav. **26**, 165010 (2009). https://doi.org/10.1088/0264-9381/26/16/165010
12. Field, S.E., Hesthaven, J.S., Lau, S.R.: Persistent junk solutions in time-domain modeling of extreme mass ratio binaries. Phys. Rev. D **81**(12), 124030 (2010)
13. Field, S.E., Gottlieb, S., Grant, Z.J., Isherwood, L.F., Khanna, G.: A GPU-accelerated mixed-precision WENO method for extremal black hole and gravitational wave physics computations. Commun. Appl. Math. Comput., 1–19 (2021). https://doi.org/10.1007/s42967-021-00129-2
14. Friedlander, F.G., Joshi, M.S., Joshi, M., Joshi, M.C.: Introduction to the Theory of Distributions. Cambridge University Press (1998)
15. Hesthaven, J., Warburton, T.: Nodal Discontinuous Galerkin Methods: Algorithms, Analysis, and Applications. Springer, Berlin (2008). https://doi.org/10.1007/978-0-387-72067-8
16. Jung, J.H., Khanna, G., Nagle, I.: A spectral collocation approximation for the radial-infall of a compact object into a Schwarzschild black hole. Int. J. Mod. Phys. C **20**(11), 1827–1848 (2009)
17. Lousto, C.O.: A time-domain fourth-order-convergent numerical algorithm to integrate black hole perturbations in the extreme-mass-ratio limit. Classical Quantum Gravity **22**(15), S543 (2005)
18. Petersson, N.A., Sjogreen, B.: Stable grid refinement and singular source discretization for seismic wave simulations. Commun. Comput. Phys. **8**(5), 1074–1110 (2010), **8**(LLNL-JRNL-419382) (2009)
19. Poisson, E., Pound, A., Vega, I.: The motion of point particles in curved spacetime. Living Rev. Relativity **14**(1), 1–190 (2011)
20. Reed, W., Hill, T.: Triangular mesh methods for the neutron transport equation. In: Conference: National Topical Meeting on Mathematical Models and Computational Techniques for Analysis of Nuclear Systems, Ann Arbor, Michigan, USA, 8 Apr 1973 LA-UR–73-479, CONF-730414–2 (1973). http://www.osti.gov/scitech/servlets/purl/4491151
21. Shearer, P.M.: Introduction to Seismology. Cambridge University Press (2019)
22. Sopuerta, C.F., Laguna, P.: Finite element computation of the gravitational radiation emitted by a pointlike object orbiting a nonrotating black hole. Phys. Rev. D **73**(4), 044028 (2006)
23. Stakgold, I., Holst, M.J.: Green's Functions and Boundary Value Problems, vol. 99. Wiley (2011)
24. Sundararajan, P.A., Khanna, G., Hughes, S.A.: Towards adiabatic waveforms for inspiral into kerr black holes: A new model of the source for the time domain perturbation equation. Phys. Rev. D **76**(10), 104005 (2007)
25. Sundararajan, P.A., Khanna, G., Hughes, S.A., Drasco, S.: Towards adiabatic waveforms for inspiral into kerr black holes. II. Dynamical sources and generic orbits. Phys. Rev. D **78**(2), 024022 (2008)
26. Tornberg, A.K., Engquist, B.: Numerical approximations of singular source terms in differential equations. J. Comput. Phys. **200**(2), 462–488 (2004)
27. Yang, Y., Shu, C.W.: Discontinuous Galerkin method for hyperbolic equations involving δ-singularities: negative-order norm error estimates and applications. Numer. Math. **124**(4), 753–781 (2013)

An Energy-Preserving High Order Method for Liouville's Equation of Geometrical Optics

R. A. M. van Gestel, M. J. H. Anthonissen, J. H. M. ten Thije Boonkkamp, and W. L. IJzerman

1 Introduction

Illumination optics deals with the design of optical systems for various applications, for instance, LED lighting [7]. Quasi-Monte Carlo ray tracing is typically employed to compute the illuminance or intensity on a target, providing insight into the effectiveness of the optical system. Quasi-Monte Carlo ray tracing, however, converges rather slowly as one over the number of rays used.

A different approach to analysing light propagation is based on a phase space description [3, 8, 12]. Here, phase space is defined as the collection of all positions and direction coordinates of rays. A point in phase space, describing a single light ray, evolves according to a Hamiltonian whenever the refractive index is smooth. In the absence of energy losses, one can define an energy density on phase space which remains constant along a light ray. The evolution of this energy density on phase space is governed by Liouville's equation of geometrical optics [10, 11].

In this work, we apply a high order discontinuous Galerkin spectral element method (DGSEM) to solve Liouville's equation. One issue arises at an optical interface, i.e., whenever a light ray strikes an optical interface its direction coordinates change discontinuously according to the laws of refraction or reflection. This manifests itself as non-local boundary conditions at an optical interface for Liouville's equation. In other words, completely different parts in phase space are

R. A. M. van Gestel (✉) · M. J. H. Anthonissen · J. H. M. ten Thije Boonkkamp
Eindhoven University of Technology, Eindhoven, The Netherlands
e-mail: r.a.m.v.gestel@tue.nl

W. L. IJzerman
Eindhoven University of Technology, Eindhoven, The Netherlands

Signify, High Tech Campus 7, Eindhoven, The Netherlands

J. M. Melenk et al. (eds.), *Spectral and High Order Methods for Partial Differential Equations ICOSAHOM 2020+1*, Lecture Notes in Computational Science and Engineering 137, https://doi.org/10.1007/978-3-031-20432-6_20

in contact with each other at an optical interface. Hence, in the DGSEM, elements are connected in a highly non-trivial fashion at the optical interface. Therefore, in this work we present a method to deal with optical interfaces while ensuring the method obeys energy conservation.

The outline of this paper is as follows. First we describe Liouville's equation in Sect. 2 and subsequently describe its discretisation and how we deal with optical interfaces in Sect. 3. Then, we will present the numerical results in Sect. 4 and end with the conclusion in Sect. 5.

2 Liouville's Equation

A light ray is described by its position $(\boldsymbol{q}, z) \in \mathbb{R}^{d+1}$ and momentum vector $(\boldsymbol{p}, p_z) \in \mathbb{R}^{d+1}$ $(d = 1, 2)$, where the momentum vector is defined as the unit direction vector multiplied by the refractive index n. The momentum vector has length $|(\boldsymbol{p}, p_z)| = n$, hence, one can write $p_z = \sigma\sqrt{n^2 - |\boldsymbol{p}|^2}$ with $\sigma \in \{-1, 0, 1\}$. To describe the evolution of a light ray we take the z-coordinate as the evolution coordinate. The ray coordinates $\boldsymbol{q}(z)$ and $\boldsymbol{p}(z)$ evolve according to Hamilton's equations [12]

$$\frac{\mathrm{d}\boldsymbol{q}}{\mathrm{d}z} = \frac{\partial h}{\partial \boldsymbol{p}}, \qquad \frac{\mathrm{d}\boldsymbol{p}}{\mathrm{d}z} = -\frac{\partial h}{\partial \boldsymbol{q}}, \tag{1a}$$

where h is the Hamiltonian given by

$$h(z, \boldsymbol{q}, \boldsymbol{p}) = -\sigma\sqrt{n(z, \boldsymbol{q})^2 - |\boldsymbol{p}|^2}, \tag{1b}$$

with $n(z, \boldsymbol{q})$ describing the refractive index field as a function of the position coordinates. In the Hamiltonian system (1) we take $\sigma = 1$ for the remainder of this work, i.e., we consider only rays moving in the positive z-direction.

At a fixed plane $z = \text{const}$, the collection of all positions $\boldsymbol{q} \in \mathbb{R}^d$ and momenta $\boldsymbol{p} \in \mathbb{R}^d$ describes the $(2d)$-dimensional phase space $\mathcal{P}$, where $d = 1$ for two-dimensional optics and $d = 2$ for three-dimensional optics. The momentum $\boldsymbol{p}$ is restricted by $|\boldsymbol{p}| \leq n$, describing a region known as Descartes' disc [12]. The flow generated by Hamilton's equations (1) describes canonical transformations on phase space [1]. A phase space volume element $\mathrm{d}q_1\mathrm{d}q_2\,\mathrm{d}p_1\mathrm{d}p_2$ remains constant under canonical transformations.

The energy of a beam of light remains constant in the absence of any losses such as absorption or scattering. In illumination optics this energy is typically the luminous flux Φ. An element of luminous flux $\mathrm{d}\Phi$ can be related to an energy density ρ on phase space by $\mathrm{d}\Phi = \rho\,\mathrm{d}\mathcal{U}$ with $\mathrm{d}\mathcal{U} = \mathrm{d}q_1\mathrm{d}q_2\,\mathrm{d}p_1\mathrm{d}p_2$. If we move along the z-axis by a distance Δz, then in the absence of losses the element of luminous flux $\mathrm{d}\Phi$ and the phase space volume element $\mathrm{d}\mathcal{U}$ remain constant, and thus ρ must remain invariant. This invariance of ρ is expressed as

$$\rho(z + \Delta z, \boldsymbol{q}(z + \Delta z), \boldsymbol{p}(z + \Delta z)) = \rho(z, \boldsymbol{q}(z), \boldsymbol{p}(z)). \tag{2}$$

In illumination optics ρ is known as the basic luminance. The basic luminance invariance (2) even holds when light is reflected or refracted. For a detailed derivation in cases involving reflection and/or refraction, we refer the reader to [2, 6].

Assuming ρ is sufficiently smooth, one can derive Liouville's equation by subtracting the right-hand side of (2) from the left-hand side and dividing by Δz and subsequently taking the limit $\Delta z \to 0$, yielding

$$\frac{\partial \rho}{\partial z} + \frac{\partial h}{\partial \boldsymbol{p}} \cdot \frac{\partial \rho}{\partial \boldsymbol{q}} - \frac{\partial h}{\partial \boldsymbol{q}} \cdot \frac{\partial \rho}{\partial \boldsymbol{p}} = 0, \tag{3}$$

where we have made use of Hamilton's equations (1). Liouville's equation (3) can be transformed into conservative form, i.e.,

$$\frac{\partial \rho}{\partial z} + \nabla \cdot \boldsymbol{f} = 0 \text{ with } \boldsymbol{f} = \rho \boldsymbol{u} = \rho \begin{pmatrix} \frac{\partial h}{\partial \boldsymbol{p}} \\ -\frac{\partial h}{\partial \boldsymbol{q}} \end{pmatrix}, \tag{4}$$

where we have used that the velocity field $\boldsymbol{u}$ is divergence-free and $\nabla = \left(\frac{\partial}{\partial \boldsymbol{q}}, \frac{\partial}{\partial \boldsymbol{p}}\right)^{\mathsf{T}}$. Whenever the Hamiltonian is discontinuous due to an optical interface, we must apply the basic luminance invariance (2) together with Snell's law and/or the law of specular reflection, which will be further discussed in Sect. 3.2.

3 Numerical Method

In the remainder of this work we restrict ourselves to two-dimensional optics, i.e., $d = 1$ and the position and momentum become scalars which we will denote with q and p, respectively. Next, we semi-discretise Liouville's equation on phase space using a DGSEM [4].

3.1 *Space Discretisation*

The two-dimensional phase space domain $\mathcal{P}$ is covered with straight-sided quadrilaterals $\Omega^k \subset \mathcal{P}$, with k the index of the element. Each quadrilateral element has four vertices, which we denote by $\boldsymbol{x}_i^k$ for $i = 1, 2, 3, 4$ and $\boldsymbol{x} = (q, p)^{\mathsf{T}}$. Each element is then mapped to a reference domain $\chi = [-1, 1]^2$ using a bilinear transformation denoted as $\boldsymbol{x}(\xi, \eta) = (q(\xi, \eta), p(\xi, \eta))^{\mathsf{T}}$ with $(\xi, \eta) \in \chi$. Liouville's equation transformed to the reference domain reads

$$\mathcal{J}\frac{\partial \rho}{\partial z} + \nabla_\xi \cdot \widetilde{f} = 0 \text{ with } \widetilde{f} = \begin{pmatrix} \frac{\partial p}{\partial \eta} & -\frac{\partial q}{\partial \eta} \\ -\frac{\partial p}{\partial \xi} & \frac{\partial q}{\partial \xi} \end{pmatrix} f, \tag{5}$$

where $\mathcal{J}$ denotes the Jacobian determinant of the transformation. The DGSEM approximates a weak form of (5), hence we multiply by a test function ϕ, apply the product rule and Gauss's theorem yielding

$$\int_\chi \mathcal{J}\frac{\partial \rho}{\partial z}\phi \, \mathrm{d}A_\xi = \int_\chi (\nabla_\xi \phi) \cdot \widetilde{f} \, \mathrm{d}A_\xi - \oint_{\partial\chi} \phi \widetilde{F} \cdot \widehat{N} \, \mathrm{d}\sigma, \tag{6}$$

where we have replaced the flux in the boundary integral with a numerical flux $\widetilde{F}$ and $\widehat{N}$ denotes the outward unit normal on the reference domain.

The DGSEM approximates the solution ρ and flux $\widetilde{f}$ by polynomials of degree N, written in a basis of Lagrange polynomials with nodes at Gauss-Legendre quadrature points. Therefore, we introduce the following approximations

$$\rho \approx \rho^{\mathrm{h}} = \sum_{i,j=0}^{N} \rho_{ij}(z)\ell_i(\xi)\ell_j(\eta), \tag{7a}$$

$$\widetilde{f} \approx \widetilde{f}^{\mathrm{h}} = \sum_{i,j=0}^{N} \widetilde{f}_{ij}(\rho_{ij}(z))\ell_i(\xi)\ell_j(\eta), \tag{7b}$$

where ℓ_i denotes the ith Lagrange polynomial of degree N and we omit $\widetilde{f}_{ij}$'s dependence on z for brevity. We take the test function equal to each basis function, i.e., $\phi(\xi, \eta) = \ell_i(\xi)\ell_j(\eta)$, and we substitute the approximations (7) in Eq. (6) and apply Gauss-Legendre quadrature to approximate the integrals, such that the quadrature nodes coincide with the nodes of the Lagrange polynomials. This results in a system of ordinary differential equations (ODEs) for the coefficients ρ_{ij}, which reads

$$\mathcal{J}_{ij}\frac{\mathrm{d}\rho_{ij}(z)}{\mathrm{d}z} = \sum_{n=0}^{N} \widehat{D}_{in}\widetilde{f}_{nj} + \sum_{m=0}^{N} \widehat{D}_{jm}\widetilde{g}_{im}$$
$$- \left[\frac{\ell_i(1)}{w_i}\widetilde{F}(1, \eta_j) - \frac{\ell_i(-1)}{w_i}\widetilde{F}(-1, \eta_j) + \frac{\ell_j(1)}{w_j}\widetilde{G}(\xi_i, 1) - \frac{\ell_j(-1)}{w_j}\widetilde{G}(\xi_i, -1)\right], \tag{8}$$

with $\mathcal{J}_{ij} = \mathcal{J}(\xi_i, \eta_j)$ the Jacobian determinant at the (ξ_i, η_j) quadrature point, $\widetilde{\boldsymbol{f}}_{ij} = (\widetilde{f}_{ij}, \widetilde{g}_{ij})^\mathsf{T}$ and

$$\widehat{D}_{ij} := \frac{w_j}{w_i}\frac{\mathrm{d}\ell_i}{\mathrm{d}\xi}(\xi_j). \tag{9}$$

Furthermore, $\widetilde{\boldsymbol{F}}(\xi, \eta) = (\widetilde{F}(\xi, \eta), \widetilde{G}(\xi, \eta))^\mathsf{T}$ denotes the numerical fluxes in Eq. (8), for which we take the upwind flux.

If the upwind flux in Eq. (8) leads across an optical interface, the law of specular reflection or Snell's law of refraction needs to be applied, which we will discuss in the next section.

3.2 *Optical Interfaces*

At an optical interface, the momentum p of a light ray changes discontinuously according to Snell's law of refraction or the law of specular reflection. Both these laws can be combined into what we refer to as Snell's function, which relates the momentum p of an incident ray to the momentum $\bar{p}$ of the outgoing ray. Let n_1 and n_2 be the refractive indices of the incident and transmitted medium, respectively. Then, a light ray with incident momentum p striking an optical interface at a point with surface unit normal $\boldsymbol{\nu} = (\nu_q, \nu_z)$, that is directed towards the incident medium, changes its momentum to $\bar{p} = \mathcal{S}(p; n_1, n_2, \boldsymbol{\nu})$. Here, $\mathcal{S}(p; n_1, n_2, \boldsymbol{\nu})$ is the so-called Snell's function, which reads [10]

$$\mathcal{S}(p; n_1, n_2, \boldsymbol{\nu}) := \begin{cases} p - (\psi + \sqrt{\delta})\nu_q & \text{if } \delta \geq 0, \\ p - 2\psi\nu_q & \text{if } \delta < 0, \end{cases} \tag{10a}$$

with the auxiliary variables δ and ψ defined by

$$\delta := n_2^2 - n_1^2 + \psi^2, \quad \psi := \begin{pmatrix} p \\ \sigma\sqrt{n_1^2 - p^2} \end{pmatrix} \cdot \begin{pmatrix} \nu_q \\ \nu_z \end{pmatrix}. \tag{10b}$$

In (10a) the case $\delta \geq 0$ describes refraction and $\delta < 0$ describes total internal reflection [10]. In what follows we will frequently use the shorthand notation $\bar{p} = \mathcal{S}(p)$. Moreover, we will also require the inverse of Snell's function $p = \mathcal{S}^{-1}(\bar{p})$, i.e., compute the momentum p such that $\bar{p} = \mathcal{S}(p)$. For refraction the inverse reads [10]

$$\mathcal{S}^{-1}(\bar{p}) = -\mathcal{S}(-\bar{p}; n_2, n_1, -\boldsymbol{\nu}). \tag{11}$$

Snell's function (10) describing the change in momentum should be combined with the basic luminance invariance (2) which results in [10]

$$\rho(z^-, q^-, p^-) = \rho(z^+, q^+, p^+), \tag{12}$$

where $p^+ = \mathcal{S}(p^-; n_1, n_2, \boldsymbol{\nu})$ and the $\pm$ superscript denotes one-sided limits towards the optical interface. Note that in imposing (12) together with Snell's function (10) we also state that a light ray is either entirely reflected or entirely refracted, i.e., there is no partial reflection as for instance in Fresnel reflections.

Consider now for simplicity a flat optical interface with $n(q) = n_1$ if $q \leq q_0$ and $n(q) = n_2$ if $q > q_0$, then with the DGSEM we will have elements with faces on the optical interface, see Fig. 1a. In Fig. 1a we consider faces L_1 and L_2 that contribute to the faces R_1, R_2, R_3 and R_4, i.e., fluxes will strike the faces L_1 and L_2 which contribute to the fluxes leaving R_1, R_2, R_3 and R_4. The total flux leaving R_i is related to the flux striking an interval $\mathcal{S}^{-1}(R_i) = \{\mathcal{S}^{-1}(\bar{p}) \, : \, \bar{p} \in R_i\}$ by

$$\int_{R_i} \rho(z^+, q_0^+, \bar{p}) a(z^+, q_0^+, \bar{p}) \, \mathrm{d}\bar{p} = \int_{\mathcal{S}^{-1}(R_i)} \rho(z^-, q_0^-, p) a(z^-, q_0^-, p) \, \mathrm{d}p, \tag{13a}$$

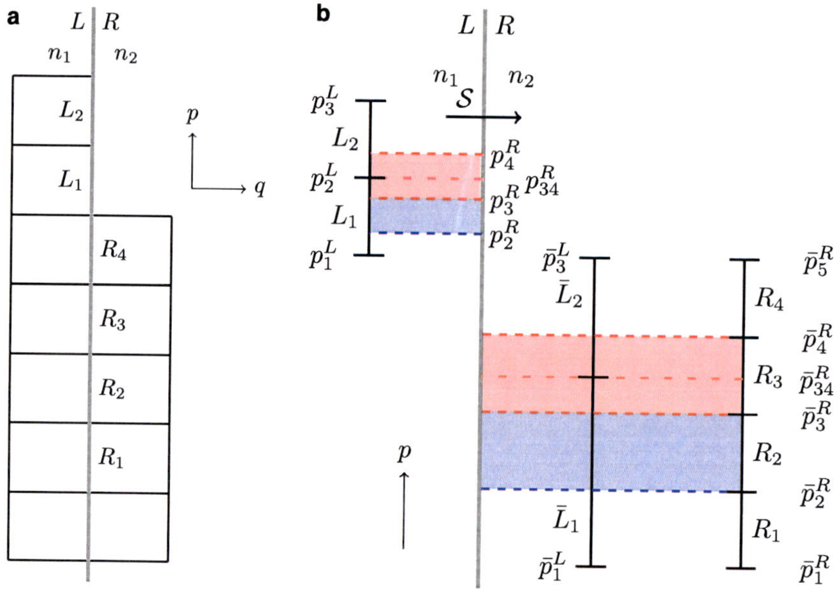

Fig. 1 (**a**) Elements in phase space connected by Snell's function. (**b**) Geometry at the optical interface

with the velocity a given by

$$a(z, q, p) = \frac{\partial h}{\partial p} = \frac{p}{\sqrt{n(z, q)^2 - p^2}}. \tag{13b}$$

Relation (13) describes an energy balance which is needed to ensure energy conservation in the DGSEM. A similar result can be derived for reflection. For more details we refer the reader to [9].

For each face R_i shown in Fig. 1a we must determine the contributing faces left of the interface. Therefore, we first transform the faces L_1 and L_2 using the law of refraction, yielding the virtual faces $\bar{L}_1 = \mathcal{S}(L_1)$ and $\bar{L}_2 = \mathcal{S}(L_2)$. These virtual faces are shown in Fig. 1b.

In the DGSEM we do not require continuity across elements, hence, we have to be careful in how we implement relation (12) whilst also ensuring energy conservation. In particular, on each face a polynomial approximation of ρ is known which can be computed by evaluating (7a) at the face. For a face L_i this polynomial depends on the momentum p and is denoted as $\rho^{L_i}(p)$. Applying the basic luminance invariance at the optical interface (12) allows us to relate ρ on the face L_i to its counterpart on the virtual face $\bar{L}_i$, by

$$\rho^{L_i}(p) = \rho^{L_i}(\mathcal{S}^{-1}(\bar{p})) = \rho^{\bar{L}_i}(\bar{p}) \quad \text{with } \bar{p} = \mathcal{S}(p). \tag{14}$$

Relation (14) together with the geometric connectivity shown in Fig. 1b allows us to describe how we compute the polynomial ρ^{R_i} on each face R_i. For instance, the polynomial ρ^{R_3} on face R_3 depends on the solution known on faces L_1 and L_2, as can be seen from Fig. 1b. The degree N polynomial $\rho^{R_3} \in \mathbb{P}_N$ must be computed from a piecewise polynomial ρ^L with the additional constraint of energy conservation, hence, we pose the problem as a constrained least-squares problem, which reads

$$\min_{\rho^{R_3} \in \mathbb{P}_N} \quad \int_{\bar{p}_3^R}^{\bar{p}_4^R} \left[\rho^{R_3}(\bar{p}) - \rho^L(\mathcal{S}^{-1}(\bar{p}))\right]^2 \mathrm{d}\bar{p}, \tag{15a}$$

$$\text{subject to} \quad \int_{\bar{p}_3^R}^{\bar{p}_4^R} F^{R_3}(\bar{p})\,\mathrm{d}\bar{p} = \int_{p_3^R}^{p_4^R} F^L(p)\,\mathrm{d}p, \tag{15b}$$

where ρ^L and F^L are piecewise polynomials given by

$$\rho^L(p) = \begin{cases} \rho^{L_1}(p) & \text{for } p \in L_1, \\ \rho^{L_2}(p) & \text{for } p \in L_2, \end{cases} \quad \text{and } F^L(p) = \begin{cases} F^{L_1}(p) & \text{for } p \in L_1, \\ F^{L_2}(p) & \text{for } p \in L_2. \end{cases} \tag{16}$$

For an interpretation of the integration limits see Fig. 1b. Equation (15b) describes the energy balance (13), where F with superscript denotes the numerical flux at a face written as a polynomial, i.e.,

$$F(p) = \sum_{j=0}^{N} F_j(\rho_j)\ell_j(\eta(p)) \quad \text{with } F_j(\rho_j) = a_j\rho_j, \tag{17}$$

where $\eta(p)$ is a linear transformation from a momentum p on the face to the reference line $[-1, 1]$, and a_j denotes the velocity coefficient.

To solve the constrained least-squares problem we write it in terms of a Lagrangian containing a Lagrange multiplier for the constraint. Subsequently we impose the requirements for a stationary point and apply Gauss-Legendre quadrature on each (part of) face resulting in a linear system for the $N + 1$ coefficients ρ_j and a Lagrange multiplier. The linear system is solved analytically to obtain the $N + 1$ coefficients ρ_j on the face R_3, from which we can compute the numerical flux that is used in the ODE system (8). For more details, see [9].

In a similar manner we apply this method to the other faces. The method can, of course, be generalised to arbitrary configurations and also works when considering total internal reflection, see [9].

4 Results

4.1 Energy-Preserving and Convergence

To illustrate that the method for optical interfaces is energy-preserving and that the DGSEM retains the $(N + 1)$th order convergence rate, we apply it to the 'bucket of water' test case introduced by van Lith et al. in [10]. In this test case, we consider the refractive index field $n(q)$ where $n(q) = n_1 = 1.4$ if $q \leq 0$ and $n(q) = n_2 = 1$ if $q > 0$. We prescribe at $z = 0$ the initial condition

$$\rho_0(q, p) := \varphi_m\left(\frac{q - q_0}{\sigma_q}\right)\left[\varphi_m\left(\frac{p - p_0}{\sigma_{p,0}}\right) + \varphi_m\left(\frac{p - p_1}{\sigma_{p,1}}\right)\right], \tag{18a}$$

with

$$\varphi_m(x) := \begin{cases} \cos^{m+1}\left(\frac{\pi}{2}x^2\right) & \text{if } |x| < 1, \\ 0 & \text{otherwise,} \end{cases} \tag{18b}$$

and for the parameters we take $m = 7$, $q_0 = -0.35$, $\sigma_q = 0.25$, $p_0 = 0.45$, $\sigma_{p,0} = 0.45$, $p_1 = \frac{1}{2}(1.3 + p_c)$ with $p_c = \sqrt{n_1^2 - n_2^2}$ and $\sigma_{p,1} = 1.3 - p_1$. For the initial condition (18a), the exact solution reads [10]

$$\rho(z, q, p) = \begin{cases} \rho_0\big(q - z\frac{p}{\sqrt{n_1^2-p^2}}, p\big) & \text{if } q < 0, p \geq 0, \\ \rho_0\big(z\frac{p}{\sqrt{n_1^2-p^2}} - q, -p\big) & \text{if } q < 0, -p_c < p < 0, \\ \rho_0\big((\delta z - z)\frac{\bar{p}}{\sqrt{n_1^2-\bar{p}^2}}, \bar{p}\big) & \text{if } q > 0, p \geq 0, \\ 0 & \text{otherwise,} \end{cases} \tag{19}$$

where $\bar{p} = -\mathcal{S}(-p; n_2, n_1, -\boldsymbol{\nu})$ with $\boldsymbol{\nu} = (-1, 0)$, and $\delta z = \frac{q}{p}\sqrt{n_2^2 - p^2}$. The region described by $\{q < 0, p \geq 0\}$ features propagation through the medium with refractive index n_1. The region $\{q < 0, -p_c < p < 0\}$ describes light that was reflected at the optical interface, and the region $\{q > 0, p \geq 0\}$ describes light that was refracted.

A numerical solution to Liouville's equation is computed using a polynomial of degree $N = 6$ and $K = 480$ rectangular elements. The ODE system (8) is integrated using an explicit low-storage fourth order Runge-Kutta method by Zingg and Chisholm [13], from $z = 0$ to $z = Z$ and $z = 2Z$ with $Z = 0.7$. The result is shown in Fig. 2. As expected, the result is discontinuous at the optical interface. Moreover, during the entire simulation a relative error in the total luminous flux, including the luminous flux striking the walls at $q = \pm 1$, of at most $1.6 \cdot 10^{-15}$ was observed, hence, the method obeys energy conservation up to machine precision.

Next, we study the convergence rate γ_{DG} for $N = 1, 2, \ldots, 6$ by computing the discretisation error e_{DG}. We take the discretisation error e_{DG} as the L^1-norm of the difference between the numerical solution and the exact solution (19), and compute the L^1-norm with an appropriate Gauss-Legendre quadrature rule. The result is shown in Table 1, where for each row the grid spacing is halved. The expected $(N + 1)$th order convergence rate is observed.

4.2 *Comparison with Ray Tracing*

We compare solving Liouville's equation with the DGSEM to quasi-Monte Carlo ray tracing. As the latter is typically performed on a one-dimensional grid, we will compute the illuminance $E(z, q)$ for $z = Z$ with both methods. For Liouville's

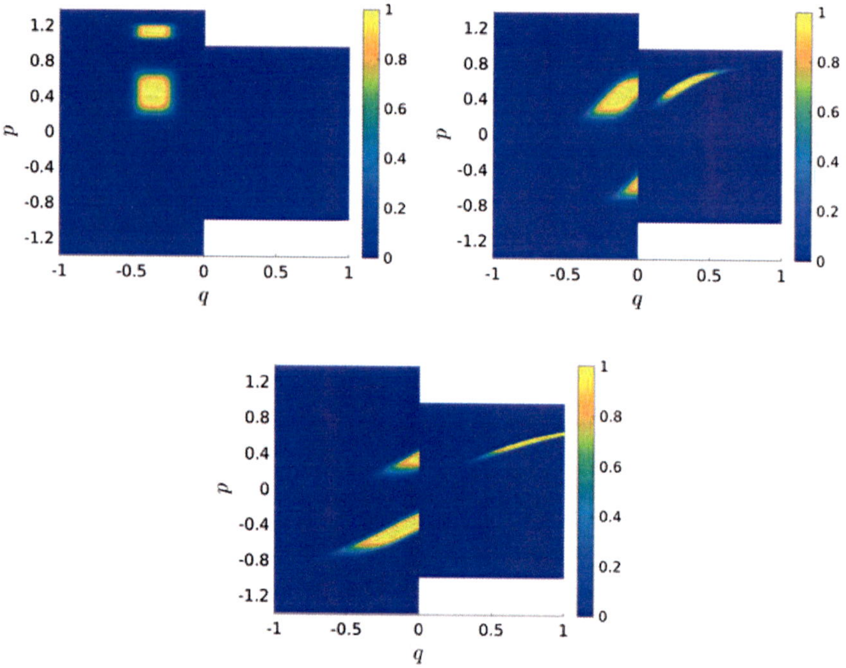

Fig. 2 Bucket of water: basic luminance distributions $\rho(z, q, p)$ at $z = 0$, $z = Z$ and $z = 2Z$ from top left to bottom middle. Parameters are $N = 6$, $K = 480$, $Z = 0.7$

Table 1 Bucket of water: convergence data

K	e_{DG}	γ_{DG}	e_{DG}	γ_{DG}	e_{DG}	γ_{DG}
	$N = 1$		$N = 2$		$N = 3$	
480	4.93e-02		1.71e-02		8.70e-03	
1920	1.82e-02	1.44	4.90e-03	1.80	1.23e-03	2.83
7680	6.25e-03	1.54	6.61e-04	2.89	8.07e-05	3.92
30,720	1.56e-03	2.00	5.82e-05	3.50	3.71e-06	4.44
	$N = 4$		$N = 5$		$N = 6$	
480	4.15e-03		2.03e-03		1.03e-03	
1920	3.55e-04	3.55	1.08e-04	4.24	3.36e-05	4.94
7680	1.17e-05	4.93	1.98e-06	5.77	3.79e-07	6.47
30,720	3.08e-07	5.24	3.10e-08	6.00	3.37e-09	6.81

equation the illuminance E is obtained by first integrating from $z = 0$ to $z = Z$ to determine ρ, followed by evaluation of the integral

$$E(z, q) = \int_{-n(z,q)}^{n(z,q)} \rho(z, q, p)\,\mathrm{d}p. \tag{20}$$

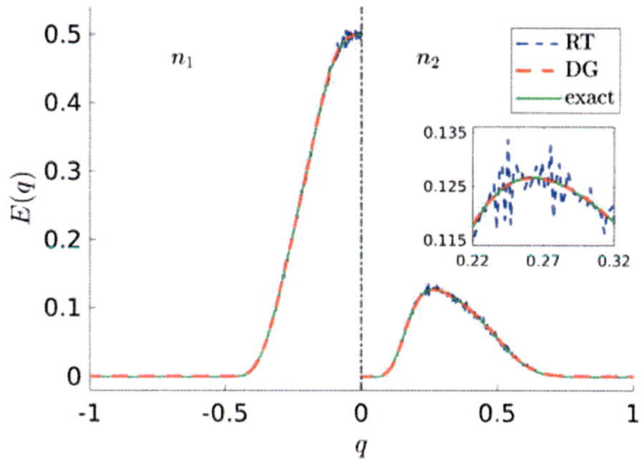

Fig. 3 Illuminance $E(q)$: ray tracing (RT) and DGSEM (DG)

For quasi-Monte Carlo ray tracing we fix the number of bins to $B = 1000$ and employ a uniform grid on $q \in [-1, 1]$. The jth bin is defined by $[Q_j, Q_{j+1}]$ with midpoint $q_j = \frac{1}{2}(Q_j + Q_{j+1})$, where

$$Q_j = (j-1)\Delta q - 1, \quad j = 1, \ldots, B+1, \tag{21}$$

and $\Delta q = 2/B$. The rays are drawn from a 2D Sobol sequence [5] using Matlab's '*sobolset*' (https://nl.mathworks.com/help/stats/sobolset.html), yielding the initial phase space coordinates (q_i, p_i) for each ray.

The illuminance E at $z = Z$ with $Z = 0.7$ is computed for the 'bucket of water' test case with initial condition (18a). The result is plotted together with the exact illuminance in Fig. 3 for both methods. Both methods take roughly the same amount of computation time using a single core. For ray tracing (RT) we use $N_{\mathrm{RT}} = 0.64 \cdot 10^6$ rays, while for DGSEM (DG) we use $K = 480$ elements and $N = 4$. From the inset in Fig. 3 it is clear that the quasi-Monte Carlo ray tracing solution is oscillatory about the exact illuminance, while the DGSEM with $N = 4$ and $K = 480$ is indistinguishable from the exact solution by eye.

To compare both methods quantitatively, we compute a discretisation error. Quasi-Monte Carlo ray tracing computes an average illuminance on each bin, hence, for a fair comparison we also average the illuminance for the DGSEM when computing the discretisation error. For the discretisation error we once again take the L^1-norm of the difference between the numerical average illuminance and the exact average illuminance.

In Fig. 4 the discretisation error as a function of computation time is plotted, for both ray tracing and the DGSEM. For ray tracing, we increase the number of rays used while keeping the number of bins fixed. For the DGSEM, we varied the polynomial degree N as well as the number of elements on phase space. From Fig. 4, one can observe that the DGSEM with $N = 1$ is already more accurate than quasi-

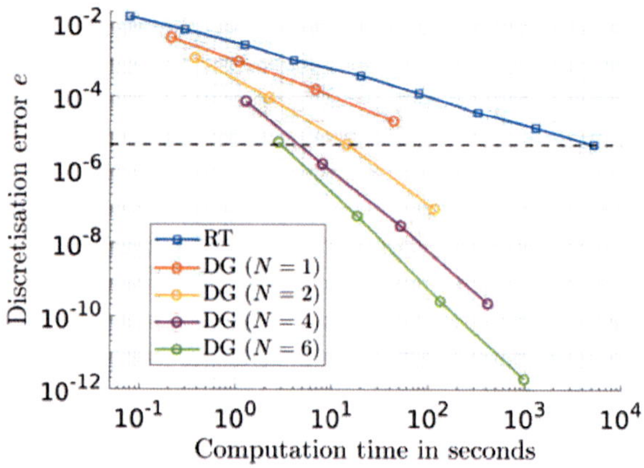

Fig. 4 Error vs computation time

Monte Carlo ray tracing for equal computation time. Most notably, is that for the discretisation error indicated with the gray dashed line, the DGSEM with $N = 6$ is over 1000 times faster than ray tracing! Hence, for this test case the DGSEM significantly outperforms quasi-Monte Carlo ray tracing.

5 Conclusion

The discretisation of Liouville's equation for two-dimensional geometrical optics using the discontinuous Galerkin spectral element method was presented. In particular, a method was presented to incorporate the non-local boundary conditions that arise at optical interfaces, such that the scheme obeys energy conservation.

In a test case, the DGSEM was verified to be energy-preserving and was shown to achieve the expected $(N + 1)$th order convergence rate. Moreover, the DGSEM was shown to significantly outperform quasi-Monte Carlo ray tracing. For instance, DGSEM with $N = 6$ was over 1000 times faster compared to ray tracing to achieve the same accuracy.

At the moment Fresnel reflections were not considered. Therefore, a next step would be to include Fresnel reflections by modifying the basic luminance invariance (12). Furthermore, we intend to extend the method to three-dimensional optics. This requires a four-dimensional phase space together with integration along the z-coordinate. Even though the computational cost increases significantly, the high convergence rates of DGSEM will be an advantage over quasi-Monte Carlo ray tracing.

Acknowledgments This work is part of the research programme NWO-TTW Perspectief with project number P15-36, which is (partly) financed by the Netherlands Organisation for Scientific Research (NWO).

References

1. Arnold, V.I.: Mathematical Methods of Classical Mechanics, vol. 60. Springer Science & Business Media (2013)
2. Chaves, J.: Introduction to Nonimaging Optics. CRC Press (2017)
3. Herkommer, A.M.: Phase space optics: an alternate approach to freeform optical systems. Optical Eng. **53**(3), 031304 (2013)
4. Kopriva, D.A.: Implementing Spectral Methods for Partial Differential Equations: Algorithms for Scientists and Engineers. Springer Science & Business Media (2009)
5. Leobacher, G., Pillichshammer, F.: Introduction to quasi-Monte Carlo Integration and Applications. Springer (2014)
6. Nicodemus, F.E.: Radiance. Am. J. Phys. **31**(5), 368–377 (1963)
7. Pelka, D.G., Patel, K.: An overview of LED applications for general illumination. In: Design of Efficient Illumination Systems, vol. 5186, pp. 15–26. International Society for Optics and Photonics (2003)
8. Rausch, D., Rommel, M., Herkommer, A.M., Talpur, T.: Illumination design for extended sources based on phase space mapping. Opt. Eng. **56**(6), 065103 (2017)
9. van Gestel, R.A.M., Anthonissen, M.J.H., ten Thije Boonkkamp, J.H.M., IJzerman, W.L.: An energy conservative hp-method for Liouville's equation of geometrical optics. J. Sci. Comput. **89**(1), 1–35 (2021)
10. van Lith, B.S., ten Thije Boonkkamp, J.H.M., IJzerman, W.L., Tukker, T.W.: A novel scheme for Liouville's equation with a discontinuous Hamiltonian and applications to geometrical optics. J. Sci. Comput. **68**(2), 739–771 (2016)
11. van Lith, B.S., ten Thije Boonkkamp, J.H.M., IJzerman, W.L.: Active flux schemes on moving meshes with applications to geometric optics. J. Comput. Phys. X **3**, 100030 (2019)
12. Wolf, K.B.: Geometric Optics on Phase Space. Springer Science & Business Media (2004)
13. Zingg, D.W., Chisholm, T.T.: Runge-Kutta methods for linear ordinary differential equations. Appl. Numer. Math. **31**(2), 227–238 (1999)

Using ℓ^1-Regularization for Shock Capturing in Discontinuous Galerkin Methods

Jan Glaubitz and Anne Gelb

1 Introduction

This work is about shock capturing in discontinuous Galerkin (DG) methods [6, 21] for solving hyperbolic conservation laws

$$\partial_t u + \partial_x f(u) = 0, \quad x \in \Omega, \; t > 0 \tag{1}$$

with adequate initial and boundary conditions. Here, f is a given flux function and u is the unknown solution. Since [33], it is known that solutions of (1) can develop spontaneous shock discontinuities, even in finite time and for smooth initial data [7, 24]. While DG methods—as well as many other methods—can achieve high orders of accuracy for smooth solutions of (1), they often lack desired robustness properties in the presence of shock discontinuities. For this reason, many shock capturing techniques have been developed over the last decades. While these are too numerous to list entirely, some examples include artificial viscosity [16, 23, 29], spectral viscosity [25, 36], modal filtering [15, 20, 26], order reduction [5, 12], WENO limiters [31], and finite volume sub-cells [35, 40]. Also see [13] and references therein.

The present contribution follows [14] in proposing ℓ^1-regularization as a new approach to shock capturing by promoting sparsity in the jump function of the numerical solution. The rationale for this is that solutions of hyperbolic conservation laws often contain only a finite number of jump discontinuities [37], which is equivalent to the jump function of the solution being sparse. Using ℓ^1 regularization, we can promote this sparsity in the "jump domain" of the exact solution for the

J. Glaubitz (✉) · A. Gelb
Department of Mathematics, Dartmouth College, Hanover, NH, USA
e-mail: Jan.Glaubitz@Dartmouth.edu; AnneGelb@Math.Dartmouth.edu

J. M. Melenk et al. (eds.), *Spectral and High Order Methods for Partial Differential Equations ICOSAHOM 2020+1*, Lecture Notes in Computational Science and Engineering 137, https://doi.org/10.1007/978-3-031-20432-6_21

numerical solution as well. Thereby, we approximate the presumably sparse jump function using linear high-order polynomial annihilation operators [3]. A similar application of ℓ^1-regularization was used in [34], though only for the Lax–Wendroff scheme and Chebyshev and Fourier spectral methods, and later also in [11]. A brief discussion of further related works can be found in [14, 34]. In contrast to [34], we integrate ℓ^1-regularization in a DG method. The underlying decomposition of the computational domain into smaller elements allows for localization of the proposed shock capturing procedure. In particular, ℓ^1-regularization is only applied in troubled elements containing spurious oscillations, and we allow for element-to-element variations in the (parameters of the) corresponding optimization problem. Finally, we note that the proposed procedure also carries over from DG to other classes of methods, with the obvious extension to discontinuous spectral element type schemes.

The rest of this contribution is organized as follows. Section 2 provides some necessary preliminaries on ℓ^1-regularization. We then describe in Sect. 3 how ℓ^1-regularization can be used for shock capturing in DG methods. In Sect. 4, we demonstrate how local conservation can be preserved when applying ℓ^1-regularization in a troubled element. Some selected numerical tests demonstrating the performance of the proposed shock capturing procedure are provided in Sect. 5. Finally, Sect. 6 offers some concluding thoughts.

2 ℓ^1-Regularization and Polynomial Annihilation

We start by collecting some necessary preliminaries on ℓ^1-regularization and polynomial annihilation operators. Furthermore, we connect ℓ^1-regularization to shock capturing for hyperbolic conservation laws.

2.1 ℓ^1-Regularization

Many applications seek to solve the *linear inverse problem*

$$\mathbf{y} = F\mathbf{x} + \boldsymbol{\nu}, \tag{2}$$

where $\mathbf{y}$ is a vector of observations, $\mathbf{x}$ is the vector of unknowns, and F is a known linear forward operator. Finally, $\boldsymbol{\nu}$ corresponds to a typically unknown noise vector. See [18, 19, 38] and references therein. It is often reasonable to assume that some linear transformation of the unknown solution $\mathbf{x}$, say $R\mathbf{x}$, is sparse. A common approach is to consider the ℓ^1-*regularized inverse problem*

$$\mathbf{x}_\lambda = \arg\min_{\mathbf{x}} \left\{ \|F\mathbf{x} - \mathbf{y}\|_2^2 + \lambda \|R\mathbf{x}\|_1 \right\}. \tag{3}$$

Here, R is referred to as the *regularization operator* and $\lambda > 0$ as the *regularization parameter*. The motivation for this approach is that the ℓ^1-norm, $\|\cdot\|_1$, serves as a convex surrogate for the ℓ^0-"norm", $\|\cdot\|_0$. Thus (3) balances data fidelity, noise, and the sparsity assumption on $R\mathbf{x}$, while still enabling efficient computations [8–10]. In our implementation, we use the alternating direction method of multipliers (ADMM) [4] to efficiently solve (3). Details can be found in [14, Section 3.4].

2.2 Connection to Numerical Hyperbolic Conservation Laws

Numerical solutions to hyperbolic conservations—at least when corresponding to a high-order method—often show spurious oscillations near (shock) discontinuities. This is sometimes explained by the Gibbs–Wilbraham phenomenon [17, 22]. Let us denote by $\mathbf{u}$ the nodal values of the solution of the conservation law (1) and by $\mathbf{u}^{\text{osc}}$ the nodal values of an oscillatory approximation to this solution. If we interpret the oscillations contained in $\mathbf{u}^{\text{osc}}$ as noise, then we may formulate the relation between $\mathbf{u}^{\text{osc}}$ and $\mathbf{u}$ as a special case of the linear inverse problem (2):

$$\mathbf{u}^{\text{osc}} = \mathbf{u} + \boldsymbol{\nu} \tag{4}$$

That is, $F = I$ for the forward operator and $\boldsymbol{\nu}$ in (4) is assumed to correspond to the spurious oscillations. Hence, we are given an oscillatory numerical solution $\mathbf{u}^{\text{osc}}$ and would like to recover the oscillation-free solution $\mathbf{u}$. Our approach is to do so by formulating this task as an ℓ^1-regularized inverse problem of the form (3). We therefore have to address in which sense the desired solution $\mathbf{u}$, or some linear transform of it, can be considered as sparse. As mentioned in the introduction, the (entropy) solution u of (1) contains only a finite number of jump discontinuities in many cases [37]. This means that the solution's jump function, $[u]$, is zero almost everywhere. Recall that the *jump function* of $u : [a, b] \to \mathbb{R}$ is defined as

$$[u](\xi) = u\left(\xi^+\right) - u\left(\xi^-\right), \quad \xi \in (a, b), \tag{5}$$

where $u\left(\xi^-\right)$ and $u\left(\xi^+\right)$ respectively denote the left- and right-hand-side limit of u at ξ. For the moment, assume that there is a linear transformation R such that $R\mathbf{u} \approx [\mathbf{u}]$ with $[\mathbf{u}]$ denoting some nodal value of the jump function $[u]$. In this case, we expect the ℓ^1-regularized inverse problem

$$\min_{\mathbf{u}} \left\{ \|\mathbf{u} - \mathbf{u}^{\text{osc}}\|_2^2 + \lambda \|R\mathbf{u}\|_1 \right\} \tag{6}$$

to give us a solution for which, on the one hand, spurious oscillations have been removed while, on the other hand, the jump function is sparse. The latter promotes the numerical solution to mimic the behavior of the exact entropy solution of (1).

2.3 Polynomial Annihilation

It remains to find a linear transformation R such that $R\mathbf{u} \approx [\mathbf{u}]$ with high accuracy. We use polynomial annihilation (PA) operators [3] to construct R. The *PA operator of order* m,

$$L_m[u](\xi) = \frac{1}{q_m(\xi)} \sum_{x_j \in S_\xi} c_j(\xi)\, u(x_j), \tag{7}$$

is designed in order to approximate the jump function $[u]$. Here

$$S_\xi = \{x_0(\xi), \ldots, x_m(\xi)\} \subset \Omega_{\text{ref}} \tag{8}$$

is a set of $m+1$ local grid points around ξ, the *annihilation coefficients* c_j are given by

$$\sum_{x_j \in S_\xi} c_j(\xi)\, p_l(x_j) = p_l^{(m)}(\xi), \quad l = 0, \ldots, m, \tag{9}$$

and $\{p_l\}_{l=0}^m$ is a basis of $\mathbb{P}_m$, the space of polynomials of degree at most m. Finally, the *normalization factor* q_m, calculated as

$$q_m(\xi) = \sum_{x_j \in S_\xi^+} c_j(\xi), \tag{10}$$

ensures convergence to the right jump strength at every discontinuity. Here S_ξ^+ denotes the set $\{x_j \in S_\xi \mid x_j \geq \xi\}$ of all local grid points to the right of ξ.

We apply the PA operator to the reference element, say $\Omega_{\text{ref}} = [-1, 1]$, of an underlying nodal DG method using $p+1$ Gauss–Lobatto points $\{\xi_k\}_{k=0}^p$. We can thus construct PA operators up to order p by allowing the sets of local grid points S_ξ to be certain subsets of the $p+1$ collocation points. In [3] it was shown that

$$L_m[u](\xi) = \begin{cases} [u](x) + O(h(\xi)) & \text{if } x_{j-1} \leq \xi, x \leq x_j, \\ O\left(h^{\min(m,k)}(\xi)\right) & \text{if } u \in C^k(I_\xi), \end{cases} \tag{11}$$

where $h(\xi) = \max\{|x_i - x_{i-1} \mid x_{i-1}, x_i \in S_\xi\}$ and I_ξ is the smallest closed interval such that $S_\xi \subset I_\xi$. Hence, L_m provides mth order convergence in regions where $u \in C^m$ and yields a first-order approximation to the jump. It should be stressed that oscillations develop around points of discontinuity as m increases. Post-processing methods, such as the minmod limiter, could reduce the impact of the oscillation, as was done in [3]. However, as long as there is sufficient resolution between two shock locations, such oscillations do not directly impact our method. This is because we use the PA operator not to detect the precise location of jump

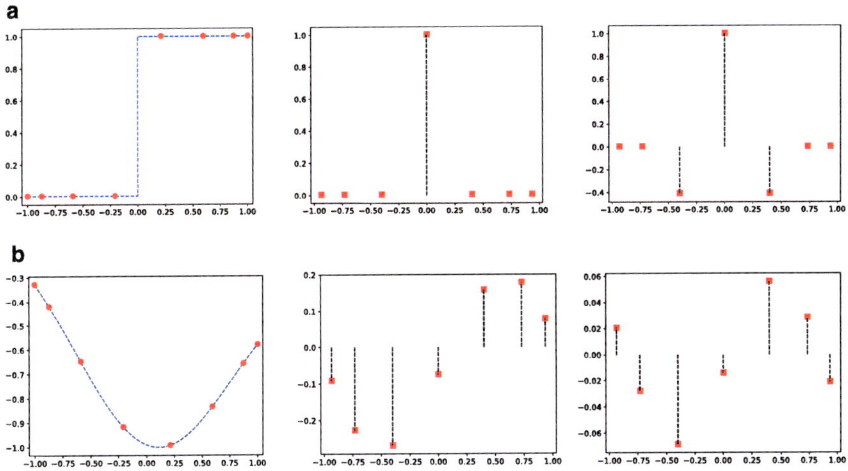

Fig. 1 (**a**) A discontinuous step function and the values of the PA operator for $m = 1, 3$. (**b**) A smooth function and the values of the PA operator for $m = 1, 3$

discontinuities but rather to enforce sparsity. Figure 1a demonstrates the PA operator for a discontinuous step function, while Fig. 1b does so for a smooth function.

3 Shock Capturing by ℓ^1-Regularization

We are now able to describe the procedure to use ℓ^1-regularization for shock capturing in nodal DG methods.

3.1 Proposed Procedure

One of the main challenges in solving nonlinear conservation laws is balancing high resolution properties and the amount of smoothing (dissipation) introduced to maintain stability, especially near shocks [16, 29, 39]. Let $u_p \in \mathbb{P}_p$ be an oscillatory numerical solution in a troubled element Ω_i and $\mathbf{u}^{\text{osc}}$ the vector of its nodal values at the grid points in Ω_i. Our procedure consists of replacing u_p by a *sparse reconstruction*:

$$\mathbf{u}^{\text{spar}} = \arg\min_{\mathbf{u}} \left\{ \|\mathbf{u} - \mathbf{u}^{\text{osc}}\|_2^2 + \lambda \|L_m \mathbf{u}\|_1 \right\}, \tag{12}$$

where $\lambda > 0$ is the regularization parameter and L_m the mth-order PA operator, in troubled elements after every (or every kth) time step by an explicit time integrator. Recall that $L_m\mathbf{u} \approx [\mathbf{u}]$ with order m in smooth regions and order 1 near discontinuities. Hence, the solution $\mathbf{u}^{\text{spar}}$ of (12) is a "sparse reconstruction" in the sense of its jump function being sparse, which mimics the assumed behavior of the exact entropy solution of (1).

3.2 Parameter Selection

ℓ^1-regularization should only be activated in troubled elements. In particular, we do not want to unnecessarily degrade the accuracy of the numerical solution in smooth regions. We thus propose to choose $\lambda = 0$ (not activate ℓ^1-regularization) in elements corresponding to smooth regions. Note that this also renders the proposed method more efficient. On the other hand, when a discontinuity is detected in an element, ℓ^1-regularization will be fully activated by choosing $\lambda = \lambda_{\max}$ in (12), which corresponds to the amount of regularization necessary to reconstruct sharp shock profiles. While no effort was made to optimize or even adapt this parameter, we found that using $\lambda_{\max} = 4 \cdot 10^2$ in all of our numerical experiments yielded good results. Choosing an appropriate λ will be the subject of future work. Between these extreme cases, $\lambda = 0$ and $\lambda = \lambda_{\max}$, we allow the regularization parameter to linearly vary and choose λ as a function of a discontinuity sensor S:

$$\lambda(S) = \begin{cases} 0 & \text{if } S \leq \kappa, \\ \lambda_{\max}(S - \kappa)/(1 - \kappa) & \text{if } \kappa < S < 1, \\ \lambda_{\max} & \text{if } S \geq 1, \end{cases} \tag{13}$$

where $\kappa \in [0, 1)$ is a problem dependent *ramp parameter*. Consequently, we obtain more accurate sparse reconstructions while still maintaining stability in regions around discontinuities. Most of the usual discontinuity sensors [23, 29] can be used to steer the regularization parameter λ according to (13). In our implementation, we used a discontinuity sensor based on comparing PA operators of different orders. See [14, Section 3.3] for more details.

Figure 2a and b demonstrate the effect of ℓ^1-regularization for an oscillatory numerical solution given by a polynomial of degree $p = 13$. While the original numerical solution contains spurious oscillations, these are removed for the sparse reconstruction without introducing a significant amount of undesired smearing. In accordance with computing the sparse reconstruction by promoting sparsity for the approximate jump functions, the values of the 3rd-order PA operator contain fewer nonzero values for the sparse reconstruction than for the original oscillatory approximation.

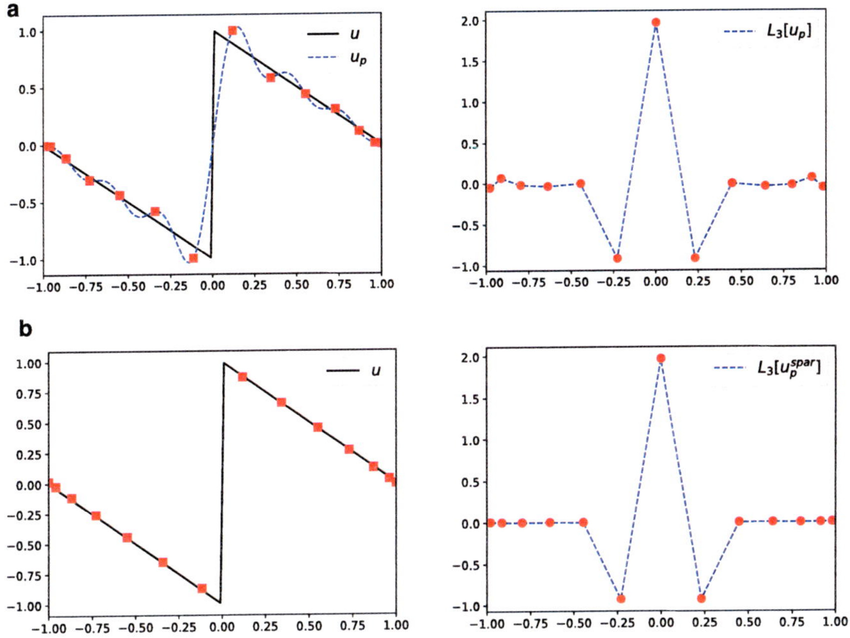

Fig. 2 (**a**) An oscillatory numerical approximation u_p to a discontinuous function u and the corresponding approximation to the jump function by a 3rd-order PA operator. (**b**) The sparse reconstruction u_p^{spar} using ℓ^1-regaulrization and the corresponding approximation to the jump function by a 3rd-order PA operator

4 Preserving Local Conservation

An essential property of DG methods for hyperbolic conservation laws is that they are locally conservative,

$$\int_{\Omega_i} u(t^{n+1}, x)\, \mathrm{d}x = \int_{\Omega_i} u(t^n, x)\, \mathrm{d}x + f^{\text{num}}|_{\partial\Omega_i}. \tag{14}$$

Any reasonable shock capturing procedure should preserve this property. In particular, in a troubled element Ω_i in which ℓ^1-regularization is applied,

$$\int_{\Omega_i} u_p^{\text{spar}}(x)\, \mathrm{d}x = \int_{\Omega_i} u_p(x)\, \mathrm{d}x \tag{15}$$

should hold. Unfortunately, (15) is violated when ℓ^1-regularization is applied too naively.

This is demonstrated in Fig. 3 for the simple discontinuous function $u(x) = \text{sign}(x - 0.5) + 1$ on $\Omega_{\text{ref}} = [-1, 1]$. Thus, we now present a simple fix for this

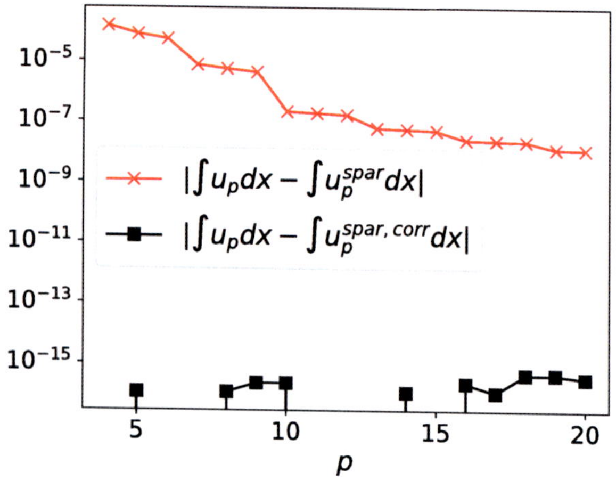

Fig. 3 Absolute difference between the mass of u_p and u_p^{spar} (red crosses) and between the mass of u_p and $u_p^{\text{spar,corr}}$ (black squares). In the cases where no (black) squares are visible the difference was below machine precision

problem. Let $\{\varphi_k\}_{k=0}^p$ be a basis of $\mathbb{P}_p$ with zero average for $k > 0$ so that

$$\int_{\Omega_{\text{ref}}} \varphi_k \, \mathrm{d}x = 0, \quad k > 0. \tag{16}$$

If we represent u_p and u_p^{spar} with respect to such a basis, then (15) is equivalent to

$$\hat{u}_0 = \hat{u}_0^{\text{spar}}, \tag{17}$$

where $\hat{u}_0$ and $\hat{u}_0^{\text{spar}}$ respectively denote the first modal coefficient of u_p and u_p^{spar} w. r. t. the basis $\{\varphi_k\}_{k=0}^p$. This observation yields the following simple strategy to ensure mass preservation when using ℓ^1-regularization.

Algorithm 1 Mass correction

1: Compute $\hat{u}_0$
2: Compute u_p^{spar} according to (12)
3: Represent u_p^{spar} w. r. t. $\{\varphi_k\}_{k=0}^p$: $u_p^{\text{spar}} = \hat{u}_0^{\text{spar}} \varphi_0 + \cdots + \hat{u}_p^{\text{spar}} \varphi_p$
4: Replace $\hat{u}_0^{\text{spar}}$ by $\hat{u}_0$

The advantage of this additional step is demonstrated in Fig. 3 as well, where the absolute difference between the mass of u_p and its sparse reconstruction with additional mass correction, denoted by $u_p^{\text{spar,corr}}$, is illustrated by (black) squares. In contrast to the sparse reconstruction without mass correction, illustrated by (red) crosses, $u_p^{\text{spar,corr}}$ is demonstrated to preserve mass nearly up to machine precision ($\approx 10^{-16}$).

5 Numerical Tests

We numerically demonstrate the application of ℓ^1-regularization (with and without mass correction) to a nodal DG method for a nonlinear system of conservation laws. Further tests can be found in [14, Section 4]. Consider the nonlinear system of conservation laws

$$\partial_t \begin{pmatrix} u_0 \\ u_1 \end{pmatrix} + \frac{1}{2}\partial_x \begin{pmatrix} u_0^2 + u_1^2 \\ 2u_0u_1 \end{pmatrix} = 0 \tag{18}$$

in the domain $\Omega = [0, 2]$ with periodic boundary conditions and smooth initial data $u_0(x, 0) = u_1(x, 0) + 1$ and

$$u_1(x, 0) = \begin{cases} e \cdot \exp\left(-\frac{r^2}{r^2-(x-0.5)^2}\right) & , \text{ if } |x - 0.5| < r, \\ 0 & , \text{ if } |x - 0.5| \geq r, \end{cases} \tag{19}$$

where $r = 0.5$. The system (18) originates from a truncated polynomial chaos approach for Burgers' equation with uncertain initial condition [27, 30]. In this context, u_0 models the expected value of the numerical solution while u_1^2 approximates the variance. Without providing the details here, we use the entropy-stable skew-symmetric nodal DG method derived in [27] to numerically solve (18).

Figure 4a and b illustrate the results of ℓ^1-regularization (with and without mass correction) for the above described test case. Thereby, $I = 100$ uniform elements, a polynomial basis of degree $p = 6$, and a 3rd-order PA operator were used. Note that while the numerical solution without ℓ^1-regularization shows heavy oscillations in both components, the numerical solution with ℓ^1-regularization provides a sharper profile. Further, it should be stressed that only ℓ^1-regularization with additional mass correction is able to capture the exact shock location. Due to missing conservation, ℓ^1-regularization without mass correction results in a slightly wrong location for the shock. Finally, the last column of Fig. 4a and b illustrates that ℓ^1-regularization (with and without mass correction) slightly increases the energy in this test case.

Remark 1 Although oscillations were almost fully removed in Fig. 2a and b, we still observe some smaller oscillations in Fig. 4b when ℓ^1-regularization is applied. This might be explained as follows: While ℓ^1-regularization was applied to a fixed snapshot of the numerical solution in Fig. 2a and b, it is applied to a time-evolving problem in Fig. 4b. Even if the shock discontinuity is perfectly captured at a certain time, oscillations will again form as time evolves due to the self-sharpening feature of shocks. ℓ^1-regularization will only become (partially) active again once the oscillations are so severe that $\lambda(S) > 0$ in (13). However, we could reduce

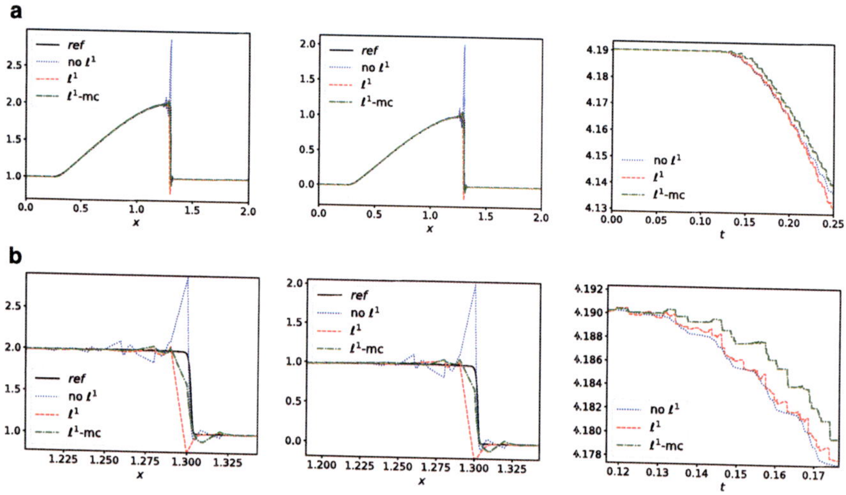

Fig. 4 (**a**) The solution components u_0 and u_1 at $t = 0.25$ and energy over time. (**b**) Zoom-in on the solution components u_0 and u_1 at $t = 0.25$ and the energy over time

the remaining small oscillations in Fig. 4b for ℓ^1-regularization by fine-tuning the discontinuity sensor S, the ramp parameter κ, and the map for the regularization parameter λ. As a second alternative, we could combine ℓ^1-regularization with another stabilization mechanism, such as artificial dissipation. Both options will be explored in future works.

6 Summary

We have presented an approach to shock capturing in nodal DG methods by ℓ^1-regularization using PA operators. Essentially, the idea was to remove spurious oscillations without introducing undesired smearing by incorporating some presumed prior knowledge about the approximated entropy solution. This prior knowledge was that the exact entropy solution is piecewise smooth and contains only a finite number of jump discontinuities, translating into a sparse jump function. The jump function was approximated using high-order PA operators, but other linear high-order edge sensors could also be used. Finally, future work might focus on incorporating energy stability (as well as other properties like TVD or positivity) by additional constraints in the minimization problem (12). In particular, it would be interesting to see if, for instance, adaptive artificial dissipation strategies [28, 32] or entropy correction [1, 2] can be used to ensure entropy stability for ℓ^1-regularization.

References

1. Abgrall, R.: A general framework to construct schemes satisfying additional conservation relations. application to entropy conservative and entropy dissipative schemes. J. Comput. Phys. **372**, 640–666 (2018)
2. Abgrall, R., Öffner, P., Ranocha, H.: Reinterpretation and extension of entropy correction terms for residual distribution and discontinuous Galerkin schemes: application to structure preserving discretization. J. Comput. Phys. 110955 (2022)
3. Archibald, R., Gelb, A., Yoon, J.: Polynomial fitting for edge detection in irregularly sampled signals and images. SIAM J. Numer. Anal. **43**(1), 259–279 (2005)
4. Boyd, S., Parikh, N., Chu, E.: Distributed Optimization and Statistical Learning via the Alternating Direction Method of Multipliers. Now Publishers Inc, Delft (2011)
5. Cockburn, B., Shu, C.W.: TVB Runge–Kutta local projection discontinuous Galerkin finite element method for conservation laws. ii. General framework. Math. Comput. **52**(186), 411–435 (1989)
6. Cockburn, B., Karniadakis, G.E., Shu, C.W.: Discontinuous Galerkin Methods: Theory, Computation and Applications, vol. 11. Springer Science & Business Media, New York (2012)
7. Dafermos, C.M.: Hyperbolic Conservation Laws in Continuum Physics, vol. 3. Springer, Berlin (2005)
8. Donoho, D.L.: Compressed sensing. IEEE Trans. Inf. Theory **52**(4), 1289–1306 (2006)
9. Eldar, Y.C., Kutyniok, G.: Compressed Sensing: Theory and Applications. Cambridge University Press, Cambridge (2012)
10. Foucart, S., Rauhut, H.: A mathematical introduction to compressive sensing. Bull. Am. Math **54**, 151–165 (2017)
11. Gelb, A., Hou, X., Li, Q.: Numerical analysis for conservation laws using ℓ_1 minimization. J. Sci. Comput. **81**(3), 1240–1265 (2019)
12. Glaubitz, J.: Shock capturing by Bernstein polynomials for scalar conservation laws. Appl. Math. Comput. **363**, 124593 (2019)
13. Glaubitz, J.: Shock Capturing and High-Order Methods for Hyperbolic Conservation Laws. Logos Verlag Berlin GmbH, Berlin (2020)
14. Glaubitz, J., Gelb, A.: High order edge sensors with ℓ^1 regularization for enhanced discontinuous Galerkin methods. SIAM J. Sci. Comput. **41**(2), A1304–A1330 (2019)
15. Glaubitz, J., Öffner, P., Sonar, T.: Application of modal filtering to a spectral difference method. Math. Comput. **87**(309), 175–207 (2018)
16. Glaubitz, J., Nogueira, A., Almeida, J., Cantão, R., Silva, C.: Smooth and compactly supported viscous sub-cell shock capturing for discontinuous Galerkin methods. J. Sci. Comput. **79**, 249–272 (2019)
17. Gottlieb, D., Shu, C.W.: On the Gibbs phenomenon and its resolution. SIAM Rev. **39**(4), 644–668 (1997)
18. Groetsch, C.W., Groetsch, C.: Inverse Problems in the Mathematical Sciences, vol. 52. Springer, Berlin (1993)
19. Hansen, P.C.: Discrete Inverse Problems: Insight and Algorithms. SIAM, Philadelphia (2010)
20. Hesthaven, J., Kirby, R.: Filtering in Legendre spectral methods. Math. Comput. **77**(263), 1425–1452 (2008)
21. Hesthaven, J.S., Warburton, T.: Nodal Discontinuous Galerkin Methods: Algorithms, Analysis, and Applications. Springer Science & Business Media, New York (2007)
22. Hewitt, E., Hewitt, R.E.: The Gibbs–Wilbraham phenomenon: an episode in Fourier analysis. Arch. Hist. Exact Sci. **21**(2), 129–160 (1979)
23. Klöckner, A., Warburton, T., Hesthaven, J.S.: Viscous shock capturing in a time-explicit discontinuous Galerkin method. Math. Model. Nat. Phenom. **6**(3), 57–83 (2011)
24. Lax, P.D.: Hyperbolic Systems of Conservation Laws and the Mathematical Theory of Shock Waves. SIAM, Philadelphia (1973)

25. Maday, Y., Kaber, S.M.O., Tadmor, E.: Legendre pseudospectral viscosity method for nonlinear conservation laws. SIAM J. Numer. Anal. **30**(2), 321–342 (1993)
26. Meister, A., Ortleb, S., Sonar, T.: Application of spectral filtering to discontinuous Galerkin methods on triangulations. Numer. Methods Partial Differ. Equ. **28**(6), 1840–1868 (2012)
27. Öffner, P., Glaubitz, J., Ranocha, H.: Stability of correction procedure via reconstruction with summation-by-parts operators for Burgers' equation using a polynomial chaos approach. ESAIM: Math. Model. Numer. Anal. **52**(6), 2215–2245 (2018)
28. Öffner, P., Glaubitz, J., Ranocha, H.: Analysis of artificial dissipation of explicit and implicit time-integration methods. Int. J. Numer. Anal. Model. **17**(3), 332–349 (2020)
29. Persson, P.O., Peraire, J.: Sub-cell shock capturing for discontinuous Galerkin methods. In: 44th AIAA Aerospace Sciences Meeting and Exhibit, p. 112 (2006)
30. Pettersson, M.P., Iaccarino, G., Nordström, J.: Polynomial Chaos Methods for Hyperbolic Partial Differential Equations: Numerical Techniques for Fluid Dynamics Problems in the Presence of Uncertainties. Springer, Berlin (2015)
31. Qiu, J., Shu, C.W.: Runge–Kutta discontinuous Galerkin method using WENO limiters. SIAM J. Sci. Comput. **26**(3), 907–929 (2005)
32. Ranocha, H., Glaubitz, J., Öffner, P., Sonar, T.: Stability of artificial dissipation and modal filtering for flux reconstruction schemes using summation-by-parts operators. Appl. Numer. Math. **128**, 1–23 (2018)
33. Riemann, B.: Über die Fortpflanzung ebener Luftwellen von endlicher Schwingungsweite. Dieterich, Mainz (1860)
34. Scarnati, T., Gelb, A., Platte, R.B.: Using ℓ_1 regularization to improve numerical partial differential equation solvers. J. Sci. Comput. 1–28 (2017)
35. Sonntag, M., Munz, C.D.: Efficient parallelization of a shock capturing for discontinuous Galerkin methods using finite volume sub-cells. J. Sci. Comput. **70**(3), 1262–1289 (2017)
36. Tadmor, E.: Shock capturing by the spectral viscosity method. Comput. Methods Appl. Mech. Eng. **80**(1–3), 197–208 (1990)
37. Tadmor, E., Tassa, T.: On the piecewise smoothness of entropy solutions to scalar conservation laws. Commun. Partial Differ. Equ. **18**(9–10), 1631–1652 (1993)
38. Vogel, C.R.: Computational Methods for Inverse Problems. SIAM, Philadelphia (2002)
39. Von Neumann, J., Richtmyer, R.D.: A method for the numerical calculation of hydrodynamic shocks. J. Appl. Phys. **21**(3), 232–237 (1950)
40. Zanotti, O., Fambri, F., Dumbser, M., Hidalgo, A.: Space–time adaptive ADER discontinuous Galerkin finite element schemes with a posteriori sub-cell finite volume limiting. Comput. Fluids **118**, 204–224 (2015)

Parallel Simulations of High-Power Optical Fiber Amplifiers

Stefan Henneking, Jacob Grosek, and Leszek Demkowicz

1 Introduction

Motivation Optical fiber amplifiers can produce highly coherent laser outputs with great efficiency. The technology has enabled advances in many engineering applications. However, at high-power operation, these fiber laser systems are susceptible to the onset of various nonlinear effects that are adverse to the beam quality of the laser. One particular challenge is mitigating the effects of heating of the silica-glass fiber. Under sufficient heat load, the fiber amplifier experiences a thermally-induced nonlinear effect called the transverse mode instability (TMI) [8, 14]. TMI is characterized by a sudden reduction of the beam coherence above a certain power threshold. Specifically, there is a transition from a stable beam to a chaotic energy transfer between the fiber's guided modes in time. The first two transverse modes of the fiber are depicted in Fig. 1. This instability is a major limitation for the average power scaling of fiber laser systems [14]. While a scientific consensus on the thermal origins of TMI has developed over the past years, finding effective mitigation strategies that do not incite other power limiting nonlinearities remains an active field of research in fiber optics.

Background In the context of studying TMI and other nonlinear effects in fibers, numerical simulations play an important role. A variety of different models are employed in fiber amplifier simulations [9, 17, 22, 24]. These models are typically

S. Henneking (✉) · L. Demkowicz
Oden Institute, The University of Texas at Austin, Austin, TX, USA
e-mail: stefan@oden.utexas.edu; leszek@oden.utexas.edu

J. Grosek
Directed Energy Directorate, Air Force Research Laboratory, Albuquerque, NM, USA
e-mail: AFRL.RDL.OrgBox@us.af.mil

J. M. Melenk et al. (eds.), *Spectral and High Order Methods for Partial Differential Equations ICOSAHOM 2020+1*, Lecture Notes in Computational Science and Engineering 137, https://doi.org/10.1007/978-3-031-20432-6_22

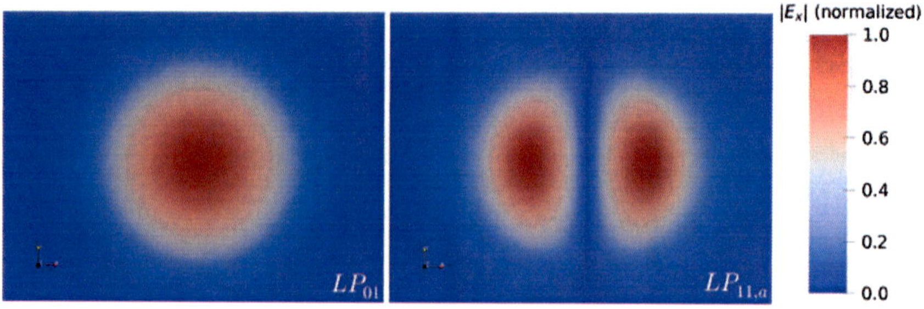

Fig. 1 Linearly polarized (LP) modes in a weakly-guiding step-index fiber. The LP_{01} mode is the fundamental mode, and the LP_{11} mode is the first higher-order mode

derived from the time-harmonic Maxwell equations. By making additional modeling assumptions, these fiber models become easier to discretize and compute than a vectorial Maxwell problem.

Because the laser light in a meter-long optical fiber has millions of wavelengths, it is extremely difficult if not computationally infeasible to resolve the wavelength scale of the propagating light for the full length. For this reason, even simplified models typically resolve a longer length scale. In the context of TMI studies, it is common to resolve only the length scale of the mode beat between the fundamental mode and higher-order modes.

The numerical simulations presented in this work stand out and are novel for two reasons:

1. The computed fiber amplifier model is a three-dimensional vectorial Maxwell model. The model avoids making assumptions typical for other models that may reduce the validity of these models, especially under common practices such as fiber coiling.
2. The large-scale simulations resolve two important length scales of the propagating light in the optical fiber: the wavelength—$\mathcal{O}(\mu\text{m})$, and the mode beat length—$\mathcal{O}(\text{mm})$. As is clear from these length scales, this effort requires resolving at least several thousand wavelengths.

Outline The paper is organized into two parts: first, a brief introduction to the fiber amplifier model, including discussion of the corresponding discontinuous Petrov–Galerkin finite element discretization; and second, a presentation of numerical results, including computational performance tests and large-scale fiber simulations to demonstrate the applicability of the model to study mode coupling phenomena. The paper concludes with a summary of the key contributions.

2 Fiber Amplifier Model

This section briefly summarizes our three-dimensional finite element model for the simulation of laser amplification and heating in an ytterbium-doped, step-index fiber amplifier. As a full vectorial finite element simulation, this model distinguishes itself from other fiber amplifier models that are typically posed as an initial value problem and make significantly more approximations.

2.1 *Ytterbium-Doped Fiber Amplifier*

Figure 2 is a schematic of a final-stage large-mode-area step-index fiber amplifier (not drawn to scale). The typical configuration includes a highly coherent *laser signal*, injected into the fiber core region, and a less coherent *pump field* launched into the inner cladding and core regions simultaneously. The core region has a slightly higher refractive index than the inner cladding region, both made primarily of fused silica, allowing for the signal to be guided along the fiber length in the core by total internal reflection. Active gain fiber amplifiers have doped core regions (e.g., ytterbium dopant). The pump and signal wavelengths are chosen such that the pump field experiences high absorption by the active dopant, while the signal field is in a regime of high emission probability of the active dopant. This way, a large percentage of the pump light can be converted into highly coherent laser signal light by the stimulated emission process. This mechanism is also referred to as *active gain*. Since the pump photons have a higher frequency than the signal photons,

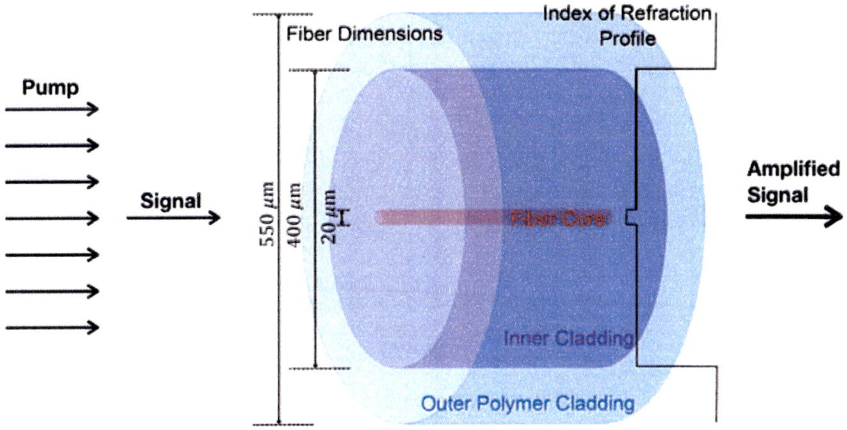

Fig. 2 Schematic of a weakly-guiding, continuous-wave, double-clad, large-mode-area, step-index fiber amplifier

some energy is lost in this process, ultimately leading to heat deposition along the fiber via this *quantum defect*.

Fiber Amplifier Model The signal and pump fields are assumed to each satisfy the time-harmonic Maxwell equations. The two Maxwell systems are coupled by a nonlinear gain function derived from ion rate equations [19], where the total dopant concentration inside the fiber core is assumed to be constant. This nonlinear system is solved by Picard iterations.

The model builds on a 3D fiber amplifier model described in [13, 18] where an artificial longitudinal scaling was employed to simulate the full signal–pump power exchange in a shortened fiber. In this effort, however, the model—now fully parallelized—is tweaked to simulate a small section of a fiber amplifier on the scale of $\mathcal{O}$(mm)–$\mathcal{O}$(cm) with very high resolution without any artificial scaling parameters. The model supports co-pumped, counter-pumped, and bidirectionally-pumped configurations.

The non-dimensionalized time-harmonic Maxwell systems for signal laser and pump field are given by

$$\nabla \times E_k = -i\omega_k H_k \,,$$
$$\nabla \times H_k = in^2\omega_k E_k - n g_0 g_k E_k \,,$$

where $k \in \{s, p\} \equiv \{\text{signal, pump}\}$, i.e. $\{E_s, H_s\}$ are the electric and magnetic field of the signal laser, and $\{E_p, H_p\}$ are the corresponding pump fields; ω_k is the angular frequency of the light, and n is the material refractive index. The gain function $g_k = g_k(I_{\{s,p\}})$ is the nonlinear coupling term of the Maxwell systems and depends on the irradiance I_k of the signal and pump fields given by $I_k = |\mathfrak{Re}\{E_k \times H_k^*\}|$ where $\mathfrak{Re}\{\cdot\}$ denotes the real part of the complex-valued vector and H_k^* is the complex conjugate of H_k. The factor g_0 is a non-dimensional gain coefficient. For a derivation of the nonlinear gain function g_k, we refer to [13].

As has been done in other models, this model couples with the time-dependent heat equation to incorporate thermally-induced effects. The diffusivity of the heat equation determines the time scale of the problem ($\mathcal{O}$(ms)). Heat deposition inside the fiber core of the optical amplifier mainly occurs due to the quantum defect. This heat load is determined through the irradiance of the signal and pump fields [10, 23].

The non-dimensionalized heat equation is given by

$$\frac{\partial(\delta T)}{\partial t} - \alpha_0 \Delta(\delta T) = Q_0 Q,$$

where α_0 and Q_0 are non-dimensional diffusivity and heat load coefficients, respectively, and δT refers to the temperature change inside the fiber. The heat source $Q = Q(I_{\{s,p\}})$ is given by

$$Q(I_{\{s,p\}}) = -\left(g_p(I_{\{s,p\}})I_p + g_s(I_{\{s,p\}})I_s\right).$$

2.2 DPG Finite Element Discretization

The discontinuous Petrov–Galerkin (DPG) method [6] yields a pre-asymptotically stable discretization. The built-in stability properties of the method come at the expense of computing *optimal test functions* [5]. In practice, this requires the use of discontinuous test spaces leading to the introduction of new unknowns on the mesh skeleton. The discretization offers additional advantages: the DPG residual, measured in the test norm, can be computed element-wise and serves as a robust error indicator which can be used for hp-adaptivity [20].

In the large-scale simulations of the fiber model, high-order elements are used with anisotropic polynomial order $(p_x, p_y, p_z) = (6, 6, 7)$ to effectively counter the numerical pollution [2, 11, 15]. The fiber geometry is approximated with isoparametric curvilinear elements: per wavelength, the geometry is discretized with two layers of (1) four prisms and four hexahedra in the fiber core, and (2) twelve hexahedra in the fiber cladding. For these high-order elements, it is essential to use fast integration techniques to accelerate the assembly of the DPG linear system [3, 16].

Let $\Omega := \Omega_t \times (0, L)$ denote the fiber domain, where $\Omega_t := \{(x, y) : x^2 + y^2 < r_{\text{clad}}\}$ is the transverse domain and L is the length of the fiber. Let $\Gamma := \partial\Omega$; define $\Gamma_r := \partial\Omega_t \times (0, L)$, $\Gamma_i := \Omega_t \times \{0\}$, $\Gamma_o := \Omega_t \times \{L\}$ as the radial, input, and output boundaries, respectively. Suppose Ω is partitioned into a set Ω_h of open disjoint elements $\{K\}_{K\in\Omega_h}$ with Lipschitz element boundaries $\{\partial K\}_{K\in\Omega_h}$.

Let $(\mathcal{U}, \hat{\mathcal{U}})$ and $\mathcal{V}$ be the trial and test space, respectively, and let $\mathcal{V}'$ be the space of antilinear functionals on $\mathcal{V}$. The DPG formulations of the Maxwell systems for signal and pump field, as well as the DPG formulation of the heat equation are each defined by a variational formulation of the form: Given $l \in \mathcal{V}'$, find $\mathfrak{u} \in \mathcal{U}$ and $\hat{\mathfrak{u}} \in \hat{\mathcal{U}}$ that satisfy

$$b(\mathfrak{u}, \mathfrak{v}) + \hat{b}(\hat{\mathfrak{u}}, \mathfrak{v}) = l(\mathfrak{v}), \quad \mathfrak{v} \in \mathcal{V}, \tag{1}$$

where b and $\hat{b}$ are sesquilinear forms on $\mathcal{U} \times \mathcal{V}$ and $\hat{\mathcal{U}} \times \mathcal{V}$, respectively.

DPG Ultraweak Maxwell Formulation The broken ultraweak Maxwell formulation [4], given by (1), is defined by the following group variables and forms:

$$\mathfrak{u} = (E_k, H_k) \in L^2(\Omega)^3 \times L^2(\Omega)^3,$$

$$\hat{\mathfrak{u}} = (\hat{E}_k, \hat{H}_k) \in \hat{\mathcal{U}}_k^E \times H^{-\frac{1}{2}}(\text{curl}, \Gamma_h),$$

$$\mathfrak{v} = (F, G) \in H(\text{curl}, \Omega_h) \times H(\text{curl}, \Omega_h),$$

$$\begin{aligned} b_k(\mathfrak{u}, \mathfrak{v}) &= (E_k, \nabla_h \times F) + i\omega_k(H_k, F) + \\ &\quad (H_k, \nabla_h \times G) - i\omega_k(n^2 E_k, G) + (n g_0 g_k(I_{\{s,p\}}) E_k, G), \\ \hat{b}_k(\hat{\mathfrak{u}}, \mathfrak{v}) &= \langle \hat{n} \times \hat{E}_k, F\rangle_{\Gamma_h} + \langle \hat{n} \times \hat{H}_k, G\rangle_{\Gamma_h}, \\ l_k(\mathfrak{v}) &= 0, \end{aligned}$$

where $k \in \{s, p\} \equiv \{\text{signal}, \text{pump}\}$, $\hat{n}$ is the outward unit normal on ∂K, and

$$\hat{\mathcal{U}}_k^E := \{\hat{q} \in H^{-\frac{1}{2}}(\text{curl}, \Gamma_h) : \hat{n} \times \hat{q} = n \times E_{0,k} \text{ on } \Gamma_i,$$
$$\hat{n} \times \hat{q} = 0 \text{ on } \Gamma_r \cup \Gamma_o\}.$$

At the fiber input, a non-homogeneous tangential electric field is prescribed—usually a combination of guided transverse modes. The radial boundary has perfect electrical conductor boundary conditions. At the fiber output, the model employs a perfectly matched layer that serves to absorb the outgoing wave and avoid artificial reflections [1]. The additional unknowns $\hat{E}, \hat{H}$ describe the tangential electromagnetic field on element boundaries on the mesh skeleton Γ_h; they are discretized as the tangential trace of $H(\text{curl})$-conforming elements. The test space is equipped with the adjoint graph norm [4].

DPG Primal Heat Formulation The DPG broken primal formulation of the heat equation with implicit Euler time-stepping [21] is given by (1) with the following definitions:

$$\mathfrak{u} = \delta T \in H^1(\Omega) : \delta T = 0 \text{ on } \Gamma_r,$$
$$\hat{\mathfrak{u}} = \hat{\sigma} \in H^{-\frac{1}{2}}(\Gamma_h) : \hat{\sigma} = 0 \text{ on } \Gamma_i \cup \Gamma_o,$$
$$\mathfrak{v} = v \in H^1(\Omega_h),$$

$$b(\mathfrak{u}, \mathfrak{v}) = (\delta T_{n+1}, v) + \delta t\, \alpha_0 (\nabla(\delta T_{n+1}), \nabla_h v),$$
$$\hat{b}(\hat{\mathfrak{u}}, \mathfrak{v}) = -\delta t\, \alpha_0 \langle \hat{\sigma}_{n+1}, v \rangle_{\Gamma_h},$$
$$l(\mathfrak{v}) = (\delta T_n, v) + \delta t\, Q_0(Q(I_{\{s,p\}}), v),$$

where $n = 0, 1, \ldots, N-1$ with time step $\delta t = t_{\max}/N$, and initial condition $\delta T_0 = 0$. The additional unknown $\hat{\sigma}_{n+1}$ describes the normal heat flux across element boundaries on the mesh skeleton Γ_h; it is discretized as the normal trace of $H(\text{div})$-conforming elements. The test space is equipped with the standard energy norm.

Parallelization The fiber amplifier model is implemented in the *hp*3D finite element (FE) software [7, 12]. The code uses a hybrid MPI/OpenMP parallelization where the computational domain is partitioned into subdomains (one per MPI rank) and each MPI process distributes its subdomain workload via OpenMP threading. The fiber amplifier domain has an elongated cylindrical shape that can be efficiently partitioned by defining subdomains via orthogonal cuts along the longitudinal axis of the fiber. The linear system is solved with a custom nested dissection solver that first eliminates degrees of freedom (DOFs) within each subdomain and then proceeds by recursively partitioning the remaining interface problem into independent subproblems that are solved in parallel. Details of the solver implementation are given in [10, Sect. 3.3].

3 Numerical Results

The numerical simulations were conducted on the *Stampede2* supercomputer at the Texas Advanced Computing Center (TACC). This section presents weak scaling results and large-scale simulations of the nonlinear fiber amplifier model with up to 24,576 cores and 512 compute nodes. All runtime results displayed in this section are based on solving one iteration of the linearized Maxwell problem for the signal laser (the pump solve is equally expensive). In the coupled Maxwell/heat model, the computational complexity is dominated by assembling and solving the Maxwell systems since the heat equation is discretized with much fewer DOFs and only solved once per time step. Therefore, the scaling results presented in this section are representative for the overall scaling of the fiber amplifier model computation.

3.1 Computational Performance

Each of *Stampede2*'s Intel Skylake (SKX) compute nodes has 48 cores. The hybrid MPI/OpenMP computation of the model performs optimally using four MPI ranks with twelve OpenMP threads each per node. Table 1 shows the setup of the weak-scaling tests performed for the Maxwell problem of the fiber model. The weak scaling analyzes inter-node (true distributed-memory) scaling with the computational performance on one compute node serving as the baseline. This baseline solves a fiber model with 32 wavelengths whereas the largest instance, computed on 512 nodes, is a fiber of 16,384 wavelengths. Usually, increasing the fiber length requires increasing the order of discretization to counter the effects of numerical pollution [11]; however, for the weak-scaling test, the workload per processor must be kept constant to obtain valid results thus the polynomial order of

Table 1 Setup of weak-scaling tests for the fiber amplifier model with uniform polynomial order $p = 6$. The smallest problem instance is a computation of 32 wavelengths on a single compute node, whereas the largest instance has 16,384 wavelengths computed on 512 nodes

Wavelengths	Nodes	Ranks	Cores	Solution DOFs
32	1	4	48	1,994,520
64	2	8	96	3,989,016
128	4	16	192	7,978,008
256	8	32	384	15,955,992
512	16	64	768	31,911,960
1024	32	128	1536	63,823,896
2048	64	256	3072	127,647,768
4096	128	512	6144	255,295,512
8192	256	1024	12,288	510,591,000
16,384	512	2048	24,576	1,021,181,976

discretization used for all fiber lengths is the same. The discretization has uniform order $p = 6$ with 20 elements in the fiber cross-section (4 prisms, 16 hexahedra) and two elements per wavelength in the longitudinal direction.

The observed parallel efficiency and runtimes are depicted in Fig. 3. Many parts of the fiber model simulation scale near-linearly, including the computation of locally optimal test functions, distributed matrix assembly, and residual computation. The linear solver consists of two parts: (1) a local subdomain solve eliminating subdomain interior DOFs, and (2) the subsequent coupled interface solve via recursive nested dissection. The first step scales near-linearly since the work is independent for every MPI process, whereas the parallel scaling of the coupled solve declines as the recursive process proceeds with additional levels. The parallel efficiency of the coupled solve (Fig. 3a) is expectedly poor because of the logarithmic increase in computation time (Fig. 3b); however, the total weak scaling efficiency still reaches 60% on 512 compute nodes since the coupled solve makes up only a relatively small portion of the total compute time.

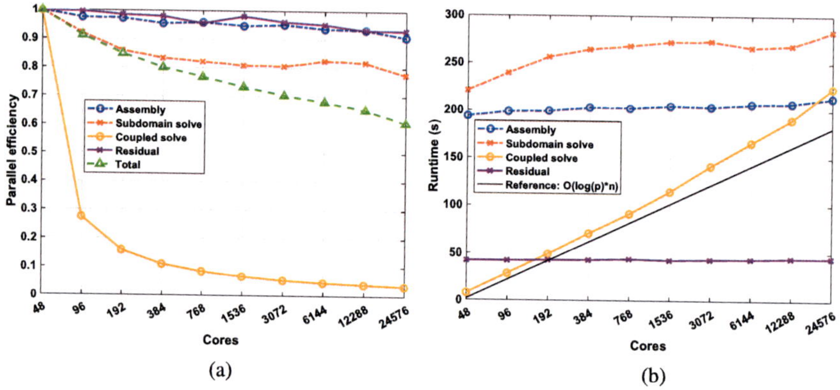

Fig. 3 Stampede2 weak-scaling results for the fiber amplifier model with uniform polynomial order $p = 6$. The implementation achieves about 60% parallel efficiency with 512 compute nodes (24,576 cores) and 2048 MPI processes. (**a**) Parallel efficiency of the main subroutines. (**b**) Runtime of each subroutine

3.2 *Large-Scale Fiber Simulations*

Finally, the model's capability for accurately solving the full mode beat length scale is demonstrated with a nonlinear mode coupling simulation for an ytterbium-doped fiber amplifier. The fiber simulation parameters are given in Table 2.

Table 2 Step-index fiber parameters

Symbol	Description	Value	Unit
r_{core}	Core radius	12.7	μm
r_{clad}	Cladding radius	127	μm
n_{core}	Refractive index in fiber core	1.4512	–
n_{clad}	Refractive index in fiber cladding	1.4500	–
NA	Core numerical aperture	0.059	–
λ_s	Signal wavelength	1064	nm
λ_p	Pump wavelength	976	nm
$V(\omega_s)$	Normalized signal frequency	4.43	–

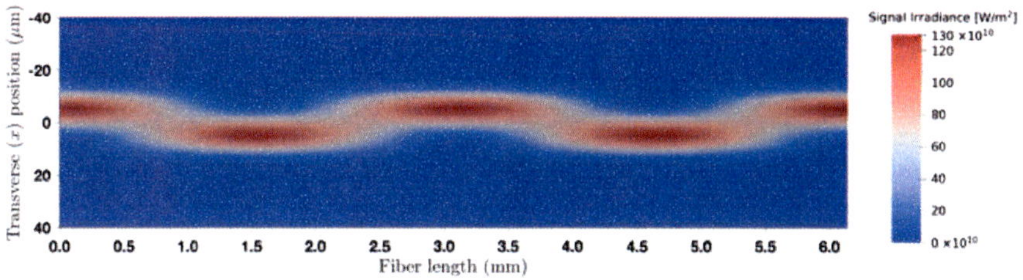

Fig. 4 Modal interference pattern of the LP_{01} and LP_{11} modes in a 6.1 mm long fiber section (approximately 8320 wavelengths). The asymmetric LP_{01}/LP_{11} mode beat has a period of ca. 4221 wavelengths (3.10 mm). The mode beating is illustrated by the periodic irradiance beating plotted in a slice orthogonal to the transverse (y) axis of the fiber

The simulation is a co-pumped configuration with the pump field modeled as a plane wave inside the fiber core. To illustrate the thermally-induced mode coupling, following a similar example in [14], the signal input is comprised of 80% LP_{01} mode and 20% LP_{11} mode. First, the simultaneous propagation of the transverse guided modes results in a modal interference pattern in the irradiance of the signal light, as depicted in Fig. 4.

Figure 5 illustrates how the modal interference pattern translates into a thermally-induced periodic grating of the material refractive index. In particular, when the higher-order mode is radially asymmetric such as the LP_{01} mode, then the corresponding index perturbation also becomes asymmetric (cf. Fig. 5c). This asymmetric periodic refractive index grating along the fiber longitudinal axis effectively couples the propagating transverse guided modes. The mode coupling results in an energy transfer between the fundamental mode (LP_{01}) and the higher-order mode (LP_{11}), as shown in Fig. 6.

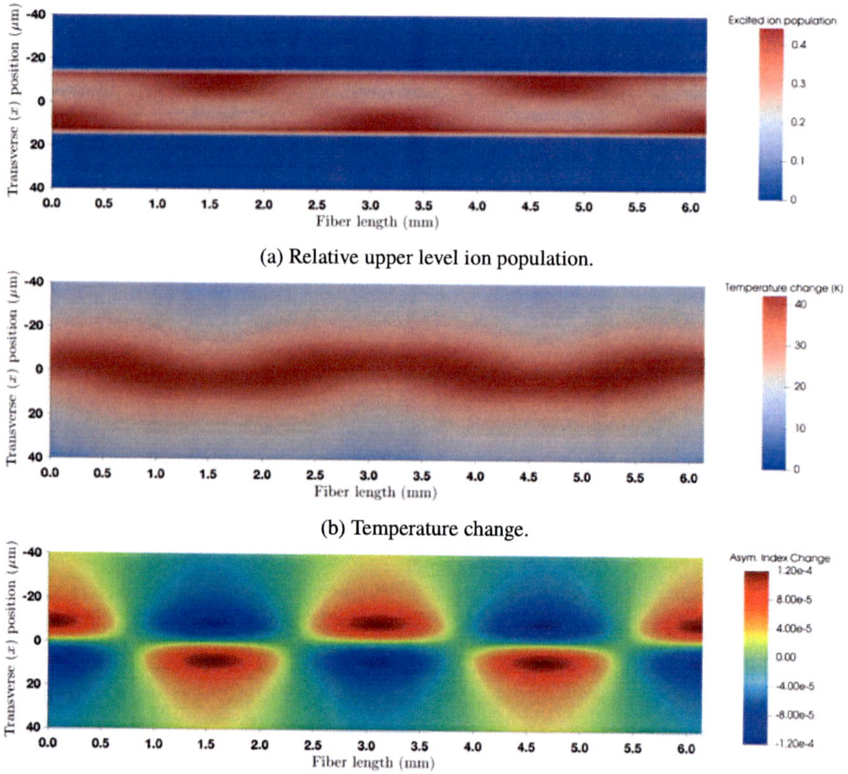

(a) Relative upper level ion population.

(b) Temperature change.

(c) Radially anti-symmetric part ofindex perturbation.

Fig. 5 Mode beating translates into a periodic asymmetric refractive index grating: (**a**) first, the beating pattern develops in the upper level ion population inside the fiber core; (**b**) then, through the quantum defect the beating occurs in the heat deposition and thus in the temperature change; and (**c**) finally, the beating pattern translates into a periodic, radially asymmetric refractive index perturbation.

4 Conclusions

This contribution presented large-scale numerical simulations of a 3D vectorial time-harmonic Maxwell model for laser amplification in optical fiber amplifiers. The model's capabilities were shown to extend to resolving both the wavelength scale and the mode beat length scale in multi-mode wave propagation. To our knowledge, this is the only vectorial Maxwell finite element implementation capable of solving optical waveguide problems with more than 10,000 wavelengths at full resolution. The numerical simulations are both computationally efficient (weak scaling) and suitable for studying short fiber sections in great detail. We believe

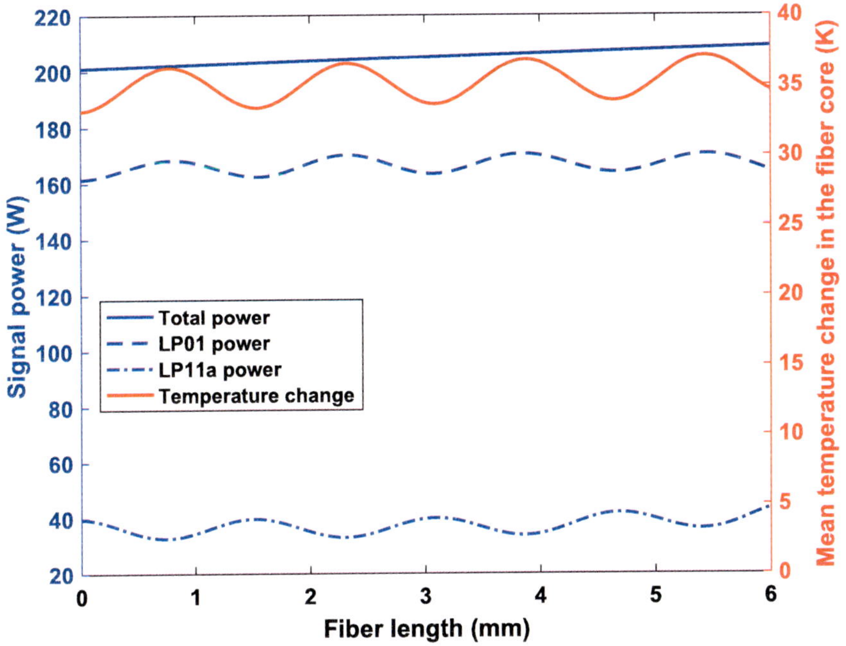

Fig. 6 Optical power and temperature change in a 6.1 mm long fiber section (approximately 8320 wavelengths). The simulation time is 7.0 ms, computed with $\delta t = 0.1$ ms (70 time steps). A grated temperature profile develops, inducing coupling between the propagating transverse guided modes

these high-fidelity simulations effectively complement widely used computationally cheaper, lower-fidelity models.

Currently, the model is limited to simulations of straight fiber configurations. Due to the practical importance of coiling the fiber around a cooling device, future work will attempt to extend the model to account for the effects of fiber bending.

Acknowledgments This work was supported by AFOSR grant no. FA9550-19-1-0237. Distribution Statement A: Approved for public release; distribution unlimited. Public Affairs release #AFRL-2021-2938.

References

1. Astaneh, A., Keith, B., Demkowicz, L.: On perfectly matched layers for discontinuous Petrov–Galerkin methods. Comput. Mech. **63**(6), 1131–1145 (2019)
2. Babuška, I., Sauter, S.: Is the pollution effect of the FEM avoidable for the Helmholtz equation considering high wave numbers? SIAM J. Numer. Anal. **34**(6), 2392–2423 (1997)

3. Badger, J., Henneking, S., Demkowicz, L.: Sum factorization for fast integration of DPG matrices on prismatic elements. Finite Elem. Anal. Des. **172**, 103385 (2020)
4. Carstensen, C., Demkowicz, L., Gopalakrishnan, J.: Breaking spaces and forms for the DPG method and applications including Maxwell equations. Comput. Math. Appl. **72**(3), 494–522 (2016)
5. Demkowicz, L., Gopalakrishnan, J.: A class of discontinuous Petrov–Galerkin methods. II: optimal test functions. Numer. Methods Partial Differ. Equ. **27**(1), 70–105 (2011)
6. Demkowicz, L., Gopalakrishnan, J.: Discontinuous Petrov–Galerkin (DPG) method. Encyclopedia of Computational Mechanics, Wiley Online Library, 2nd edn, pp. 1–15 (2017)
7. Demkowicz, L., Kurtz, J., Pardo, D., Paszyński, M., Rachowicz, W., Zdunek, A.: Computing with hp Finite Elements. II. Frontiers: Three Dimensional Elliptic and Maxwell Problems with Applications. Chapman & Hall/CRC, Boca Raton (2007)
8. Eidam, T., Wirth, C., Jauregui, C., Stutzki, F., Jansen, F., Otto, H., Schmidt, O., Schreiber, T., Limpert, J., Tünnermann, A.: Experimental observations of the threshold-like onset of mode instabilities in high power fiber amplifiers. Opt. Express **19**(14), 13218–13224 (2011)
9. Goswami, T., Grosek, J., Gopalakrishnan, J.: Simulations of single-and two-tone Tm-doped optical fiber laser amplifiers. Opt. Express **29**(8), 12599–12615 (2021)
10. Henneking, S.: A scalable hp-adaptive finite element software with applications in fiber optics. Ph.D. Thesis, The University of Texas at Austin (2021)
11. Henneking, S., Demkowicz, L.: A numerical study of the pollution error and DPG adaptivity for long waveguide simulations. Comput. Math. Appl. **95**, 85–100 (2021)
12. Henneking, S., Demkowicz, L.: hp3D User Manual. arXiv preprint arXiv:2207.12211 (2022)
13. Henneking, S., Grosek, J., Demkowicz, L.: Model and computational advancements to full vectorial Maxwell model for studying fiber amplifiers. Comput. Math. Appl. **85**, 30–41 (2021)
14. Jauregui, C., Stihler, C., Limpert, J.: Transverse mode instability. Adv. Opt. Photonics **12**(2), 429–484 (2020)
15. Melenk, J., Sauter, S.: Wavenumber-explicit hp-FEM analysis for Maxwell's equations with transparent boundary conditions. Found. Comput. Math. 1–117 (2020)
16. Mora, J., Demkowicz, L.: Fast integration of DPG matrices based on sum factorization for all the energy spaces. Comput. Methods Appl. Math. **19**(3), 523–555 (2019)
17. Naderi, S., Dajani, I., Madden, T., Robin, C.: Investigations of modal instabilities in fiber amplifiers through detailed numerical simulations. Opt. Express **21**(13), 16111–16129 (2013)
18. Nagaraj, S., Grosek, J., Petrides, S., Demkowicz, L., Mora, J.: A 3D DPG Maxwell approach to nonlinear Raman gain in fiber laser amplifiers. J. Comput. Phys. **2**, 100002 (2019)
19. Pask, H., Carman, R., Hanna, D., Tropper, A., Mackechnie, C., Barber, P., Dawes, J.: Ytterbium-doped silica fiber lasers: versatile sources for the 1-1.2 μm region. IEEE J. Sel. Top. Quantum Electron. **1**(1), 2–13 (1995)
20. Petrides, S., Demkowicz, L.: An adaptive multigrid solver for DPG methods with applications in linear acoustics and electromagnetics. Comput. Math. Appl. **87**, 12–26 (2021)
21. Roberts, N., Henneking, S.: Time-stepping DPG formulations for the heat equation. Comput. Math. Appl. **95**, 242–255 (2021)
22. Saitoh, K., Koshiba, M.: Full-vectorial finite element beam propagation method with perfectly matched layers for anisotropic optical waveguides. J. Light. Technol. **19**(3), 405–413 (2001)
23. Smith, A., Smith, J.: Mode instability thresholds for Tm-doped fiber amplifiers pumped at 790 nm. Opt. Express **24**(2), 975–992 (2016)
24. Ward, B.: Modeling of transient modal instability in fiber amplifiers. Opt. Express **21**(10), 12053–12067 (2013)

Reduced Order Modeling for Spectral Element Methods: Current Developments in Nektar++ and Further Perspectives

Martin W. Hess, Andrea Lario, Gianmarco Mengaldo, and Gianluigi Rozza

1 Introduction

The spectral element method (SEM) [12] is a discretization technique for partial differential equations (PDEs) that uses high-order polynomials on a tessellation of elements constituting the computational domain. Unlike other numerical discretizations, such as traditional finite element, finite difference and finite volume methods, SEM provides exponential convergence properties to the PDE solution and allows for tunable numerical properties, such as tunable diffusion and dispersion errors. This flexibility has been extensively leveraged to construct high-fidelity simulation capabilities [15] in the context of both continuous Galerkin (CG) [16, 19], and discontinuous Galerkin (DG) formulations [5, 13, 14, 14, 17, 18, 20, 21, 28]. The former (CG) constructs a numerical discretization enforcing the solution to be continuous between elements. The latter (DG) enforces the numerical fluxes (and not the solution) to be continuous between elements. The use of SEM for high-fidelity simulations provides a pathway to next generation computational tools for engineering analysis and design, in fluid dynamics and other sectors. This offers the opportunity to devise improved reduced order models that may be used for real-time design and control purposes.

In this paper, with present recent efforts that were undertaken to develop reduced order modeling (ROM) [4, 7, 8, 10, 26] approaches within SEM, both for CG and DG methods. These efforts were conveyed in the implementation of ROM into the

M. W. Hess (✉) · A. Lario · G. Rozza
SISSA mathLab, International School for Advanced Studies, Trieste, Italy
e-mail: mhess@sissa.it; alario@sissa.it; grozza@sissa.it

G. Mengaldo
National University of Singapore (NUS), Singapore, Singapore
e-mail: mpegim@nus.edu.sg

J. M. Melenk et al. (eds.), *Spectral and High Order Methods for Partial Differential Equations ICOSAHOM 2020+1*, Lecture Notes in Computational Science and Engineering 137, https://doi.org/10.1007/978-3-031-20432-6_23

spectral/hp element library Nektar++[1] [3, 22]. In particular, we show the global assembly of the PDE solution and how the model reduction relates to the full-order solver for both CG and DG. The different solver modules of Nektar++ share basic functionalities such as the geometry description and quadrature rules, among others. This also holds true for the Nektar++ incompressible and compressible solver modules. The former uses a CG discretization, while the latter uses DG. As a consequence, the model reduction codes for incompressible and compressible flow simulations are independent from each other. The implementation of ROM methods in Nektar++ is available in the open-source model reduction software ITHACA-SEM.[2]

The open-source software ITHACA-SEM currently has the capability to generate POD-based ROMs for 2D incompressible Navier-Stokes equations with parametric variation in geometry and/or kinematic viscosity. The parametric dependency on geometry parameters is assumed to be affine and the user can specify the affine form in a header file. After another compilation of ITHACA-SEM, the affine form is then available. It can thus serve as a head start for a developer seeking to work in ROMs with the SEM and also to a practitioner within the boundaries mentioned. ITHACA-SEM is delivered and compiled with the Nektar++ master branch, which is periodically merged into the code. A few test cases are part of the Nektar++ unit test and additional examples are available.

The paper is organized as follows. In Sect. 2, we present the CG method and the ROM approach implemented. In Sect. 3, we detail the DG method and the associated ROM method implemented. In Sect. 4, we draw some conclusions and future perspectives.

2 Continuous Galerkin: Incompressible Flow Simulations

2.1 Overview

Let Ω denote the spatial computational domain. Incompressible, viscous fluid motion in the domain Ω over a time interval $(0, T)$ is governed by the incompressible *Navier-Stokes* equations

$$\frac{\partial u}{\partial t} + u \cdot \nabla u = -\nabla p + \nu \Delta u + f, \tag{1}$$

$$\nabla \cdot u = 0. \tag{2}$$

[1] www.nektar.info.

[2] https://mathlab.sissa.it/ITHACA-SEM.

where u is the fluid velocity, p is the pressure, ν is the kinematic viscosity, and f is a body forcing term. The following boundary and initial conditions

$$u = d \quad \text{on } \Gamma_D \times (0, T), \tag{3}$$

$$\nabla u \cdot n = g \quad \text{on } \Gamma_N \times (0, T), \tag{4}$$

$$u = u_0 \quad \text{in } \Omega \times 0, \tag{5}$$

fully define the problem (1)–(2), with d, g and u_0 given and $\partial\Omega = \Gamma_D \cup \Gamma_N$, $\Gamma_D \cap \Gamma_N = \emptyset$. The *Reynolds* number Re depends on the kinematic viscosity ν through the characteristic velocity U and characteristic length L as

$$Re = \frac{UL}{\nu}. \tag{6}$$

As an example of the inner workings of Nektar++, the computation of steady states is explained in detail, which will used for the model reduction in the next section. The steady states are solutions where $\frac{\partial u}{\partial t} = 0$ holds.

For the ROM in the next section, a parametric variation of the viscosity ν will be assumed. This corresponds to a variable Reynolds number via the relation (6) and allows to compute the flow for various Reynolds numbers. To compute steady states for varying viscosity ν, a solution $u(\nu_0)$ for a parameter value ν_0 is used as an initial guess within a fixed point iteration to obtain the steady state solution $u(\nu_1)$ at a parameter value ν_1. This is repeated for ν_2 and so on. In this way, the parameter range of interest can be explored iteratively with a fixed point iteration.

The *Oseen*-iteration is a secant modulus fixed-point iteration with a linear rate of convergence. Given the current iterate (or initial condition) u^k, the linear system

$$-\nu\Delta u + (u^k \cdot \nabla)u + \nabla p = f \text{ in } \Omega, \tag{7}$$

$$\nabla \cdot u = 0 \text{ in } \Omega, \tag{8}$$

$$u = d \quad \text{on } \Gamma_D, \tag{9}$$

$$\nabla u \cdot n = g \quad \text{on } \Gamma_N, \tag{10}$$

is solved for the next iterate $u^{k+1} = u$. A usual stopping criterion is that the relative change between iterates in the L^2 or H^1 norm falls below a given tolerance. The initial solution $u^0(\nu_0)$ is computed by time-advancement of (1)–(2) from zero initial conditions at a parameter value ν_0. From this starting point, solutions on the whole parameter domain can be found.

The discretized system solved in each step of the *Oseen*-iteration is decomposed as (11) as

$$\begin{bmatrix} A & -D_{bnd}^T & B \\ -D_{bnd} & 0 & -D_{int} \\ \tilde{B}^T & -D_{int}^T & C \end{bmatrix} \begin{bmatrix} v_{bnd} \\ p \\ v_{int} \end{bmatrix} = \begin{bmatrix} f_{bnd} \\ 0 \\ f_{int} \end{bmatrix} \tag{11}$$

where v_{bnd} and v_{int} denote velocity degrees of freedom on the boundary and in the interior, respectively. The forcing terms f_{bnd} and f_{int} refer to the boundary and interior, respectively. The matrix A assembles the boundary-boundary terms, B the boundary-interior terms, $\tilde{B}$ the interior-boundary terms and C assembles the interior-interior terms of elemental velocity *ansatz* functions. In a *Stokes* system, it holds that $B = \tilde{B}^T$, but this is not the case in the *Oseen* equation, since the linearization term $(u^k \cdot \nabla)u$ is present in (7). The matrices D_{bnd} and D_{int} provide the pressure-velocity boundary and pressure-velocity interior couplings, respectively.

The linear system (11) is assembled in local degrees of freedom, leading to block matrices A, B, $\tilde{B}$, C, D_{bnd} and D_{int}, with each block referring to one spectral element. Thus, the system is singular in this form. To solve the system, the local degrees of freedom can be gathered into the global degrees of freedom, but here a multi-level static condensation is employed. See also the documentation and source code of *Nektar++*.

2.2 *Reduced Order Modeling for Continuous Galerkin Methods*

The reduced order model (ROM) needs to approximate the full order solutions accurately over the parameter domain of interest. The reduced basis (RB) model reduction uses a projection onto a low order space of snapshot solutions (i.e., full order solutions) and an offline-online decomposition to facilitate computational efficiency. A set of snapshots is computed over a coarse sample of the parameter domain and used to define a projection space U of size N using the standard solver provided by Nektar++. The POD computes a singular value decomposition (SVD) of the snapshot solutions to 99.9% of the most dominant POD modes. This defines a projection matrix $U \in \mathbb{R}^{N_\delta \times N}$. The software package *Eigen* is used in ITHACA-SEM to compute the SVD.

The offline-online decomposition allows for fast input-output solves, because they are independent of the original model size N_δ. It is an important part of efficient reduced order modeling, but since the static condensation includes the inversion of the parameter-dependent matrix C, the projection is applied to the system (11). Alternatively, also some degrees of freedom can be gathered, see [7]. In the offline phase, snapshot solutions have been gathered over the parameter domain, which now serve as a projection space to define the reduced order setting. To have fast reduced order solves, the offline-online decomposition expands (11) in the parameter of interest and stores the parameter independent projections as small-sized matrices of the order $N \times N$. Since during the *Oseen*-iteration each matrix is dependent on the previous iterate, the submatrices corresponding to each basis function are assembled and then formed online with the reduced basis coordinate representation of the current iterate. This is analogous to reduced order assembly of the nonlinear term in the *Navier-Stokes* equations. For more details and applications, see [9].

Particular care must be taken regarding the different levels of accuracy of a function within Nektar++. There exist three levels, the global degrees of freedom, the local degrees of freedom, where degrees of freedom on the boundaries between spectral elements are present multiple times and the physical degrees of freedom, which correspond to quadrature points. The nonlinear terms need to be evaluated on the level of physical degrees of freedom for the increased accuracy but the projection takes place in the local degrees of freedom. This correspondingly requires two separate sets of projection bases.

2.2.1 A Numerical Example

Consider a variable kinematic viscosity in the interval $\nu \in [0.15; 10]$ for a channel flow as depicted in Figs. 1 and 2. A parabolic inflow profile is prescribed on the left wall at $y = 0$, a natural outflow boundary is prescribed at $y = 8$ and the other walls are no-slip boundaries. The polynomial order of the ansatz functions is chosen as 11. The relative error between full order solutions upon increasing the polynomial order is about 0.05%. The Figs. 1 and 2 show the extreme cases considered here for a very small viscosity and very large viscosity.

There is a bifurcation occurring nearby for some $\nu < \nu^{\min} = 0.15$ as investigated in [23]. Since the bifurcation point represents a singularity, the convergence speed

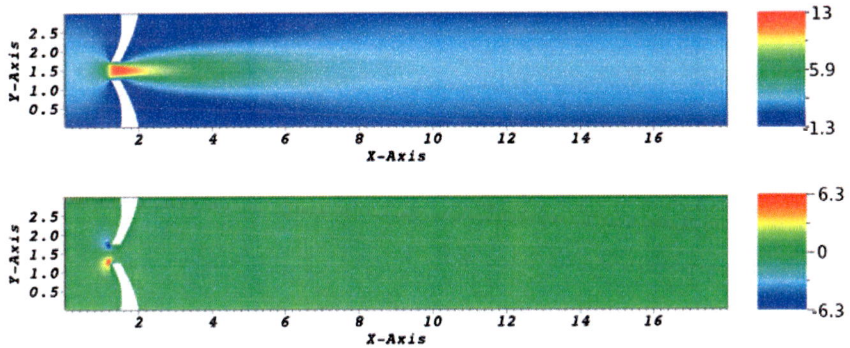

Fig. 1 Full order, steady-state solution for $\nu = 0.15$: velocity in x-direction (top) and y-direction (bottom)

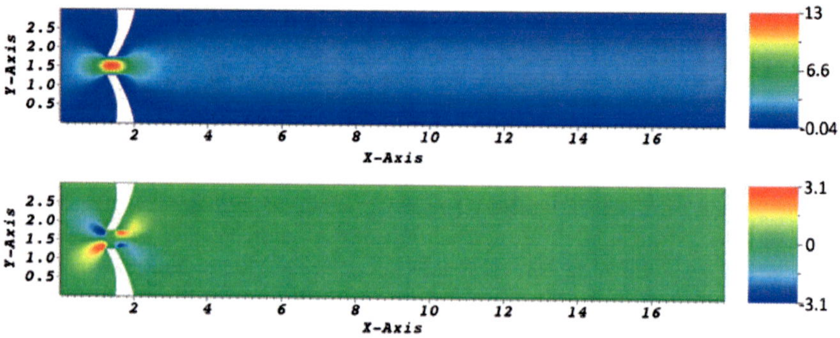

Fig. 2 Full order, steady-state solution for $\nu = 10$: velocity in x-direction (top) and y-direction (bottom)

of the fixed-point iteration is very low close to the bifurcation. The tolerance for the fixed-point iteration has thus been set 1e−4 for a change among two iterates.

Sample solutions are computed at 22 values of ν in the interval of interest [0.15; 10]. The POD is computed and the POD energy reaches a threshold of 99% with 2 modes and a threshold of 99.99% with 6 modes. Ideally, the exponential decay in POD energy should translate into an exponential decay in relative approximation error. Figure 3 shows the mean and maximum relative $L^2(\Omega)$ error in the velocity as an increasing number of POD modes is considered. A mean relative error of 1% is reached with 5 basis functions and a maximum relative error of 1% is reached with 6 basis functions.

The error plateaus at about 1e−4, which is because the fixed-point iteration is not more accurate than that. When the problem is such that the fixed-point iteration can work with a higher accuracy, then this will also be recovered in the ROM. The offline time took 46s and the online evaluation at all parameter values and over all basis sizes took 1s on a workstation with an i7-6700.

2.2.2 Parametric Variation in Geometry

Numerical examples where a parametrized geometry is considered can be found in [8] and [9]. In case of an affine parameter dependency, this parameter dependency can be made explicit in the system matrix as

$$A(\mu)x = \left(\sum_{i=1}^{Q} \Theta_i(\mu) A_i \right) x = b, \tag{12}$$

given a parameter vector μ with scalar functions Θ_i and parameter independent matrices A_i. The matrix A corresponds to the matrix (11) in the full order assembly. The precise transformations to arrive at the form (12) can be found in

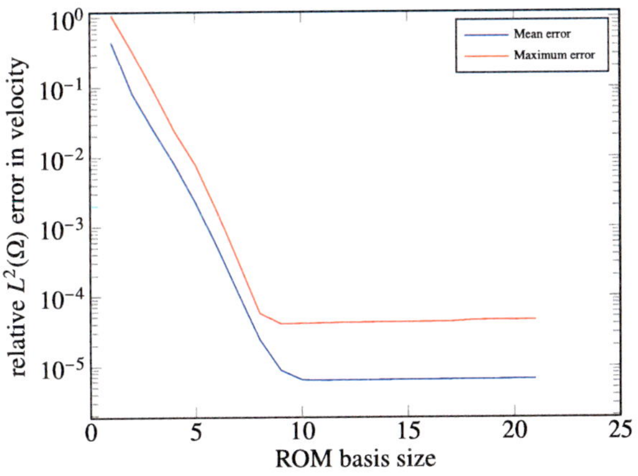

Fig. 3 Mean and maximum relative $L^2(\Omega)$ error in the velocity with increasing basis size

[24]. ITHACA-SEM allows the user to specify the form (12) in a header file which is then used to achieve the offline-online decomposition by exploiting

$$V^T A(\mu) V x_r = \left(\sum_{i=1}^{Q} \Theta_i(\mu) V^T A_i V \right) x_r = V^T b, \tag{13}$$

given a projection matrix V. The terms $V^T A_i V$ are computed in the offline phase and are then available online as small scale matrices, which form the reduced order system by assembling the sum for a parameter of interest μ. The approximation to the full-order solution x is then given by

$$x \approx V x_r. \tag{14}$$

In case of a parametric variation in geometry which is non-affine, the *empirical interpolation method* (EIM) is usually used, see [6]. The EIM approximates an affine form, such as (12). To do that, a few degrees of freedom are identified, which allow to recover the whole system matrix in an affine form. Its speed-up depends on the ability to quickly compute a few entries of the system matrix. This is typically the case if the *ansatz* functions have a small support such as in a finite element method (FEM). But in the SEM the *ansatz* functions have a much larger support, such that this speed-up is not possible and the applicability of the EIM is limited. This was investigated in [8].

3 Discontinuous Galerkin: Compressible Flow Simulations

3.1 Overview

The DG-based solver implemented in Nektar++ is used for the prediction of compressible flows in which density variations are not negligible. This implies that the zero-divergence condition of the velocity fields no longer holds; the resulting set of equations follows:

$$\frac{\partial \rho}{\partial t} + \nabla \cdot (\rho u) = 0 \tag{15}$$
$$\frac{\partial u}{\partial t} + \nabla(\rho u \times u) = -\nabla p + \nabla \cdot \sigma + \rho f,$$
$$\frac{\partial E}{\partial t} + \nabla \cdot (uE) + \nabla \cdot (up) = \nabla \cdot (u \cdot \sigma) + k\nabla T \quad ,$$

where ρ is the density, E is the total energy, and T the temperature. Moreover σ indicates the viscous dissipation tensor:

$$\sigma = 2\nu\rho\left(S_{ij} + \frac{1}{3}\nabla \cdot u\delta_{ij}\right) \quad , \tag{16}$$

being the strain tensor S_{ij} defined as the symmetric part of the gradient of the velocity u. To close the problem, proper initial and boundary conditions must be provided:

$$U_i = d \quad \text{on } \Gamma_D \times (0, T), \tag{17}$$
$$\nabla U_i \cdot n = g \quad \text{on } \Gamma_N \times (0, T),$$
$$U_i = U_{i0} \quad \text{in } \Omega \times 0,$$

being $U = U_1, U_2, U_3 = \rho, u, T$.

In detail the DG scheme is a high order method that, differently from the continuous Galerkin approach, does not impose the continuity of the solution between contiguous elements. In order to keep the consistency of the solution, proper flux must be exchanged and in general flux splitting schemes are devised by the ones developed in the Finite Volume Method framework. In detail, in this work the HLLC flux splitting scheme was used. The HLLC differs from the more common HLL scheme because it is able to reconstruct the contact surface of a Riemann problem. For further details, one can refer to [29]. Time integration was performed with an explicit fourth order Runge-Kutta marching scheme, chosen because of its low dissipative properties. A more comprehensive description of the full order method implemented in Nektar++ can be found in [3].

3.2 *Reduced Order Modeling for Discontinuous Galerkin Methods*

As in the previous case, the ROM for compressible flows is structured according to an online/offline paradigm. During the offline phase, once that N snapshots have been generated with the aid of the full order solver implemented in Nektar++ and gathered in a database, the dominant dynamics are extracted through a SVD procedure. The computational speed up of the ROMs relies on the fact that only the first $N_m < N$ most energetic modes are retained when the building of the bases is carried on, being N the number of degree of freedom of the full order solver. The Eq. (16) are linearized and the operators are projected on the reduced basis once that a suitable inner product has been identified. ROMs for compressible flows are challenging since an energy-based inner product is not defined and the evolution of ROMs does not satisfy the energy equation. In this work a stabilization matrix inspired by the symmetry matrix introduced by Barone et al. [2] is used for recovering part of the stability of the reduced order system.

The introduction of the stabilization matrix aims to improve, beyond the stability, also the controllability of the system, e.g. moving the unstable (positive) eigenvectors closer to zero. This has important consequences when Eigenvalues Replacement (ER) techniques are employed for stabilizing the system since smaller spaces have to be explored. The first method based on Eigenvalues Replacement can be found in Balajewicz et al. [1], who propose to achieve the stability of the reduced system through a minimal subspace rotation, thus by introducing viscosity. In this work the ER is based on a Swarm Particle Optimization in order to find the combination of eigenvalues which minimize the distance between the predicted coefficients a_{ROM} and the ones a_{FOM} obtained by projecting the snapshots onto the reduced basis [25]. Moreover one requires that the total power of the system W decreases with time. Hence:

$$min_{\lambda} \sum_{i}^{N_s} ||a^k_{ROM} - a^k_{FOM}||^2_2 \quad s.t. \quad W(t) < 0 \quad , \tag{18}$$

being N_s the number of snapshots and the total power computed as $E \approx a^2_{ROM}$. To include the constraints related to the total power, Eq. (18) can be rewritten as follows:

$$\min_{\lambda} \sum_{i}^{N_s} ||a^k_{ROM} - a^k_{FOM}||^2_2 + c1(\alpha + c_2) \quad , \tag{19}$$

where c_1 and $c2$ are constants, α is the coefficient of the linear regression which gives the best fit of $W(t) = \alpha t + \beta$. In general c_1 is a constant which represent a penalization term and it has to be determined each time, while c_2 is a small positive number, that in the present work is evaluated as:

$$c_2 = \max(10^{-5}, \alpha_{FOM}) \quad , \tag{20}$$

where α_{FOM} is the angular coefficient of the linear regression which best approximates the total power computed starting from the coefficients obtained from the full order snapshots. The time integration is performed with an explicit fourth order Runge-Kutta scheme in order to be consistent with the full order solver.

3.3 Results

A NACA 0012 airfoil exposed to a flow field with an angle of attack of 5° is considered as a test case for applying the reduction strategy described in the previous sections. The unperturbed flow field is characterized by a Mach number of 0.5 and standard sea level air conditions are used for initializing the flow variables. The full order snapshots have been obtained using a two-dimensional computational grid whose main dimensions were 40 times the airfoil chord length and it was composed by approximately 4000 discrete elements. Moreover, a polynomial order P equal to four was used for the simulation. Figure 4 shows the distributions of pressure coefficients along the x-axis obtained by using two distinct polynomial

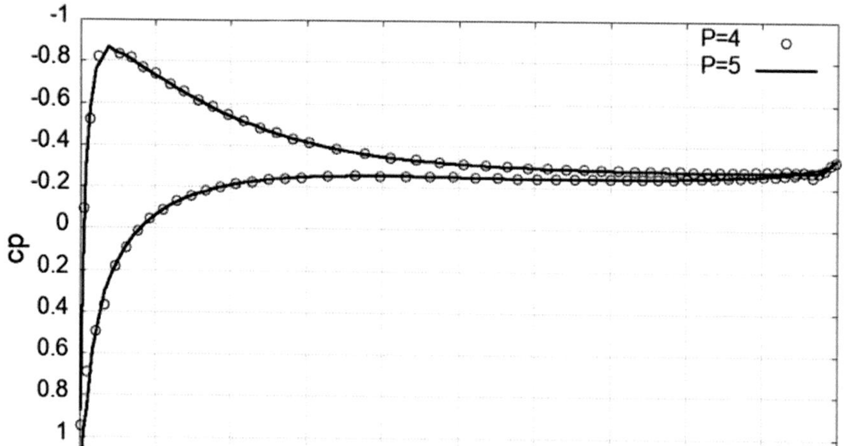

Fig. 4 Time-averaged pressure coefficients over the airfoil obtained with two different polynomial orders: $P = 4$ (empty dots), $P = 5$ (solid line)

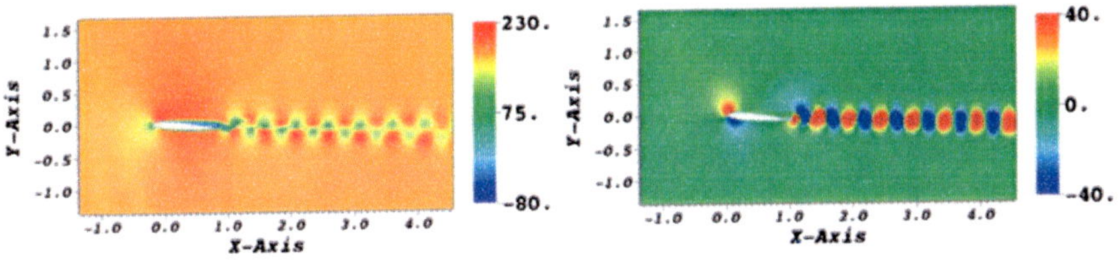

Fig. 5 Flow field for Re = 7500: velocity along the x-axis (left) and y-axis (right)

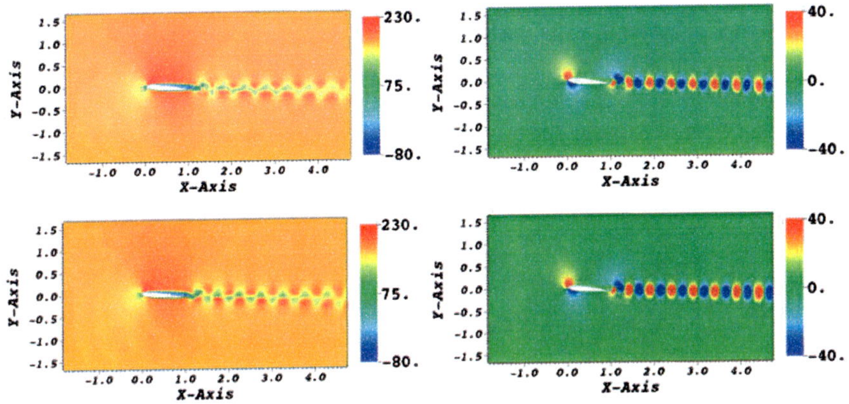

Fig. 6 Velocity along the x-axis (left) and y-axis (right): comparison between a simple stabilization strategy (top) and one reinforced with an eigenvalue replacement (bottom)

orders (four and five); the two solutions do not differ significantly, thus indicating the convergence of the numerical model for $P = 4$. The resulting distribution of velocity components along the x-axis and y-axis obtained with the full order solver for a Reynolds number equal to 7500 is shown in Fig. 5.

For each flow realization, 60 snapshots were saved and 8 modes were generated. Two distinct reconstruction strategies were adopted: in the first case the linearized governing equations were projected on the reduced basis and the reduced matrices were computed; in the second case the eigenvalues of these matrices were further optimized with the aid of the PSO-strategy previously described. Reconstructed flow fields for Re = 7500 are reported in Fig. 6; as one can observe the additional step contribute to enhance the stability and this can be seen in particular for the u component of the velocity vector. The computational effort required by the online phase is approximately 28 seconds, on the same machine the FOM solution requires approximately 5 hours, given the small time step needed to guarantee the stability of the solution.

4 Conclusions

In this paper, we presented the first steps towards intrusive ROM capabilities for SEM. In particular, we highlighted the implementation of offline-online ROM decomposition strategies for both CG and DG, in the context of incompressible and compressible flow problems, respectively. These steps constitutes the basis for further ROM developments for SEM, that is of particular interest for high-fidelity flow simulations in complex geometries, such as explored for finite volume methods in [11, 27]. In detail next steps will be focused on the development of non-intrusive Reduced Order Models based on Machine Learning techniques and on the stabilization on non-linear compressible ROMs. Indeed, non-intrusive methods present some advantages with respect to intrusive ones, among the others they are naturally stable and more user friendly to deal with.

Acknowledgments We acknowledge the support by European Union Funding for Research and Innovation - Horizon 2020 Program - in the framework of European Research Council Executive Agency: Consolidator Grant H2020 ERC CoG 2015 AROMA-CFD project 681447 "Advanced Reduced Order Methods with Applications in Computational Fluid Dynamics" (P.I.Prof. Gianluigi Rozza). This work was partially supported by MIUR (Italian ministry for university and research) through FARE-X-AROMA-CFD project, P.I. Prof. Gianluigi Rozza. Gianmarco Mengaldo acknowledges support from NUS startup grant R-265-000-A36-133.

References

1. Balajewicz, M., Tezaur, I., Dowell, E.: Minimal subspace rotation on the Stiefel manifold for stabilization and enhancement of projection-based reduced order models for the compressible Navier–Stokes equations. J. Comput. Phys. **321**, 224–241 (2016)
2. Barone, M.F., Kalashnikova, I., Segalman, D.J., Thornquist, H.K.: Stable Galerkin reduced order models for linearized compressible flow. J. Comput. Phys. **228**(6), 1932–1946 (2009)
3. Cantwell, C.D., Moxey, D., Comerford, A., Bolis, A., Rocco, G., Mengaldo, G., De Grazia, D., Yakovlev, S., Lombard, J.E., Ekelschot, D., et al.: Nektar++: An open-source spectral/hp element framework. Comput. Phys. Commun. **192**, 205–219 (2015)
4. Deville, M.O., Fischer, P.F., Mund, E.H.: High-Order Methods for Incompressible Fluid Flow. Cambridge Monographs on Applied and Computational Mathematics. Cambridge University Press, Cambridge (2002). https://doi.org/10.1017/CBO9780511546792
5. Fernandez, P., Moura, R.C., Mengaldo, G., Peraire, J.: Non-modal analysis of spectral element methods: towards accurate and robust large-eddy simulations. Comput. Methods Appl. Mech. Eng. **346**, 43–62 (2019)
6. Grepl, M.A., Maday, Y., Nguyen, N.C., Patera, A.T.: Efficient reduced-basis treatment of nonaffine and nonlinear partial differential equations. ESAIM: Math. Model. Numer. Anal. **41**(3), 575–605 (2007). https://doi.org/10.1051/m2an:2007031
7. Hess, M.W., Rozza, G.: A spectral element reduced basis method in parametric CFD. In: Numerical Mathematics and Advanced Applications—ENUMATH 2017. Lecture Notes in Computational Science and Engineering, vol. 126, pp. 693–701. Springer, Cham (2019). https://doi.org/10.1007/978-3-319-96415-7_64

8. Hess, M.W., Quaini, A., Rozza, G.: Reduced basis model order reduction for Navier-Stokes equations in domains with walls of varying curvature. Int. J. Comput. Fluid Dyn. **34**(2), 119–126 (2020). https://doi.org/10.1080/10618562.2019.1645328
9. Hess, M.W., Quaini, A., Rozza, G.: A spectral element reduced basis method for Navier–Stokes equations with geometric variations. In: Sherwin, S.J., Moxey, D., Peiró, J., Vincent, P.E., Schwab, C. (eds.) Spectral and High Order Methods for Partial Differential Equations ICOSAHOM 2018, pp. 561–571. Springer International Publishing, Cham (2020)
10. Hesthaven, J., Rozza, G., Stamm, B.: Certified Reduced Basis Methods for Parametrized Partial Differential Equations. Springer Briefs in Mathematics. Springer, Cham (2015)
11. Hijazi, S., Stabile, G., Mola, A., Rozza, G.: Data-driven POD-Galerkin reduced order model for turbulent flows. J. Comput. Phys. **416**, 109513 (2020). https://doi.org/10.1016/j.jcp.2020.109513
12. Karniadakis, G., Sherwin, S.: Spectral/hp Element Methods for CFD. Numerical Mathematics and Scientific Computation. Oxford University Press, Oxford (2005)
13. Mengaldo, G.: Discontinuous spectral/hp element methods: development, analysis and applications to compressible flows. Ph.D. Thesis, Imperial College London (2015)
14. Mengaldo, G., Moura, R., Giralda, B., Peiró, J., Sherwin, S.: Spatial eigensolution analysis of discontinuous galerkin schemes with practical insights for under-resolved computations and implicit les. Comput. Fluids **169**, 349–364 (2018)
15. Mengaldo, G., Moxey, D., Turner, M., Moura, R.C., Jassim, A., Taylor, M., Peiro, J., Sherwin, S.: Industry-relevant implicit large-eddy simulation of a high-performance road car via spectral/hp element methods. SIAM Rev. **63**(4), 723–755 (2021)
16. Moura, R.C., Sherwin, S.J., Peiró, J.: Eigensolution analysis of spectral/hp continuous galerkin approximations to advection–diffusion problems: insights into spectral vanishing viscosity. J. Comput. Phys. **307**, 401–422 (2016)
17. Moura, R.C., Mengaldo, G., Peiró, J., Sherwin, S.J.: An les setting for dg-based implicit les with insights on dissipation and robustness. In: Spectral and High Order Methods for Partial Differential Equations ICOSAHOM 2016, pp. 161–173. Springer, Cham (2017)
18. Moura, R.C., Mengaldo, G., Peiró, J., Sherwin, S.J.: On the eddy-resolving capability of high-order discontinuous Galerkin approaches to implicit les/under-resolved dns of Euler turbulence. J. Comput. Phys. **330**, 615–623 (2017)
19. Moura, R.C., Aman, M., Peiró, J., Sherwin, S.J.: Spatial eigenanalysis of spectral/hp continuous Galerkin schemes and their stabilisation via dg-mimicking spectral vanishing viscosity for high reynolds number flows. J. Comput. Phys. **406**, 109112 (2020)
20. Moura, R.C., Fernandez, P., Mengaldo, G., Sherwin, S.J.: Viscous diffusion effects in the eigenanalysis of (hybridisable) dg methods. In: Spectral and High Order Methods for Partial Differential Equations ICOSAHOM 2018, pp. 371–382. Springer, Cham (2020)
21. Moxey, D., Cantwell, C., Mengaldo, G., Serson, D., Ekelschot, D., Peiró, J., Sherwin, S., Kirby, R.: Towards p-adaptive spectral/hp element methods for modelling industrial flows. In: Spectral and High Order Methods for Partial Differential Equations Icosahom 2016, pp. 63–79. Springer, Cham (2017)
22. Moxey, D., Cantwell, C.D., Bao, Y., Cassinelli, A., Castiglioni, G., Chun, S., Juda, E., Kazemi, E., Lackhove, K., Marcon, J., et al.: Nektar++: enhancing the capability and application of high-fidelity spectral/hp element methods. Comput. Phys. Commun. **249**, 107110 (2020)
23. Pitton, G., Quaini, A., Rozza, G.: Computational reduction strategies for the detection of steady bifurcations in incompressible fluid-dynamics: Applications to Coanda effect in cardiology. J. Comput. Phys. **344**, 534–557 (2017). https://doi.org/10.1016/j.jcp.2017.05.010. https://www.sciencedirect.com/science/article/pii/S0021999117303790
24. Quarteroni, A., Rozza, G.: Numerical solution of parametrized Navier–Stokes equations by reduced basis methods. Numer. Methods Partial Differ. Equ. **23**(4), 923–948 (2007). https://doi.org/10.1002/num.20249. https://onlinelibrary.wiley.com/doi/abs/10.1002/num.20249
25. Rezaian, E., Wei, M.: Eigenvalue reassignment by particle swarm optimization toward stability and accuracy in nonlinear reduced-order models. In: 2018 Fluid Dynamics Conference, p. 3095 (2018)

26. Schilders, W.H., van der Vorst, H.A., Rommes, J. (eds.): Model Order Reduction: Theory, Research Aspects and Applications. Mathematics in Industry. Springer, Berlin (2008). https://doi.org/10.1007/978-3-540-78841-6
27. Stabile, G., Zancanaro, M., Rozza, G.: Efficient geometrical parametrization for finite-volume-based reduced order methods. Int. J. Numer. Methods Eng. **121**(12), 2655–2682 (2020). https://doi.org/10.1002/nme.6324
28. Tonicello, N., Moura, R.C., Lodato, G., Mengaldo, G.: Fully-discrete spatial eigenanalysis of discontinuous spectral element methods: insights into well-resolved and under-resolved vortical flows. Preprint (2021). arXiv:2111.13891
29. Toro, E.F., Spruce, M., Speares, W.: Restoration of the contact surface in the hll-Riemann solver. Shock Waves **4**(1), 25–34 (1994)

Interface Discontinuities in Spectral-Element Simulations with Adaptive Mesh Refinement

Daniele Massaro, Adam Peplinski, and Philipp Schlatter

1 Introduction

Direct numerical simulations (DNS) are increasingly used to study not only canonical flows but also more complex geometries. To increase the Reynolds number further towards industrially relevant values and resolve all the turbulent scales down to dissipative ones, adaptive mesh refinement (AMR) becomes important [17]. This technique plays a pivotal role when an apriori knowledge of the solution is not known. It offers more flexibility than a static mesh, increasing the mesh resolution only when it is required [12]. In AMR we have two main ingredients: the error measurement and the refinement strategy. In the open-source CFD code Nek5000 [4] our group has implemented and extensively tested two different error measurements [12–14]. On the one hand, the spectral error indicator (SEI) [10]; on the other, the adjoint error estimator (AEE) [1]. Nek5000 is a highly scalable and portable code based on the spectral element method (SEM) [15], which offers minimal dissipation and dispersion, with spectral accuracy. The high scalability of the code is owed to the vectorization at the element level via tensor product representations and a matrix-free implementation at the global solver level. This is also the case when AMR is used [16]. The h-refinement strategy is adopted, splitting the elements isotropically and introducing hanging nodes at the interfaces. We rely on external libraries such as p4est [2] and parMETIS (or parRSB) [7] for dynamic management of octrees and the parallel handling partitioning respectively. The continuity constraint at the non-conforming (or non-conformal) interface is imposed by interpolating the solution from the side with the lower resolution (parent

D. Massaro (✉) · A. Peplinski · P. Schlatter
SimEx/FLOW, Engineering Mechanics, KTH Royal Institute of Technology, Stockholm, Sweden
e-mail: dmassaro@kth.se; adam@mech.kth.se; pschlatt@mech.kth.se

J. M. Melenk et al. (eds.), *Spectral and High Order Methods for Partial Differential Equations ICOSAHOM 2020+1*, Lecture Notes in Computational Science and Engineering 137, https://doi.org/10.1007/978-3-031-20432-6_24

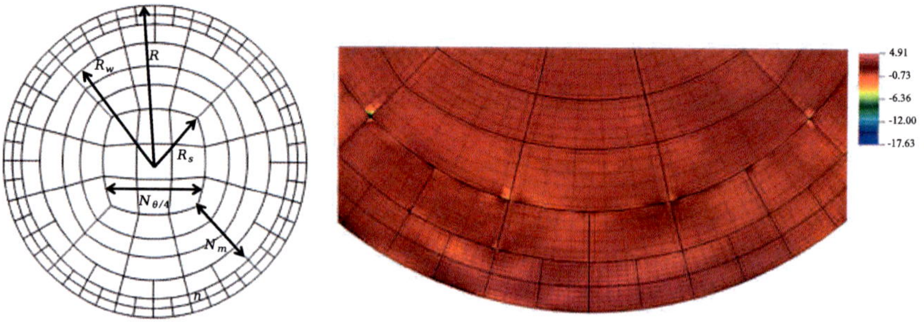

Fig. 1 On the left, the design of the coarse mesh with 2 refinement levels at the wall. We need to specify the geometrical parameters, the number of refinements at the wall and the refinement sequence. On the right, the second-order derivative of the V velocity component w.r.t. to y, being y the vertical direction in the Cartesian reference system. $\partial V^2/\partial y^2$ is averaged in time and axial direction. The intersections of the solid black and light grey lines identify the element boundaries and the GLL points' location respectively

element) onto the side with the higher resolution (children elements). No mortar elements are used [8].

In recent applications, we have used AMR for turbulent flows at moderately high Reynolds number, e.g. the flow around a NACA 4412 airfoil [18] or 3D not uniform cylinders [9]. Nevertheless, the possibility to apply AMR in internal wall-bounded flows has not been explored. We do not expect to get a saving as large as for open-bounded flows, but it can still be significant. To answer this question, we consider the turbulent flow in a straight pipe with periodic boundary conditions. The flow requires high resolution at the wall, where the dissipative scales are located, but does the centre require such resolution too? The conformal mesh forces us to keep the same resolution at each r location in the azimuthal and axial directions, while in the radial direction coarsening towards the centre is possible, see Fig. 1. Hereby, a significant saving is expected to be measured with a non-conformal approach.

Our study also aims at investigating the discontinuities which arise at the non-conforming interfaces. For derivative terms, this is unavoidable since the Galerkin method only ensures C_0 continuity, albeit converging spectrally fast. Visually, we observe wiggles appearing in each term which involves at least first-order derivate. The question we answer here is: does the hanging node at the parent-children face amplify this phenomenon? In addition, does the non-conformal interface introduce a local or global effect? If local, can any local filtering reduce it? The pipe flow simulation is an excellent test case to address this problem, as explained below.

In Sect. 2 the flow case is introduced. In Sect. 3 we present the strategy which has been pursued, a possible non-conformal filtering and a predictive mathematical model for larger Reynolds cases. Eventually, in Sect. 4 the main conclusions are presented and a future outlook is given.

2 Problem Description

We consider the three-dimensional turbulent flow in a straight pipe as a test case. The friction Reynolds number is increased from $Re_\tau = 180$ to $Re_\tau = 550$, where $Re_\tau = u_\tau R/\nu$, u_τ is the friction velocity, R the pipe radius and ν the kinematic viscosity. Regarding spatial and temporal discretisation, the polynomial order for the Lagrangian interpolants and Courant–Friedrich–Lewy (CFL) are set according to the reference study [3]. In both cases the $P_N - P_{N-2}$ formulation is used, with Gauss–Lobatto–Legendre (GLL) and Gauss–Legendre (GL) points for the velocity and pressure fields respectively. The time discretization is performed via third-order implicit backward differentiation (BDF), while the advective term is treated explicitly using the convective form and an extrapolation (EXT) formula. With the same simulation settings, the actual saving in the number of elements can be measured.

To analyse the AMR influence on the wiggles appearing at the interfaces we need an indicator as simple as possible and robust. We thus look at turbulent kinetic energy (TKE) budgets, i.e. the terms present in the transport equation for turbulent kinetic energy k,

$$\frac{\partial k}{\partial t} = P^k + \epsilon^k + D^k + T^k - \Pi^{k,s} + \Pi^{k,t} - C^k, \tag{1}$$

where each term is obtained by summing up the diagonal components of the respective tensor, e.g. $P^k = 1/2(P_{xx} + P_{yy} + P_{zz})$, U is the mean velocity and u is the velocity fluctuation. The second-order tensors are defined as

$$P_{ij} = -\overline{u_i u_k}\frac{\partial U_j}{\partial x_k} - \overline{u_j u_k}\frac{\partial U_i}{\partial x_k}, \qquad \epsilon_{ij} = -2\nu\overline{\frac{\partial u_i}{\partial x_k}\frac{\partial u_j}{\partial x_k}},$$

$$D_{ij} = \nu\frac{\partial^2}{\partial x_k^2}\overline{u_i u_j}, \qquad T_{ij} = -\frac{\partial}{\partial x_k}\overline{u_i u_j u_k}, \qquad C_{ij} = U_k\frac{\partial}{\partial x_k}\overline{u_i u_j},$$

$$\Pi^s_{ij} = -\frac{1}{\rho}\left(\overline{p\frac{\partial u_i}{\partial x_j}} + \overline{p\frac{\partial u_j}{\partial x_i}}\right), \qquad \Pi^t_{ij} = -\frac{1}{\rho}\left(\frac{\partial}{\partial x_j}\overline{p u_i} + \frac{\partial}{\partial x_i}\overline{p u_j}\right),$$

and denoted as the production P_{ij}, the pseudo-dissipation ϵ_{ij}, the viscous diffusion D_{ij}, the turbulent transport T_{ij}, the pressure strain Π^s_{ij}, the pressure convection Π^t_{ij} and the convection C_{ij}. The TKE budgets represent a fixed point in a statistical analysis of the turbulent flow. In addition, each term involves first or second-order derivatives and it is possible to average them in the homogeneous and periodic directions, i.e azimuthal and axial respectively (other than in time). In the end, each statistical quantity is reduced to be dependent only on the radial coordinate. This makes the TKE budget a suitable 1D "wiggles-indicator", pointing out any discontinuities. We complete numerous simulations testing different configurations and obtaining fully converged statistics for each of these might have become extremely expensive [19]. Aware that not fully temporally converged statistics are presented, we rather focus on spatial discontinues.

2.1 *Turbulent Pipe Flow*

The initial mesh with radius $R = 1$ and length $L = 8$ is shown in Fig. 1. The case is shorter than [3], where $L = 25$, and the length reduction is justified by a large number of tests carried out. We can easily extrapolate the equivalent number of elements for a longer pipe to compare the two cases and measure the actual saving. The flow in the axial direction is driven by a pressure gradient, which is adjusted dynamically by the time-integration scheme to ensure a constant mass flux. Each simulation is initialised with a travelling wave in azimuthal and streamwise direction initial condition (IC) to quickly go through transition. Periodic boundary conditions are set at the inlet/outlet ($\boldsymbol{u}(x, y, z) = \boldsymbol{u}(x, y, z + L)$) and no-slip condition at the wall. Let us now define some geometrical parameters for the initial configuration:

- R_w defines the radial distance of the external conformal layer.
- R_s defines the half diagonal of the square central part.
- N_θ is the number of elements in the azimuthal direction. As a consequence the elements in the central part results $(N_\theta/4)^2$.
- N_m is the number of elements in the middle layer, i.e. the region between the external layer and the central part.
- N_z is the number of elements in the axial direction.

In addition, there are AMR parameters such as the number of refinements at the wall n and the refinement sequence. The minimum sequence with $n = 1$ leads to a refinement level in the layers from the wall towards the centre as $1 - 1 - 0$. Considering $n = 2$ several sequences are possible, for example $2 - 2 - 1 - 0$ (the sequential one) or $2 - 2 - 1 - 1 - 1 - 0$ (with extended $n = 1$ layer).

3 Numerical Tests and Results

In this section, we present the carried-out tests and the main results. The original goal was to design a mesh that satisfies DNS requirements at the wall and coarsen it towards the centre. In a first-round, we manually set the number of wall refinements. The resulting mesh is under-resolved with discontinuities not just at the parent-children interface. Inspired by Ref. [5], we do not suppress them but rather investigate the non-conformality effect. Eventually, we use an error-driven approach to design the mesh, testing a non-conforming filter as well.

3.1 Manual Refinement

To design the initial mesh, some geometrical quantities have to be set, i.e. the radius $R = 1$ and length $L = 8$. On the contrary, we need to establish the discretisation parameters properly to minimise the number of elements and respect the wall resolution requirements. We implement an algorithm which performs a parametric study for different R_w values, taking as inputs Re_τ, R, L, p and checking if the resolution requirements are satisfied. The wall requirements read:

$$\Delta r^+ = \frac{R_w}{2^n} \cdot f \cdot Re_\tau < 5, \qquad y^+_{\mathrm{GLL1}} \approx 0.5 \tag{2}$$

$$\Delta\theta^+ = \frac{2\pi R}{2^n \cdot N_\theta} \cdot f \cdot Re_\tau < 5 \tag{3}$$

$$\Delta z^+ = \frac{L}{2^n \cdot N_z} \cdot f \cdot Re_\tau < 10 \tag{4}$$

where Δr^+, $\Delta\theta^+$ and Δz^+ are the maximum grid spacing between GLL points in the radial, azimuthal and streamwise respectively. They are expressed in plus unit, i.e. normalized with the friction length. The position of the first GLL point is $y_{\mathrm{GLL1}} \approx 0.5$ and f is the maximum spacing factor among GLL points within one element for a given polynomial order p (for $p = 7$, $f \approx 0.2$).

In the first attempt, the refinement sequence is fixed $(n - (n-1) - \ldots - 0)$ since we do not use our AMR framework in an adaptive way, but we only set the wall refinement level manually. To keep the resolution in the other regions under control, we assume isotropic turbulence and we look at the Kolmogorov length scale η. For each direction, the maximum grid spacing is compared to the equivalent Kolmogorov length scale at that location. Alternatively, the cubic root of the volume grid spacing is compared to η. At different Reynolds numbers, the data from [3] are used to compute the Kolmogorov scales. The procedure is as follows:

1. Provide input paramaters: Re_τ, R, L and p.
2. Perform a parametric study for different values and combinations of N_z, N_θ, N_m, R_s and R_w.
3. Check the radial DNS wall resolution (2), if it is not respected the parameters combination is discarded.
4. Check the azimuthal DNS wall resolution (3), if it is not respected the parameters combination is discarded.
5. Check the axial DNS wall resolution (4), if it is not respected the parameters combination is discarded.
6. If all the previous DNS wall requirements are respected, verify the resolution in the other internal layers. We guarantee that $\Delta r^+/\eta$, $\Delta\theta^+/\eta$ and $\Delta z^+/\eta$ are less than 10. An alternative and relaxed condition is $(\Delta r^+ \cdot \Delta\theta^+ \cdot \Delta z^+)^{1/3}/\eta < 10$.

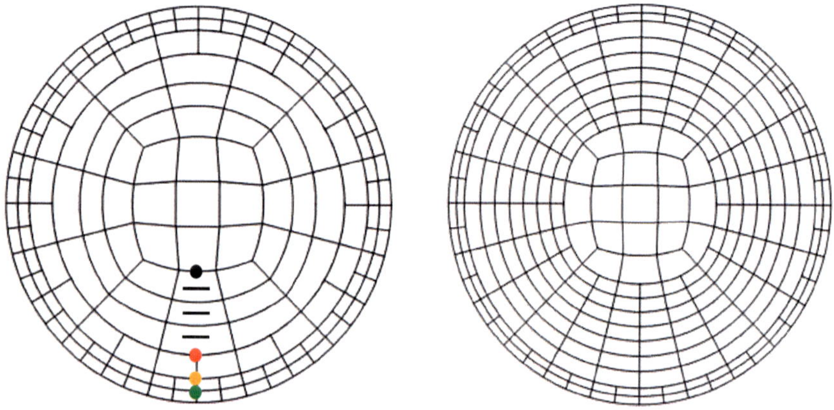

Fig. 2 Comparison between the mesh design via manual refinement (left) and error-driven approach (right) at $Re_\tau = 180$. The green, yellow and red points indicate the n-n, n-$(n-1)$ and n-$(n-2)$ interfaces respectively; where $n = 2$ is the wall refinement level. The black bars report the layers where the refinement level $n = 1$ has been extended to observe whether it is a local or global effect. In this latter case, the hanging node of the $1-0$ interface is shifted in the black point location

The first attempt is shown in Fig. 2-left. The mean velocity profile for the streamwise component shows an excellent agreement with the DNS reference data by El Khoury et al. [3], see Fig. 3. When we look at our "wiggles indicator", i.e. the TKE budgets, some wiggles at different interfaces appear. A small discontinuity is present at the first conformal layer (green point) for the viscous diffusion term. Improving the resolution in that region, the discontinuities get smaller up to completely disappear. In the case Fig. 2-left, on purpose we perform a relaxation in the radial resolution, setting $y_{\mathrm{GLL1}} \approx 0.7$. In this way, we show that the issue is not only related to the non-conformal interface. The other two wiggles, yellow and red points in 3-top right, are located at the n-$(n-1)$ and $(n-1)$-0 boundaries, respectively.

First, we try to understand if this is a local or global effect. To observe if the internal resolution affects the non-conformal wiggles, we extend the refinement level $n-1$ towards the centre. The layers marked with a black line in Fig. 2-left have been refined with a level equal to 1. As a consequence, a non-conformal layer is introduced at the black point location. The wiggles at the yellow and red locations are not affected by a higher internal level of refinement, pointing out that this is a local phenomenon. Moreover, the non-conformal layer at the black location does not introduce any visible discontinuities, since the mesh is sufficiently fine there. This is the first evidence that the hanging nodes do not introduce further instabilities when we have a proper resolution. To verify such observation, we also consider the conformal mesh previously used by El Khoury et al. [3], adding one level of refinement in the first two layers. Again, the non-conforming layer does not introduce discontinuities for well-resolved meshes. In Fig. 1-right the second-order

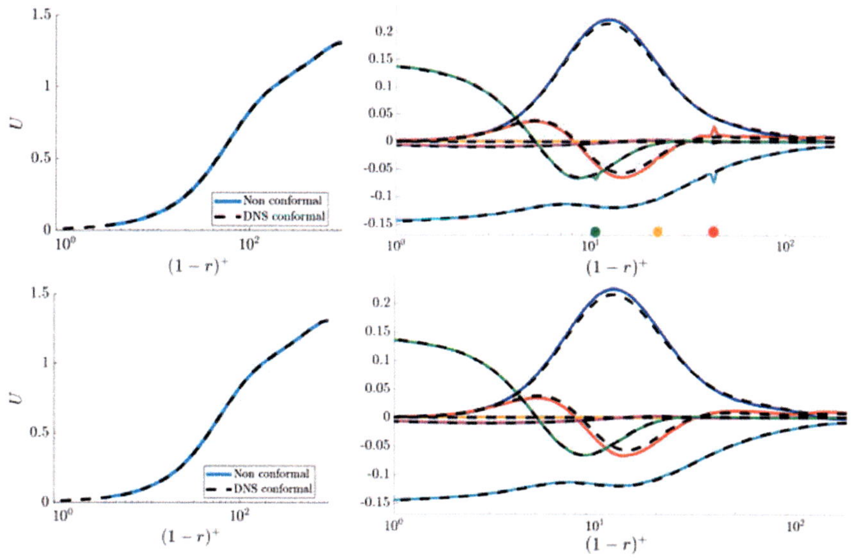

Fig. 3 On the first row, for the manually designed mesh (see Fig. 2-left) we compare the mean axial velocity U (left) and the TKE budgets (right) with the reference DNS data by El Khoury et al. [3] (black dashed lines). The green, yellow and red points indicate the n-n, n-$(n-1)$ and n-$(n-2)$ interfaces respectively; where $n = 2$ is the wall refinement level. On the second row, the error-driven designed mesh (see Fig. 2-right) is considered. In the TKE budgets, we have the production term P^k in ——— , the turbulent transport T^k in ——— , the viscous diffusion D^k in ——— , the pseudo-dissipation ϵ^k in ——— , the pressure convection Π_t^k in ——— and the pressure strain Π_s^k in ———

derivate of the v velocity component w.r.t. y (averaged in time and axial direction) confirms our observation. The spikes are not located at the hanging node location, but rather at conformal vertices.

This was the first important result, which confirms the robustness of our code and allows us to exploit AMR even in such geometries. The only requirement is to guarantee an adequate resolution in each region. In order to quantify what "adequate" means exactly, we need to rely on an error-driven approach.

3.2 Error-Driven Refinement

The manual refinement approach does not control the solution error, dangerously leading to an under-refined mesh where interface jumps appear. Thus, we decide to introduce the spectral error indicator for driving the refinement procedure.

The SEI was introduced in Refs. [10, 11] and relies on the local properties of the solution. According to SEM theory, we can consider an exponential decay of the spectral coefficients if the solution is sufficiently smooth. Based on this assumption, the spectral error indicator evaluates the L^2-norm of the error between the exact solution and its spectral expansion at order N. The total error is decomposed into two contributions: a truncation error due to the finite number of coefficients in the spectral expansion and a second error due to quadrature. Considering a 1D problem, $u(x)$ is the exact solution to a system of partial differential equations and u_N is an approximate spectral-element solution with polynomial order N. The expansion of $u(x)$ on a reference element in terms of the Legendre polynomials reads $u(x) = \sum_{k=0}^{\infty} \hat{u}_k L_k(x)$, where $\hat{u}_k$ are the associated spectral coefficients and $L_k(x)$ is the Legendre polynomial of order k. The error indicator $\varepsilon_{\text{ind}} = \|u - u_N\|_{L^2}$ is expressed as:

$$\varepsilon_{\text{ind}} = \left(\int_N^\infty \frac{\hat{u}(k)^2}{\frac{2k+1}{2}} \mathrm{d}k + \frac{\hat{u}_N^2}{\frac{2N+1}{2}} \right)^{\frac{1}{2}} \tag{5}$$

assuming exponential decay for the spectral coefficients of the form $\hat{u}(k) \approx c \exp(-\sigma k)$. The parameters c and σ are obtained via interpolating in a linear least-squares sense the $\log(\hat{u}_k)$ w.r.t. k for $k \leq N$. As it was shown in a previous work [13], these indicators are best suited to track flow features with high gradients, such as shear layers or fluid-wall interaction region.

For the pipe at $Re_\tau = 180$ the comparison between the manual and error-driven mesh is shown in Fig. 2. Controlling the error via SEI the mesh is automatically designed with adequate resolution in the full domain. Moreover, we reduce by 20% the overall computational cost using fewer GLL points. For $Re_\tau = 550$ the TKE budgets are shown in Fig. 4; the mean velocity profile shows an excellent agreement as well (not shown here). This latter result is promising since the mesh has only half of the elements used by Ref. [3].

For the sake of completeness, we have also implemented and tested a local filter acting only at the non-conformal interfaces. The modal filter is defined as:

$$\boldsymbol{u}_c^{n+1} = \boldsymbol{u}_c^* - \chi H(\boldsymbol{u}_c^*) \Delta t \tag{6}$$

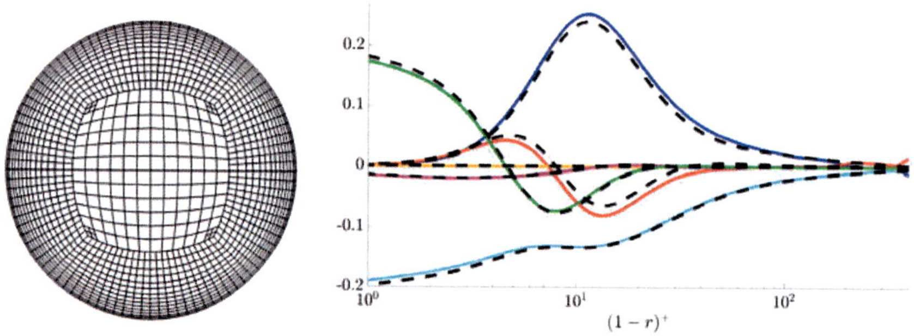

Fig. 4 On the left, the cross-section of the non-conformal mesh at $Re_\tau = 550$. On the right, the comparison between the TKE budgets for the current case (coloured lines) and the DNS reference data [3] (black-dashed lines)

where $\boldsymbol{u}_c$ is the solution on the children GLL points, $.^*$ refers to the unfiltered solution and H is the step-function which defines the high-pass filtered velocity field $H(\boldsymbol{u}_c^*)$. It acts as low-pass filtering, cutting a certain percentage of the highest modes. The effect of this non-conformal filtering has been observed on the solution fields in previous studies, see [6, 12, 13, 18]; but here we use it to smooth spikes in the derivatives. We can apply the filter uniformly on all the GLL points of the children-elements or linearly activate it towards the boundaries. Moreover, we tune the stiffness and the width of the filtering function. To maintain the divergence-free condition, we do not cut explicitly the highest modes. Instead, we subtract the highest modes properly scaled by χ. The width χ has to be large enough to guarantee that we remove energy from the highest wavenumbers, but not too much to avoid unphysical effects.

Regarding the solution field, the filtering has shown good capabilities in avoiding numerical instabilities at the interfaces [12]. However, it did not alleviate the measured discontinuities in the derivate terms. In future studies, we could test different post-processing or in-situ filters, but this is out of the goal of the current manuscript.

3.3 Predictive Model

We conduct the simulations for a range of Reynolds numbers which goes from $Re_\tau = 180$ to $Re_\tau = 550$. It is worth figuring out what computational saving could be obtained as Re_τ increases.We thus derive a predictive model for estimating the expected computational saving. The model is based on empirical observations from the previous tests. Here, we point out some common SEI indications:

- the non-conformal hanging nodes wherever located do not introduce any discontinuities, as long as the mesh is well-resolved.
- to reasonably compare with the conformal case [3] we need to guarantee DNS resolution requirements at the wall, but how far shall we propagate such resolution internally? As general guidelines, the SEI points out that the wall resolution needs to be extended up to $y^+ \approx 100$. Testing different N_θ and N_z, we see that this is always the case for the azimuthal and axial resolutions. Regarding the radial one, at the wall there is a maximum grid spacing between GLL points $\Delta r^+ \approx 2.5$ (for $p = 7$) because we enforce the first GLL point at $y^+ \approx 0.5$. This high resolution does not need to be propagated up to $y^+ \approx 100$ and the refinement level $n-1$, which ensures $\Delta r^+ \approx 5$, is sufficient. Nevertheless, we refine isotropically, so coarsening in only one direction is impossible. In our predictive model, we assume that the initial mesh is designed to take these aspects into account (setting R_w properly) and propagate the $n-1$ level up to $y^+ = 100$.
- to design the mesh we need to set a minimum number of parameters: geometrical (L, R, R_w) and AMR related, e.g. the number of refinement n and the

Table 1 The table compares the results obtained for the Reynolds numbers $Re_\tau = 180, 360, 550$ and the estimations by the model (7). In addition, the predictions for $Re_\tau = 1000, 2000$ are shown. From the left to right we report: the friction Reynolds number; the number of GLL points for the AMR-designed meshes, the model prediction and the reference DNS [3]; the actual and predicted saving; the actual and predicted wall refinement level

Re_τ	n_{GLL}	$n_{\text{GLL}e}$	$n_{\text{GLL}r}$	s	s_e	n	n_e
180	$14.57 \cdot 10^6$	$21.660 \cdot 10^6$	$18.67 \cdot 10^6$	0.219	–	1	1
360	$91.55 \cdot 10^6$	$122.53 \cdot 10^6$	$121.4 \cdot 10^6$	0.271	0.252	1	1
550	$209.41 \cdot 10^6$	$302.28 \cdot 10^6$	$437.00 \cdot 10^6$	0.508	0.308	1	1
1000	–	$963.03 \cdot 10^6$	$2.184 \cdot 10^9$	–	0.559	–	2
2000	–	$3.9232 \cdot 10^9$	$1.296 \cdot 10^{10}$	–	0.672	–	3

sequence. While L and R are fixed, and R_w and n are computed to minimize the number of elements; the refinement sequence has to be assumed known to enable predictions at a higher Reynolds number. We always get two levels n at the wall and then the level $n-1$ is maintained up to $y^+ = 100$. Beyond this region, a sequential coarsening is considered.

Based on these observations, we define a predictive model for higher Reynolds numbers:

$$n_{el} = N_z\left(2\left(2^{n(d-1)}\right)N_\theta + 2^{(n-1)(d-1)}\tilde{n}N_\theta + \sum_{i=n-2}^{0} 2^{i(d-1)}N_\theta + (N_\theta/4)^2\right) \quad (7)$$

where d is the case dimension ($d = 3$) and $\tilde{n}$ is the number of layers with refinement $n-1$ needed to satisfy mesh resolution up to $y^+ \approx 100$. Within the round brackets, the first term keeps track of the two near-wall layers with refinement n and the second one counts the $\tilde{n}$ layers up to $y^+ \approx 100$. The third term keeps the score of the elements in the sequential coarsening sequence from $n-2$ to $n = 0$ and the last one evaluates the central region.

The model predicts in a excellent way the required refinement parameters (n_e and $n_{\text{GLL}e}$) at larger Reynolds number, see Table 1. It fails at $Re_\tau = 180$, where it overestimates the maximum number of GLL points $n_{\text{GLL}e}$. Comparing with the reference data $n_{\text{GLL}r}$, we observe the actual and the estimated saving, respectively, s and s_e. Moreover, the required refinement level at the wall n_e is in excellent agreement. According to this model, we appear not to be able to save more than 65% even at higher Reynolds numbers.

4 Conclusions

The study investigates the discontinuities in the derivatives at the non-conforming interfaces of spectral elements. Considering the turbulent flow in a straight pipe, we test manual and error-driven approaches relying on the turbulent kinetic energy budgets as a "wiggles indicator". Moreover, we also measure the number of elements saved through adaptive mesh refinement.

Referring to Ref. [3], the same simulation setup is considered, but with a shorter pipe ($L = 8$ instead of $L = 25$). The effect of pipe length on the observed quantities has been quantified and taken into account. This limitation has been necessary since we conduct a large number of tests. Both conforming and non-conforming meshes show jumps in the derivate terms as long as the solution field is under resolved. The hanging node present when AMR is enabled does not introduce further instabilities. This is the first important outcome of our study, which ensures the robustness of our AMR implementation. In addition, the effect is local. We are thus encouraged to apply filters to suppress the interface discontinuities. A possible high pass filter acting on the solution field is tested without success. Eventually, we also derive a predictive model to estimate the computational saving at higher Reynolds numbers. According to the indications provided by the spectral error indicator, we always expect a saving of more than a factor 2. The importance of relying on an error-driven approach is fundamental. Indeed, ensuring only the standard DNS resolution requirements at the wall turned out to be insufficient for avoiding interface discontinuities.

The study shows a significant saving via adaptive mesh refinement for internal flows. On the one hand, for the refinement method, the possibility to implement a non-isotropic refinement is attractive, since some directions demand more resolution than others, i.e. the radial one. On the other hand, promising results encourage us to use the spectral error indicator to design meshes for even more complicated wall-bounded flows, e.g. bent pipes. This latter is a case where we expect a larger saving in the overall number of elements.

Acknowledgments We acknowledge that the results of this research have been achieved using the DECI resource ARCHER2 based in UK at the University of Edinburgh with support from the PRACE aisbl.

References

1. Bangerth, W., Rannacher, R.: Adaptive Finite Element Methods for Differential Equations, Springer, Basel (2002)
2. Burstedde, C., Wilcox, L., Ghattas, O.: p4est: scalable algorithms for parallel adaptive mesh refinement on forests of octrees. SIAM J. Sci. Comput. **33**(3), 1103–1133 (2011)
3. El Khoury, G., Schlatter, P., Noorani, A., Fischer, P., Brethouwer, G., Johansson, A.: Direct numerical simulation of turbulent pipe flow at moderately high Reynolds numbers. Flow Turbulence Combust. **91**, 475–495 (2013)
4. Fischer, P., Lottes, J., Kerkemeier, S.: Nek5000: open source spectral element CFD solver (2008). https://nek5000.mcs.anl.gov/
5. Gresho, P., Lee, R.: Don't suppress the wiggles-they're telling you something! Comput. Fluids **9**, 223–253 (1981)
6. Johnson, N., Bareford, M., Cantwell, C., Nielsen, A., Jansson, N., Offermans, N., Gong, J., Peplinksi, A., Jammy, S., Zhang, J., Dick, B., Vogler, P., Sherwin, S., Vymazal, M., Moxey, D.:

Final report on the exaflow algorithms, energy efficiency and IO strategies. Tech. Rep. D2.4, ExaFLOW Consortium, European Commission (2018)
7. Karypis, G., Kumar, V.: MeTis: unstructured graph partitioning and sparse matrix ordering system, version 4.0 (2009). http://www.cs.umn.edu/metis
8. Kruse, G.W.: Parallel nonconforming spectral element solution of the incompressible Navier–Stokes equations in three dimensions. Ph.D. Thesis, Brown University (1997)
9. Massaro, D., Peplinski, A., Schlatter, P.: Direct numerical simulation of turbulent flow around 3D stepped cylinder with adaptive mesh refinement. In: Twelfth International Symposium on Turbulence and Shear Flow Phenomena - TSFP12 (2022)
10. Mavriplis, C.: Nonconforming discretizations and a posteriori error estimators for adaptive spectral element techniques. Ph.D. Thesis, Massachusetts Institute of Technology (1989)
11. Mavriplis, C.: A posteriori error estimators for adaptive spectral element techniques. In: Proceedings of the Eighth GAMM-Conference on Numerical Methods in Fluid Mechanics. Notes on Numerical Fluid Mechanics, pp. 333–342 (1990)
12. Offermans, N.: Aspects of adaptive mesh refinment in the spectral element method. Ph.D. Thesis, Royal Institute of Technlogy (2019)
13. Offermans, N., Peplinski, A., Marin, O., Schlatter, P.: Adaptive mesh refinement for steady flows in Nek5000. Comput. Fluids **197**, 104352 (2020)
14. Offermans, N., Massaro, D., Peplinski, A., Schlatter, P.: Error-driven adaptive mesh refinement for unsteady turbulent flows in spectral-element simulations. Comput. Fluids **251**, 105736 (2022)
15. Patera, A.: A spectral element method for fluid dynamics: laminar flow in a channel expansion. J. Comput. Phys. **54**(3), 468–488 (1984)
16. Peplinski, A., Offermans, N., Fischer, P., Schlatter, P.: Non-conforming elements in Nek5000: pressure preconditioning and parallel performance. In: Proceedings of Spectral and High Order Methods for Partial Differential Equations ICOSAHOM 2018. Lecture Notes in Computational Science and Engineering, vol. 134 (2020)
17. Slotnick, J., Khodadoust, A., Alonso, J., Darmofal, D., Gropp, W., Lurie, E., Mavriplis, C.: CFD vision 2030 study: a path to revolutionary computational aeroscience. Report by National Aeronautics and Space Administration (2014)
18. Tanarro, A., Mallor, F., Offermans, N., Peplinski, A., Vinuesa, R., Schlatter, P.: Enabling adaptive mesh refinement for spectral-element simulations of turbulence around wing sections. Flow Turbulence Combust. **105**, 415–436 (2020)
19. Vinuesa, R., Prus, C., Schlatter, P., Nagib, H.: Convergence of numerical simulations of turbulent wall-bounded flows and mean cross-flow structure of rectangular ducts. Meccanica **51**, 3025–3042 (2016)

Fully Discrete Entropy Conserving/Stable Discontinuous Galerkin Discretization of the Euler Equations in Entropy Variables

Alessandra Nigro, Andrea Crivellini, and Alessandro Colombo

1 Introduction

A faithful representation of entropy evolution is a very important aspect and a desirable property to perform high-fidelity iLES/uDNS simulations of turbulent flows. To this end, in recent years, several novel entropy conserving [4, 9, 10, 15, 17, 18, 20] and entropy stable [4, 5, 9–11, 15, 19] numerical fluxes have been developed with the aim to ensure a variation of entropy equal to zero and to preserve its positivity, respectively. In particular, the entropy stable numerical fluxes are obtained adding to the entropy conservative ones, that are typically centered fluxes, some dissipation that enforces the production of entropy. Semi-discrete entropy conservative and entropy stable numerical schemes have been investigated in the literature in several papers, but there are few papers which include an extensive analysis of the effect of the time integration [6, 13, 16, 20], and the question of the entropy stability of explicit and implicit methods is still an open question [13]. Nevertheless, even if entropy conserving/stable schemes in space are used to perform unsteady simulations, there is a non-negligible impact of the time integration process on these properties, i.e. even if entropy conserving/stable properties are fulfilled at a discrete level for the spatial approximation, if the time integration scheme is not able to respect these properties and relatively large time step sizes are used, the time

A. Nigro (✉) · A. Crivellini
Marche Polytechnic University, Department of Industrial Engineering and Mathematical Sciences, Ancona, Italy
e-mail: a.nigro@univpm.it; a.crivellini@@univpm.it

A. Colombo
Università degli Studi di Bergamo, Department of Engineering and Applied Sciences, Bergamo, Italy
e-mail: alessandro.colombo@unibg.it

J. M. Melenk et al. (eds.), *Spectral and High Order Methods for Partial Differential Equations ICOSAHOM 2020+1*, Lecture Notes in Computational Science and Engineering 137, https://doi.org/10.1007/978-3-031-20432-6_25

integration error overwhelm the spatial one and the conserving/stable properties of entropy cannot be fulfilled.

The scope of this paper is to present a fully discrete entropy conserving/stable dG discretization in entropy variables for unsteady inviscid flows. In particular, will be investigated the accuracy and the entropy conservation/stability properties of several numerical fluxes [4, 12, 15] and of the entropy conserving generalized Crank-Nicolson scheme [16, 20].

The rest of the paper is organized as follows. Section 2 reviews the main building blocks of the spatial and temporal discretizations. In Sect. 3 are shown the numerical results performed on two inviscid unsteady test-cases: the isentropic vortex convection problem and the double shear layer. Finally, concluding remarks are reported at the end of the article.

2 Spatial and Temporal Discretization

The Euler equations of compressible flows can be written, in terms of conservative variable, as:

$$\frac{\partial \mathbf{q}}{\partial t} + \nabla \cdot \mathbf{F}(\mathbf{q}) = \mathbf{0}, \tag{1}$$

where $\mathbf{q} = (\rho, \rho u_1, \rho u_2, \rho u_3, \rho E)^T$ is the set of conservative variables and $\mathbf{F}(\mathbf{q}) = (\mathbf{f}_1, \mathbf{f}_2, \mathbf{f}_3)$ is the inviscid flux vector, which components are:

$$\mathbf{f}_j = \begin{pmatrix} \rho u_j \\ \rho u_1 u_j + p\delta_{1j} \\ \rho u_2 u_j + p\delta_{2j} \\ \rho u_3 u_j + p\delta_{3j} \\ \rho H u_j \end{pmatrix} \quad \text{with} \quad j = 1, 2, 3. \tag{2}$$

In these equations ρ is the fluid density, p the pressure, u_j the components of the velocity vector $\mathbf{u}$, E and H the total energy and enthalpy for unit mass, with $H = E + p/\rho$.

In order to fulfil the entropy conservation/stable properties, the Euler equations are expressed in terms of entropy variables, which leads to the symmetrized Euler equations:

$$\mathbf{P}(\mathbf{v}) \frac{\partial \mathbf{v}}{\partial t} + \nabla \cdot \mathbf{F}(\mathbf{v}) = \mathbf{0}, \tag{3}$$

where $\mathbf{P}(\mathbf{v}) = \partial \mathbf{q}(\mathbf{v})/\partial \mathbf{v}$ is the matrix that takes into account of the change of variables from the conservative set to the entropy set, this last one defined as:

$$\mathbf{v} = \left(\frac{\gamma - s}{\gamma - 1} - \frac{\rho}{2p} |\mathbf{u}|^2, \ \frac{\rho u_1}{p}, \ \frac{\rho u_2}{p}, \ \frac{\rho u_3}{p} - \frac{\rho}{p} \right)^{\top}, \tag{4}$$

where $\gamma = c_p/c_v$ is the ratio of gas specific heats and $s = \ln(p/\rho^{\gamma})$ is the physical entropy.

To discretize Eq. (3) we consider $\mathcal{K}_h = \{K\}$ a mesh of the domain Ω composed by non-overlapping elements K, and we define a discrete polynomial space in physical coordinates:

$$\mathbb{P}_d^k(\mathcal{K}_h) \stackrel{\text{def}}{=} \left\{ v_h \in L^2(\Omega) \mid v_{h|K} \in \mathbb{P}_d^k(K), \ \forall K \in \mathcal{K}_h \right\}, \tag{5}$$

where d is the number of geometrical dimensions, k is a non-negative integer and $\mathbb{P}_d^k(K)$ denotes the restriction to K of the polynomial functions of d variables and total degree $\leq k$. By multiplying Eq. (3) by an arbitrary smooth test function $\mathbf{w} = \{w_1, \ldots, w_{2+d}\}$, and integrating by parts, we obtain the weak formulation:

$$\int_{\Omega} \mathbf{w} \cdot \left(\mathbf{P}(\mathbf{v}) \frac{\partial \mathbf{v}}{\partial t} \right) d\mathbf{x} - \int_{\Omega} \nabla \mathbf{w} : \mathbf{F}(\mathbf{v}) d\mathbf{x} + \int_{\partial \Omega} \mathbf{w} \otimes \mathbf{n} : \mathbf{F}(\mathbf{v}) d\sigma = \mathbf{0}, \tag{6}$$

where $\mathbf{n}$ is the unit vector normal to the boundary. In Eq. (6) the solution $\mathbf{v}$ and the test function $\mathbf{w}$ are replaced with a finite element approximation $\mathbf{v}_h$ and a discrete test function $\mathbf{w}_h$, respectively, where $\mathbf{v}_h$ and $\mathbf{w}_h$ belong to the discrete space $\mathbf{V}_h \stackrel{\text{def}}{=} [\mathbb{P}_d^k(\mathcal{K}_h)]^{2+d}$. Each component $v_{h,j}$, $j = 1, \ldots, 2+d$, of the numerical solution $\mathbf{v}_h \in \mathbf{V}_h$ can be expressed, in terms of the elements of the global vector $\mathbf{V}$ of unknown degrees of freedom, as $q_{h,j} = \phi_l V_{j,l}$, $l = 1, \ldots, N_{dof} = \binom{k-1}{d}$, $\forall K \in \mathcal{K}_h$, where ϕ_l belongs to the set of orthogonal and hierarchical basis functions [2]. We define the set $\mathcal{F}_h$ of the mesh faces $\mathcal{F}_h \stackrel{\text{def}}{=} \mathcal{F}_h^i \cup \mathcal{F}_h^b$, where $\mathcal{F}_h^b$ collects the faces located on the boundary of Ω_h and for any $F \in \mathcal{F}_h^i$ there exist two elements $K^+, K^- \in \mathcal{K}_h$ such that $F \in \partial K^+ \cap \partial K^-$. Moreover, for all $F \in \mathcal{F}_h$, $\mathbf{n}_F = \{n_i\}$ with $i = 1, \ldots, d$, is the univocally defined normal unit vector pointing from K^+ to K^-. As a function $w_h \in \mathbb{P}_d^k(\mathcal{K}_h)$ is double valued over an internal face $F \in \mathcal{F}_h^i$, we introduce the jump trace operator $[\![w_h]\!] \stackrel{\text{def}}{=} w_{h|K^+} - w_{h|K^-}$, that acts componentwise when applied to a vector.

The dG discretization of the Euler equations consists in seeking, for $j = 1, \ldots, 2+d$, the elements of $\mathbf{V}$ such that:

$$\sum_{K \in \mathcal{K}_h} \int_K \phi_i P_{j,k}(\mathbf{v}_h)\, \phi_l \frac{dV_{k,l}}{dt} d\mathbf{x} - \sum_{K \in \mathcal{K}_h} \int_K \frac{\partial \phi_i}{\partial x_n} F_{j,n}(\mathbf{v}_h) d\mathbf{x} \\ + \sum_{F \in \mathcal{F}_h} \int_F [\![\phi_i]\!]\, \widehat{F}_j\left(\mathbf{v}_h^{\pm}, \mathbf{n}_F\right) d\mathbf{x} = \mathbf{0}, \tag{7}$$

for $i = 1, \ldots, N_{dof}$, where repeated indices imply summation over the ranges $k = 1, \ldots, 2+d$, $l = 1, \ldots, N_{dof}$ and $n = 1, \ldots, d$. At each face $F \in \mathcal{F}_h$ the convective physical flux $\mathbf{F}$ in normal direction is replaced by a proper numerical function $\widehat{\mathbf{F}}$, to compute which any of the numerical flux functions considered in the finite volume method can be used.

In the present work several entropy conserving and entropy stable numerical fluxes have been tested. The entropy conserving fluxes investigated are: (i) the entropy-conservative (EC) flux of Ismail and Roe [15]; (ii) the kinetic energy preserving and entropy-conservative (KEEC) flux of Chandrashekar [4]. The entropy stable fluxes investigated are: (i) a Rusanov-type dissipation (ESRS); (ii) a Roe-type dissipation (ESRO); (iii) the entropy-consistent dissipation of Roe (ESEC), named EC1 in [15]; (iv) a Godunov flux (ERS) that compute the convective flux thanks to the exact solution of the Riemann problem [12].

The semi-discrete system of equations (7) can be written for a single element K in the following compact form:

$$\int_K \phi_i \mathbf{P}(\mathbf{v}_h) \frac{d\mathbf{v}_h}{dt} d\mathbf{x} = \mathbf{r}(\mathbf{v}_h) \tag{8}$$

The time discretization of this equation with a standard Crank-Nicolson scheme [8] leads to the following system of algebraic equations:

$$\int_K \phi_i \mathbf{P}\left(\mathbf{v}_h^{n+1/2}\right) \frac{\mathbf{v}_h^{n+1} - \mathbf{v}_h^n}{\Delta t} d\mathbf{x} = \mathbf{r}\left(\mathbf{v}_h^{n+1/2}\right) \tag{9}$$

where $\mathbf{v}_h^{n+1/2} = 1/2(\mathbf{v}_h^{n+1} + \mathbf{v}_h^n)$. Differently, the entropy conserving generalised Crank-Nicolson, presented in [13, 16, 20] for a finite volume method and conservative variables, considers an intermediate state $\mathbf{v}_h^{n+1/2}$, defined as:

$$\mathbf{v}_h^{n+1/2} = \int_{-1/2}^{1/2} \mathbf{v}\left(\frac{\mathbf{q}_h^n + \mathbf{q}_h^{n+1}}{2} + \zeta \Delta \mathbf{q}_h\right) d\zeta, \quad \text{with } \Delta \mathbf{q}_h = \mathbf{q}_h^{n+1} - \mathbf{q}_h^n, \tag{10}$$

on the base of which the solution is advanced in time, in the context of the finite volume method, as follows:

$$\frac{\mathbf{q}_h^{n+1} - \mathbf{q}_h^n}{\Delta t} = \mathbf{r}\left(\mathbf{v}_h^{n+1/2}\right) \tag{11}$$

It is in fact possible to demonstrate that the above equations is equivalent to:

$$\frac{S_h^{n+1} - S_h^n}{\Delta t} = \mathbf{v}_h^{n+1/2} \cdot \mathbf{r}\left(\mathbf{v}_h^{n+1/2}\right) \quad \text{with} \quad S_h = -\rho_h s_h, \tag{12}$$

that ensure the entropy conservation property in time. Inspired by this work, in our framework the solution is advanced in time $\forall K$ as:

$$\int_K \phi_i \frac{\mathbf{q}_h^{n+1} - \mathbf{q}_h^n}{\Delta t} d\mathbf{x} = \mathbf{r}\left(\mathbf{v}_h^{n+1/2}\right) \tag{13}$$

where the intermediate state is computed as:

$$\int_K \phi_i \mathbf{v}_h^{n+1/2} d\mathbf{x} = \int_K \phi_i \int_{-1/2}^{1/2} \mathbf{v}\left(\frac{\mathbf{q}_h^n + \mathbf{q}_h^{n+1}}{2} + \zeta \Delta \mathbf{q}_h\right) d\zeta d\mathbf{x}, \tag{14}$$

and $\mathbf{q}_h^n$ as:

$$\int_K \phi_i \mathbf{q}_h^n d\mathbf{x} = \int_K \phi_i \mathbf{q}\left(\mathbf{v}_h^n\right) d\mathbf{x}, \tag{15}$$

thus, $\mathbf{q}_h^n$ is the projection on the dG space of the conservative variables computed from the entropy ones. It can be proven that the above equations are equivalent to:

$$\int_K \frac{S_h^{n+1} - S_h^n}{\Delta t} d\mathbf{x} = r_s\left(\mathbf{v}_h^{n+1/2}\right), \tag{16}$$

where r_s represents the spatial discretization of the equation governing the evolution of S_h which is known to conserve the total entropy, for a dG discretization, when entropy variables and entropy conservative fluxes are used. However, to fulfill the entropy conservation property, an exact evaluation of the integrals is required, which results in an *over-integration* procedure. In other words, we use a Gauss integration formula characterised by a high order of exactness $(4k + 3)$ in order to almost nullify the integration error. We refer to this second order accurate method as a "Generalized Crank-Nicolson" (GCN) following the Tadmor [20] nomenclature. Note that Eq. (14) also requires a numerical integration, which is performed here with a very high-order Gauss-integration formula at least equal to 27. The implementation of a non linear solution algorithm for Eq. (13) is not straightforwards and it is often considered "impractical". Here, we solved these difficulties evaluating the Jacobian of $\mathbf{r}(\mathbf{v}_h^{n+1/2})$ at the intermediate state $\mathbf{v}_h^{n+1/2} = 1/2(\mathbf{v}_h^n + \mathbf{v}_h^{n+1})$ instead of the one defined in Eq. (14). The approach results in an inexact Newton non-linear solution algorithm, nevertheless we never encountered convergence problems. The accuracy and the conservation properties of this time integration scheme will be investigated in the next section and compared with the ones obtained with a standard Crank-Nicolson (SCN) scheme [8].

For further details about the dG method [3], the entropy conserving/stable fluxes investigated [4, 12, 15], the GCN entropy conserving time integration [13, 16, 20] and its implementation within the described framework [7], the interest reader can refer to cited works.

3 Numerical Results

In this section are shown the results of two bi-dimensional inviscid test-cases: the convection of an isentropic vortex [14, 21, 22] and the double shear layer [5].

Several temporal and spatial refinement studies have been performed to evaluate the accuracy and the conservation properties of the GCN scheme and of the entropy conserving/stable numerical fluxes. To this end, the following errors have been defined:

$$\eta(\circ, \bullet_{ref}) = (\Omega_h)^{-1/2} \,||\, \circ - \bullet_{ref} \,||_{L^2}, \tag{17}$$

$$\varepsilon(\circ, \bullet_{ref}) = \Omega_h^{-1} \left(\int_{\Omega_h} \circ \, d\mathbf{x} - \int_{\Omega_h} \bullet_{ref} \, d\mathbf{x} \right), \tag{18}$$

where $\circ$ and $\bullet_{ref}$ are the numerical and the reference solutions, respectively, the last one equal to the L^2 projection of the initial solution on the dG polynomial space. Note in fact that for the inviscid isentropic vortex the initial flow field is equal to the exact solution, since the simulations are performed for one period of vortex revolution. For the double shear layer the reference solutions of the investigated variables, i.e. kinetic energy and enstrophy, are considered equal to their initial values since this test case has been computed for a very low Mach number, and therefore the deviation of kinetic energy with respect to the initial value, due to the compressibility effects, is very small.

In this work, to solve the resulting non-linear algebraic systems arising from the time discretization process, a Newton-Krilov method is used, which is preconditioned with an ILU(0) decomposition applied on each mesh partition, exploiting the algorithms present in the PETSc library [1]. The non-linear solver tolerance was set to 10^{-14} and the linear one to 10^{-2}, furthermore, when an entropy conserving numerical flux is used, the *over-integration* technique has been adopted in order to ensure the entropy conservation property.

3.1 The Isentropic Vortex Convection Problem

The inviscid isentropic convecting vortex is a well known unsteady test case that is often used to assess the accuracy of time integration methods. Its initial flow field is:

$$
\begin{aligned}
u_1 &= u_{1\infty} - \frac{\alpha}{2\pi}(y - y_0)\, e^{\phi(1-r^2)}, \\
u_2 &= u_{2\infty} + \frac{\alpha}{2\pi}(x - x_0)\, e^{\phi(1-r^2)}, \\
T &= T_\infty - \frac{\alpha^2(\gamma - 1)}{16\phi\gamma\pi^2} e^{2\phi(1-r^2)},
\end{aligned}
\tag{19}
$$

and the pressure is prescribed assuming the isentropic flow condition. The free-stream non dimensional conditions are $u_{1\infty} = u_{2\infty} = \sqrt{\gamma}$, with $\gamma = 1.4$, $T_\infty = p_\infty = 1$. Furthermore, $r = \sqrt{(x - x_0)^2 + (y - y_0)^2}$ is the distance of a generic point of the computational domain of coordinates (x, y) with respect to the vortex center, with coordinates (x_0, y_0). ϕ and α are equal to $= 1/2$ and 5, respectively. The computational domain is a square of side 10 discretized with several equi-spaced cartesian grids. Period boundary conditions are set at both the x and y directions, and the simulations are performed up to a final time equal to one period of vortex revolution.

To verify the formal order of accuracy of the GCN time integration method and its entropy conservation property, a time refinement study has been performed using the EC flux. The results of this study are shown in Fig. 1, together with the ones obtained

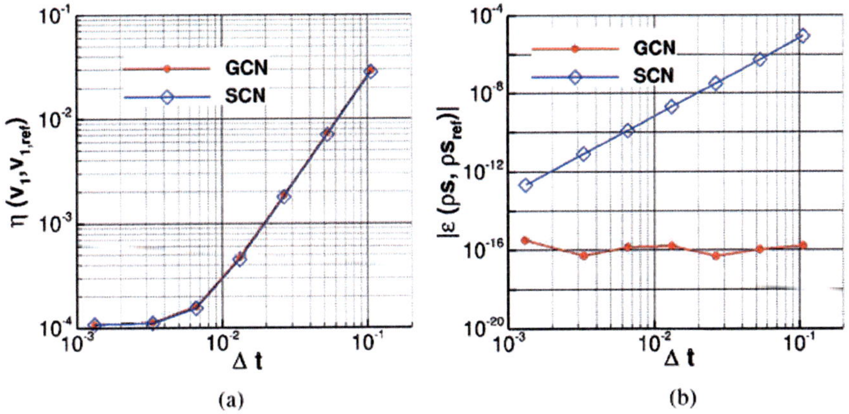

Fig. 1 Time refinement study performed with a cartesian grid composed by 50×50 elements, $k = 3$ dG approximation and the EC flux. In the plots the results obtained by using the GCN and the SCN schemes are reported. **(a)** η errors of v_1. **(b)** ε errors of ρs

using the SCN scheme as comparison. The plot on the left shows the convergence of the η error, see Eq. (17), computed for the first entropy variable only, v_1, since similar results were obtained for the other entropy variables. The slopes of the curves related to the GCN and the SCN schemes confirm their designed order of accuracy equal to 2. Furthermore, the plot shows that for a given time step size the two time integration schemes achieve the same accuracy. The plot on the right of the figure, which report the absolute values of ε error for the ρs quantity, see Eq. (18), confirms the entropy conservation property of the GCN scheme, since, independently from the time step size used, the ε error is around the order of the machine precision. The SCN scheme, as expected, shows instead different values of this error depending on the time step sizes used, with a convergence rate equal to 4. In fact, due to the definition of ε error, the convergence rate measured in this case mainly refers to dissipation error only.

In Fig. 2 are reported the results of another time refinement study, performed to highlight the entropy conserving/stable properties of several numerical fluxes. By looking at the plot on the left, it is possible to conclude that all the entropy stable numerical fluxes here investigated show, like the entropy conserving ones, roughly the same plateau value, that is of order 10^{-4} for the entropy stable fluxes and 10^{-3} for the entropy conservative ones. The plot on the right confirms the entropy conservation property of the GCN time integration scheme, since the ε errors are around the machine precision, independently from the time step size used, when it is coupled with an entropy conserving flux, and are always positive (of the order of 10^{-7}) when it is coupled with an entropy stable flux.

Figure 3 shows the results of a spatial accuracy test, performed with the EC and the ESRO numerical fluxes. In this case a very small time step size has been used for each simulations in order to evaluate the spatial accuracy. As reported in the plots,

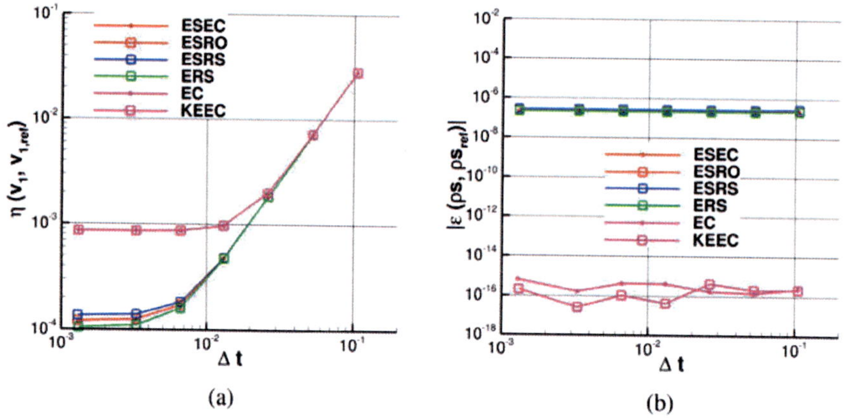

Fig. 2 Time refinement study performed with a cartesian grid composed by 25 × 25 elements, $k = 3$ dG approximation and several entropy conserving/stable fluxes. In the plots the results obtained by using the GCN scheme are reported. **(a)** η errors of v_1. **(b)** ε errors of ρs

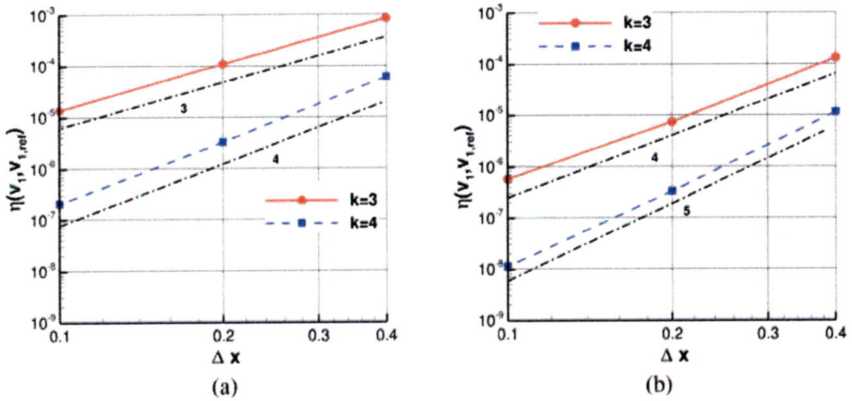

Fig. 3 Spatial refinement study performed on three grids (25 × 25, 50 × 50 and 100 × 100), $k = 3$ and $k = 4$ dG approximations, EC and ESRO numerical fluxes. (**a**) η errors of v_1 with EC. (**b**) η errors of v_1 with ESRO

both the odd ($k = 3$) and even ($k = 4$) dG approximations have a convergence rate equal only to k when the EC flux is used and equal to $k + 1$ when the ESRO flux is employed. In the plots only the results for the first entropy variable are reported since the same convergence rates are obtained for the other entropy variables.

3.2 *The Inviscid Double Shear Layer*

The initial flow field of this test case is:

$$u_1 = \begin{cases} U_\infty \tanh\left[(y - \pi/2)/\delta_1\right] & \text{if } y \leq \pi, \\ U_\infty \tanh\left[(3\pi/2 - y)/\delta_1\right] & \text{if } y > \pi, \end{cases}$$
$$u_2 = U_\infty \delta_2 \sin(x)$$
$$p = 1 \quad \text{and} \quad T = 1 \tag{20}$$

where $U_\infty = M_\infty \sqrt{\gamma}$, with $\gamma = 1.4$, $\delta_1 = \pi/15$, $\delta_2 = 0.05$, and, in order to obtain a nearly incompressible flow, $M_\infty = 0.01$. The numerical setup is as follows: the computational domain is a square of side 2π discretized with a 8×8 uniform cartesian grid and periodic boundary conditions are imposed in both the x and y directions. The simulations are performed up to a final time $T = 80$, which is a value ten times larger than the one usually used in literature [5].

The results obtained by using the GCN time integration scheme with different time step sizes, the KEEC numerical flux and several dG approximations are reported in Fig. 4. In particular, at the top row of the figure it is reported the ε error

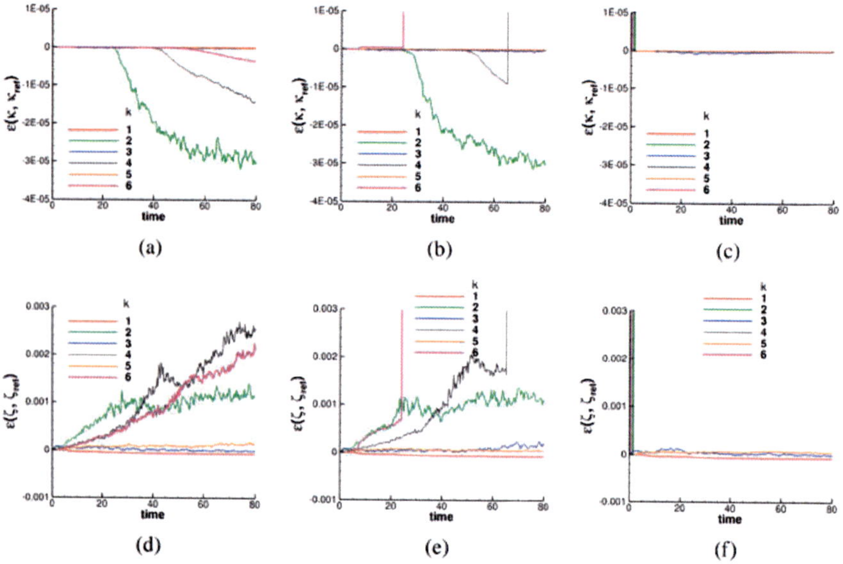

Fig. 4 Simulations performed with a cartesian grid composed by 8 × 8 elements, up to $k = 6$ dG approximations, the KEEC entropy conserving flux and final time $T = 80$. In the plots are shown the time evolution of ε errors computed for kinetic energy (top row) and enstrophy (bottom row) obtained by using the GCN scheme and several time step sizes. **(a)** $\Delta t = T/800$. **(b)** $\Delta t = T/400$. **(c)** $\Delta t = T/200$. **(d)** $\Delta t = T/800$. **(e)** $\Delta t = T/400$. **(f)** $\Delta t = T/200$

computed for the kinetic energy κ, defined as $\kappa = \frac{1}{2}\rho u_i u_i$ and $i = 1, 2$. At the bottom row of the figure it is shown the ε error related to the enstrophy ζ, defined as $\zeta = \omega^2$, where ω is the vorticity. The plots on the left ($\Delta t = T/800$) show an evident odd-even effect of the dG approximation [6]. All the even approximations display decreasing values of kinetic energy and increasing values of enstrophy, while all the odd approximations show better preservation property of both these variables. The odd-even effect is confirmed even by the worst stability properties shown by all the even approximations. By looking at the figure, in fact, for $\Delta t = T/400$ it was not possible to obtain a convergent solution for $k = 6$ and $k = 4$, and, for $\Delta t = T/200$, it was possible to obtain a convergent solution only for the odd approximations.

Figure 5 shows the results performed by using the GCN time integration scheme, the entropy stable flux ESEC and $k = 5$ and $k = 6$ dG approximations. Overall, it is possible to note that in this case there is no odd-even effect of the dG approximation. Furthermore, the preservation properties of the investigated variables are independent from the time step size used, thus confirming the entropy conservation property of the GCN time integration scheme.

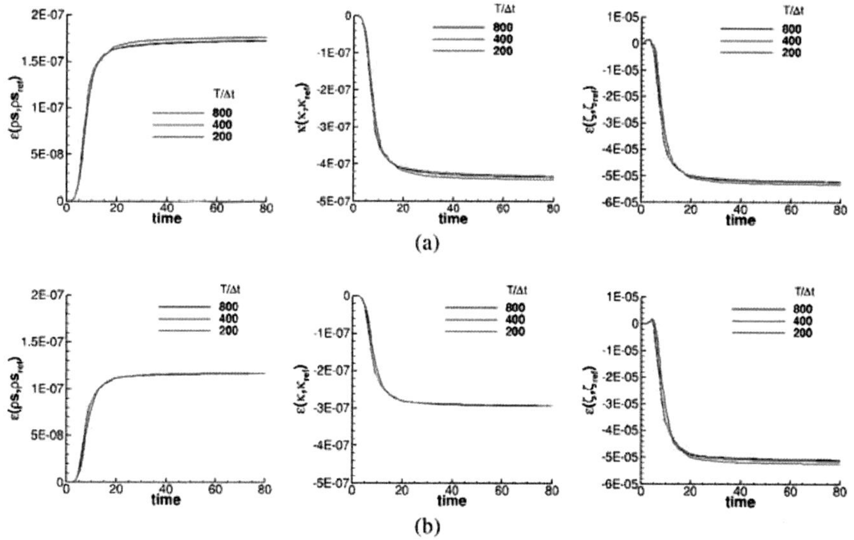

Fig. 5 Simulations performed with a cartesian grid composed by 8×8 elements, the ESEC entropy stable flux, final time $T = 80$ and $k = 5$ (top row) and $k = 6$ (bottom row) dG approximations. In the plots are shown the time evolutions of ε errors computed for ρs, kinetic energy κ and enstrophy ζ obtained by using the GCN scheme and several time step sizes. (**a**) dG approximation $k = 5$. (**b**) dG approximation $k = 6$

4 Conclusions

In this article has been presented a fully discrete implicit high-order accurate discontinuous Galerkin (dG) entropy conserving/stable scheme for inviscid flows. The accuracy and the conservation properties of the numerical fluxes and of the time integration scheme used have been evaluated for two unsteady flows, performing, for both the test cases, several temporal and spatial refinement studies. The temporal refinement studies have confirmed the designed order of convergence of two for the generalized Crank-Nicolson scheme, used for time integration, and its entropy conservation property. The spatial refinement study has put in evidence a sub-optimal order of convergence of the dG approximation when using the entropy conserving numerical fluxes. Furthermore, the results presented for the double shear layers and a *long-time* simulation, have shown a clear odd/even effect of the dG approximation when an entropy conserving flux is used. The even approximations show very inaccurate kinetic energy and enstrophy and are less stable than the odd ones. On the contrary, when an entropy stable flux is used, there is no odd-even effect, and the results confirm the entropy conserving property of the generalized Crank-Nicolson scheme.

References

1. Balay, S., et al.: PETSc web page. http://www.mcs.anl.gov/petsc
2. Bassi, F., Botti, L., Colombo, A., Di Pietro, D.A., Tesini, P.: On the flexibility of agglomeration based physical space discontinuous Galerkin discretizations. J. Comput. Phys. **231**(1), 45–65 (2012)
3. Bassi, F., Botti, L., Colombo, A., Crivellini, A., Ghidoni, A., Massa, F.: On the development of an implicit high-order discontinuous Galerkin method for DNS and implicit LES of turbulent flows. Eur. J. Mech.-B/Fluids **55**, 367–379 (2016)
4. Chandrashekar, P.: Kinetic energy preserving and entropy stable finite volume schemes for compressible Euler and Navier-Stokes equations. Commun. Comput. Phys. **14**(5), 1252–1286 (2013)
5. Chen, S., Yan, C., Lou, S., Lin, B.: An improved entropy-consistent Euler flux in low Mach number. J. Comput. Sci. **27**, 271–283 (2018)
6. Colombo, A., Crivellini, A., Nigro, A.: On the entropy conserving/stable implicit DG discretization of the Euler equations in entropy variables. Comput. Fluids **232**, 105198 (2021)
7. Colombo, A., Crivellini, A., Nigro, A.: Entropy conserving implicit time integration in a discontinuous Galerkin solver in entropy variables. J. Comput. Phys. **472**, 111683 (2023)
8. Crank, J., Nicolson, P.: A practical method for numerical evaluation of solutions of partial differential equations of the heat conduction type. Math. Proc. Cambridge Philos. Soc. **43**(1), 50–67 (1947)
9. Fisher, T.C., Carpenter, M.H.: High-order entropy stable finite difference schemes for nonlinear conservation laws: finite domains. Technical Report NASA/TM-2013- 217971 (2013)
10. Fisher, T.C., Carpenter, M.H.: High-order entropy stable finite difference schemes for nonlinear conservation laws: finite domains. J. Comput. Phys. **252**, 518–557 (2013)
11. Fjordholm, U.S., Mishra, S., Tadmor, E.: Arbitrarily high-order accurate entropy stable essentially nonoscillatory schemes for systems of conservation laws. SIAM J. Numer. Anal. **50**(2), 544–573 (2012)
12. Gottlieb, J., Groth, C.: Assessment of riemann solvers for unsteady one-dimensional inviscid flows of perfect gases. J. Comput. Phys. **78**(2), 437–458 (1988)
13. Gouasmi, A., Murman, S.M., Duraisamy, K.: Entropy conservative schemes and the receding flow problem. J. Sci. Comput. **78**(2), 971–994 (2019)
14. Hu, C., Shu, C.: Weighted essentially non-oscillatory schemes on triangular meshes. J. Comput. Phys. **150**, 97–127 (1999)
15. Ismail, F., Roe, P.: Affordable, entropy-consistent Euler flux functions II: entropy production at shocks. J. Comput. Phys. **228**(15), 5410–5436 (2009)
16. LeFloch, P.G., Mercier, J.M., Rohde, C.: Fully discrete, entropy conservative schemes of arbitrary order. SIAM J. Numer. Anal. **40**(5), 1968–1992 (2003)
17. Ranocha, H.: Entropy conserving and kinetic energy preserving numerical methods for the Euler equations using summation-by-parts operators. In: Sherwin, S.J., Moxey, D., Peiró, J., Vincent, P.E., Schwab, C. (eds.) Spectral and High Order Methods for Partial Differential Equations ICOSAHOM 2018, pp. 525–535. Springer, Cham (2020)
18. Ranocha, H., Gassner, G.J.: Preventing pressure oscillations does not fix local linear stability issues of entropy-based split-form high-order schemes. Commun. Appl. Math. Comput. **4**, 880–903 (2022)
19. Ray, D., Chandrashekar, P.: An entropy stable finite volume scheme for the two dimensional Navier-Stokes equations on triangular grids. Appl. Math. Comput. **314**, 257–286 (2017)
20. Tadmor, E.: Entropy stability theory for difference approximations of nonlinear conservation laws and related time-dependent problems. Acta Numer. **12**, 451–512 (2003)
21. Wang, L., Mavriplis, D.: Implicit solution of the unsteady euler equations for high-order accurate discontinuous Galerkin discretizations. J. Comput. Phys. **225**, 1994–2015 (2007)
22. Yee, H., Sandham, N., Djomehri, M.: Low dissipative high order shock-capturing methods using characteristic-based filters. J. Comput. Phys. **150**, 199–238 (1999)

Inexact IETI-DP for Conforming Isogeometric Multi-Patch Discretizations

Rainer Schneckenleitner and Stefan Takacs

1 Introduction

We are interested in a fast solver for linear systems that are obtained from the discretization of boundary value problems using Isogeometric Analysis (IgA; [5]) schemes. We consider computational domains that are composed of multiple non-overlapping patches, for which FETI-DP type algorithms are a canonical choice. Adaptations of FETI-DP, introduced in [2], have already been made to IgA, see, e.g., [3, 7]. This approach is sometimes called Dual-Primal Isogeometric Tearing and Interconnecting (IETI-DP) method. Recently, a convergence analysis for IETI-DP methods that is explicit in the grid sizes, the patch diameters, the spline degree and other parameters like the smoothness of the splines within the patches or the number of patches was carried out for a conforming Galerkin IgA discretization, see [10]. There, the authors considered a Schur complement IETI-DP method, where the subdomain problems are solved with sparse direct solvers. In case of large subdomain problems, direct solvers slow down the overall algorithm and require a lot of memory resources. The saddle point formulation of IETI-DP allows the use of inexact local solvers. The successful use of inexact solvers for FETI-DP has already been demonstrated in [4, 6]. In this paper, we use the fast diagonalization (FD) method introduced in [9] to construct solvers for the local subproblems. We show that the inexact IETI-DP version satisfies the same condition number bound as the IETI-DP solver from [10].

R. Schneckenleitner (✉)
Electrical Engineering Unit, Tampere University, Tampere, Finland
e-mail: rainer.schneckenleitner@tuni.fi

S. Takacs
Institute of Computational Mathematics, Johannes Kepler University Linz, Linz, Austria
e-mail: stefan.takacs@numa.uni-linz.ac.at

J. M. Melenk et al. (eds.), *Spectral and High Order Methods for Partial Differential Equations ICOSAHOM 2020+1*, Lecture Notes in Computational Science and Engineering 137, https://doi.org/10.1007/978-3-031-20432-6_26

The structure of the paper is as follows. Section 2 is devoted to the introduction of the model problem and the IETI-DP solver. In Sect. 3, we give a condition number estimate of the preconditioned system. Numerical results are presented in Sect. 4.

2 Model Problem and Its Solution

Let $\Omega \subset \mathbb{R}^2$ be an open and bounded Lipschitz domain with boundary $\partial\Omega$ and $f \in L_2(\Omega)$ be a given source function. We consider the following model problem: Find $u \in H_0^1(\Omega)$ such that

$$\int_\Omega \nabla u \cdot \nabla v \,\mathrm{d}x = \int_\Omega f v \,\mathrm{d}x \qquad \text{for all } v \in H_0^1(\Omega). \tag{1}$$

We assume that Ω is composed of K non-overlapping patches $\Omega^{(k)}$ that are parameterized with geometry mappings

$$G_k : \widehat{\Omega} := (0,1)^d \to \Omega^{(k)} := G_k(\widehat{\Omega}),$$

where any two patches with non-empty intersection share either a common vertex or a common edge, cf. [10, Ass. 2]. Additionally, we assume that the number of patches sharing a vertex is uniformly bounded, cf. [10, Ass. 3]. Moreover, we assume that there is a constant C_G such that

$$\|\nabla G_k\|_{L_\infty(\widehat{\Omega})} \le C_G\, H_k \quad \text{and} \quad \|(\nabla G_k)^{-1}\|_{L_\infty(\widehat{\Omega})} \le C_G\, H_k^{-1}, \tag{2}$$

where $H_k := \mathrm{diam}(\Omega^{(k)})$ is the patch size, see [10, Ass. 1]. The local discretization spaces on the parameter domain $\widehat{\Omega}$ are tensor-product B-splines spaces of degree p with bases obtained using the Cox-de Boor formula. We assume that these spaces are based on quasi-uniform grids with sizes $\widehat{h}_k$, see [10, Ass. 4]. The local discretization spaces on the physical patches $\Omega^{(k)}$ are obtained by the pull-back principle. The quantity $h_k := H_k \widehat{h}_k$ measures the grid size on the physical domain. We assume that the geometry mappings as well as the discretizations agree on all interfaces between patches, cf. [10, Ass. 5]. So, we are able to set up a fully matching discretization with the function space

$$V = \{v \in H_0^1(\Omega) : v \circ G_k \text{ is a B-spline function}\} \cap C(\Omega).$$

The corresponding discrete problem is obtained by restricting the variational problem (1) to this space.

In the following, we introduce the IETI-DP solver. The patches from the definition of the computational domain provide a canonical choice of substructures which we use to set up the solver. By assembling the variational problem (1) on the patches separately, we obtain yet uncoupled local systems

$$A^{(k)}\underline{u}^{(k)} = \underline{f}^{(k)} \quad \text{for} \quad k = 1, \dots, K,$$

where $A^{(k)}$ is the local stiffness matrix and $\underline{f}^{(k)}$ is the local source vector.

Following the DP approach, we need to select a set of primal degrees of freedom. We restrict ourselves to choosing the function values at the corners of the patches as primal degrees of freedom, see [10, Alg. A]. By splitting of the degrees of freedom into the corner values (index C) and the remaining degrees of freedom (index Δ), we obtain

$$A^{(k)} = \begin{pmatrix} A^{(k)}_{CC} & A^{(k)}_{C\Delta} \\ A^{(k)}_{\Delta C} & A^{(k)}_{\Delta\Delta} \end{pmatrix}, \quad \underline{u}^{(k)} = \begin{pmatrix} \underline{u}^{(k)}_{C} \\ \underline{u}^{(k)}_{\Delta} \end{pmatrix}, \quad \underline{f}^{(k)} = \begin{pmatrix} \underline{f}^{(k)}_{C} \\ \underline{f}^{(k)}_{\Delta} \end{pmatrix}. \tag{3}$$

As local spaces, we choose the patch-local functions where the corner values vanish, so the still uncoupled local systems are

$$A^{(k)}_{\Delta\Delta}\underline{u}^{(k)}_{\Delta\Delta} = \underline{f}^{(k)}_{\Delta\Delta} \quad \text{for} \quad k = 1, \dots, K. \tag{4}$$

Moreover, we need a primal space spanned by energy minimizing basis functions that form a nodal basis for the primal degrees of freedom. This basis is represented by the matrix $\Psi^{(k)}$ that is characterized by the linear system

$$\underbrace{\begin{pmatrix} A^{(k)} & (C^{(k)})^\top \\ C^{(k)} & \end{pmatrix}}_{=:\,\widetilde{\mathcal{A}}^{(k)}} \begin{pmatrix} \Psi^{(k)} \\ \Delta^{(k)} \end{pmatrix} = \begin{pmatrix} 0 \\ R^{(k)}_c \end{pmatrix}, \tag{5}$$

where $C^{(k)} = (C^{(k)}_C, C^{(k)}_\Delta) = (I, 0)$ is a matrix that represents the evaluation of the primal degrees of freedom and $R^{(k)}_c$ is a binary local-to-global mapping that identifies the local ordering of the primal degrees of freedom and their global ordering. We compute $\Psi^{(k)}$ by solving (5) using a MINRES solver, preconditioned with

$$\widetilde{P}^{(k)} = \begin{pmatrix} (\widehat{A}^{(k)}_M)^{-1} & \\ & (C^{(k)}(\widehat{A}^{(k)}_M)^{-1}C^{(k)})^{-1} \end{pmatrix},$$

where

$$\widehat{A}^{(k)}_M := \widehat{A}^{(k)} + \gamma_k \widehat{M}^{(k)}\underline{e}^{(k)}_h(\underline{e}^{(k)}_h)^\top \widehat{M}^{(k)},$$

$\widehat{A}^{(k)}$ is the stiffness matrix on the parameter domain (obtained from discretizing $\int_{\widehat{\Omega}} \nabla\widehat{u} \cdot \nabla\widehat{v}\,\mathrm{d}x$), $\widehat{M}^{(k)}$ is the analogously defined mass matrix and $\underline{e}^{(k)}_h$ represents the constant function with value 1. We choose $\gamma_k = 1$ if $\Omega^{(k)}$ does not contribute to the Dirichlet boundary $\partial\Omega$ and $\gamma_k = 0$ otherwise. The application of $(\widehat{A}_M^{(k)})^{-1}$ is realized with the FD method, see [9]. After the computation of all matrices $\Psi^{(k)}$, we obtain the global primal basis representation matrix Ψ by canonical mappings.

To ensure continuity, we introduce jump matrices $B^{(k)}$, where the condition $\sum_{k=1}^K B^{(k)}\underline{u}^{(k)} = 0$ holds if and only if the basis functions are continuous between the patches (except the continuity at the corners), in the usual way, see [10, Sect. 3]. The matrices $B_\Delta^{(k)}$ are obtained from $B^{(k)}$ again by eliminating the entries corresponding to the primal degrees of freedom.

The overall IETI-DP saddle point system is obtained by coupling the local systems (4) and the primal system using the jump matrices and reads as follows:

$$\begin{pmatrix} A_{\Delta\Delta}^{(1)} & & & & (B_\Delta^{(1)})^\top \\ & \ddots & & & \vdots \\ & & A_{\Delta\Delta}^{(K)} & & (B_\Delta^{(K)})^\top \\ & & & A_\Pi & B_\Pi^\top \\ B_\Delta^{(1)} & \cdots & B_\Delta^{(K)} & B_\Pi & 0 \end{pmatrix} \begin{pmatrix} \underline{u}_\Delta^{(1)} \\ \vdots \\ \underline{u}_\Delta^{(K)} \\ \underline{u}_\Pi \\ \underline{\lambda} \end{pmatrix} = \begin{pmatrix} \underline{f}_\Delta^{(1)} \\ \vdots \\ \underline{f}_\Delta^{(K)} \\ \underline{f}_\Pi \\ 0 \end{pmatrix}, \tag{6}$$

where

$$A_\Pi := \sum_{k=1}^K (\Psi^{(k)})^\top A^{(k)} \Psi^{(k)}, \quad \underline{f}_\Pi := \Psi^\top \underline{f} \quad \text{and} \quad B_\Pi := \sum_{k=1}^K B^{(k)} \Psi^{(k)}.$$

We solve (6) with a MINRES solver, preconditioned with

$$\mathcal{P} := \operatorname{diag}\,(P, A_\Pi^{-1}, \widehat{M}_{\text{sD}}), \tag{7}$$

where $P := \operatorname{diag}\,(P^{(1)}, \cdots, P^{(K)})$ and $P^{(k)} := Q^{(k)}(\widehat{A}_M^{(k)})^{-1}(Q^{(k)})^\top$ and $Q^{(k)} := ((A_{\Delta\Delta}^{(k)})^{-1}A_{\Delta C}^{(k)}, I)$ is the $A^{(k)}$-orthogonal projection from the local function space into the space of functions with vanishing corner values; its entries can be extracted from $\Psi^{(k)} = (I, -((A_{\Delta\Delta}^{(k)})^{-1}A_{\Delta C}^{(k)})^\top)^\top R_c$. The matrix $\widehat{A}_M^{(k)}$ is defined as above and realized using the FD method. A_Π^{-1} is realized using a direct solver. For the setup of the inexact scaled Dirichlet preconditioner $\widehat{M}_{\text{sD}}$, we define analogously to (3) a splitting into basis functions vanishing at the interfaces (index I) and remaining basis functions (index Γ). Then, we define

$$\widehat{M}_{\text{sD}} := B_\Gamma D^{-1} \widehat{S} D^{-1} B_\Gamma^\top,$$

where $\widehat{S} := \operatorname{diag}(\widehat{S}^{(1)}, \ldots, \widehat{S}^{(K)})$ with $\widehat{S}^{(k)} := \widehat{A}_{\Gamma\Gamma}^{(k)} - \widehat{A}_{\Gamma I}^{(k)}(\widehat{A}_{II}^{(k)})^{-1}\widehat{A}_{I\Gamma}^{(k)}$. The diagonal matrix D is based on the principle of multiplicity scaling, cf. [10]. The application of $(\widehat{A}_{II}^{(k)})^{-1}$ is also realized using the FD method.

3 Analysis of the Method

3.1 Analysis of the Main Iteration

First, we show that the chosen preconditioners for the local problems are optimal up to constants.

Lemma 1 *The preconditioner* $P^{(k)} = Q^{(k)}(\widehat{A}_M^{(k)})^{-1}(Q^{(k)})^\top$ *is spectrally equivalent to the matrix* $(A_{\Delta\Delta}^{(k)})^{-1}$, *i.e.,* $P^{(k)} \eqsim (A_{\Delta\Delta}^{(k)})^{-1}$.

Proof We show the equivalent statement $A_{\Delta\Delta}^{(k)} P^{(k)} A_{\Delta\Delta}^{(k)} \eqsim A_{\Delta\Delta}^{(k)}$. Since we assume (2), [10, Lemma 4.13] yields $A^{(k)} \eqsim \widehat{A}^{(k)}$. From this observation, the desired result follows immediately if $\partial\Omega^{(k)} \cap \partial\Omega \neq \emptyset$. For the case $\partial\Omega^{(k)} \cap \partial\Omega = \emptyset$, we define $\underline{e}_h$ be the vector representing the constant function with value 1. Due to the definitions of $P^{(k)}$, the $A^{(k)}$-orthogonal projection $Q^{(k)}$, we get using the norm equivalence, [10, Lemma 4.13], that

$$
\begin{aligned}
A_{\Delta\Delta}^{(k)} P^{(k)} A_{\Delta\Delta}^{(k)} &= A_{\Delta\Delta}^{(k)} Q^{(k)} (\widehat{A}^{(k)} + \widehat{M}^{(k)} \underline{e}_h \underline{e}_h^\top \widehat{M}^{(k)})^{-1} (Q^{(k)})^\top A_{\Delta\Delta}^{(k)} \\
&= \begin{pmatrix} A_{\Delta C}^{(k)} & A_{\Delta\Delta}^{(k)} \end{pmatrix} (\widehat{A}^{(k)} + \widehat{M}^{(k)} \underline{e}_h \underline{e}_h^\top \widehat{M}^{(k)})^{-1} \begin{pmatrix} A_{C\Delta}^{(k)} \\ A_{\Delta\Delta}^{(k)} \end{pmatrix} \\
&\eqsim \begin{pmatrix} 0 & I \end{pmatrix} A^{(k)} (A^{(k)} + X)^{-1} A^{(k)} \begin{pmatrix} 0 \\ I \end{pmatrix} \\
&= \underbrace{\begin{pmatrix} 0 & I \end{pmatrix} A^{(k)} \begin{pmatrix} 0 \\ I \end{pmatrix}}_{= A_{\Delta\Delta}^{(k)}} - \begin{pmatrix} 0 & I \end{pmatrix} A^{(k)} (A^{(k)} + X)^{-1} X \begin{pmatrix} 0 \\ I \end{pmatrix},
\end{aligned}
$$

where $X := H_k^{-4} M^{(k)} \underline{e}_h \underline{e}_h^\top M^{(k)}$ and $M^{(k)}$ is the mass matrix on the physical patch. From $(A^{(k)} + X)\underline{e}_h = 0 + H_k^{-4} M^{(k)} \underline{e}_h \underline{e}_h^\top M^{(k)} \underline{e}_h = |\Omega^{(k)}| H_k^{-4} M^{(k)} \underline{e}_h$ we obtain $(A^{(k)} + X)^{-1} M^{(k)} \underline{e}_h = |\Omega^{(k)}|^{-1} H_k^4 \underline{e}_h$. In total, we obtain

$$
A_{\Delta\Delta}^{(k)} P^{(k)} A_{\Delta\Delta}^{(k)} \eqsim A_{\Delta\Delta}^{(k)} - |\Omega^{(k)}|^{-1} H_k^4 \begin{pmatrix} 0 & I \end{pmatrix} \underbrace{A^{(k)} \underline{e}_h}_{=0} \underline{e}_h^\top M^{(k)} \begin{pmatrix} 0 \\ I \end{pmatrix},
$$

which finishes the proof.

Let $B_\Delta := (B_\Delta^{(1)}, \ldots, B_\Delta^{(K)})$. Using algebraic reformulations and Lemma 1, it follows that the approximate Schur complement matrix $\widehat{F} = B_\Delta P B_\Delta^\top + B_\Pi A_\Pi^{-1} B_\Pi^\top$ is spectrally equivalent to the IETI-DP matrix F from [10].

An analogous statements holds for the scaled Dirichlet preconditioner $M_{\mathrm{sD}} = B_\Gamma D^{-1} S D^{-1} B_\Gamma^\top$ and $\widehat{M}_{\mathrm{sD}} = B_\Gamma D^{-1} \widehat{S} D^{-1} B_\Gamma^\top$ by using the fact that $\widehat{S}^{(k)}$ and $S^{(k)}$

represent discrete harmonic extensions, i.e., the extensions that minimize the energy norm and the norm equivalence, [10, Lemma 4.13], again applicable because of (2).

The following theorem shows that the use of our proposed preconditioner P does not degrade the qualitative behavior of the condition number estimate from [10, Theorem 4.1] for the overall system.

Theorem 2 *Under the presented assumptions, the condition number of the system (6) preconditioned with (7) is bounded by*

$$C\,p\left(1+\log p+\max_{k=1,\dots,K}\log\frac{H_k}{h_k}\right)^2,$$

where the constant C only depends on the constant C_G, the quasi-uniformity constant (see [10, Ass. 4]) and the maximum number of patches sharing a vertex (see [10, Ass. 3]).

Proof All constants in this theorem are positive and only depend on the abovementioned constants. Using [8, Theorem 22] and [10, Theorem 4.1], we have

$$\sigma(M_{\text{sD}}F)\subseteq\left[1,\,p\left(1+\log p+\max_{k=1,\dots,K}\log\frac{H_k}{h_k}\right)^2\overline{\sigma}_1\right]$$

for some constant $\overline{\sigma}_1$. Using the equivalences $F\eqsim\widehat{F}$ and $M_{\text{sD}}\eqsim\widehat{M}_{\text{sD}}$, we obtain

$$\sigma(\widehat{M}_{\text{sD}}\widehat{F})\subseteq\left[\underline{\sigma}_2,\,p\left(1+\log p+\max_{k=1,\dots,K}\log\frac{H_k}{h_k}\right)^2\overline{\sigma}_2\right] \tag{8}$$

for constants $\underline{\sigma}_2$ and $\overline{\sigma}_2$. Using Lemma 1, we get

$$\sigma(PA_{\Delta\Delta})\subseteq[\underline{\sigma}_3,\overline{\sigma}_3], \tag{9}$$

where $A_{\Delta\Delta}:=\text{diag}\,(A^{(1)}_{\Delta\Delta},\dots,A^{(K)}_{\Delta\Delta})$, with some constants $\underline{\sigma}_3$ and $\overline{\sigma}_3$. The theorem of Brezzi, cf. [1], in combination with (8) and (9) concludes the proof.

3.2 Analysis of the System for the Primal Basis

In this subsection, we show a condition number result that guarantees that the local MINRES solvers required to obtain the local bases $\Psi^{(k)}$ can be realized with a computational cost that does not exceed the complexity of the main iteration. First, we need to show an auxiliary lemma.

Lemma 3 *Let $\widehat{u}$ be a B-spline function of degree p over a quasi uniform grid on $\widehat{\Omega}$ with grid size $\widehat{h}$ and assume that $\widehat{u}(0) = 0$. Then,*

$$\left(\int_{\widehat{\Omega}} \widehat{u}(x)\mathrm{d}x\right)^2 \lesssim \left(1 + \log p + \log \frac{1}{\widehat{h}}\right) |\widehat{u}|^2_{H^1(\widehat{\Omega})}.$$

Proof We decompose $\widehat{u} = \widehat{u}_0 + c$ such that $\int_{\widehat{\Omega}} \widehat{u}_0(x)\,\mathrm{d}x = 0$ and $c = \int_{\widehat{\Omega}} \widehat{u}(x)\,\mathrm{d}x \in \mathbb{R}$. [10, Lemma 4.14] yields

$$|\widehat{u}_0(0)|^2 \lesssim \left(1 + \log p + \log \frac{1}{\widehat{h}}\right) \|\widehat{u}_0\|^2_{H^1(\widehat{\Omega})}.$$

Using Poincaré's inequality, we obtain

$$|\widehat{u}_0(0)|^2 \lesssim \left(1 + \log p + \log \frac{1}{\widehat{h}}\right) |\widehat{u}_0|^2_{H^1(\widehat{\Omega})} = \left(1 + \log p + \log \frac{1}{\widehat{h}}\right) |\widehat{u}|^2_{H^1(\widehat{\Omega})}.$$

$\widehat{u}(0) = 0$ implies $|\widehat{u}_0(0)| = |c| = |\int_{\widehat{\Omega}} \widehat{u}(x)\,\mathrm{d}x|$, which finishes the proof.

Theorem 4 *For $k = 1, \ldots, K$, the condition number of $\widetilde{P}^{(k)}\widetilde{\mathcal{A}}^{(k)}$ is bounded by*

$$C\left(1 + \log p + \log \frac{H_k}{h_k}\right),$$

where the constant C only depends on the constant C_G and the quasi-uniformity constant (see [10, Ass. 4]).

Proof We prove this equivalence by showing that the assumptions of Brezzi's theorem [1] are satisfied. By the norm equivalence [10, Lemma 4.13], we obtain $A^{(k)} \eqsim \widehat{A}^{(k)}$. In the kernel of $C^{(k)}$, we use Lemma 3 to obtain the equivalence $A^{(k)} \eqsim \widehat{A}^{(k)} \eqsim \widehat{A}^{(k)}_M$. This kernel coercivity and the boundedness assumptions on $A^{(k)}$ are satisfied. The remaining inf-sup and boundedness conditions are trivially fulfilled due to the use of a Schur complement preconditioner. Hence, the statement of the theorem follows.

4 Numerical Results

In this section, we show numerical results of the proposed inexact IETI-DP method for the computational domains as given in Fig. 1. The first domain is a quarter annulus consisting of 32 patches and the second one is the Yeti-footprint decomposed into 84 patches.

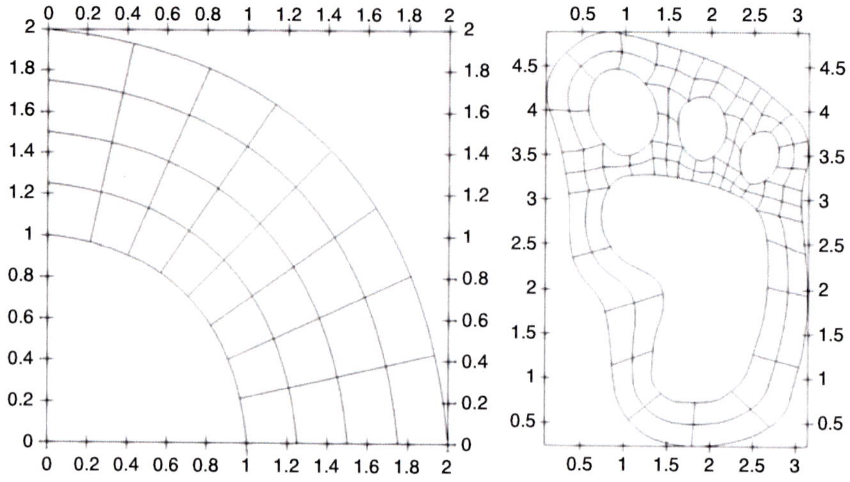

Fig. 1 Computational domains; quarter annulus (left); Yeti-footprint (right)

We consider the model problem

$$-\Delta u(x, y) = 2\pi^2 \sin(\pi x) \sin(\pi y) \quad \text{for} \quad (x, y) \in \Omega$$

with homogeneous Dirichlet boundary conditions. Within each patch, we use B-splines of degree p and maximum smoothness C^{p-1}. The coarsest discretization space ($r = 0$) is the space of patchwise global polynomials, only the more rectangular patches of the Yeti-footprint have one inner knot on each of the longer sides of the patches. The subsequent refinements $r = 1, 2, \ldots$ are obtained via uniform refinement steps.

All numerical experiments have been carried out using the C++ library G+Smo,[1] the CPU times have been recorded on the Radon1[2] Cluster in Linz.

We compare three different IETI-DP solvers: the proposed solver as introduced in Sect. 2 (=MFD), a IETI-DP solver for the saddle point system (6) without the primal degrees of freedom eliminated with direct solvers for the local subproblems (=MLU), and the Schur complement based approach as introduced in [10] (=CGLU). We use MINRES as outer solver in the cases MFD and MLU and conjugate gradient as outer solver for the case CGLU. For MLU and CGLU, we use sparse direct LU solvers from the Pardiso project[3] for the local subproblems and for computing the bases $\Psi^{(k)}$. We start all numerical experiments with zero initial guess and stop the iterations if the ℓ_2-norm of the residual vector is reduced

---

[1] https://github.com/gismo/gismo.

[2] https://www.ricam.oeaw.ac.at/hpc/.

[3] https://www.pardiso-project.org/.

Table 1 Alg. A; $p = 5$; timings in sec.; quarter annulus

	r	Ψ	Θ_S	Θ_A	Solving	Total	It
MFD	6	1.2	0.1	0.4	4.2	5.5	71
MLU		0.8	7.6	21.2	43.0	51.4	37
CGLU		0.8	7.6	9.5	17.0	15.4	15
MFD	7	8.0	0.3	4.9	25.0	33.3	80
MLU		4.0	42.3	106.4	216.0	262.3	39
CGLU		4.0	42.3	45.4	81.0	127.3	15
MFD	8	35.5	2.0	26.9	126.3	163.8	88
MLU		18.1	243.6	503.8	1015.0	1276.7	41
CGLU		18.0	242.1	225.3	412.0	672.1	17

Table 2 Alg. A; $p = 8$; timings in sec.; quarter annulus

	r	Ψ	Θ_S	Θ_A	Solving	Total	It
MFD	6	2.5	0.2	0.5	8.2	10.9	76
MLU		1.2	17.9	33.8	72.0	91.1	39
CGLU		1.2	17.8	14.4	26.0	45.0	15
MFD	7	15.1	0.4	5.4	41.0	56.5	84
MLU		5.8	105.0	163.3	343.0	453.8	41
CGLU		5.8	105.0	73.8	134.0	244.8	17
MFD	8	60.1	2.3	29.9	200.0	262.4	94
MLU		OoM					
CGLU		OoM					

by a factor of 10^{-6} compared to the ℓ_2-norm of the right-hand side vector. For MFD, the primal bases $\Psi^{(k)}$ are solved with MINRES up to an accuracy of 10^{-8}.

In the Tables 1 and 2, we present the timings of the algorithms on the quarter annulus domain. The time required for computing the primal basis Ψ is indicated with the same letter. The accumulated setup and application times of the different local preconditioners for all patches K are indicated by Θ_S and Θ_A, respectively. Moreover, we present the overall solving times and the number of iterations. It required by the main Krylov space solver.

We observe that the solving and total times required by MFD are three to five times smaller compared to the other methods MLU and CGLU. MFD is much faster than the other methods despite the fact that the required number of iterations are up to approximately six times larger. One disadvantage of MFD is the computation of the primal basis Ψ. The tables show a larger computation time when using MFD. This is a weakness of the classical preconditioned MINRES method when applied to problems with multiple right-hand sides. In general, we have to solve systems with four right-hand sides to compute $\Psi^{(k)}$. Moreover, we see that the setup and application of the FD based preconditioner is much faster compared to the factorization of the matrices $\widetilde{A}^{(k)}$ and the application of these factorizations. Table 2 shows another advantage of the MFD method. Since its memory footprint

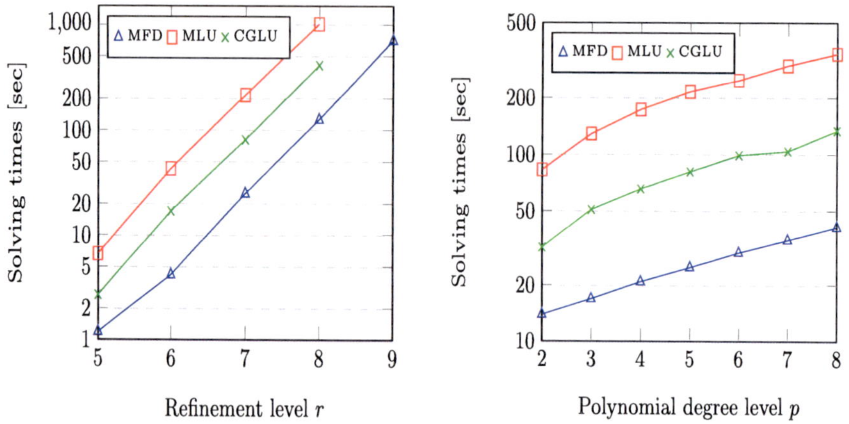

Fig. 2 Solving times for $p = 5$ (left) and $r = 7$ (right); MFD (blue lines); MLU (red lines); CGLU (green lines); quarter annulus

is smaller, it also provides a solution vector to the considered linear system for the refinement level $r = 8$. The plots in Fig. 2 visualize solving times of the IETI-DP solvers on the quarter annulus domain. We mark the performance of MFD with blue lines and triangles, MLU with red lines and squares and the performance of CGLU is indicated with green lines and crosses. In both graphs, we observe that MFD is the fastest algorithm. In the left plot, we see that for spline degree $p = 5$, the solving times increase rather linearly with respect to the number of unknowns. Moreover, the left graph shows that MFD computes the solution for the linear system even for refinement level $r = 9$ ($\approx$ 8.5M dofs). In the right graph, we present the solving times with respect to the spline degree for refinement level $r = 7$. The solving times for the three IETI-DP solvers increase about linearly with the spline degree.

The plots in Fig. 3 show solving times of the IETI-DP solvers on the Yeti-footprint. We marked the performance of the different IETI-DP solvers as above in the experiments on the quarter annulus. The plot on the left shows the increase of the solving time with respect to the refinement level with polynomial degree $p = 3$ and the plot on the right shows the increase of the solving time with respect to the polynomial degree, where we have fixed the refinement level to $r = 6$. As for the quarter annulus, we see that MFD is superior compared to MLU and CGLU also on the Yeti-footprint with respect to the solving times and the smaller memory footprint of MFD allows us to consider larger problems. In both plots, we observe similar growth rates of the solving time for all three solvers as in Fig. 2. In the Tables 3 and 4, we present and compare the required timings for the polynomial degrees $p = 3$ and $p = 7$ for different refinement levels for the Yeti-footprint.

To conclude, we presented a fast IETI-DP method which allows the incorporation of inexact solvers for the local subproblems while maintaining the condition number bound as established in [10]. It is beneficial both because of its smaller memory footprint and its faster convergence for the model problems.

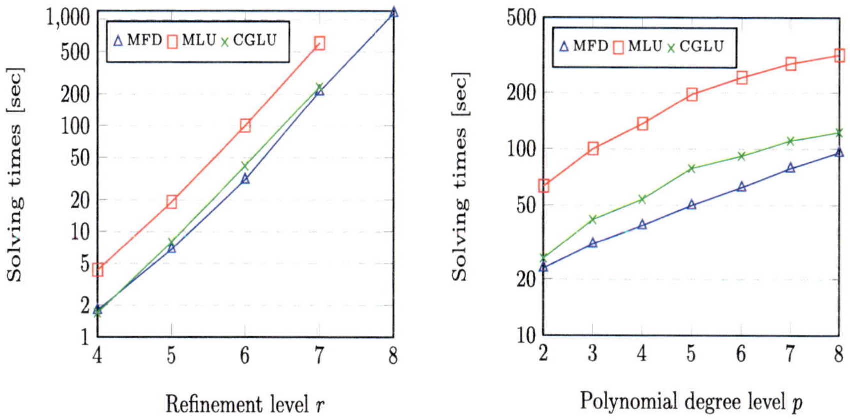

Fig. 3 Solving times for $p = 3$ (left) and $r = 6$ (right); MFD (blue lines); MLU (red lines); CGLU (green lines); Yeti-footprint

Table 3 Alg. A; $p = 3$; timings in sec.; Yeti-footprint

	r	Ψ	Θ_S	Θ_A	Solving	Total	It
MFD	5	0.7	0.1	0.8	6.8	7.6	213
MLU		0.3	2.3	9.4	19.0	21.6	45
CGLU		0.3	2.3	4.5	8.0	10.6	20
MFD	6	3.4	0.2	5.0	31.0	34.6	242
MLU		1.2	10.5	49.6	100.0	111.7	51
CGLU		1.2	10.4	23.1	42.0	53.6	22
MFD	7	21.8	1.2	52.7	212.0	235.0	274
MLU		6.8	50.3	295.2	600.5	657.6	55
CGLU		6.8	50.1	126.1	234.0	290.9	22

Table 4 Alg. A; $p = 7$; timings in sec.; Yeti-footprint

	r	Ψ	Θ_S	Θ_A	Solving	Total	It
MFD	5	2.2	0.2	1.2	20.0	22.4	249
MLU		0.6	10.5	23.5	48.0	59.1	51
CGLU		0.6	10.4	10.9	19.0	30.0	22
MFD	6	9.5	0.4	5.9	79.0	88.9	282
MLU		3.0	44.8	140.5	286.0	333.8	57
CGLU		3.0	44.8	60.6	111.0	158.8	23
MFD	7	54.2	1.5	63.2	414.0	469.7	309
MLU		14.1	265.3	739.1	1521.0	1800.4	61
CGLU		13.8	261.7	304.4	570.0	845.5	25

Acknowledgments The first author was supported by the Austrian Science Fund (FWF): S117-03 and W1214-04. Also, the second author has received support from the Austrian Science Fund (FWF): P31048.

References

1. Brezzi, F.: On the existence, uniqueness and approximation of saddle-point problems arising from lagrangian multipliers. ESAIM Math. Model. Numer. Anal. **8**(R2), 129–151 (1974)
2. Farhat, C., Lesoinne, M., Tallec, P.L., Pierson, K., Rixen, D.: FETI-DP: a dual-primal unified FETI method I: a faster alternative to the two-level FETI method. Int. J. Numer. Methods Eng. **50**, 1523–1544 (2001)
3. Hofer, C., Langer, U.: Dual-primal isogeometric tearing and interconnecting solvers for multipatch continuous and discontinuous Galerkin IgA equations. Proc. Appl. Math. Mech. **16**(1), 747–748 (2016)
4. Hofer, C., Langer, U., Takacs, S.: Inexact dual-primal isogeometric tearing and interconnecting methods. In: Bjørstad, P.E., Brenner, S.C., Halpern, L., Kim, H.H., Kornhuber, R., Rahman, T., Widlund, O.B. (eds.) Domain Decomposition Methods in Science and Engineering XXIV, pp. 393–403. Springer, Berlin (2018)
5. Hughes, T.J.R., Cottrell, J.A., Bazilevs, Y.: Isogeometric analysis: CAD, finite elements, NURBS, exact geometry and mesh refinement. Comput. Methods Appl. Mech. Eng. **194**(39–41), 4135–4195 (2005)
6. Klawonn, A., Rheinbach, O.: Inexact FETI-DP methods. Int. J. Numer. Methods Eng. **69**(2), 284–307 (2007)
7. Kleiss, S., Pechstein, C., Jüttler, B., Tomar, S.: IETI-isogeometric tearing and interconnecting. Comput. Methods Appl. Mech. Eng. **247–248**, 201–215 (2012)
8. Mandel, J., Dohrmann, C.R., Tezaur, R.: An algebraic theory for primal and dual substructuring methods by constraints. Appl. Numer. Math. **54**(2), 167–193 (2005)
9. Sangalli, G., Tani, M.: Isogeometric preconditioners based on fast solvers for the Sylvester equation. SIAM J. Sci. Comput. **38**(6), A3644–A3671 (2016)
10. Schneckenleitner, R., Takacs, S.: Condition number bounds for IETI-DP methods that are explicit in h and p. Math. Models Methods Appl. Sci. **30**(11), 2067–2103 (2020)

Split Form ALE DG Methods for the Euler Equations: Entropy Stability and Kinetic Energy Dissipation

Gero Schnücke, Gregor J. Gassner, and Nico Krais

1 Introduction

Numerical simulations related to the equations of gas dynamics, e.g. the Euler and Navier-Stokes equations, with high order methods, e.g. discontinuous Galerkin (DG) [4, 12, 18] methods, compact finite difference methods [24] or summation-by-parts (SBP) finite difference methods [29], can be affected by aliasing errors due to the strong non-linearity of the flux functions [1, 16]. These errors are generated in the volume integrals and cannot be sufficiently controlled with the dissipation of the numerical surface fluxes alone. The DG spectral element method (DGSEM), e.g., [18], is constructed with local tensor-product Lagrange polynomial basis functions computed from Legendre-Gauss-Lobatto (LGL) points. The collocation of interpolation and quadrature nodes is used in the spatial discretization. This approach ensures that the derivative matrix in the DGSEM provides a SBP operator [8]. A SBP operator gives a discrete analogue of the integration-by-parts formula [6, 8, 22] and thus ideas from the continuous entropy or kinetic energy stability analysis can be mimicked at the discrete level. In this sense, the SBP property is a powerful tool to construct the numerical approximation in a way that aliasing

G. Schnücke (✉)
Faculty of Mathematics and Computer Science, University of Jena, Jena, Germany
e-mail: gero.schnuecke@uni-jena.de

G. J. Gassner
Department for Mathematics/Computer Science, Center for Data and Simulation Science, University of Cologne, Cologne, Germany
e-mail: ggassner@uni-koeln.de

N. Krais
Institute of Aerodynamics and Gas Dynamics (IAG), University of Stuttgart, Stuttgart, Germany
e-mail: krais@iag.uni-stuttgart.de

J. M. Melenk et al. (eds.), *Spectral and High Order Methods for Partial Differential Equations ICOSAHOM 2020+1*, Lecture Notes in Computational Science and Engineering 137, https://doi.org/10.1007/978-3-031-20432-6_27

issues are avoided. Since continuous analysis is mimicked, these methods satisfy by construction discrete analogues for entropy conditions or relations connected to the kinetic energy.

A framework to construct high order entropy conserving (EC) schemes in periodic domains has been given by LeFloch et al. [23]. Fisher and Carpenter [7] combined this approach with SBP operators and proved that two-point EC fluxes can be used to construct high order schemes when the derivative approximations in space are SBP operators. Gassner et al. [9, 10] showed that skew-symmetric-like (split form) DGSEM formulations for the Euler equations can be discretely recovered when specific numerical volume fluxes in the flux form volume integral of Fisher and Carpenter are chosen. In particular, the work from Gassner et al. provides a framework to construct general split form and kinetic energy preserving (KEP) DGSEM methods for the Euler equations. Moreover, It is formally possible to construct ES and kinetic energy dissipating (KED) DGSEM, when a two-point ES or KED flux function is used as surface flux in the DG method, the derivative matrix satisfies the SBP property and a two-point EC flux or KEP flux function is used as volume flux in the flux form volume integral of Fisher and Carpenter.

On the other hand numerical simulations of turbulent flows require adaptive discretizations, where the resolution is increased in regions with large spatial variations. The r-adaptive method involves the re-distribution of the mesh nodes in regions of rapid variation of the solution [36]. A r-adaptive method can be only used when the effect of mesh movement is accounted in the discretization. This can be done by an arbitrary Lagrangian Eulerian (ALE) approach [5]. Some ALE DG methods have been developed in [19, 25, 27, 28, 30]. In [21, 33] provable EC/KEP or ES/KEP split form ALE DGSEM have been constructed.

The goal of this paper is to give a review of the split form ALE DGSEM. The structure of the present work is as follows: First, the ALE transformation of the three dimensional Euler equations is given. Then, the approaches [10, 33] are briefly reviewed and provable EC/KEP ALE DGSEM for the Euler equations on curved moving elements are constructed. Finally, the constructed methods are numerically investigated for the Taylor-Green vortex (TGV) problem [34].

2 The Euler Equations

The three dimensional compressible Euler equations are written in compact block vector form

$$\frac{\partial \mathbf{u}}{\partial t} + \nabla_x \cdot \overset{\leftrightarrow}{\mathbf{f}} (\mathbf{u}) = 0. \tag{1}$$

These equations are considered on a time-dependent domain $\Omega(t) \subseteq \mathbb{R}^3$, $t \in \mathbb{R}_{\geq 0}$, with suitable initial data and boundary conditions. The state vector is given by

$$\mathbf{u} = \left[\rho, \rho\mathbf{u}, \rho e + \frac{1}{2}\rho\, |\mathbf{u}|^2\right]^T, \tag{2}$$

where ρ is the mass density, $\mathbf{u} = [u_1, u_2, u_3]^T$ the velocity, p is the pressure and the specific internal energy is $e = p/(\gamma - 1)\rho$, $\gamma > 1$. The advective block flux $\overleftrightarrow{\mathbf{f}} = \left[\mathbf{f}_1^T, \mathbf{f}_2^T, \mathbf{f}_3^T\right]^T$ consists of the flux vectors for $\iota = 1, 2, 3$ given by

$$\begin{aligned} \mathbf{f}_\iota^1 &= \rho u_\iota, \\ \mathbf{f}_\iota^{\upsilon+1} &= \rho u_\iota u_\upsilon + p\delta_{\iota\upsilon}, \qquad \upsilon = 1, 2, 3, \\ \mathbf{f}_\iota^5 &= \rho u_\iota \left[e + \frac{1}{2}|\mathbf{u}|^2\right] + p u_\iota, \end{aligned} \tag{3}$$

where $\delta_{\iota\upsilon}$ is the Kronecker delta.

2.1 The ALE Transformation

The domain $\Omega(t)$ is divided in K time-dependent non-overlapping curved elements $e_k(t)$, $k = 1, \ldots, K$. Each element $e_k(t)$ is mapped onto a reference element $\mathrm{E} = [-1, 1]^3$ by a time-dependent mapping

$$\mathbf{x}(t) = \boldsymbol{\chi}(\tau, \boldsymbol{\xi}), \qquad \boldsymbol{\xi} = \left(\xi^1, \xi^2, \xi^3\right)^T \in \mathrm{E}, \qquad \tau \in \mathbb{R}_{\geq 0}. \tag{4}$$

This mapping provides the covariant basis vectors $\mathbf{a}_i := \frac{\partial \boldsymbol{\chi}}{\partial \xi^i}$, $i = 1, 2, 3$, and the volume weighted contravariant vectors

$$J\mathbf{a}^\alpha = \mathbf{a}_\beta \times \mathbf{a}_\gamma, \quad (\alpha, \beta, \gamma) \text{ cyclic}, \quad J := \det \nabla\boldsymbol{\chi}, \tag{5}$$

where $\alpha, \beta \in \{1, 2, 3\}$ and $\nabla\boldsymbol{\chi}$ is the Jacobian matrix of (4). The same computation as in [21, 33] leads to the following ALE formulation for the Euler equation (1) in a time-dependent element $e_k(t)$:

$$\sum_{\alpha=1}^{3} \frac{\partial J\mathbf{a}^\alpha}{\partial \xi^\alpha} = 0. \tag{6a}$$

$$\frac{\partial J}{\partial \tau} = \nabla_\xi \cdot \tilde{\boldsymbol{\nu}}, \tag{6b}$$

$$\frac{\partial J\mathbf{u}}{\partial \tau} + \nabla_\xi \cdot \overleftrightarrow{\tilde{\mathbf{g}}} = 0. \tag{6c}$$

The quantity $\boldsymbol{\nu} = \left[\nu_1, \nu_2, \nu_3\right]^T$ is the grid velocity field and the contravariant flux vectors are given by

$$\tilde{\boldsymbol{\nu}}_\iota = J\mathbf{a}^\iota \cdot \boldsymbol{\nu}, \qquad \overleftrightarrow{\tilde{\mathbf{g}}}_\iota = J\mathbf{a}^\iota \cdot \overleftrightarrow{\mathbf{f}} - \left(J\mathbf{a}^\iota \cdot \boldsymbol{\nu}\right)\mathbf{u}, \qquad \iota = 1, 2, 3. \tag{7}$$

Remark 1 Equation (6a) contains the metric identities for the contravariant vectors and Eq. (6b) is the geometric conservation law. In the discretization, care must be taken to ensure that these equations are *discretely satisfied*. Otherwise, the stability proofs do not work and the method might be unstable as a consequence.

2.2 Entropy Analysis in Three Dimensions

The Euler equations are equipped with the entropy/entropy flux pair

$$s = -\frac{\rho\varsigma}{\gamma - 1}, \qquad f_\iota^s = -\frac{\rho\varsigma u_\iota}{\gamma - 1}, \qquad \iota = 1, 2, 3, \tag{8}$$

where $\varsigma = \log\left(p\rho^{-\gamma}\right)$ is the thermodynamic entropy of the fluid (cf. Harten [11]). The entropy variables are given by

$$\frac{\partial s}{\partial \mathbf{u}} = \left[\frac{\gamma - \varsigma}{\gamma - 1} - \frac{\rho}{2p}|\mathbf{u}|^2, \frac{\rho\mathbf{u}}{p}, -\frac{\rho}{p}\right]^T =: \mathbf{w}, \tag{9}$$

and the entropy/entropy flux pair satisfies the following identities

$$\mathbf{w}^T \frac{\partial (J\mathbf{u})}{\partial \tau} = \frac{\partial (Js)}{\partial \tau} + \left(\nabla_\xi \cdot \tilde{\mathbf{v}}\right)\left(\mathbf{w}^T \mathbf{u} - s\right), \tag{10}$$

$$\mathbf{w}^T \left(\nabla_\xi \cdot \overleftrightarrow{\tilde{\mathbf{g}}}\right) = \nabla_\xi \cdot \left(\tilde{\mathbf{f}}^s - \tilde{\mathbf{v}} s\right) - \left(\nabla_\xi \cdot \tilde{\mathbf{v}}\right)\left(\mathbf{w}^T \mathbf{u} - s\right). \tag{11}$$

These identities provide for smooth solutions of the Euler equations the integral equation

$$\frac{\partial}{\partial \tau}\int_E Js\, d\boldsymbol{\xi} = -\int_{\partial E}\left(\tilde{\mathbf{f}}^s - \tilde{\mathbf{v}} s\right)^T \hat{n}\, dS. \tag{12}$$

2.3 Analysis of the Kinetic Energy Evolution

The kinetic energy is given by

$$\Bbbk := \frac{1}{2}\rho\,|\mathbf{u}|^2 \qquad \mathbf{v} := \frac{\partial \Bbbk}{\partial \mathbf{u}} = \left[-\frac{1}{2}|\mathbf{u}|^2, \mathbf{u}, 0\right]^T. \tag{13}$$

Similar to an entropy/entropy flux pair, the following equations are satisfied

$$\mathbf{v}^T \frac{\partial (J\mathbf{u})}{\partial \tau} = \frac{\partial (J\Bbbk)}{\partial \tau}, \qquad \mathbf{v}^T \nabla_\xi \cdot \overleftrightarrow{\tilde{\mathbf{g}}} = \nabla_\xi \cdot \left[\left(\tilde{\mathbf{u}} - \tilde{\boldsymbol{\nu}}\right) \Bbbk \right] + \left[\nabla_\xi p\right] \cdot \tilde{\mathbf{u}}.$$

Thus, it follows for smooth solutions of the Euler equations

$$\frac{\partial}{\partial \tau} \int_{\mathrm{E}} J\Bbbk \, d\boldsymbol{\xi} = - \int_{\mathrm{E}} \left[\nabla_\xi p\right] \cdot \tilde{\mathbf{u}} \, d\boldsymbol{\xi} - \int_{\partial \mathrm{E}} \left[\left(\tilde{\mathbf{u}} - \tilde{\boldsymbol{\nu}}\right) \Bbbk \right]^T \hat{n} \, dS. \tag{14}$$

3 Discontinuous Galerkin Spectral Element Method

3.1 *Building Blocks for the Spectral Element Approximation*

The spectral element approximation is based on a nodal approach with Lagrange basis functions $\left\{\ell_j\left(\cdot\right)\right\}_{j=0}^N$ constructed from LGL points $\{\xi_i\}_{i=0}^N$. On the reference element $\mathrm{E} = [-1, 1]^3$ the solution and fluxes of the system (1) are approximated by tensor product Lagrange polynomials of degree N, e.g.,

$$J\left(\xi^1, \xi^2, \xi^3, t\right) \approx \mathscr{J}\left(\xi^1, \xi^2, \xi^3, t\right) := \sum_{i,j,k=0}^{N} \mathscr{J}_{ijk}\left(t\right) \ell_i\left(\xi^1\right) \ell_j\left(\xi^2\right) \ell_k\left(\xi^3\right), \tag{15a}$$

$$\mathbf{u}\left(\xi^1, \xi^2, \xi^3, t\right) \approx \mathbf{U}\left(\xi^1, \xi^2, \xi^3, t\right) := \sum_{i,j,k=0}^{N} \mathbf{U}_{ijk}\left(t\right) \ell_i\left(\xi^1\right) \ell_j\left(\xi^2\right) \ell_k\left(\xi^3\right). \tag{15b}$$

The space of tensor product polynomials with three dimensional domain is denoted by $\mathbb{P}^N\left(\mathrm{E}\right)$. In addition, we define the space

$$\mathbb{P}^N\left(\mathrm{E}, \mathbb{R}^5\right) := \left\{\boldsymbol{\varphi} \mid \boldsymbol{\varphi} = \left[\varphi_1, \varphi_2, \varphi_3, \varphi_4, \varphi_5\right]^T, \quad \varphi_i \in \mathbb{P}^N\left(\mathrm{E}\right), \quad i = 1, 2, 3, 4, 5\right\}. \tag{16}$$

From now on, polynomial approximations are highlighted by capital letters, e.g., $\mathbf{U}$ is an approximation for the state vector $\mathbf{u}$ and $\mathbf{F}_j$, $j = 1, 2, 3$, are approximations for the fluxes $\mathbf{f}_j$, $j = 1, 2, 3$. The approximation for the determinant J of the Jacobian matrix $\nabla_\xi \chi$ is highlighted by $\mathcal{J}$. Furthermore, the interpolation operator for a function $\mathbf{g}$ is given by

$$\mathbb{I}_N(\mathbf{g})\left(\xi^1,\xi^2,\xi^3\right) = \sum_{i,j,k=0}^{N} \mathbf{g}_{ijk}\ell_i\left(\xi^1\right)\ell_j\left(\xi^2\right)\ell_k\left(\xi^3\right), \tag{17}$$

where $\mathbf{g}_{ijk} := \mathbf{g}\left(\xi_i^1,\xi_j^2,\xi_k^3\right)$ and $\left\{\xi_i^1\right\}_{i=0}^N$, $\left\{\xi_i^2\right\}_{i=0}^N$, $\left\{\xi_i^3\right\}_{i=0}^N$ are sets of LGL points. Derivatives are approximated by exact differentiation of the polynomial interpolants. It is worth to mention that differentiation and interpolation only commute if there are no interpolation errors [2, 18].

Remark 2 Gassner [8] showed that the DGSEM approximation with LGL points for interpolation and quadrature gives a SBP operator $\mathbf{Q} = \mathbf{MD}$ with the mass matrix $\mathbf{M}$ and the derivative matrix $\mathbf{D}$. The mass matrix and the derivative matrix have the entries

$$\mathbf{M}_{ij} = \omega_i\delta_{ij}, \qquad \mathbf{D}_{ij} = \ell_j'\left(\xi_i\right), \qquad i,j=0,\dots,N. \tag{18}$$

A SBP operator satisfies the property [22]

$$\mathbf{Q}+\mathbf{Q}^T = \mathbf{B}, \quad \mathbf{B} = \operatorname{diag}\left(-1,0,\dots,0,1\right). \tag{19}$$

3.1.1 Discrete Integrals

Integrals are approximated by a tensor product extension of a $2N-1$ accurate LGL quadrature formula. Hence, interpolation and quadrature nodes are collocated. For two functions $\mathbf{f}$ and $\mathbf{g}$ the following bracket notation

$$\langle \mathbf{f},\mathbf{g}\rangle_N := \sum_{i=0}^{N}\sum_{j=0}^{N}\sum_{k=0}^{N} \omega_i\omega_j\omega_k \mathbf{f}_{ijk}^T\mathbf{g}_{ijk} = \sum_{i,j,k=0}^{N} \omega_{ijk}\mathbf{f}_{ijk}^T\mathbf{g}_{ijk}, \tag{20}$$

is used. Here denotes ω_i, $i=0,\dots,N$, the LGL quadrature weights and ξ_i, $i=0,\dots,N$, are LGL points. Furthermore, for a block vector $\overleftrightarrow{\mathbf{F}}$ and test functions $\boldsymbol{\varphi} \in \mathbb{P}^N\left(\mathrm{E},\mathbb{R}^5\right)$, we define the discrete surface integral

$$\int_{\partial \mathrm{E},N} \boldsymbol{\varphi}^T\left\{\overleftrightarrow{\mathbf{F}}\cdot\hat{n}\right\} dS = \sum_{\iota=1}^{3}\int_{\partial\mathrm{E},N} \boldsymbol{\varphi}^T\mathbf{F}_\iota\hat{n}^\iota\, dS, \tag{21}$$

where $\hat{n} = \left[\hat{n}^1, \hat{n}^2, \hat{n}^3\right]^T$ is the unit outward normal at the faces of the reference element E and

$$\begin{aligned}
\int_{\partial E, N} \boldsymbol{\varphi}^T \mathbf{F}_1 \hat{n}^1 \, dS &:= \sum_{j,k=0}^{N} \omega_j \omega_k \left(\boldsymbol{\varphi}_{Njk}^T \left(\mathbf{F}_1\right)_{Njk} - \boldsymbol{\varphi}_{0jk}^T \left(\mathbf{F}_1\right)_{0jk} \right), \\
\int_{\partial E, N} \boldsymbol{\varphi}^T \mathbf{F}_2 \hat{n}^2 \, dS &:= \sum_{i,k=0}^{N} \omega_i \omega_k \left(\boldsymbol{\varphi}_{iNk}^T \left(\mathbf{F}_2\right)_{iNk} - \boldsymbol{\varphi}_{i0k}^T \left(\mathbf{F}_2\right)_{i0k} \right), \\
\int_{\partial E, N} \boldsymbol{\varphi}^T \mathbf{F}_3 \hat{n}^3 \, dS &:= \sum_{i,j=0}^{N} \omega_i \omega_j \left(\boldsymbol{\varphi}_{ijN}^T \left(\mathbf{F}_3\right)_{ijN} - \boldsymbol{\varphi}_{ij0}^T \left(\mathbf{F}_3\right)_{ij0} \right).
\end{aligned} \tag{22}$$

3.1.2 Discrete Metric Identities

The contravariant coordinate vectors need to be discretized in such a way that the metric identities (6a) are satisfied on the discrete level, too. Kopriva [17] introduced the conservative curl form to approximate the metric terms. In this approach the coefficients of the volume weighted contravariant coordinate vectors $J\mathbf{a}^\alpha$, $\alpha = 1, 2, 3$, are computed by

$$J\mathbf{a}_\beta^\alpha \approx \mathbb{J}\mathbf{a}_\beta^\alpha := -\hat{x}_\alpha \cdot \nabla_\xi \times \left(\mathbb{I}_N\left(\chi_\gamma \nabla_\xi \chi_\delta\right)\right), \quad \alpha, \beta = 1, 2, 3, \quad (\beta, \gamma, \delta) \text{ cyclic.} \tag{23}$$

Here $\boldsymbol{\chi} = [\chi_1, \chi_2, \chi_3]^T$ represents the mapping from the physical element to the reference element and $\hat{x}_\alpha$ is the unit vector in the α-th Cartesian coordinate direction. Since the contravariant coordinate vectors are discretized by (23), it holds

$$\sum_{m=0}^{N} \left(\mathbf{D}_{im} \left(\mathbb{J}a_\beta^1\right)_{mjk} + \mathbf{D}_{jm} \left(\mathbb{J}a_\beta^2\right)_{imk} + \mathbf{D}_{km} \left(\mathbb{J}a_\beta^3\right)_{ijm} \right) = 0, \quad \begin{cases} \beta = 1, 2, 3, \\ i, j, k = 0, \ldots, N, \end{cases} \tag{24}$$

This equation is the discrete version of the metric identities (6a).

3.2 *Standard ALE DGSEM*

The construction of the standard ALE DGSEM for the Euler equations follows by the same steps as in [19, 27]. The solution $\mathbf{u}$ and the Jacobian J are replaced by (15b) and (15a). The advective block flux are approximated by the interpolation operator (17). Then the Eq. (6b) is multiplied with test functions $\varphi \in \mathbb{P}^N(\mathrm{E})$ and the transformed Euler equations with $\boldsymbol{\varphi} \in \mathbb{P}^N\left(\mathrm{E}, \mathbb{R}^5\right)$. The resulting equations are integrated and integration-by-parts is used. The integrals in the variational form are approximated with the tensor product LGL quadrature. At the element interfaces, numerical surface fluxes $\tilde{\boldsymbol{\nu}}^*$ and $\overset{\leftrightarrow}{\tilde{\mathbf{G}}}{}^*$ are inserted. Next, the SBP property (19) is used. Finally, the standard ALE DGSEM strong form is given by: For all $\varphi \in \mathbb{P}^N(\mathrm{E})$ and for all $\boldsymbol{\varphi} \in \mathbb{P}^N\left(\mathrm{E}, \mathbb{R}^5\right)$ apply

$$\left\langle \frac{\partial \mathcal{J}}{\partial \tau}, \varphi \right\rangle_N - \left\langle \nabla_\xi \cdot \mathbb{I}_N(\tilde{\boldsymbol{\nu}}), \varphi \right\rangle_N - \int_{\partial \mathrm{E}, N} \varphi \left(\tilde{\nu}_{\hat{n}}^* - \tilde{\nu}_{\hat{n}}\right) dS = 0, \tag{25a}$$

$$\left\langle \frac{\partial \left(\mathcal{J}\mathbf{U}\right)}{\partial \tau}, \boldsymbol{\varphi} \right\rangle_N + \left\langle \nabla_\xi \cdot \mathbb{I}_N\left(\overset{\leftrightarrow}{\tilde{\mathbf{g}}}\right), \boldsymbol{\varphi} \right\rangle_N + \int_{\partial \mathrm{E}, N} \boldsymbol{\varphi}^T \left(\tilde{\mathbf{G}}_{\hat{n}}^* - \tilde{\mathbf{G}}_{\hat{n}}\right) dS = 0. \tag{25b}$$

The quantity $\tilde{\nu}_{\hat{n}}$ in (25a) and the flux $\tilde{\mathbf{G}}_{\hat{n}}$ in (25b) are defined by

$$\tilde{\nu}_{\hat{n}} = \left(\hat{s}\mathbf{n}\right) \cdot \boldsymbol{\nu} = \sum_{\iota=1}^{3} \hat{n}^\iota \left(\mathbb{J}a_1^\iota \nu_1 + \mathbb{J}a_2^\iota \nu_2 + \mathbb{J}a_3^\iota \nu_3\right), \tag{26}$$

$$\tilde{\mathbf{G}}_{\hat{n}} = \left(\hat{s}\mathbf{n}\right) \cdot \overset{\leftrightarrow}{\mathbf{G}} = \sum_{\iota=1}^{3} \hat{n}^\iota \left(\mathbb{J}a_1^\iota \mathbf{G}_1 + \mathbb{J}a_2^\iota \mathbf{G}_2 + \mathbb{J}a_3^\iota \mathbf{G}_3\right). \tag{27}$$

The surface numerical fluxes in (25) are defined by

$$\tilde{\nu}_{\hat{n}}^* = \hat{s}\left(n_1 \{\{\nu_1\}\} + n_2 \{\{\nu_2\}\} + n_3 \{\{\nu_3\}\}\right), \tag{28}$$

$$\mathbf{G}_{\hat{n}}^* = \hat{s}\left(n_1 \mathbf{G}_1^* + n_2 \mathbf{G}_2^* + n_3 \mathbf{G}_3^*\right). \tag{29}$$

Here the functions $\tilde{\mathbf{G}}_j^*$, $j = 1, 2, 3$ are consistent Cartesian numerical fluxes. Moreover, the quantities

$$[\![*]\!] := (*)^+ - (*)^-, \quad \text{and} \quad \{\{*\}\} := \frac{1}{2}\left[(*)^+ + (*)^-\right], \tag{30}$$

are the orientated jump and the average operator at the interfaces. The notation "−" and "+" is used to denote the value of states between two elements at the LGL nodes.

3.3 Split Form ALE DGSEM

The terms $\nabla_\xi \cdot \mathbb{I}_N(\tilde{\boldsymbol{\nu}})$ and $\nabla_\xi \cdot \mathbb{I}_N\left(\overset{\leftrightarrow}{\tilde{\mathbf{g}}}\right)$ cause aliasing errors in the standard strong form (25). In [9, 10] the split form DGSEM approach was introduced. In this approach a special derivative operator is introduced to replace the terms that cause aliasing errors. In [21, 33] the following derivative operator were introduced:

$$\begin{aligned}\mathbb{D}_N \cdot \tilde{\boldsymbol{\nu}}^{\#}_{ijk} := \sum_{m=0}^{N} \quad & 2\,\mathbf{D}_{im}\left(\{\!\{\boldsymbol{\nu}\}\!\}_{(i,m)jk} \cdot \{\!\{\mathbb{J}\mathbf{a}^1\}\!\}_{(i,m)jk}\right) \\ & + 2\,\mathbf{D}_{jm}\left(\{\!\{\boldsymbol{\nu}\}\!\}_{i(j,m)k} \cdot \{\!\{\mathbb{J}\mathbf{a}^2\}\!\}_{i(j,m)k}\right) \\ & + 2\,\mathbf{D}_{km}\left(\{\!\{\boldsymbol{\nu}\}\!\}_{ij(m,k)} \cdot \{\!\{\mathbb{J}\mathbf{a}^3\}\!\}_{ij(m,k)}\right),\end{aligned} \tag{31}$$

$$\begin{aligned}\mathbb{D}_N \cdot \overset{\leftrightarrow}{\tilde{\mathbf{G}}}{}^{\#}_{ijk} := \sum_{m=0}^{N} \quad & 2\,\mathbf{D}_{im}\left(\overset{\leftrightarrow}{\mathbf{G}}{}^{\#}\left(\boldsymbol{\nu}_{ijk}, \boldsymbol{\nu}_{mjk}, \mathbf{U}_{ijk}, \mathbf{U}_{mjk}\right) \cdot \{\!\{\mathbb{J}\mathbf{a}^1\}\!\}_{(i,m)jk}\right) \\ & + 2\,\mathbf{D}_{jm}\left(\overset{\leftrightarrow}{\mathbf{G}}{}^{\#}\left(\boldsymbol{\nu}_{ijk}, \boldsymbol{\nu}_{imk}, \mathbf{U}_{ijk}, \mathbf{U}_{imk}\right) \cdot \{\!\{\mathbb{J}\mathbf{a}^2\}\!\}_{i(j,m)k}\right) \\ & + 2\,\mathbf{D}_{km}\left(\overset{\leftrightarrow}{\mathbf{G}}{}^{\#}\left(\boldsymbol{\nu}_{ijk}, \boldsymbol{\nu}_{ijm}, \mathbf{U}_{ijk}, \mathbf{U}_{ijm}\right) \cdot \{\!\{\mathbb{J}\mathbf{a}^3{}_{ij(k,m)}\}\!\}\right),\end{aligned} \tag{32}$$

where the volume averages are given by $\{\!\{*\}\!\}_{(i,m)jk} := \frac{1}{2}\left[(*)_{ijk} + (*)_{mjk}\right]$. The functions $\tilde{\mathbf{G}}^{\#}_{\iota}$, $\iota = 1, 2, 3$, in (32) are numerical volume fluxes. It is important that these volume fluxes are consistent and symmetric such that for all $i, j, k, m = 0, \ldots, N$ holds

$$\begin{aligned}\mathbf{G}^{\#}_{\iota}\left(\boldsymbol{\nu}_{ijk}, \boldsymbol{\nu}_{mjk}, \mathbf{U}_{ijk}, \mathbf{U}_{ijk}\right) &= \mathbf{F}_{\iota}\left(\mathbf{U}_{ijk}\right) - \{\!\{\boldsymbol{\nu}_{\iota}\}\!\}_{(i,m)jk}\mathbf{U}, \qquad \iota = 1, 2, 3, \\ \mathbf{G}^{\#}_{\iota}\left(\boldsymbol{\nu}_{ijk}, \boldsymbol{\nu}_{mjk}, \mathbf{U}_{ijk}, \mathbf{U}_{mjk}\right) &= \mathbf{G}^{\#}_{\iota}\left(\boldsymbol{\nu}_{mjk}, \boldsymbol{\nu}_{ijk}, \mathbf{U}_{mjk}, \mathbf{U}_{ijk}\right), \qquad \iota = 1, 2, 3,\end{aligned}$$

where $\mathbf{F}_{\iota}\left(\mathbf{U}_{ijk}\right)$ is the advective Euler flux (3) evaluated at the LGL points. Then, for each element the semi-discrete split form ALE DGSEM for the Euler equations is represented on the reference element E by: For all $\varphi \in \mathbb{P}^N(\mathrm{E})$ and for all $\boldsymbol{\varphi} \in \mathbb{P}^N\left(\mathrm{E}, \mathbb{R}^5\right)$ apply

$$\left\langle \frac{\partial \mathscr{J}}{\partial \tau}, \varphi \right\rangle_N - \left\langle \mathbb{D}_N \cdot \tilde{\boldsymbol{\nu}}^{\#}, \varphi \right\rangle_N - \int_{\partial \mathrm{E}, N} \varphi\left(\tilde{\nu}^{*}_{\hat{n}} - \tilde{\nu}_{\hat{n}}\right) dS = 0, \tag{33a}$$

$$\left\langle \frac{\partial \left(\mathscr{J}\mathbf{U}\right)}{\partial \tau}, \boldsymbol{\varphi} \right\rangle_N + \left\langle \mathbb{D}_N \cdot \overset{\leftrightarrow}{\tilde{\mathbf{G}}}^{\#}, \boldsymbol{\varphi} \right\rangle_N + \int_{\partial E, N} \boldsymbol{\varphi}^T \left(\tilde{\mathbf{G}}^*_{\hat{n}} - \tilde{\mathbf{G}}_{\hat{n}} \right) dS = 0. \tag{33b}$$

3.4 EC/KEP Split form ALE DGSEM

We call a numerical method an EC scheme, if the method satisfies a discrete version of the Eq. (12), likewise a method is called a KEP scheme, if the method satisfies a discrete version of the Eq. (14). In [21, 33] it has been proven that the split form ALE DGSEM (33) becomes an EC scheme, when the Cartesian surface flux functions $\tilde{\mathbf{G}}^*_\iota$, $\iota = 1, 2, 3$, satisfy the discrete entropy conditions

$$[\![\mathbf{w}]\!]^T \mathbf{G}^*_\iota = [\![\rho u_\iota]\!] - \{\{\nu_\iota\}\}[\![\rho]\!], \qquad \iota = 1, 2, 3, \tag{34}$$

and the volume flux functions $\tilde{\mathbf{G}}^{\#}_\iota$, $\iota = 1, 2, 3$, in the derivative projection operator (32) satisfy in the LGL points for $i, j, k, m = 0, \ldots, N$, the discrete entropy conditions:

$$\begin{aligned}
[\![\mathbf{W}]\!]^T_{(i,m)jk} \mathbf{G}^{\#}_1 \left(\boldsymbol{\nu}_{ijk}, \boldsymbol{\nu}_{mjk}, \mathbf{U}_{ijk}, \mathbf{U}_{mjk}\right) &= [\![\rho \mathbf{u}]\!]_{(i,m)jk} - \{\{\boldsymbol{\nu}\}\}_{(i,m)jk}[\![\rho]\!]_{(i,m)jk}, \\
[\![\mathbf{W}]\!]^T_{i(j,m)k} \mathbf{G}^{\#}_2 \left(\boldsymbol{\nu}_{ijk}, \boldsymbol{\nu}_{imk}, \mathbf{U}_{ijk}, \mathbf{U}_{imk}\right) &= [\![\rho \mathbf{u}]\!]_{i(j,m)k} - \{\{\boldsymbol{\nu}\}\}_{i(j,m)k}[\![\rho]\!]_{i(j,m)k}, \\
[\![\mathbf{W}]\!]^T_{ij(k,m)} \mathbf{G}^{\#}_3 \left(\boldsymbol{\nu}_{ijk}, \boldsymbol{\nu}_{ijm}, \mathbf{U}_{ijk}, \mathbf{U}_{ijm}\right) &= [\![\rho \mathbf{u}]\!]_{ij(k,m)} - \{\{\boldsymbol{\nu}\}\}_{ij(k,m)}[\![\rho]\!]_{ij(k,m)}.
\end{aligned} \tag{35}$$

These entropy conditions are generalizations of Tadmor's condition [35] to construct two point EC flux functions on a static mesh. Moreover, in [21] it is proven that the split form ALE DGSEM (33) becomes a KEP scheme, when the Cartesian surface flux functions $\tilde{\mathbf{G}}^*_\iota$ and the volume flux functions $\mathbf{G}^{\#}_\iota$, $\iota = 1, 2, 3$, in the derivative projection operator (32) satisfy

$$\begin{aligned}
\mathbf{G}^{\upsilon+1,*}_\iota &= \mathbf{G}^{1,*}_\iota \{\{u_\upsilon\}\} + \{\{p\}\}^\star \delta_{\iota\upsilon}, \\
\mathbf{G}^{\upsilon+1,\#}_\iota &= \mathbf{G}^{1,\#}_\iota \{\{u_\upsilon\}\} + \{\{p\}\} \delta_{\iota\upsilon},
\end{aligned} \qquad \upsilon = 1, 2, 3, \tag{36}$$

where $\{\{p\}\}^\star$ can be any consistent numerical trace approximation of the pressure. These entropy conditions are generalizations of Jameson's [13] condition to construct two point KEP flux functions on a static mesh.

Remark 3 In general, the Euler equations are describing irreversible processes with discontinuous solutions. Hence, it cannot be expected that an EC/KEP split form ALE DGSEM provides a physical meaningful discretization for the Euler equations. Therefore, for practical simulation an artificial dissipation term is added to the surface flux functions. In [33, Appendix C.3] a suitable matrix dissipation operator is given. The construction of such a matrix dissipation operator is based on the fact that there are block diagonal scaling matrices such that the Hessian matrix of an entropy for the Euler equations can be represented by scaled right eigenvector matrices (cf. Merriam [26]). This matrix dissipation operator leads to an ES/KED split form ALE DGSEM.

4 Numerical Results

The numerical computation in this section have been done with the open source high order DG solver FLEXI[1] (cf. [20]). In the following, experimental convergence rates as well as the KEP and EC property of the different split forms (choice of the two point volume flux) are investigated. The following two point flux functions are considered: Chandrashekar (CH) [3], Kennedy and Gruber (KG) [14], Kuya, Totani and Kawai (KTK), Pirozzoli (PI) [31], Ranocha (RA) [32]. The RK4(3)5[2R+] explicit time stepping scheme is used (cf. [15]). Moreover, the time evolution of the grid points $\mathbf{x}$ distribution is described by the equation

$$\mathbf{x}(t) = \mathbf{x}(0) + 0.05L\sin(2\pi t)\sin\left(\frac{2\pi}{L}x_1(0)\right)\sin\left(\frac{2\pi}{L}x_2(0)\right)\sin\left(\frac{2\pi}{L}x_3(0)\right), \tag{37}$$

where $\mathbf{x}(t=0)$ is the position of the grid point in the non-deformed configuration of the mesh, and L is the length of the domain.

4.1 Experimental Convergence Rates

In this section, the convergence behavior under mesh refinement of the split form ALE DGSEM is assessed for different flux functions using the method of manufactured solutions. The results are used to verify that the proposed methodology retains its high order accuracy on moving grids. For the following simulations, we assume a solution of the form

$$\begin{aligned} \rho(\mathbf{x},t) &= 2 + 0.1\sin(\pi(x_1 + x_2 + x_3 - 0.6t)), \\ \rho u_\iota(\mathbf{x},t) &= 2 + 0.1\sin(\pi(x_1 + x_2 + x_3 - 0.6t)), \qquad \iota = 1,2,3, \\ E(\mathbf{x},t) &= [2 + 0.1\sin(\pi(x_1 + x_2 + x_3 - 0.6t))]^2, \end{aligned} \tag{38}$$

[1] http://www.flexi-project.org.

with total energy $E = e + \frac{1}{2}\rho\,|\mathbf{u}|^2$ on the domain $[-1, 1]^3$ and compute the residual when (38) is inserted into the Euler equations. The resulting terms are then used as sources for the simulations, and are discretized as a solution independent part of the computation. The simulations are performed on initially Cartesian grids with an increasing number of elements

$$K = 2^3, 4^3, 8^3, 16^3, 32^3. \tag{39}$$

All meshes are undergoing a forced periodic motion, corresponding to a standing wave. The simulation is advanced until final time $T = 5$. Since the stability region of the explicit time integration scheme is restricted by the Courant-Friedrichs-Lewy (CFL) condition, the allowable time step Δt is computed as in [4]

$$\frac{\Delta t}{\min\limits_{1 \le \alpha \le K} |h_\alpha(t^n)|} \le \frac{C_{\mathrm{CFL}}}{(2N+1)\,\lambda_{\max}}, \tag{40}$$

where $h_\alpha(t)$ is the size of element $e_\alpha(t)$, λ_{max} the fastest signal velocity of the Euler equations and C_{CFL} is set to $C_{\mathrm{CFL}} = 0.1$. The surface fluxes are used with Roe-type dissipation terms, which can be found in [33, Appendix C.3]. Figure 1 shows the results obtained for the manufactured solution test cases. The quantity used for comparison is the L^2 norm of the error in the density ρ, when compared with the manufactured solution (38).

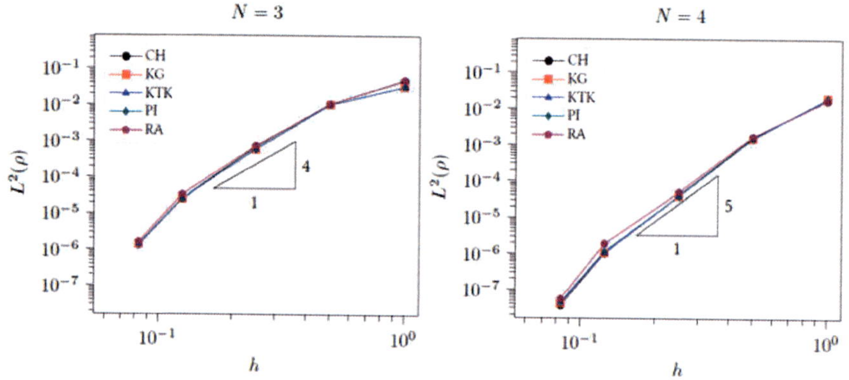

Fig. 1 Experimental convergence rates for different flux functions. Manufactured solution test case with standing wave mesh movement, for both $N = 3$ and $N = 4$. Shown are the L^2 error norms of the density ρ over the mesh size h

4.2 *Numerical Validation of the Entropy and Kinetic Energy Analysis*

The TGV problem [34] is considered to investigate the KEP and EC property of the different split forms. This is an important canonical problem for laminar-turbulent transition and turbulent flows. The flow is defined on the domain $\Omega = [0, 2\pi]^3$ with periodic boundary conditions on all sides, and its initial conditions are given by

$$
\begin{aligned}
\rho(\mathbf{x}, 0) &= 1, \\
\mathbf{u}(\mathbf{x}, 0) &= [\sin(x_1)\cos(x_2)\cos(x_3), -\cos(x_1)\sin(x_2)\cos(x_3), 0]^T, \\
p(\mathbf{x}, 0) &= p_0 + \frac{1}{16}(\cos(2x_1) + \cos(2x_2))(\cos(2x_3) + 2),
\end{aligned}
\tag{41}
$$

where p_0 is chosen such that the Mach number becomes $\mathrm{Ma} = \frac{1}{\sqrt{\gamma p_0}} = 0.1$. The flow is evolved until final time $T = 13$, which is past the critical point for stability reached at $t \approx 9$, where the maximum of turbulent dissipation occurs. The mesh is initially Cartesian with $K = 16^3$ elements. The behavior of the discrete error for the temporal evolution of the global entropy and global kinetic energy are investigated. These quantities are given by:

$$
\Delta_{\mathbb{S}}(T) = \mathbb{S}(T) - \mathbb{S}(0), \qquad \Delta_{\mathbb{K}}(T) = \mathbb{K}(T) - \mathbb{K}(0),
$$

$$
\mathbb{S}(\tau) := \sum_{k=1}^{K} \langle \mathbb{I}_N(s(\tau)), \mathcal{J}(\tau) \rangle_N \approx \int_E s(\tau) J(\tau)\, d\boldsymbol{\xi}, \qquad \forall \tau \in [0, T],
$$

$$
\mathbb{K}(\tau) := \sum_{k=1}^{K} \langle \mathbb{I}_N(\Bbbk(\tau)), \mathcal{J}(\tau) \rangle_N \approx \int_E \Bbbk(\tau) J(\tau)\, d\boldsymbol{\xi}, \qquad \forall \tau \in [0, T],
$$

It is important to note that no additional dissipation terms are added to the surface fluxes, all dissipative effects must be attributed to the numerical scheme. This allows to judge the conservation and preservation for all considered flux formulations. We note that the kinetic energy is not a conserved quantity for compressible flows, but as the Mach number is very small, the pressure contribution in the momentum equation decreases and the kinetic energy is almost conserved. The numerical results are plotted in Fig. 2.

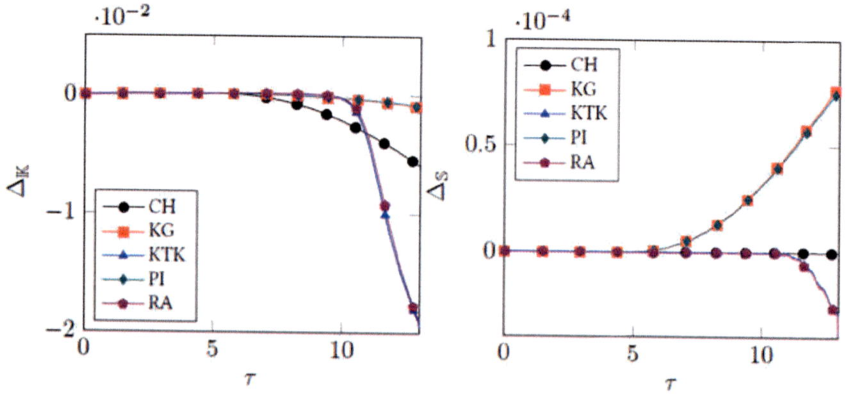

Fig. 2 The temporal evolution of the error in kinetic energy $\Delta_{\mathbb{K}}$ and entropy $\Delta_{\mathbb{S}}$ during the TGV test case for various split form configurations with $N = 3$

5 Conclusions

In this work, the framework to construct split form ALE DGSEM to solve the three dimensional Euler equations on a moving domain has been presented. The split form DG framework [9, 10] has been used to avoid aliasing in the discretization of the volume integrals. This approach allows to construct EC/KEP split form ALE DGSEM, when a two-point EC or KED flux function is used as surface flux in the DG method and a two-point EC flux or KEP flux function in the derivative operator (32). However, it should be mentioned that the approach in this work and the associated analysis is for semi-discrete EC/ES or KEP/KED high order DGSEM and the construction of these methods rely on the assumption that the discrete density and pressure are positive. In general, this might not always be the case for high order methods, since the positivity of the discrete density and pressure can be affected by spurious oscillations in the numerical solution.
The TGV test problem was used to investigate the split form ALE DGSEM numerical. This test case was chosen, since (i) it has simple periodic boundary conditions, removing the influence of e.g. walls, (ii) is challenging for the stability of the scheme, especially if the inviscid case is investigated, and (iii) it is analytically isentropic if viscous effects are neglected, allowing to confirm the EC or KEP property of the methods.

Acknowledgments This research is supported by the European Research Council (ERC) under the European Union's Eights Framework Program Horizon 2020 with the research project Extreme, ERC grant agreement no. 714487. The authors gratefully acknowledge the support and the computing time on "Hazel Hen" provided by the High-Performance Computing Center Stuttgart (HLRS) through the project hpcdg".

References

1. Blaisdell, G.A., Spyropoulos, E.T., Qin, J.H.: The effect of the formulation of nonlinear terms on aliasing errors in spectral methods. Appl. Numer. Math. **21**(3), 207–219 (1996)
2. Canuto, C., Hussaini, M.Y., Quarteroni, A., Zang, T.A.: Spectral methods. In: Fundamentals in Single Domains. Springer, Berlin (2006)
3. Chandrashekar, P.: Kinetic energy preserving and entropy stable finite volume schemes for compressible Euler and Navier-Stokes equations. Commun. Comput. Phys. **14**(5), 1252–1286 (2006)
4. Cockburn, B., Shu, C.-W.: Runge-Kutta discontinuous galerkin methods for convection-dominated problems. J. Sci. Comput. **16**(3), 173–261 (2001)
5. Donea, J., Huerta, A., Ponthot, J.-P., Rodríguez-Ferran, A.: Arbitrary Lagrangian-Eulerian methods. In: Encyclopedia of Computational Mechanics, 2nd edn., pp.1–32 (2017)
6. Fernández, D.C.D.R., Hicken, J.E., Zingg, D.W.: Review of summation-by-parts operators with simultaneous approximation terms for the numerical solution of partial differential equations. Comput. Fluids **95**, 171–196 (2014)
7. Fisher, T.C., Carpenter, M.H.: High-order entropy stable finite difference schemes for nonlinear conservation laws: finite domains. J. Comput. Phys. **252**, 518–557 (2013)
8. Gassner, G.J.: A skew-symmetric discontinuous Galerkin spectral element discretization and its relation to SBP-SAT finite difference methods. SIAM J. Sci. Comput. **35**(3), A1233–A1253 (2013)
9. Gassner, G.J., Winters, A.R., Kopriva, D.A.: Split form nodal discontinuous Galerkin schemes with summation-by-parts property for the compressible Euler equations. J. Comput. Phys. **327**, 39–66 (2016)
10. Gassner, G.J., Winters, A.R., Hindenlang, F.J., Kopriva, D.A.: The BR1 scheme is stable for the compressible Navier-Stokes equations. J. Sci. Comput. **77**(1), 154–200 (2018)
11. Harten, A.: On the symmetric form of systems of conservation laws with entropy. J. Comput. Phys. **49**, 151–164 (1983)
12. Hesthaven, J.S., Warburton, T.: Nodal Discontinuous Galerkin Methods: Algorithms, Analysis, and Applications. Springer, Berlin (2007)
13. Jameson, A.: Formulation of kinetic energy preserving conservative schemes for gas dynamics and direct numerical simulation of one-dimensional viscous compressible flow in a shock tube using entropy and kinetic energy preserving schemes. J. Sci. Comput. **34**(2), 188–208 (2008)
14. Kennedy, C.A., Gruber, A.: Reduced aliasing formulations of the convective terms within the Navier–Stokes equations for a compressible fluid. J. Comput. Phys. **227**(3), 1676–1700 (2008)
15. Kennedy, C.A., Carpenter, M.H., Lewis, R.M.: Low-storage, explicit Runge-Kutta schemes for the compressible Navier-Stokes equations. Appl. Numer. Math. **35**(3), 177–219 (2000)
16. Kirby, R.M., Karniadakis, G.E.: De-aliasing on non-uniform grids: algorithms and applications. J. Comput. Phys. **191**(1), 249–264 (2003)
17. Kopriva, D.A.: Metric identities and the discontinuous spectral element method on curvilinear meshes. J. Sci. Comput. **26**(3), 301 (2006)
18. Kopriva, D.A.: Implementing Spectral Methods for Partial Differential Equations: Algorithms for Scientists and Engineers. Springer, Berlin (2009)
19. Kopriva, D.A., Winters, A.R., Bohm, M., Gassner, G.J.: A provably stable discontinuous Galerkin spectral element approximation for moving hexahedral meshes. Comput. Fluids **139**, 148–160 (2016)
20. Krais, N., Beck, A., Bolemann, T., Frank, H., Flad, D., Gassner, G., Hindenlang, F., Hoffmann, M., Kuhn, T., Sonntag, M., Munz, C.-D.: FLEXI: a high order discontinuous Galerkin framework for hyperbolic-parabolic conservation laws (2019). arXiv:1910.02858
21. Krais, N., Schnücke, G., Bolemann, T., Gassner, G.J.: Split form ALE discontinuous Galerkin methods with applications to under-resolved turbulent low-Mach number flows. J. Comput. Phys. **421**, 109726 (2020)

22. Kreiss, H.-O., Oliger, J.: Comparison of accurate methods for the integration of hyperbolic equations. Tellus **24**(3), 199–215 (1972)
23. Lefloch, P.G., Mercier, J.-M., Rohde, C.: Fully discrete, entropy conservative schemes of arbitrary order. SIAM J. Numer. Anal. **40**(5), 1968–1992 (2002)
24. Lele, S.K.: Compact finite difference schemes with spectral-like resolution. J. Comput. Phys. **103**(1), 16–42 (1992)
25. Lomtev, I., Kirby, R.M., Karniadakis, G.E.: A discontinuous Galerkin ALE method for compressible viscous flows in moving domains. J. Comput. Phys. **155**(1), 128–159 (1999)
26. Merriam, M.L.: Towards a rigorous approach to artificial dissipation. National Aeronautics and Space Administration, Moffett Field. Ames Research Center (1989)
27. Minoli, C.A.A., Kopriva, D.A.: Discontinuous Galerkin spectral element approximations on moving meshes. J. Comput. Phys. **230**(5), 1876–1902 (2011)
28. Nguyen, V.-T.: An arbitrary Lagrangian–Eulerian discontinuous Galerkin method for simulations of flows over variable geometries. J. Fluids Struct. **26**(2), 312–329 (2010)
29. Nordström, J., Carpenter, M.H.: Boundary and interface conditions for high-order finite-difference methods applied to the Euler and Navier-Stokes equations. J. Comput. Phys. **148**(2), 621–645 (1999)
30. Persson, P.-O., Bonet, J., Peraire, J.: Discontinuous Galerkin solution of the Navier-Stokes equations on deformable domains. Comput. Methods Appl. Mech. Eng. **198**(17–20), 1585–1595 (2009)
31. Pirozzoli, S.: Generalized conservative approximations of split convective derivative operators. J. Comput. Phys. **229**(19), 7180–7190 (2010)
32. Ranocha, H.: Generalised summation-by-parts operators and entropy stability of numerical methods for hyperbolic balance laws. Cuvillier Verlag, Göttingen (2018)
33. Schnücke, G., Krais, N., Bolemann, T., Gassner, G.J.: Entropy stable discontinuous Galerkin schemes on moving meshes for hyperbolic conservation laws. J. Sci. Comput. **82**, 1–42 (2020)
34. Shu, C.-W., Don, W.-S., Gottlieb, D., Schilling, O., Jameson, L.: Numerical convergence study of nearly incompressible, inviscid Taylor-Green vortex flow. J. Sci. Comput. **24**(1), 1–27 (2005)
35. Tadmor, E.: Entropy stability theory for difference approximations of nonlinear conservation laws and related time-dependent problems. Acta Numer. **12**, 451–512 (2003)
36. Tang, T.: Moving mesh methods for computational fluid dynamics. Contemp. Math. **383**(8), 141–173 (2005)

Explicit Modal Discontinuous Galerkin Approximations for Three-Dimensional Electronic Boltzmann Transport Equation

Satyvir Singh and Marco Battiato

1 Introduction

Over the last decades, the studies of several unique properties in low dimensional materials have sparked an interest in ultrafast dynamics [1, 2]. However, several experimental investigations are carried out to explore the underlying physics, the theoretical and computational efforts in this regard have been hampered by numerous obstacles [3]. The most difficult task is to effectively characterize the laser interactions, scatterings, and transport of numerous quasi-particles (electrons, phonons, excitons) [4]. Furthermore, because the majority of the intriguing effects occur in the nonequilibrium domain, a complete description of the time-varying nonequilibrium population increases the complexity. To overcome these complexities, the real nonequilibrium materials requires advanced computational modelling.

The Boltzmann transport equation (BTE) has already been used to describe the physics of gases, plasmas, and semiconductors, and has proven to be quite effective in dealing with these complexities [5]. The BTE is widely recognized as the most accurate model of incoherent transport and scattering between different types of quasi-particles in materials. Because of the large dimensionality of the system, the high order of the scattering operator, conservation rules, and time-step restriction, the development of an efficient and accurate numerical approach for solving the BTE system is a difficult task. As a result of these significant obstacles, the development of an effective numerical technique necessitates a careful balance between computing complexity and accuracy.

S. Singh (✉) · M. Battiato
School of Physical and Mathematical Sciences, Nanyang Technological University, Singapore, Singapore
e-mail: satyvir.singh@ntu.edu.sg; marco.battiato@ntu.edu.sg

J. M. Melenk et al. (eds.), *Spectral and High Order Methods for Partial Differential Equations ICOSAHOM 2020+1*, Lecture Notes in Computational Science and Engineering 137, https://doi.org/10.1007/978-3-031-20432-6_28

In literature, two types of numerical techniques are well-known for solving the BTE: stochastic and deterministic approaches. In former approach, the particle-based stochastic methods, such as the direct simulation Monte Carlo, are widely used because of computational limitations and the high dimensionality of the system [6]. However, the stochastic approach has some drawbacks, such as computational cost and statistical noise. While, the deterministic methods basically use the discretization approach for solving the BTE on the phase space. Among various deterministic methods, the discontinuous Galerkin (DG) approximations have been a popular method for solving kinetic equations in gases and quasi-particles in recent years [7–11]. The DG approximations, which combine the common features of Finite Element and Finite Volume methods, have been successfully applied to a wide range of applications in gas dynamics, shock dynamics, computational biology, and many more [12–14].

The main of this study is to develop an explicit modal DG approximation based on hexahedral element for solving three-dimensional electronic BTE under nonequilibrium conditions. A simple collisional model—relaxation time approximation, is used to handle the complex scattering operator. The structure of the paper is the following: in Sect. 2, the formulation of three-dimensional electronic BTE is briefly discussed. In Sect. 3, the numerical procedure is discussed in depth. In Sect. 4, the numerical results, including accuracy test and dynamics of BTE are presented. Finally, in Sect. 5, conclusions remarks with further development in line of the current study.

2 Governing Equations

In ultrafast dynamics, the electron evolution is represented by a distribution function $f(\mathbf{x}, \mathbf{k}, t)$, where $\mathbf{x} \in \Re^3$ is the position in real space, $\mathbf{k} \in \Re^3$ is the momentum vector and $t > 0$ is the time. This distribution function is assumed to satisfy the electronic BTE [7–9]

$$\frac{\partial f}{\partial t} + \frac{1}{\hbar}\nabla_{\mathbf{k}}\epsilon(\mathbf{k}) \cdot \frac{\partial f}{\partial \mathbf{x}} - \frac{e}{\hbar}\Big[\mathbf{E}(\mathbf{x}, t) + \frac{1}{\hbar}\nabla_{\mathbf{k}}\epsilon(\mathbf{k}) \times \mathbf{B}(\mathbf{x}, t)\Big] \cdot \frac{\partial f}{\partial \mathbf{k}} = Q(f). \tag{1}$$

Here, $Q(f)$ denotes the Boltzmann collision integral of the interaction between two electrons. Here e is the absolute value of electron charge, $\hbar$ is the reduced Planck constant, $\mathbf{E}(\mathbf{x}, t)$ is the macroscopic electric field and $\mathbf{B}(\mathbf{x}, t)$ is the magnetic field. With the specified initial condition $f(\mathbf{x}, \mathbf{k}, 0) = f_0(\mathbf{x}, \mathbf{k})$, the Eq. (1) must be solved for phase space coordinates $(\mathbf{x}, \mathbf{k}) \in \Omega_{\mathbf{x}} \times \Omega_{\mathbf{k}} = \Omega$ and time $t \in \big[0, t_f\big]$. The domain of momentum vector $\mathbf{k}$ i.e. $\Omega_{\mathbf{k}}$ is the first Brillouin zone (BZ). Here, we consider the electrons as uniform and homogeneous with $\frac{\partial f}{\partial \mathbf{x}} = 0$. Then electronic BTE (1) becomes

$$\frac{\partial f}{\partial t} - \frac{e}{\hbar}\Big[\mathbf{E}(\mathbf{x}, t) + \frac{1}{\hbar}\nabla_{\mathbf{k}}\epsilon(\mathbf{k}) \times \mathbf{B}(\mathbf{x}, t)\Big] \cdot \frac{\partial f}{\partial \mathbf{k}} = Q(f). \tag{2}$$

Here, we are focusing on three dimensional materials with a magnetic field. In the three-dimensional momentum-space, the electronic BTE (2) is expressed as

$$\frac{\partial f}{\partial t} + v_k^x \frac{\partial f}{\partial k_x} + v_k^y \frac{\partial f}{\partial k_y} + v_k^z \frac{\partial f}{\partial k_z} = Q(f), \tag{3}$$

where, $(k_x, k_y, k_z) \in \Omega_k^x \times \Omega_k^y \times \Omega_k^z = \Omega_{\mathbf{k}}$ are the components of the momentum vector $\mathbf{k}$. The v_k^x, v_k^y, and v_k^z denote the components of momentum velocity $\mathbf{v}_k$ in k_x, k_y and k_z directions, respectively.

$$\begin{aligned} v_k^x &= -\frac{e}{\hbar}\left[E_x + \frac{\partial \epsilon}{\partial k_y}\frac{B_z}{\hbar} - \frac{\partial \epsilon}{\partial k_z}\frac{B_y}{\hbar}\right], \\ v_k^y &= -\frac{e}{\hbar}\left[E_y + \frac{\partial \epsilon}{\partial k_z}\frac{B_x}{\hbar} - \frac{\partial \epsilon}{\partial k_x}\frac{B_z}{\hbar}\right], \\ v_k^z &= -\frac{e}{\hbar}\left[E_z + \frac{\partial \epsilon}{\partial k_x}\frac{B_y}{\hbar} - \frac{\partial \epsilon}{\partial k_y}\frac{B_x}{\hbar}\right]. \end{aligned} \tag{4}$$

Here (E_x, E_y, E_z), and (B_x, B_y, B_z) are the components of the applied electric field $\mathbf{E}$, and the magnetic field $\mathbf{B}$, respectively in k_x, k_y and k_z directions. A simple collisional model−relaxation time approximation − is used here for the scattering operator.

$$Q(f) = -\frac{f(\mathbf{k}, t) - f_{eq}(\mathbf{k})}{\tau}, \tag{5}$$

where, $f_{eq}(\mathbf{k})$ represents an equilibrium electronic distribution, while the parameter τ denotes the relaxation time. At equilibrium, without applied fields or temperature gradients, the $f_{eq}(\mathbf{k})$ is given by Fermi-Dirac distribution function,

$$f_{eq}(\mathbf{k}) = \frac{1}{1 + e^{\beta(\epsilon(\mathbf{k}) - \mu)}}. \tag{6}$$

with $\beta = 1/k_B T$ being the inverse temperature, k_B the Boltzmann constant, T the absolute temperature, and μ the chemical potential. In present study, the electric field $\mathbf{E}$ is examined as the time independent constant, but it is in general not required to be so. The numerical method allows for any type of arbitrary energy dispersion relation $\epsilon(\mathbf{k})$. Here we show numerical results for a tight-binding anysotropic energy dispersion relation defined as

$$\epsilon(k_x, k_y, k_z) = A_1 \cos\left(\frac{2\pi k_x}{L_x}\right) + A_2 \cos\left(\frac{2\pi k_y}{L_y}\right) + A_3 \cos\left(\frac{2\pi k_z}{L_z}\right). \tag{7}$$

In expression (7), the A_1, A_2, A_3 are the arbitrary constants and the L_x, L_y and L_z are the domain sizes of Ω_k^x, Ω_k^y and Ω_k^z. The schematic of this energy dispersion

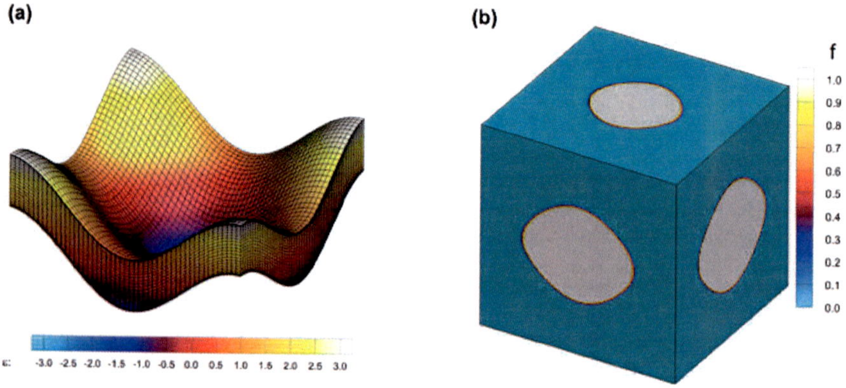

Fig. 1 (**a**) Schematic contours of the energy dispersion relation, and (**b**) the momentum resolved distribution function at thermal equilibrium

relation within the first Brillouin zone and the momentum resolved distribution function at thermal equilibrium is plotted in Fig. 1.

3 Numerical Procedure

The three-dimensional electronic BTE (3) described in the previous section are discretized using an explicit modal discontinuous Galerkin (DG) approximation. For this purpose, the hexahedron elements are used, which is the tensor product of grids in k_x, k_y, and k_z directions. Let $\{\mathcal{T}\}$ be a family of partitions of the computational domain $\Omega_{\mathbf{k}} = \Omega_k^x \times \Omega_k^y \times \Omega_k^z$ which is partitioned into $N_x \times N_y \times N_z$ uniform and non-overlapping hexahedron elements as

$$\begin{aligned} k_x^L &= k_x^{1/2} < k_x^{3/2} < \cdots < k_x^{i+1/2} < k_x^{N_x+1/2} = k_x^R, \\ k_y^L &= k_y^{1/2} < k_y^{3/2} < \cdots < k_y^{j+1/2} < k_y^{N_y+1/2} = k_y^R, \\ k_z^L &= k_z^{1/2} < k_z^{3/2} < \cdots < k_z^{j+1/2} < k_z^{N_z+1/2} = k_z^R. \end{aligned} \tag{8}$$

After then, a three-dimensional Cartesian element is defined here as

$$\begin{aligned} &\mathcal{T} := \{T_{ijk} = I_i \times J_j \times K_k, \quad 1 \le i \le N_x, 1 \le j \le N_y, 1 \le k \le N_k\}, \\ &I_i := \left[k_x^{i-1/2}, k_x^{i+1/2}\right], \quad \forall i = 1 \cdots N_x; \\ &I_j := \left[k_y^{j-1/2}, k_y^{j+1/2}\right], \quad \forall j = 1 \cdots N_y; \\ &I_k := \left[k_z^{k-1/2}, k_z^{k+1/2}\right], \quad \forall k = 1 \cdots N_z. \end{aligned} \tag{9}$$

Here, $k_x^i = \frac{1}{2}(k_x^{i-1/2} + k_x^{i+1/2})$, $k_y^j = \frac{1}{2}(k_y^{j-1/2} + k_y^{j+1/2})$, and $k_z^k = \frac{1}{2}(k_k^{k-1/2} + k_z^{k+1/2})$ are the center points of the cells. For the Ω_k^x-domain, we introduce the piecewise polynomial space of the functions $\xi : \Omega_k^x \longmapsto \Re$ as

$$Z_h^l = \{\xi \in L_2(\Omega_k^x) : \quad \xi \mid_{\Omega_k^x} \in \mathbb{P}^l(I_i), \quad i = 1, \cdots N_x\}, \tag{10}$$

where $L_2(\Omega_k^x)$ is the space function of the squared Lebesgue integral over the domain Ω_k^x and $\mathbb{P}^l(I_i)$ denotes the space of polynomial functions of degree at most l. For the domain $\Omega_{\mathbf{k}} = \Omega_k^x \times \Omega_k^y \times \Omega_k^z$, we consider $\varphi : \Omega_{\mathbf{k}} \longmapsto \Re$,

$$V_h^l = \{b \in L_2(\Omega_{\mathbf{k}}) : \quad b \mid_{\Omega_{\mathbf{k}}} \in \mathbb{Q}^l(T_{ijk})\}, \tag{11}$$

where $\mathbb{Q}^l(T_{ijk}) = \mathbb{P}^l(I_i) \otimes \mathbb{P}^l(J_j) \otimes \mathbb{P}^l(J_k)$ denotes the space of polynomials of degree at most l on T_{ijk}. The number of degrees of freedom of $\mathbb{Q}^l(T_{ijk})$ is $N_p = (l+1)(l+2)(l+3)/6$. The numerical solution of the electronic BTE (3) is approximated by $f_h \in V_h^l$ in the local element T_{ijk},

$$f_h(\mathbf{k}, t) = \sum_{i=1}^{N_p} \hat{f}_h^i(t) b_i(\mathbf{k}), \quad \mathbf{k} \in T_{ijk}, \tag{12}$$

where $\hat{f}_h^i$ is the local degree of freedom, and $b_i(\mathbf{k})$ is the piecewise polynomial function. In the present study, the three-dimensional scaled Legendre polynomials are used, which are defined as

$$\begin{aligned} b_k(\xi, \eta, \zeta) &= \psi_i(\xi) \otimes \psi_j(\eta) \otimes \psi_k(\zeta), \\ \psi_i(\xi) &= \frac{2^i (i!)^2}{(2i)!} P_i^{0,0}(\xi), \quad -1 \le \xi \le 1, \\ \psi_j(\eta) &= \frac{2^j (j!)^2}{(2j)!} P_j^{0,0}(\eta), \quad -1 \le \eta \le 1, \\ \psi_k(\zeta) &= \frac{2^k (k!)^2}{(2k)!} P_k^{0,0}(\zeta), \quad -1 \le \zeta \le 1 \end{aligned} \tag{13}$$

Here, $P_n^{0,0}$ denotes the Legendre polynomial. Now, the exact solution of electronic BTE (3) is replaced by the corresponding numeric approximation defined in Eq. (12), and then, is multiplied with a test function $b_h(\mathbf{k})$, integrating by parts over

$T_{ijk} \in \mathcal{T}$. Taking $b_h \in V_h^l(\mathcal{T})$ and $f_h(t,.,.) \in V_h^l(\mathcal{T}), \forall t \in [0, t_f]$, we obtain

$$\begin{aligned}
&\frac{\partial}{\partial t}\iint\limits_{T_{ijk}} f_h b_h dk_x dk_y dk_z - \iint\limits_{T_{ijk}} v_k^x f_h \cdot \nabla_{k_x} b_h dk_x dk_y dk_z + \\
&\int\limits_{\partial_{k_x} T_{ijk}} \mathbf{n} \cdot \widehat{v_k^x f_h} b_h d\nu dk_y dk_z - \iint\limits_{T_{ijk}} v_k^y f_h \cdot \nabla_{k_y} b_h dk_x dk_y dk_z + \\
&\int\limits_{\partial_{k_y} T_{ijk}} \mathbf{n} \cdot \widehat{v_k^y f_h} b_h dk_x d\varsigma dk_z - \iint\limits_{T_{ijk}} v_k^z f_h \cdot \nabla_{k_z} b_h dk_x dk_y + \\
&\int\limits_{\partial_{k_z} T_{ijk}} \mathbf{n} \cdot \widehat{v_k^z f_h} b_h dk_x dk_y d\sigma = - \iint\limits_{T_{ijk}} \Big(\frac{f - f_{eq}}{\tau}\Big) b_h dk_x dk_y dk_z.
\end{aligned} \tag{14}$$

where $\partial_{k_x} T_{ijk}$, $\partial_{k_y} T_{ijk}$, and $\partial_{k_z} T_{ijk}$ are the boundaries of the local element T_{ijk}. The symbol $\mathbf{n}$ denotes the outward pointing normal vector on the boundary of the element T_{ijk}. The hat symbol denotes the so-called numerical fluxes and the symbols ν, ς, and σ denote the integration variables on $\partial_{k_x} T_{ijk}$, $\partial_{k_y} T_{ijk}$, and $\partial_{k_y} T_{ijk}$, respectively. For the numerical flux , the upwind scheme is employed at elemental interface. The number of quadrature points necessary for $l-$th order finite element space depends on the type of quadrature rule employed in the numerical process [9]. In the present study, the volume and surface integrations appearing in Eq. (14) are computed by Gauss-Legendre quadrature rule with $(2l + 1)$ quadrature points to ensure accuracy [9].

After that, by collecting all the DG contribution simultaneously, the DG formulation for three-dimensional electronic BTE (3) delivered a semi-discrete ordinary differential equation in time for each element as

$$\frac{df_h}{dt} = \mathbf{M}^{-1}\mathbf{L}(f_h), \tag{15}$$

where $\mathbf{M}^{-1}$ represents the inverse of orthogonal mass matrix and $\mathbf{L}(f_h)$ is the residual function. In present study, an explicit third-order accurate SSP RK method is adopted for time integration [14],

$$\begin{aligned}
f_h^{(1)} &= f_h^n + \Delta t \mathbf{M}^{-1}\mathbf{L}(f_h^n), \\
f_h^{(2)} &= \frac{3}{4} f_h^n + \frac{1}{4} f_h^{(1)} + \frac{1}{4}\Delta t \mathbf{M}^{-1}\mathbf{L}(f_h^{(1)}), \\
f_h^{n+1} &= \frac{1}{3} f_h^n + \frac{2}{3} f_h^{(2)} + \frac{2}{3}\Delta t \mathbf{M}^{-1}\mathbf{L}(f_h^{(2)}),
\end{aligned} \tag{16}$$

where $\mathbf{L}(f_h^n)$ represents a numerical approximation of the solution at time t_n and Δt is the time step.

4 Results and Discussion

4.1 Accuracy Test

Here, a three-dimensional linear scalar problem is considered to evaluate the code and estimate the order of accuracy for the present modal DG approximation. This problem is taken into account because it is the one for which these numerical methods were first devised, and its exact solution provides a simple and smooth analytical formulation.

$$\begin{aligned} &\frac{\partial f}{\partial t} + v_k^x \frac{\partial f}{\partial k_x} + v_k^y \frac{\partial f}{\partial k_y} + v_k^z \frac{\partial f}{\partial k_z} = 0, \\ &f(k_x, k_y, k_z, 0) = f_0(k_x, k_y, k_z) = \sin(\pi(k_x + k_y + k_z)). \end{aligned} \tag{17}$$

The computational domain $\Omega_\mathbf{k} = [0, 2]^3$ is considered and periodic boundary conditions are imposed everywhere. The characteristic speed coefficients are set to be $v_k^x = v_k^y = v_k^z = 1$. The initial conditions and the characteristic speed coefficients are set such that the isolines of the solution and the direction of the flow do not coincide with the orientation of the mesh. The final time of computation is set to $t_f = 1$ sec. The exact solution of this problem is given by $f(k_x, k_y, k_z, t) = f_0(k_x - v_k^x t, k_y - v_k^y t, k_z - v_k^z t)$. A series of four different successively refined hexahedral meshes is adopted: $N_x \times N_y \times N_z = 10 \times 10 \times 10, 20 \times 20 \times 20, 40 \times 40 \times 40,$ and 80×80 with piecewise polynomial order $l = 1, 2$. The numerical result via third-order DG scheme with 40×40 meshes as shown in Fig. 2a and observe high resolution in this example. Figure 2b illustrates a comparison between the analytical solution and the third-order DG solution which are extracted along diagonal-line from the computational domain. As seen from result, the present scheme shows a very good agreement with the analytical solution. In Table 1, the discrete L_2 errors and rate of convergence of the present numerical scheme are computed. It is clear that the current numerical scheme delivers the requisite order of accuracy.

4.2 Dynamics of Three-Dimensional Electronic BTE

Here, we present the numerical results for the dynamic description of the three-dimensional electronic BTE from equilibrium to nonequilibrium regime. Emphasis is placed on the time evolution of distribution function and role of electric fields. For numerical experiments, the following parameters are considered here: $\mathbf{E} =$

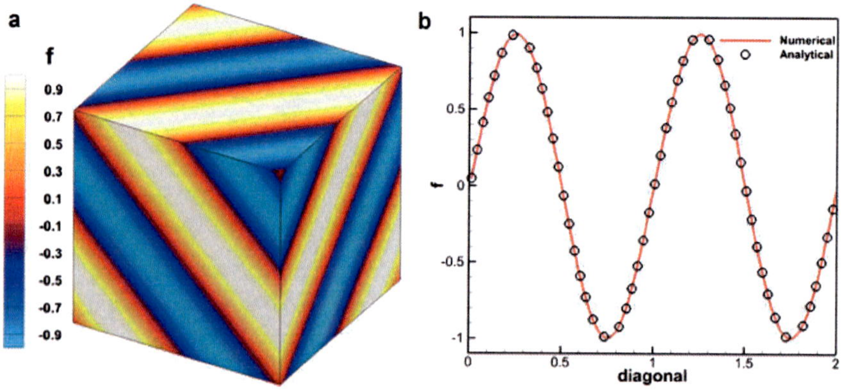

Fig. 2 Three-dimensional linear scalar equation: (**a**) solution contours, and (**b**) comparison between present DG and analytical solutions along the diagonal-line with 40×40 meshes at $t_f = 1$ sec

Table 1 Convergence study of the three-dimensional linear scalar problem

l	$N_x \times N_y \times N_z$	L_2 error	Rate of convergence
1	$10 \times 10 \times 10$	1.400×10^{-2}	–
	$20 \times 20 \times 20$	3.609×10^{-3}	1.96
	$40 \times 40 \times 40$	7.379×10^{-4}	2.29
	$80 \times 80 \times 80$	1.707×10^{-4}	2.11
2	$10 \times 10 \times 10$	1.702×10^{-2}	–
	$20 \times 20 \times 20$	3.031×10^{-3}	2.49
	$40 \times 40 \times 40$	4.962×10^{-4}	2.61
	$80 \times 80 \times 80$	7.312×10^{-4}	2.76

$(E_x, E_y, E_z) = 1 \sim 4 \times 10^4$ V/cm; $\mathbf{B} = (B_x, B_y, B_z) = 0.1$ T; $e/\hbar = 1.54 \times 10^{-4}$; $\mu = 0.0$; $\beta = 50$; and $\tau = 1$. A polynomial expansion of second-order accuracy is used to approximate solutions in the finite element space. All the computations are carried out on $70 \times 70 \times 70$ meshes and the periodic boundary conditions are enforced in all boundaries. For the numerical simulations, the final time for all steady state solution is set as $t_f = 7$ ps.

In Figs. 3 and 4, an extensive investigation on the time evolution of the distribution function for the 3D electronic BTE solution is carried out with $\mathbf{E} = 5000$ V/cm. Because the Lorentz force is orthogonal to the electron's velocity, no change is seen when a magnetic field is introduced to an initial condition of thermal equilibrium in the absence of an electric field. Furthermore, when the Fermi-Dirac distribution is employed as the initial condition, all points along those level lines have the same population, indicating that the electron population is unaffected by an external magnetic field. When an external electric field is added, however, this scenario alters rapidly. The electric field along the (1, 1, 1) direction accelerates electron in the opposite direction, as seen in Figs. 3 and 4.

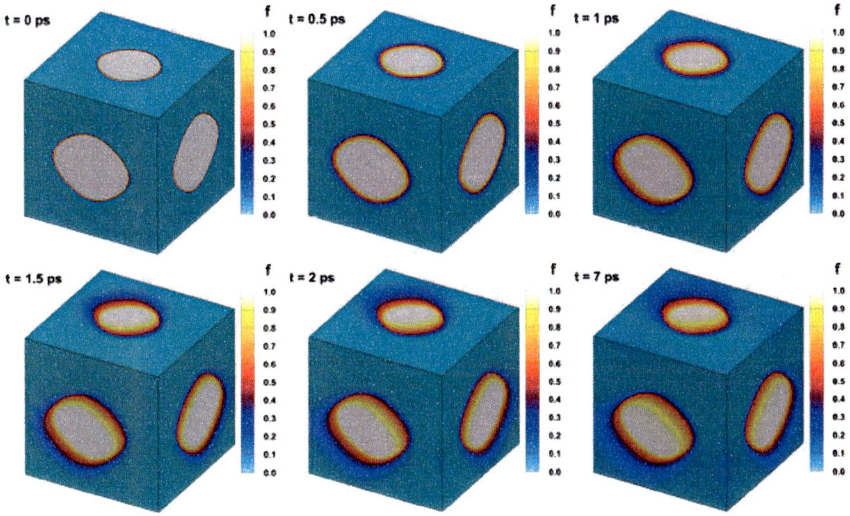

Fig. 3 Time evolution of 3D electronic BTE dynamics: contours of distribution function with $\mathbf{E} = 5 \times 10^3$ V/cm, $\mathbf{B} = 0.1$ T, $\beta = 50$, $\mu = 0$ and $\tau = 1$

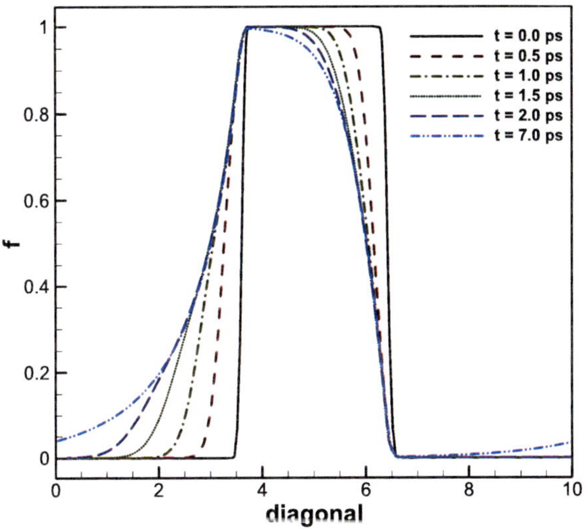

Fig. 4 Time evolution of 3D electronic BTE dynamics: profiles of distribution function along the diagonal-line with $\mathbf{E} = 5 \times 10^3$ V/cm, $\mathbf{B} = 0.1$ T, $\beta = 50$, $\mu = 0$ and $\tau = 1$

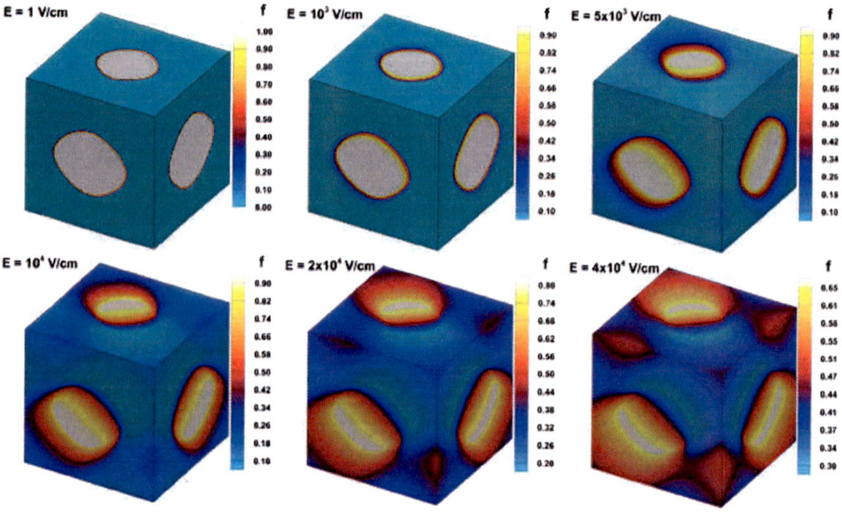

Fig. 5 Effect of electric field on electronic BTE dynamics: contours of distribution function with $\mathbf{B} = 0.1$ T, $\beta = 50$, $\mu = 0$ and $\tau = 1$

In the electronic BTE investigation, strongly nonequilibrium dynamics are considered as the most appealing and relevant phenomena. Therefore, six different electric fields $\mathbf{E} = 1, 10^3, 5 \times 10^3, 10^4, 2 \times 10^4$ and 4×10^4 V/m are selected To demonstrate the effects of electric field on BTE dynamics. The 3D contours and corresponding 1D cross-section profiles of the electronic distribution for different electric fields are illustrated in Figs. 5 and 6. At $\mathbf{E} = 1$ V/cm, a rigid shift in Fermi surface is observed, and the distribution function stays close to the equilibrium state. In this situation, a steady state solution is obtained after a time of the order of the relaxation time. At $\mathbf{E} = 10^3$ V/cm, the equilibrium distribution function begins to diverge significantly from the steady state distribution. The distribution function, on the other hand, is no longer a simple rigid shift in the Fermi surface, but rather a significant smoothening of the Fermi edge, even at low temperatures. At higher electric field $\mathbf{E} = 5 \times 10^3$ V/cm, the system switches to a damped Bloch oscillation regime. These Bloch oscillations are triggered by significant accelerations in the distribution function, which drive electrons past the Brillouin zone's border. The sharp profile of the electronic distribution is gradually dampened due to the existence of the scattering operator, and the system enters a nonequilibrium steady state. Interestingly, The distribution function crosses the Brillouin zone faster at very high electric field values i.e. $\mathbf{E} = 4 \times 10^4$ V/cm, before reaching a steady state. The steady state population tends to become nearly constant in the direction of the applied electric field at extremely high fields, as one would expect [8].

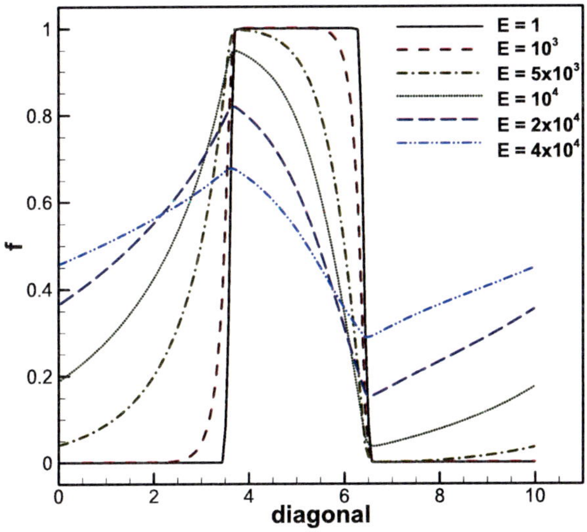

Fig. 6 Effect of electric field on electronic BTE dynamics: profiles of distribution function with $\mathbf{B} = 0.1$ T, $\beta = 50$, $\mu = 0$ and $\tau = 1$

5 Conclusion

The objective of this study was to develop an efficient numerical algorithm to simulate the three-dimensional electronic Boltzmann transport equation (BTE) in equilibrium to nonequilibrium regimes, within a single framework. We developed a three-dimensional explicit modal discontinuous Galerkin approximations based on hexahedral elements to solve the electronic BTE in conjunction with the relaxation time approximation. In this scheme, the hierarchical basis functions based on orthogonal scaled Legendre polynomials for hexahedral elements are used. The Gaussian quadrature rule was adopted for evaluating surface and volume integration. The upwind scheme was used for handling the numerical flux function, while, an explicit third-order accurate SSP-RK scheme was used for temporal discretization. To verify the order of accuracy of the numerical scheme, we solved a three-dimensional linear scalar problem. An extensive range of numerical simulations was conducted to investigate the effects of physical parameters on electronic distribution function. The numerical results shows that the proposed scheme performs equally well and consistently over a very wide range of parameters and regimes in treating the ultrafast dynamics. This work will be expanded in the future to include some further validation studies with analytical and experimental data, as well as an investigation of electrical and Hall conductivity in nonequilibrium conditions to further explore ultrafast physics.

Acknowledgments The authors would like to acknowledge the financial support of the NAP-SUG grant program funded by the Nanyang Technological University, Singapore.

References

1. Beaurepaire, E., Merle, J.C., Daunois, A., Bigot, J.Y.: Ultrafast spin dynamics in ferromagnetic nickel. Phys. Rev. Lett. **76**, 4250–4253 (1996)
2. Lindemann, M., Xu, G., Pusch, T., Michalzik, R., Hofmann, M.R., Žutić, I., Gerhardt, N.C.: Ultrafast spin-lasers. Nature **568**, 212–215 (2019)
3. Hertel, T., Knoesel, E., Wolf, M., Ertl, G.: Ultrafast electron dynamics at Cu (111): response of an electron gas to optical excitation. Phys. Rev. Lett. **76**, 535–538 (1996)
4. Battiato, M., Held, K.: Ultrafast and gigantic spin injection in semiconductors. Phys. Rev. Lett. **116**, 196601 (2016)
5. Boltzmann, L.: Sitzungsber. Kaiserlichen Akad. Wiss. **66**, 275 (1872)
6. Jungemann, C., Meinerzhagen, B.: Hierarchical Device Simulation: The Monte-Carlo Perspective. Springer, Berlin (2003)
7. Singh, S., Battiato, M.: Effect of strong electric fields on material responses: the Bloch oscillation resonance in high field conductivities. Materials **13**, 1070 (2020)
8. Singh, S., Battiato, M.: Strongly out-of-equilibrium simulations for electron Boltzmann transport equation using modal discontinuous Galerkin approach. Int. J. Appl. Comput. Math. **6**, 133 (2020)
9. Singh, S., Battiato, M.: An explicit modal discontinuous Galerkin method for Boltzmann transport equation under electronic nonequilibrium conditions. Comput. Fluids **224**, 104972 (2021)
10. Wadgaonkar, I., Jain, R., Battiato, M.: Numerical scheme for the far-out-of-equilibrium time-dependent Boltzmann collision operator: 1D second-degree momentum discretisation and adaptive time stepping. Comput. Phys. Commun. **263**, 107863 (2021)
11. Wadgaonkar, I., Wais, M., Battiato, M.: Numerical solver for the out-of-equilibrium time dependent Boltzmann collision operator: application to 2D materials. Comput. Phys. Commun. **271**, 108207 (2022)
12. Singh, S.: Numerical investigation of thermal non-equilibrium effects of diatomic and polyatomic gases on the shock-accelerated square light bubble using a mixed-type modal discontinuous Galerkin method. Int. J. Heat Mass Transfer **179**, 121708 (2021)
13. Singh, S.: Mixed-type discontinuous Galerkin approach for solving the generalized FitzHugh–Nagumo reaction–diffusion model. Int. J. Appl. Comput. **7**, 207 (2021)
14. Singh, S., Karchani, A., Chourushi, T., Myong, R.S.: A three-dimensional modal discontinuous Galerkin method for the second-order Boltzmann-Curtiss-based constitutive model of rarefied and microscale gas flows. J. Comput. Phys. **457**, 111052 (2022)

High Order Compact Central Spatial Discretization Under the Framework of Entropy Split Methods

Björn Sjögreen and H. C. Yee

1 Introduction and Background

Compact (Padé) spatial finite discretizations (or implicit spatial schemes) were developed over five decades ago for computational fluid dynamics (CFD) [1, 10]. It is well known in the computational turbulence community that high order compact schemes are methods of choice for many incompressible, nearly incompressible and low speed compressible turbulent/acoustic flows for DNS (direction numerical simulation) and LES (large eddy simulation) computations. This is largely due to the fact that, in a linear analysis, the compact scheme has the advantage of requiring a low number of grid points per wavelength and exhibits low aliasing error. In the presence of multiscale shock interactions, however, this desired property of high order compact schemes seems to have diminished in both the gas dynamic and MHD test cases that Yee and Sjögreen demonstrated in 2008 [19]. There is an added disadvantage of compact schemes in the parallel computation environment for large scale complex multi-dimensional turbulence flows. As summarized in Fang et al. [6] quote: "Significant effort has been devoted to the parallelisation of compact schemes. In general, there are two categories of methods for dealing with a parallel compact scheme. The first group focuses on the method of solving a linear system in parallel, using either pipeline methods, alternating direction methods, parallel diagonal dominant algorithms or line-relaxation methods. The second group essentially decouples compact schemes to enable them to be solved independently

B. Sjögreen
Multid Analyses AB, Gothenburg, Sweden
e-mail: bjorn.sjogreen@multid.se

H. C. Yee (✉)
NASA Ames Research Center, Moffett Field, CA, USA
e-mail: Helen.M.Yee@nasa.gov

J. M. Melenk et al. (eds.), *Spectral and High Order Methods for Partial Differential Equations ICOSAHOM 2020+1*, Lecture Notes in Computational Science and Engineering 137, https://doi.org/10.1007/978-3-031-20432-6_29

on each processor. Consequently, the computational field can be partitioned with classic domain decomposition methods and each subdomain is solved independently on a processor, in common with most CFD solvers using explicit schemes." One of the shortcomings as shown in [12] is that compact schemes display bias in the direction of convection using the domain decomposition for parallel computations of multi-dimensional problems. This bias in the direction of convection, often produces numerical instability near the inflow and severely damps the solution, always near the outflow. New more complicated compact schemes proposed in Fang et al. and Sengupta et al. have been specifically designed for parallel computing using domain decomposition methods in order to minimize the bias in the direction of convection.

Putting aside the aforementioned CPU intensity and complexity of parallelization of compact schemes for multi-dimensional applications, the objective of this research is to examine the performance of the Yee et al. and Sjögreen and Yee high order entropy split methods [14–16, 18, 20–23] by employing the high order compact scheme. Under the entropy split method framework, this study shows that stability is greatly improved in the use of the compact schemes without the need of low pass high order compact filters to stabilize long time integration of flows, e.g., DNS (direction numerical simulation) and LES (large eddy simulation) computations. For the numerical experiments an eighth-order compact scheme was compared with the explicit scheme of the same order. Both schemes have fourth-order accurate summation-by-parts (SBP) boundary closure at the domain boundaries. See [17] for explicit central scheme SBP boundary closure. Furthermore, good performance when using the entropy split approach with the compact scheme as base discretization together with the dissipative portion of a seventh-order WENO nonlinear filter for the equations of magnetohydrodynamics (MHD) was demonstrated. The following present the notation, a truncation errors comparison and the formulation of the entropy split approach using compact schemes for the gas dynamics and MHD equations. Section 2 compares the stability and accuracy of the compact scheme approach presented in this Section with the explicit spatial scheme of the same order.

Consider a one-dimensional domain of length L and $0 < x < L$. The domain is divided by uniform N grid points and grid spacings h with $x_j = (j-1)h$, $j = 1, \ldots, N$. Consider also a continuous function $u(x)$ with $0 < x < L$ and let u_j, be the approximation of the function value $u(x_j)$ $\quad j = 1, \ldots, N$. Here, explicit centered finite difference operators are denoted by D. Fully written out with coefficients

$$Du_j = \frac{1}{h}\sum_{k=1}^{s} \alpha_k (u_{j+k} - u_{j-k}),$$

where the stencil width is $2s + 1$. The optimal order of accuracy with a $2s + 1$ point wide stencil is $p = 2s$. Implicit (or compact) finite difference operators are denoted by C. Written with coefficients, these are

$$\gamma_0(Cu)_j + \sum_{k=1}^{r} \gamma_k((Cu)_{j+k} + (Cu)_{j-k}) = \frac{1}{h}\sum_{k=1}^{s} \beta_k(u_{j+k} - u_{j-k}). \tag{1}$$

Hence, the stencil width is $2r + 1$ on thc implicit side and $2s + 1$ on the explicit part. The optimal order of accuracy for C is $p = 2r + 2s$.

One of the reasons to use the more complicated compact difference operators is that the truncation error is smaller than for the explicit operator of the same order. The pth-order accurate explicit operator has truncation error

$$D_p u(x_j) - u_x(x_j) = c_p^{(e)} h^p u^{(p+1)} + O(h^{p+2}),$$

and the pth-order accurate compact operator has truncation error

$$C_p(u(x_j) - u_x(x_j) = c_p^{(i)} h^p u^{(p+1)} + O(h^{p+2}).$$

Table 1 compares the error constants $c_p^{(e)}$ and $c_p^{(i)}$ of the optimal order operators for $p = 4, 6, 8$. The second column shows the explicit stencil size, s, and the third column shows the compact operator's implicit and explicit stencil sizes (r, s). The last column, giving the ratio $c_p^{(e)}/c_i^{(i)}$, clearly shows that the compact operators have smaller error constants.

Near boundaries, the operators need to be modified to keep the stencils inside the domain $1 \le j \le N$. When defined at all points in the bounded domain, the operators can be written as matrices acting on grid function vectors, $\mathbf{u} = (u_1, \ldots, u_N)$. For the explicit operators this is denoted by $D\mathbf{u}$. For the implicit operators $C\mathbf{u} = P^{-1}Q\mathbf{u}$, where P is formed from the elements γ_k of (1) and Q is defined in terms of β_k.

1.1 SBP Boundary Closure

Introduce a discrete scalar product on $0 \le x \le L$ by

$$(u, v)_h = \sum_{i=1}^{N}\sum_{j=1}^{N} hH_{i,j}u_i v_j,$$

Table 1 Error constants for optimal order explicit and implicit difference operators

p	s	(r, s)	$c^{(e)}$	$c^{(i)}$	$c^{(e)}/c^{(i)}$
4	2	(1,1)	1/30	1/210	7
6	3	(1,2)	1/140	1/2100	15
8	4	(2,2)	1/630	1/44100	70

where the norm weights $H_{i,j}$ form a symmetric positive definite matrix, H. If the matrix is diagonal (with positive elements), the standard weighted scalar product obtained is

$$(u, v)_h = \sum_{j=1}^{N} h H_j u_j v_j.$$

SBP approximations use boundary operators and a norm H that are determined to satisfy

$$(u, Dv)_h = -(Du, v)_h + u_N v_N - u_1 v_1.$$

SBP approximations are well-known for explicit operators [17].

In the SBP construction non-diagonal norms, or full norms, allow higher boundary accuracy thanks to the larger number of degrees of freedom involved. However, full norms have some drawbacks in comparison with diagonal norms. First, unlike diagonal norms, imposing boundary conditions by projection is not equivalent to simple injection. Secondly, the operator to move a variable coefficient $a(x_j)$ between arguments $(u, av)_h = (au, v)_h$ is in general not valid with full norms.

SBP for compact operators has been considered in the past but is less well-developed than SBP for explicit operators. Construction of SBP compact operators is done by determining boundary modifications of matrices H, P, and Q in such a way that they satisfy the SBP condition. There are many ways to determine these operators. One convenient simplification is to let H proportional to P with P positive definite, because then there are only the two matrices P and Q to determine instead of the three matrices H, P and Q, thereby reducing the number of unknowns. However, this is a full norm operator with the drawbacks pointed out above. Furthermore, it was pointed out in [8] that taking H proportional to P can give an additional restriction on the attainable order of accuracy of the boundary operators.

For problems with periodic boundary conditions, no boundary modification of the operators is necessary and the property

$$(u, Cv)_h = -(Cu, v)_h$$

holds for both explicit and implicit operators in the standard scalar product where H is the identity matrix.

We here use the eighth-order compact operator with fourth-order SBP boundary closure given in [2]. The coefficients are

$$\gamma_0 = 6/7, \quad \gamma_1 = 3/28, \quad \gamma_2 = -3/70, \quad \gamma_3 = 1/140 \tag{2}$$

$$\beta_1 = 3/4, \quad \beta_2 = -3/20, \quad \beta_3 = 1/60 \tag{3}$$

Note, this operator is not of optimal order, since the stencil size is wider than necessary for an eighth-order scheme. The truncation error in the interior has error constant $c_8^{(i)} = 1/2520$ which is a factor 4 smaller than the explicit optimal order eight-order operator. This operator uses the full norm obtained by setting $H = P$. The SBP norms and boundary operators are given in the Appendix.

The explicit eighth-order operator has $s = 4$ with coefficients

$$\alpha_1 = 4/5 \quad \alpha_2 = -1/5 \quad \alpha_3 = 4/105 \quad \alpha_4 = -1/280 \tag{4}$$

away from boundaries.

Comparison of Truncation Errors One objective of this study is to assess the influence of SBP boundary operators on the overall accuracy of the schemes. The accuracy advantage for the implicit schemes demonstrated by Table 1 does not take into account boundary modification of the operators. Since compact schemes are global methods, potentially, when solving a time dependent partial differential equation, errors from the boundary modified part of the operators could dominate the total solution error leading to a different conclusion. Unlike explicit spatial schemes which are local schemes, the boundary modified part of the operators only dominate the solution error locally.

Figure 1 compares numerically computed truncation errors for the explicit eighth-order SBP-modified operator (red color) with the compact eighth-order SBP-modified operator (3) (blue color). The operators are evaluated on the grid function x^5 on a grid with 60 points. Since the boundary operators are fourth-order accurate, the truncation error of the form $ch^4u^{(5)}$ will evaluate to $ch^4/120$. The eighth-order

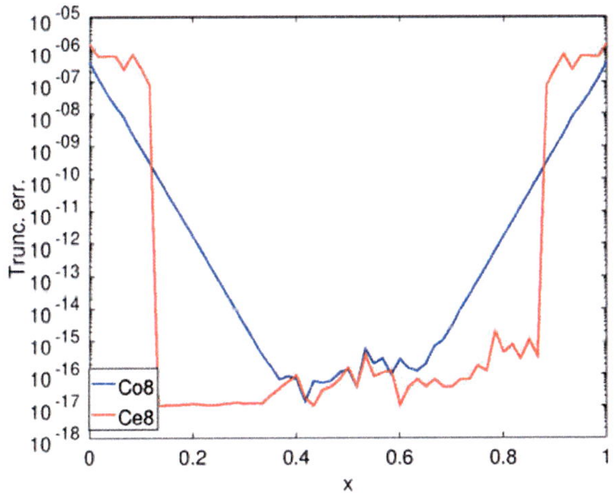

Fig. 1 Truncation errors of eighth order SBP modified operators, compact (blue) and explicit (red)

interior operators evaluate the derivative of x^5 exactly, and only contributes with arithmetic round-off errors.

The explicit operator is boundary modified at the first eighth and the last eighth points of the domain. The non-zero truncation error at these points is clearly seen in Fig. 1. Away from boundaries the error is close to zero as expected.

For the compact operator, the application of P^{-1} will spread the boundary errors into a layer of decreasing truncation errors. The layer has a fixed width in terms of grid points, here, about 20 grid points. It becomes steeper if the grid is refined. The truncation is larger for the explicit operator at the first (and last) eighth grid points, but the compact operator has a wider layer of boundary errors. This test example indicated that, for a global spatial compact scheme with non-periodic BC, any mishandling of the numerical boundary scheme treatment will contaminate the accuracy and stability of the interior domain.

1.2 Application to the Equations of Gas Dynamics and MHD

The equations of magnetohydrodynamics (MHD) are the system of conservation laws

$$\mathbf{q}_t + \mathbf{f}_x + \mathbf{g}_y + \mathbf{h}_z = \mathbf{e}\mathrm{div}\mathbf{B}. \tag{5}$$

The conserved variables are $\mathbf{q} = (\rho\ \rho u\ \rho v\ \rho w\ e\ B_1\ B_2\ B_3)^T$. where ρ is the density, $(\rho u,\ \rho v,\ \rho w)$ the momentum in the three coordinate directions, e is the total energy and, B_1, B_2, and B_3 are the magnetic field components in the x-, y-, and z-directions.

The vector in the right hand side is $\mathbf{e} = (0\ B_1\ B_2\ B_3\ \mathbf{u}^T\mathbf{B}\ u\ v\ w)^T$, where we denote $\mathbf{u} = (u\ v\ w)$ and $\mathbf{B} = (B_1\ B_2\ B_3)$. The flux in the $(k_1\ k_2\ k_3)$ direction is given by

$$\hat{f} = \begin{pmatrix} \rho\hat{u} \\ \rho u\hat{u} + k_1(p + \frac{1}{2}|\mathbf{B}|^2) - \hat{B}B_1 \\ \rho v\hat{u} + k_2(p + \frac{1}{2}|\mathbf{B}|^2) - \hat{B}B_2 \\ \rho w\hat{u} + k_3(p + \frac{1}{2}|\mathbf{B}|^2) - \hat{B}B_3 \\ \hat{u}(e + p + \frac{1}{2}|\mathbf{B}|^2) - \hat{B}\mathbf{u}^T\mathbf{B} \\ \hat{u}B_1 - \hat{B}u \\ \hat{u}B_2 - \hat{B}v \\ \hat{u}B_3 - \hat{B}w \end{pmatrix},$$

where $\hat{u} = k_1 u + k_2 v + k_3 w$ and $\hat{B} = k_1 B_1 + k_2 B_2 + k_3 B_3$. The pressure is given by

$$p = (\gamma - 1)(e - \frac{1}{2}\rho|\mathbf{u}|^2 - \frac{1}{2}|\mathbf{B}|^2). \tag{6}$$

The equations allow entropies of the form

$$E_H = -\frac{\gamma + \alpha}{\gamma - 1}\rho(p\rho^{-\gamma})^{\frac{1}{\alpha+\gamma}}. \tag{7}$$

where α is a parameter satisfying $\alpha < -\gamma$. Associated with the entropy are the entropy variables, $\mathbf{v} = \nabla_{\mathbf{q}} E_H$.

The entropy split form of the Eq. (5) is obtained by splitting the flux derivatives into conservative and non-conservative parts according to

$$\mathbf{q}_t + \frac{\beta}{\beta+1}(\mathbf{f}_x^{(x)} + \mathbf{f}_y^{(y)} + \mathbf{f}_z^{(z)}) + \\ \frac{1}{\beta+1}(A^{(x)}\mathbf{v}_x + A^{(y)}\mathbf{v}_y + A^{(z)}\mathbf{v}_z) + \frac{\beta}{\beta+1}\mathbf{e}\mathrm{div}\mathbf{B} = \mathbf{0}, \tag{8}$$

where parameter β is related to α in (7) by $\beta = (\alpha + \gamma)/(1 - \gamma)$ [21, 23].

The matrices are defined by

$$A^{(x)} = \mathbf{f}_{\mathbf{v}}^{(x)} + C^{(x)} \tag{9}$$

and similarly for the other coordinate directions. The matrices $C^{(x)}$, $C^{(y)}$, and $C^{(z)}$ are obtained by rewriting the divergence term with derivatives of the entropy variables as

$$\mathbf{e}\mathrm{div}\mathbf{B} = C^{(x)}\mathbf{v}_x + C^{(y)}\mathbf{v}_y + C^{(z)}\mathbf{v}_z.$$

The semi-discrete entropy split approximation of the one-dimensional version of (5) is

$$\frac{d}{dt}\mathbf{q}_j + \frac{\beta}{\beta+1}D\mathbf{f}_j + \frac{1}{\beta+1}A_j D\mathbf{v}_j + \frac{\beta}{\beta+1}\mathbf{e}_j D B_1 = 0, \quad j = 1, \ldots, N, \tag{10}$$

where D denotes an arbitrary linear difference operator. In the numerical experiments D will be selected as an explicit SBP operator, D, or a compact SBP operator, C. The parameter $\beta > 0$ determines the fraction of conservative vs. non-conservative part. (Note: Negative β is non-physical [23]). The MHD equations can be converted back to gas dynamics equations by setting all $\mathbf{B}$ vectors to zero.

2 Numerical Examples

In this section, we first compare the propagation of errors between the compact and explicit schemes using the SBP boundary operator for non-periodic boundaries. Then we compare the performance of the two schemes for two ideal MHD test cases. All computations use a fourth-order Runge-Kutta time discretization. Denote the eighth-order spatial explicit centered scheme by Ce and the eighth-order spatial compact scheme described above by Co.

Linear Advection Test Case The equation $u_t + u_x = 0$ is solved on the domain $0 < x < 2$. The initial data is

$$u_0(x) = \begin{cases} \phi(x - 1/2) & 1/2 < x < 3/2 \\ 0 & \text{otherwise} \end{cases},$$

where

$$\phi(x) = Ke^{\frac{1}{(1-x)x}}$$

with $K = e^4$. The initial data $u_0(x)$ is shown in Fig. 2. To study the propagation of errors from the SBP boundary operator, we perform two numerical experiments. First, solve the problem with periodic boundary conditions to time 2. Second, use the SBP boundary operators at the boundaries and impose data $g(t) = u_0(t)$ on the left boundary $x = 0$. Solving to time 2 will lead to the same solution as with periodic boundary conditions.

The CFL number was set to 0.1, which is determined to make spatial errors dominate over time discretization errors. Because time dependent data are imposed on the Runge-Kutta stages, special care was taken to make sure boundary accuracy in time is maintained at fourth-order [3]. The numerical experiments use injection of boundary data, which lead to a stable method. However, to guarantee stability for general problems, boundary data should be imposed by projection in the SBP scalar product associated with the SBP full norm. The errors on successively refined grids, going from $h = 0.1$ to $h = 3.125 \times 10^{-3}$, are plotted in Fig. 2. The middle subplot shows the maximum norm errors, and the right subplot shows the L^2 norm errors. The compact SBP modified eighth-order scheme, in blue color, is consistently smaller than the error of the explicit eighth-order SBP modified scheme in red color. The convergence rate as the grid is refined is fifth-order for both methods, as expected. The ratio of the maximum norm error between the two methods is between 4 and 5 for the four finest grids, which is close to the ratio for the interior truncation errors. The cross-over point is at $h = 0.05$ where both schemes give similar errors.

Because of the non-diagonal SBP norm used by the compact scheme, one could argue that it would be a more fair comparison if the boundary closure of the explicit operator also is allowed to have a non-diagonal SBP norm. In that case the order

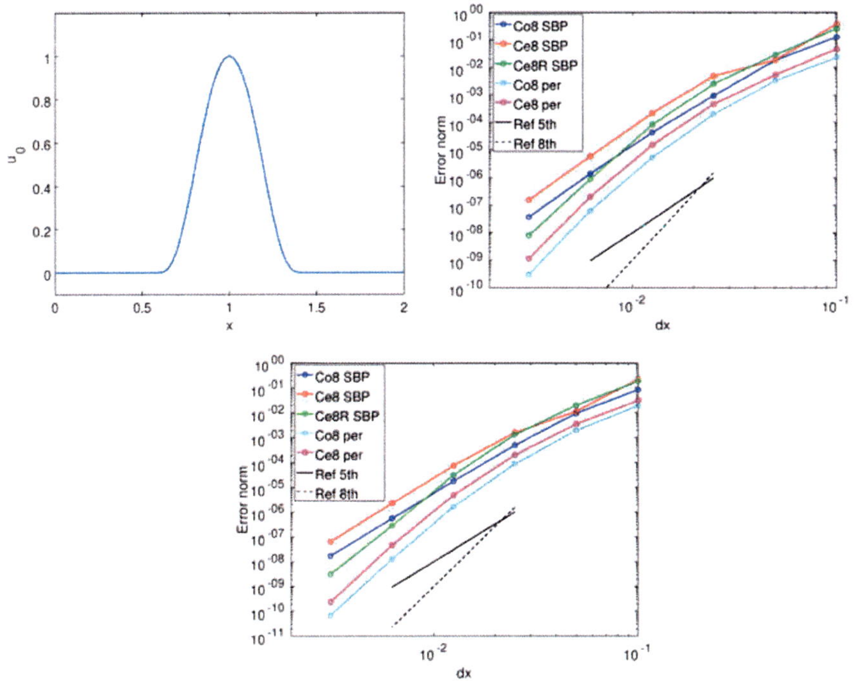

Fig. 2 Linear advection: initial data (left), Norm of solution error at $t = 2$. Maximum norm (middle) and L^2 norm (right)

of accuracy on the boundary could be brought up to seven for the explicit scheme. We tried an explicit SBP full norm operator with seventh-order accurate boundary closure, but because the boundary data are enforced by injection in the solver, the scheme was not stable. Instead a SBP restricted full norm operator with sixth-order accurate boundary closure was used, with results shown in green color (label "Ce8R SBP") in Fig. 2. If the grid is fine enough, the higher order boundary operator will lead to a smaller error than both the compact and explicit operators with fourth-order accurate boundary operators. The computations with periodic boundary conditions give smaller errors, but the convergence rate is only approaching eighth-order between the two finest grids. The ratio between explicit and compact schemes errors is again to the advantage of the compact scheme, but by a somewhat smaller margin than in the non-periodic case. The above comparison represents only the result of solving a linear constant coefficient case with smooth initial data. Next is the comparison between the compact and explicit schemes for a well-known difficult MHD test case with a slow compound wave, a contact discontinuity and a slow shock wave.

Brio-Wu Riemann Problem Test Case The Brio-Wu Riemann problem is a well-known difficult test problem to evaluate the accuracy of shock-capturing methods for equations of MHD. The problem is solved on a domain $0 < x < 1$ with initial states

$$(\rho_L, u_L, p_L, B^{(x)}, B^{(y)}, B^{(z)}) = (1, 0, 1, 0.75, 1, 0)$$

on $x < 0.5$, and

$$(\rho_R, u_R, p_R, B^{(x)}, B^{(y)}, B^{(z)}) = (0.125, 0, 0.1, 0.75, -1, 0)$$

on $x > 0.5$. The solution consists of two fast rarefaction waves, a slow compound wave, a contact discontinuity and a slow shock wave. According to the Flash code [7] and many published work, e.g., [4], standard high order shock-capturing methods exhibit oscillatory solutions, e.g., PPM (piecewise parabolic method of Colella and Woodward [5]) and third- to fifth-order WENO (WENO3 - WENO5). Most researchers resorted to use first order or very diffusive 2nd-order methods with local Lax-Freidrichs (LLF) or variants of the HLL-type numerical fluxes [9, 11]. The oscillations increase with an increase of B_y; the reason being that a stronger B_y introduces a more transverse effect that resists shock propagation in the x direction causing the shock to move slowly.

The Brio-Wu test case was solved to time 0.1 on 801 uniform grid points. In order to handle discontinuities, the solution was post-proccessed (nonlinear filter) by the dissipative portion of a seventh-order WENO (WENO7) guided by a multiresolution wavelet flow sensor to provide the amount of numerical dissipation to be used after each full time step of the Runge-Kutta method. See [13] for details on the shock capturing nonlinear filter. The same filter was applied to both the explicit and the compact schemes. Figure 3 compares the density at time 0.1 by the entropy split scheme based on explicit difference operator (CeES) with the entropy split scheme based on implicit (compact) operators (CoES). The three different values 1.5, 2.5, and 25.5 of the entropy splitting parameter β are shown. Note that the

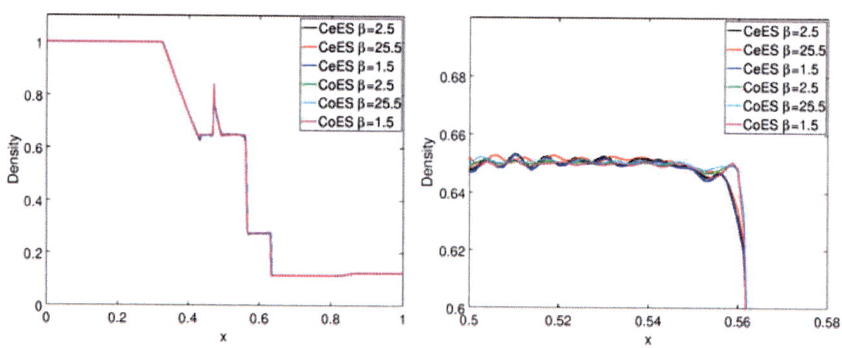

Fig. 3 Density at $t = 0.1$ of Brio-Wu Riemann problem. Right subfigure is a close up segment from the left subfigure

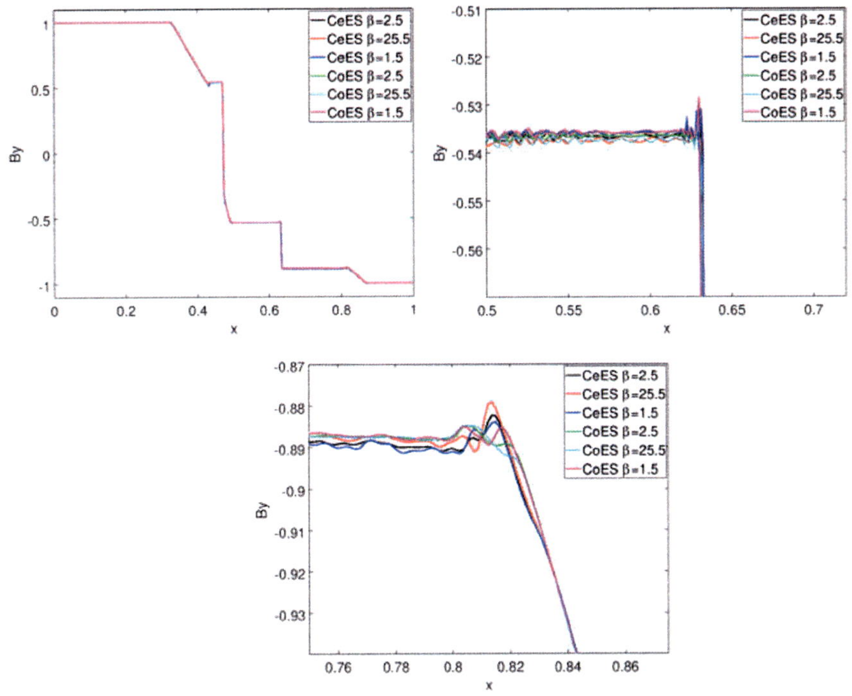

Fig. 4 Magnetic field y-component at $t = 0.1$ of Brio-Wu Riemann problem. Bottom subfigures are two close up segments from the top subfigure

bigger the $\beta > 0$, the smaller the non-conservative portion of the entropy split inviscid flux derivative is used at the expense of a lesser benefit in improving numerical instability without resorting to added numerical dissipation, especially for problems with stronger discontinuities [14–16, 23]. Similarly, Figs. 4 and 5 show the computed y-component of the magnetic field and the pressure, respectively.

For this test case, the highly oscillatory solutions using standard higher than 2nd-order shock-capturing methods have been greatly suppressed by both the compact and explicit scheme in the frame work of entropy split and nonlinear filter approaches. A study using the WENO7 with the diffusive LLF numerical flux scheme for two grids, 801 and 10,001 showed that small numerical oscillations are still present even when using the finest grid. Due to a space limitation, figures are not shown. See [4, 7, 9, 11] for early study and Yee and Sjögreen [21] for higher order structure-preserving methods comparison.

Alfvén Wave Test Case This two-dimensional MHD test case advects an Alfvén wave with phase $\phi = 2\pi(x\cos\alpha + y\sin\alpha + t)$. The exact solution is given by

$$\rho = 1 \quad p = 1 \quad \mathbf{v} = A(-\sin\alpha\sin\phi,\ \cos\alpha\sin\phi,\ \cos\phi) \tag{11}$$

$$\mathbf{B} = (\cos\alpha - A\sin\alpha\sin\phi,\ \sin\alpha + A\cos\alpha\sin\phi,\ A\cos\phi)\ . \tag{12}$$

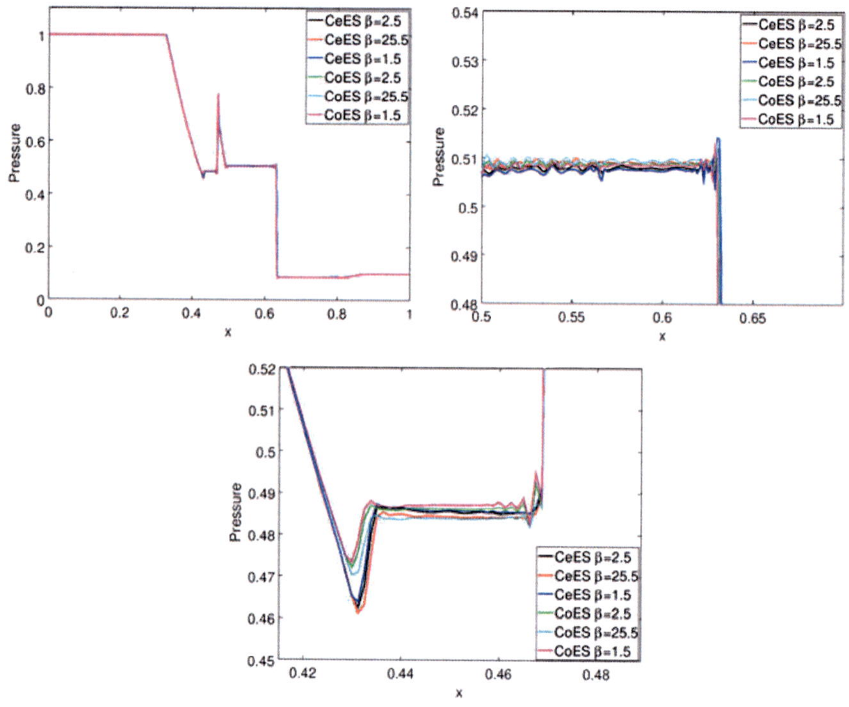

Fig. 5 Pressure at $t = 0.1$ of Brio-Wu Riemann problem. Bottom subfigures are two close up segments from the top subfigure

We consider a wave traveling in the direction $\mathbf{n} = (\cos\alpha, \ \sin\alpha, \ 0)$, where $\alpha = \pi/6$, with periodic boundary conditions. For this computation $A = 0.1$. The computational domain is of size 1.3546×2.1915, discretized by a grid of spacing $\Delta x = \Delta y = 0.022$. The problem is solved up to time 200, corresponding to approximately 70,000 time steps.

Figure 6 shows the maximum error vs. time for six different eight-order methods. The standard explicit approximation (4), shown in black, maintains stability to around time 75, whereas the solution by the compact approximation (3), shown in red, blows up near time 40. This is consistent with the observations in [19], where a numerical example showed that a compact approximation produced more spurious oscillations than an explicit approximation of the same order of accuracy.

The remaining four curves in Fig. 6 show results using the entropy split approximation (10) with explicit (CeES) or compact difference operators (CoES) for $\beta = 1$ and $\beta = 2.5$. The error stays small for the longest time for CeES at $\beta = 2.5$ and for CoES at $\beta = 1$.

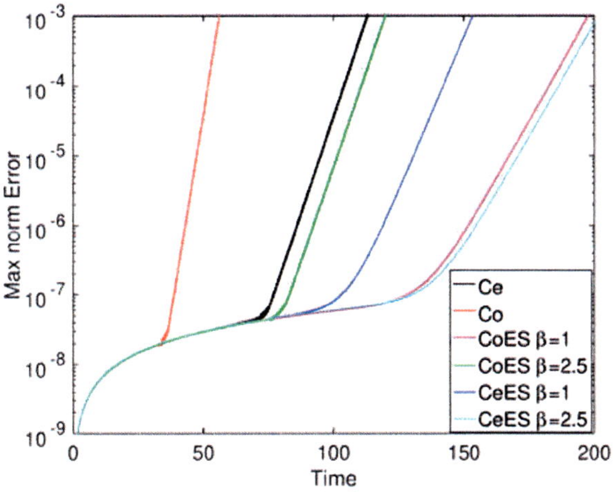

Fig. 6 Alfvén wave, maximum norm error vs. time for Ce, CeES(β = 1), CeES(β = 2.5), Co, CoES(β = 1) and CoES(β = 2.5)

3 Summary

Investigation on the performance of an eighth-order compact finite difference spatial scheme showed that its accuracy remains competitive comparing with explicit finite difference spatial scheme of the same order when the respective SBP boundary closures are used. Numerical experiments conducted on the compact scheme also illustrated to work well when used in an entropy split approximation of the equations of MHD without the need for low pass high order linear compact filter. Furthermore, the application of the entropy split compact scheme to a difficult MHD test case showed an accurate discontinuity capturing in the context of our nonlinear filter approach in conjunction with a multi-resolution wavelet flow sensor to control the amount of shock-capturing dissipation developed in [13]. The shortcoming of the compact scheme is more CPU intensive, and less friendly in the parallel computation environment than the explicit central scheme counterpart and yet at the same time exhibit no visible gain in numerical stability and accuracy than the explicit central scheme counterpart. Compact schemes are global methods. For non-periodic problems, any mishandling of the numerical boundary scheme treatment will contaminate the accuracy and stability of the interior domain, unlike explicit spatial schemes which are local schemes the error due the numerical boundary treatment is local. Future studies on complex DNS and LES flow problems are needed to explore the aforementioned results, including the development of SBP boundary closure for compact discretizations using diagonal norms.

Appendix

We here summarize the stencil coefficients of the eighth-order SBP compact operator given in [2]. Away from boundaries, the stencil coefficients are

$$\gamma_0 = 120, \ \ \gamma_1 = 15, \ \ \gamma_2 = -6, \ \ \gamma_3 = 1 \tag{13}$$

$$\beta_1 = 105, \ \ \beta_2 = -21, \ \ \beta_3 = 7/3. \tag{14}$$

These are the same coefficients as in (3), but scaled differently. The matrix representation of P near the left boundary is of the form

$$P = \begin{pmatrix} x & x & x & x & x & 0 & 0 & 0 & 0 \\ x & x & x & x & x & 0 & 0 & 0 & 0 \\ x & x & x & x & x & 1 & 0 & 0 & 0 \\ x & x & x & x & x & -6 & 1 & 0 & 0 \\ x & x & x & x & x & 15 & -6 & 1 & 0 \\ 0 & 0 & 1 & -6 & 15 & 120 & 15 & -6 & 1 \\ & & & & \ldots & \ldots & \ldots & \ldots & \ldots \end{pmatrix}$$

with the upper left 5×5 matrix being

$$\hat{P} = \begin{pmatrix} \frac{21547}{729} & \frac{20560}{729} & -\frac{12734}{729} & 2 & 1 \\ \hat{p}_{12} & \frac{3971857}{23328} & -\frac{300533}{23328} & \frac{445379}{23328} & -\frac{21263}{2592} \\ \hat{p}_{13} & \hat{p}_{23} & \frac{273185}{2592} & \frac{368689}{23328} & -\frac{43069}{23328} \\ \hat{p}_{14} & \hat{p}_{24} & \hat{p}_{34} & \frac{2828713}{23328} & \frac{293771}{23328} \\ \hat{p}_{15} & \hat{p}_{25} & \hat{p}_{35} & \hat{p}_{45} & \frac{2821361}{23328} \end{pmatrix}.$$

The SBP norm is a scaled version of P, $H = \frac{1}{140}P$. The matrix representation of Q near the left boundary is of the form

$$Q = \begin{pmatrix} x & x & x & x & x & 0 & 0 & 0 & 0 \\ x & x & x & x & x & 0 & 0 & 0 & 0 \\ x & x & x & x & x & \frac{7}{3} & 0 & 0 & 0 \\ x & x & x & x & x & -21 & \frac{7}{3} & 0 & 0 \\ x & x & x & x & x & 105 & -21 & \frac{7}{3} & 0 \\ 0 & 0 & -\frac{7}{3} & 21 & -105 & 0 & 105 & -21 & \frac{7}{3} \\ & & & & \ldots & \ldots & \ldots & \ldots & \ldots \end{pmatrix}$$

with the upper left 5×5 matrix being

$$\hat{Q} = \begin{pmatrix} -70 & \frac{2863}{27} & -\frac{11267}{243} & \frac{2753}{243} & -1 \\ -\hat{q}_{12} & 0 & \frac{28561}{243} & -\frac{2395}{324} & -\frac{3991}{972} \\ -\hat{q}_{13} & -\hat{q}_{23} & 0 & \frac{77245}{972} & -\frac{10337}{972} \\ -\hat{q}_{14} & -\hat{q}_{24} & -\hat{q}_{34} & 0 & \frac{2756}{27} \\ -\hat{q}_{15} & -\hat{q}_{25} & -\hat{q}_{35} & -\hat{q}_{45} & 0 \end{pmatrix}.$$

References

1. Adam Y.: Highly accurate compact implicit methods and boundary conditions. J Comput Phys. **24**, 10–22 (1977)
2. Chang, C.-C., Chen, M.-H., Teng, C.-T.: High-order summation-by-parts implicit difference operators for wave problems: presentation. https://www.math.ncku.edu.tw/~mhchen/HOFD/HOSBP_slide.pdf
3. Carpenter, M.H., Gottlieb, D., Arbabanel, S., Don, W.-S.: The theoretical accuracy of Runge-Kutta time discretizations for the initial boundary value problem: a study of the boundary error. SIAM J. Sci. Comput. **16**, 1241–1252 (1995)
4. Castro, M.J., Gallardo, J.M., Marquina, A.: Jacobian-free incomplete riemann solvers. In: Proceedings of Theory, Numerics and Application Problems I, Aachen, pp. 292–307 (2016)
5. Colella, P., Woodward, P.R.: The piecewise parabolic method (PPM) for gas-dynamical simulations. J. Comput. Phys. **54**, 174–201 (1984)
6. Fang, J., Gao, F., Moulinec, C., Emerson, D.R.: An improved parallel compact scheme for domain-decoupled simulation of turbulence. Int. J. Numer. Methods Fluids **90**(10), 479–500. https://doi.org/10.1002/fld.4731
7. Flash Manual - UserManual.wiki, University of Chicago (2021)
8. Gottlieb, D., Gustafsson, B., Olsson, P., Strand B.: On the superconvergence of Galerkin methods for hyperbolic IBVP. SIAM J. Numer. Anal. **33**, 1778–1796 (1996)
9. Gurski, K.F.: An HLLC-type approximate riemann solver for ideal magnetohydrodynamics. SIAM J. Sci. Comput. **25**, 2165–2187 (2004)
10. Hirsh, R.S.: Higher order accurate difference solutions of fluid mechanics problems by a compact differencing eechnique. J. Comput. Phys. **19**, 90–109 (1975)
11. Li, S.: An HLLC Riemann solver for magneto-hydrodynamics. J. Comput. Phys. **203**(1), 344–357 (2005)
12. Sengupta, T.K., Dipankar, A., Rao, K.: A new compact scheme for parallel computing using domain decomposition. J. Comput. Phys. **220**, 654–677 (2007)
13. Sjögreen, B., Yee, H.C.: Multiresolution wavelet based adaptive numerical dissipation control for high order methods. J. Sci. Comput. **20**, 211–255 (2004)
14. Sjögreen, B., Yee, H.C.: High order entropy conservative central schemes for wide ranges of compressible gas dynamics and MHD flows. J. Comput. Phys. **364**, 153–185 (2018)
15. Sjögreen, B., Yee, H.C.: Entropy stable method for the euler equations revisited: central differencing via entropy splitting and SBP. J. Sci. Comput. **81**, 1359–1385 (2019). https://doi.org/10.1007/s10915-019-01013-1
16. Sjögreen, B., Yee, H.C.: Construction of conservative numerical fluxes for the entropy split method. Commun. Appl. Math. Comput. (2021). https://doi.org/10.1.1007/s42967-020-00111-4

17. Strand, B.: Summation by parts for finite difference approximations for d/dx. J. Comput. Phys. **110**, 47–97 (1996)
18. Yee, H.C., Sjögreen, B.: Development of low dissipative high order filter schemes for multiscale Navier-Stokes and MHD systems. J. Comput. Phys. **225**, 910–934 (2007)
19. Yee, H.C., Sjögreen, B.: Adaptive filtering and limiting in compact high order methods for multiscale gas dynamics and MHD systems. Comput. Fluid **37**, 593–619 (2008)
20. Yee, H.C., Sjögreen, B.: High order filter methods for wide range of compressible flow speeds. In: Proceedings of the ICOSAHOM09, June 22–26, 2009, Trondheim (2009)
21. Yee, H.C., Sjögreen, B.: Comparative study on a variety of structure-preserving high order spatial discretizations with the entropy split methods for MHD. In: Proceedings of the ICOSAHOM21, July 12–16, 2021 (2021)
22. Yee, H.C., Sandham, N.D., Djomehri, M.J.: Low-dissipative high order shock-capturing methods using characteristic-based filters. J. Comput. Phys. **150**, 199–238 (1999)
23. Yee, H.C., Vinokur, M., Djomehri, M.J.: Entropy splitting and numerical dissipation. J. Comput. Phys. **162**, 33–81 (2000)

High Order Solution to Exterior 3D Wave Equation by the Method of Difference Potentials

Fouché Frantz Smith, Semyon Tsynkov, and Eli Turkel

1 Introduction

Consider the acoustic wave equation

$$\partial_t^2 u - c^2(\mathbf{x})\Delta u = F(\mathbf{x}, t), \quad (\mathbf{x}, t) \in \mathbb{R}^3 \times (0, T] \tag{1}$$

with zero initial data and a source compactly supported in space. We assume there exists a bounded connected set Υ such that $c(\mathbf{x}) = c_\infty > 0$ whenever $\mathbf{x} \notin \Upsilon$. To solve (1) numerically, we truncate the unbounded domain $\mathbb{R}^3$ with a spherical artificial outer boundary and set a pth order artificial boundary condition (ABC):

$$\partial_t^2 u - c^2(\mathbf{x})\Delta u = F(\mathbf{x}, t), \quad (\mathbf{x}, t) \in \Omega \times (0, T] \tag{2a}$$

$$B_p u = 0 \text{ on } \Gamma \times (0, T] \tag{2b}$$

where $\Omega = \{\mathbf{x} \mid \|\mathbf{x}\|_2 \le R\}$ such that $\Omega \supset \overline{\Upsilon \cup \mathrm{supp}(F)}$ and $\Gamma = \partial\Omega$. As p increases, the ABC (2b) is expected to better approximate (1). The specific form of the ABC (2b) will be introduced later (see Sect. 3).

F. F. Smith (✉) · S. Tsynkov
Department of Mathematics, North Carolina State University, Raleigh, NC, USA
e-mail: ffsmith@ncsu.edu; tsynkov@math.ncsu.edu

E. Turkel
School of Mathematical Sciences, Tel-Aviv University, Tel Aviv, Israel
e-mail: turkel@tauex.tau.ac.il

J. M. Melenk et al. (eds.), *Spectral and High Order Methods for Partial Differential Equations ICOSAHOM 2020+1*, Lecture Notes in Computational Science and Engineering 137, https://doi.org/10.1007/978-3-031-20432-6_30

2 Fourth Order Compact Scheme (FOCS)

We construct a fourth order discretization of the exterior acoustic wave equation using a $3 \times 3 \times 3$ stencil in space and three levels in time. Given the uniform time step τ, applying the θ-scheme [9] to (1) produces a one parameter family of EPDEs on the upper time level

$$(\Delta - \kappa^2)u^{n+1} = f^{n+1} \triangleq (2 - 1/\theta) f^n - f^{n-1} - \kappa^2 u^n/\theta - \left(F^{n+1} + (1/\theta - 2)F^n + F^{n-1}\right)/c^2 \quad (3)$$

where $\kappa^2 = (\theta\tau^2 c^2(\mathbf{x}))^{-1}$. When $\theta = 1/12$, (3) is fourth order accurate in t. Applying an "equation based" compact finite difference scheme[1] to (3) yields a fourth order compact scheme (FOCS)

$$\mathbf{L}_h[\kappa^2]u^{n+1}_{i,j,k} = h^2 f^{n+1}_{\mathbf{R}} \triangleq h^2 \mathbf{R}_h f^{n+1}_{i,j,k} = h^2 \left(2f^{n+1}_{i,j,k}/3 + f^{n+1}_{ss}/36 + f^{n+1}_{sc}/72\right). \quad (4)$$

The LHS operator in (4) is given by

$$\mathbf{L}_h[\kappa^2]u_{i,j,k} \triangleq -4u_{i,j,k} + u_{ss}/3 + u_{sc}/6 - h^2 \left(2(\kappa^2 u)_{i,j,k}/3 + (\kappa^2 u)_{ss}/36 + (\kappa^2 u)_{sc}/72\right)$$

where $u_{ss} = \sum_{\substack{-1 \le \bar{i},\bar{j},\bar{k} \le 1 \\ |\bar{i}|+|\bar{j}|+|\bar{k}|=1}} u_{i+\bar{i},j+\bar{j},k+\bar{k}}$ and $u_{sc} = \sum_{\substack{-1 \le \bar{i},\bar{j},\bar{k} \le 1 \\ |\bar{i}|+|\bar{j}|+|\bar{k}|=2}} u_{i+\bar{i},j+\bar{j},k+\bar{k}}$. The CFL number, $\lambda(\mathbf{x}) = c(\mathbf{x})\tau/h$ should satisfy $\max_{\mathbf{x}} \lambda(\mathbf{x}) \le \sqrt{5/8}$ to guarantee stability. For a complete derivation of the FOCS, the CFL number, and numerical validation over a cubic domain see our previous results in [21].

3 Method of Difference Potentials (MDP)

We apply the θ-scheme (Sect. 2) to the truncated problem (2) and choose the radiation boundary condition [7] for (2b). This ABC introduces new artificial variables v_j on the outer artificial boundary. We then define a sequence of equations (one for each v_j) along the outer boundary. The number of v_j determines the accuracy of the ABC as a function of the size (diameter) of Ω. Thus, problem (2) becomes

$$(\Delta - \kappa^2)u^{n+1} = f^{n+1}, \quad \mathbf{x} \in \Omega \quad (5a)$$

$$(\partial_t/c_\infty + \partial_r + 1/R)u^{n+1} = v^{n+1}_1, \quad \mathbf{x} \in \Gamma \quad (5b)$$

$$\left(\partial_t/c_\infty + j/R\right)v^{n+1}_j = (j(j-1)+\Delta_{\theta,\varphi})v^{n+1}_{j-1}/4R^2 + v^{n+1}_{j+1}, \quad \mathbf{x} \in \Gamma,\ j = 1, \ldots, p \quad (5c)$$

[1] Start with a central difference formula including the leading error term. Second, differentiate the governing equation to replace the high order derivatives contained in the leading error term. Third, approximate this expression with central differences. See [21] for detail.

where $v_0^{n+1} = 2u^{n+1}$ and $v_{p+1}^{n+1} = 0$. We refer to (5b)–(5c) as NRBC(p). We chose this ABC because it works in spherical geometry where we have formulated the elliptic PDE and it is easy to implement for any given p. Note that in [11], we used a different approach and defined the high order BGT ABCs (time harmonic case) directly.

Advancing the time marching scheme amounts to solving the EPDE (5) at every time step. The method of difference potentials (MDP) utilizes the uniformly discretized FOCS, but has the capacity to handle the curvilinear geometry of Ω. Difference potentials can be considered discrete counterparts to Calderon-Seeley potentials which reduce a given PDE to an equivalent pseudo-differential equation on the boundary of its domain. The MDP embeds the BVP (5) into a simple cubic auxiliary domain $\Omega^0 \supset \Omega$ while the Calderon-Seeley potentials are approximated with difference potentials constructed from the discrete solution operator to the EPDE on the auxiliary domain. Since the equation is positive definite and the auxiliary domain is a cube, a geometric multigrid method can compute the discrete solution operator in $O(N \log N)$ operations and achieve optimal (multigrid) convergence rates [22]. In addition, the MDP uses a spectral representation of the boundary condition on Γ. See the monograph [18] for more on the MDP and [1, 3–6, 10, 14, 19] for various applications of the MDP.

In Sect. 3.1, we define some constructs pertinent to the MDP. In Sect. 3.2, we introduce the Boundary Equation with Projection (BEP) and the governing theorem which shows the relationship between the solution to the BEP and the solution to (5) on Ω. In Sect. 3.3, we show how to solve the BEP.

3.1 Preliminaries

The following constructs are necessary to solve the EPDE (5) at every time step.

- **Grid Sets**: Let $\mathbb{N}^0/\mathbb{M}^0$ denote the uniform mesh of Ω^0 including/excluding the boundary nodes, $\Gamma^0 = \mathbb{N}^0 \setminus \mathbb{M}^0$, $\mathbb{M}^+ = \mathbb{M}^0 \cap \bar{\Omega}$, and $\mathbb{M}^- = \mathbb{M}^0 \setminus \mathbb{M}^+$. For any $(x_i, y_j, z_k) \in \mathbb{M}^0$, let $\mathcal{N}_{i,j,k} = \{(x_{i+\bar{i}}, y_{j+\bar{j}}, z_{k+\bar{k}}) \mid |\bar{i}|, |\bar{j}|, |\bar{k}| \le 1,\ 1 \le |\bar{i}| + |\bar{j}| + |\bar{k}| \le 2\}$. Then let $\mathbb{N}^{\pm} = \{\cup \mathcal{N}_{i,j,k} \mid (x_i, y_j, z_k) \in \mathbb{M}^{\pm}\}$. Finally, the discrete boundary $\gamma = \mathbb{N}^+ \cap \mathbb{N}^-$ are nodes which straddle the continuous boundary Γ. Choose Ω^0 large enough so $\gamma \cap \Gamma^0 = \emptyset$.
- **Auxiliary Problem (AP):** Given the discrete AP $\mathbf{L}_h[\kappa^2]w = g$ in $\mathbb{M}^0$ and $w = 0$ on Γ^0, where the RHS g on $\mathbb{M}^0$ can be arbitrary, the *solution operator* $\mathbf{G}_h$ (Green's operator) produces the unique solution $w = \mathbf{G}_h g$ to the discrete AP.
- **Difference Potential:** A *density*, v_γ, is a grid function supported on γ. The *difference potential* with density v_γ is $v_{\mathbb{N}^+} = \mathbf{P}_{\mathbb{N}^+} v_\gamma = w - \mathbf{G}_h\big(\mathbf{L}_h[\kappa^2]w|_{\mathbb{M}^+}\big)$, where $v_\gamma = w|_\gamma$ and the *difference projection* is $\mathbf{P}_\gamma v_\gamma = \mathbf{P}_{\mathbb{N}^+} v_\gamma|_\gamma$.

3.2 Boundary Equation with Projection (BEP)

The following theorem (Theorem 1.1.7 [18] or Proposition 3.4 [23]) gives us a representation of the solution to the BVP (5) using the difference potential.

Theorem 1 (Boundary Equations of Calderón-Seely Type) *Consider the discrete EPDE*

$$\mathbf{L}_h[\kappa^2]u_{\mathbb{N}^+}^{n+1} = h^2\tilde{f}_{\mathbf{R}}^{n+1} = \begin{cases} 0, & \mathbf{x}_h \in \mathbb{M}^- \\ h^2 f_{\mathbf{R}}^{n+1}, & \mathbf{x}_h \in \mathbb{M}^+ \end{cases} \tag{6}$$

where the RHS is assembled according to (3)*. The density* u_γ^{n+1} *coincides with the trace of a solution to* (6) *on* γ: $u_\gamma^{n+1} = u_{\mathbb{N}^+}^{n+1}|_\gamma$, *if and only if it satisfies the inhomogeneous BEP*

$$u_\gamma^{n+1} = \mathbf{P}_\gamma u_\gamma^{n+1} + \mathbf{G}_h \tilde{f}_{\mathbf{R}}^{n+1}|_\gamma. \tag{7}$$

If the above holds, $u_{\mathbb{N}^+}^{n+1}$ *is given by the generalized Green's formula:*

$$u_{\mathbb{N}^+}^{n+1} = \mathbf{P}_{\mathbb{N}^+} u_\gamma^{n+1} + \mathbf{G}_h \tilde{f}_{\mathbf{R}}^{n+1}|_{\mathbb{N}^+}. \tag{8}$$

Note that Theorem 1 doesn't make any explicit reference to the boundary condition on Γ. It shows how the trace of the solution, $u_\gamma^{n+1} = u_{\mathbb{N}^+}^{n+1}|_\Gamma$, can be substituted into the generalized Green's Formula (8) to solve the discrete EPDE (6). However, the boundary condition on Γ is necessary to construct u_γ^{n+1}.

3.3 Solving the Boundary Equation with Projection

The density u_γ^{n+1} is obtained from $\boldsymbol{\xi}_\Gamma^{n+1} = \left(u^{n+1}|_\Gamma, \partial_r u^{n+1}|_\Gamma\right)$ by applying an affine operator called the extension operator (see Appendix 1). It is derived using a combination of Taylor's Theorem about the continuous boundary Γ and a spectral representation of $\boldsymbol{\xi}_\Gamma^{n+1}$ in terms of spherical harmonics (recall, Ω is a sphere). From Theorem 1, the trace of the discrete solution must satisfy the inhomogeneous BEP (7). Consequently, substituting (21) into (7) yields the linear system

$$\mathbf{Q}_0\mathbf{c}_0^{n+1} + \mathbf{Q}_1\mathbf{c}_1^{n+1} = -\mathbf{G}_h \tilde{f}_{\mathbf{R}}^{n+1}|_{\mathbb{N}^+} - \mathbf{q}_I^{n+1}. \tag{9}$$

$\mathbf{Q}_i \in \mathbb{C}^{|\gamma|\times(1+L)^2}$ in (9) is the result of applying $\mathbf{P}_\gamma - \mathbf{I}_\gamma$ to each of the columns of $\mathbf{A}_i \in \mathbb{C}^{|\gamma|\times(1+L)^2}$ (defined in (21), Appendix 1), $\mathbf{q}_I^{n+1} = (\mathbf{P}_\gamma - \mathbf{I}_\gamma)\mathbf{Ex}_I^{n+1}$ ($\mathbf{Ex}_I^{n+1}$ defined in (18)), and $\mathbf{c}_i^{n+1} \in \mathbb{C}^{(1+L)^2}$ (defined in (22)) are the Fourier coefficients of $\partial_r^i u^{n+1}|_\Gamma$. $\mathbf{Q}_0$ and $\mathbf{Q}_1$ require $(1+L)^2$ calls of the solution operator $\mathbf{G}_h$. However,

$\mathbf{P}_{\mathbb{N}^+}(\alpha_{lm}^{(i)} Y_l^m) = (-1)^m \mathbf{P}_{\mathbb{N}^+}(\alpha_{lm}^{(i)} Y_l^m)^*$ [14, Eq. (53)] reduces the number of times we need to call $\mathbf{G}_h$ by nearly one half. The matrices $\mathbf{Q}_0$ and $\mathbf{Q}_1$ only need to be computed once as they don't depend on time. Since we assume the solution to the exterior problem (1) is smooth[2] the total number of columns $(1+L)^2 \ll |\gamma|$. The spectral form of NRBC(p) provides the additional equations necessary to solve for both unknowns in (9). Accordingly, (5b)–(5c) satisfies (see [7, eqs. (40) and (45)])

$$(\partial_t/c_\infty + \partial_r + 1/R)u = \sum_{l=1}^{\infty} \sum_{m=-l}^{l} (\mathbf{v}_{lm} \cdot \mathbf{e}_1) Y_l^m(\theta, \varphi)$$

$$d\mathbf{v}_{lm}/dt = \mathbf{A}_l \mathbf{v}_{lm} - l(l+1) c_\infty \langle u, Y_l^m \rangle \mathbf{e}_1 / 2R^2$$

where $\mathbf{v}_{lm} = [v_{lm}^{(1)}, \ldots, v_{lm}^{(p)}]^T \in \mathbb{C}^p$, $\mathbf{e}_1 = [1, 0, \ldots, 0]^T \in \mathbb{R}^p$, the tridiagonal matrix $\mathbf{A}_l$ contains $-\frac{c_\infty}{R}[1, \ldots, p]^T \in \mathbb{R}^p$ along the diagonal, $c_\infty[1, \ldots, 1]^T \in \mathbb{R}^{p-1}$ along the superdiagonal, and $\frac{c_\infty}{4R^2}[2(1) - l(l+1), \ldots, p(p-1) - l(l+1)]^T \in \mathbb{R}^{p-1}$ along the subdiagonal, and the weighted inner product is defined in (19). We truncate the auxiliary variables of NRBC(p) using $l = 0, \ldots, L$ and $m = -l, \ldots, l$ to be consistent with the truncation of the extension operator (see Appendix 1). First, we discretize the spectral boundary condition in time. If we discretize the ODE $dw/dt = g$ with the fourth order linear multistep method backwards differentiation formula (BDF4) $w^{n+1} + \sum_{j=1}^{4} a_j w^{n+1-j} = b_0 \tau g^{n+1}$ [15], replace u and $\partial_r u$ with the expansion (20), use orthogonality of the spherical harmonics (19), and combine like terms, we obtain a system of equations for the Fourier coefficients:

$$-\sum_{j=1}^{4} a_i \left\langle u^{n+1-j}, Y_0^0 \right\rangle = (1 + \tau c_\infty b_0/R) \left\langle u^{n+1}, Y_0^0 \right\rangle + \tau c_\infty b_0 \left\langle \partial_r u, Y_0^0 \right\rangle \quad (10a)$$

$$-\sum_{j=1}^{4} a_i \mathbf{\Phi}_{lm}^{n+1-j} = \hat{\mathbf{A}}_l \mathbf{\Phi}_{lm}^{n+1} + \left\langle \partial_r u^{n+1}, Y_l^m \right\rangle \mathbf{b}_l, \; l = 1, \ldots, L, |m| \le l, \quad (10b)$$

$\mathbf{\Phi}_{lm} = [\langle u, Y_l^m \rangle, \mathbf{v}_{lm}]^T \in \mathbb{C}^{p+1}$, $\mathbf{b}_l = [\tau c_\infty b_0, 0, \ldots, 0]^T \in \mathbb{R}^{p+1}$, and the tridiagonal matrix

$$\hat{\mathbf{A}}_l = \left(\begin{array}{c|c} (1 + \tau c_\infty b_0/R) & -\tau b_0 c_\infty \mathbf{e}_1^T \\ \hline \tau b_0 l(l+1) c_\infty \mathbf{e}_1 / 2R^2 & \mathbf{I}_{p \times p} - \tau b_0 \mathbf{A}_l \end{array} \right) \in \mathbb{R}^{p+1 \times p+1} \quad (11)$$

[2] Let $g \in C^{k,\gamma}(\mathbb{S}^2)$ for some $k \ge 0$ and $\gamma \in (0, 1]$ such that $k + \gamma > 1/2$. Then there exists $C > 0$ such that $\|g - \mathcal{Q}_L g\|_\infty \le C/L^{k+\gamma-1/2}$ where $\mathcal{Q}_L g$ is the orthogonal projection of g on the space of polynomials of degree $\le L$ on $\mathbb{S}^2$ [2, Corollary 4.14].

Rearranging (10) yields:

$$-\sum_{j=1}^{4}\frac{a_i}{1+\tau c_\infty b_0/R}\langle u^{n+1-j}, Y_0^0\rangle = \langle u^{n+1}, Y_0^0\rangle + \frac{\tau c_\infty b_0}{1+\tau c_\infty b_0/R}\langle \partial_r u^{n+1}, Y_0^0\rangle \tag{12a}$$

$$-\hat{\mathbf{A}}_l^{-1}\left(\sum_{j=1}^{4} a_i \mathbf{\Phi}_{lm}^{n+1-j}\right) = \mathbf{\Phi}_{lm}^{n+1} + \langle \partial_r u^{n+1}, Y_l^m\rangle \hat{\mathbf{A}}_l^{-1}\mathbf{b}_l, l = 1, \dots, L, |m| \le l. \tag{12b}$$

Now we are in a position to substitute the spectral form of NRBC(p) into (9). Since (9) doesn't depend on the auxiliary variables $\mathbf{v}_{lm}^{n+1}$ we extract the equations in (12) which only include the Fourier coefficients of $(u^{n+1}|_\Gamma, \partial_r u^{n+1}|_\Gamma)$. If we take the first component of (12b) for $l = 1, \dots, L$ and $m = -l, \dots, l$ along with (12a) and arrange them as a linear system where the lm-th component is placed in row $l(l+1)+m+1$, then

$$\mathbf{c}_0^{n+1} + \mathbf{M}\mathbf{c}_1^{n+1} = \mathbf{c}_{\mathrm{NRBC}(p)}^{n+1} \tag{13}$$

where

$$\mathbf{M} = \mathrm{diag}\left(\frac{\tau c_\infty b_0}{1+\tau c_\infty b_0/R}, \hat{\mathbf{A}}_1^{-1}\mathbf{b}_1 \cdot \tilde{\mathbf{e}}_1, \dots, \hat{\mathbf{A}}_L^{-1}\mathbf{b}_L \cdot \tilde{\mathbf{e}}_1\right)$$

and

$$\mathbf{c}_{\mathrm{NRBC}(p)}^{n+1} = -\sum_{j=1}^{4} a_j \left(\frac{\langle u^{n+1-j}, Y_0^0\rangle}{1+\tau c_\infty b_0/R}, \hat{\mathbf{A}}_1^{-1}\mathbf{\Phi}_{1,-1}^{n+1-j} \cdot \tilde{\mathbf{e}}_1, \dots, \hat{\mathbf{A}}_L^{-1}\mathbf{\Phi}_{L,L}^{n+1-j} \cdot \tilde{\mathbf{e}}_1\right)^T$$

and $\tilde{\mathbf{e}}_1 = [1, 0, \dots, 0]^T \in \mathbb{R}^{p+1}$. Substituting (13) into (9) yields:

$$(\mathbf{Q}_1 - \mathbf{Q}_0\mathbf{M})\mathbf{c}_1 = -\mathbf{Q}_0\mathbf{c}_{\mathrm{NRBC}(p)}^{n+1} - \mathbf{G}_h \tilde{f}_{\mathbf{R}}^{n+1}|_{\mathbb{N}^+} - \mathbf{q}_I^{n+1}. \tag{14}$$

The overdetermined system (14) is solved for $\boldsymbol{c}_1^{n+1}$ in the sense of least squares using QR factorization. Then $\boldsymbol{c}_0^{n+1}$ is computed with (13). The NRBC(p) auxiliary variables $\mathbf{v}_{lm}^{n+1}$ are computed from (12b) for $l = 1, \dots, L$ and $m = -l, \dots, l$. Now that the Fourier coefficients are all known we can use the extension operator (21) to compute the density u_γ^{n+1} which we substitute into the generalized Green's formula (8) to compute the solution $u_{\mathbb{N}^+}^{n+1}$ and advance the time marching scheme.

4 Numerical Simulations

Table 1 defines the components of our test problems (see Appendix 2). For **T1**, the direction of propagation is normal to Γ which corresponds to the maximum absorption. For **T2**, the off-center location of the center of the source implies the solution no longer meets Γ at a right angle. For **T3**, the off center location of the source and differentiation defined by means of the multi-index $\boldsymbol{\alpha}$ generate even more asymmetry than **T2**. In addition, the source term provides continuous output for all $t > 0$. Finally, **T4** and **T5** repeat **T1** and **T2** respectively with variable speed. In all our simulations $R = 1.5$, $\Omega^0 = [-2, 2]^3$, $L = 18$, and the termination criteria is $\|r_h^{(i)}\| \leq 10^{-12}\|r_h^{(0)}\| + 10^{-12}$ where the residual of the ith iteration is $r_h^{(i)} = g - \boldsymbol{L}_h u_h^{(i)}$. For our multigrid method we use the V(1,1) cycle per iteration using full-weighting as the restriction operator and tri-cubic interpolation as the prolongation operator.

Figure 1 qualitatively demonstrates why high order ABCs are necessary. A grid refinement analysis would fail to demonstrate fourth order accuracy since the plots from several grids overlap when $t \in (4, 5)$. Increasing the order of the NRBC clearly improves the absorption since the gap between dash and dashed-dotted (or dashed-dotted and dotted) plots increases over the interval when $t \in (4, 5)$. However, the order isn't high enough to eliminate the reflection error on each and every grid.

To perform a full-fledged grid refinement analysis we run our simulations with NRBC(6). Table 2 shows fourth order convergence for all test problems defined in Table 1. Recall that, there are two sources of error. The first source of error is the reflection error due to replacing (1) with (2), which decreases as the order of NRBC(p) increases. The second source of error is the discretization error from the FOCS which decreases as the grid is refined. An ABC of a sufficiently high order guarantees the reflection errors are so much smaller than the discretization error that refining the grid effectively produces fourth order convergence. There exists a sufficiently fine grid where refining the grid no longer improves the overall error as seen Fig. 1. Fortunately, a high order ABC decreases the "floor" of this threshold value.

Table 1 Components of the Test solution and source term defined in Appendix 2

Name	$\boldsymbol{\alpha}$	R_0	m	$\mathbf{x}_0$	t_0	$S(t)$	c^2
T1	(0, 0, 0)	1.5	7	(0, 0, 0)	3.0	$5(1 - 12t^2)\exp(-6t^2)$	1
T2	(0, 0, 0)	1.0	7	(15/100, 15/100, 15/100)	4.0	$\sin(8t)\exp(-6t^2)$	1
T3	(1, 1, 0)	0.75	12	(0, 0, 1/4)	0.2	$\left(\sin^{11}(\pi t/5) + \frac{1}{\sqrt{3}}\sin^{11}(\pi t/5\sqrt{2})\right)\chi_{(0,\infty)}$	1
T4	(0, 0, 0)	1.5	7	(0, 0, 0)	3.0	$5(1 - 12t^2)\exp(-6t^2)$	$4/5 + \exp(-20\|\mathbf{x}\|_2^2)/5$
T5	(0, 0, 0)	1.0	7	(15/100, 15/100, 15/100)	4.0	$\sin(8t)\exp(-6t^2)$	$4/5 + \exp(-20\|\mathbf{x}\|_2^2)/5$

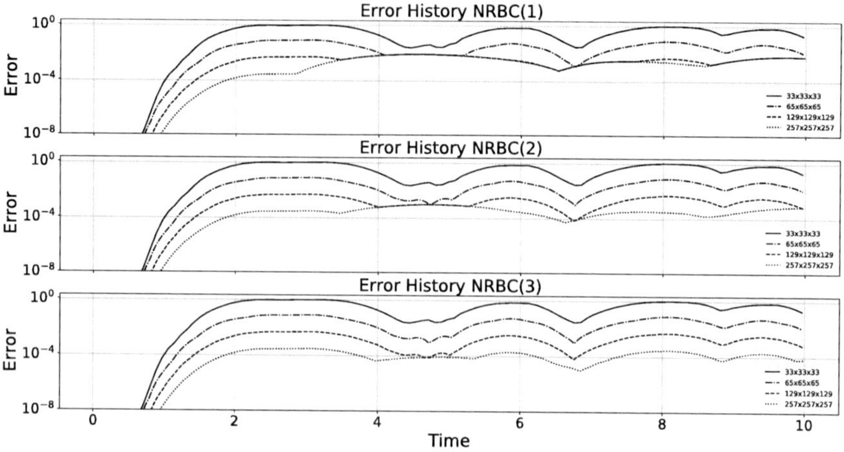

Fig. 1 Error history for **T3** for low order NRBC with respect to the infinity norm

Table 2 Error (∞-norm on $\mathbb{M}^+$) when using NRBC(6), $\tau = 9h\sqrt{5/8}/10$, and $T = 10.0$

				CPUTIME (s)		Avg. # multigrid iterations		
	h	Error	Rate	get $\mathbf{Q}_0, \mathbf{Q}_1$	Time marching	$\mathbf{G}_h \tilde{f}_{\mathbf{R}}^{n+1}$	$\mathbf{q}_I^{n+1}$	$\mathbf{P}_{\mathbb{N}^+} u_\gamma^{n+1}$
T1	4/33	9.55e-01	–	10.1	56.7	4.28	3.96	3.96
	4/65	3.75e-02	4.67	138.7	916.2	4.09	3.71	3.66
	4/129	2.24e-03	4.07	1328.7	12984.4	3.69	3.41	3.37
	4/257	1.37e-04	4.03	10524.03	195352.92	3.41	3.20	3.11
T2	4/33	7.92e-01	–	9.9	25.1	3.79	3.53	3.52
	4/65	5.10e-02	3.96	135.0	393.9	3.73	3.38	3.25
	4/129	2.73e-03	4.22	1343.9	5720.4	3.38	3.13	3.06
	4/257	1.55e-04	4.05	10572.62	79376.43	3.18	3.02	2.88
T3	4/33	9.33e-01	–	10.1	23.0	4.64	4.36	4.39
	4/65	7.34e-02	3.67	140.1	392.1	4.64	4.31	4.32
	4/129	4.61e-03	3.99	1346.2	5577.1	4.64	4.29	4.29
	4/257	2.84e-04	4.02	10514.64	80684.77	4.63	4.29	4.28
T4	4/33	1.85e+01	–	10.8	26.78	4.96	4.49	4.35
	4/65	7.09e-02	4.71	165.37	431.53	4.54	4.14	4.03
	4/129	3.84e-03	4.21	1571.36	6266.19	4.40	3.97	3.80
	4/257	2.31e-04	4.06	11931.39	90038.02	4.02	3.73	3.59
T5	4/33	1.10e+00	–	10.78	17.28	4.13	3.98	3.73
	4/65	1.06e-01	3.38	161.66	260.71	4.02	3.71	3.55
	4/129	4.90e-03	4.43	1454.40	3439.12	3.95	3.50	3.37
	4/257	2.87e-04	4.09	11821.86	49433.50	3.60	3.36	3.22

5 Concluding Remarks and Future Work

We derived a high order scheme for (1) using a combination of the θ-scheme, a compact finite difference scheme in space, multigrid, the MDP, and a high order ABC. Our numerical examples demonstrate that if the order of the ABC is *sufficiently* high, refining the grid *effectively* produces fourth order accuracy.

A perfectly matched layer (PML) is another approach to truncating unbounded problems. A PML is a layer surrounding the domain which rapidly attenuates any incoming waves. Modern PMLs are derived by transforming the governing equation in time to the frequency space, using a complex coordinate transformation, performing some algebraic manipulation, and transforming back to the time space. The PML modified governing equation resembles the original governing equation with some additional parameters and some auxiliary equations supported in the PML. Discretizing PMLs with high order accuracy while maintaining long term stability is challenging. Instead, we derived a FOCS for a Cartesian sponge layer [8], which behaves similarly to a PML without the additional auxiliary equations, but offers less effective absorption. We were able to demonstrate fourth order convergence provided the sponge layer was sufficiently thick. Otherwise the convergence rate would stall. This mirrors the results shown in Sect. 4.

In the future, we will solve the three dimensional wave scattering problem about a spherical body adapting the current high order MDP scheme as a foundation. We will close the unbounded problem with NRBC(p) or the newly developed sponge layer [8], construct an auxiliary problem which contains the scattering region and the above closure, then define a similar BEP to solve the scattering problem.

Appendix 1: Extension Operator

Consider the pair $(\mathbf{x}_h, \tilde{\mathbf{x}}_h) \in (\gamma, \Gamma)$ where $\tilde{\mathbf{x}}_h$ is the orthogonal projection of $\mathbf{x}_h$ onto Γ. By Taylor's Theorem

$$\mathbf{Ex}(\xi^{n+1}) = \left(u^{n+1} + \varrho \partial_r u^{n+1} + \sum_{j=2}^{4} \frac{\varrho^j}{j!} \partial_r^j u^{n+1}\right)|_{\tilde{x}_h} + O(\varrho^5) \tag{15}$$

where $\varrho = |\mathbf{x}_h - \tilde{\mathbf{x}}_h|$ if $\mathbf{x}_h \notin \Omega$ or $\varrho = -|\mathbf{x}_h - \tilde{\mathbf{x}}_h|$ if $\mathbf{x}_h \in \Omega$. We choose a fifth order extension operator even though we only desire fourth order accuracy. According to Reznik's Theorem [16, 17], a sixth order extension operator is sufficient for maintaining fourth order accuracy for a second order PDE discretized with fourth order accuracy. However, Reznik's Theorem isn't always necessary. For example, [3] uses a fourth order MDP scheme for the 2D acoustic wave equation using a fourth order extension operator. Dirichlet boundary conditions maintained fourth order accuracy, but Neumann boundary conditions dropped to third order accuracy. Thus, we opted for a fifth order accurate extension operator. Differentiating the acoustic wave equation (1) in spherical coordinates (i.e. $\Delta u = \partial_r^2 u + \frac{2}{r}\partial_r u + \Delta_{\theta,\varphi} u / r^2$)

yields the high order derivatives:

$$\partial_r^2 u = -\left((2/R)\partial_r u + \Delta_{\theta,\varphi} u/R^2\right) + (1/c_\infty^2)\partial_t^2 u \tag{16a}$$

$$\partial_r^3 u = \partial_t^2\partial_r u/c_\infty^2 - 2\partial_t^2 u/(c_\infty^2 R) + (6-\Delta_{\theta,\varphi})\partial_r u/R^2 + 4\Delta_{\theta,\varphi} u/R^3 \tag{16b}$$

$$\begin{aligned}\partial_r^4 u = {} & (8-\Delta_{\theta,\varphi})\partial_t^2 u/(c_\infty^2 R^2) - 2\partial_t^2\partial_r u/(c_\infty^2 R) + \partial_t^2\partial_r^2 u/c_\infty^2 + (\Delta_{\theta,\varphi}^2 - 18\Delta_{\theta,\varphi})u/R^4 \\ & + (8\Delta_{\theta,\varphi} - 24)\partial_r u/R^3\end{aligned} \tag{16c}$$

since the source term is zero on Γ and the speed $c(\boldsymbol{x}) = c_\infty$ on Γ. Replacing the time derivatives of (16) with the one sided difference scheme in time $\partial_r^i\partial_t^2 u^{n+1} = d_i^{n+1}\partial_r^i u^{n+1} + \sum_{j=0}^{n_i} d_i^{n-j}\partial_r^i u^{n-j} + O(\tau^{n_i})$, which preserves fifth order accuracy since $\tau = O(\varrho)$, and substituting the expressions into (15) leads to

$$\begin{aligned}\mathbf{Ex}(\boldsymbol{\xi}^{n+1}) = {} & \Big[1 + \varrho^2\big(d_0^{n+1}/c_\infty^2 - \Delta_{\theta,\varphi}/R^2\big)/2 + \varrho^3\big(-2d_0^{n+1}/(c_\infty^2 R) + 4\Delta_{\theta,\varphi}/R^3\big)/6 + \varrho^4\big(d_0^{n+1}d_2^{n+1}/c_\infty^4 \\ & + (\Delta_{\theta,\varphi}^2 - 18\Delta_{\theta,\varphi})/R^4 + (8d_0^{n+1} - (d_0^{n+1} + d_2^{n+1})\Delta_{\theta,\varphi})/(c_\infty^2 R^2)\big)/24\Big]u^{n+1} + \Big[\varrho - \varrho^2/R \\ & + \varrho^3\big((6-\Delta_{\theta,\varphi})/R^2 + d_1^{n+1}/c_\infty^2\big)/6 - \varrho^4\big(2(d_2^{n+1} + d_1^{n+1})/(c_\infty^2 R) + (24 - 8\Delta_{\theta,\varphi})/R^3\big)/24\Big]\partial_r u^{n+1} + \mathbf{Ex}_I^{n+1}\end{aligned} \tag{17}$$

where the inhomogeneous term is given by

$$\begin{aligned}\mathbf{Ex}_I^{n+1} = {} & \Big[\varrho^2/(2c_\infty^2) - \varrho^3/(3c_\infty^2 R) + \varrho^4\big((8-\Delta_{\theta,\varphi})/(c_\infty^2 R^2) + d_2^{n+1}/c_\infty^4\big)/24\Big]\sum_{j=0}^{n_0} d_0^{n-j} u^{n-j} \\ & + \Big[\varrho^3/(6c_\infty^2) - \varrho^4/(12c_\infty^2 R)\Big]\sum_{j=0}^{n_1} d_1^{n-j}\partial_r u^{n-j} + (\varrho^4/(24c_\infty^2))\sum_{j=0}^{n_2} d_2^{n-j}\partial_r^2 u^{n-j}.\end{aligned} \tag{18}$$

Finally, we introduce the spectral form of the trace $\boldsymbol{\xi}_\Gamma^{n+1} = (u^{n+1}|_\Gamma, \partial_r u^{n+1}|_\Gamma)$ using spherical harmonics as our basis functions. The spherical harmonics are eigenfunctions of the Laplace-Beltrami operator, i.e., $\Delta_{\theta,\varphi} Y_l^m(\theta,\varphi) = -l(l+1)Y_l^m(\theta,\varphi)$ on the unit sphere $\mathbb{S}^2$ [2, Section 3.3] and form an orthonormal basis on the sphere of radius R centered at the origin with respect to the weighted inner product [2, eq. (4.6)]:

$$\langle v, w\rangle = \int_0^{2\pi}\int_0^{\pi} v(R,\theta,\varphi)w(R,\theta,\varphi)^* \sin(\theta)\mathrm{d}\theta\mathrm{d}\varphi \tag{19}$$

The derivatives are given by

$$\partial_r^k u \approx \sum_{l=0}^{L}\sum_{m=-l}^{l}\langle\partial_r^k u, Y_l^m\rangle Y_l^m(\theta,\varphi), \quad \text{on } \Gamma, \quad k = 0, 1, \ldots \tag{20}$$

and then the extension operator (17) becomes

$$u_\gamma^{n+1} = \mathbf{Ex}(\boldsymbol{\xi}^{n+1}) = A_0\mathbf{c}_0^{n+1} + A_1\mathbf{c}_1^{n+1} + \mathbf{Ex}_I^{n+1} \tag{21}$$

The matrices are given by

$$\mathbf{A}_i = \left[\alpha_{0,0}^{(i)}Y_0^0 \ \alpha_{1,-1}^{(i)}Y_1^{-1} \ \alpha_{1,0}^{(i)}Y_1^0 \ \alpha_{1,1}^{(i)}Y_1^1 \ \dots \ \alpha_{L,L}^{(i)}Y_L^L\right], \quad i \in \{0, 1\}$$

where the coefficients are

$$\begin{aligned}\alpha_{lm}^{(0)} &= 1 + \varrho^2\left(d_0^{n+1}/c_\infty^2 + l(l+1)/R^2\right)/2 - \varrho^3\left(2d_0^{n+1}/(c_\infty^2 R) + 4l(l+1)/R^3\right)/6\\ &+ \varrho^4\left(l(l+1)(l(l+1)+18)/R^4 + d_0^{n+1}d_2^{n+1}/c_\infty^4 + (8d_0^{n+1} + (d_0^{n+1} + d_2^{n+1})l(l+1))/(c_\infty^2 R^2)\right)/24\\ \alpha_{lm}^{(1)} &= \varrho - \varrho^2/R + \varrho^3\left((6 + l(l+1))/R^2 + d_1^{n+1}/c_\infty^2\right)/6 - \varrho^4\left(2(d_2^{n+1} + d_1^{n+1})/(c_\infty^2 R) + (8l(l+1)+24)/R^3\right)/24\end{aligned}$$

and the Fourier coefficients are

$$\mathbf{c}_i^n = \left[\langle \partial_r^i u^n, Y_0^0 \rangle \ \langle \partial_r^i u^n, Y_1^{-1} \rangle \ \dots \ \langle \partial_r^i u^n, Y_L^L \rangle\right]^T, \quad i \in \{0, 1, 2\} \tag{22}$$

Substituting the expansions (20) into (18) produces the spectral from of $\mathbf{Ex}_I^{n+1}$. To compute the spherical harmonics numerically, see [12]. Given $g \in C(\mathbb{S})$, the Fourier coefficients $\langle g, Y_l^m \rangle$ for $l = 0, 1, \dots, L$ and $m = -l, \dots, l$ can be computed fast using the spherical harmonic transform described in [20].

Appendix 2: Generating Test Solutions

The function

$$u(\mathbf{x}, t) = S(t - t_0 - r/c)/(4\pi r) \quad \text{with} \quad r = \|\mathbf{x} - \mathbf{x}_0\|_2 \tag{23}$$

solves the wave equation (1), with constant speed c, driven by the point source term $F(\mathbf{x}, t) = \delta_{\mathbf{x}_0}(\mathbf{x})S(t)$. We construct a family of smooth test solutions similar to (23) employing the same strategy as [13, Section 8.2]. Define the test solution

$$u^{\text{Test}}(\mathbf{x}, t) = D^\alpha\left(\phi_m(r/R_0)S(t - t_0 - r/c)/(4\pi r)\right) \tag{24}$$

where $D^\alpha = \partial^{|\alpha|}/\partial x^{\alpha_1}\partial y^{\alpha_2}\partial z^{\alpha_3}$, and the smooth step function is defined as follows

$$\phi_m(r) = \begin{cases} r^{m+1}\sum_{k=0}^{m}\binom{m+k}{m}\binom{2m+1}{m-k}(-r)^k & \text{if } 0 \le r < 1\\ 1 & \text{if } r \ge 1\end{cases}$$

Since $\phi_m(r/R_0)/r = O(r^m)$ as $r \downarrow 0$, (24) vanishes at $\mathbf{x}_0$ provided $m > |\boldsymbol{\alpha}|$. By construction, (24) equals (23) on the complement of the sphere of radius R_0 centered at $\mathbf{x}_0$ when $|\boldsymbol{\alpha}| = 0$, and satisfies the acoustic wave equation (1) where the source term is given by

$$F^{\text{Test}}(\mathbf{x}, t) \triangleq D^{\boldsymbol{\alpha}}\left(c\phi_m'(r/R_0)S'(t-t_0-r/c)/(2\pi R_0 r) - c^2\phi_m''(r/R_0)S(t-t_0-r/c)/(4\pi R_0^2 r)\right) \tag{25}$$

The source term (25) is compactly supported in space in the ball of radius R_0 centered at $\mathbf{x}_0$. The translation t_0 is chosen sufficiently large so that the initial data are zero (up to machine precision) assuming that S is a smooth rapidly decaying function or compactly supported.

To generate a family of test solutions with variable speed, consider a function of the form $\phi_m(r/R_0)S(t-t_0-r/c(\mathbf{x}))/(4\pi r)$, then follow the same procedure as outlined above. The resulting source term will be supported on the union of Υ and the sphere of radius R_0 centered at $\mathbf{x}_0$.

Acknowledgments Work supported by the US Army Research Office (ARO) under grant W911NF-16-1-0115 and the US–Israel Binational Science Foundation (BSF) under grants 2014048 and 2020128.

References

1. Albright, J., Epshteyn, Y., Steffen, K.R.: High-order accurate difference potentials methods for parabolic problems. Appl. Numer. Math. **93**, 87–106 (2015)
2. Atkinson, K., Han, W.: Spherical Harmonics and Approximations on the Unit Sphere: An Introduction. Lecture Notes in Mathematics. Springer Berlin Heidelberg, Berlin, Heidelberg (2012)
3. Britt, S., Tsynkov, S., Turkel, E.: Numerical solution of the wave equation with variable wave speed on nonconforming domains by high-order difference potentials. J. Comput. Phys. **354**, 26–42 (2018)
4. Epshteyn, Y.: Upwind-difference potentials method for Patlak-Keller-Segel chemotaxis model. J. Sci. Comput. **53**(3), 689–713 (2012)
5. Epshteyn, Y., Xia, Q.: Efficient numerical algorithms based on difference potentials for chemotaxis systems in 3d. J. Sci. Comput. **80**(1), 26–59 (2019)
6. Epshteyn, Y., Xia, Q.: Difference potentials method for models with dynamic boundary conditions and bulk-surface problems. Adv. Comput. Math. **46**(5), 1–39 (2020)
7. Huan, R., Thompson, L.L.: Accurate radiation boundary conditions for the time-dependent wave equation on unbounded domains. Int. J. Numer. Methods Eng. **47**(9), 1569–1603 (2000). https://doi.org/10.1002/(SICI)1097-0207(20000330)47:9<1569::AID-NME845>3.0.CO;2-9
8. Kahana, A., Smith, F., Turkel, E., Tsynkov, S.: A high order compact time/space finite difference scheme for the 2D and 3D wave equation with a damping layer. J. Comput. Phys. **460**, Paper No. 111161 (2022)
9. Liang, H., Liu, M.Z., Lv, W.: Stability of θ-schemes in the numerical solution of a partial differential equation with piecewise continuous arguments. Appl. Math. Lett. **23**(2), 198–206 (2010)

10. Medvinsky, M., Tsynkov, S., Turkel, E.: The method of difference potentials for the Helmholtz equation using compact high order schemes. J. Sci. Comput. **53**(1), 150–193 (2012)
11. Medvinsky, M., Tsynkov, S., Turkel, E.: Direct implementation of high order BGT artificial boundary conditions. J. Comput. Phys. **376**, 98–128 (2019)
12. Mohlenkamp, M.J.: A user's guide to spherical harmonics (2010)
13. Petropavlovsky, S., Tsynkov, S.: Non-deteriorating time domain numerical algorithms for Maxwell's electrodynamics. J. Comput. Phys. **336**, 1–35 (2017)
14. Petropavlovsky, S., Tsynkov, S., Turkel, E.: A method of boundary equations for unsteady hyperbolic problems in 3d. J. Comput. Phys. **365**, 294–323 (2018)
15. Ralston, A., Rabinowitz, P.: A First Course in Numerical Analysis. Courier Corporation (2001)
16. Reznik, A.A.: Approximation of surface potentials of elliptic operators by difference potentials. Soviet Math. Dokl. **25**(2), 543–545 (1982)
17. Reznik, A.A.: Approximation of the surface potentials of elliptic operators by difference potentials and solution of boundary-value problems. Ph.D. thesis, Moscow Institute for Physics and Technology, Moscow, USSR (1983). [in Russian]
18. Ryaben'kii, V.S.: Method of Difference Potentials and Its Applications, Springer Series in Computational Mathematics, vol. 30. Springer, Berlin (2002)
19. Ryaben'kii, V., Torgashov, V.: An iteration-free approach to solving the navier–stokes equations by implicit finite difference schemes in the vorticity-stream function formulation. J. Sci. Comput. **81**(3), 1136–1149 (2019)
20. Schaeffer, N.: Efficient spherical harmonic transforms aimed at pseudospectral numerical simulations. Geochem. Geophys. Geosyst. **14**(3), 751–758 (2013)
21. Smith, F., Tsynkov, S., Turkel, E.: Compact high order accurate schemes for the three dimensional wave equation. J. Sci. Comput. **81**(3), 1181–1209 (2019)
22. Trottenberg, U., Schuller, A.: Multigrid. Academic Press, Orlando (2001)
23. Utyuzhnikov, S.V.: Generalized Calderón-Ryaben'kii's potentials. IMA J. Appl. Math. **74**(1), 128–148 (2009)

Dual-Primal Isogeometric Tearing and Interconnecting Methods for the Stokes Problem

Jarle Sogn and Stefan Takacs

1 Introduction

We explore fast solvers for linear systems that arise from the discretization of the Stokes problem in two space dimensions using Isogeometric Analysis (IgA; [2, 6]). We consider multipatch domains, that is, the computational domains consists of multiple non-overlapping patches. For such domains, FETI-DP methods (introduced in [3]) are a canonical choice. FETI-DP was first adapted to IgA in [8] and named the *Dual-Primal Isogeometric Tearing and Interconnecting* (IETI-DP) method. Several IETI-DP solvers for second-order elliptic boundary value problems have already been explored, see, e.g., [4, 5] and the more recent convergence analysis [11], which is, besides grid sizes h_k and the patch diameters H_k, also explicit in the spline degree p and spline smoothness, namely the condition number of the preconditioned system is bounded by a constant times

$$p\left(1+\log p+\max_k \log \frac{H_k}{h_k}\right)^2. \tag{1}$$

In [10], a FETI-DP like solver has been applied to an isogeometric Taylor-Hood element for a single-patch domain. There, the substructures used for the setup of the solvers are non-overlapping parts of the considered patch. The isogeometric

J. Sogn (✉)
Johann Radon Institute for Computational and Applied Mathematics, Austrian Academy of Sciences, Linz, Austria
e-mail: jarle.sogn@ricam.oeaw.ac.at

S. Takacs
Institute of Computational Mathematics, Johannes Kepler University Linz, Linz, Austria
e-mail: stefan.takacs@numa.uni-linz.ac.at

J. M. Melenk et al. (eds.), *Spectral and High Order Methods for Partial Differential Equations ICOSAHOM 2020+1*, Lecture Notes in Computational Science and Engineering 137, https://doi.org/10.1007/978-3-031-20432-6_31

Taylor-Hood element uses spline degree $p + 1$ for velocity and spline degree p for pressure and smoothness $p - 1$ for both of them. The derived solver maintains this smoothness. In [10], a solver has also been proposed for the elasticity problem for quasi incompressible materials. These results have recently been extended in [14]. In the finite element literature, there are a few results on FETI-DP methods for the Stokes problem, see, e.g., [7, 9, 12] for the case of two dimensions and [13] for the case of three dimensions.

We consider IETI-DP solvers for multi-patch domains, where each patch is a substructure. Concerning the coupling between the patches, we go a slightly different way and use the minimum smoothness requirements that are feasible in order to obtain a conforming discretization. This means that the pressure space is discontinuous across the interfaces between the patches. For the velocity, we only impose continuity between the patches. Within the patches, we use the aforementioned isogeometric Taylor-Hood scheme. Several challenges arise when one aims to extend the IETI-DP solvers to the Stokes problem. The first is that the resulting linear system is not positive definite, it rather has a saddle point structure. The second challenge is that in case of Dirichlet conditions, the differential operator has a nonzero nullspace. The nullspace consists of all constant pressure modes. This is normally remedied by restricting the solution space such that the average pressure is zero. When one wants to apply a scaled Dirichlet preconditioner, such a condition would be required locally. We obtain such a local condition by adding the patchwise averages of the pressure to the space of primal degrees of freedom. In the case of the Stokes equations, the condition numbers of the preconditioned system observed in the numerical results are bounded by (1), analogous to those obtained for the Poisson equation.

The remainder of this paper is organized as follows. In Sect. 2, we formulate the problem and introduce a few possible ways of realizing the IETI-DP solver. In Sect. 3, we present a numerical results. Finally, we draw some conclusions in Sect. 4.

2 Problem Formulation and IETI-DP

As model problem, we consider the Stokes problem: Let $\Omega \subset \mathbb{R}^2$ be a bounded Lipschitz domain. For a given right-hand side $\mathbf{f} \in [L(\Omega)]^2$, find $(\mathbf{u}, p) \in \left[H_0^1(\Omega)\right]^2 \times L_0^2(\Omega)$ such that

$$\begin{aligned} (\nabla \mathbf{u}, \nabla \mathbf{v})_{L^2(\Omega)} + (p, \operatorname{div} \mathbf{v})_{L^2(\Omega)} &= (\mathbf{f}, \mathbf{v})_{L^2(\Omega)} && \forall\, \mathbf{v} \in \left[H_0^1(\Omega)\right]^2, \\ (\operatorname{div} \mathbf{u}, q)_{L^2(\Omega)} &= 0 && \forall\, q \in L_0^2(\Omega). \end{aligned} \tag{2}$$

We assume that the domain Ω is composed of K non-overlapping patches $\Omega^{(k)}$ which are parameterized with geometry mappings $\mathbf{G}_k : (0, 1)^2 \to \Omega^{(k)}$ such that $\Omega^{(k)} = \mathbf{G}_k((0,1)^2)$. We assume that both the Jacobian $\nabla \mathbf{G}_k$ and its inverse $(\nabla \mathbf{G}_k)^{-1}$ are almost everywhere uniformly bounded (cf. [11, Ass. 1]).

As local discretization spaces on the parameter domain $(0,1)^2$, we use tensor-product B-splines, which we denote by $S_{p,\alpha}$, where p is the spline degree and α is the smoothness, this is, such that the splines are α times continuously differentiable. So, $\alpha = p-1$ corresponds to splines of maximum smoothness. For the Stokes equations, we need a inf-sup stable discretization space. We use the generalized Taylor-Hood space (cf. [1]), which we transfer to the physical domain $\Omega^{(k)}$ using the pull-back principle. The spaces on the individual patches are given by

$$\begin{aligned} \mathbf{V}^{(k)} &:= \left\{ \mathbf{u} : \mathbf{u} \circ \mathbf{G}_k \in [S_{p+1,\alpha}]^2 \text{ and } \mathbf{u}|_{\partial\Omega\cap\partial\Omega^{(k)}} = 0 \right\}, \\ Q^{(k)} &:= \{ q : q \circ \mathbf{G}_k \in S_{p,\alpha} \}, \end{aligned} \tag{3}$$

where $\mathbf{u}|_{\partial\Omega\cap\partial\Omega^{(k)}}$ denotes the restriction of $\mathbf{u}$ to $\partial\Omega \cap \partial\Omega^{(k)}$ (trace operator).

The overall space for the velocity is $[H_0^1(\Omega)]^2$, therefore a conforming discretization needs to be continuous. Thus, we assume that the spaces are fully matching, this is, for all interfaces $\Gamma^{(k,\ell)}$ between two patches, the following statement holds true: For every basis function $\mathbf{u} \in \mathbf{V}^{(k)}$ with $\mathbf{u}|_{\Gamma^{(k,\ell)}} \neq 0$, there is a basis function $\mathbf{v} \in \mathbf{V}^{(\ell)}$ such that $\mathbf{u}|_{\Gamma^{(k,\ell)}} = \mathbf{v}|_{\Gamma^{(k,\ell)}}$ (cf. [11, Ass. 5]). This allows us to define the overall discretization space for the velocity via $\mathbf{V} := \{\mathbf{u} \in [H_0^1(\Omega)]^2 : \mathbf{u}|_{\Omega^{(k)}} \in \mathbf{V}^{(k)}\}$. For the pressure, continuity is not required in order to obtain a conforming discretization. We define $Q := \{q \in L_0^2(\Omega) : q|_{\Omega^{(k)}} \in Q^{(k)}\}$, where $L_0^2(\Omega)$ is the space of functions on $L^2(\Omega)$ with vanishing mean value.

By using the spaces from (3) for a patch-wise assembling of the problem (2), we obtain local (still uncoupled) linear systems

$$A^{(k)}\underline{\mathbf{x}}^{(k)} = \begin{pmatrix} K^{(k)} & (D^{(k)})^\top \\ D^{(k)} & 0 \end{pmatrix} \begin{pmatrix} \underline{\mathbf{u}}^{(k)} \\ \underline{p}^{(k)} \end{pmatrix} = \begin{pmatrix} \underline{\mathbf{f}}^{(k)} \\ 0 \end{pmatrix} = \underline{\mathbf{b}}^{(k)},$$

where the matrices $K^{(k)}$ and $D^{(k)}$ and the vector $\underline{\mathbf{f}}^{(k)}$ are obtained by evaluating the terms $(\nabla \cdot, \nabla \cdot)_{L^2(\Omega^{(k)})}$, $(\cdot, \operatorname{div} \cdot)_{L^2(\Omega^{(k)})}$ and $(\mathbf{f}, \cdot)_{L^2(\Omega^{(k)})}$, respectively, for the basis functions for the spaces $\mathbf{V}^{(k)}$ and/or $Q^{(k)}$. An underlined quantity, like $\underline{\mathbf{u}}^{(k)}$, represents the coefficient vector of the corresponding function $\mathbf{u}^{(k)}$ with respect to the chosen basis.

Next, we split the degrees of freedom for the velocity variable into those associated with the interfaces, denoted $\underline{\mathbf{u}}_\Gamma$, and the remaining degrees of freedom, associated with basis functions vanishing on the interface, denoted $\underline{\mathbf{u}}_\mathrm{I}$. Using this splitting, we obtain

$$A^{(k)}\underline{\mathbf{x}}^{(k)} = \begin{pmatrix} K_{\Gamma\Gamma}^{(k)} & K_{\Gamma\mathrm{I}}^{(k)} & (D_\Gamma^{(k)})^\top \\ K_{\mathrm{I}\Gamma}^{(k)} & K_{\mathrm{II}}^{(k)} & (D_\mathrm{I}^{(k)})^\top \\ D_\Gamma^{(k)} & D_\mathrm{I}^{(k)} & 0 \end{pmatrix} \begin{pmatrix} \underline{\mathbf{u}}_\Gamma^{(k)} \\ \underline{\mathbf{u}}_\mathrm{I}^{(k)} \\ \underline{p}^{(k)} \end{pmatrix} = \begin{pmatrix} \underline{\mathbf{f}}_\Gamma^{(k)} \\ \underline{\mathbf{f}}_\mathrm{I}^{(k)} \\ 0 \end{pmatrix} = \underline{\mathbf{b}}^{(k)}.$$

We introduce primal degrees of freedom, which is done in order to guarantee that the system matrices of the patch-local problems are non-singular. As in the case of the Poisson problem, the patch-local systems of patches that do not contribute to the Dirichlet boundary represent problems with pure Neumann boundary conditions. On these patches, the null space of $A^{(k)}$ corresponds to constant velocity modes. We deal with this problem by strongly enforcing the continuity of the primal degrees of freedom. We consider three variants:

- Variant Π_{c}: Each velocity component on each of the corners.
- Variant Π_{ce}: Each velocity component on each of the corners and the average of each velocity component on each of the edges.
- Variant Π_{cn}: Each velocity component on each of the corners and the average of the normal component of the velocity on each of the edges.

In each of the local problems, we require that these primal degrees of freedom vanish. This constraint is represented by the condition $C_v^{(k)}\underline{\mathbf{u}}_\Gamma^{(k)} = 0$.

Moreover, we introduce primal degrees of freedom for the pressure. For the scaled Dirichlet preconditioner, it is necessary that the local Dirichlet problems are uniquely solvable. However, the Stokes system with Dirichlet boundary conditions has a non-trivial null space, which consists of the constant pressure modes. We also choose the averages of the pressure on the individual patches as primal degrees of freedom. For each of the local problems we require that this primal degree of freedom vanishes. This constraint is represented by the condition $C_p^{(k)}\underline{p}^{(k)} = 0$.

The local systems read as follows

$$\bar{A}^{(k)}\underline{\bar{\mathbf{x}}}^{(k)} = \begin{pmatrix} A^{(k)} & (C^{(k)})^\top \\ C^{(k)} & 0 \end{pmatrix} \begin{pmatrix} \underline{\mathbf{x}}^{(k)} \\ \underline{\mu}^{(k)} \end{pmatrix}$$

$$= \left(\begin{array}{ccc:cc} K_{\Gamma\Gamma}^{(k)} & K_{\Gamma\mathrm{I}}^{(k)} & (D_\Gamma^{(k)})^\top & 0 & C_v^\top \\ K_{\mathrm{I}\Gamma}^{(k)} & K_{\mathrm{II}}^{(k)} & (D_\mathrm{I}^{(k)})^\top & 0 & 0 \\ D_\Gamma^{(k)} & D_\mathrm{I}^{(k)} & 0 & (C_p^{(k)})^\top & 0 \\ \hdashline 0 & 0 & C_p^{(k)} & 0 & 0 \\ C_v^{(k)} & 0 & 0 & 0 & 0 \end{array}\right) \begin{pmatrix} \underline{\mathbf{u}}_\Gamma^{(k)} \\ \underline{\mathbf{u}}_\mathrm{I}^{(k)} \\ \underline{p}^{(k)} \\ \hdashline \mu_p^{(k)} \\ \underline{\mu}_v^{(k)} \end{pmatrix} = \begin{pmatrix} \underline{\mathbf{f}}_\Gamma^{(k)} \\ \underline{\mathbf{f}}_\mathrm{I}^{(k)} \\ 0 \\ \hdashline 0 \\ 0 \end{pmatrix} = \underline{\bar{\mathbf{b}}}^{(k)}, \tag{4}$$

where $\mu_p^{(k)}$ and $\underline{\mu}_v^{(k)}$ are the Lagrangian multipliers corresponding to the primal degrees of freedom.

The continuity of the velocity variables (except the corner values) between the patches is enforced weakly using a constraint that is represented by the boolean matrix $\bar{B}^{(k)} = (B_\Gamma^{(k)}, 0, 0, 0, 0)$. The velocity is continuous if $\sum_{k=1}^K \bar{B}^{(k)}\underline{\bar{\mathbf{x}}}^{(k)} = 0$ or, equivalently, $\sum_{k=1}^K B_\Gamma^{(k)}\underline{\mathbf{u}}_\Gamma^{(k)} = 0$.

Finally, we consider the primal problem, this is, the global problem for the primal degrees of freedom. We use a nodal basis for the primal degrees of freedom which is $A^{(k)}$-orthogonal to the remaining degrees of freedom on each patch. This basis is represented with respect to the patch-local bases by the matrices $\Psi^{(k)}$, which are the solutions to

$$\begin{pmatrix} A^{(k)} & (C^{(k)})^\top \\ C^{(k)} & 0 \end{pmatrix} \begin{pmatrix} \Psi^{(k)} \\ \mathrm{M}^{(k)} \end{pmatrix} = \begin{pmatrix} 0 \\ R_C^{(k)} \end{pmatrix},$$

where $R_C^{(k)}$ is a boolean matrix, which represents the mapping between a patch-local ordering of the primal degrees of freedom and the global ordering of the primal degrees of freedom. Using the matrices $\Psi^{(k)}$, we define the system matrix, jump matrix and right-hand side for the primal problem as

$$A_\Pi := \sum_{k=1}^{K} (\Psi^{(k)})^\top A^{(k)} \Psi^{(k)}, \quad B_\Pi := \sum_{k=1}^{K} B^{(k)} \Psi^{(k)} \quad \text{and} \quad \mathbf{b}_\Pi := \sum_{k=1}^{K} (\Psi^{(k)})^\top \underline{\mathbf{b}}^{(k)}.$$

Before we can write up the global system, we need to revisit the condition that the average of the pressure vanishes, this is, $p \in L_0^2(\Omega)$. Our choice of the primal degrees of freedom ensures that the average of the pressure vanishes for all patch-local problems. Since the patch-wise constant pressure modes are primal, they form part of the primal system. In order to obtain the unique solvability of the primal system, we introduce a constant multiplier that guarantees that the global average of the pressure vanishes. This condition is enforced by adding a vector $\mathbf{p}$ representing this constraint. So, we extend the primal system as follows

$$\bar{A}_\Pi := \begin{pmatrix} A_\Pi & \mathbf{p}^\top \\ \mathbf{p} & 0 \end{pmatrix}, \quad \bar{B}_\Pi := \begin{pmatrix} B_\Pi, & 0 \end{pmatrix}, \quad \underline{\bar{\mathbf{x}}}_\Pi := \begin{pmatrix} \underline{\mathbf{x}}_\Pi \\ \mu_0 \end{pmatrix} \quad \text{and} \quad \underline{\bar{\mathbf{b}}}_\Pi := \begin{pmatrix} \underline{\mathbf{b}}_\Pi \\ 0 \end{pmatrix}, \tag{5}$$

where μ_0 is a new Lagrangian multiplier associated with the additional constraint.

By coupling the patch local systems and the primal system, we obtain the IETI-DP saddle point system, which characterizes the solution and which reads as follows

$$\begin{pmatrix} \bar{A}^{(1)} & & & & (\bar{B}^{(1)})^\top \\ & \ddots & & & \vdots \\ & & \bar{A}^{(K)} & & (\bar{B}^{(K)})^\top \\ & & & \bar{A}_\Pi & \bar{B}_\Pi^\top \\ \bar{B}^{(1)} & \dots & \bar{B}^{(K)} & \bar{B}_\Pi & 0 \end{pmatrix} \begin{pmatrix} \underline{\bar{\mathbf{x}}}^{(1)} \\ \vdots \\ \underline{\bar{\mathbf{x}}}^{(K)} \\ \underline{\bar{\mathbf{x}}}_\Pi \\ \underline{\lambda} \end{pmatrix} = \begin{pmatrix} \underline{\bar{\mathbf{b}}}^{(1)} \\ \vdots \\ \underline{\bar{\mathbf{b}}}^{(K)} \\ \underline{\bar{\mathbf{b}}}_\Pi \\ 0 \end{pmatrix}. \tag{6}$$

We solve this linear system as follows. We first define the Schur-complement $\bar{F}$ and the corresponding right-hand side $\underline{g}$ by

$$\bar{F} := \bar{B}_\Pi \bar{A}_\Pi^{-1} \bar{B}_\Pi^\top + \sum_{k=1}^{K} \bar{B}^{(k)} (\bar{A}^{(k)})^{-1} (\bar{B}^{(k)})^\top, \ \underline{g} := \bar{B}_\Pi \bar{A}_\Pi^{-1} \underline{\bar{\mathbf{b}}}_\Pi + \sum_{k=1}^{K} \bar{B}^{(k)} (\bar{A}^{(k)})^{-1} \underline{\bar{\mathbf{b}}}^{(k)},$$

where we note that the symmetric and positive definite matrix $\bar{F}$ is never computed. Products of the form $\bar{F}\underline{\lambda}$ can be easily computed by solving the primal problem and the patch-local problems with right-hand sides $\bar{B}_\Pi^\top\underline{\lambda}$ and $(\bar{B}^{(k)})^\top\underline{\lambda}$, respectively. Consequently, we solve the linear system

$$\bar{F}\underline{\lambda} = \underline{g} \tag{7}$$

using a preconditioned conjugate gradient solver. After the computation of $\underline{\lambda}$, the local solutions are recovered using (6):

$$\bar{\underline{\mathbf{x}}}^{(k)} = (\bar{A}^{(k)})^{-1}(\bar{\underline{\mathbf{b}}}^{(k)} - (\bar{B}^{(k)})^\top\underline{\lambda}) \quad \text{and} \quad \bar{\underline{\mathbf{x}}}_\Pi = \bar{A}_\Pi^{-1}(\bar{\underline{\mathbf{b}}}_\Pi - \bar{B}_\Pi^\top\underline{\lambda}).$$

The local solution $\underline{\mathbf{x}}^{(k)} = (\underline{\mathbf{u}}^{(k)}, \underline{p}^{(k)})$ is then obtained by $\underline{\mathbf{x}}^{(k)} = \begin{pmatrix} I, & 0 \end{pmatrix} \bar{\underline{\mathbf{x}}}^{(k)} + \Psi^{(k)}\begin{pmatrix} I, & 0 \end{pmatrix} \bar{\underline{\mathbf{x}}}_\Pi$, where the matrices $(I, 0)$ eliminate the rows corresponding to the Lagrangian multipliers μ_p, $\underline{\mu}_v$ and μ_0, cf. (4) and (5).

Finally, we have to discuss the choice of the preconditioner for the conjugate gradient solver. We consider two variants of a scaled Dirichlet preconditioner. The first variant follows the standard construction principles and is based on the equation of interest, which is the Stokes equation. We choose

$$M_{\text{sD},1} := \sum_{k=1}^{K} B_\Gamma^{(k)}(D^{(k)})^{-1}S_{\Gamma,1}^{(k)}(D^{(k)})^{-1}(B_\Gamma^{(k)})^\top,$$

where the corresponding Schur complement is given by

$$S_{\Gamma,1}^{(k)} := K_{\Gamma\Gamma}^{(k)} - \begin{pmatrix} K_{\Gamma\mathrm{I}}^{(k)} & (D_\Gamma^{(k)})^\top & 0 \end{pmatrix} \begin{pmatrix} K_{\mathrm{II}}^{(k)} & (D_\mathrm{I}^{(k)})^\top & 0 \\ D_\mathrm{I}^{(k)} & 0 & (C_p^{(k)})^\top \\ 0 & C_p^{(k)} & 0 \end{pmatrix}^{-1} \begin{pmatrix} K_{\mathrm{I}\Gamma}^{(k)} \\ D_\Gamma^{(k)} \\ 0 \end{pmatrix}$$

and $D^{(k)} := 2I$ is set up based on the principle of multiplicity scaling. Often, the primal degrees of freedom do not affect the definition of the Schur complement used for the scaled Dirichlet preconditioner. Note that our definition includes the constraint that guarantees that the pressure average vanishes. This is required in order to make the local system matrix solvable.

The second variant is motivated by the observation that the scaled Dirichlet preconditioner is usually set up in order to realize a $H^{1/2}$-scalar product. To archive this goal, we simply choose

$$M_{\text{sD},2} := \sum_{k=1}^{K} B_\Gamma^{(k)}(D^{(k)})^{-1}S_{\Gamma,2}^{(k)}(D^{(k)})^{-1}(B_\Gamma^{(k)})^\top,$$

where

$$S_{\Gamma,2}^{(k)} := K_{\Gamma\Gamma}^{(k)} - K_{\Gamma\mathrm{I}}^{(k)} (K_{\mathrm{II}}^{(k)})^{-1} K_{\mathrm{I}\Gamma}^{(k)},$$

which corresponds to a vector-valued Poisson problem.

3 Numerical Experiments

We consider the Stokes problem (2) with the right-hand side function

$$\mathbf{f}(x, y) = (-\pi \cos(\pi x) - 2\pi^2 \sin(\pi x) \cos(\pi y),\ 2\pi^2 \cos(\pi x) \sin(\pi y))$$

and the inhomogeneous Dirichlet boundary conditions

$$\mathbf{u}(x, y) = (-\sin(\pi x) \cos(\pi y),\ \cos(\pi x) \sin(\pi y)) \quad \text{for} \quad (x, y) \in \partial\Omega.$$

We consider three computational domains: a quarter annulus (64 patches), the Yeti-footprint (84 patches) and the unit square (64 patches), see Fig. 1. For the quarter annulus and the unit square, the patch-local function spaces on the coarsest grid level ($\ell = 0$) consist of global polynomials only, this is, there are no inner knots. For the Yeti-footprint, the patch-local function spaces on the long and thin patches at the bottom consist of two elements, obtained by uniformly refining the two longer sides of the patch. For the remaining patches of the Yeti-footprint, the function spaces consist of global polynomials only. The grid levels $\ell = 1, 2, \ldots$ are obtained by ℓ uniform refinement levels. In either case, the grid size on the parameter domain is $\widehat{h} = 2^{-\ell}$. Within each patch, we have splines of degree $p+1$ and smoothness $p-1$ for the velocity and splines of degree p and smoothness $p-1$ for the pressure. As

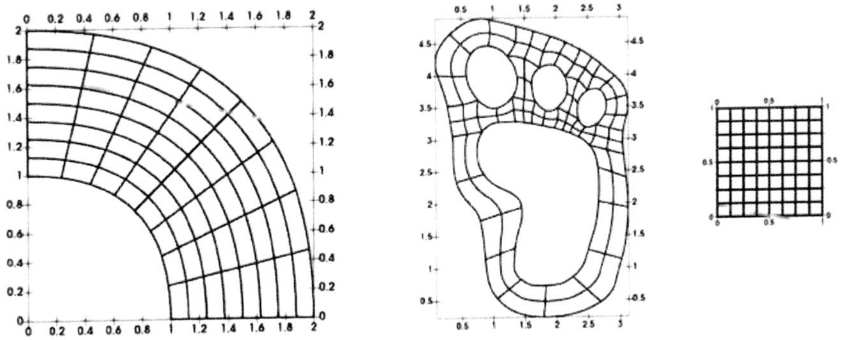

Fig. 1 Computational domains: Quarter annulus (left); Yeti-footprint (middle); unit square (right)

Table 1 Iteration numbers for quarter annulus using Π_c and $M_{sD,1}$ (left), $M_{sD,2}$ (right)

$\ell \setminus p$	2	3	4	5	6
2	43	42	43	43	38
3	47	48	48	48	44
4	53	51	55	53	52
5	56	60	61	58	58

$\ell \setminus p$	2	3	4	5	6
2	28	27	28	27	25
3	29	31	30	30	29
4	34	34	35	33	32
5	35	37	38	36	35

Table 2 Iteration numbers for quarter annulus using Π_{ce} and $M_{sD,1}$ (left), $M_{sD,2}$ (right)

$\ell \setminus p$	2	3	4	5	6
2	16	16	16	15	15
3	17	17	17	17	16
4	18	18	18	18	18
5	20	20	20	19	19

$\ell \setminus p$	2	3	4	5	6
2	11	12	11	11	11
3	13	13	13	12	12
4	14	14	14	13	13
5	15	15	15	15	14

Table 3 Iteration numbers for quarter annulus using Π_{cn} and $M_{sD,1}$ (left), $M_{sD,2}$ (right)

$\ell \setminus p$	2	3	4	5	6
2	22	23	22	22	22
3	25	25	25	24	24
4	26	27	27	26	25
5	29	29	28	28	27

$\ell \setminus p$	2	3	4	5	6
2	17	17	17	17	16
3	18	19	19	18	18
4	20	20	20	20	19
5	22	22	22	21	21

outlined in the previous section, the coupling between the patches is continuous for the velocity and discontinuous for the pressure.

In the following, we discuss the convergence of the preconditioned conjugate gradient solver, used to solve (7). We start the conjugate gradient solver with a random initial guess and stop the iteration when the Euclidean norm of the residual vector is reduced by a factor of 10^{-6} compared to the Euclidean norm of the initial residual vector. The local linear systems that need to be solved in the pre-processing steps (computation of $\underline{g}$) and during the main iteration are realized using a sparse LU-solver. All experiments have been implemented using the G+smo library[1] and have been performed on the Radon1 cluster[2] in Linz.

In Tables 1, 2, and 3, we present the results for the quarter annulus domain. In these tables, we indicate the number of iterations of the preconditioned conjugate gradient solver required to reach the desired threshold. In the tables on the left side, we present the results obtained using the Stokes based scaled Dirichlet preconditioner $M_{sD,1}$. The tables on the right side show the results obtained using the Poisson based scaled Dirichlet preconditioner $M_{sD,2}$. Tables 1, 2, and 3 show the results for the primal degrees of freedom Π_c, Π_{ce} and Π_{cn}, respectively. We observe that only choosing the corner values as primal degrees of freedom (variant

[1] https://github.com/gismo/gismo.

[2] https://www.ricam.oeaw.ac.at/hpc/.

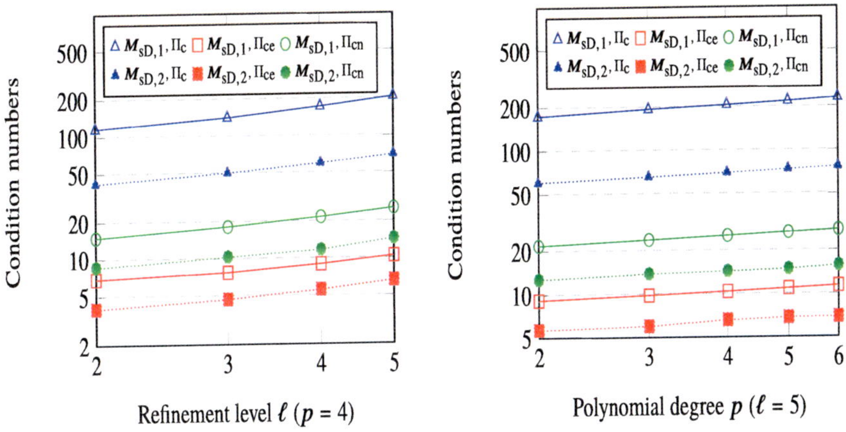

Fig. 2 Condition numbers for the quarter annulus

Π_c) leads to the largest iteration numbers. The variant Π_{ce}, which has a slightly larger primal problem than Π_{cn}, leads to the smallest iteration numbers. In all cases, the dependence of the condition number on the spline degree p and the grid size $h_k \sim H_k 2^{-\ell}$ seems to be consistent with (1). $M_{sD,2}$ outperforms $M_{sD,1}$ in all cases. Note that the local problems required to be solved to realize $M_{sD,2}$ are smaller and they involve symmetric and positive definite matrices. So, $M_{sD,2}$ can also be realized more efficiently than $M_{sD,1}$.

We have estimated the condition numbers of the preconditioned system with the conjugate gradient solver. We present these results in Fig. 2. The left diagram shows the dependence on the refinement level ℓ for a fixed choice of the spline degree $p = 4$. In the right diagram, one can see the dependence on the spline degree p for a fixed choice of the refinement level $\ell = 5$. In both cases, one can observe a mild dependence. The convergence of the solver for the variants Π_{ce} and Π_{cn} is close to what one would estimate from the condition number, that is, the convergence rate is only slightly smaller than

$$\frac{\sqrt{\kappa} - 1}{\sqrt{\kappa} + 1},$$

where κ is the condition number of the preconditioned system. If we consider the variant Π_c, the difference is much more significant, that is, the true iteration numbers are much smaller than the estimated condition number would suggest.

The next domain to consider is the Yeti-footprint. Note that the inf-sup stability highly depends on the shape of the domain. Long and thin channels, which constitute the Yeti-footprint, are known to lead to small inf-sup constants. In Tables 4, 5, and 6 and in Fig. 3, we present the results for the Yeti-footprint. The iteration counts and the condition numbers are only slightly larger than for the quarter annulus. The largest difference is obtained for the choice Π_c, particularly if the condition number

Table 4 Iteration numbers for Yeti-footprint using Π_c and $M_{sD,1}$ (left), $M_{sD,2}$ (right)

$\ell \setminus p$	2	3	4	5	6
2	143	150	133	132	126
3	172	175	146	183	176
4	191	201	203	151	174
5	223	245	238	240	234

$\ell \setminus p$	2	3	4	5	6
2	76	78	67	70	63
3	85	86	73	85	78
4	94	97	95	74	72
5	101	112	106	109	103

Table 5 Iteration numbers for Yeti-footprint using Π_{ce} and $M_{sD,1}$ (left), $M_{sD,2}$ (right)

$\ell \setminus p$	2	3	4	5	6
2	21	22	22	23	23
3	25	27	24	28	26
4	29	30	30	30	31
5	31	34	33	33	35

$\ell \setminus p$	2	3	4	5	6
2	14	14	14	14	13
3	15	16	15	16	15
4	17	18	16	16	17
5	19	19	19	18	18

Table 6 Iteration numbers for Yeti-footprint using Π_{cn} and $M_{sD,1}$ (left), $M_{sD,2}$ (right)

$\ell \setminus p$	2	3	4	5	6
2	20	21	22	22	23
3	23	25	26	26	26
4	27	28	28	28	29
5	30	31	31	31	32

$\ell \setminus p$	2	3	4	5	6
2	16	17	16	16	16
3	18	18	18	18	17
4	20	20	20	19	19
5	22	22	22	21	20

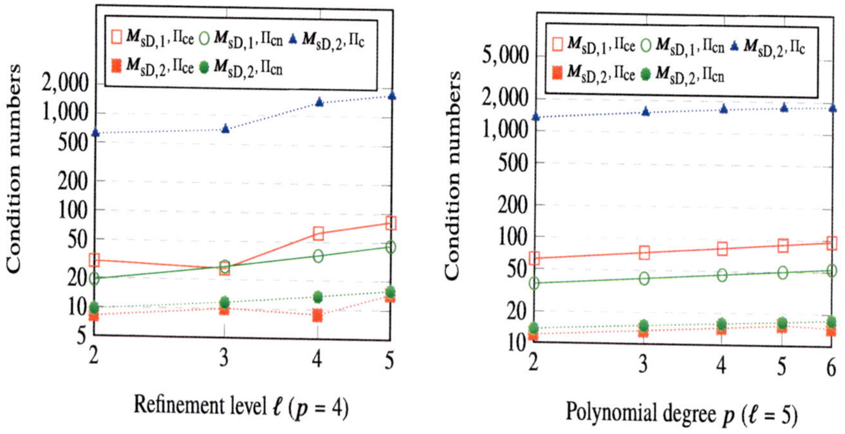

Fig. 3 Condition numbers for the Yeti-footprint

is considered. If the preconditioner $M_{sD,1}$ is used, the condition number estimates gave a "not a number" results in many cases. Consequently, they were not added to the diagram. Concerning the dependence on the grid size and the spline degree, the choice of the primal degrees of freedom and the choice of the preconditioner, we observe qualitatively the same results as for the quarter annulus.

Table 7 Iteration numbers for unit square using Π_c and $M_{sD,1}$ (left), $M_{sD,2}$ (right)

$\ell \setminus p$	2	3	4	5	6	$\ell \setminus p$	2	3	4	5	6
2	28	29	30	28	27	2	21	21	22	21	20
3	32	32	33	32	31	3	23	24	24	24	23
4	37	37	37	36	35	4	27	26	27	26	25
5	40	43	42	42	39	5	27	30	30	29	28

Table 8 Iteration numbers for unit square using Π_{ce} and $M_{sD,1}$ (left), $M_{sD,2}$ (right)

$\ell \setminus p$	2	3	4	5	6	$\ell \setminus p$	2	3	4	5	6
2	14	14	14	14	14	2	18	18	18	19	18
3	15	16	16	16	15	3	20	21	21	21	20
4	17	17	17	17	17	4	22	23	22	22	22
5	18	18	18	18	18	5	24	25	25	24	23

Table 9 Iteration numbers for unit square using Π_{cn} and $M_{sD,1}$ (left), $M_{sD,2}$ (right)

$\ell \setminus p$	2	3	4	5	6	$\ell \setminus p$	2	3	4	5	6
2	18	18	18	19	18	2	14	15	15	15	14
3	20	21	21	21	20	3	16	16	16	16	16
4	22	23	22	22	22	4	18	18	18	18	17
5	24	25	25	24	23	5	19	20	20	20	19

Finally, we present results for the unit square in order to better see the effect of the geometry transformation. In Tables 7, 8, and 9 and in Fig. 4, one can see the results for this domain. In any case, the iteration counts and the condition numbers are smaller than those for the non-trivial domains. This difference is only mild for the choices Π_{ce} and Π_{cn}, however for the choice Π_c, this difference is quite large. When one only compares the results for the unit square, one would conclude that choice Π_c would be acceptable as well. The results for the other computational domains however show that Π_c is significantly inferior.

4 Conclusions and Final Remarks

We solved the Stokes equations, discretized using multipatch IgA, by means of ITEI-DP solvers. Even though the Stokes system is indefinite, the reduced problem is symmetric positive definite, which we can solve efficiently using a preconditioned conjugate gradient solver. Two scaled Dirichlet preconditioners were tested. The simpler one, which is based on the vector-valued Poisson problem and thus easier to realize in practice, is numerically superior. Concerning the choice of the primal degrees of freedom, we observe that it is worthwhile to include edge averages of the velocity value (either for both components or only for the normal component).

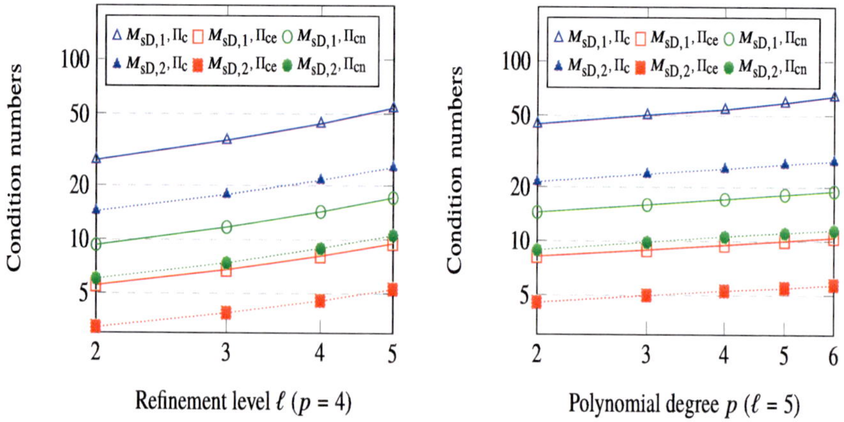

Fig. 4 Condition numbers for the unit square

This observation cannot be made if only the unit square is considered. Thus, it is necessary to test ITEI-DP methods on non-trivial domains.

Convergence analysis will be discussed in a forthcoming paper.

Acknowledgments This work was supported by the Austrian Science Fund (FWF): P31048. This support is gratefully acknowledged.

References

1. Bressan, A., Sangalli, G.: Isogeometric discretizations of the Stokes problem: stability analysis by the macroelement technique. IMA J. Numer. Anal. **33**(2), 629–651 (2013)
2. da Veiga, L.B., Buffa, A., Sangalli, G., Vázquez, R.: Mathematical analysis of variational isogeometric methods. Acta Numer. **23**, 157–287 (2014)
3. Farhat, C., Lesoinne, M., LeTallec, P., Pierson, K., Rixen, D.: FETI-DP: a dual–primal unified FETI method—part I: A faster alternative to the two-level FETI method. Int. J. Numer. Methods Eng. **50**(7), 1523–1544 (2001)
4. Hofer, C., Langer, U.: Dual-primal isogeometric tearing and interconnecting solvers for multipatch dG-IgA equations. Comput. Methods Appl. Mech. Eng. **316**, 2–21 (2017)
5. Hofer, C., Langer, U.: Dual-primal isogeometric tearing and interconnecting methods. In: Contributions to Partial Differential Equations and Applications, pp. 273–296. Springer (2019)
6. Hughes, T.J., Cottrell, J.A., Bazilevs, Y.: Isogeometric analysis: CAD, finite elements, NURBS, exact geometry and mesh refinement. Comput. Methods Appl. Mech. Eng. **194**(39–41), 4135–4195 (2005)
7. Kim, H.H., Lee, C.O., Park, E.H.: A FETI-DP formulation for the Stokes problem without primal pressure components. SIAM J. Numer. Anal. **47**(6), 4142–4162 (2010)
8. Kleiss, S.K., Pechstein, C., Jüttler, B., Tomar, S.: IETI–isogeometric tearing and interconnecting. Comput. Methods Appl. Mech. Eng. **247**, 201–215 (2012)
9. Li, J.: A dual-primal FETI method for incompressible Stokes equations. Numer. Math. **102**(2), 257–275 (2005)

10. Pavarino, L.f., Scacchi, S.: Isogeometric block FETI-DP preconditioners for the Stokes and mixed linear elasticity systems. Comput. Methods Appl. Mech. Eng. **310**, 694–710 (2016)
11. Schneckenleitner, R., Takacs, S.: Condition number bounds for IETI-DP methods that are explicit in h and p. Math. Models Methods Appl. Sci. **30**(11), 2067–2103 (2020)
12. Tu, X., Li, J.: A unified dual-primal finite element tearing and interconnecting approach for incompressible Stokes equations. Int. J. Numer. Methods Eng. **94**(2), 128–149 (2013)
13. Tu, X., Li, J.: A FETI-DP type domain decomposition algorithm for three-dimensional incompressible Stokes equations. SIAM J. Numer. Anal. **53**(2), 720–742 (2015)
14. Widlund, O., Zampini, S., Scacchi, S., Pavarino, L.F.: Block FETI–DP/BDDC preconditioners for mixed isogeometric discretizations of three-dimensional almost incompressible elasticity. Math. Comput. **90**(330), 1773–1797 (2021)

Mimetic Relaxation Runge Kutta Methods

A. Srinivasan and J. E. Castillo

1 Introduction

Hyperbolic PDEs can be generalized as the initial value problem

$$u_t + \mathcal{F}(t, u) = 0, \quad u(t = 0) = u_0 \tag{1}$$

The numerical integration of Eq. (1) can be performed using the Runge Kutta (RK) methods. The explicit RK schemes possess the advantage of ease of implementation, while achieving up to sixth-order accuracy [1]. The traditional explicit RK methods do not preserve numerical energy [2] and result in the violation of energy conservation at each time step of integration. Energy-preserving RK methods do exist for Hamiltonian systems [3], but are primarily implicit methods.

An s-stage RK method with real coefficients $a_{s1}, a_{s2}, \ldots, b_1, \ldots, b_s, c_2, \ldots, c_s$ of the Butcher tableau

$$\begin{array}{c|c} c & A \\ \hline & b \end{array}$$

with n denoting the spatial and time step discretizations, is given by

$$\mathbf{u}^{n+1} - \mathbf{u}^n + \Delta t \sum_{i=1}^{s} b_i \, \mathbf{k}_i^n, \quad \text{where} \tag{2}$$

A. Srinivasan (✉) · J. E. Castillo
San Diego State University, San Diego, CA, USA
e-mail: asrinivasan0709@sdsu.edu; jcastillo@sdsu.edu

J. M. Melenk et al. (eds.), *Spectral and High Order Methods for Partial Differential Equations ICOSAHOM 2020+1*, Lecture Notes in Computational Science and Engineering 137, https://doi.org/10.1007/978-3-031-20432-6_32

$$\mathbf{k}_s^n = \mathbf{u}^n + \Delta t \sum_{j=1}^{i-1} f(t^n + c_j\ \Delta t, a_{ij}\mathbf{k}_j^n), \quad i = 1, 2, \ldots, s \tag{3}$$

The problem (1) is non-dissipative and therefore energy is conserved. The following section presents an overview of the Mimetic spatial discretization and Relaxation RK temporal schemes that ensure conservation of numerical energy.

2 Overview of Mimetic and Relaxation RK Methods

Mimetic spatial discretization methods [4] work on a staggered grid. The discrete operators DIV and $GRAD$ mimic their continuum counterparts of divergence and gradient, and are therefore said to be faithful to the physics of the problem. The discrete Mimetic DIV and $GRAD$ coefficients satisfy the extended Gauss' divergence theorem

$$\int_{\Omega} (\nabla \cdot \mathbf{v})\, f\, \mathrm{d}V + \int_{\Omega} \mathbf{v} \cdot (\nabla f)\, \mathrm{d}V = \int_{\partial\Omega} (f\mathbf{v}) \cdot \mathbf{n}\, \mathrm{d}S \tag{4}$$

The Castillo-Grone Mimetic methods [5] achieve even order of accuracy at the boundaries and interiors of the grids for both the DIV and $GRAD$ matrices. The methods achieve high orders of accuracy (up to sixth order) while satisfying the generalized inner-product norm. The discrete equivalent of the extended Gauss' divergence theorem is given by

$$\langle \hat{\mathbf{D}} v, \hat{f} \rangle_Q + \langle \mathbf{G}^T v, \hat{f} \rangle_P = \langle \mathbf{B} v, \hat{f} \rangle, \tag{5}$$

where $\hat{\mathbf{D}}$ and $\mathbf{G}$ are the Mimetic DIV and $GRAD$ operators, $\mathbf{B}$ the boundary operator and Q and P the quadrature matrices. In 1D, this becomes the equivalent of integration by parts. The Castillo-Grone Mimetic coefficients satisfy a global conservation law and are therefore structure-preserving. The quadrature matrices are diagonal with positive weights, and therefore guarantee positivity of solutions.

The Mimetic spatial discretizations result in energy-stable numerical solutions of PDEs. We illustrate this for the 1D advection equation $u_t + \nabla \cdot u = 0$ with periodic Dirichlet boundary conditions, $u(t, 0) = u(t, 1) = 0, x \in [0, 1]$. The energy method [6] can be implemented for the advection equation by multiplying it with the solution u and integrating over the domain to obtain

$$\int_{\Omega} u\, u_t\, \mathrm{d}V + \int_{\Omega} u\, \nabla \cdot u\, \mathrm{d}V = 0 \tag{6}$$

The first term in eq. (6) becomes $\frac{1}{2}\frac{\mathrm{d}}{\mathrm{d}t}\int_{\Omega} u^2 \mathrm{d}V = \frac{1}{2}\frac{\mathrm{d}}{\mathrm{d}t}||u||^2$. The second term can be represented in terms of the extended Gauss' divergence theorem as

$$\int_{\Omega} u \, \nabla \cdot u \, \mathrm{d}V = \int_{\Omega} \nabla \cdot (u^2) \mathrm{d}V = \frac{1}{2}\int_{\partial\Omega} u^2 \, \mathrm{d}S \tag{7}$$

After applying the boundary conditions, we obtain $\frac{\mathrm{d}}{\mathrm{d}t}||u||^2 = -u^2|_0^1$. This is an initial value problem ordinary differential equation and the solution depends continuously on the initial and boundary conditions. The Mimetic discretization method therefore leads to a well-posed and energy-stable numerical solution of the advection equation.

Relaxation RK (RRK) methods [7, 8] ensure numerical energy conservation at each time step of integration by introducing the relaxation parameter γ. The step-update formula, Eq. (2), is modified as

$$\mathbf{u}_{\gamma}^{n+1} = \mathbf{u}^n + \gamma^n \, \Delta t \sum_{i=1}^{s} b_i \, \mathbf{k}_i^n \tag{8}$$

The relaxation parameter γ^n is calculated such that the energy norm is satisfied at each time step. Thus,

$$||\mathbf{u}_{\gamma}^{n+1}|| \, \leq ||\mathbf{u}^n|| \tag{9}$$

γ^n is calculated numerically by utilizing the energy functional at each time step. The RRK methods therefore ensure conservation of numerical energy at each time step. As noted earlier, the Mimetic methods guarantee positivity of solutions on the spatial discretization. The Mimetic RRK methods are therefore well suited for the numerical solution of hyperbolic PDEs. We illustrate some numerical examples of this technique in the next section.

3 Numerical Results

The Matlab codes for the numerical examples presented in this section can be accessed via https://github.com/asrinivasan0709/ICOSAHOM2021. The details of the implementation are presented in report [9].

3.1 Example 1

The one dimensional advection equation with a constant velocity a is given by

$$u_t + \nabla \cdot (au) = 0 \tag{10}$$

The spatial discretization for the divergence operator $\nabla\cdot$ is obtained using the fourth and sixth order Mimetic DIV operators. Since the DIV is defined on the cell centers, an interpolant is required at each time step to map the data from the centers to the nodes. The fourth and sixth order finite difference flux discretization of Wicker and Skamarock [10] was used as a comparison. The time discretization was performed using the fourth and sixth order RRK methods. The fourth order scheme implemented was the classical RK method, while the sixth order scheme was the Verner's method [1]. The numerical study was performed on a one dimensional grid $x \in [-5, 5]$ and $a = 1$, with a Gaussian initial condition

$$u(x, 0) = \frac{1}{\sqrt{2\pi\sigma}} exp\left(-\frac{(x-\mu)^2}{2\sigma}\right), \quad \mu = 0,\ \sigma = 0.15 \tag{11}$$

Figure 1 shows the convergence rates calculated at time $t = 1s$, with the maximum error norm calculated as the difference between the exact and numerical

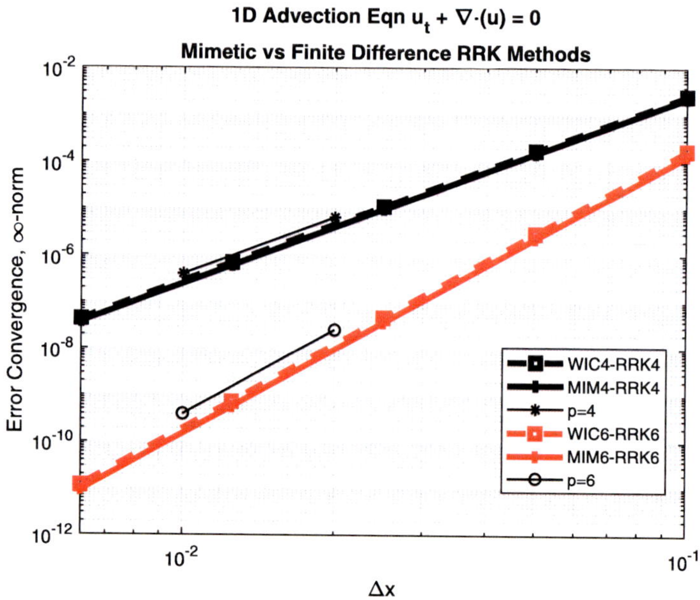

Fig. 1 Spatial convergence rates for the 1D advection equation

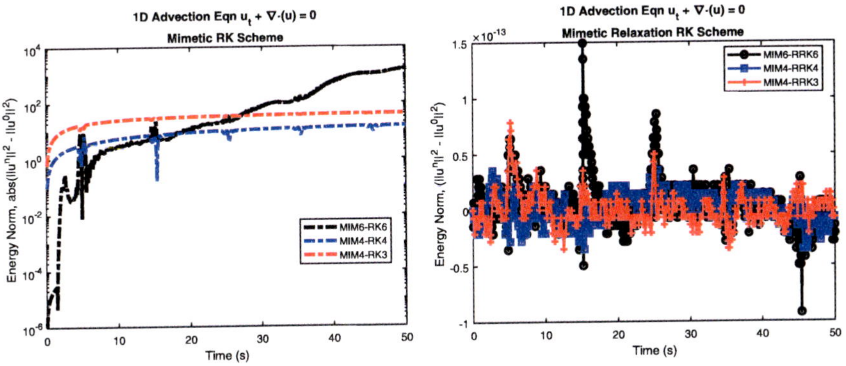

Fig. 2 Energy evolution using the standard RK methods (left) and the Relaxation RK methods (right). The Mimetic-RRK methods preserve energy to machine numeric precision

solutions. Both the Mimetic and the finite difference methods achieve the desired order of accuracy, with the Mimetic discretization achieving a slightly lower error norm.

The energy functional for the advection equation is given by

$$E = \int u^2 \mathrm{d}x \tag{12}$$

Figure 2 shows the energy evolution for the advection equation, computed using the standard RK methods (left plot of Fig. 2) and the Relaxation RK method on the right. The standard RK scheme results in an asymptotically increasing numerical energy with respect to the initial condition, while the Relaxation scheme conserves energy to machine numeric precision.

Figure 3 shows the numerical solution obtained using three different schemes:

- fourth order Mimetic and third order Runge Kutta (legends MIM4-RK3 & MIM4-RRK3),
- fourth order Mimetic and Runge Kutta (legends MIM4-RK4 & MIM4-RRK4),
- sixth order Mimetic and Runge Kutta (legends MIM6-RK6 & MIM6-RRK6).

The time integrations were performed using the standard and the Relaxation Runge Kutta methods. It can be observed that the MIM4-RRK3 scheme captures the peak of the gaussian pulse more closely than its RK3 counterpart. The MIM4-RK4 and MIM4-RRK4 schemes exhibit low frequency oscillations at the leading edge of the gaussian wave. The sixth order RK scheme exhibits numerical instability in the solution. The RRK scheme more closely resembles the true solution, albeit with high frequency oscillations at the leading edge of the wave.

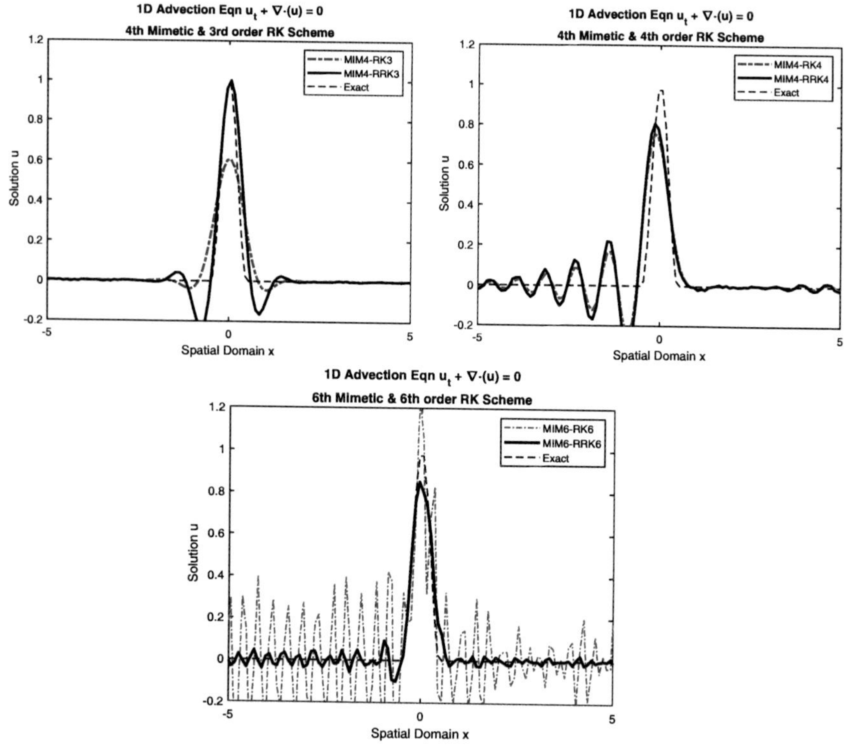

Fig. 3 Comparison of numerical versus exact solutions for the schemes considered in example 1

3.2 *Example 2: System of Hyperbolic Equations*

The numerical solution of the 1D wave equation $u_{tt} = a^2 \nabla \cdot \nabla u$ is investigated in this example. The wave equation can be represented as a system of equations that are first order in time, such that $u_t = v$, $v_t = a^2 \nabla \cdot \nabla u$. It is also possible to discretize the spatial domain using only first order spatial operators. However, the Laplacian operator has been chosen in this case in an attempt to demonstrate the applicability with the Mimetic Gradient operator. The conserved energy (analytical and numerical) for this system of equations is given by

$$E = \int \left(v^2 + (\nabla u)^2\right) \mathrm{d}x \quad \text{and} \quad E_n = ||v||^2 + ||\nabla u||^2 \tag{13}$$

The spatial discretization was performed using the fourth order Mimetic method, implemented on a grid $x \in [-5, 5]$ with wave velocity $a = 1$ and periodic boundary conditions. The gaussian input from Eq. (11) was used as initial condition for $u(x, 0)$, and $v(x, 0) = 0$. The fourth order temporal schemes implemented were

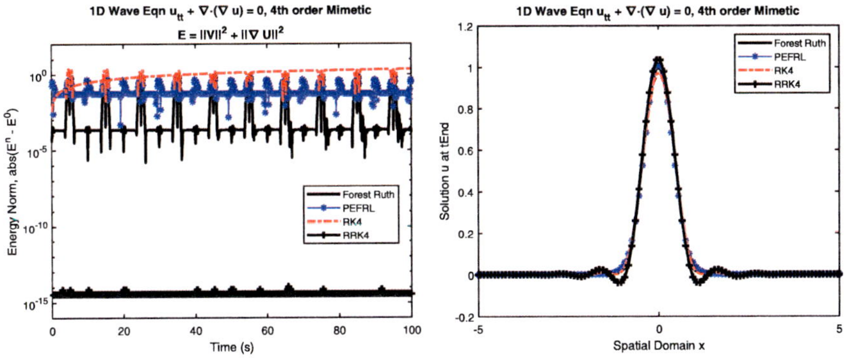

Fig. 4 Energy evolution and numerical solution for wave equation (example 2) with CFL condition of 0.6 for Forest Ruth and 1.20 for the PEFRL, RK4 and RRK4 schemes

the Forest Ruth, [11], Position Extended Forest Ruth Algorithm (PEFRL, [12]), RK4 and RRK4.

The plot on the left of Fig. 4 shows the energy evolution for the four methods. The numerical energy for the Forest Ruth and PEFRL schemes is conserved over time, and the average change in energy with respect to the initial energy at time $t = 0$ remains constant. However, the calculated energy shows numerical oscillations at every periodic reflection from the boundaries. This is reflected by the spikes in the energy at periodic intervals. In contrast, the energy of the RK4 scheme shows an asymptotically increasing trend over time. The RRK4 scheme conserves energy to close to machine numeric precision. Figure 4 (right) shows the numerical solution at time $t = 100$ s for the four schemes. The Mimetic-RRK4 scheme provides a numerical solution that is on par with the other three schemes, with the added benefit of conserving numerical energy to machine precision.

Figure 5 shows the energy evolution (left) and the numerical solution (right) obtained for the RK4 and RRK4 schemes with a CFL condition of 1.22. The RK4 scheme introduces numerical errors (exhibited as spurious oscillations in the numerical solution) as evidenced by the sharp increase in energy around time $t = 90$ s. In contrast, the energy-conserving Mimetic-Relaxation scheme continues to adequately capture the numerical solution as shown in the plot on the right.

3.3 Example 3: Non-linear Burgers' Equation

The non-linear inviscid Burgers' equation in conservation form

$$u_t + \nabla \cdot \left(\frac{u^2}{2}\right) = 0 \tag{14}$$

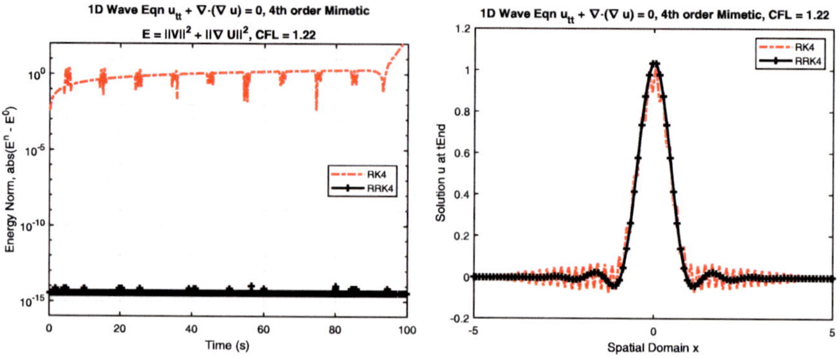

Fig. 5 Energy evolution and numerical solution for the wave equation (example 2) with CFL condition of 1.22

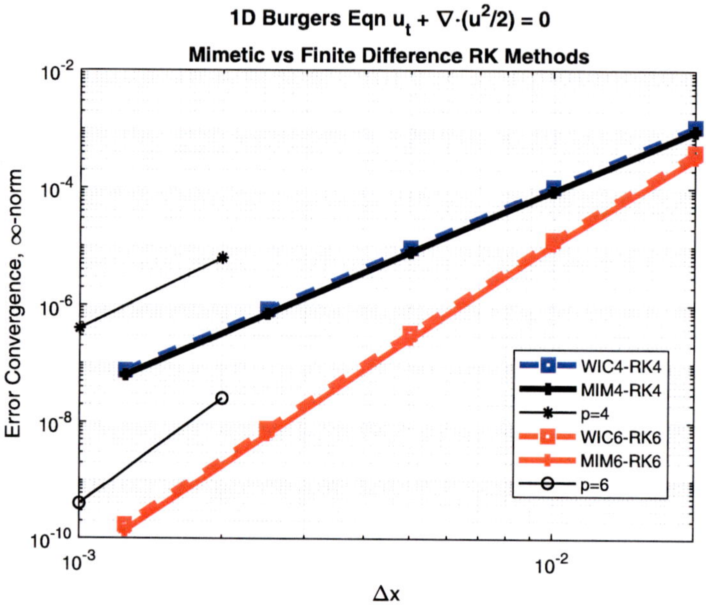

Fig. 6 Spatial convergence rates for the inviscid Burgers' equation

was solved using the discretization methods outlined in example 1. The initial condition was $u(x, 0) = sin(\pi x)$ in the domain $x \in [0, 1]$. The analytical and numerical solutions were calculated using a CFL condition of 0.1 at time $t = 0.1s$ prior to the occurrence of the shock. Figure 6 shows the convergence rates obtained using the RK4 and RK6 schemes, using the fourth and sixth order Mimetic and finite difference spatial discretizations. The Mimetic spatial convergence rates achieve the desired order of accuracy for each method. It can also be noted that the errors as

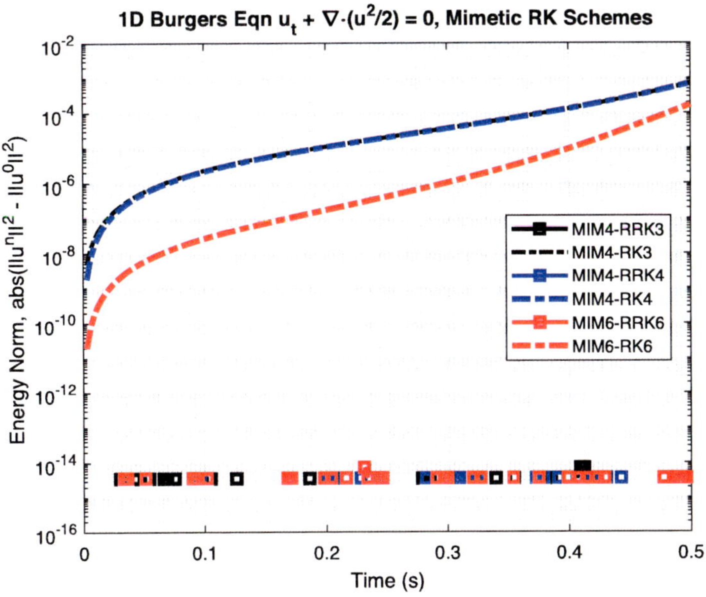

Fig. 7 Energy evolution of the Burgers' equation

obtained from the Mimetic schemes are comparable to, or if not slightly better than those obtained from the finite difference flux discretization schemes.

Figure 7 shows the energy evolution for the standard RK schemes, contrasted with the energy conserving RRK schemes. As expected, the energy is conserved to machine numeric precision with the Mimetic-Relaxation schemes.

Figure 8 shows the numerical solutions at time $t = 0.1s$. It can be observed that the Relaxation schemes produce identical results as that of the standard RK schemes with Mimetic spatial discretization.

4 Conclusion

The implementation of the Mimetic Relaxation Runge Kutta methods for hyperbolic PDEs has been presented in this paper. The Mimetic methods are structure preserving and ensure positivity of solutions. The RRK time integration scheme ensures that the numerical energy is conserved at each time step of integration. The Mimetic-RRK methods are thus well suited for hyperbolic PDEs. Both linear and non-linear numerical examples have been shown to illustrate the convergence and numerical accuracy of the schemes for hyperbolic PDEs.

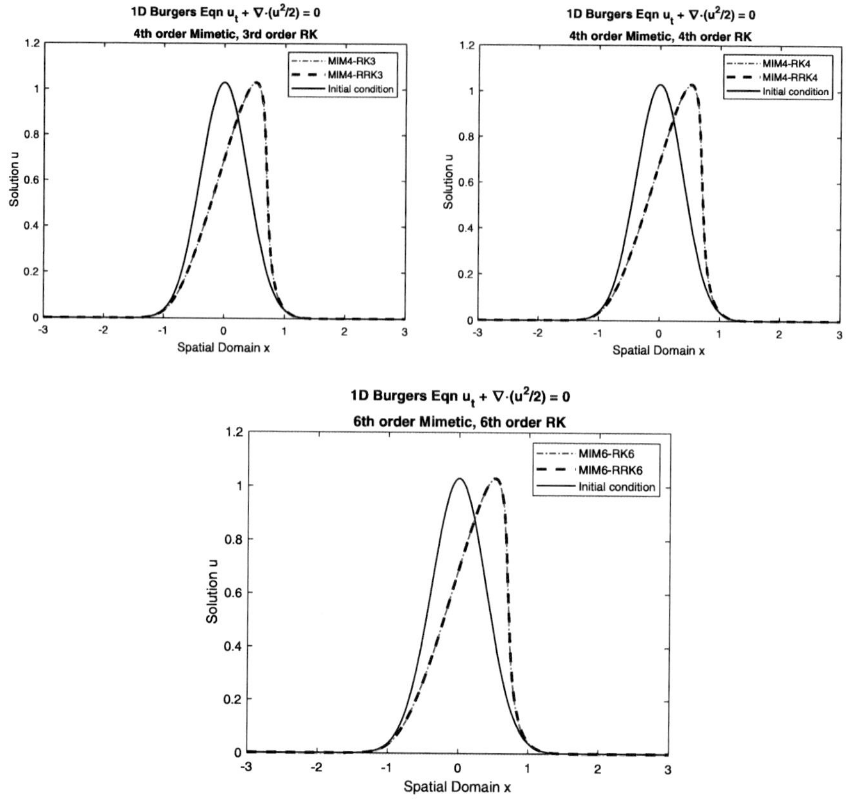

Fig. 8 Comparison of numerical solutions for the Burgers' equation

Acknowledgments We would like to thank David Ketcheson for the helpful discussions and the feedback on the subject of this paper.

References

1. Hairer, E., Norsett, S., Wanner, G.: Solving Ordinary Differential Equations I: Nonstiff Problems, Springer Series In Computational Mathematics, vol. 8 (1993)
2. Butcher, J.C.: Numerical Methods For Ordinary Differential Equations, 2nd edn. Wiley (2008)
3. Hairer, E., Lubich, C., Wanner, G.: Geometric Numerical Integration: Structure-preserving Algorithms for Ordinary Differential Equations, vol. 31. Springer Science & Business Media (2006)
4. Castillo, J., Miranda, G.: Mimetic Discretization Methods. CRC Press (2013)
5. Castillo, J., Grone, R.: A Matrix Analysis Approach to Higher-Order Approximations for Divergence and Gradients Satisfying a Global Conservation Law. SIAM J. Matrix Anal. Appl. **25**(1), 128–142 (2003)

6. Friedrich, L., Fernández, L., Winters, D., Gassner, A., Zingg, G., Hicken, D., Hicken, J.: Conservative and stable degree preserving SBP finite difference operators for non-conforming meshes. J. Sci. Comput. (2018)
7. Ketcheson, D., Relaxation Runge Kutta methods: conservation and stability for inner-product norms. SIAM J. Numer. Anal. **57**(6), 2850–2870 (2019)
8. Ranocha, H., Sayyari, M., Dalcin, L., Parsani, M., Ketcheson, D.: Relaxation Runge Kutta methods: fully discrete explicit entropy-stable schemes for the compressible Euler and Navier Stokes equations. SIAM J. Sci. Comput. **42**, A612 (2020)
9. Srinivasan, A., Corbino, J., Castillo, J.: Implementation of mimetic relaxation Runge Kutta methods. CSRC Research Report CSRCR2022-01. http://www.csrc.sdsu.edu/csrc/research_reports/CSRCR2022-01.pdf
10. Wicker, L., Skamarock, W., Time-splitting methods for elastic models using forward time schemes. Monthly Weather Review - MON WEATHER REV. 130 (2002)
11. Forest, E., Ruth, R., Fourth order symplectic integration. Physica D **43**, 105 (1990)
12. Omelyan, I., Mryglod, I., Folk, R.: Optimized Forest-Ruth- and Suzuki-like algorithms for integration of motion in many-body systems. Comput. Phys. Commun. (2002)

DoD Stabilization for Higher-Order Advection in Two Dimensions

Florian Streitbürger, Gunnar Birke, Christian Engwer, and Sandra May

1 Introduction

Modern simulations often require to mesh complex geometries. One approach that is particularly suited for this purpose are embedded boundary meshes. One simply cuts the geometry out of a structured background mesh, resulting in *cut cells* along the boundary of the embedded object. Cut cells have different shapes and can become arbitrarily small. In the context of solving time-dependent hyperbolic conservation laws this causes the *small cell problem*: for standard explicit time stepping, the scheme is not stable on small cut cells when the time step is chosen with respect to the larger background cells.

Existing solution approaches in a finite volume regime are typically bound to at most second order, see for example the flux redistribution method [6, 7], the h-box method [5, 16], the mixed explicit-implicit scheme [18], the dimensionally split approach [13, 17], or the state redistribution (SRD) method [4]. An exception is the extension of the active flux method to cut cell meshes [15], which aims for third order.

For discontinuous Galerkin (DG) schemes it is significantly easier to achieve higher order. The development of DG schemes that overcome the small cell problem has only started very recently. Some work relies on cell merging, e.g. [20], other work on algorithmic solution approaches such as the usage of a ghost penalty term

F. Streitbürger (✉) · S. May
TU Dortmund University, Dortmund, Germany
e-mail: florian.streitbuerger@math.tu-dortmund.de; sandra.may@math.tu-dortmund.de

G. Birke · C. Engwer
University of Münster, Münster, Germany
e-mail: g_birk01@uni-muenster.de; christian.engwer@uni-muenster.de

J. M. Melenk et al. (eds.), *Spectral and High Order Methods for Partial Differential Equations ICOSAHOM 2020+1*, Lecture Notes in Computational Science and Engineering 137, https://doi.org/10.1007/978-3-031-20432-6_33

as done by Fu and Kreiss [11] or the extension of the SRD method to a DG setting by Giuliani [12].

Another approach, proposed previously by the authors, is the *Domain-of-Dependence* (DoD) stabilization. In this approach, a penalty term is added on small cut cells that restores the proper domains of dependence in the neighborhood of small cut cells and therefore makes standard explicit time stepping stable again. In [10] we first introduced the DoD stabilization for linear advection in 1d and 2d for linear polynomials only. In [19], we extend the stabilization in 1d to non-linear systems and higher order. For the extension to higher order in 1d we found that it is necessary to add an extra term in the stabilization, which adjusts the mass distribution within inflow neighbors of small cut cells. With this term it is possible to show an L^2 stability result for the semi-discrete setting (keeping the time continuous) in 1d [19].

In this contribution, we partially extend the 1d results from [19] to 2d by solving linear advection with higher order polynomials. For the case of a planar ramp geometry we show an L^2 stability result for the semi-discrete setting. We will also provide corresponding numerical results. These results show the expected convergence orders of $p+1$ for polynomial degree p in the L^1 norm. In the L^∞ norm, we observe a slight decay, resulting in convergence orders between $p+\frac{1}{2}$ and $p+1$.

2 Problem Setup

Within the scope of this work, we will focus on the 2d linear advection equation

$$u_t + \langle \beta, \nabla u \rangle = 0 \qquad \text{in } \Omega \times (0, T), \tag{1}$$

$$u = g \qquad \text{on } \partial\Omega^{\text{in}} \times (0, T), \tag{2}$$

$$u = u_0 \qquad \text{on } \Omega \times \{t = 0\}. \tag{3}$$

We denote by Ω an open, connected domain in $\mathbb{R}^2$ and by $\partial\Omega$ its boundary. The inflow boundary is defined as $\partial\Omega^{\text{in}} := \{x \in \partial\Omega : \langle \beta(x), n(x) \rangle < 0\}$ with $n \in \mathbb{R}^2$ being the outer unit normal vector on $\partial\Omega$. Analogously, we define $\partial\Omega^{\text{out}} := \{x \in \partial\Omega : \langle \beta(x), n(x) \rangle > 0\}$. Moreover, $\beta \in \mathbb{R}^2$ is the velocity field and $\langle \cdot, \cdot \rangle$ the standard scalar product in $\mathbb{R}^2$.

For simplicity and brevity of presentation, we will focus on the case of a ramp geometry in this work. We assume $\beta \in \mathbb{R}^2$ to be an incompressible velocity field that is parallel to the ramp. The geometry setup and mesh creation is sketched in Fig. 1. We refer to the internal and external faces of our mesh $\mathcal{M}_h$ as

$$\Gamma_{\text{int}} = \left\{ e_{E_1,E_2} = \partial E_1 \cap \partial E_2 \;\middle|\; E_1, E_2 \in \mathcal{M}_h \text{ and } E_1 \neq E_2 \text{ and } |e_{E_1,E_2}| > 0 \right\},$$

$$\Gamma_{\text{ext}} = \{ e_E = \partial E \cap \partial\Omega \mid E \in \mathcal{M}_h \text{ and } |e_E| > 0 \},$$

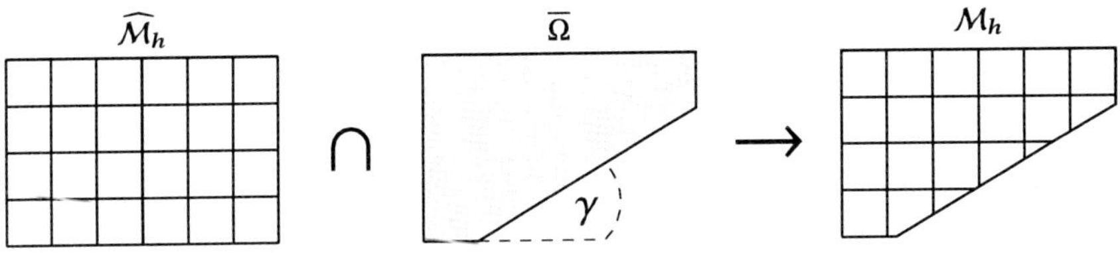

Fig. 1 Construction of cut cell mesh $\mathcal{M}_h$ for the case of a ramp geometry: We intersect the structured background mesh $\widehat{\mathcal{M}}_h$ of a larger rectangular domain $\widehat{\Omega} \supset \Omega$ with the domain $\overline{\Omega}$. Here, Ω has a ramp geometry described by the angle γ. This results in cut cells $E = \widehat{E} \cap \overline{\Omega}$ along the ramp, where $\widehat{E} \in \widehat{\mathcal{M}}_h$ is an element of the background mesh

with $|e|$ denoting the length of an edge e. We then further split Γ_{ext} in $\Gamma_{\text{ext,Cart}}$ and $\Gamma_{\text{ext,ramp}}$: $\Gamma_{\text{ext,Cart}}$ contains all Cartesian boundary faces and $\Gamma_{\text{ext,ramp}}$ contains the boundary faces along the ramp that were created by cutting out the ramp geometry.

On the partition $\mathcal{M}_h$, we define the discrete function space $\mathcal{V}_h^p(\mathcal{M}_h) \subset L^2(\Omega)$ by

$$\mathcal{V}_h^p(\mathcal{M}_h) = \left\{ v_h \in L^2(\Omega) \mid \forall E \in \mathcal{M}_h,\, v_h|_E \in P^p(E) \right\},$$

where P^p denotes the polynomial space of degree p. We use the same number of degrees of freedom and the same polynomial degree on cut cells as on regular elements. In our tests below we use a scaled and centered monomial basis.

On a face e between two adjacent cells E_1 and E_2, i.e., $e = \partial E_1 \cap \partial E_2$, we define the scalar-valued *average* as

$$\{u_h\} = \frac{1}{2}(u_h|_{E_1} + u_h|_{E_2}),$$

and the *jump* to be vector-valued as

$$[\![u_h]\!] := u_h|_{E_1} n_{E_1} + u_h|_{E_2} n_{E_2},$$

with n_{E_i} denoting the outer unit normal vector of cell E_i, $i = 1, 2$. With these prerequisites we can now define the scheme that we use to solve (1).

We use a method of lines approach: we first discretize in space and then in time. The unstabilized semi-discrete scheme is given by: Find $u_h(t) \in \mathcal{V}_h^p(\mathcal{M}_h)$ such that

$$(d_t u_h(t), w_h)_{L^2} + a_h^{\text{upw}}(u_h(t), w_h) + l_h(w_h) = 0 \quad \forall w_h \in \mathcal{V}_h^p(\mathcal{M}_h), \tag{4}$$

with

$$a_h^{\text{upw}}(u_h, w_h) := -\sum_{E\in\mathcal{M}_h} \int_E u_h \langle \beta, \nabla w_h\rangle \,\mathrm{d}x + \sum_{e\in\Gamma_{\text{ext}}} \int_e \langle \beta, n\rangle^{\oplus} u_h w_h \mathrm{d}s$$
$$+ \sum_{e\in\Gamma_{\text{int}}} \int_e \left(\{u_h\} \langle \beta, [\![w_h]\!]\rangle + \frac{1}{2} |\langle \beta, n_e\rangle| \langle [\![u_h]\!], [\![w_h]\!]\rangle \right) \mathrm{d}s,$$

$$l_h(w_h) := -\sum_{e\in\Gamma_{\text{ext}}} \int_e \langle \beta, n\rangle^{\ominus} g \, w_h \mathrm{d}s.$$

Here, $(\cdot,\cdot)_{L^2}$ denotes the standard scalar product in $L^2(\Omega)$ and $n_e \in \mathbb{R}^2$ is a unit normal on a face e (of arbitrary but fixed orientation). We define the negative and positive parts of $x \in \mathbb{R}$ as $x^{\ominus} := \frac{|x|-x}{2}$ and $x^{\oplus} := \frac{|x|+x}{2}$. Note that the standard upwind flux is used in the definition of a_h^{upw}.

The proposed stabilization modifies the *space* discretization, whereas in time we are free to use a time stepping scheme of our choice. We will use standard explicit strong stability preserving (SSP) Runge Kutta (RK) schemes [14].

3 Stabilization Terms

The stabilization is designed as an additional term J_h that is added to the semi-discrete formulation in (4). The DoD stabilized semi-discrete scheme is then given by: Find $u_h(t) \in \mathcal{V}_h^p(\mathcal{M}_h)$ such that for all $w_h \in \mathcal{V}_h^p(\mathcal{M}_h)$

$$(d_t u_h(t), w_h)_{L^2} + a_h^{\text{upw}}(u_h(t), w_h) + J_h(u_h(t), w_h) + l_h(w_h) = 0. \tag{5}$$

The stabilization term J_h is given by

$$J_h(u_h, w_h) = J_h^0(u_h, w_h) + J_h^1(u_h, w_h) = \sum_{E\in\mathcal{I}} \left(J_h^{0,E}(u_h, w_h) + J_h^{1,E}(u_h, w_h) \right).$$

We define $J_h^{0,E}$ and $J_h^{1,E}$ in detail below. The set $\mathcal{I}$ denotes the set of small cut cells that need stabilization. For a planar cut in 2d, there are 3-sided, 4-sided, and 5-sided cut cells. In [10], we have shown (see Lemma 3.5), that for the considered setup, it is sufficient to stabilize *triangular cut cells* only. For a triangular cut cell E_{cut} in our setup, each edge has a different boundary condition, see Fig. 2:

- On the *boundary edge* e_{bdy} we have a no-flow boundary condition as the flow is parallel to the ramp.
- Out of the two remaining edges, one edge is the *inflow edge* e_{in}, which is characterized by $\langle \beta, n_{E_{\text{cut}}}\rangle < 0$.

Fig. 2 Triangular cut cell

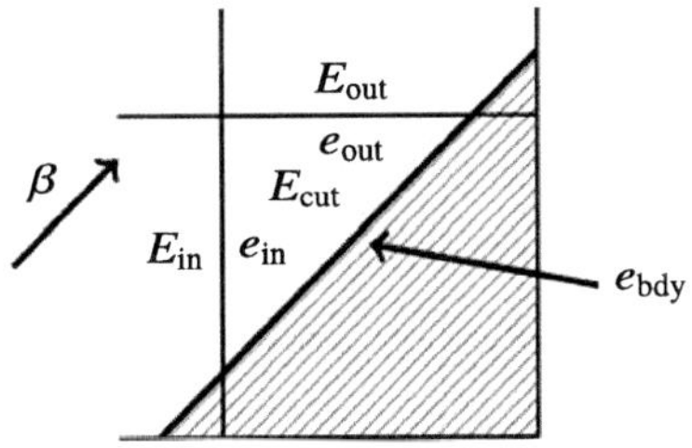

- The remaining edge is the *outflow edge* e_{out}, which is characterized by $\langle \beta, n_{E_{\text{cut}}} \rangle \geq 0$.

Thus we can uniquely define an inflow neighbor E_{in} and an outflow neighbor E_{out} for a triangular cut cell E_{cut}. The assumption of a no-penetration boundary condition on the cut boundary is a natural condition to consider. We exploit in the design of J_h that there is exactly one inflow neighbor and one outflow neighbor for the cells that we stabilize. We are currently working on the extension of the stabilization to inflow or outflow boundary conditions on the cut boundary.

If the time step is chosen according to the size of the larger background cells and does not respect the size of small cut cells, then, physically, mass passes within one time step from the inflow cell E_{in} through the small cut cell E_{cut} into the outflow cell E_{out}. The idea behind the DoD stabilization is to make this possible by directly passing part of the mass that enters E_{cut} from E_{in} into the outflow neighbor E_{out}. This way we restore the domain of dependence of the outflow neighbor E_{out} and make sure that the small cut cell E_{cut} only keeps as much mass as it can hold. For the latter one we define the concept of *capacity* below.

In order to create this flux of information between the inflow neighbor and the outflow neighbor of E_{cut}, we introduce an extension operator: The operator $\mathcal{L}_{E'}^{\text{ext}}$ extends a function $u_h \in \mathcal{V}_h^p$ from a cell $E' \in \mathcal{M}_h$ to the whole domain Ω. This simply corresponds to evaluating a polynomial outside its original support. In particular, we will evaluate the polynomial solution defined on cell E_{in} in cell $\overline{E_{\text{cut}}}$. We will refer to this both as $\mathcal{L}_{E_{\text{in}}}^{\text{ext}}(u_h)(x)$, $x \in \overline{E_{\text{cut}}}$, and simply as $u_{E_{\text{in}}}(x)$ to ease notation.

With these prerequisites we can now define J_h^0 and J_h^1. Generally, the terms of the DoD stabilization target two different goals:

- J_h^0 aims for redistributing the mass *among* the small cut cells and their neighbors appropriately. It therefore consists of *cell interface* terms.
- J_h^1 aims for redistributing the mass *within* the small cut cells and their neighbors appropriately. It therefore consists of *volume* terms.

The term $J_h^{0,E}$ is given by

$$J_h^{0,E}(u_h, w_h) = \eta_E \int_{e_{\text{out}}} (\mathcal{L}_{E_{\text{in}}}^{\text{ext}}(u_h) - u_h) \langle \beta, [\![w_h]\!] \rangle \, ds, \tag{6}$$

with the stabilization parameter η_E defined below. Note that we only redistribute mass across outflow edges of small cut cells and that we use the extended solution of the inflow neighbor to determine the size of the correction. The term $J_h^{1,E}$ is given by

$$J_h^{1,E}(u_h, w_h) = \eta_E \int_E (\mathcal{L}_{E_{\text{in}}}^{\text{ext}}(u_h) - u_h) \langle \beta, \mathcal{L}_{E_{\text{in}}}^{\text{ext}}(\nabla w_h) - \nabla w_h \rangle \, \mathrm{d}x. \tag{7}$$

This term is designed to adjust the mass distribution primarily within the small cut cell E and secondarily within its neighbor. Note that we apply the extension operator to both the discrete solution and the test function from inflow neighbor E_{in}. In [10], where we only considered piecewise linear polynomials, we proposed a different formulation of $J_h^{1,E}$, which did not contain the term $\mathcal{L}_{E_{\text{in}}}^{\text{ext}}(\nabla w_h)$. In 1d [19], we found that when going to higher order one can run into instabilities without this extra term. In addition, with this augmented definition of $J_h^{1,E}$ we are able to show an L^2 stability result for the semi-discrete scheme, which we will present below.

Both stabilization terms are scaled with the stabilization parameter η_E. We set $\eta_E = 1 - \alpha_{E,1/(2p+1)}$ with the capacity $\alpha_{E,\omega}$ and p being the polynomial degree of the discrete function space. We define the capacity of a cut cell E, see [10], as

$$\alpha_{E,\omega} := \min\left(\omega \frac{|E|}{\Delta t \int_{\partial E} \langle \beta, n_E \rangle^{\ominus} \, \mathrm{d}s}, 1\right), \qquad \omega \in (0, 1].$$

For $\omega = 1$, the capacity estimates the fraction of the inflow that is allowed to flow into the cut cell E and stay there without producing overshoot. Note that by definition $0 \leq \eta_E \leq 1$.

3.1 L^2 *Stability for Semi-discrete Scheme*

In the following we will show an L^2 stability result for the stabilized semi-discrete scheme for an arbitrary polynomial degree p. Generally, the L^2 stability result for the considered ramp setup is influenced by the inflow and outflow across $\partial\Omega^{\text{in}}$ and $\partial\Omega^{\text{out}}$ during the time $(0, T)$. Note that only Cartesian faces $e \in \Gamma_{\text{ext,Cart}}$ are contained in $\partial\Omega^{\text{in}} \cup \partial\Omega^{\text{out}}$ as we have a no-flow boundary condition for faces $e \in \Gamma_{\text{ext,ramp}}$ along the ramp. Our goal here is to show that L^2 stability still holds true for the stabilized scheme *with cut cells being present*, and not to analyze the influence of the inflow and outflow on the L^2 stability. We will therefore for simplicity assume that the solution has compact support inside $\hat{\Omega}$ during the considered time interval $(0, T)$ and does not intersect the Cartesian boundary, i.e. $\text{supp}(u) \cap (\partial\Omega^{\text{in}} \cup \partial\Omega^{\text{out}}) = \emptyset$, which implies that we have a homogeneous right hand side during the whole time frame $(0, T)$ and in particular that there is no in- or outflow.

Theorem 1 *Consider the advection equation (1) for the setup of a ramp with incompressible velocity field $\beta = (\beta_1, \beta_2)^T$ parallel to the ramp. Let the solution u have compact support during the considered time interval $(0, T)$ and $supp(u) \cap (\partial\Omega^{in} \cup \partial\Omega^{out}) = \emptyset$. Let $u_h(t)$, with $u_h(t) \in \mathcal{V}_h^p$ for any fixed t, be the solution to the stabilized semi-discrete problem (5). Then, under the assumption of exact integration, the semi-discrete solution satisfies for all $t \in (0, T)$*

$$\|u_h(t)\|_{L^2(\Omega)} \le \|u_h(0)\|_{L^2(\Omega)} .$$

Proof Setting $w_h = u_h(t)$ in (5) and ignoring boundary contributions with respect to $\partial\Omega^{\text{in}}$, we get

$$(d_t u_h(t), u_h(t))_{L^2} + a_h^{\text{upw}}(u_h(t), u_h(t)) + J_h(u_h(t), u_h(t)) = 0.$$

Integration of the first term in time yields

$$\begin{aligned}\int_0^t (d_\tau u_h(\tau), u_h(\tau))_{L^2}\, \mathrm{d}\tau &= \int_0^t \frac{d}{d\tau}\frac{1}{2}\|u_h(\tau)\|^2_{L^2(\Omega)}\, \mathrm{d}\tau\\ &= \frac{1}{2}\|u_h(t)\|^2_{L^2(\Omega)} - \frac{1}{2}\|u_h(0)\|^2_{L^2(\Omega)},\end{aligned}$$

and it remains to show that for any fixed t

$$a_h^{\text{upw}}(u_h(t), u_h(t)) + J_h(u_h(t), u_h(t)) \ge 0.$$

We will first discuss a_h^{upw} and then J_h. (We will drop the explicit time dependence in the following for brevity.)

By definition of a_h^{upw} and ignoring outflow across $\partial\Omega^{\text{out}}$, there holds

$$\begin{aligned}a_h^{\text{upw}}(u_h, u_h) = &- \sum_{E\in\mathcal{M}_h} \int_E u_h \langle\beta, \nabla u_h\rangle\, \mathrm{d}x\\ &+ \sum_{e\in\Gamma_{\text{int}}} \int_e \left(\{\!\{u_h\}\!\} \langle\beta, [\![u_h]\!]\rangle + \frac{1}{2}|\langle\beta, n_e\rangle| \left\langle [\![u_h]\!], [\![u_h]\!]\right\rangle \right) \mathrm{d}s.\end{aligned}$$

For the integral term we rewrite (exploiting $\nabla \cdot \beta = 0$)

$$-\int_E u_h \langle\beta, \nabla u_h\rangle\, \mathrm{d}x = -\int_E \nabla \cdot \left(\frac{1}{2}\beta u_h^2\right) \mathrm{d}x = -\int_{\partial E} \left(\frac{1}{2}\beta u_h^2\right) \cdot n\, \mathrm{d}s.$$

Fig. 3 Setup for Cartesian cells

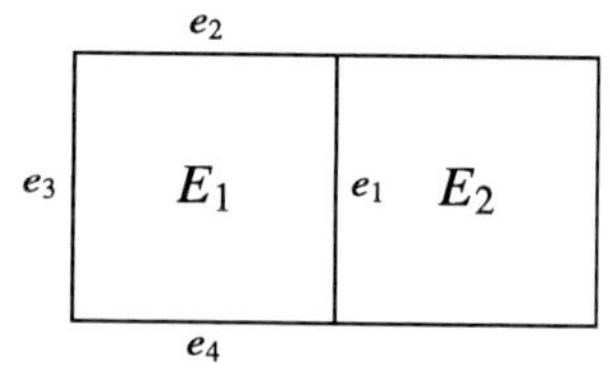

Let us first consider a standard Cartesian cell E_1 with edges as shown in Fig. 3. Then, for $\beta = (\beta_1, \beta_2)^T$ and assuming without loss of generality that $\beta_1, \beta_2 \geq 0$, there holds

$$-\int_{E_1} u_h \langle \beta, \nabla u_h \rangle \, \mathrm{d}x = -\int_{e_1} \frac{1}{2}\beta_1 u_h^2 \mathrm{d}s - \int_{e_2} \frac{1}{2}\beta_2 u_h^2 \mathrm{d}s + \int_{e_3} \frac{1}{2}\beta_1 u_h^2 \mathrm{d}s + \int_{e_4} \frac{1}{2}\beta_2 u_h^2 \mathrm{d}s.$$

For the edge terms in a_h^{upw}, let us consider the internal edge e_1, connecting E_1 and E_2. Then, (using from now on the notation $u_{E'}$ to indicate that we evaluate the discrete solution from cell E', potentially outside of its original support)

$$\begin{aligned}
&\int_{e_1} \left(\{\!\!\{u_h\}\!\!\} \langle \beta, [\![u_h]\!] \rangle + \frac{1}{2} |\langle \beta, n_{e_1} \rangle| \langle [\![u_h]\!], [\![u_h]\!] \rangle \right) \mathrm{d}s \\
&= \int_{e_1} \left(\frac{1}{2}\beta_1 (u_{E_1} + u_{E_2})(u_{E_1} - u_{E_2}) + \frac{1}{2}\beta_1 (u_{E_1} - u_{E_2})^2 \right) \mathrm{d}s \\
&= \int_{e_1} \beta_1 \left((u_{E_1})^2 - u_{E_1} u_{E_2} \right) \mathrm{d}s.
\end{aligned}$$

Combining this with the corresponding contributions for edge e_1 from the volume terms from cells E_1 and E_2, we get

$$\begin{aligned}
-\int_{e_1} \frac{1}{2}\beta_1 (u_{E_1})^2 \, \mathrm{d}s + \int_{e_1} \frac{1}{2}\beta_1 (u_{E_2})^2 \, \mathrm{d}s& \\
+ \int_{e_1} \left(\{\!\!\{u_h\}\!\!\} \langle \beta, [\![u_h]\!] \rangle + \frac{1}{2} |\langle \beta, n_{e_1} \rangle| \langle [\![u_h]\!], [\![u_h]\!] \rangle \right) \mathrm{d}s& \\
= \int_{e_1} \left(\frac{1}{2}\beta_1 (u_{E_1})^2 - \beta_1 u_{E_1} u_{E_2} + \frac{1}{2}\beta_1 (u_{E_2})^2 \right) \mathrm{d}s& \\
= \int_{e_1} \frac{1}{2}\beta_1 \left(u_{E_1} - u_{E_2}\right)^2 \mathrm{d}s.&
\end{aligned}$$

Let us now add the cut cells. For the small triangular cut cell E_{cut} with the notation from Fig. 2, we get with $\beta = (\beta_1, \beta_1)^T$ and assuming that $\beta_1, \beta_2 \geq 0$

$$-\int_{E_{\text{cut}}} u_h \langle \beta, \nabla u_h \rangle \, dx = -\int_{\partial E_{\text{cut}}} \left(\frac{1}{2}\beta u_h^2\right) \cdot n \, ds = -\int_{e_{\text{out}}} \frac{1}{2}\beta_2 u_h^2 \, ds + \int_{e_{\text{in}}} \frac{1}{2}\beta_1 u_h^2 \, ds.$$

Therefore, taking the boundary term in a_h^{upw} into account as well as the contribution from the volume term of cell E_{in}, we get for the edge e_{in}

$$\begin{aligned}
-\int_{e_{\text{in}}} \frac{1}{2}\beta_1 (u_{E_{\text{in}}})^2 \, ds + &\int_{e_{\text{in}}} \frac{1}{2}\beta_1 (u_{E_{\text{cut}}})^2 \, ds \\
&+ \int_{e_{\text{in}}} \left(\{u_h\} \langle \beta, [\![u_h]\!] \rangle + \frac{1}{2} |\langle \beta, n_{e_{\text{in}}} \rangle| \langle [\![u_h]\!], [\![u_h]\!] \rangle \right) ds \\
&= \int_{e_{\text{in}}} \left(\frac{1}{2}\beta_1 (u_{E_{\text{in}}})^2 - \beta_1 u_{E_{\text{in}}} u_{E_{\text{cut}}} + \frac{1}{2}\beta_1 (u_{E_{\text{cut}}})^2 \right) ds \\
&= \int_{e_{\text{in}}} \frac{1}{2}\beta_1 \left(u_{E_{\text{in}}} - u_{E_{\text{cut}}}\right)^2 ds.
\end{aligned}$$

We obtain a similar term for edge e_{out}, involving solutions from cells E_{cut} and E_{out}. Therefore, ignoring boundary contributions across $\partial\Omega^{\text{in}} \cup \partial\Omega^{\text{out}}$ due to the assumption of compact support, there holds

$$a_h^{\text{upw}}(u_h, u_h) = \sum_{e \in \Gamma_{\text{int}}} \frac{1}{2} \int_e |\langle \beta, n_e \rangle| \langle [\![u_h]\!], [\![u_h]\!] \rangle \, ds. \tag{8}$$

Therefore, without the stabilization term J_h, there holds L^2 stability.

Let us now add the stabilization term

$$J_h(u_h, u_h) = \sum_{E \in \mathcal{I}} J_h^{0,E}(u_h, u_h) + J_h^{1,E}(u_h, u_h).$$

We only stabilize small triangular cells of type E_{cut}. There holds

$$\begin{aligned}
J_h^{0,E_{\text{cut}}}(u_h, u_h) &= \eta_{E_{\text{cut}}} \int_{e_{\text{out}}} (u_{E_{\text{in}}} - u_{E_{\text{cut}}}) \langle \beta, [\![u_h]\!] \rangle \, ds \\
&= \eta_{E_{\text{cut}}} \int_{e_{\text{out}}} \beta_2 (u_{E_{\text{in}}} - u_{E_{\text{cut}}})(u_{E_{\text{cut}}} - u_{E_{\text{out}}}) ds \\
&= \eta_{E_{\text{cut}}} \int_{e_{\text{out}}} \beta_2 \left(u_{E_{\text{in}}} u_{E_{\text{cut}}} - u_{E_{\text{in}}} u_{E_{\text{out}}} - (u_{E_{\text{cut}}})^2 + u_{E_{\text{cut}}} u_{E_{\text{out}}} \right) ds.
\end{aligned}$$

We now consider $J_h^{1,E_{\text{cut}}}$ given by

$$J_h^{1,E_{\text{cut}}}(u_h, u_h) = \eta_{E_{\text{cut}}} \int_{E_{\text{cut}}} (u_{E_{\text{in}}} - u_{E_{\text{cut}}}) \left\langle \beta, \nabla u_{E_{\text{in}}} - \nabla u_{E_{\text{cut}}} \right\rangle \mathrm{d}x.$$

With $\beta = (\beta_1, \beta_2)^T$ and $\nabla \cdot \beta = 0$, there holds

$$\begin{aligned} J_h^{1,E_{\text{cut}}}(u_h, u_h) &= \eta_{E_{\text{cut}}} \int_{E_{\text{cut}}} \nabla \cdot \left(\frac{1}{2} \beta (u_{E_{\text{in}}} - u_{E_{\text{cut}}})^2 \right) \mathrm{d}x \\ &= \eta_{E_{\text{cut}}} \int_{\partial E_{\text{cut}}} \left(\frac{1}{2} \beta (u_{E_{\text{in}}} - u_{E_{\text{cut}}})^2 \right) \cdot n \,\mathrm{d}s \\ &= \eta_{E_{\text{cut}}} \int_{e_{\text{out}}} \left(\frac{1}{2} \beta_2 (u_{E_{\text{in}}} - u_{E_{\text{cut}}})^2 \right) \mathrm{d}s - \eta_{E_{\text{cut}}} \int_{e_{\text{in}}} \left(\frac{1}{2} \beta_1 (u_{E_{\text{in}}} - u_{E_{\text{cut}}})^2 \right) \mathrm{d}s. \end{aligned}$$

As $0 \leq \eta_{E_{\text{cut}}} \leq 1$, the negative term over the edge e_{in} can be compensated with the edge term $\int_{e_{\text{in}}} \beta_1 \left(\frac{1}{2} (u_{E_{\text{in}}} - u_{E_{\text{cut}}})^2 \right) \mathrm{d}s$ from a_h^{upw} in (8). For the edge e_{out}, we collect all terms from $J_h^{0,E_{\text{cut}}}$ and $J_h^{1,E_{\text{cut}}}$ to get

$$\begin{aligned} &\eta_{E_{\text{cut}}} \int_{e_{\text{out}}} \beta_2 \Big(u_{E_{\text{in}}} u_{E_{\text{cut}}} - u_{E_{\text{in}}} u_{E_{\text{out}}} - (u_{E_{\text{cut}}})^2 + u_{E_{\text{cut}}} u_{E_{\text{out}}} + \frac{1}{2} (u_{E_{\text{in}}} - u_{E_{\text{cut}}})^2 \Big) \mathrm{d}s \\ &= \eta_{E_{\text{cut}}} \int_{e_{\text{out}}} \beta_2 \left(\frac{1}{2} (u_{E_{\text{in}}})^2 - \frac{1}{2} (u_{E_{\text{cut}}})^2 - u_{E_{\text{in}}} u_{E_{\text{out}}} + u_{E_{\text{cut}}} u_{E_{\text{out}}} \right) \mathrm{d}s \\ &= \eta_{E_{\text{cut}}} \int_{e_{\text{out}}} \frac{1}{2} \beta_2 (u_{E_{\text{in}}} - u_{E_{\text{out}}})^2 \,\mathrm{d}s - \eta_{E_{\text{cut}}} \int_{e_{\text{out}}} \frac{1}{2} \beta_2 (u_{E_{\text{cut}}} - u_{E_{\text{out}}})^2 \,\mathrm{d}s. \end{aligned}$$

The right term in the last line involves the standard jump over edge e_{out} and (same as for edge e_{in}) can be compensated with its positive counterpart in the sum in (8). The first term in the last line consists of a new extended jump involving the difference of the solution of cell E_{in} and the solution of cell E_{out}, both evaluated on the outflow edge e_{out}. This concludes the proof. □

4 Numerical Results

In this section, we present numerical results for the linear advection equation in 2d using higher order polynomials for the ramp setup introduced above for different angles γ, see Fig. 1. Following the test setup from [10] we choose $\hat{\Omega} = (0, 1)^2$ and start the ramp at $x = 0.2001$. For the definition of the initial data, we use a rotated and shifted coordinate system $(\hat{x}, \hat{y})$ that we derive from the standard Cartesian coordinate system (x, y) by

$$\begin{pmatrix} \hat{x} \\ \hat{y} \end{pmatrix} = \begin{pmatrix} \cos\gamma & \sin\gamma \\ -\sin\gamma & \cos\gamma \end{pmatrix} \cdot \begin{pmatrix} x - 0.2001 \\ y \end{pmatrix}. \tag{9}$$

This newly described coordinate system is defined in such a way that the $\hat{x}$-direction is parallel and the $\hat{y}$-direction is orthogonal to the ramp. In this coordinate system, the velocity field β and the smooth initial data are given by

$$\beta(\hat{x}, \hat{y}) = (2 - \hat{y}) \begin{pmatrix} 1 \\ 0 \end{pmatrix}, \quad u_0(\hat{x}, \hat{y}) = \sin\left(\frac{\sqrt{2}\pi\hat{x}}{1 - 0.2001}\right).$$

The velocity field β is parallel to the ramp with decreasing speed for increasing distance. We derive the inflow conditions on $\partial\Omega_{\text{in}}$ from the exact solution. We compute the discrete solution at time $T = 0.3$ using polynomials of degrees $p = 1, 2, 3$. In time we use an SSP RK scheme of the same order as the space discretization. We compute the time step Δt by

$$\Delta t \leq 0.4 \frac{1}{2p + 1} \frac{h}{\|\beta\|}.$$

Here, $h = 1/N$ with N being the number of cells in x- and y-direction on $\hat{\Omega}$.

The implementation is based on the DUNE [2, 3] framework, the cut-cell DG extension `dune-udg` package [1, 8] and its integration with `dune-pdelab`. The geometry is represented as a discrete level set function, using vertex values. Based on this representation the cut cells and their corresponding quadrature rules are constructed via the TPMC library [9].

In Fig. 4, we show convergence results for ramp angles of $\gamma = 25°$ and $\gamma = 45°$ in the L^1 and L^∞ norm. In the L^1 norm we observe convergence orders that are (roughly) $p + 1$ for polynomials of degree p for both angles. In the L^∞ norm, the results are between $p + \frac{1}{2}$ and $p + 1$. This is overall consistent with the findings of Giuliani [12] who reports for the annulus test for the L^∞ error orders between $p + \frac{1}{2}$ and $p + 1$ for polynomials of degrees $p = 1, \ldots, 5$.

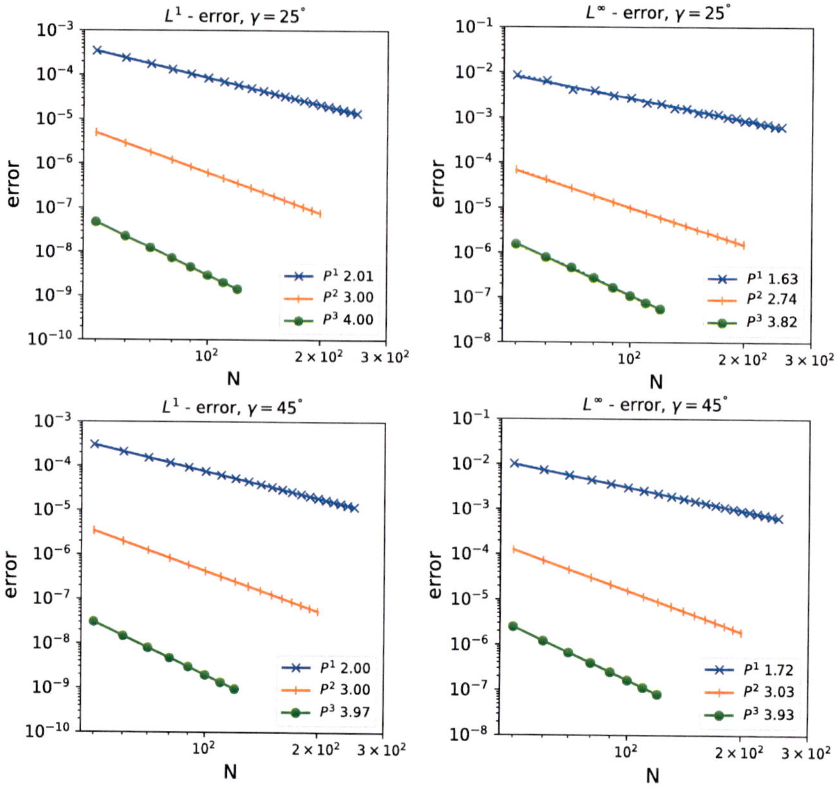

Fig. 4 Convergence orders in L^1 and L^∞ norm for the error at time $T = 0.3$ for a ramp geometry with $\gamma = 25°$ and $\gamma = 45°$ and different polynomial degrees $p = 1, 2, 3$

5 Conclusion

In this contribution, we introduce the formulation of the DoD stabilization for the linear advection equation for higher order polynomials. Compared to [10], where we only considered linear polynomials, we have augmented the penalty term $J_h^{1,E}$ to also involve the extended test function of the inflow neighbor of a small cut cell. For this new formulation, we show an L^2 stability result for the semi-discrete stabilized scheme for the ramp geometry. We also provide numerical results for a smooth test function, showing convergence rates between $p+\frac{1}{2}$ and $p+1$ for polynomial degree p. In the future, we plan to extend the stabilization to non-linear problems in 2d.

Acknowledgments This work was partially supported by the Deutsche Forschungsgemeinschaft (DFG, German Research Foundation) under the project number 439956613 (HyperCut) under the contracts MA 7773/4-1 and EN 1042/5-1 and under Germany's Excellence Strategy EXC 2044-390685587, Mathematics Münster: Dynamics-Geometry-Structure.

References

1. Bastian, P., Engwer, C.: An unfitted finite element method using discontinuous Galerkin. Internat. J. Numer. Methods Eng. **79**, 1557–1576 (2009)
2. Bastian, P., Blatt, M., Dedner, A., Engwer, C., Klöfkorn, R., Kornhuber, R., Ohlberger, M., Sander, O.: A generic grid interface for parallel and adaptive scientific computing. Part II: Implementation and tests in DUNE. Computing **82**(2–3), 121–138 (2008). https://doi.org/10.1007/s00607-008-0004-9
3. Bastian, P., Blatt, M., Dedner, A., Engwer, C., Klöfkorn, R., Ohlberger, M., Sander, O.: A generic grid interface for parallel and adaptive scientific computing. Part I: Abstract framework. Computing **82**(2–3), 103–119 (2008). https://doi.org/10.1007/s00607-008-0003-x
4. Berger, M., Giuliani, A.: A state redistribution algorithm for finite volume schemes on cut cell meshes. J. Comput. Phys. **428** (2021). https://doi.org/10.1016/j.jcp.2020.109820
5. Berger, M., Helzel, C.: A simplified h-box method for embedded boundary grids. SIAM J. Sci. Comput. **34**(2), A861–A888 (2012)
6. Chern, I.L., Colella, P.: A conservative front tracking method for hyperbolic conservation laws. Tech. rep., Lawrence Livermore National Laboratory, Livermore, CA (1987). Preprint UCRL-97200
7. Colella, P., Graves, D.T., Keen, B.J., Modiano, D.: A Cartesian grid embedded boundary method for hyperbolic conservation laws. J. Comput. Phys. **211**(1), 347–366 (2006)
8. Engwer, C., Heimann, F.: Dune-udg: a cut-cell framework for unfitted discontinuous Galerkin methods. In: Advances in DUNE, pp. 89–100. Springer (2012)
9. Engwer, C., Nüßing, A.: Geometric reconstruction of implicitly defined surfaces and domains with topological guarantees. ACM Trans. Math. Softw. (TOMS) **44**(2), 14 (2017)
10. Engwer, C., May, S., Nüßing, A., Streitbürger, F.: A stabilized DG cut cell method for discretizing the linear transport equation. SIAM J. Sci. Comput. **42**(6), A3677–A3703 (2020)
11. Fu, P., Kreiss, G.: High order cut discontinuous Galerkin methods for hyperbolic conservation laws in one space dimension. SIAM J. Sci. Comput. **43**(4), A2404–A2424 (2021)
12. Giuliani, A.: A two-dimensional stabilized discontinuous Galerkin method on curvilinear embedded boundary grids (2021). https://epubs.siam.org/doi/abs/10.1137/21M1396277
13. Gokhale, N., Nikiforakis, N., Klein, R.: A dimensionally split Cartesian cut cell method for hyperbolic conservation laws. J. Comput. Phys. **364**, 186–208 (2018)
14. Gottlieb, S., Shu, C.W., Tadmor, E.: Strong stability-preserving high-order time discretization methods. SIAM Rev. **43**(1), 89–112 (2001)
15. Helzel, C., Kerkmann, D.: An active flux method for cut cell grids. In: Klöfkorn, R., Keilegavlen, E., Radu, A., Fuhrmann, J. (eds.) Finite Volumes for Complex Applications IX - Methods, Theoretical Aspects, Examples, pp. 507–515. Springer International Publishing (2020)
16. Helzel, C., Berger, M., LeVeque, R.: A high-resolution rotated grid method for conservation laws with embedded geometries. SIAM J. Sci. Comput. **26**(3), 785–809 (2005)
17. Klein, R., Bates, K.R., Nikiforakis, N.: Well-balanced compressible cut-cell simulation of atmospheric flow. Philos. Trans. R. Soc. A **367**, 4559–4575 (2009)
18. May, S., Berger, M.: An explicit implicit scheme for cut cells in embedded boundary meshes. J. Sci. Comput. **71**, 919–943 (2017)

19. May, S., Streitbürger, F.: DoD stabilization for non-linear hyperbolic conservation laws on cut cell meshes in one dimension. Appl. Math. Comput. **419**, 126854 (2022)
20. Schoeder, S., Sticko, S., Kreiss, G., Kronbichler, M.: High-order cut discontinuous Galerkin methods with local time stepping for acoustics. Internat. J. Numer. Methods Eng. **121**(13), 2979–3003 (2020)

Space-Time Error Control Using a Partition-of-Unity Dual-Weighted Residual Method Applied to Low Mach Number Combustion

Jan P. Thiele and Thomas Wick

1 Introduction

This work is devoted to space-time goal-oriented a posteriori error control. Such space-time schemes for error estimation and adaptivity in time, space, or both, are of current interest with various applications in parabolic problems [14], incompressible Navier-Stokes equations [3, 4], dynamic Signorini, obstacle, and hyperbolic problems [5, 6, 12] and fluid-structure interaction [9, 10].

Our method is based on prior work, in which the dual-weighted error estimator is realized within a weak formulation using a partition-of-unity [13]. We note that another weak realization is achieved with the so-called filtering approach [7], which was already applied in [3, 4, 14] to space-time error control and adaptivity.

However, the extension of the partition-of-unity (PU) localization to space-time settings has not yet been established in the published literature. We provide a detailed algorithmic derivation of the error estimator and discuss the important ingredients. As numerical example, we consider a nonlinear low mach number combustion problem. A key part is the backward-in-time running adjoint problem. One difficulty is that, due to Galerkin orthogonality, the adjoint problem must contain higher order information (see [2]) in the primal error part and vice versa in the adjoint error part. Recently, for stationary settings a new class of algorithms could be established by using interpolation information in a smart way [8]. For our proposed space-time setting we investigate the performance by simply choosing different combinations for the spatial finite elements, such as low-order $cG(1)/cG(1)$ and high-order equal elements $cG(2)/cG(2)$ as well as the natural approach $cG(1)/cG(2)$ (low order primal and higher order adjoint for the

J. P. Thiele (✉) · T. Wick
Leibniz University Hannover, Hanover, Germany
e-mail: thiele@ifam.uni-hannover.de; thomas.wick@ifam.uni-hannover.de

J. M. Melenk et al. (eds.), *Spectral and High Order Methods for Partial Differential Equations ICOSAHOM 2020+1*, Lecture Notes in Computational Science and Engineering 137, https://doi.org/10.1007/978-3-031-20432-6_34

primal error part). These choices are investigated with respect to their convergence properties and evaluation of the effectivity indices. We will, however, limit ourselves to the variation of finite element orders in space and use the equal order approach for the temporal discretization. We also notice that some preliminary results on space-time adaptivity with the PU-DWR method are published in [15].

The outline of this work is as follows: In Sect. 2, the low mach number combustion equations are introduced and their weak formulation is provided. Next, in Sect. 3, the discretization with finite elements is described. In the main Sect. 4 a space-time PU-DWR error estimator is derived in detail. Finally, in Sect. 5 some numerical experiments are reported, that show the performance of our developments. The code for these simulations is based on our extension of the package `dwr-diffusion` [11] to solve nonlinear problems. The package itself uses `deal.II` [1] as the finite element library. We conclude our work in Sect. 6.

2 The Low Mach Number Combustion Equations

The nonlinear parabolic problem we want to investigate describes a combustion reaction under the low Mach number hypothesis. Under that hypothesis the dimensionless temperature θ and the concentration of the combustible species Y are not influencing the fluid velocity field.

For the special but important case of $v = 0$ all convection terms vanish and θ and Y are only influenced by diffusion and by the reaction mechanism in which Y combusts and raises the temperature. For constant diffusion coefficients we arrive at the following set of equations

$$\partial_t \theta - \Delta\theta = \omega(\theta, Y) \quad \text{in } \Omega \times (0, T), \tag{1}$$

$$\partial_t Y - \frac{1}{Le}\Delta Y = -\omega(\theta, Y) \quad \text{in } \Omega \times (0, T), \tag{2}$$

where the reaction is described by Arrhenius law

$$\omega(\theta, Y) = \frac{\beta^2}{2Le} Y e^{\frac{\beta(\theta-1)}{1+\alpha(\theta-1)}}. \tag{3}$$

The parameters are the Lewis number $Le = 1$, the gas expansion $\alpha = 0.8$ and the nondimensional activation energy $\beta = 10$.

Part of the boundary $\Gamma_R \subset \Gamma := \partial\Omega$ will be cooled. This can be described by a Robin condition $\partial_n\theta = -k\theta$. Following the standard procedure, we obtain the following space-time variational formulation of our problem. Find $u = (\theta, Y)$ such that

$$A(u, \phi) = (\partial_t\theta, \phi^\theta) + (\nabla\theta, \nabla\phi^\theta) + (\partial_t Y, \phi^Y) + (\nabla Y, \nabla\phi^Y) \tag{4}$$

$$+(\omega(\theta, Y), \phi^Y - \phi^\theta) + \int_0^T \int_{\Gamma_R} k\theta\phi^\theta \,\mathrm{d}s\mathrm{d}t = 0 =: F(\phi)\ \forall\phi = (\phi^\theta, \phi^Y)$$

where (f, g) describes the space-time scalar product defined as

$$(f, g) := \int_0^T \int_\Omega fg \mathrm{d}x \mathrm{d}t.$$

As the homogeneous Neumann condition on Γ_N is a natural condition it does not appear in the variational formulation. The inhomogeneous Dirichlet conditions on Γ_D are imposed as usual and inserted into the finite element spaces.

3 Discretization

As we want to use different finite elements for the temporal and the spatial test- and trial functions we will start by partitioning $[0, T]$ into M subintervals $I_n = (t_{n-1}, t_n]$, with the discretization parameter $k = t_n - t_{n-1}$. In time we will use piecewise constant discontinuous elements $\phi_k(t) \in dG(0)$. To be able to use different refined meshes over time, so called dynamic meshes, we will discretize Ω on each subinterval by a triangulation $\mathcal{T}_h^n$. Using quadrilaterals (in two-dimensional configurations) for the spatial triangulation, we can use continuous finite element functions $\phi_h(x) \in cG(p)$ of order p as test functions and trial functions. The fully discrete equations on each subinterval are then obtained by using $\phi kh \in dG(0)cG(p)$ as test functions and trial functions in (4). For a more detailed look at the discretization and the corresponding finite element spaces see [14].

4 Space-Time PU-DWR Error Estimation

Denoting our quantity of interest by the goal functional $J(u)$, we obtain the Lagrange functional for minimizing the error in said quantity as

$$\mathcal{L}(u, z) = J(u) + F(z) - A(u, z). \tag{5}$$

As a first order optimality condition we obtain the KKT (Karush-Kuhn-Tucker) system and with it an auxiliary adjoint problem. In summary, we then have

$$\mathcal{L}'_u(u)(\psi, z) = J'_u(u)(\psi) - A'_u(u)(\psi, z) \stackrel{!}{=} 0 \text{ (adjoint problem)} \tag{6}$$

$$\mathcal{L}'_z(z)(u, \phi) = F(\phi) - A(u, \phi) \stackrel{!}{=} 0 \text{ (primal problem)} \tag{7}$$

Note that for nonlinear problems $A(u, z)$ is a semilinear form that is linear in z and that $F(z)$ is always a linear form. Thus, the directional derivative in direction ϕ w. r. t. to z, i. e. $A'_z(z)(u, \phi)$ coincides with $A(u, \phi)$. The same holds for linear goal functionals and linear problems resulting in the dual problem $A(\psi, z) = J(\psi)$. Also note that the adjoint problem obtained by this derivation applies the temporal derivative to the test function. To rectify this, a partial integration in time is applied to $(\partial_t \psi, z)$, yielding $(\psi, -\partial_t z)$. This results in a problem that runs backwards in time and has a final condition instead of an initial condition.

4.1 Error Estimation

Following Proposition (2.1) in [2] we obtain the error representation

$$J(u) - J(u_{kh}) = \frac{1}{2}\mathcal{L}'_z(z_{kh})(u_{kh}, z - z_{kh}) + \frac{1}{2}\mathcal{L}'_u(u_{kh})(u - u_{kh}, z_{kh}) + \mathcal{R}, \tag{8}$$

where $\mathcal{R}$ is a higher order remainder term. In many cases it is sufficient to approximate the error by only computing the primal residual i.e.

$$\begin{aligned} J(u) - J(u_{kh}) &\approx \mathcal{L}'_z(z_{kh})(u_{kh}, z - z_{kh}) \\ &= F(z - z_{kh}) - A(u_{kh}, z - z_{kh}) =: \rho_{kh}(u_{kh}, z - z_{kh}), \end{aligned} \tag{9}$$

which is also called primal error estimator. Subsequently the second term in the error representation is called adjoint error estimator $\rho^*_{kh}(u_{kh})(u - u_{kh}, z_{kh})$. Introducing the semidiscrete solutions u_k and z_k which are still continuous in space the primal error estimator can be further split into a temporal estimator ρ_k and a spatial estimator ρ_h

$$\begin{aligned} J(u) - J(u_{kh}) &= [J(u) - J(u_k)] + [J(u_k) - J(u_{kh})] \\ &\approx \rho_k(u_k, z - z_k) + \rho_h(u_{kh}, z_k - z_{kh}). \end{aligned} \tag{10}$$

4.2 Practical Evaluation

As the exact solutions u and z are unknown, we have to further approximate them to calculate the error estimators. For the temporal primal estimator, we will construct a piecewise linear solution $i_k^{(1)} z$ on each grid point by linear interpolation between the piecewise constant solutions z_n and z_{n-1} in the interval I_n.

For the spatial estimator we will look at three different approaches. The simplest approach is calculating u_{kh} with $dG(0)cG(1)$ elements and z_{kh} with $dG(0)cG(2)$ elements. Then, we assume z_{kh} to be the approximation of the exact solution and interpolate it down into $cG(1)$ in space obtaining $i_h^{(2,1)} z_{kh}$. This interpolation should be included in most finite element packages.

For also calculating the dual estimator we also need an approximation for u, which can be obtained by approximating u_{kh} and z_{kh} with $dG(0)cG(2)$ elements. Using the same interpolation as before we can approximate the discrete solutions as $\tilde{u}_{kh} := i_h^{(2,1)} u_{kh}$ and $\tilde{z}_{kh} := i_h^{(2,1)} z_{kh}$, while the exact solutions are approximated by u_{kh} and z_{kh}.

As this approach can be quite memory intensive, both u_{kh} and z_{kh} can be solved using $dG(0)cG(1)$ elements. The approximation for z can then be obtained by combining neighbouring $cG(1)$ elements into one large $cG(2)$ patch with the operator $i_{2h}^{(2)}$. The operator and the requirements for the mesh are described in [7]. Using those interpolations we obtain the following primal estimators for the different approaches

$$\eta_k^{cG(1)/cG(1)} = F(i_k^{(1)} z_{kh} - z_{kh}) - A(u_{kh}, i_k^{(1)} z_{kh} - z_{kh}), \tag{11}$$

$$\eta_h^{cG(1)/cG(1)} = F(i_{2h}^{(2)} z_{kh} - z_{kh}) - A(u_{kh}, i_{2h}^{(2)} z_{kh} - z_{kh}), \tag{12}$$

for $u_{kh} \in dG(0)cG(1)$ and $z_{kh} \in dG(0)cG(1)$,

$$\eta_k^{cG(1)/cG(2)} = F(i_k^{(1)} i_h^{(2,1)} z_{kh} - i_h^{(2,1)} z_{kh}) - A(u_{kh}, i_k^{(1)} i_h^{(2,1)} z_{kh} - i_h^{(2,1)} z_{kh}), \tag{13}$$

$$\eta_h^{cG(1)/cG(2)} = F(z_{kh} - i_h^{(2,1)} z_{kh}) - A(u_{kh}, z_{kh} - i_h^{(2,1)} z_{kh}), \tag{14}$$

for $u_{kh} \in dG(0)cG(1)$ and $z_{kh} \in dG(0)cG(2)$,

$$\eta_k^{cG(2)/cG(2)} = F(i_k^{(1)} \tilde{z}_{kh} - \tilde{z}_{kh}) - A(\tilde{u}_{kh}, i_k^{(1)} \tilde{z}_{kh} - \tilde{z}_{kh}), \tag{15}$$

$$\eta_h^{cG(2)/cG(2)} = F(z_{kh} - \tilde{z}_{kh}) - A(\tilde{u}_{kh}, z_{kh} - \tilde{z}_{kh}), \tag{16}$$

for $u_{kh} \in dG(0)cG(2)$ and $z_{kh} \in dG(0)cG(2)$.

The corresponding dual estimators are obtained by the same interpolation operators, but applied to the primal solution u_{kh} and inserted into the adjoint problem.

$$\eta_k^{*cG(1)/cG(1)} = J_u'(u_{kh})(i_k^{(1)} u_{kh} - u_{kh}) - A_u'(u_{kh})(i_k^{(1)} u_{kh} - u_{kh}, z_{kh}) \tag{17}$$

$$\eta_h^{*cG(1)/cG(1)} = J_u'(u_{kh})(i_{2h}^{(2)} u_{kh} - u_{kh}) - A_u'(u_{kh})(i_{2h}^{(2)} u_{kh} - u_{kh}, z_{kh}) \tag{18}$$

$$\eta_k^{*cG(1)/cG(2)} = J_u'(u_{kh})(i_k^{(1)} u_{kh} - u_{kh}) - A_u'(u_{kh})(i_k^{(1)} u_{kh} - u_{kh}, i_h^{(2,1)} z_{kh}) \tag{19}$$

$$\eta_h^{*cG(1)/cG(2)} = J_u'(u_{kh})(i_{2h}^{(2)} u_{kh} - u_{kh}) - A_u'(u_{kh})(i_{2h}^{(2)} u_{kh} - u_{kh}, i_h^{(2,1)} z_{kh}) \tag{20}$$

$$\eta_k^{*cG(2)/cG(2)} = J'_u(\tilde{u}_{kh})(i_k^{(1)}\tilde{u}_{kh} - \tilde{u}_{kh}) - A'_u(\tilde{u}_{kh})(i_k^{(1)}\tilde{u}_{kh} - \tilde{u}_{kh}, \tilde{z}_{kh}), \tag{21}$$

$$\eta_h^{*cG(2)/cG(2)} = J'_u(\tilde{u}_{kh})(u_{kh} - \tilde{u}_{kh}) - A'_u(\tilde{u}_{kh})(u_{kh} - \tilde{u}_{kh}, \tilde{z}_{kh}). \tag{22}$$

4.3 Variational PU Localization

For use in adaptive refinement we need to obtain indicators η_K^n or η_i^n for each cell K or DoF i on the time interval I_n, such that

$$\eta = \sum_{n=1}^{M} \sum_{K \in \mathcal{T}_h^n} \eta_K^n = \sum_{n=1}^{M} \sum_{i \in \mathcal{T}_h} \eta_i^n. \tag{23}$$

We propose a DoF-wise partition of unity (PU) χ_i^n, with

$$\sum_{n=1}^{M} \sum_{i \in \mathcal{T}_h} \chi_i^n \equiv 1, \tag{24}$$

the simplest choice is $\chi_i^n \in dG(0)cG(1)$. Effectively, this leads to a spatial PU per time step, that is identical to the approach of [13] for stationary problems. The estimators are obtained by multiplying the directions in the derivatives of the Lagrangian with the PU, which leads to the localization of the original error representation (8):

$$[J(u) - J(u_{kh})]_i^n := \frac{1}{2}\mathcal{L}'_z(z_{kh})(u_{kh}, (z - z_{kh})\chi_i^n) + \mathcal{L}'_u(u_{kh})((u - u_{kh})\chi_i^n, z_{kh}), \tag{25}$$

$$J(u) - J(u_{kh}) = \mathcal{R} + \sum_{n=1}^{M} \sum_{i \in \mathcal{T}_h} [J(u) - J(u_{kh})]_i^n. \tag{26}$$

Finally, inserting the PU into the estimators described in the previous subsection yields the error indicators for each space-time DoF.

5 Numerical Example

In our numerical example, we solve the equations described in Sect. 2 on the geometry shown in Fig. 1. Here, the left edge of the domain Γ_D is kept at a constant temperature $\theta = 1$ without any combustible species $Y = 0$. The recessed area

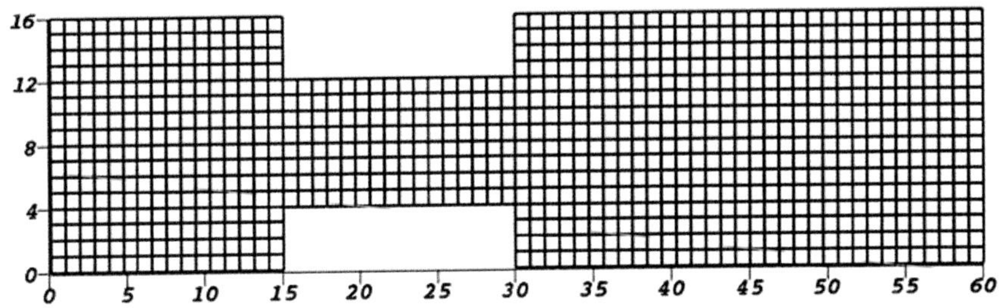

Fig. 1 Initial grid with $N = 896$ cells and $1970(cG(1)^2)$ and $7522(cG(2)^2)$ degrees of freedom

between $x = 15$ and $x = 30$ is the cooled Robin boundary Γ_R, with $\partial_n\theta = -0.1\theta$ and $\partial_n Y = 0$. On the rest of the boundary Γ_N homogeneous Neumann conditions are set.

The initial conditions are described by

$$\theta_0(x, z) = \begin{cases} 1, & x \leq 9, \\ e^{9-x}, & x > 9 \end{cases} \tag{27}$$

$$Y_0(x, z) = \begin{cases} 0, & x \leq 9, \\ 1 - e^{Le(9-x)}, & x > 9 \end{cases} \tag{28}$$

The functional of interest is the space-time averaged reaction rate

$$J(u) = \frac{1}{T|\Omega|} \int_0^T \int_\Omega \omega(\theta, Y) \mathrm{d}x \, \mathrm{d}t, \tag{29}$$

with final time $T = 60$.

The initial grid is solved with $k = 0.234375$, resulting in $M = 256$ time intervals.

5.1 Comparison of Error Estimators

To compare the estimators without influence of the adaptivity, the following simulations were done with global refinement in space and time. Tables 1, 2, and 3 show the results for the error $J(u) - J(u_{kh})$ in comparison with the primal, adjoint and full estimators respectively. Since we use a different number of degrees of freedom for each approach, we decided to base the comparison on the number of time steps M and the number of spatial cells N. As a rough approximation the solution $(\theta, Y) \in cG(1)^2$ needs $2N$ and the solution $(\theta, Y) \in cG(2)^2$ needs $8N$ spatial degrees of freedom. As a $cG(1)$ representation of the solution is inserted

Table 1 Primal estimators for different global refinement levels

M	N	$J(u) - J(u_{kh})$	$\eta_{kh}^{cG(1)/cG(1)}$	$\eta_{kh}^{cG(1)/cG(2)}$	$\eta_{kh}^{cG(2)/cG(2)}$
256	896	1.07197741e−02	1.92058654e−03	7.61264827e+06	5.30861697e−04
512	3584	2.48965242e−03	6.43743177e−04	4.08892798e−01	2.34439679e−04
1024	14336	5.67024544e−04	2.61386699e−04	1.32470408e−03	2.02825765e−04
2048	57344	1.11743216e−04	1.08270274e−04	1.62070260e−04	1.02680294e−04

Table 2 Adjoint/dual estimators for different global refinement levels

M	N	$J(u) - J(u_{kh})$	$\eta_{kh}^{cG(1)/cG(1)}$	$\eta_{kh}^{cG(1)/cG(2)}$	$\eta_{kh}^{cG(2)/cG(2)}$
256	896	1.07197741e−02	1.19648379e−03	1.82646766e+08	1.21421097e−03
512	3584	2.48965242e−03	8.33534627e−04	2.79437497e+00	7.95366447e−04
1024	14,336	5.67024544e−04	4.25107914e−04	4.21358177e−03	4.21309997e−04
2048	57,344	1.11743216e−04	2.74717107e−04	3.83350535e−04	1.70568548e−04

Table 3 Full estimators for different global refinement levels

M	N	$J(u) - J(u_{kh})$	$\eta_{kh}^{cG(1)/cG(1)}$	$\eta_{kh}^{cG(1)/cG(2)}$	$\eta_{kh}^{cG(2)/cG(2)}$
256	896	1.07197741e−02	1.37575686e−03	9.20610356e+07	8.20354657e−04
512	3584	2.48965242e−03	5.59566023e−04	1.60163388e+00	5.14903063e−04
1024	14336	5.67024544e−04	3.32992866e−04	2.54117001e−03	2.74961514e−04
2048	57344	1.11743216e−04	1.77760869e−04	2.55543805e−04	1.25485484e−04

as u_{kh} in (8), the error $J(u) - J(u_{kh})$ is evaluated in either the $cG(1)/cG(1)$ or the $cG(1)/cG(2)$ case. Note that the error would be larger for the projection of the $cG(2)$ solution into $cG(1)$ for the $cG(2)/cG(2)$ approach, as that interpolation simply uses the values at the vertex DoFs with the respective $cG(1)$ basis functions and discards all other parts of the solution.

Comparing the results over all tables, we see that the equal order approaches perform relatively similar and better than the mixed order approach. Especially on lower refinement levels the $cG(1)/cG(2)$ results are orders of magnitude above the actual error. On closer inspection the adjoint solutions get larger with each time step, which leads to the estimator being dominated by the indicators on the first few time intervals. As the codes for solving the $cG(2)$ adjoint problems for $cG(2)/cG(2)$ and $cG(1)/cG(2)$ basically only differ in the u_{kh} that is inserted in the assembly of the right hand side and the nonlinear part of the matrix, we surmise that the errors from inserting a lower order solution get amplified with each time step. This would also explain why this approach does not fail for stationary problems even on coarse meshes. In previous simulations we also saw that pairing the $cG(1)/cG(2)$ approach with solving the adjoint with $cG(1)$ elements in time led to worse results than the temporal equal order approach, even for the linear heat equation with the

L_2 error as functional of interest. For adaptivity on dynamic meshes this is of course a considerable problem, as it is advisable to start with a coarse mesh to only capture the local behaviour of the solution/functional at each time interval.

When comparing the tables with each other, we see that for this problem the primal error estimator on itself performs better than the adjoint error estimator and is comparable to the full estimator. Overall, the $cG(1)/cG(1)$ approach is preferable as it is considerably cheaper to calculate compared to the $cG(2)/cG(2)$ approach for which multiple linear systems need to be solved with $\approx 8N$ unknowns (one solve for the adjoint and several solves for the primal Newton solver steps) instead of $\approx 2N$ unknowns. Additionally, the primal solution vectors have to be kept either in RAM or on hard disk as the adjoint is solved backwards in time, so the $cG(2)/cG(2)$ leads to a considerable increase in memory or storage demand. From a computational standpoint one can see why the $cG(1)/cG(2)$ approach would be a nice tradeoff between accuracy and memory demand as only a single linear solve per time step has to performed on the larger set of unknowns.

5.2 *Adaptive Results*

When comparing the different estimators in the $cG(1)/cG(1)$ approach the primal estimator is closest to the actual error, so we decided to use this estimator as a basis for an adaptive simulation. As we have DoF-based indicators we compute cell-wise indicators to use build-in refinement strategies in `deal.II`. These are obtained by simply adding the four spatial indicators of the cell vertices. As a refinement strategy we chose fixed fraction marking in which the indicators are sorted and the $x\%$ of cells with the largest indicators are marked for refinement. The same strategy is applied to the time intervals for which the indicators are calculated as the sum over all temporal DoF-indicators on the corresponding spatial triangulation. As fractions we chose 50% for the temporal and 33% for the spatial refinement which leads to roughly $1.5M$ time intervals and $2N$ spatial cells per time interval compared to $2M$ and $4N$ for global refinement. Figure 2 shows that the exact error converges faster for adaptive refinement, when comparing the number of primal DoFs.

To see if our novel localization approach works well in capturing the local behaviour of the goal functional, Figs. 3 and 4 show the evolution of the reaction rate ω over $[0, T]$ and the corresponding meshes. In all timesteps the combustion reaction is captured well by the fine cells.Additionally, for time steps after the flame passed the cooled rods, there is also some refinement around the sharp corners, which is to be expected.

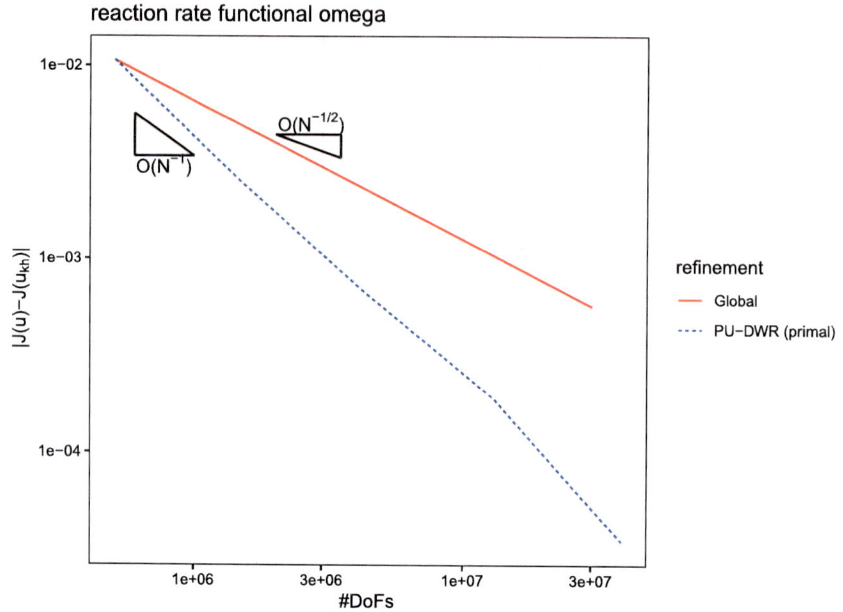

Fig. 2 Comparison of the actual errors for global refinement vs. adaptive refinement with the primal $cG(1)/cG(1)$ estimator with marking 50% of the time intervals and 33% of the spatial cells on each interval for refinement

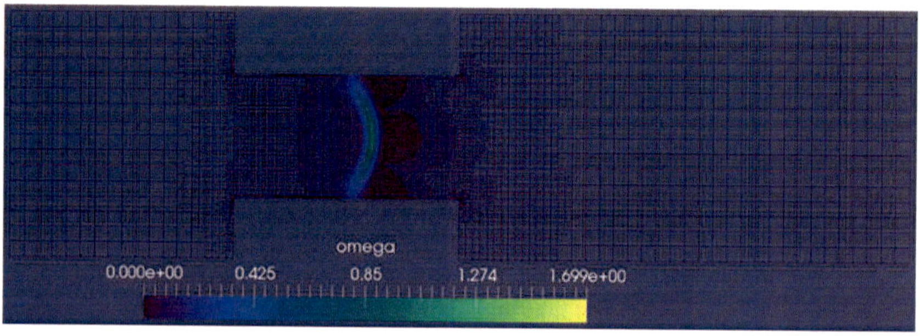

Fig. 3 reaction rate ω at $t = 20$

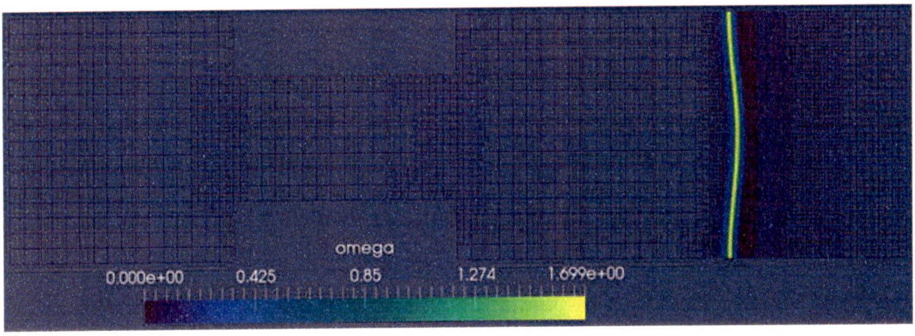

Fig. 4 reaction rate ω at $t = 60$

6 Conclusions

In this work, we developed a space-time goal-oriented a posteriori error estimator using a newly developed partition-of-unity dual-weighted residual localization. As model problem we considered a nonlinear low Mach number combustion problem. Specific emphasis was on different space-time finite element combinations for the primal and adjoint subproblems. Therein, we detected a better performance for equal-order combinations of $cG(1)/cG(1)$ and $cG(2)/cG(2)$ type in comparison to a $cG(1)/cG(2)$ finite element. The reason has not yet been fully understood by us and needs further future investigations whether algorithmic or mathematical problems are the reason. Finally, some illustrations of locally adaptive meshes show that the error indicators obtained by our proposed method yield excellent findings in terms of resolving the local flame front.

References

1. Arndt, D., Bangerth, W., Blais, B., Clevenger, T.C., Fehling, M., Grayver, A.V., Heister, T., Heltai, L., Kronbichler, M., Maier, M., Munch, P., Pelteret, J.P., Rastak, R., Tomas, I., Turcksin, B., Wang, Z., Wells, D.: The deal.II library, Version 9.2. J. Numer. Math. **28**(3), 131–146 (2020)
2. Becker, R., Rannacher, R.: An optimal control approach to a posteriori error estimation in finite element methods. Acta Numer. **10**, 1–102 (2001)
3. Besier, M.: Adaptive finite element methods for computing nonstationary incompressible flows. Ph.D. thesis, University of Heidelberg (2009)
4. Besier, M., Rannacher, R.: Goal-oriented space-time adaptivity in the finite element galerkin method for the computation of nonstationary incompressible flow. Int. J. Numer. Methods Fluids **70**, 1139–1166 (2012)
5. Blum, H., Rademacher, A., Schröder, A.: Space adaptive finite element methods for dynamic obstacle problems. Electron. Trans. Numer. Anal. **32**, 162–172 (2008)
6. Blum, H., Rademacher, A., Schröder, A.: Space adaptive finite element methods for dynamic signorini problems. Comput. Mech. **44**(4), 481–491 (2009)
7. Braack, M., Ern, A.: A posteriori control of modeling errors and discretization errors. Multiscale Model. Simul. **1**(2), 221–238 (2003)

8. Endtmayer, B., Langer, U., Wick, T.: Reliability and efficiency of DWR-type a posteriori error estimates with smart sensitivity weight recovering. Comput. Methods Appl. Math. **21**(2), 351 (2021)
9. Failer, L.: Optimal control of time-dependent nonlinear fluid-structure interaction. Ph.D. thesis, Technical University Munich (2017)
10. Failer, L., Wick, T.: Adaptive time-step control for nonlinear fluid-structure interaction. J. Comput. Phys. **366**, 448–477 (2018)
11. Köcher, U., Bruchhäuser, M.P., Bause, M.: Efficient and scalable data structures and algorithms for goal-oriented adaptivity of space–time FEM codes. SoftwareX **10**, 100239 (2019)
12. Rademacher, A.: Adaptive finite element methods for nonlinear hyperbolic problems of second order. Ph.D. thesis, Technische Universität Dortmund (2009)
13. Richter, T., Wick, T.: Variational localizations of the dual weighted residual estimator. J. Comput. Appl. Math. **279**(0), 192–208 (2015)
14. Schmich, M., Vexler, B.: Adaptivity with dynamic meshes for space-time finite element discretizations of parabolic equations. SIAM J. Sci. Comput. **30**(1), 369–393 (2008)
15. Thiele, J., Wick, T.: Space-time PU-DWR error control and adaptivity for the heat equation. Proc. Appl. Math. Mech. **21**(1), e202100174 (2021)

Eigenspectral Analysis of Preconditioners in an Adaptive Compressible Flow Solver

Zhen-Guo Yan, Yu Pan, Joaquim Peiró, and Spencer J. Sherwin

1 Introduction

There is a growing need for a better understanding of complex flow phenomena in areas such as turbo-machinery [5, 16], in which the flow phenomena are intrinsically unsteady, anisotropic and across multiple scales [1, 16]. To fulfill the demand for enhanced accuracy, high-fidelity methods are being developed rapidly in the computational fluid dynamics (CFD) community, which directly resolve the flow structures that are difficult to model [4]. High-order numerical methods, such as DG, are good candidates for high-fidelity simulations [5, 17]. However, improving efficiency is one of the key issues to be addressed before the DG methods can be widely used in industrial high-fidelity simulations [1, 14, 17].

One of the main obstacles limiting the efficiency of high-order simulations with explicit time integration methods is the severe time step restrictions [8]. In this case, developing implicit time integration methods to relax the time step restrictions has the potential to largely improve their efficiency. Researchers have studied different aspects in developing efficient implicit time integration methods including efficient temporal discretization schemes, approximations of the Jacobian matrix, and efficient linear system solvers [2, 3, 6, 9, 12, 13, 19]. Preconditioning is one of the factors that significantly affect the efficiency of implicit solvers. Preconditioners

Z.-G. Yan (✉)
State Key Laboratory of Aerodynamics, Mianyang, People's Republic of China

Department of Aeronautics, Imperial College London, London, UK
e-mail: zgyan@skla.cardc.cn; yanzhg@mail.ustc.edu.cn

Y. Pan · J. Peiró · S. J. Sherwin
Department of Aeronautics, Imperial College London, London, UK
e-mail: yu.pan16@imperial.ac.uk

J. M. Melenk et al. (eds.), *Spectral and High Order Methods for Partial Differential Equations ICOSAHOM 2020+1*, Lecture Notes in Computational Science and Engineering 137, https://doi.org/10.1007/978-3-031-20432-6_35

based on iterative methods, such as Jacobi and Gauss-Seidel iterative method, and factorizations of the Jacobian matrix, such as incomplete LU factorization, have been developed [2, 3]. Alternatively, a p-multigrid solver can be used as an efficient linear solver [10].

Unfortunately, there is not a general answer to assess the performance of a preconditioner [11]. For symmetric systems, it can be proved that the convergence rate is linked with the eigenvalue distributions [7]. In another word, a system with preconditioned matrices of only a few tightly clustered eigenvalues is typically easier to solve. However, when solving the nonsymmetric linear system, there is no universal theory to guide the design of a good preconditioner [11]. If the preconditioned matrix is not far from normal, the cluster of eigenvalues could still be a good indicator of an efficient preconditioning strategy. Other spectral properties like condition number can also indicate the stiffness of a preconditioned linear equation system.

To study the feasibility of eigenspectral analysis in assessing preconditioners and compare the effects of different preconditioners, eigenspectral analysis is performed for preconditioners based on Jacobi iteration, Gauss-Seidel iteration, ILU factorization, and p-multigrid. The study is conducted based on an implicit-in-time compressible flow solver with an adaptive time stepping that balances the different discretization errors [13, 19]. The studies based on a low Reynolds number lid-driven flow show that the eigenspectral analysis is still a tool that can give qualitative guidance in assessing preconditioning effects. In addition, a comparison of the different preconditioners is conducted.

A brief introduction of the adaptive compressible flow solver is presented in Sect. 2. Then, eigenspectral analysis of different preconditioners is conducted in Sect. 3. At last, concluding remarks are summarized in Sect. 4.

2 Adaptive Compressible Flow Solver

The non-dimensional governing compressible Navier-Stokes equations, representing conservation of mass, momentum, and energy, can be written in an abridged form as

$$\frac{\partial \mathbf{U}}{\partial t} = L\left(\mathbf{U}\right) = -\boldsymbol{\nabla} \cdot \mathbf{H}\,; \qquad \mathbf{H}_i = \mathbf{F}_i(\mathbf{U}) - \mathbf{G}_i(\mathbf{U}, \nabla \mathbf{U}), \, , i = 1, 2, \ldots, d, \tag{1}$$

where L is the analytical nonlinear spatial operator, $\mathbf{U} = [\rho, \rho u_1 \ldots \rho u_d, E]^T$ is the vector of conservative variables, d is the number of dimensions of the problem. $\mathbf{F} = \mathbf{F}(\mathbf{U})$ represents the inviscid flux and $\mathbf{G} = \mathbf{G}(\mathbf{U}, \nabla\mathbf{U})$ represents the viscous flux, which can be found in detail in [13, 19].

Section 2.1 presents the spatial discretization methods. The implicit time integration method is presented in Sect. 2.2. Finally, a balanced time stepping method is presented in Sect. 2.3.

2.1 Spatial Discretization

The spatial operator in Eq. (1) is discretized using the DG method, in which the computational domain Ω is divided into N_e non-overlapping elements, Ω_e. The space of test functions is defined as piecewise polynomial spaces of degree P on each element. The weak form of Eq. (1) is obtained by multiplying it by the test function ϕ_p and performing integration by parts,

$$\int_{\Omega_e} \frac{\partial \mathbf{U}}{\partial t}\phi_p d\Omega_e = \int_{\Omega_e} \nabla\phi_p \cdot \mathbf{H} d\Omega_e - \int_{\Gamma_e} \phi_p \mathbf{H}^n d\Gamma_e, \tag{2}$$

where Γ_e is the element boundary, $\mathbf{H}^n = \mathbf{F}^n - \mathbf{G}^n$, $\mathbf{F}^n = \mathbf{F}\cdot\mathbf{n}$, and $\mathbf{G}^n = \mathbf{G}\cdot\mathbf{n}$ are boundary fluxes on the elemental outward normal direction ($\mathbf{n}$).

The vector of conservative variables $\mathbf{U}$ is approximated in the same polynomial space as the test functions, where $\mathbf{u}_q(t)$ is the qth coefficient of the basis (or trial) function $\phi_q(\mathbf{x})$, and N_P is the total degrees of freedom (DoFs) in the element. The flow variable values on the quadrature points are calculated using a backward transformation with $\mathbf{B}$ the backward transformation matrix. A semi-discrete equation for the vector $\mathbf{U}_\delta$ can be written as

$$\frac{\partial \mathbf{U}_\delta}{\partial t} = L_\delta(\mathbf{U}_\delta), \tag{3}$$

where

$$L_\delta(\mathbf{U}_\delta(\mathbf{u})) = \mathbf{B}\mathcal{L}_\delta(\mathbf{u}), \mathbf{U}_\delta = \mathbf{B}\mathbf{u}. \tag{4}$$

The weak DG scheme for the advection term is complete as long as a Riemann numerical flux is used to calculate the normal flux. Currently, the implicit solver is based on the compact IP (interior penalty) method since it is easier to derive its Jacobian matrix. The interior penalty methods can be found in detail in references [13, 19].

2.2 Implicit Time Integration

The time derivative is approximated using an explicit first stage singly diagonal implicit Runge-Kutta (ESDIRK) method [13]. The implicit stages of the ESDIRK methods can be written as a nonlinear system

$$\mathbf{N}(\mathbf{u}^{(i)}) = \mathbf{u}^{(i)} - \mathbf{s}^{(i)} - \Delta t a_{ii}\mathcal{L}_\delta\left(\mathbf{u}^{(i)}\right) = \mathbf{0}, \tag{5}$$

where $\mathbf{s}$ is a source term based on information from the previous stages, a_{ii} is a coefficient of the ESDIRK scheme and Δt is the time step. This system is iteratively solved by the Newton's method [9]. When the Newton residual is smaller than a convergence tolerance of the Newton iterations τ, i.e.

$$\|\mathbf{N}(\mathbf{v}^k)\| \leq \tau, \tag{6}$$

$\mathbf{u}_{it}^{(i)} = \mathbf{v}^k$ is regarded as the approximate solution of the nonlinear system. A proper choice of τ is important because it determines the magnitude of the iterative error introduced by the iterative solver [12].

2.3 Balanced Adaptive Time Stepping

The leading term of the spatial truncation and temporal errors can be expressed as

$$\mathbf{E}_s(\mathbf{U}_\delta) = L_\delta(\mathbf{U}_\delta) - L(\mathbf{U}) = C_s D_s(\mathbf{U}) h^{P+1} + O(h^{P+2}), \tag{7}$$

$$\mathbf{E}_t^{n+1} = \mathbf{U}^{n+1} - \hat{\mathbf{U}}^{n+1} = C_t D_t(\mathbf{U}) \Delta t^{N+1} + O(\Delta t^{N+2}), \tag{8}$$

where C_s and C_t are coefficients related to the discretization scheme but independent of $\mathbf{U}$, $D_s(\mathbf{U})$ and $D_t(\mathbf{U})$ are spatial and temporal derivative terms, and P and N are the polynomial order of the DG basis functions and order of accuracy of the ESDIRK, respectively.

As discussed in reference [13], the leading term of the error generated within one time step is

$$\begin{aligned} \mathbf{U}_{it}^{n+1} - \mathbf{U}(t^{n+1}) \approx & C_t D_t(\mathbf{U})(\Delta t)^{N+1} + \Delta t C_s \bar{D}_s(\mathbf{U}) h^{P+1} + \Delta t \bar{\mathbf{E}}_{it} \\ = & \mathbf{E}_t^{n+1} + \Delta t \bar{\mathbf{E}}_s + \Delta t \bar{\mathbf{E}}_{it} \\ = & \mathbf{B}\left(\mathbf{e}_t^{n+1} + \Delta t \bar{\mathbf{e}}_s + \Delta t \bar{\mathbf{e}}_{it}\right), \end{aligned} \tag{9}$$

where we have used the overbar to denote summations of the form $\bar{\psi} = \sum_{i=1}^{S} b_i \psi^{(i)}$, which is a weighted average of ψ because $\sum_{i=1}^{S} b_i = 1$.

An adaptive time stepping strategy is developed based on the idea of balancing different errors. The time step and Newton tolerance are adapted so that the spatial error is dominating and further decreasing the temporal and iteration errors will not obviously improve the results. The adaptive time step Δt^{n+1} and Newton tolerance τ are calculated by

$$\Delta t_{e,m}^{n+1} = \Delta t^n \left(\frac{\beta \Delta t \|\bar{\mathbf{E}}_s\|_{e,m} + 1.5^N \epsilon_m}{\|\mathbf{E}_t\|_{e,m} + \epsilon_m} \right)^{1/N}, \epsilon_m = 10^{-12} \mathbf{U}_{it,m}^{\text{rms}}, \tag{10}$$

$$\Delta t^{n+1} = \min(\Delta t_m^{n+1}),\ \Delta t_m^{n+1} = \sum_{e=1}^{N_e} \|\mathbf{E}_t\|_{e,m}^r \Delta t_{e,m}^{n+1} / \sum_{e=1}^{N_e} \|\mathbf{E}_t\|_{e,m}^r, \tag{11}$$

$$\tau = \eta \|\mathbf{e}_t\|,\ \eta = 0.1. \tag{12}$$

As shown in our previous work [13], the proposed balanced adaptive time stepping maintains temporal accuracy while is close to the most efficient parameter choice in all the cases tested. It shows strong universality in unsteady/steady, under-resolved/well-resolved problems tested without the need of tuning the parameters.

3 Eigenspectral Analysis of Different Preconditioners

The linear problem in each Newton iteration is solved using a restarted generalized minimal residual method (GMRES) [15]. Instead of solving the linear system directly, one can solve instead the preconditioned linear system

$$\left(\frac{\partial \mathbf{N}}{\partial \mathbf{u}}\mathbf{P}^{-1}\right)\left(\mathbf{P} \bigtriangleup \bar{\mathbf{u}}^l\right) = -\mathbf{N}\left(\bar{\mathbf{u}}^l\right), \tag{13}$$

where $\mathbf{P}$ is the preconditioning matrix.

The preconditioners studied in this section are the block Jacobi iterative preconditioner (BJac), block Gauss-Seidel iterative preconditioner (BGS), block ILU preconditioner (BILU), and p-multigrid preconditioner. The generation and usage of these preconditioners are based on the numerically generated Jacobian matrix using finite difference approximations. This eigenspectral analysis is tested in a small-scale Lid-driven flow to help us understand the performance of preconditioners.

3.1 Different Preconditioners

A standard linear equation system is denoted as

$$\mathbf{A}\mathbf{y} = \mathbf{c}. \tag{14}$$

The matrix $\mathbf{A}$ is split into several parts

$$\mathbf{A} = \mathbf{L} + \mathbf{D} + \mathbf{U}, \tag{15}$$

where $\mathbf{L}$, $\mathbf{D}$ and $\mathbf{U}$ are the block lower triangular part, block diagonal part and block upper triangular part of $\mathbf{A}$.

Block Jacobi Iterative Preconditioner

The block Jacobi iterative preconditioner is implemented in an iterative way

$$\mathbf{y}^z = \mathbf{D}^{-1}(\mathbf{c} - (\mathbf{L} + \mathbf{U})\mathbf{y}^{z-1}),\ z = 1, 2 \ldots, Z \tag{16}$$

Block Gauss-Seidel Iterative Preconditioner

Different from the Jacobi iterative method, a Gauss-Seidel iterative method is adopted here.

$$\mathbf{y}^z = \mathbf{D}^{-1}(\mathbf{c} - \mathbf{L}\mathbf{y}^z - \mathbf{U}\mathbf{y}^{z-1}),\ z = 1, 2 \ldots, Z \tag{17}$$

Block ILU Preconditioner

The BILU preconditioner performs an incomplete LU factorization to the coefficient matrix **A** and uses it as a preconditioner. The process of the zero fill-in ILU factorization can be expressed as follows.

For $i = 2, \ldots, N_e$
    For $k = 1, \ldots, i - 1$ and $(i, k) \in S(\mathbf{A})$
    Calculate the block matrix $\mathbf{A}_{i,k} = \mathbf{A}_{i,k}/\mathbf{A}_{k,k}$
    For $j = k + 1, \ldots, N_e$ and $(i, j) \in S(\mathbf{A})$
        Calculate the block matrix $\mathbf{A}_{i,j} = \mathbf{A}_{i,j} - \mathbf{A}_{i,k} \times \mathbf{A}_{k,j}$
    End
  End
End

Block *p*-Multigrid Preconditioner

The process can be summarized as 'smooth the high-frequency error in the high-order system and solve the equation in the lower-order system'. The first step is to pre-smooth the solutions at the current level using smoothers like BJac or BGS.

$$\mathbf{y}_1^{Z_1} = \mathbf{D}^{-1}(\mathbf{c} - (\mathbf{L} + \mathbf{U})\mathbf{y}_1^{Z_1 - 1}). \tag{18}$$

Obtaining the solution after Z_1 iterations, the equation residual vector $\mathbf{r} = \mathbf{c} - \mathbf{A}\mathbf{y}_1^{Z_1}$ at a high-level system is projected to a lower-level resolution

(continued)

$$\mathbf{r}^l = \tilde{\mathbf{R}}(\mathbf{c} - \mathbf{A}\mathbf{y}_1^{Z_1}), \tag{19}$$

where the restriction matrix $\tilde{\mathbf{R}}$ is a projection matrix to map the residual to the lower level. The superscript l denotes the lower level.

The defect equation at the lower-level system is resolved to obtain the corrected error $\mathbf{e}^l$.

$$\mathbf{A}^l\mathbf{e}^l = \mathbf{r}^l, \tag{20}$$

where $\mathbf{A}^l$ is the operation matrix for the defect equation. It is related to the projection matrices $\tilde{\mathbf{R}}$ and $\tilde{\mathbf{P}}$ (mentioned below).

The error is then interpolated to correct high-level approximations $\mathbf{y}_1^{Z_1}$

$$\mathbf{y}_2 = \mathbf{y}_1^{Z_1} + \tilde{\mathbf{P}}\mathbf{e}^l, \tag{21}$$

where the prolongation matrix $\tilde{\mathbf{P}}$ is a projection matrix to map the error back to the high-level system.

3.2 Eigenspectral Analysis

The lid-driven cavity problem is calculated in a 2D square domain, where all the boundaries are walls. The top wall moves at u=1m/s in x-direction while the other three walls are stationary. The simulations are run at a low Reynolds number of Re $= 5$ as shown in Fig. 1. The eigenspectral analysis is based on a flow field, in which the residual $L_\delta(\mathbf{U}_\delta)$ is reduced by a factor of 10^{-9}.

To do eigenspectral analysis, we compare two properties, the condition number and the eigenvalue distributions. The condition number is defined as the ratio between the maximum and the minimum eigenvalues. The condition number of unpreconditioned Jacobian matrix is 2.794×10^3 and the eigenvalue distributions are shown in Fig. 1. The simulation is hard to converge without preconditioning [18].

In the first test, we compare the performance of different BJac preconditioners. The number in the bracket represents the number of smoothing iterations. The simulation using BJac(1) does not converge, the result of which is not shown here. Figure 2a demonstrates that many eigenvalues of BJac(1) are distributed around the circle with its center at (1,0) and radius=1. The eigenvalues of BJac(3), and BJac(5) tend to cluster in a smaller circle and are distributed near the center. The condition number after BJac(1), BJac(3), and BJac(5) preconditioning are 107, 35, and 21, respectively. Lastly, the simulations with BJac(3) and BJac(5) take 187 and 181 GMRES iterations to converge. The changes in eigenvalue distributions and condition numbers are consistent with the improvement of convergence after preconditioning, which indicates the eigenspectral analysis correctly reflects the preconditioning effects of different preconditioners in this case.

The second test compares the performance of different BGS preconditioners. The simulation using BGS(1) also does not converge. Different from the eigenvalue distributions of BJac preconditioners, Fig. 2b shows the eigenvalues of BGS preconditioned matrices are asymmetrically distributed in this test case. The condition number after BGS(1), BGS(3), and BGS(5) preconditioning are 47, 15, and 8, respectively, which are smaller than their counterparts of BJac preconditioners. Lastly, the simulations with BGS(3) and BGS(5) take 174 and 127 GMRES iterations to converge, which are also consistent with the changes in eigenspectral properties.

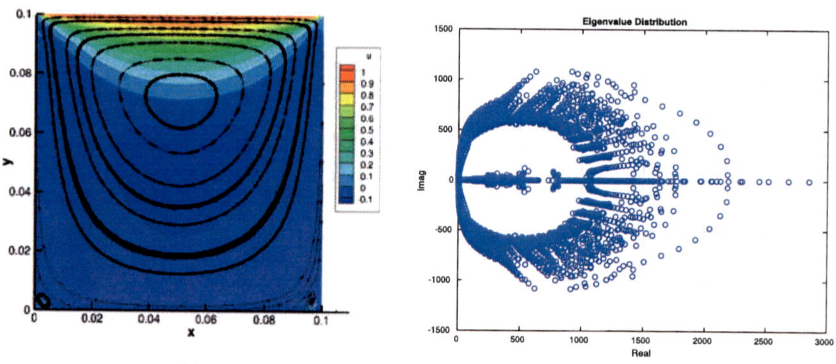

(a) (b)

Fig. 1 Lid-driven cavity flow: Steady-state field (Re = 5) and eigenvalue distributions of the unpreconditioned Jacobian matrix. (**a**) Field. (**b**) Eigenvalue distributions

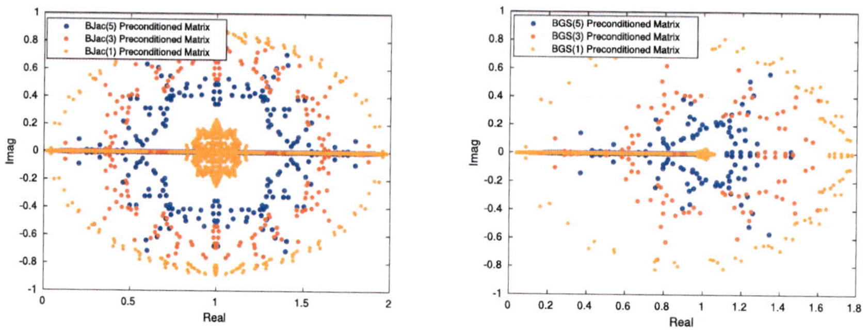

(a) (b)

Fig. 2 Eigenvalue distributions of the block Jacobi and Gauss-Seidel preconditioned matrix. (**a**) BJac. (**b**) BGS

The third test compares the performance of p-multigrid preconditioners of different levels. The smoothers for the p-multigrid preconditioners are the BGS solvers. The simulations using p-multigrid of Level2(3), Level3(5) and Level4(7) take 174, 129 and 101 GMRES iterations to converge, in which the number in the bracket represents the total iteration number and 'Level' represents the number of levels for a multi-level p-multigrid preconditioner. For example, Level3(5) is a 3 levels p-multigrid preconditioner, where the highest level resolution is at $P = 3$, the second level is at $P = 2$ and the lowest level is at $P = 1$. The number 5 in Level3(5) means there are in total 5 smoothing iterations, consisting of 1 pre-smooth in the highest level, 1 pre-smooth in the second level, 1 smoothing in the lowest level, 1 post-smooth in the second level, 1 post-smooth in the highest level. Compared with the performance BGS(5) preconditioner using 127 GMRES iterations, Level3(5)'s performance is a little inferior to it with 129 iterations. The smoother performance at a low level still needs to be optimized to achieve the same performance as high-level smoother. But the low-level resolution saves computational costs. Lastly, the condition numbers after Level2, Level3 and Level4 p-multigrid preconditioning are 15, 8 and 6.

In the fourth test, the simulation using BILU takes 139 GMRES iterations to converge. The condition number after BILU preconditioning is 31. The eigenvalue distributions of the Jacobian matrix after BILU preconditioning are not compared separately. Figure 3b compares the distributions of the BILU preconditioner with other types of preconditioners to demonstrate some interesting details.

To summarize, for the same type of preconditioner, when we apply more iterations on the relevant preconditioners such as BJac, and BGS, the system converges faster. The clustering of eigenvalues is consistent with better convergence property as expected. However, for different preconditioners, they cluster in different directions and there is no obvious rule between different types of preconditioners. When the smoothing number is the same, BGS performs better

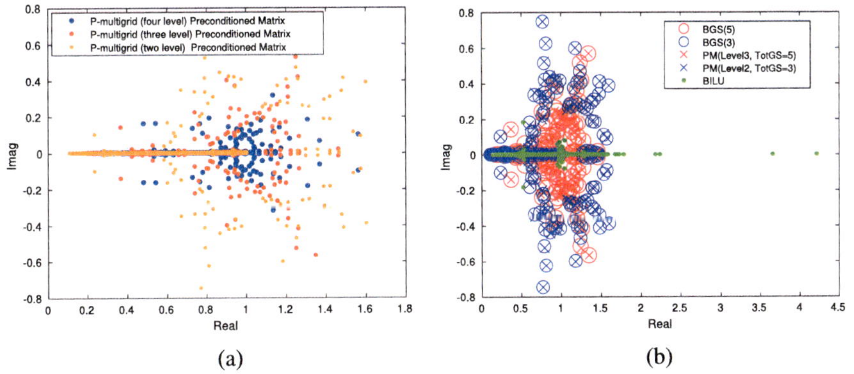

Fig. 3 Eigenvalue distributions of the BILU preconditioned matrix and comparison of different preconditioners. (**a**) p-multigrid. (**b**) Comparisons

than BJac. Because in each smoothing iteration, BGS can influence more elements and smooth errors in a wider computational domain. The BILU preconditioner does not perform well compared with BGS(3) in this case. Comparing Figs. 2 and 3a, the eigenvalue distributions of p-multigrid are quite similar to the linear solver used for smoothing in each level. To be clear, as shown in Fig. 3b, when p-multigrid preconditioner and BGS use the same number of iterations, their eigenvalue distributions are nearly the same. The condition number after preconditioning is even the same in this test. For example, the value of condition number after BGS(5) and Level3(5) preconditioning are both 8. The eigenspectral properties are very similar to the p-multigrid preconditioner and it's smoother. It, therefore, hints that an efficient smoother plays an important role in p-multigrid preconditioner. Most importantly, this property indicates that even though p-multigrid preconditioner does not obviously improve the convergence property compared with its original smoothing method (BGS) in this test when the total smoothing number is the same, the computational cost can be potentially reduced since part of smoothing iterations of p-multigrid are at lower levels.

Extra eigenspectral analysis based on the converged field of the Re = 100 Lid-driven case has been conducted and similar conclusions can be verified.

3.3 *Numerical Tests*

The preconditioned implicit solver is compared to an explicit solver through cases of 2D flow over a circular cylinder with M = 0.3 at two different Reynolds numbers, Re = 1200 and Re = 40. A second-order DG is adopted. The implicit solver utilizes a third-order DIRK (DIRK3) scheme with BJac(7) preconditioner, while a third-order Runge-Kutta (RK3) scheme is adopted for the explicit solver.

In the case with Re = 1200, a St number of 0.243 is obtained. Time steps of $\Delta t = 5 \times 10^{-2}$ and $\Delta t = 5 \times 10^{-5}$ are used for the implicit and explicit solvers, respectively. The corresponding computational CPU time is 5323 and 83 seconds for the explicit and implicit methods, respectively. The implicit solver is around 64 times faster than the explicit one. Figure 4a shows the variation of density at the history point $(2r, 0)$ within one period.

In the case with Re = 40, the implicit solver and the explicit solver are run using their respective largest time step $\Delta t = 10^{-2}$ and $\Delta t = 10^{-7}$, with equally CFL = 100 and CFL = 0.001. The history of residual is shown in Fig. 4b. Compared with the explicit solver, the implicit solver speeds up by approximately two orders of magnitude.

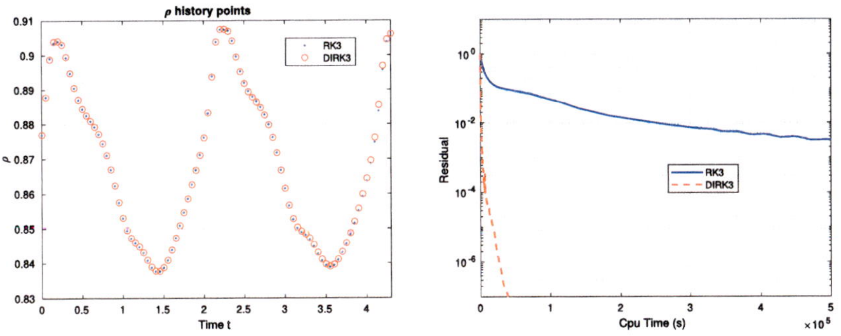

(a) (b)

Fig. 4 Comparison of implicit and explicit solver in 2D flow over a circular cylinder. (**a**) ρ variation within one period for Re = 1200. (**b**) Residual history for Re = 40

4 Conclusion

To further speed up the efficiency of an implicit-in-time compressible DG solver with adaptive time stepping, preconditioners based on Jacobi iteration, Gauss-Seidel iteration, incomplete LU factorization, and p-multigrid have been compared using eigenspectral analysis and numerical experiments. The results show that the condition number and clustering of eigenvalues can correctly reflect the preconditioning effects and can give qualitative guidance in assessing different preconditioners. The BILU preconditioner shows better performance in clustering the eigenvalues. Different from h-multigrid, which accelerates simulations by promoting information exchange and reducing the total iteration needed, the p-multigrid method has no obvious effects in reducing the iteration (smoothing) number. However, it still has the potential to accelerate simulations by reducing iteration costs.

Acknowledgments This study was supported by the National Natural Science Foundation of China (Grant Nos.11902344, 12172375), the National Numerical Wind-tunnel project, the foundation of the State Key Laboratory of Aerodynamics (Grant No. SKLA2019010101). The study has been supported by EPSRC grant (EP/R029423/1) and UK Turbulence Consortium grant (EP/R029326/1). We would like to gratefully acknowledge access to the computing facilities provided by the Imperial College Research Computing Service (DOI: https://doi.org/10.14469/hpc/2232).

References

1. Brunet, V., Croner, E., Minot, A., Laborderie, J.D., Lippinois, E., Richard, S., Boussuge, J.F., Dombard, J., Duchaine, F., Gicquel, L., Poinsot, T., Puigt, G., Staffelbach, G., Segui, L., Vermorel, O., Villedieu, N., Cagnone, J.S., Hillewaert, K., Rasquin, M., Lartigue, G., Moureau, V., Couaillier, V., Martin, E., Plata, M.d.l.L., Gouez, J.M.L., Renac, F.: Comparison of various

CFD codes for LES simulations of turbomachinery: from inviscid vortex convection to multistage compressor. In: ASME Turbo Expo 2018: Turbomachinery Technical Conference and Exposition, pp. V02CT42A013–V02CT42A013. American Society of Mechanical Engineers, New York (2018)
2. Diosady, L., Darmofal, D.: Discontinuous Galerkin solutions of the Navier-Stokes equations using linear multigrid preconditioning. In: 18th AIAA Computational Fluid Dynamics Conference, p. 3942 (2007)
3. Ghai, A., Lu, C., Jiao, X.: A comparison of preconditioned Krylov subspace methods for large-scale nonsymmetric linear systems. Numer. Linear Algebra Appl. **26**(1), e2215 (2019)
4. Grinstein, F.F., Margolin, L.G., Rider, W.J.: Implicit Large Eddy Simulation: Computing Turbulent Fluid Dynamics. Cambridge University Press, Cambridge (2007)
5. Hillewaert, K.: Development of the discontinuous Galerkin method for high-resolution, large scale CFD and acoustics in industrial geometries. Ph.D. Thesis, Université Catholique de Louvain (2013). Google-Books-ID: IBJij3jtUCQC
6. Hillewaert, K.: Development of the discontinuous Galerkin method for high-resolution, large scale CFD and acoustics in industrial geometries. Presses Universitaires de Louvain, Louvain-la-Neuve (2013)
7. Hogben, L.: Handbook of Linear Algebra. CRC Press, Boca Raton (2006)
8. Karniadakis, G., Sherwin, S.: Spectral/hp Element Methods for Computational Fluid Dynamics, 2nd edn. Oxford Science Publications, Oxford (2013)
9. Knoll, D.A., Keyes, D.E.: Jacobian-free Newton-Krylov methods: a survey of approaches and applications. J. Comput. Phys. **193**(2), 357–397 (2004)
10. Luo, H., Baum, J.D., Löhner, R.: A p-multigrid discontinuous Galerkin method for the Euler equations on unstructured grids. J. Comput. Phys. **211**(2), 767–783 (2006)
11. Nachtigal, N.M., Reddy, S.C., Trefethen, L.N.: How fast are nonsymmetric matrix iterations? SIAM J. Matrix Anal. Appl. **13**(3), 778–795 (1992)
12. Noventa, G., Massa, F., Rebay, S., Bassi, F., Ghidoni, A.: Robustness and efficiency of an implicit time-adaptive discontinuous Galerkin solver for unsteady flows. Comput. Fluids **204**, 104529 (2020)
13. Pan, Y., Yan, Z.G., Peiró, J., Sherwin, S.J.: Development of a balanced adaptive time-stepping strategy based on an implicit JFNK-DG compressible flow solver. Commun. Appl. Math. Comput. **4**, 728–757 (2022)
14. Persson, P.O.: Scalable parallel Newton-Krylov solvers for discontinuous Galerkin discretizations. In: 47th AIAA Aerospace Sciences Meeting including The New Horizons Forum and Aerospace Exposition, p. 606. Denver, Colorado (2009)
15. Saad, Y., Schultz, M.H.: GMRES: a generalized minimal residual algorithm for solving nonsymmetric linear systems. SIAM J. Sci. Stat. Comput. **7**(3), 856–869 (1986)
16. Tucker, P.G.: Progress in aerospace sciences computation of unsteady turbomachinery flows: part 1 progress and challenges. Progress Aerosp. Sci. **47**(7), 522–545 (2011)
17. Wang, Z.J., Fidkowski, K., Abgrall, R., Bassi, F., Caraeni, D., Cary, A., Deconinck, H., Hartmann, R., Hillewaert, K., Huynh, H.T., Kroll, N., May, G., Persson, P.O., van Leer, B., Visbal, M.: High-order CFD methods: current status and perspective. Int. J. Numer. Methods Fluids **72**, 811–845 (2013)
18. Wathen, A.J.: Preconditioning. Acta Numer. **24**, 329–376 (2015)
19. Yan, Z.G., Pan, Y., Castiglioni, G., Hillewaert, K., Peiró, J., Moxey, D., Sherwin, S.J.: Nektar++: design and implementation of an implicit, spectral/hp element, compressible flow solver using a Jacobian-free Newton Krylov approach. Comput. Math. Appl. **81**, 351–372 (2021)

Comparative Study on a Variety of Structure-Preserving High Order Spatial Discretizations with the Entropy Split Methods for MHD

H. C. Yee and Björn Sjögreen

1 Introduction

Temporal and spatially finite discretizations that conserve fundamental properties of the chosen governing equations are of vital importance for reliable and accurate simulations of fluid flows as not all standard finite discretizations conserve fundamental properties of the governing equations. This type of temporal, spatial, and/or both finite discretizations that conserve certain physical property (properties) of the governing equations, are most often referred to, in a broad sense, as structure-preserving numerical methods. See e.g., [18]. In the applied mathematics circle for computational fluid dynamics (CFD) with mathematical approaches, entropy conserving and stable methods have been flourishing emerging research for the last two decades [10, 14, 28, 30–32]. On the other hand, in the applied CFD circle, the importance of developing structure-preserving finite discretization methods that conserve some or all fundamental properties of the governing equations (mass, momentum and physical entropy conserving, positivity of density and pressure, and kinetic energy preserving, etc.) have been on-going for the last 40 years. See e.g., [1, 3, 8, 11, 14, 25, 28–30, 41]. The present study concentrates on high order spatial structure-preserving numerical methods that are entropy conserving, momentum conserving or kinetic energy preserving, and/or a combination of two of these properties. The majority of these high order methods belong to the class of skew-symmetric splitting of the inviscid flux derivatives. In particular, the comparative studies concentrate on

H. C. Yee (✉)
NASA Ames Research Center, Moffett Field, CA, USA
e-mail: Helen.M.Yee@nasa.gov

B. Sjögreen
Multid Analyses AB, Gothenburg, Sweden
e-mail: bjorn.sjogreen@multid.se

J. M. Melenk et al. (eds.), *Spectral and High Order Methods for Partial Differential Equations ICOSAHOM 2020+1*, Lecture Notes in Computational Science and Engineering 137, https://doi.org/10.1007/978-3-031-20432-6_36

Tadmor-type of discrete entropy conserving [32], momentum conserving [8], kinetic energy preserving [7, 15], Yee et al., Sjogreen and Yee [21, 25, 28–30, 39, 41] and their combination of these physical-preserving methods. All of these considered methods are not only preserve certain physical properties of the chosen governing equations but are also known to either improve numerical stability, and/or minimize aliasing errors in long time integration in turbulent flow simulations without the aid of added numerical dissipation for selected flow types. Temporal structure-preserving high order time discretization is beyond the scope of this investigation. The comparison assumed the fourth-order Runge-Kutta time-discretizations using fine enough time steps to not interfere with the comparative study.

Sjögreen and Yee [28, 30] proved that the high order entropy split methods based on Harten's entropy function [14] are entropy stable for central differencing with SBP operators for both periodic and non-periodic boundary conditions for nonlinear thermally-perfect gas dynamics Euler equations. Harten's entropy function also can be part of the Tadmor-type entropy conserving method [25, 28]. The objective of this paper is to use the same Harten entropy function for the ideal MHD governing equation set and examine its performance. Test cases are included with comparative studies among other structure-preserving methods. For the Brio-Wu challenging test case [4], contrary to previous published work using low order diffusive shock-capturing methods to remove highly oscillatory computations, it is noted that here we were able to use very high order methods with Roe-type of approximate Riemann solver to obtain highly accurate stable solutions.

Due to a space limitation, the derivation part of the mathematical formulations are omitted. An expanded version of this short article with more detail formulation and extensive comparison with grid refinement can be found in [40].

2 The Symmetrizable Ideal MHD Equations [12, 26, 35]

$$\mathbf{q}_t + \mathbf{f}_x + \mathbf{g}_y + \mathbf{h}_z = \mathbf{e}\,\mathrm{div}\,\mathbf{B} \tag{1}$$

where the conserved variables are $\mathbf{q} = (\rho\ \rho u\ \rho v\ \rho w\ e\ B_1\ B_2\ B_3)^T$. The magnetic field components in the x-, y-, and z-directions are denoted by B_1, B_2, and B_3.

The vector in the right hand side is $\mathbf{e} = (0\ B_1\ B_2\ B_3\ \mathbf{u}^T\mathbf{B}\ u\ v\ w)^T$, where we denote $\mathbf{u} = (u\ v\ w)$ and $\mathbf{B} = (B_1\ B_2\ B_3)$. The flux in the $(k_1\ k_2\ k_3)$ direction is given by

$$\hat{f} = \begin{pmatrix} \rho\hat{u} \\ \rho u\hat{u} + k_1(p + \frac{1}{2}|\mathbf{B}|^2) - \hat{B}B_1 \\ \rho v\hat{u} + k_2(p + \frac{1}{2}|\mathbf{B}|^2) - \hat{B}B_2 \\ \rho w\hat{u} + k_3(p + \frac{1}{2}|\mathbf{B}|^2) - \hat{B}B_3 \\ \hat{u}(e + p + \frac{1}{2}|\mathbf{B}|^2) - \hat{B}\mathbf{u}^T\mathbf{B} \\ \hat{u}B_1 - \hat{B}u \\ \hat{u}B_2 - \hat{B}v \\ \hat{u}B_3 - \hat{B}w \end{pmatrix}$$

where $\hat{u} = k_1 u + k_2 v + k_3 w$ and $\hat{B} = k_1 B_1 + k_2 B_2 + k_3 B_3$. The pressure is given by

$$p = (\gamma - 1)(e - \frac{1}{2}\rho|\mathbf{u}|^2 - \frac{1}{2}|\mathbf{B}|^2) \tag{2}$$

An entropy is a convex function $E = E(\mathbf{q})$ such that smooth solutions of the Euler equations for gas dynamics satisfy the additional scalar conservation law

$$E_t + F_x + G_y + H_z = 0 \tag{3}$$

where the entropy fluxes, F, G, and H are related to the Euler fluxes. In the $x-direction$, the entropy flux is related to f_x by $E_{\mathbf{q}}\mathbf{f}_x = F_x$. Similarly for the y- and z-directions. The entropy variables are defined by

$$\mathbf{v} = E_{\mathbf{q}}(\mathbf{q}),$$

which is a well-defined change of variables, $\mathbf{v} = \mathbf{v}(\mathbf{q})$, due to the convexity of $E(\mathbf{q})$.

There are two entropies for the Euler equations that are used in entropy based numerical methods. Harten[14] considered the class of entropies

$$E_H = -\frac{\gamma + \alpha}{\gamma - 1}\rho(p\rho^{-\gamma})^{\frac{1}{\alpha+\gamma}}, \tag{4}$$

where α is a parameter. To ensure that E_H is convex, i.e., that the matrix $(E_H)_{\mathbf{q},\mathbf{q}}$ is positive definite, α is required to satisfy $\alpha > 0$ or $\alpha < -\gamma$. The logarithmic entropy by Tadmor [32],

$$E_L = -\rho \log(p\rho^{-\gamma}), \tag{5}$$

is another commonly used entropy. Entropy conserving (EC) schemes are numerical discretizations for which the semi-discrete counterpart of (3) holds.

The entropies (5) and (4) indicated above are valid entropies also for (1) (with p defined by (2)). Hence, the entropy conservation law (3) can be derived from (1), and the second derivatives, $E_{\mathbf{q},\mathbf{q}}$, is a positive definite matrix.

The kinetic energy, $e_k = \frac{1}{2}\rho(u^2 + v^2 + w^2)$, satisfies the equation

$$(e_k)_t + (ue_k)_x + (ve_k)_y + (we_k)_z + up_x + vp_y + wp_z = 0, \tag{6}$$

which can be derived from the form for the gas dynamics [40].

Kinetic energy preserving (KEP) discretizations are numerical discretization for which a semi-discrete counterpart of (6) holds. The exact form of the discrete (6) will be described shortly.

Semi-discretizations of (1) will be of the form

$$\frac{d\mathbf{q}_j}{dt} + \frac{\mathbf{h}_{j+1/2} - \mathbf{h}_{j-1/2}}{\Delta x} = \mathbf{e}_j D\hat{B}_j \tag{7}$$

where we consider two-point fluxes $\mathbf{h}_{j+1/2} = \mathbf{h}(\mathbf{q}_j, \mathbf{q}_{j+1})$, and D is the second order accurate $D\hat{B}_j = (\hat{B}_{j+1} - \hat{B}_{j-1})/(2\Delta x)$.

Method (7) conserves entropy if

$$(\mathbf{v}_{j+1} - \mathbf{v}_j)^T \mathbf{h}_{j+1/2} = \psi_{j+1} - \psi_j + \frac{1}{2}(\mathbf{v}_{j+1}^T \mathbf{e}_{j+1} + \mathbf{v}_j^T \mathbf{e}_j)(\hat{B}_{j+1} - \hat{B}_j). \tag{8}$$

See [10, 33].

Nomenclature for Spatial Discretization with Various Structure-Preserving Properties The considered structure-preserving properties with the following nomenclature for spatial discretizations are:

- EC: Entropy conserving using the Tadmor entropy function $E_L = -\rho \log(p\rho^{-\gamma})$ or the Harten class of entropy functions $E_H = -\frac{\gamma+\alpha}{\gamma-1} \rho(p\rho^{-\gamma})^{\frac{1}{\alpha+\gamma}}$ [14, 32].
- ECLOG: Tadmor-type entropy conserving method using the Tadmor entropy function.
- ES: Skew-symmetric splitting of the inviscid flux derivative that is entropy conserving and stable using the Harten entropy function [14] and the generalized energy norm with summation-by-parts (SBP) [27, 28, 41].
- DS: Momentum conserving Ducros et al. skew-symmetric split of the inviscid flux derivative [8].
- KGP: Kennedy-Gruber-Pirozzoli (KGP) skew-symmetric splitting of the inviscid flux derivative that is kinetic energy preserving [7, 15, 19].
- ESDS: Entropy split (ES) with Ducros et al. splitting [28].
- ESSW: Entropy split (ES) with Ducros et al. splitting but switch to regular central near discontinuities [30].
- ECLOGKP: Tadmor-type entropy conserving method using the Tadmor entropy function with Ranocha's kinetic energy preserving modification [20].
- ECHKP: Tadmor-type entropy conserving using the Harten entropy function. It turned out that this method in its base form also satisfies Ranocha's kinetic energy preservation condition; so there is only one variant for this method [20, 25].
- DSKP: Ducros Split with KEP.

All of the ten methods are conservative methods except the entropy split methods ES, ESDS and ESSW. From our previous studies [28–30], EC and ECLOG, ECLOGKP are the most CPU intensive methods among the 10 methods. It is approximately twice the CPU per time step than the ES, ESDS and ESSW methods. DS is the least CPU intensive. The CPU comparison is based on one single processor and eight processors as well as the operations count per grid point and per time step.

2.1 ECLOG for MHD

This Tadmor-type entropy conserving method using the Tadmor entropy function conserves the entropy (5). The logarithmic average of the density is defined by $E_L = -\rho \log(p\rho^{-\gamma})$ The numerical flux vector is defined by

$$h^{(1)} = \rho^{ln}\{\hat{u}\} \tag{9}$$

$$h^{(2)} = \rho^{ln}\{\hat{u}\}\{u\} + k_1\left(\frac{\{\rho\}}{\{\rho/p\}} + \frac{1}{2}(\{B_1^2\} + \{B_2^2\} + \{B_3^2\})\right) - \{\hat{B}\}\{B_1\} \tag{10}$$

$$h^{(3)} = \rho^{ln}\{\hat{u}\}\{v\} + k_2\left(\frac{\{\rho\}}{\{\rho/p\}} + \frac{1}{2}(\{B_1^2\} + \{B_2^2\} + \{B_3^2\})\right) - \{\hat{B}\}\{B_2\} \tag{11}$$

$$h^{(4)} = \rho^{ln}\{\hat{u}\}\{w\} + k_3\left(\frac{\{\rho\}}{\{\rho/p\}} + \frac{1}{2}(\{B_1^2\} + \{B_2^2\} + \{B_3^2\})\right) - \{\hat{B}\}\{B_3\} \tag{12}$$

$$h^{(6)} = \frac{\{\rho\hat{u}/p\}}{\{\rho/p\}}\{B_1\} - \{\hat{B}\}\frac{\{\rho u/p\}}{\{\rho/p\}} \tag{13}$$

$$h^{(7)} = \frac{\{\rho\hat{u}/p\}}{\{\rho/p\}}\{B_2\} - \{\hat{B}\}\frac{\{\rho v/p\}}{\{\rho/p\}} \tag{14}$$

$$h^{(8)} = \frac{\{\rho\hat{u}/p\}}{\{\rho/p\}}\{B_3\} - \{\hat{B}\}\frac{\{\rho w/p\}}{\{\rho/p\}} \tag{15}$$

and

$$\begin{aligned} h^{(5)} = &\frac{1}{\gamma-1}\frac{\rho^{ln}}{(\rho/p)^{ln}}\{\hat{u}\} + \frac{\{\rho\}}{\{\rho/p\}}\{\hat{u}\} - \frac{1}{2}\rho^{ln}\{\hat{u}\}(\{u^2\} + \{v^2\} + \{w^2\}) \\ &+ \rho^{ln}\{\hat{u}\}(\{u\}^2 + \{v\}^2 + \{w\}^2) + \frac{\{\rho\hat{u}/p\}}{\{\rho/p\}}(\{B_1\}^2 + \{B_2\}^2 + \{B_3\}^2) \\ &- \{\hat{B}\}\left(\frac{\{\rho u/p\}}{\{\rho/p\}}\{B_1\} + \frac{\{\rho v/p\}}{\{\rho/p\}}\{B_2\} + \frac{\{\rho w/p\}}{\{\rho/p\}}\{B_3\}\right) \end{aligned} \tag{16}$$

The logarithmic average of ρ/p required in the energy flux is defined by with ρ/p replacing ρ.

2.2 ECLOGKP for MHD

This Tadmor-type entropy conserving method using the Tadmor entropy function E_L with Ranocha's kinetic energy preserving modification is entropy conserving and kinetic energy preserving.

The numerical flux vector is defined by

$$h^{(1)} = \rho^{ln}\{\hat{u}\} \tag{17}$$

$$h^{(2)} = \rho^{ln}\{\hat{u}\}\{u\} + k_1\left(\{p\} + \frac{1}{2}(\{B_1^2\} + \{B_2^2\} + \{B_3^2\})\right) - \{\hat{B}\}\{B_1\} \tag{18}$$

$$h^{(3)} = \rho^{ln}\{\hat{u}\}\{v\} + k_2\left(\{p\} + \frac{1}{2}(\{B_1^2\} + \{B_2^2\} + \{B_3^2\})\right) - \{\hat{B}\}\{B_2\} \tag{19}$$

$$h^{(4)} = \rho^{ln}\{\hat{u}\}\{w\} + k_3\left(\{p\} + \frac{1}{2}(\{B_1^2\} + \{B_2^2\} + \{B_3^2\})\right) - \{\hat{B}\}\{B_3\} \tag{20}$$

$$h^{(6)} = \frac{\{\rho\hat{u}/p\}}{\{\rho/p\}}\{B_1\} - \{\hat{B}\}\frac{\{\rho u/p\}}{\{\rho/p\}} \tag{21}$$

$$h^{(7)} = \frac{\{\rho\hat{u}/p\}}{\{\rho/p\}}\{B_2\} - \{\hat{B}\}\frac{\{\rho v/p\}}{\{\rho/p\}} \tag{22}$$

$$h^{(8)} = \frac{\{\rho\hat{u}/p\}}{\{\rho/p\}}\{B_3\} - \{\hat{B}\}\frac{\{\rho w/p\}}{\{\rho/p\}} \tag{23}$$

and

$$\begin{aligned} h^{(5)} = {} & \frac{1}{\gamma - 1}\frac{\rho^{ln}}{(\rho/p)^{ln}}\{\hat{u}\} + \{p\}\{\hat{u}\} - \frac{1}{2}\rho^{ln}\{\hat{u}\}(\{u^2\} + \{v^2\} + \{w^2\}) \\ & + \rho^{ln}\{\hat{u}\}(\{u\}^2 + \{v\}^2 + \{w\}^2) - \frac{1}{4}(p_{j+1} - p_j)(\hat{u}_{j+1} - \hat{u}_j) \\ & + \frac{\{\rho\hat{u}/p\}}{\{\rho/p\}}(\{B_1\}^2 + \{B_2\}^2 + \{B_3\}^2) \\ & - \{\hat{B}\}\left(\frac{\{\rho u/p\}}{\{\rho/p\}}\{B_1\} + \frac{\{\rho v/p\}}{\{\rho/p\}}\{B_2\} + \frac{\{\rho w/p\}}{\{\rho/p\}}\{B_3\}\right) \end{aligned} \tag{24}$$

The logarithmic average of ρ/p required in the energy flux is defined by with ρ/p replacing ρ.

2.3 ECHKP for MHD

This Tadmor-type entropy conserving method using the Harten entropy function E_H with Ranocha's kinetic energy preserving modification is entropy conserving and kinetic energy preserving. The numerical two-point flux is defined by

$$h^{(1)} = \frac{\{z\hat{u}\}}{\{z^{-\gamma/\alpha}\}\{p^{(1-\gamma-\alpha)/\alpha}\}_e}$$

$$h^{(2)} = h^{(1)}\{u\} + k_1\left(\{p\} + \frac{1}{2}(\{B_1^2\} + \{B_2^2\} + \{B_3^2\})\right) - \{\hat{B}\}\{B_1\}$$

$$h^{(3)} = h^{(1)}\{v\} + k_2\left(\{p\} + \frac{1}{2}(\{B_1^2\} + \{B_2^2\} + \{B_3^2\})\right) - \{\hat{B}\}\{B_2\}$$

$$h^{(4)} = h^{(1)}\{w\} + k_3\left(\{p\} + \frac{1}{2}(\{B_1^2\} + \{B_2^2\} + \{B_3^2\})\right) - \{\hat{B}\}\{B_3\}$$

$$h^{(6)} = \frac{\{z\hat{u}\}}{\{z\}}\{B_1\} - \frac{\{zu\}}{\{z\}}\{\hat{B}\}$$

$$h^{(7)} = \frac{\{z\hat{u}\}}{\{z\}}\{B_2\} - \frac{\{zv\}}{\{z\}}\{\hat{B}\}$$

$$h^{(8)} = \frac{\{z\hat{u}\}}{\{z\}}\{B_3\} - \frac{\{zw\}}{\{z\}}\{\hat{B}\}$$

$$\begin{aligned} h^{(5)} = h^{(1)}(&\frac{\gamma}{\gamma-1}\{p^{-(\gamma-1)/\alpha}\}\{z^{-\frac{\gamma+\alpha}{\alpha}}\}_e \\ &+ \{u\}^2 + \{v\}^2 + \{w\}^2 - \frac{1}{2}\{u^2+v^2+w^2\}) \\ &+ \frac{\{z\hat{u}\}}{\{z\}}\left(\{B_1\}^2 + \{B_2\}^2 + \{B_3\}^2\right) \\ &- \{\hat{B}\}\left(\frac{\{zu\}}{\{z\}}\{B_1\} + \frac{\{zv\}}{\{z\}}\{B_2\} + \frac{\{zw\}}{\{z\}}\{B_3\}\right) \end{aligned}$$

where z is defined by $z = \frac{\rho}{p}(p\rho^{-\gamma})^{\frac{1}{\alpha+\gamma}}$ The exponential average is defined by $\{q^\beta\}_e = (q_{j+1}^{\beta+1} - q_j^{\beta+1})/((\beta+1)(q_{j+1} - q_j))$

2.4 *Ducros Split (DS) for MHD*

The Ducros et al. skew-symmetric splitting of the inviscid flux derivative is a momentum conserving method for central spatial discretization. See Sj"ogreen and Yee [29] for the MHD formulation.

In [29], some different generalizations of the DS to MHD were developed and studied. The method used here, denoted DS2 in [29], is defined as follows.

$$h^{(1)} = \{\rho\}\{\hat{u}\} \tag{25}$$

$$h^{(2)} = \{\rho u\}\{\hat{u}\} + k_1\left(\{p\} + \frac{1}{2}(\{B_1^2\} + \{B_2^2\} + \{B_3^2\})\right) - \{\hat{B}\}\{B_1\} \tag{26}$$

$$h^{(3)} = \{\rho v\}\{\hat{u}\} + k_2\left(\{p\} + \frac{1}{2}(\{B_1^2\} + \{B_2^2\} + \{B_3^2\})\right) - \{\hat{B}\}\{B_2\} \tag{27}$$

$$h^{(4)} = \{\rho w\}\{\hat{u}\} + k_3\left(\{p\} + \frac{1}{2}(\{B_1^2\} + \{B_2^2\} + \{B_3^2\})\right) - \{\hat{B}\}\{B_3\} \tag{28}$$

$$h^{(5)} = \left(\{e+p\} + \frac{1}{2}(\{B_1^2\} + \{B_2^2\} + \{B_3^2\})\right)\{\hat{u}\}$$

$$- \{\hat{B}\}(\{uB_1\} + \{vB_2\} + \{wB_3\}) \tag{29}$$

$$h^{(6)} = \{\hat{u}\}\{B_1\} - \{u\}\{\hat{B}\} \tag{30}$$

$$h^{(7)} = \{\hat{u}\}\{B_2\} - \{v\}\{\hat{B}\} \tag{31}$$

$$h^{(8)} = \{\hat{u}\}\{B_3\} - \{w\}\{\hat{B}\} \tag{32}$$

2.5 Ducros Split with KEP for MHD (DSKP)

The numerical flux vector for this Ducros et al. skew-symmetric splitting with kinetic energy preservation for the MHD is defined by

$$h^{(1)} = \{\rho\hat{u}\} \tag{33}$$

$$h^{(2)} = \{\rho\hat{u}\}\{u\} + k_1\left(\{p\} + \frac{1}{2}(\{B_1^2\} + \{B_2^2\} + \{B_3^2\}\right) - \{B\}\{B_1\} \tag{34}$$

$$h^{(3)} = \{\rho\hat{u}\}\{v\} + k_2\left(\{p\} + \frac{1}{2}(\{B_1^2\} + \{B_2^2\} + \{B_3^2\}\right) - \{B\}\{B_2\} \tag{35}$$

$$h^{(4)} = \{\rho\hat{u}\}\{w\} + k_3\left(\{p\} + \frac{1}{2}(\{B_1^2\} + \{B_2^2\} + \{B_3^2\}\right) - \{B\}\{B_3\} \tag{36}$$

$$h^{(5)} = \{\rho\hat{u}\}\{\frac{e+p}{\rho} + \frac{1}{2\rho}(B_1^2 + B_2^2 + B_3^2)\} - \{\hat{B}\}\{\mathbf{u}\}^T\{\mathbf{B}\} \tag{37}$$

$$h^{(6)} = \{\hat{u}\}\{B_1\} - \{\hat{B}\}\{u\} \tag{38}$$

$$h^{(7)} = \{\hat{u}\}\{B_2\} - \{\hat{B}\}\{v\} \tag{39}$$

$$h^{(8)} = \{\hat{u}\}\{B_3\} - \{\hat{B}\}\{w\} \tag{40}$$

2.6 KGP Split for MHD

The Kennedy-Gruber-Pirozzoli (KGP) skew-symmetric splitting of the inviscid flux derivative that is kinetic energy preserving is generalized to the equations of MHD by adding the MHD terms of a form similar DS. This gives the numerical flux vector

$$h^{(1)} = \{\rho\}\{\hat{u}\} \tag{41}$$

$$h^{(2)} = \{\rho\}\{\hat{u}\}\{u\} + k_1\{p + \frac{1}{2}(B_1^2 + B_2^2 + B_3^2)\} - \{\hat{B}\}\{B_1\} \tag{42}$$

$$h^{(3)} = \{\rho\}\{\hat{u}\}\{v\} + k_2\{p + \frac{1}{2}(B_1^2 + B_2^2 + B_3^2)\} - \{\hat{B}\}\{B_2\} \tag{43}$$

$$h^{(4)} = \{\rho\}\{\hat{u}\}\{w\} + k_3\{p + \frac{1}{2}(B_1^2 + B_2^2 + B_3^2)\} - \{\hat{B}\}\{B_3\} \tag{44}$$

$$h^{(5)} = \{\rho\}\{\hat{u}\}\left(\{(e+p)/\rho\} + \{\frac{1}{2\rho}((B_1^2 + B_2^2 + B_3^2))\}\right) - \tag{45}$$

$$\{\hat{B}\}(\{u\}\{B_1\} + \{v\}\{B_2\} + \{w\}\{B_3\}) \tag{46}$$

$$h^{(6)} = \{\hat{u}\}\{B_1\} - \{\hat{B}\}\{u\} \tag{47}$$

$$h^{(7)} = \{\hat{u}\}\{B_2\} - \{\hat{B}\}\{v\} \tag{48}$$

$$h^{(8)} = \{\hat{u}\}\{B_3\} - \{\hat{B}\}\{w\}. \tag{49}$$

2.7 *Entropy Split Scheme (ES) for MHD*

To generalize the entropy split method for nonlinear thermally perfect gas dynamics Euler equations to MHD, Eqs. (1) are split in conservative and non-conservative parts as

$$\mathbf{q}_t + \frac{\beta}{\beta+1}(\mathbf{f}_x^{(x)} + \mathbf{f}_y^{(y)} + \mathbf{f}_z^{(z)}) + \frac{1}{\beta+1}(A^{(x)}\mathbf{v}_x + A^{(y)}\mathbf{v}_y + A^{(z)}\mathbf{v}_z) + \frac{\beta}{\beta+1}\mathbf{e}\,\mathrm{div}\,\mathbf{B} = \mathbf{0}. \tag{50}$$

The matrices are defined by

$$A^{(x)} = \mathbf{f}_{\mathbf{v}}^{(x)} + C^{(x)} \tag{51}$$

and similarly for the other coordinate directions. The matrices $C^{(x)}$, $C^{(y)}$, and $C^{(z)}$ are obtained by rewriting the divergence term as

$$\mathbf{e}\,\mathrm{div}\,\mathbf{B} = C^{(x)}\mathbf{v}_x + C^{(y)}\mathbf{v}_y + C^{(z)}\mathbf{v}_z.$$

The entropy variables are derived from the entropy (4) and are given by

$$\mathbf{v} = \frac{\rho}{p}s^{\frac{1}{\alpha+\gamma}}(-\frac{\alpha}{\gamma-1}\frac{p}{\rho} - \frac{1}{2}|\mathbf{u}|^2\ u\ v\ w\ -1\ B_1\ B_2\ B_3)^T$$

The matrices $A^{(x)}$, $A^{(y)}$, and $A^{(z)}$ are symmetric.

The semi-discrete entropy split approximation of the one-dimensional version of (1) is

$$\frac{d}{dt}\mathbf{q}_j + \frac{\beta}{\beta+1}D\mathbf{f}_j + \frac{1}{\beta+1}A_j D\mathbf{v}_j + \frac{\beta}{\beta+1}\mathbf{e}_j D B_1 = 0, \quad j = 1, \ldots, N, \tag{52}$$

where D is a linear finite difference operator, e.g., $\mathbf{f}_x(x_j) \approx Df_j = \frac{1}{\Delta x}\sum_{k=1}^{p}\alpha_k(\mathbf{f}_{j+k} - \mathbf{f}_{j-k})$ and $\beta > 0$ is a parameter related to α in (4) by $\beta = (\alpha+\gamma)/(1-\gamma)$.

This scheme is formally defined in the same way as in the case of the gas dynamics equations. However, because the MHD flux function is not homogeneous, entropy conservation cannot be proved in the same way as for gas dynamics. The appendix in the expanded version of this articles gives the elements of the matrices of the form (51) that appear in the entropy split MHD equations (50). The formulas are given for the matrix in the direction $(k_1,\ k_2,\ k_3)$. See [40] for details. A conservative version of the ES is denoted by ESSW. It is an entropy split method with Ducros splitting but switched to regular central near discontinuities [30].

3 Numerical Experiments

Two test cases are considered, an Alfven wave, and a challenging Brio-Wu Riemann problem. The problems were solved by the eight-order accurate versions of the schemes EC, ECLOG, ECLOGKP, ECHKP, KGP, ES, ESDS, ESSW, DS and DSKP, described in Sect. 2. The entropy split schemes using $\beta = 1, 2, 2.5, 25.5$ were considered. Grid refinement was also performed. For a detailed comparison, see the expanded version [40]. Here, due to a space limitation, only seven of the method for fine grid using $\beta = 1$ for the Alfven wave test case and $\beta = 25.5$ for the Brio-Wu test case will be shown. A fourth-order accurate Runge-Kutta method was used in time.

3.1 2D Alfvén Wave

We select an Alfvén wave (a traveling transverse wave) with phase $\phi = 2\pi(x\cos\alpha + y\sin\alpha + t)$, and given by

$$\rho = 1 \quad p = 1 \quad \mathbf{v} = A(-\sin\alpha\sin\phi,\ \cos\alpha\sin\phi,\ cos\phi) \tag{53}$$

$$\mathbf{B} = (\cos\alpha - A\sin\alpha\sin\phi,\ \sin\alpha + A\cos\alpha\sin\phi,\ Acos\phi)\ . \tag{54}$$

We will consider a two-dimensional wave traveling in the direction $\mathbf{n} = (\cos\alpha,\ \sin\alpha,\ 0)$, where $\alpha = 30°$, with periodic boundary conditions. For this

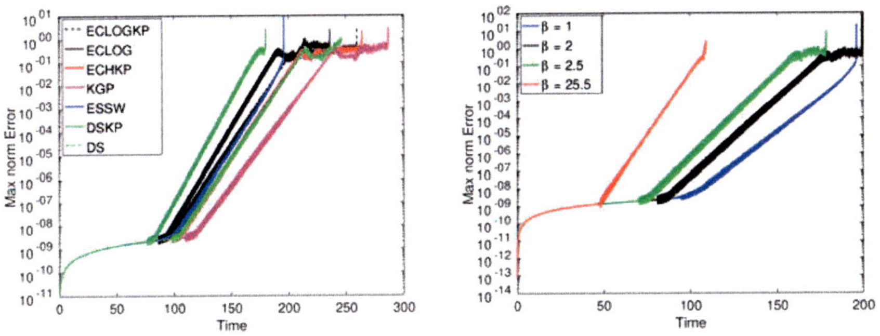

Fig. 1 Alfvén wave, maximum norm error vs. time: Comparison among seven structure-preserving methods (left) and the maximum error vs. β (right)

computation $A = 0.1$. The computational domain is of size 1.3546×2.1915, discretized by a grid of spacing $\Delta x = \Delta y = 0.022$. A grid refinement study was also perform using $\Delta x = \Delta y = 0.011$. The flow is solved up to time 300 until all methods eventually diverge, corresponding to approximately 100,000 time steps. All of the ten methods indicated on the nomenclature in the introduction section were examined.

Figure 1 shows the time evolution of the maximum norm error over the computational domain for seven methods using the fine grid. The error is maintained at a very low level up to time around 130 for all the methods. Furthermore, at times before 130, the different numerical schemes produce errors that are indistinguishable. At time 130 the error of the DSKP scheme start to grow faster than the errors of other methods. After time 160, non-linear instabilities starts to become significant, and some difference in error between the methods can be observed. Beyond time 275, all methods diverge. For this shock-free test case, a smaller $\beta = 1$ by the ES method is more accurate and stable for a much longer time integration than $\beta > 1$. For physical argument, it is preferred to use $\beta = 1$ or greater. Figure not shown, a two grid refinement were performed on the mass conservation error between the ES and ESSW methods. Both methods conservative mass for similar time duration of integration.

3.2 *1D MHD Brio-Wu Riemann Problem [4]*

For problem with complex shock waves, we selected the Wu-Brio test case as it is challenging for standard shock-capturing methods. The initial state vectors are

$$(\rho_L, u_L, p_L, B^{(x)}, B^{(y)}, B^{(z)}) = (1, 0, 1, 0.75, 1, 0) \tag{55}$$

$$(\rho_R, u_R, p_R, B^{(x)}, B^{(y)}, B^{(z)}) = (0.125, 0, 0.1, 0.75, -1, 0) \tag{56}$$

It involves two fast rarefaction waves, a slow compound wave, a contact discontinuity and a slow shock wave. According to the Flash code [9] and many published work, e.g., [5], standard high order shock-capturing methods exhibit the oscillatory solutions, e.g., PPM (piecewise parabolic method of Collela and Woodward [6] and third- to fifth-order WENO (WENO3–WENO5). Most researchers resorted to use first order or very diffusive 2nd-order methods with local Lax-Friedrichs (LLF) or variants of the HLL-type numerical fluxes [13, 17]. The oscillations increase with an increase of B_y; the reason being that stronger B_y introduces more transverse effect that resists shock propagation in the x direction causing the shock to move slowly.

For the current simulations, this test case was solved on the domain $0 \leq x \leq 1$ to time 0.1, using 801 uniformly distributed grid points. A grid refinement study was also performed using a 1601 uniform grid.

The solution was computed by the base schemes ECLOG, ECLOGKP, ECHKP, KGP, ES, DSKP, and DS. In order to suppress spurious oscillations around discontinuities, after the completion of the full Runge-Kutta time step, the computed solution is filtered by the dissipative portion of the WENO7 scheme premultiplied by a wavelet flow sensor after each time step. See Yee and Sjögreen, and Sjögreen and Yee [18, 20, 22, 23, 36, 38] for the nonlinear filter formulation. The Gallice form of the Roe-type approximate Riemann solver [35, 37] was used for the shock capturing nonlinear filter step. Contrary to previous published work, it is noted that here we were able to use a very high order method with Roe-type of approximate Riemann solver to obtain highly accurate stable solutions. The highly spurious oscillatory solutions by standard high order shock-capturing methods are greatly reduced by the considered seven methods.

The density is shown in Fig. 2, the pressure in Fig. 3, and the y-direction magnetic field in Fig. 4 for the B_y. Comparing the present simulations with published work, the oscillatory solutions are almost diminished by the considered seven methods.

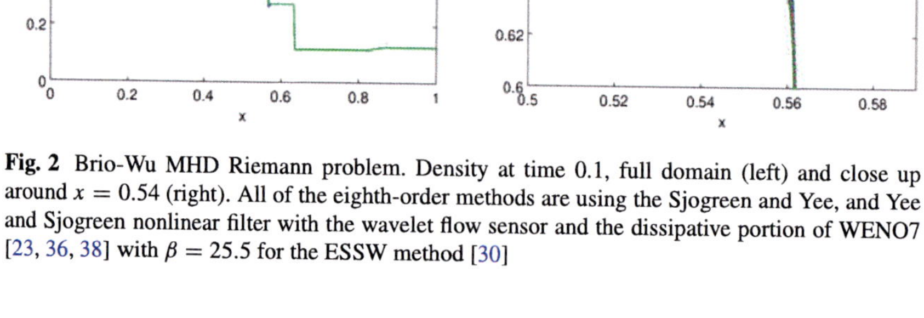

Fig. 2 Brio-Wu MHD Riemann problem. Density at time 0.1, full domain (left) and close up around $x = 0.54$ (right). All of the eighth-order methods are using the Sjogreen and Yee, and Yee and Sjogreen nonlinear filter with the wavelet flow sensor and the dissipative portion of WENO7 [23, 36, 38] with $\beta = 25.5$ for the ESSW method [30]

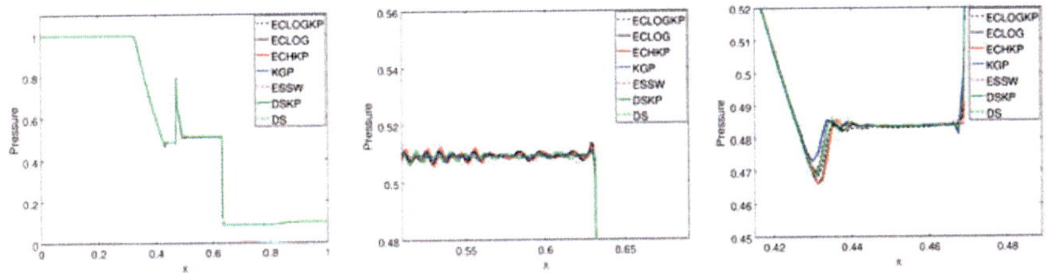

Fig. 3 Brio-Wu MHD Riemann problem. Pressure at time 0.1, full domain (left), close up around $x = 0.45$ (middle), and around $x = 0.60$ (right). All of the methods are using the Sjogreen and Yee, and Yee and Sjogreen nonlinear filter with the wavelet flow sensor and the dissipative portion of WENO7 [2, 16, 22–24, 36, 38] with $\beta = 25.5$ for the ESSW method [30]

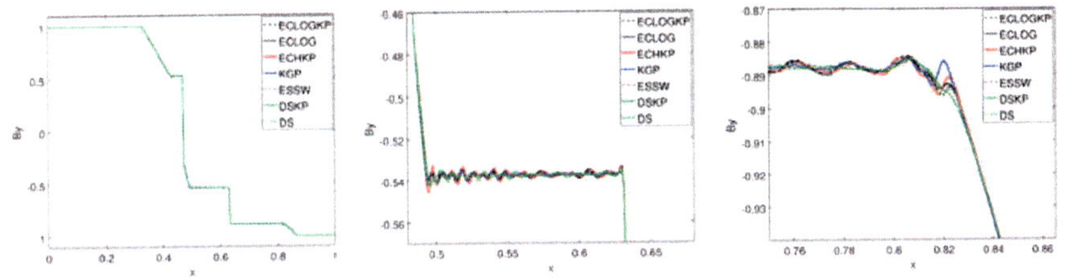

Fig. 4 Brio-Wu MHD Riemann problem. $B^{(y)}$ at time 0.1, full domain (left), close up around $x = 0.51$ (middle), and around $x = 0.80$ (right). All of the methods are using the Sjogreen and Yee, and Yee and Sjogreen nonlinear filter with the wavelet flow sensor and the dissipative portion of WENO7 [23, 36, 38] with $\beta = 25.5$ for the ESSW method [30]

Summary In summary, for the two chosen test cases, the relative stability and accuracy are comparable among the KGP-type, DS-type split methods and the ES and ESSW methods. For the difficult Brio-Wu test case, unlike most published work reported in the literature, the current study indicated that stability and high accuracy were obtained using very high order spatial discretization using the Roe-type approximate Riemann solver. The numerical instability and highly spurious oscillatory solutions by standard high order shock-capturing methods are drastically reduced by the ES and ESSW methods. With the current studies and from our previous studies [28–30, 39], EC and ECLOG, ECLOGKP are the most CPU intensive methods among the 10 methods. It is approximately twice the CPU per time step than the ES, ESDS and ESSW methods and yet exhibit similar resolutions. DS is the least CPU intensive. The CPU comparison is based on one single processor and eight processors as well as the operations count per grid point and per time step.

References

1. Arakawa, A.: Computational design for long-term numerical integration of the equations of fluid motion: two-dimensional incompressible flow, Part I. J. Comput. Phys. **1**, 119–143 (1966)
2. Balsara, D., Shu, Chi-Wang: monotonicity preserving weighted essentially non-oscillatory schemes with increasingly high order of accuracy. J. Comput. Phys. **160**, 405–452 (2000)
3. Blaisdell, G.A., Spyropoulos, E.T., Qin, J.H.: The effect of the formulation of nonlinear terms on aliasing errors in spectral methods. Appl. Num. Math. **21**, 207–219 (1996)
4. Brio, M., Wu, C.C.: An upwind differencing scheme for the equations of ideal magnetohydrodynamics. J. Comput. Phys. **75**, 400–422 (1988)
5. Castro, M.J., Gallardo, J. M., Marquina, A.: Jacobian-free incomplete Riemann solvers. In: Proceedings of Theory, Numerics and Application Problems I: Aachen Germany, pp. 292–307 (2016)
6. Colella, P., Woodward, P.R.: The piecewise parabolic method (PPM) for gas-dynamical simulations. J. Comput. Phys. **54**, 174–201 (1984)
7. Coppola, G., Capuano, F., Pirozzoli, S., de Luca, L.: Numerically stable formulations of convective terms for turbulent compressible flows. J. Comput. Phys. (2019). https://doi.org/10.1016/j.jcp.2019.01.007
8. Ducros, F., Laporte, F., Soulères, T., Guinot, V., Moinat, P., Caruelle, B.: High-order fluxes for conservative skew-symmetric-like schemes in structured meshes: application to compressible flows. J. Comput. Phys. **161**, 114–139 (2000)
9. Flash Manual - UserManual.wiki. University of Chicago (2021)
10. Gassner, G., Winters, A.R.: A novel robust strategy for discontinuous Galerkin methods in computational fluid mechanics: Why? When? What? Where? Front. Phys. **8**, 500690 (2021). https://doi.org/10.3389/fphy.2020.500690
11. Gerritsen, M, Olsson, P.: Designing an efficient solution strategy for fluid flows. I. A stable high order finite difference scheme and sharp shock resolution for the Euler equations. J. Comput. Phys. **129**, 245–262 (1996)
12. Godunov, S.K.: Symmetric form of the equations of magnetohydrodynamics. Numer. Methods Mech. Continuum Medium **13**(1), 26–34 (1972)
13. Gurski, K.F.: An HLLC-type approximate Riemann solver for ideal magnetohydrodynamics. SIAM J. Sci. Comput. **25**, 2165–2187 (2004)
14. Harten, A: On the symmetric form of systems for conservation laws with entropy: J. Comput. Phys. **49**, 151 (1983)
15. Kennedy, C.A., Gruber, A.: Reduced aliasing formulations of the convective terms within the Navier-Stokes equations. J. Comput. Phys. **227**, 1676–1700 (2008)
16. Kritsuk, A.G., Kotov, D.V., Sjögreen, B., Yee, H.C.: High order nonlinear filter methods for subsonic turbulence simulation with stochastic forcing. J. Comput. Phys. **431**, 1676–1700, 110118 (2021). https://doi.org/10.1016/j.jcp.2021.110118
17. Li, S.,: An HLLC Riemann solver for magneto-hydrodynamics. J. Comput. Phys. **203**(1), 344–357 (2005)
18. Palha, A., Gerritsma, M.: A mass energy, enstrophy and vorticity conserving (MEEVC) mimetic spectral element discretization for the 2D incompressible Navier-Stokes equations. https://doi.org/10.1016/j.jcp.2016.10.009
19. Pirozzoli, S.: Generalized conservative approximations of split convective derivative operators. J. Comput. Phys. **219**, 7180–7190 (2010)
20. Ranocha. H.: Entropy conserving and kinetic energy preserving numerical methods for the Euler equations using summation-by-parts operators. In: Proceedings of the ICOSAHOM-2018, Imperial College, London, UK, July 9–13 (2018)
21. Sandham, N.D., Li, Q., Yee, H.C.: Entropy splitting for high-order numerical simulation of compressible turbulence. J. Comput. Phys. **23**, 307–322 (2002)
22. Shu, C.-W., Osher, S.J.: Efficient implementation of essentially non-oscillatory shock capturing schemes. J. Comput. Phys. **83**, 32–78 (1989)

23. Sjögreen, B., Yee, H.C.: Multiresolution wavelet based adaptive numerical dissipation control for high order methods. J. Sci. Comput. **20**, 211–255 (2004)
24. Sjögreen, B., Yee, H.C.: On skew-symmetric splitting and entropy conservation schemes for the Euler equations. In: Proceedings of ENUMATH09, June 29–July 2. Uppsala University, Sweden (2009)
25. Sjögreen, B., Yee, H.C.: High order entropy conserving central schemes for wide ranges of compressible gas dynamics and MHD flows. J. Comput. Phys., **364**, 153–185 (2018)
26. Sjögreen, B., Yee, H.C.: High order entropy conservative central schemes for wide ranges of compressible gas dynamics and MHD flows. J. Comput. Phys. **364**, 153–185 (2018)
27. Sjögreen, B., Yee, H.C.: An entropy stable method revisited: central differencing via entropy splitting and SBP. In: Proceeding of Proceeding of ICOSAHOM-2018, July 9–13, 2018. Imperial College, London (2018)
28. Sjögreen, B., Yee, H.C.: Entropy stable method for the euler equations revisited: central differencing via entropy splitting and SBP. J. Sci. Comput. **81**, 1359–1385 (2019). https://doi.org/10.1007/s10915-019-01013-1
29. Sjögreen, B., Yee, H.C.: Skew-symmetric splitting for multiscale gas dynamics and MHD turbulence flows. Extended version of Proceedings of ASTRONUM-2016, June 6–10, 2016, Monterey, CA, USA, J. Sci. Comput. **83**, 43 (2020)
30. Sjögreen, B., Yee, H.C.: Construction of conservative numerical fluxes for the entropy split method. Speical Issue of Commun. Appl. Math. Comput. (CAMC) (2021)
31. Sjögreen, B., Yee, H.C., Kotov, D.V., Kritsuk, A.G.: Skew-symmetric splitting for multiscale gas dynamics and MHD turbulence flows. J. Sci. Comput. **83**, 1–43 (2020)
32. Tadmor, E.: Entropy stability theory for difference approximations of nonlinear conservation laws and related time-dependent problems. Acta Numer. **12**, 451–512 (2003)
33. Winters, A.R., Gassner, G.J.: Affordable, entropy conserving and entropy stable flux functions for the ideal MHD equations. J. Comput. Phys. **304**, 72–108 (2016)
34. Yee, H.C., Sandham, N.D., Djomehri, M.J.: Low-dissipative high order shock-capturing methods using characteristic-based filters. J. Comput. Phys. **150**, 199–238 (1999)
35. Yee, H.C., Sjögreen, B.: Efficient low dissipative high order schemes for multiscale MHD flows, II: Minimization of div **B** numerical error. J. Sci. Comput. **29**(1) (2006). https://doi.org/10.1007/s10915-005-9004-5
36. Yee, H.C., Sjögreen, B.: Development of low dissipative high order filter schemes for multiscale Navier-Stokes and MHD systems. J. Comput. Phys. **225**, 910–934 (2007)
37. Yee, H.C., Sjögreen, B.: Adaptive filtering and limiting in compact high order methods for multiscale gas dynamics and MHD systems. Comput. Fluid **37**, 593–619 (2008)
38. Yee, H.C., Sjögreen, B.: High order filter methods for wide range of compressible flow speeds. In: Proceedings of the ICOSAHOM09, June 22–26, 2009, Trondheim, Norway (2009)
39. Yee, H.C., Sjögreen, B.: On entropy conservation and kinetic energy preservation methods. In: Proceedings of the ICOSAHOM-2019, July 1–5, 2019, Paris, France (2019)
40. Yee, H.C., Sjögreen, B.: Recent development of entropy split methods for gas dynamics and MHD. J. Appl. Math. Comput. (2022). Submitted
41. Yee, H.C., Vinokur, M., Djomehri, M.J.: Entropy splitting and numerical dissipation. J. Comput. Phys. **162**, 33–81 (2000)

High-Order Discretisations and Efficient Direct Space-Time Finite Element Solvers for Parabolic Initial-Boundary Value Problems

Marco Zank

1 Introduction

Traditional approaches for approximating solutions of parabolic initial-boundary value problems are usually based on semi-discretisations in space and time, where the discretisation in space and time is split accordingly. In contrast to these approaches, we consider space-time methods, which discretise time-dependent partial differential equations without separating the temporal and spatial directions. In this work, the model problem is the homogeneous Dirichlet problem for the heat equation,

$$\left.\begin{aligned} \partial_t u(x,t) - \Delta_x u(x,t) &= f(x,t) && \text{for } (x,t) \in Q = \Omega \times (0,T), \\ u(x,t) &= 0 && \text{for } (x,t) \in \Sigma = \partial\Omega \times [0,T], \\ u(x,0) &= 0 && \text{for } x \in \Omega, \end{aligned}\right\} \tag{1}$$

where $\Omega = (0, L)$ is an interval for $d = 1$, or Ω is polygonal for $d = 2$, or Ω is polyhedral for $d = 3$, $T > 0$ is a terminal time and f is a given right-hand side. The underlying variational setting is crucial for the space-time approximation. To derive a space-time variational formulation, we define the space-time Sobolev spaces $H^{1,1/2}_{0;0,}(Q) := H^{1/2}_{0,}(0,T; L^2(\Omega)) \cap L^2(0,T; H^1_0(\Omega))$, $H^{1,1/2}_{0;,0}(Q) := H^{1/2}_{,0}(0,T; L^2(\Omega)) \cap L^2(0,T; H^1_0(\Omega))$, see [14, 16] for more details, with the Hilbertian norms $\|\cdot\|_{H^{1,1/2}_{0;0,}(Q)} := \sqrt{\|\cdot\|^2_{H^{1/2}_{0,}(0,T,L^2(\Omega))} + \|\nabla_x \cdot\|^2_{L^2(Q)}}$,

M. Zank (✉)
Universität Wien, Wien, Austria
e-mail: marco.zank@univie.ac.at

J. M. Melenk et al. (eds.), *Spectral and High Order Methods for Partial Differential Equations ICOSAHOM 2020+1*, Lecture Notes in Computational Science and Engineering 137, https://doi.org/10.1007/978-3-031-20432-6_37

$\| \cdot \|_{H^{1,1/2}_{0;,0}(Q)} := \sqrt{\| \cdot \|^2_{H^{1/2}_{,0}(0,T;L^2(\Omega))} + \|\nabla_x \cdot \|^2_{L^2(Q)}}$. Here, the Bochner–Sobolev space with initial condition and its Bochner–Sobolev norm are given by

$$H^{1/2}_{0,}(0,T;L^2(\Omega)) := \left\{ v \in H^{1/2}(0,T;L^2(\Omega)) \colon \|v\|_{H^{1/2}_{0,}(0,T;L^2(\Omega))} < \infty \right\},$$

$$\|v\|_{H^{1/2}_{0,}(0,T;L^2(\Omega))} := \left(\|v\|^2_{H^{1/2}(0,T;L^2(\Omega))} + \int_0^T \|v(\cdot,t)\|^2_{L^2(\Omega)} t^{-1} \mathrm{d}t \right)^{1/2}.$$

For $s \in [0,1], s \neq 1/2$, the space $H^s_{0,}(0,T;L^2(\Omega))$ with initial condition and the space $H^{1/2}_{,0}(0,T;L^2(\Omega))$ with terminal condition are introduced analogously, see [16, Chapter 2]. With the bilinear form $a(\cdot,\cdot)\colon H^{1,1/2}_{0;0,}(Q) \times H^{1,1/2}_{0;,0}(Q) \to \mathbb{R}$, defined by

$$a(u,w) := \langle \partial_t u, w \rangle_Q + \langle \nabla_x u, \nabla_x w \rangle_{L^2(Q)}, \quad u \in H^{1,1/2}_{0;0,}(Q), \quad w \in H^{1,1/2}_{0;,0}(Q),$$

the space-time variational formulation of (1) is to find $u \in H^{1,1/2}_{0;0,}(Q)$ such that

$$\forall w \in H^{1,1/2}_{0;,0}(Q)\colon \quad a(u,w) = \langle f, w \rangle_Q \tag{2}$$

with f given as an element of the dual space $[H^{1,1/2}_{0;,0}(Q)]'$, where $\langle \cdot,\cdot \rangle_Q$ denotes the duality pairing in $[H^{1,1/2}_{0;,0}(Q)]'$ and $H^{1,1/2}_{0;,0}(Q)$ as extension of the inner product in $L^2(Q)$. We have the following existence and uniqueness theorem [16, Theorem 3.4.19].

Theorem 1 *Let $f \in [H^{1,1/2}_{0;,0}(Q)]'$ be given. Then, the variational formulation (2) has a unique solution $u \in H^{1,1/2}_{0;0,}(Q)$, satisfying $\|u\|_{H^{1,1/2}_{0;0,}(Q)} \le C \|f\|_{[H^{1,1/2}_{0;,0}(Q)]'}$ with a constant $C > 0$. Furthermore, the solution operator is an isomorphism.*

A conforming tensor-product space-time discretisation of (2) with piecewise polynomial, continuous ansatz and test functions is not stable in general, independent of a CFL condition $h_t \le C h_x$ with a constant $C > 0$, where h_t and h_x are the mesh sizes in time and space, see [16, Section 3.3]. To overcome these stability issues, we define the modified Hilbert transformation $\mathcal{H}_T\colon L^2(Q) \to L^2(Q)$ by

$$(\mathcal{H}_T w)(x,t) := \sum_{i=1}^{\infty} \sum_{k=0}^{\infty} w_{i,k} \cos\left(\left(\frac{\pi}{2} + k\pi \right) \frac{t}{T} \right) \phi_i(x), \quad (x,t) \in Q, \tag{3}$$

where the given function $w \in L^2(Q)$ is represented by the Fourier series

$$w(x,t) = \sum_{i=1}^{\infty} \sum_{k=0}^{\infty} w_{i,k} \sin\left(\left(\frac{\pi}{2} + k\pi \right) \frac{t}{T} \right) \phi_i(x), \quad (x,t) \in Q,$$

with the eigenfunctions $\phi_i \in H_0^1(\Omega)$ and eigenvalues $\mu_i \in \mathbb{R}$, satisfying $-\Delta_x \phi_i = \mu_i \phi_i$ in Ω, $\phi_i = 0$ on $\partial\Omega$, $\|\phi_i\|_{L^2(\Omega)} = 1$, $i \in \mathbb{N}$. This approach was introduced recently in [14] and [16, Section 3.4]. Note that the novel transformation $\mathcal{H}_T$ acts only on the finite interval $(0, T)$, whereas analogous considerations of an infinite time interval $(0, \infty)$ with the classical Hilbert transformation are investigated in [3, 5, 9]. The most important properties of $\mathcal{H}_T$ are summarised in the following, see [13–17]. The map $\mathcal{H}_T: H_{0;0,}^{1,1/2}(Q) \to H_{0;\,,0}^{1,1/2}(Q)$ is norm preserving and bijective. Moreover, the relations

$$\forall v \in H_{0,}^{1/2}(0, T; L^2(\Omega)): \quad \langle \partial_t v, \mathcal{H}_T v \rangle_Q \simeq \|v\|^2_{H_{0,}^{1/2}(0,T;L^2(\Omega))}, \tag{4}$$

$$\forall s \in (0, 1]: \quad \forall v \in H_{0,}^{s}(0, T; L^2(\Omega)), v \neq 0: \quad \langle v, \mathcal{H}_T v \rangle_{L^2(Q)} > 0 \tag{5}$$

hold true. With the modified Hilbert transformation $\mathcal{H}_T$, the variational formulation (2) is equivalent to find $u \in H_{0;0,}^{1,1/2}(Q)$ such that

$$\forall v \in H_{0;0,}^{1,1/2}(Q): \ a(u, \mathcal{H}_T v) = \langle f, \mathcal{H}_T v \rangle_Q. \tag{6}$$

Hence, unique solvability of the variational formulation (6) follows from the unique solvability of (2).

Next, we consider conforming discretisations of the space-time variational formulation (6). For some conforming space-time finite element space $\mathcal{V}_h \subset H_{0;0,}^{1,1/2}(Q)$, the discrete variational formulation of (6) is to find $u_h \in \mathcal{V}_h$ such that

$$\forall v_h \in \mathcal{V}_h: \ a(u_h, \mathcal{H}_T v_h) = \langle f, \mathcal{H}_T v_h \rangle_Q. \tag{7}$$

Note that ansatz and test spaces are equal. In [16, Theorem 3.4.20], the unique solvability and unconditional stability of the discrete problem (7) for any conforming space-time finite element space $\mathcal{V}_h \subset H_{0;0,}^{1,1/2}(Q)$ is proven under the assumption $f \in [H_{,0}^{1/2}(0, T; L^2(\Omega))]'$, i.e., no CFL condition is required. Using the choice $\mathcal{V}_h = Q_{h,0}^p(Q) \subset H_{0;0,}^{1,1/2}(Q)$, i.e., a tensor-product space-time ansatz space of piecewise polynomial, continuous functions with polynomial degree p in the spatial and temporal variables, see Sect. 3 for the notations, results in a huge global linear system

$$K_h \underline{u} = \underline{f}, \tag{8}$$

where the system matrix K_h admits a representation as a sum of Kronecker products. In [8], efficient direct space-time solvers for the linear system (8) are considered and analysed for the case of a polynomial degree $p = 1$, i.e., piecewise multilinear, globally continuous functions. In this work, we generalise the ideas in [8] from the linear case to the higher-order case. In greater detail, we state the resulting algorithm of the space-time solver for arbitrary polynomial degrees $p \in \mathbb{N}$ using the so-called Bartels–Stewart method [1] with real-Schur decomposition and analyse the complexity of the proposed algorithm. Alternatively, the global linear system (8) is solved by (preconditioned) iterative solvers. This topic will be considered in future work. Concerning preconditioned iterative solvers applied to different numerical approaches for parabolic equations, we refer to [7] and references there.

The rest of the paper is organised as follows: In Sect. 2, properties of the Kronecker product are summarised and in Sect. 3, notations of the used finite element spaces are fixed. Section 4 introduces the space-time finite element method. The new algorithm is investigated in Sect. 5. Numerical examples for a one-dimensional spatial domain and piecewise polynomials of higher-order are presented in Sect. 6. Finally, we draw some conclusions in Sect. 7.

2 Kronecker Product

In this section, some basic properties of the Kronecker product are stated, see, e.g., [12]. Let $A, C \in \mathbb{R}^{N_A \times N_A}$, $B, D \in \mathbb{R}^{N_B \times N_B}$ and $X \in \mathbb{R}^{N_A \times N_B}$ be given matrices for $N_A, N_B \in \mathbb{N}$. The matrix $A \otimes B \in \mathbb{R}^{N_A \cdot N_B \times N_A \cdot N_B}$ is the Kronecker product defined in the usual way. Furthermore, the vectorisation of a matrix converts the matrix into a column vector, i.e., we define

$$\begin{aligned}\operatorname{vec}(X) = (X[1,1], X[2,1], \ldots, X[N_A,1], X[1,2], \ldots, X[N_A,N_B])^\top \\ \in \mathbb{R}^{N_A N_B \times 1}.\end{aligned}$$

Conversely, for a given vector $\underline{v} \in \mathbb{R}^{N_A \cdot N_B}$, define the matrix $V = \begin{pmatrix}\underline{v}_1 & \underline{v}_2 & \cdots & \underline{v}_{N_B}\end{pmatrix} \in \mathbb{R}^{N_A \times N_B}$ with $\underline{v}_i \in \mathbb{R}^{N_A}$ given by $\underline{v}_i[k] = \underline{v}[(i-1)N_A + k]$ for $k = 1, \ldots, N_A$, $i = 1, \ldots, N_B$, i.e., $\operatorname{vec}(V) = \underline{v}$. Then, the relation

$$(B^\top \otimes A)\underline{v} = (B^\top \otimes A)\operatorname{vec}(V) = \operatorname{vec}(AVB) \tag{9}$$

holds true. We use the following properties of the Kronecker product $(A \otimes B)^\top = A^\top \otimes B^\top$ and $(A \otimes B)^{-1} = A^{-1} \otimes B^{-1}$, where in the second relation the matrices are assumed to be regular, and the mixed-product property $(A \otimes B)(C \otimes D) = (AC) \otimes (BD)$.

3 Space-Time Finite Element Spaces

In this section, notations of the used finite element spaces are stated. For this purpose, let the bounded Lipschitz domain $\Omega \subset \mathbb{R}^d$ be an interval $\Omega = (0, L)$ for $d = 1$, or polygonal for $d = 2$, or polyhedral for $d = 3$. For a tensor-product ansatz, we consider admissible decompositions $\overline{Q} = \overline{\Omega} \times [0, T] = \bigcup_{k=1}^{N_x} \overline{\omega_k} \times \bigcup_{\ell=1}^{N_t} [t_{\ell-1}, t_\ell]$ with $N = N_x \cdot N_t$ space-time elements, where the time intervals $\tau_\ell := (t_{\ell-1}, t_\ell)$ with mesh sizes $h_{t,\ell} = t_\ell - t_{\ell-1}$ for $\ell = 1, \ldots, N_t$ are defined via the decomposition

$$0 = t_0 < t_1 < t_2 < \cdots < t_{N_t-1} < t_{N_t} = T \tag{10}$$

of the time interval $(0, T)$. The maximal and the minimal time mesh sizes are denoted by $h_t := h_{t,\max} := \max_\ell h_{t,\ell}$ and $h_{t,\min} := \min_\ell h_{t,\ell}$, respectively. For the spatial domain Ω, we consider a shape-regular sequence $(\mathcal{T}_\nu)_{\nu \in \mathbb{N}}$ of admissible decompositions $\mathcal{T}_\nu := \{\omega_k \subset \mathbb{R}^d : k = 1, \ldots, N_x\}$ of Ω into finite elements $\omega_k \subset \mathbb{R}^d$ with mesh sizes $h_{x,k}$, the maximal mesh size $h_x := h_{x,\max} := \max_k h_{x,k}$ and the minimal mesh size $h_{x,\min} := \min_k h_{x,k}$. The spatial elements ω_k are intervals for $d = 1$, triangles for $d = 2$, and tetrahedra for $d = 3$.

Next, for a fixed polynomial degree $p \in \mathbb{N}$, we introduce the tensor-product finite element spaces $Q_h^p(Q) := S_{h_x}^p(\Omega) \otimes S_{h_t}^p(0, T)$ and, satisfying initial and boundary conditions, $Q_{h,0}^p(Q) := S_{h_x,0}^p(\Omega) \otimes S_{h_t;0,}^p(0, T) \subset H_{0;0,}^{1,1/2}(Q)$ of piecewise polynomial, continuous functions, i.e., $S_{h_x,0}^p(\Omega) := S_{h_x}^p(\Omega) \cap H_0^1(\Omega)$ and $S_{h_t;0,}^p(0, T) := S_{h_t}^p(0, T) \cap H_{0,}^1(0, T)$, where $v \in H_{0,}^1(0, T)$ satisfies $v(0) = 0$. Here, $\otimes$ is the Hilbert tensor-product, and

$$S_{h_t}^p(0, T) := \left\{ v_{h_t} \in C[0, T] : \forall \ell \in \{1, \ldots, N_t\} \colon v_{h_t|\overline{\tau_\ell}} \in \mathbb{P}^p(\overline{\tau_\ell}) \right\}$$

denotes the space of piecewise polynomial, continuous functions on intervals, where $\mathbb{P}^p(A)$ is the space of polynomials on a subset $A \subset \mathbb{R}^d$ of global degree at most p. The temporal basis functions of $S_{h_t}^p(0, T)$ are denoted by φ_n^p for $n = 0, \ldots, pN_t$ with $\varphi_0^p(0) = 1$ and $\varphi_n^p(0) = 0$ for $n = 1, \ldots, pN_t$, i.e., $S_{h_t}^p(0, T) = \operatorname{span}\{\varphi_n^p\}_{n=0}^{pN_t}$ and $S_{h_t;0,}^p(0, T) = \operatorname{span}\{\varphi_n^p\}_{n=1}^{pN_t}$. Analogously,

$$S_{h_x}^p(\Omega) := \left\{ v_{h_x} \in C(\overline{\Omega}) : \forall \omega \in \mathcal{T}_\nu \colon v_{h_x|\overline{\omega}} \in \mathbb{P}^p(\overline{\omega}) \right\}$$

is the space of piecewise polynomial, continuous functions on intervals ($d = 1$), triangles ($d = 2$), or tetrahedra ($d = 3$). The spatial basis functions ψ_j^p, $j = 1, \ldots, M_x, M_x + 1, \ldots, \widetilde{M}_x$, of the space $S_{h_x}^p(\Omega)$ are ordered such that $\psi_{j|\partial\Omega}^p = 0$ for $j = 1, \ldots, M_x$ and $\psi_{j|\partial\Omega}^p \neq 0$ for $j = M_x + 1, \ldots, \widetilde{M}_x$, i.e., $S_{h_x}^p(\Omega) = \operatorname{span}\{\psi_j^p\}_{j=1}^{\widetilde{M}_x}$ and $S_{h_x,0}^p(\Omega) = \operatorname{span}\{\psi_j^p\}_{j=1}^{M_x}$. With these notations, the number of the degrees of freedom of the space-time finite element space $Q_{h,0}^p(Q)$ is given by $\mathrm{dof} = pN_t \cdot M_x$.

4 Space-Time Finite Element Method

In this section, we investigate the conforming space-time finite element method (7) with continuous ansatz and test functions $\mathcal{V}_h = Q^p_{h,0}(Q)$, which are piecewise polynomials of arbitrary polynomial degree $p \in \mathbb{N}$ with respect to the spatial variable and the temporal variable. For this purpose, we fix a polynomial degree $p \in \mathbb{N}$. For $f \in L^2(Q)$, we consider the variational formulation to find $u_h \in Q^p_{h,0}(Q)$ such that

$$\forall v_h \in Q^p_{h,0}(Q): \quad a(u_h, \mathcal{H}_T v_h) = \langle f, \mathcal{H}_T v_h \rangle_{L^2(Q)}, \tag{11}$$

which is uniquely solvable and unconditionally stable due to [16, Theorem 3.4.20]. Additionally, the variational formulation (11) fulfils the space-time error estimates

$$\|u - u_h\|_{L^2(Q)} \le c\, h^{p+1}, \qquad |u - u_h|_{H^1(Q)} \le c\, h^p, \tag{12}$$

$$\|u - u_h\|_{H^{1/2}_{0,}(0,T;L^2(\Omega))} \le c\, h^{p+1/2} \tag{13}$$

with $h = \max\{h_t, h_x\}$ and with a constant $c > 0$ for a sufficiently smooth solution $u \in H^{1,1/2}_{0;0,}(Q)$ of (2) and a sufficiently regular domain Ω, where, for the $H^1(Q)$ error estimate (12), the sequence $(\mathcal{T}_\nu)_{\nu\in\mathbb{N}}$ of decompositions of Ω is additionally assumed to be globally quasi-uniform, see [14, 16] for details.

For an easier implementation, we approximate in (11) the right-hand side $L^2(Q) \ni f \approx \Pi^p_h f \in Q^p_h(Q)$, where Π^p_h is the $L^2(Q)$ projection onto $Q^p_h(Q)$. Thus, we consider the perturbed variational formulation to find $\widetilde{u}_h \in Q^p_{h,0}(Q)$ such that

$$\forall v_h \in Q^p_{h,0}(Q): \quad a(\widetilde{u}_h, \mathcal{H}_T v_h) = \langle \Pi^p_h f, \mathcal{H}_T v_h \rangle_{L^2(Q)}. \tag{14}$$

Note that, for the approximation $\Pi^p_h f \approx f$, the space-time error estimates (12), (13) are not spoilt. The discrete variational formulation (14) is equivalent to the global linear system

$$K_h \underline{u} = \underline{\widetilde{F}}^{\mathcal{H}_T} \tag{15}$$

with the system matrix $K_h = A^{\mathcal{H}_T}_{h_t} \otimes M_{h_x} + M^{\mathcal{H}_T}_{h_t} \otimes A_{h_x} \in \mathbb{R}^{pN_t \cdot M_x \times pN_t \cdot M_x}$, where $\otimes$ is the Kronecker product, see Sect. 2. Here, $M_{h_x} \in \mathbb{R}^{M_x \times M_x}$ and $A_{h_x} \in \mathbb{R}^{M_x \times M_x}$ denote spatial mass and stiffness matrices given by $M_{h_x}[i,j] = \langle \psi^p_j, \psi^p_i \rangle_{L^2(\Omega)}$, $A_{h_x}[i,j] = \langle \nabla_x \psi^p_j, \nabla_x \psi^p_i \rangle_{L^2(\Omega)}$ for $i, j = 1, \dots, M_x$. The temporal matrices $M^{\mathcal{H}_T}_{h_t} \in \mathbb{R}^{pN_t \times pN_t}$ and $A^{\mathcal{H}_T}_{h_t} \in \mathbb{R}^{pN_t \times pN_t}$ are defined by

$$M^{\mathcal{H}_T}_{h_t}[\ell, k] := \langle \varphi^p_k, \mathcal{H}_T \varphi^p_\ell \rangle_{L^2(0,T)}, \quad A^{\mathcal{H}_T}_{h_t}[\ell, k] := \langle \partial_t \varphi^p_k, \mathcal{H}_T \varphi^p_\ell \rangle_{L^2(0,T)} \tag{16}$$

for $\ell, k = 1, \ldots, pN_t$. In (16), the modified Hilbert transformation, acting on solely time-dependent functions, is defined as in (3) when the spatial part is omitted. Due to properties of the modified Hilbert transformation $\mathcal{H}_T$, the matrix $A_{h_t}^{\mathcal{H}_T}$ is dense, symmetric and positive definite, see (4), whereas the matrix $M_{h_t}^{\mathcal{H}_T}$ is dense, nonsymmetric and positive definite, see (5). Additionally, the vector of the right-hand side in (15) is given by $\underline{\widetilde{F}}^{\mathcal{H}_T} := \left(\underline{\widetilde{f}}_1, \ldots, \underline{\widetilde{f}}_{pN_t}\right)^\top \in \mathbb{R}^{pN_t \cdot M_x}$ with the vectors $\underline{\widetilde{f}}_k \in \mathbb{R}^{M_x}$, $k = 1, \ldots, pN_t$, where

$$\underline{\widetilde{f}}_k[i] := \langle \Pi_h^p f, \psi_i^p \mathcal{H}_T \varphi_k^p \rangle_{L^2(Q)} = \sum_{j=1}^{\widetilde{M}_x} \sum_{\ell=0}^{pN_t} f_{j,\ell} \langle \psi_j^p, \psi_i^p \rangle_{L^2(\Omega)} \langle \varphi_\ell^p, \mathcal{H}_T \varphi_k^p \rangle_{L^2(0,T)},$$

$i = 1, \ldots, M_x$, with the $L^2(Q)$ projection $\Pi_h^p f = \sum_{j=1}^{\widetilde{M}_x} \sum_{\ell=0}^{pN_t} f_{j,\ell} \psi_j^p \varphi_\ell^p$. Using the property (9), the coefficients $f_{j,\ell} \in \mathbb{R}$ of $\Pi_h^p f$ are given by $F = \widetilde{M}_{h_x}^{-1} F_{\Pi_h^p} \widetilde{M}_{h_t}^{-1} \in \mathbb{R}^{\widetilde{M}_x \times (pN_t+1)}$ with $f_{j,\ell} = F[j,\ell]$, $j = 1, \ldots, \widetilde{M}_x$, $\ell = 0, \ldots, pN_t$. Here, the mass matrices and the right-hand side of the projection are defined by $\widetilde{M}_{h_x}[i,j] := \langle \psi_j^p, \psi_i^p \rangle_{L^2(\Omega)}$, $\widetilde{M}_{h_t}[k,\ell] := \langle \varphi_\ell^p, \varphi_k^p \rangle_{L^2(0,T)}$ and $F_{\Pi_h^p}[j,\ell] := \langle f, \psi_j^p \varphi_\ell^p \rangle_{L^2(Q)}$ with indices $i, j = 1, \ldots, \widetilde{M}_x$, and $k, \ell = 0, \ldots, pN_t$. Thus, we have

$$\underline{\widetilde{f}}_k[i] = \left(\widehat{M}_{h_x} \widetilde{M}_{h_x}^{-1} F_{\Pi_h^p} \widetilde{M}_{h_t}^{-1} \left(\widehat{M}_{h_t}^{\mathcal{H}_T}\right)^\top\right)[i,k], \quad k = 1, \ldots, pN_t, \; i = 1, \ldots, M_x,$$

with the matrices $\widehat{M}_{h_t}^{\mathcal{H}_T}[k,\ell] := \langle \varphi_\ell^p, \mathcal{H}_T \varphi_k^p \rangle_{L^2(0,T)}$, $\widehat{M}_{h_x}[i,j] := \widetilde{M}_{h_x}[i,j]$ for $\ell = 0, \ldots, pN_t$, $k = 1, \ldots, pN_t$, $i = 1, \ldots, M_x$, $j = 1, \ldots, \widetilde{M}_x$.

5 Space-Time Solver

In this section, for the large-scale space-time system (15), we investigate an efficient solver based on the Bartels–Stewart method with real-Schur decomposition [1, 6, 12]. The proposed solver is a generalisation of the solver developed in [8, Subsection 4.2] from the linear case to the higher-order case. In greater detail, for a fixed polynomial degree $p \in \mathbb{N}$, we state solvers for the global linear system

$$(A_{h_t}^{\mathcal{H}_T} \otimes M_{h_x} + M_{h_t}^{\mathcal{H}_T} \otimes A_{h_x}) \underline{u} = \underline{\widetilde{F}}^{\mathcal{H}_T} \tag{17}$$

given in (15) with the symmetric, positive definite matrices $M_{h_x} \in \mathbb{R}^{M_x \times M_x}$, $A_{h_x} \in \mathbb{R}^{M_x \times M_x}$, $A_{h_t}^{\mathcal{H}_T} \in \mathbb{R}^{pN_t \times pN_t}$ and the nonsymmetric, positive definite matrix

$M_{h_t}^{\mathcal{H}_T} \in \mathbb{R}^{pN_t \times pN_t}$. Since (17) is a (generalised) Sylvester equation, we can apply the Bartels–Stewart method [1] with real-Schur decomposition to solve (17), see also [6, 12]. For this purpose, the matrix pencil $(M_{h_t}^{\mathcal{H}_T}, A_{h_t}^{\mathcal{H}_T})$ is decomposed in the form

$$(A_{h_t}^{\mathcal{H}_T})^{-1} M_{h_t}^{\mathcal{H}_T} = Q_t R_t Q_t^\top \tag{18}$$

with the orthogonal matrix $Q_t \in \mathbb{R}^{pN_t \times pN_t}$ and the upper quasi-triangular matrix $R_t \in \mathbb{R}^{pN_t \times pN_t}$, where the diagonals of R_t have 2×2 and 1×1 blocks, corresponding to complex and real eigenvalues of the matrix $(A_{h_t}^{\mathcal{H}_T})^{-1} M_{h_t}^{\mathcal{H}_T}$. In greater detail, to each real eigenvalue $\lambda \in \mathbb{R}$, we can relate a 1×1 block given as $(\lambda) \in \mathbb{R}^{1\times 1}$. The complex eigenvalues occur in conjugate pairs. Thus, each conjugate pair $\alpha \pm \iota\beta$ corresponds to a 2×2 block $\begin{pmatrix} \alpha & b_1 \\ b_2 & \alpha \end{pmatrix} \in \mathbb{R}^{2\times 2}$, satisfying $|\beta| = \sqrt{|b_1 b_2|} > 0$ with b_1 and b_2 having different signs.

Next, we state the solution $\underline{u}$ of (17) in terms of the temporal matrices Q_t, R_t and the spatial matrices M_{h_x}, A_{h_x}. For this purpose, we define $Y_t := (A_{h_t}^{\mathcal{H}_T} Q_t)^{-1}$, which leads to the representations $A_{h_t}^{\mathcal{H}_T} = Y_t^{-1} Q_t^\top$ and $M_{h_t}^{\mathcal{H}_T} = A_{h_t}^{\mathcal{H}_T} Q_t R_t Q_t^\top = Y_t^{-1} R_t Q_t^\top$. Thus, the global linear system (17) is equivalent to solving

$$(Y_t^{-1} \otimes I_{M_x})(I_{pN_t} \otimes M_{h_x} + R_t \otimes A_{h_x})(Q_t^\top \otimes I_{M_x})\underline{u} = \underline{\widetilde{F}}^{\mathcal{H}_T}$$

with the identity matrices $I_{pN_t} \in \mathbb{R}^{pN_t \times pN_t}$ and $I_{M_x} \in \mathbb{R}^{M_x \times M_x}$. With the properties of the Kronecker product, the solution of (17) is given by

$$\underline{u} = (Q_t \otimes I_{M_x})(I_{pN_t} \otimes M_{h_x} + R_t \otimes A_{h_x})^{-1}(Y_t \otimes I_{M_x})\underline{\widetilde{F}}^{\mathcal{H}_T}. \tag{19}$$

Next, we describe the calculation of the solution $\underline{u}$ using the representation (19). The first step in (19) is the calculation of the vector

$$\underline{g} = (\underline{g}_1, \underline{g}_2, \ldots, \underline{g}_{pN_t})^\top = (Y_t \otimes I_{M_x})\underline{\widetilde{F}}^{\mathcal{H}_T} = \operatorname{vec}\big(\hat{F}(A_{h_t}^{\mathcal{H}_T})^{-1} Q_t\big) \in \mathbb{R}^{pN_t \cdot M_x} \tag{20}$$

with a matrix $\hat{F} \in \mathbb{R}^{M_x \times pN_t}$ due to (9), satisfying $\operatorname{vec}(\hat{F}) = \underline{\widetilde{F}}^{\mathcal{H}_T}$, where $\underline{g}_\ell \in \mathbb{R}^{M_x}$, $\ell = 1, \ldots, pN_t$. The second step in (19) is to solve the linear system

$$(I_{pN_t} \otimes M_{h_x} + R_t \otimes A_{h_x})\underline{z} = \underline{g} \tag{21}$$

for the vector $\underline{z} = (\underline{z}_1, \underline{z}_2, \ldots, \underline{z}_{pN_t})^\top \in \mathbb{R}^{pN_t \cdot M_x}$, where $\underline{z}_\ell \in \mathbb{R}^{M_x}$, $\ell = 1, \ldots, pN_t$. The third step in (19) is the calculation of the desired unknown

$$\underline{u} = (Q_t \otimes I_{M_x})\underline{z} = \operatorname{vec}\left(Z Q_t^\top\right) \in \mathbb{R}^{pN_t \cdot M_x} \tag{22}$$

with a matrix $Z \in \mathbb{R}^{M_x \times pN_t}$ corresponding to the relation (9), satisfying $\operatorname{vec}(Z) = \underline{z}$.

It remains to solve the linear system (21), which has a special triangular structure. Thus, this system can be solved by a backward substitution technique, which we describe in the following in more detail. Let $k \in \{1, \ldots, pN_t - 1\}$ be such that $\underline{z}_{k+1}, \ldots, \underline{z}_{pN_t}$ are already computed, or let $k = pN_t$, where we set $\sum_{j=pN_t+1}^{pN_t}(\cdot) := 0$. Then, two cases occur, as the diagonals of R_t have 2×2 and 1×1 blocks:

1. In the case of a 1×1 block of R_t, i.e., $k = 1$ or $R_t[k, k-1] = 0$, the linear system

$$(M_{h_x} + R_t[k,k]A_{h_x})\underline{z}_k = \underline{g}_k - \sum_{j=k+1}^{pN_t} R_t[k,j]A_{h_x}\underline{z}_j \tag{23}$$

has to be solved for $\underline{z}_k$.

2. In the case of a 2×2 block of R_t, i.e., $k > 1$ and $R_t[k, k-1] \neq 0$, the linear system

$$\begin{pmatrix} |b_2|\left(M_{h_x} + \alpha A_{h_x}\right) & -b_1|b_2|A_{h_x} \\ |b_1|b_2 A_{h_x} & -|b_1|\left(M_{h_x} + \alpha A_{h_x}\right) \end{pmatrix} \begin{pmatrix} \underline{z}_{k-1} \\ -\underline{z}_k \end{pmatrix} = \begin{pmatrix} |b_2|\left(\underline{g}_{k-1} - \sum_{j=k+1}^{pN_t} R_t[k-1,j]A_{h_x}\underline{z}_j\right) \\ |b_1|\left(\underline{g}_k - \sum_{j=k+1}^{pN_t} R_t[k,j]A_{h_x}\underline{z}_j\right) \end{pmatrix} \tag{24}$$

with $\alpha = R_t[k-1,k-1] = R_t[k,k]$, $b_1 = R_t[k-1,k] \neq 0$, $b_2 = R_t[k,k-1] \neq 0$ has to be solved for $\underline{z}_{k-1}$ and $\underline{z}_k$.

The linear systems (23) and (24) show that the behaviour of the eigenvalues of the matrix $(A_{h_t}^{\mathcal{H}_T})^{-1}M_{h_t}^{\mathcal{H}_T}$ is essential for their solver. So, let $\lambda_1, \ldots, \lambda_{pN_t} \in \mathbb{C}$ be the eigenvalues of the matrix $(A_{h_t}^{\mathcal{H}_T})^{-1}M_{h_t}^{\mathcal{H}_T} \in \mathbb{R}^{pN_t \times pN_t}$. Due to the properties of $\mathcal{H}_T$, the matrices $M_{h_t}^{\mathcal{H}_T}$ and $A_{h_t}^{\mathcal{H}_T}$ are positive definite, and $A_{h_t}^{\mathcal{H}_T}$ is symmetric. Thus, [7, Lemma 3.2] yields that the real parts of the eigenvalues satisfy

$$\forall \ell \in \{1, \ldots, pN_t\}: \quad \Re(\lambda_\ell) > 0, \tag{25}$$

independent of the polynomial degree p and the time mesh (10). With the property (25), the system matrix of the linear system (23) is symmetric and positive definite, since $R_t[k,k] > 0$ is a real eigenvalue of the matrix $(A_{h_t}^{\mathcal{H}_T})^{-1} M_{h_t}^{\mathcal{H}_T}$. The system matrix of the linear system (24) is symmetric, but indefinite due to the property $b_1|b_2| = -|b_1|b_2$. Note that the linear system (24) is uniquely solvable, since multiplying the second equation in (24) by -1 leads to a nonsymmetric, but positive definite system matrix, due to the property (25) leading to $\alpha = R_t[k-1,k-1] = R_t[k,k] > 0$. The spatial linear systems (23) and (24) can be solved by (preconditioned) iterative solvers or by direct solvers. In this work, we consider sparse direct solvers only, see, e.g., [2, 10, 11]. The resulting algorithm of the Bartels–Stewart method with real-Schur decomposition is summarised as follows:

1. Compute the real-Schur decomposition $(A_{h_t}^{\mathcal{H}_T})^{-1} M_{h_t}^{\mathcal{H}_T} = Q_t R_t Q_t^\top$ in (18).
2. Solve $\underline{g} = (\underline{g}_1, \underline{g}_2, \ldots, \underline{g}_{pN_t})^\top = (Y_t \otimes I_{M_x}) \underline{\widetilde{F}}^{\mathcal{H}_T} = \text{vec}\left(\hat{F}(A_{h_t}^{\mathcal{H}_T})^{-1} Q_t\right)$ in (20).
3. Set $k = pN_t$ and compute $\underline{z} = (\underline{z}_1, \underline{z}_2, \ldots, \underline{z}_{pN_t})^\top$ sequentially
   - in the case $k = 1$ or $R_t[k, k-1] = 0$, by solving (23) for $\underline{z}_k$, and setting $k = k - 1$,
   - or otherwise, by solving (24) for $(\underline{z}_{k-1}, -\underline{z}_k)^\top$ and setting $k = k - 2$.
4. Compute the matrix-vector product $\underline{u} = (Q_t \otimes I_{M_x})\underline{z} = \text{vec}\left(Z Q_t^\top\right)$ in (22).

In the remainder of this section, we state the computational cost and the memory requirement of the Bartels–Stewart method with real-Schur decomposition, which is investigated in the same way as in [8, Subsection 4.2.1]. The overall computational cost and memory consumption of the Bartels–Stewart method with real-Schur decomposition are

$$O(p^3 N_t^3 + M_x p^2 N_t^2 + C_C(M_x) \cdot pN_t) \quad \text{and} \quad O(p^2 N_t^2 + M_x pN_t + C_S(M_x)),$$

where solving a spatial linear system of the form (23) or (24) requires $O(C_C(M_x))$ operations and $O(C_S(M_x))$ storage with the cost function $C_C(\cdot)$ and the storage function $C_S(\cdot)$ defined by the spatial solver for the corresponding linear systems. Note that the calculations related to the term $M_x p^2 N_t^2$ are few matrix multiplications, which are parallelisable and can be written as highly efficient BLAS-3 operations.

For the case of a uniform refinement strategy in temporal and spatial direction, i.e., N_t doubles and M_x grows by a factor $O(2^d)$ in each refinement step, the number of the degrees of freedom dof $= pN_t \cdot M_x$ increases by a factor $O(2^{d+1})$. Hence, we

Table 1 Summary of the complexity of the Bartels–Stewart method for a uniform refinement strategy, using a sparse direct solver for spatial structured grids

d	Computations	Memory
1	$O(\text{dof}^{3/2} + p^2\text{dof})$	$O(\text{dof} + p\,\text{dof}^{1/2})$
2	$O(\text{dof}^{4/3} + p^4\text{dof})$	$O(\text{dof} + p^2\text{dof}^{2/3})$
3	$O(\text{dof}^{7/4} + p^6\text{dof})$	$O(\text{dof} + p^3\text{dof}^{3/4})$

have $pN_t \sim M_x^{1/d} \sim \text{dof}^{1/(d+1)}$, which results in the complexity in Table 1, when the spatial direct solver satisfies

$$C_C(M_x) = \begin{cases} p^2 M_x, & d=1, \\ M_x^{3/2} + p^4 M_x, & d=2, \\ M_x^2 + p^6 M_x, & d=3, \end{cases} \quad \text{and} \quad C_S(M_x) = \begin{cases} p\, M_x, & d=1, \\ M_x \ln \frac{M_x}{p^2} + p^2 M_x, & d=2, \\ M_x^{4/3} + p^3 M_x, & d=3, \end{cases}$$

see [2, 11] for the p dependence on structured grids. Note that M_x depends also on the polynomial degree p, see, e.g., [4, (19.11a) and (19.11b)] for $d = 2$ or $d = 3$, and for $d = 1$, $M_x = pN_x - 1$ with N_x spatial elements.

6 Numerical Examples

In this section, numerical examples for the space-time finite element method (14) are given. For this purpose, we consider the parabolic initial-boundary value problem (1) in the one-dimensional spatial domain $\Omega := (0, 1)$ with the terminal time $T = 10$, i.e., the rectangular space-time domain $Q = \Omega \times (0, T) = (0, 1) \times (0, 10)$. As exact solution, we choose the smooth function $u(x, t) = \left(e^{-tx} - 1\right) \sin(\pi x) \cos(4tx)$ for $(x, t) \in Q$. The spatial domain $\Omega = (0, 1)$ is decomposed into nonuniform elements with the vertices

$$x_0 = 0, \quad x_1 = 1/4, \quad x_2 = 1, \tag{26}$$

whereas the temporal domain $(0, T) = (0, 10)$ is decomposed into nonuniform elements with the vertices

$$t_0 = 0, \quad t_1 = T/8, \quad t_2 = T/4, \quad t_3 = T. \tag{27}$$

We apply a uniform refinement strategy for the meshes (26), (27), and we choose the polynomial degrees $p = 1$ or $p = 2$ or $p = 6$. The assembling of the matrices $A_{h_t}^{\mathcal{H}_T}$ and $M_{h_t}^{\mathcal{H}_T}$ is done as proposed in [15, Subsection 3.1] (or alternatively, [17, Subsection 2.2]), where we use standard Lagrange finite elements

Table 2 Numerical results of the Galerkin finite element discretisation (14) for $p = 1$ for the smooth function u for a uniform refinement strategy with the initial meshes (26), (27)

dof	$h_{x,\max}$	$h_{x,\min}$	$h_{t,\max}$	$h_{t,\min}$	$\|u - \widetilde{u}_h\|_{H^{1/2}_{0,}(0,T;L^2(\Omega))}$	eoc	Solving
24384	0.011719	0.003906	0.117188	0.019531	5.816e−02	1.5	0.1
97920	0.005859	0.001953	0.058594	0.009766	2.055e−02	1.5	0.3
392448	0.002930	0.000977	0.029297	0.004883	7.264e−03	1.5	1.1
1571328	0.001465	0.000488	0.014648	0.002441	2.568e−03	1.5	6.1
6288384	0.000732	0.000244	0.007324	0.001221	9.079e−04	1.5	51.1
25159680	0.000366	0.000122	0.003662	0.000610	3.210e−04	1.5	455.3
100651008	0.000183	0.000061	0.001831	0.000305	1.135e−04	1.5	3438.1

Table 3 Numerical results of the Galerkin finite element discretisation (14) for $p = 2$ for the smooth function u for a uniform refinement strategy with the initial meshes (26), (27)

dof	$h_{x,\max}$	$h_{x,\min}$	$h_{t,\max}$	$h_{t,\min}$	$\|u - \widetilde{u}_h\|_{H^{1/2}_{0,}(0,T;L^2(\Omega))}$	eoc	Solving
24384	0.023438	0.007812	0.234375	0.039062	1.169e−02	2.5	0.1
97920	0.011719	0.003906	0.117188	0.019531	2.042e−03	2.5	0.3
392448	0.005859	0.001953	0.058594	0.009766	3.587e−04	2.5	1.4
1571328	0.002930	0.000977	0.029297	0.004883	6.329e−05	2.5	6.5
6288384	0.001465	0.000488	0.014648	0.002441	1.118e−05	2.5	56.1
25159680	0.000732	0.000244	0.007324	0.001221	1.976e−06	2.5	470.3
100651008	0.000366	0.000122	0.003662	0.000610	3.494e−07	2.5	3407.3

[4, Subsection 6.3]. The appearing integrals to compute the related right-hand side of (15) are calculated by using high-order quadrature rules. To solve the global linear system (15), all steps of the Bartels–Stewart method with real-Schur decomposition of Sect. 5 are realised in MATLAB, including its backslash operator for solving linear systems. Due to the interpolation inequality, $\sqrt{\|\cdot\|_{L^2(Q)}\|\partial_t\cdot\|_{L^2(Q)}}$ is an upper bound for $C\|\cdot\|_{H^{1/2}_{0,}(0,T;L^2(\Omega))}$ with $C > 0$ and is used as approximation of $\|\cdot\|_{H^{1/2}_{0,}(0,T;L^2(\Omega))}$ for all numerical experiments in this section. All calculations presented were performed on a PC with two Intel Xeon E5-2687W v4 CPUs 3.00 GHz with 512 GB main memory. Last, recall that the number of the degrees of freedom is dof $= pN_t \cdot M_x$.

For $p = 1$, $p = 2$ and $p = 6$, the results for the smooth solution u are stated in Tables 2, 3 and 4, respectively, where we illustrate that the space-time finite element method (14) is unconditionally stable and the convergence rates with respect to the space-time norm $\|\cdot\|_{H^{1/2}_{0,}(0,T;L^2(\Omega))}$ are as expected, see (13). The last column of Tables 2, 3 and 4 states the computation times in seconds of the Bartels–Stewart method with real-Schur decomposition, where the assembling time of the involved matrices and right-hand sides is not included. We observe that the calculating time for the last level in Tables 2, 3 and 4 grows with factors 7.6 for $p = 1$, 7.2 for $p = 2$ and 4.3 for $p = 6$, respectively, which are smaller than the factor 8 resulting from $O(\mathrm{dof}^{3/2})$ in Table 1. Moreover, a comparison of Tables 2, 3 and 4 shows

Table 4 Numerical results of the Galerkin finite element discretisation (14) for $p = 6$ for the smooth function u for a uniform refinement strategy with the initial meshes (26), (27)

dof	$h_{x,\max}$	$h_{x,\min}$	$h_{t,\max}$	$h_{t,\min}$	$\|u - \widetilde{u}_h\|_{H^{1/2}_{0,}(0,T;L^2(\Omega))}$	eoc	Solving
198	0.750000	0.250000	7.500000	1.250000	1.747e+00	–	≈ 0.0
828	0.375000	0.125000	3.750000	0.625000	6.887e−01	1.3	≈ 0.0
3384	0.187500	0.062500	1.875000	0.312500	2.107e−02	5.0	≈ 0.0
13680	0.093750	0.031250	0.937500	0.156250	3.075e−04	6.1	≈ 0.0
55008	0.046875	0.015625	0.468750	0.078125	3.334e−06	6.5	0.2
220608	0.023438	0.007812	0.234375	0.039062	3.505e−08	6.6	0.8
883584	0.011719	0.003906	0.117188	0.019531	3.926e−10	6.5	3.4

that a polynomial degree $p > 1$ is advisable since the numbers of the degrees of freedom and computation times are much lower for $p > 1$ than for $p = 1$ when a fixed accuracy is desired. For example, we need 455.3 seconds for dof = 25159680 degrees of freedom for $p = 1$, 1.4 seconds for dof = 392448 degrees of freedom for $p = 2$ and ≈ 0.0 seconds for dof = 13680 degrees of freedom for $p = 6$ to receive the error in $\|\cdot\|_{H^{1/2}_{0,}(0,T;L^2(\Omega))}$ within a comparable range.

7 Conclusions

In this work, we generalised the efficient direct solvers for the global linear system arising from the space-time Galerkin finite element discretisation of parabolic initial-boundary value problems to arbitrary polynomial degrees. The developed algorithm is based on the Bartels–Stewart method. We gave complexity estimates for this algorithm. Moreover, we presented numerical experiments for a one-dimensional spatial domain for different polynomial degrees, where the spatial subproblems were solved by sparse direct solvers. These numerical results confirmed the unconditional stability, the space-time error estimates and the efficient applicability of the higher-order space-time approach in connection with the proposed direct space-time solver.

References

1. Bartels, R.H., Stewart, G.W.: Algorithm 432: Solution of the matrix equation $AX + XB = C$. Commun. ACM **15**, 820–826 (1972)
2. Collier, N., Pardo, D., Paszynski, M., Calo, V.M.: Computational complexity and memory usage for multi-frontal direct solvers in structured mesh finite elements [cs.NA] (2012). arXiv:1204.1718, arXiv.org
3. Devaud, D.: Petrov–Galerkin space-time *hp*-approximation of parabolic equations in $H^{1/2}$. IMA J. Numer. Anal. **40**(4), 2717–2745 (2020)

4. Ern, A., Guermond, J.L.: Finite Elements I: Approximation and Interpolation, vol. 72. Springer, Cham (2020)
5. Fontes, M.: Initial-boundary value problems for parabolic equations. Ann. Acad. Sci. Fenn. Math. **34**(2), 583–605 (2009)
6. Gardiner, J.D., Laub, A.J., Amato, J.J., Moler, C.B.: Solution of the Sylvester matrix equation $AXB^T + CXD^T = E$. ACM Trans. Math. Softw. **18**(2), 223–231 (1992)
7. Hofer, C., Langer, U., Neumüller, M., Schneckenleitner, R.: Parallel and robust preconditioning for space-time isogeometric analysis of parabolic evolution problems. SIAM J. Sci. Comput. **41**(3), A1793–A1821 (2019)
8. Langer, U., Zank, M.: Efficient direct space-time finite element solvers for parabolic initial-boundary value problems in anisotropic Sobolev spaces. SIAM J. Sci. Comput. **43**(4), A2714–A2736 (2021)
9. Larsson, S., Schwab, C.: Compressive space-time Galerkin discretizations of parabolic partial differential equations [math.NA] (2015). arXiv:1501.04514, arXiv.org
10. Martinsson, P.G.: Fast Direct Solvers for Elliptic PDEs, vol. 96. SIAM, Philadelphia (2020)
11. Pardo, D., Paszynski, M., Collier, N., Alvarez, J., Dalcin, L., Calo, V.M.: A survey on direct solvers for Galerkin methods. SeMA J. **57**, 107–134 (2012)
12. Simoncini, V.: Computational methods for linear matrix equations. SIAM Rev. **58**(3), 377–441 (2016)
13. Steinbach, O., Missoni, A.: A note on a modified Hilbert transform. Applicable Anal., 1–8 (2022). https://doi.org/10.1080/00036811.2022.2030725
14. Steinbach, O., Zank, M.: Coercive space-time finite element methods for initial boundary value problems. Electron. Trans. Numer. Anal. **52**, 154–194 (2020)
15. Steinbach, O., Zank, M.: A note on the efficient evaluation of a modified Hilbert transformation. J. Numer. Math. **29**(1), 47–61 (2021)
16. Zank, M.: Inf-Sup Stable Space-Time Methods for Time-Dependent Partial Differential Equations. Monographic Series TU Graz: Computation in Engineering and Science, vol. 36. Verlag der Technischen Universität Graz, Graz (2020)
17. Zank, M.: An exact realization of a modified Hilbert transformation for space-time methods for parabolic evolution equations. Comput. Methods Appl. Math. **21**(2), 479–496 (2021)

A Subcell Limiting Based on High-Order Compact Nonuniform Nonlinear Weighted Schemes for High-Order CPR Method

Huajun Zhu, Zhen-Guo Yan, Feiran Jia, and Guo-Quan Shi

1 Introduction

High-order methods have been widely used in large eddy simulations (LES) and direct numerical simulations (DNS) of turbulent flows, computational aeroacoustics (CAA) and shock-induced separation flows [4, 10, 20]. Among high-order methods, high-order correction procedure via reconstruction method (CPR) are compact, highly parallelizable, efficient for high-performance computing and applicable to complex unstructured meshes [9, 10, 19, 20]. For Euler equations, since solution may contain discontinuities, numerical methods need to be designed carefully to capture shocks effectively. There exist different strategies to deal with spurious oscillations of finite element (FE) method containing DG method and CPR method, such as adding artificial viscosity [5, 13], limiting high-order solution distribution [1, 11, 12, 14, 22], and developing a hybrid method [2, 3, 6]. This paper will investigate subcell limiting strategy, which is developing a hybrid method. Many authors do lots of works in developing this strategy [6–8, 18].

Recently, a priori subcell limiting approach based on compact nonuniform nonlinear weighted schemes (abbr. CNNW) was first proposed by Zhu et al. in [24] for CPR method with Legendre-Gauss solution points. A CPR scheme based on subcell CNNW limiting (abbr. CPR-CNNW) is a hybrid scheme. Both high-order and low-order CNNW can be applied to solve troubled cells. Various numerical

H. Zhu · Z.-G. Yan (✉) · G.-Q. Shi
State Key Laboratory of Aerodynamics, Mianyang, China
e-mail: zgyan@skla.cardc.cn; yanzhg@mail.ustc.edu.cn

F. Jia
State Key Laboratory of Aerodynamics, Mianyang, China

School of Power and Energy, Northwesten Polytechnical University, Xi'an, China

J. M. Melenk et al. (eds.), *Spectral and High Order Methods for Partial Differential Equations ICOSAHOM 2020+1*, Lecture Notes in Computational Science and Engineering 137, https://doi.org/10.1007/978-3-031-20432-6_38

experiments for linear wave equations and Euler equations are conducted to show the good properties of the CPR-CNNW in high resolution, good robustness in shock capturing and satisfying discrete conservation law. And one CPR-CNNW was generalized to unstructured meshes in [17]. This paper will further investigate one kind of CPR-CNNW which only use high-order CNNW for solving troubled cells. One-dimensional Shu-Osher problem, two-dimensional Riemann problem, and shock-vortex interaction problem are simulated to show the effectiveness of the subcell high-order CNNW limiting of CPR in keeping high resolution and shock capturing robustness.

2 Review of High-Order CPR and High-Order CNNW

2.1 High-Order CPR

Correction procedure via reconstruction (CPR) method was originally proposed by Huynh as flux reconstruction (FR) for structured grids [9] and then was generalized to unstructured grids by Wang et al.[19]. Here we give a brief review of the CPR method. For more details we refer to papers[9, 17, 19, 23, 24].

Consider conservation law in physical space

$$\frac{\partial \mathbf{U}}{\partial t} + \nabla \cdot \mathbf{F}(\mathbf{U}) = \mathbf{0}, \tag{1}$$

where $\mathbf{U}$ is the conservative variable vector, and $\mathbf{F}$ is the inviscid flux vector. After transformation to the computational space, the conservation law (1) in the two-dimensional case becomes

$$\frac{\partial \hat{\mathbf{U}}}{\partial t} + \frac{\partial \hat{\mathbf{F}}}{\partial \xi} + \frac{\partial \hat{\mathbf{G}}}{\partial \eta} = \mathbf{0}, \tag{2}$$

where $\hat{\mathbf{U}} = J\mathbf{U}$, $\hat{\mathbf{F}} = \mathbf{F}\widehat{\xi}_x + \mathbf{G}\widehat{\xi}_y$, $\hat{\mathbf{G}} = \mathbf{F}\widehat{\eta}_x + \mathbf{G}\widehat{\eta}_y$. Here $\widehat{\xi}_x$, $\widehat{\xi}_y$, $\widehat{\eta}_x$ and $\widehat{\eta}_y$ are grid metrics, and J is the Jacobian. For more details we refer to [9, 19, 23].

In the CPR method, the conservation variable vector inside one element is approximated by polynomials, for example the degree K Lagrange interpolation polynomial. Then, the Lagrange Polynomial (LP) approach is applied to approximate the second term and the third term in (2),

$$\begin{aligned} \widehat{\mathbf{F}}_{i,j}(\xi,\eta) &= \sum_{l=1}^{K+2}\sum_{m=1}^{K+1} \widehat{\mathbf{F}}_{i,j,l,m} L_l(\xi) L_m(\eta), \\ \widehat{\mathbf{G}}_{i,j}(\xi,\eta) &= \sum_{l=1}^{K+1}\sum_{m=1}^{K+2} \widehat{\mathbf{F}}_{i,j,l,m} L_l(\xi) L_m(\eta). \end{aligned} \tag{3}$$

The flux derivatives in (2) can be computed. Therefore, the nodal values of the state variable **U** at the solution points are updated by the following equations

$$\frac{\partial \widehat{\mathbf{U}}_{i,j,l,m}}{\partial t} + \frac{\partial \widehat{\mathbf{F}}_{i,j}(\xi_l, \eta_m)}{\partial \xi} + \frac{\partial \widehat{\mathbf{G}}_{i,j}(\xi_l, \eta_m)}{\partial \eta} + \delta_{i,j}(\xi_l, \eta_m) = 0, \tag{4}$$

where $1 \leq l, m \leq K + 1$ and

$$\begin{aligned}\delta_{i,j}(\xi_l, \eta_m) = \\ &\left[\overline{F}_{i,j}(-1, \eta_m) - \widehat{F}_{i,j}(-1, \eta_m)\right] g'_L(\xi_l) + \left[\overline{F}_{i,j}(1, \eta_m) - \widehat{F}_{i,j}(1, \eta_m)\right] g'_R(\xi_l) \\ &+ \left[\overline{G}_{i,j}(\xi_l, -1) - \widehat{G}_{i,j}(\xi_l, -1)\right] g'_L(\eta_m) + \left[\overline{G}_{i,j}(\xi_l, 1) - \widehat{G}_{i,j}(\xi_l, 1)\right] g'_R(\eta_m).\end{aligned}$$

Here $\delta_{i,j}(\xi, \eta)$ is a correction flux polynomial, $g_L(\xi)$ and $g_R(\xi)$ are both the degree $K + 1$ polynomials called correction functions. $\overline{F}$ and $\overline{G}$ are the common fluxes.

In this paper, for the CPR method we choose the Legendre-Gauss (LG) points as solution points and the Legendre-Gauss-Lobatto (LGL) points as flux points, and the correction functions take the Radau polynomials. These are the same as in [24]. In this paper, we also mainly consider 5th-order CPR (CPR5) with $K = 4$.

2.2 High-Order CNNW

A high-order compact nonuniform nonlinear weighted (CNNW) schemes was developed for subcell limiting of high-order CPR method in [24]. Here we give a brief review of fifth-order CNNW (C5NNW5). The solution points and flux points are the same as CPR and a CPR cell will be split into CNNW subcells according to the location of the flux points, as shown in Fig. 1. A fifth-order CNNW was constructed by using a 5th-order nonuniform nonlinear weighted (NNW) interpolation and a 5th-order compact flux difference operator. Riemann fluxes are used for each fluxes. The flux derivatives are computed by 5th-order compact flux difference operator of the Lagrange polynomial at the solution points

$$\frac{\partial \widetilde{F}(\xi)}{\partial \xi}|_{i,sp_m} = \sum_{l=1}^{6} u_{m,l} \overline{F}_{i,fp_l},$$

where $\widetilde{F}(\xi)$ is Lagrange polynomial

$$\widetilde{F}(\xi) = \sum_{l=1}^{6} \overline{F}_{i,fp_l} L_l(\xi),$$

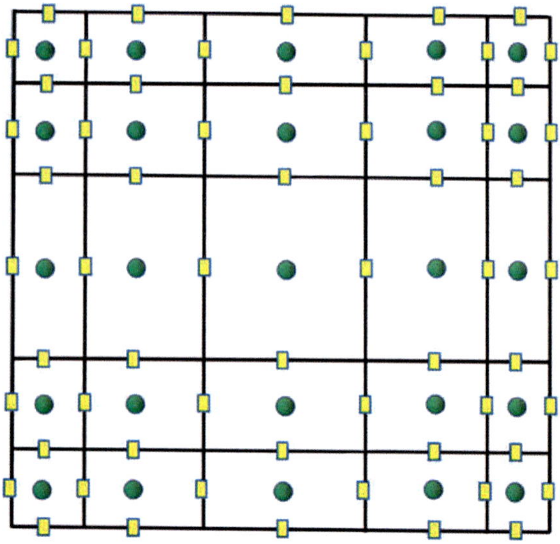

Fig. 1 Solution points and flux points for CNNW5 subcells of CPR5 ($K = 4$)

$\overline{F}_{i,fp_l}$ is the Riemann flux at the LGL flux points $\overline{F}_{i,fp_l}(u^L_{i,fp_l}, u^R_{i,fp_l})$, and $\{L_l(\xi), l = 1, 2, \cdots, 6\}$ are the Lagrange basis. Here u^L_{i,fp_l} and u^R_{i,fp_l} are obtained by 5th-order NNW interpolation. The new simple smoothness indicator in [24] is used.

The difference between the high-order CNNW and the high-order CPR is that CNNW uses nonlinear interpolation based on solution points of the cell and its neighbor cells, uses Riemann fluxes for each flux points, and does not use correction functions.

3 A Priori Subcell CNNW Limiting Approach for CPR Method

Firstly, a troubled cell indicator is used to detect troubled cells which may have discontinuities. Then, the troubled cells are split into subcells and computed by the CNNW schemes while the other cells are computed by the CPR scheme.

3.1 Troubled Cell Indicator

To find troubled cells, we consider the highest modal decay indicator based on the "extended" stencil (MDHE indicator) in [17, 24], which follows the indicator proposed in [8] and the idea presented by Persson and Peraire [13].

For the fifth-order CPR scheme with $K = 4$, the indicator of the ith cell is calculated based on the stencil with seven points $\{\epsilon_L, \epsilon_{i,1}, \epsilon_{i,2}, \epsilon_{i,3}, \epsilon_{i,4}, \epsilon_{i,5}, \epsilon_R\}$ to consider the jump at cell interfaces, where $\epsilon_{i,1}, \epsilon_{i,2}, \epsilon_{i,3}, \epsilon_{i,4}, \epsilon_{i,5}$ are the quantity $\epsilon = \rho p$ at five solution points, $\epsilon_L = aver(\epsilon_{i-1,5}, \epsilon_{i,1})$ and $\epsilon_R = aver(\epsilon_{i,5}, \epsilon_{i+1,1})$ are Roe average values at cell interfaces. Here $aver(\epsilon_1, \epsilon_2) = aver(\rho_1, \rho_2) \cdot aver(p_1, p_2)$ and $aver(\cdot, \cdot)$ is the Roe average function. Firstly, the representation of the quantity $\epsilon = \rho p$ with Lagrange interpolation polynomials of degree N is transformed to a modal representation with Legendre interpolation polynomials. Secondly, the maximum of proportion of the highest modes and proportion of the second highest mode to the total energy of the Legendre interpolation polynomial is calculated as

$$EI = \max\left(\frac{m_N^2}{\sum_{j=0}^{N} m_j^2}, \frac{m_{N-1}^2}{\sum_{j=0}^{N-1} m_j^2}\right), \tag{5}$$

where $N = K + 2$ and $\{m_j | j = 0, 1, \cdots, N\}$ are the modal coefficients. We take a threshold value

$$c(a) = a \cdot 10^{-1.8((K+2)+1)^{1/4}}, \tag{6}$$

which is similar as those in [8, 17, 24]. Different a will be considered in numerical tests. If $EI \geq c(a)$, the element is denoted as a troubled cell.

3.2 Subcell Limiting Based on CNNW

After troubled cell detection, the troubled cells are decomposed into subcells and solved by CNNW, while the other cells are computed by CPR. Thus, a CPR scheme based on subcell CNNW limiting (abbr. CPR-CNNW) in [24] is a hybrid scheme. CNNW has the same solution points as CPR, which make the CPR based on subcell CNNW limiting approach has some merits. The hybrid CPR-CNNW scheme (abbr. HCCS) can be expressed as

$$\text{HCCS} = \begin{cases} \text{CPR5}, & 0 \leq EI \leq S_1, \\ \text{C5NNW5}, & S_1 < EI \leq S_2, \\ \text{C2NNW5}, & S_2 < EI \leq S_3, \\ \text{C2NNW2}, & S_3 < EI \leq 1. \end{cases} \tag{7}$$

with the partition vector $\mathbf{dv} = (S_1, S_2, S_3)$. By controlling the vector $\mathbf{dv}$, the CPR-CNNW scheme (7) can contain some of the four schemes and can be denoted by HCCS(d_1, d_2, d_3, d_4) with d_1, d_2, d_3, d_4 marking status of CPR, C5NNW5, C2NNW5, C2NNW2 correspondingly. Here $d_i = 1$ means that the correspond-

Table 1 Hybrid CPR-CNNW schemes

HCCS(d_1, d_2, d_3, d_4)	$\mathbf{dv} = (S1, S2, S3)$	Containing schemes
HCCS(1,1,0,0)	$0<S1<S2=S3=1$	CPR5,C5NNW5
HCCS(1,0,0,1)	$0<S1=S2=S3<1$	CPR5,C2NNW2
HCCS(1,1,1,1)	$0<S1<S2<S3<1$	CPR5,C5NNW5,C2NNW5,C2NNW2

ing scheme is contained by the hybrid scheme, otherwise it is not included. Hybrid schemes HCCS(1, 1, 1, 1) and HCCS(1, 0, 0, 1) were tested in [24] and HCCS(1, 0, 0, 1) has been generalized to unstructured meshes in [17]. In this paper, we will mainly test HCCS(1, 1, 0, 0) which only contains CPR and C5NNW5, as shown in Table 1.

Riemann fluxes at the interface of CPR and C5NNW5 are computed based on the two values of one side from CPR cell and the other from C5NNW5.

4 Numerical Investigation

In this section, three test cases are adopted for the Euler equations to illustrate the properties of HCCS.

The threshold value in the MDHE indicator is set to $c(a) = a \cdot 10^{-1.8(6+1)^{1/4}}$. HCCS(1,1,0,0) with $\mathbf{dv} = (c(a), 1.0, 1.0)$ under different a is tested. We consider three cases $a = 0.5$, $a = 0.05$ and $a = 0.005$. Z-weights with $\varepsilon = 10^{-10}$ are used to compute nonlinear weights based characteristic variables as default in the NNW procedure. The explicit third-order TVD Runge-Kutta scheme is used for time integration.

4.1 *Shu-Osher Problem*

The Shu-Osher problem with initial conditions

$$(\rho, u, p) = \begin{cases} (3.857143, 2.629369, 10.333333), & -5 \leq x < -4, \\ (1.0 + 0.2\sin(5x), 0, 1.0), & -4 \leq x < 5, \end{cases}$$

is solved till $t = 1.8$ with 80 cells and $DoFs = 400$.

HCCS(1,1,0,0) with $\mathbf{dv} = (c(a), 1.0, 1.0)$ is applied to solve this problem. For comparison, HCCS(1,0,0,1) with $\mathbf{dv} = (c(a), c(a), c(a))$ is also tested. The Lax-Friedrichs flux is adopted to compute common fluxes. Troubled cells of HCCS(1,1,0,0) at $t = 1.8$ are 3, 9, 21 for $a = 0.5$, $a = 0.05$ and $a = 0.005$, correspondingly. And troubled cells of HCCS(1,0,0,1) at $t = 1.8$ are 3, 7, 19 for $a = 0.5$, $a = 0.05$ and $a = 0.005$, correspondingly. From Fig. 2c and d we can

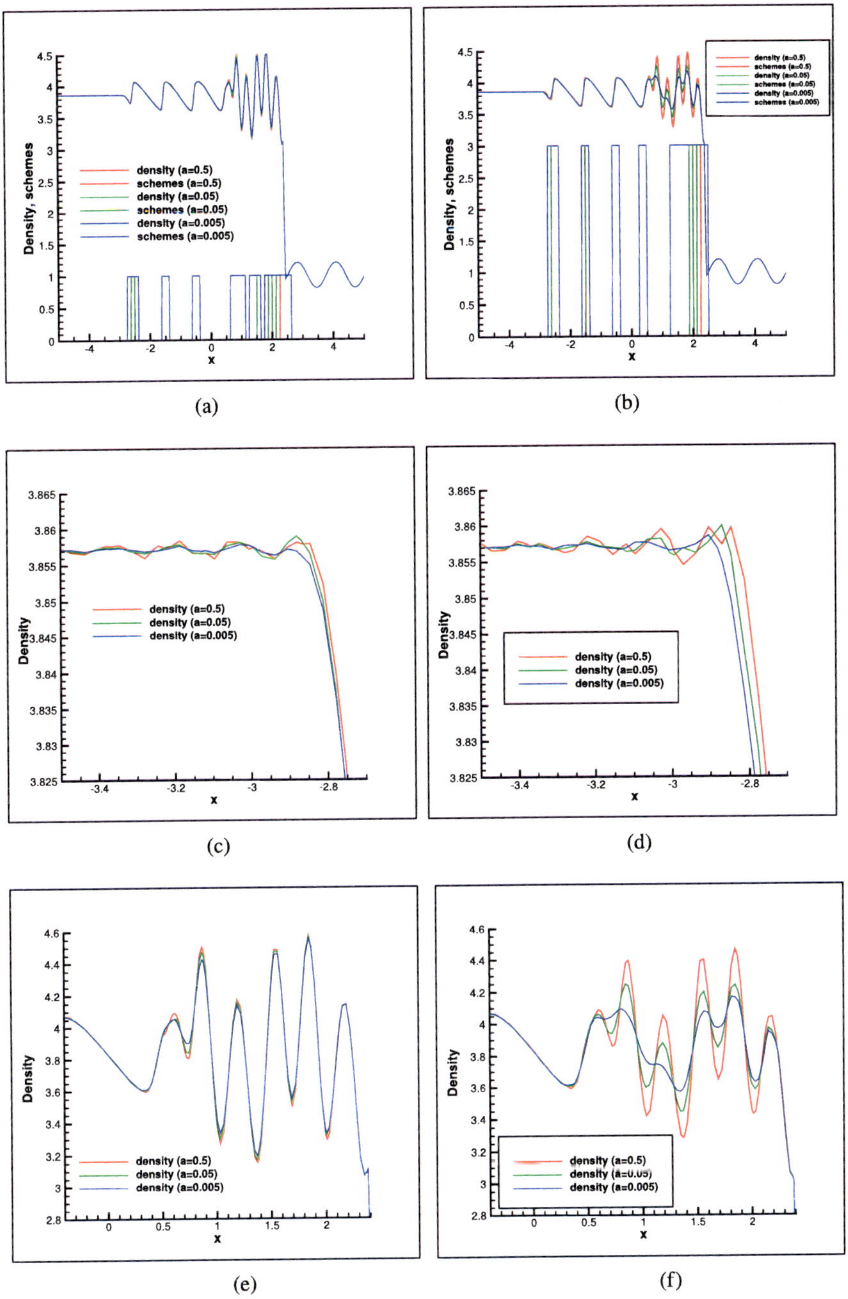

Fig. 2 1D Shu-Osher problem solved by hybrid CPR-CNNW schemes HCCS(1,1,0,0) (left plots) and HCCS(1,0,0,1) (right plots). (**a**) Density on the area [−5.00, 5.00]. (**b**) Density on the area [−5.00, 5.00]. (**c**) Density on the area [−3.50, −2.70]. (**d**) Density on the area [−3.50, −2.70]. (**e**) Density on the area [−0.4, 2.40]. (**f**) Density on the area [−0.4, 2.40]

see that the magnitude of oscillations near $x = -2.8$ is smallest for $a = 0.005$ and biggest for $a = 0.5$. In addition, Fig. 2e and f show that $a = 0.005$ has lowest resolution while $a = 0.5$ has highest resolution. Comparing Fig. 2e and f, we can see that HCCS(1,1,0,0) has much higher resolution than HCCS(1,0,0,1). HCCS(1,1,0,0) can still obtain similar resolution during increasing troubled cells (by increasing a) while the resolution of HCCS(1,0,0,1) decrease dramatically. This results indicate that it is better to introduce high-order shock capturing schemes in subcell limiting to keep high resolution.

4.2 2D Riemann Problem

The hybrid CPR-CNNW scheme is applied to solve 2D Riemann problem proposed by Schulz-Rinne [16]. This problem was also solved in [24]. The Lax-Friedrichs flux is adopted to compute common fluxes. The computational domain $[0, 1] \times [0, 1]$ is divided into four quadrants by two lines $x = 0.8$ and $y = 0.8$ and the initial constant states on the four quadrants are

$$(\rho, u, v, p) = \begin{cases} \mathbf{V}_1 = (1.500, 0, 0, 1.500), & 0.8 \le x \le 1.0,\ \ 0.8 \le y \le 1.0, \\ \mathbf{V}_2 = (0.5323, 1.206, 0, 0.3), & 0.0 \le x < 0.8,\ \ 0.8 \le y \le 1.0, \\ \mathbf{V}_3 = (0.138, 1.206, 1.206, 0.029), & 0.0 \le x < 0.8,\ \ 0.0 \le y < 0.8, \\ \mathbf{V}_4 = (0.5323, 0, 1.206, 0.3), & 0.8 < x \le 1.0,\ \ 0.0 \le y < 0.8. \end{cases}$$

HCCS(1,1,0,0) is adopted to solve this problem till $T = 0.8$. The results in Fig. 3 show that HCCS(1,1,0,0) can still obtain similar results during increasing troubled cells (by increasing a). Again the magnitude of oscillations is smallest for $a = 0.005$ and biggest for $a = 0.5$, which indicates that increasing troubled cells will have some effects in decreasing small oscillations.

4.3 Shock-Vortex Interaction

The hybrid CPR-CNNW scheme is adopted to solve shock-vortex interaction, which is originally proposed by Rault et al. [15] and is one of the benchmark problems of the 5th International Workshop on High Order CFD Methods (HiOCFD5). The test case involves a complex flow pattern with both smooth features and discontinuous waves and has become a benchmark for a high-order numerical scheme [6, 17, 21, 24]. The computational domain is $[0, 2] \times [0, 1]$. The initial conditions are given by a stationary normal shock wave placed at $x = 0.5$ with shock Mach number M_s and by a vortex placed at $(x_c, y_c) = (0.5, 0.5)$ with the strength M_v. In this paper, $M_s = 1.5$ and $M_v = 0.9$ are taken, which is the same as that in [15]. A

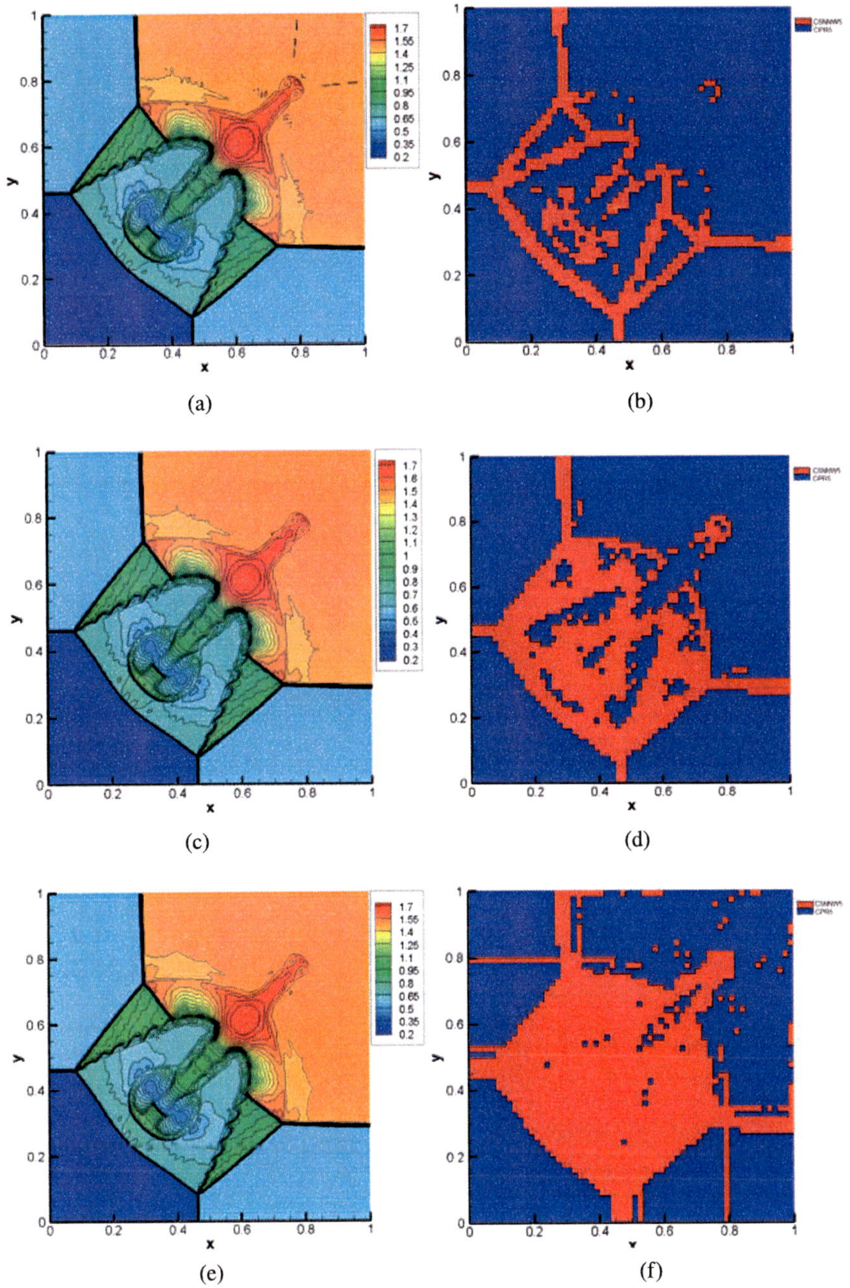

Fig. 3 Density distribution from 0.2 to 1.7 with 31 contours and cell distribution for 2D Riemann problem solved by HCCS(1,1,0,0) with $a = 0.5$, $a = 0.05$, $a = 0.005$. (60×60 grid, $DOFs = 300 \times 300$, $T = 0.8$). (**a**) Density distribution, $a = 0.5$. (**b**) Cell distribution, $a = 0.5$. (**c**) Density distribution, $a = 0.05$. (**d**) Cell distribution, $a = 0.05$. (**e**) Density distribution, $a = 0.005$. (**f**) cell distribution, $a = 0.005$

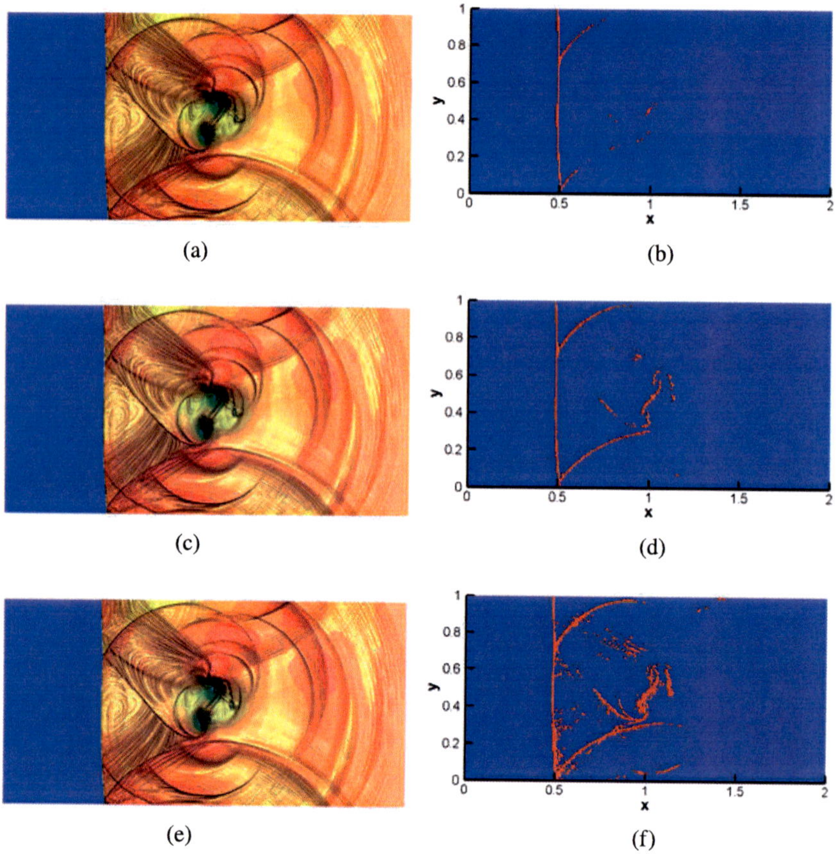

Fig. 4 Shock-vortex interaction problem solved by HCCS(1,1,0,0) ($h = 1/140$, $DoFs = 1400 \times 700$, $T = 0.7$). (**a**) Distribution of density ($a = 0.5$). (**b**) Different cells ($a = 0.5$). (**c**) Distribution of density ($a = 0.05$) (**d**) Different cells ($a = 0.05$). (**e**) Distribution of density ($a = 0.005$). (**f**) Different cells ($a = 0.005$)

computational grid with $h = 1/140$ and $DoFs = 1400 \times 700 = 980000$ is used. Roe flux is adopted to compute common fluxes.

HCCS(1,1,0,0) is applied to solve this problem. The distribution of the mass density and of different cells at time $T = 0.7$ are shown in Fig. 4. Density along two lines are also shown in Fig. 5. The results show that HCCS(1,1,0,0) can capture shock wave as well as smooth vortex features. In addition, HCCS(1,1,0,0) can still obtain similar results during increasing troubled cells (by increasing a). From Fig. 5 we can see that $a = 0.5$ has the highest resolution in capturing vortical structures while the magnitude of oscillations is smallest for $a = 0.005$ and biggest for $a = 0.5$.

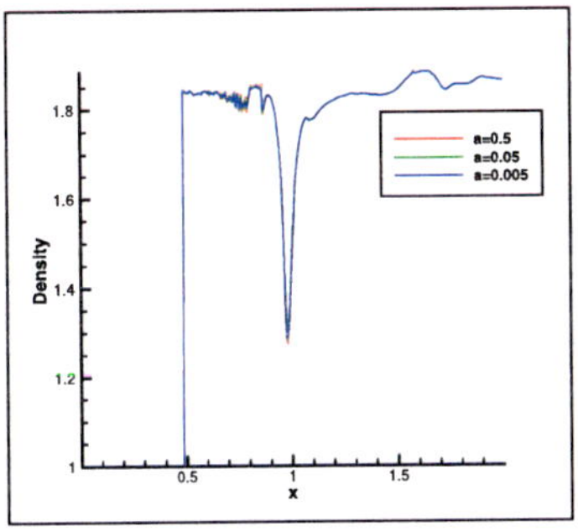

(a)

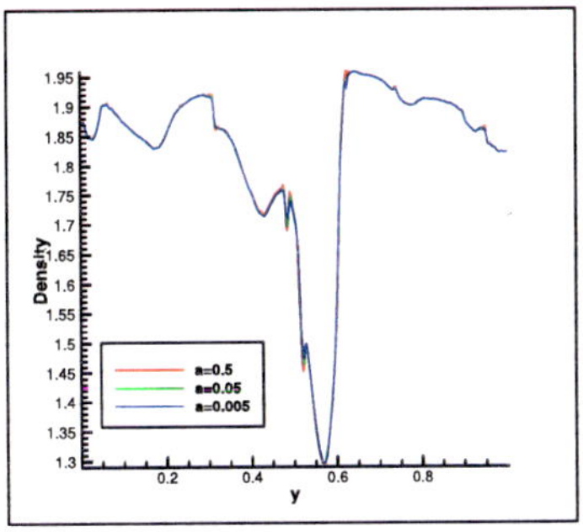

(b)

Fig. 5 Results along two lines of HCCS(1,1,0,0) in solving shock-vortex interaction ($h = 1/140$, $DoFs = 1400 \times 700$, $T = 0.7$). (**a**) Density along Line ($y = 0.4$). (**b**) Density along Line ($x = 1.05$)

5 Concluding Remarks

In this paper, a special CPR-CNNW is further investigated to show its capability in shock capturing and keeping high resolution. Numerical investigations show that the CPR with subcell fifth-order CNNW limiting has higher resolution than subcell second-order CNNW limiting. The CPR-CNNW can keep high resolution during increasing troubled cells. Increasing troubled cells will have some effects in decreasing small oscillations. The scheme combines the advantages of high-order CPR schemes in smooth regions with the robustness and accuracy of high-order CNNW for shock capturing. The limiting approach can be easily applied to DG or other kind of FE method.

Acknowledgments This study was supported by National Natural Science Foundation of China (Grant Nos.12172375, 11902344, 11572342), National Numerical Wind-tunnel project, the foundation of State Key Laboratory of Aerodynamics (Grant No. SKLA2019010101).

References

1. Balsara, D., Meyer, C., Dumbser, M.: A sub-cell based indicator for troubled zones in RKDG schemes and a novel class of hybrid RKDG+HWENO schemes. J. Comput. Phys. **226**, 586–620 (2007)
2. Baumann, C., Oden, J.: A discontinuous hp finite element method for the Euler and Navier-Stokes equations. Int. J. Numer. Methods Fluids. **31**, 79–95 (1999)
3. Burbeau, A., Sagaut, P., Bruneau, C.: A problem-independent limiter for high-order Runge-Kutta discontinuous Galerkin methods. J. Comput. Phys. **169**(1), 111–150 (2001)
4. Deng, X., Mao, M., Tu, G., Zhang, H., Zhang, Y.: High-order and high accurate CFD methods and their applications for complex grid problems. Commun. Comput. Phys. **11**, 1081–1102 (2012)

5. Discacciati, N., Hesthaven, J., Ray, D.: Cotrolling oscillations in high-order discontinuous Galerkin schemes using artificial viscosity tuned by neural networks. J. Comput. Phys. **409**, 109304 (2020)
6. Dumbser, M., Zanotti, O., Loubere, R., Diot, S.: A posteriori subcell limiting of the discontinuous Galerkin finite element method for hyperbolic conservation laws. J. Comput. Phys. **278**, 47–75 (2014)
7. Guo, J., Zhu, H., Yan, Z., Tang, L., Song, S.: High-order hybrid WCNS-CPR scheme for shock capturing of conservation laws. Int. J. Aerosp. Eng. 8825445 (2020)
8. Hennemann, S., Rueda-Ramírez, A., Hindenlang, F., Gassner, G.: A provably entropy stable subcell shock capturing approach for high order split form DG for the compressible Euler equations. J. Comput. Phys. **426**, 109935 (2021)
9. Huynh, H.T.: A flux reconstruction approach to high-order schemes including discontinuous Galerkin methods. In: 18th AIAA Computational Fluid Dynamics Conference, 2007-4079 (2007)
10. Huynh, H., Wang, Z., Vincent, P.: High-order methods for computational fluid dynamics: a brief review of compact differential formulations on unstructured grids. Comput. Fluids **98**, 209–220 (2014)
11. Li, W., Wandand, Q., Ren, Y.X.: A p-weighted limiter for the discontinuous Galerkin method on one-dimensional and two-dimensional triangular grids. J. Comput. Phys. **407**, 109246 (2020)
12. Park, J., Kim, C.: Hierarchical multi-dimensional limiting strategy for correction procedure via reconstruction. J. Comput. Phys. **308**, 57–80 (2016)
13. Persson, P., Peraire, J.: Subcell shock capturing for discontinuous Galerkin methods. In: Proceedings of the 44th AIAA Aerospace Science Meeting and Exhibit (2006)
14. Qiu, J., Shu, C.W.: Hermite WENO schemes and their application as limiters for Runge-Kutta discontinuous Galerkin method: one-dimensional case. J. Comput. Phys. **193**, 115–135 (2004)
15. Rault, A., Chiavassa, G., Donat, R.: Shock-vortex interactions at high Mach numbers. J. Sci. Comput. **19**, 347–371 (2003)
16. Schulz-Rinne, C.: Numerical solution of the Riemann problem for two-dimensional gas dynamics. J. Sci. Comput. **14**, 1394–1414 (1993)
17. Shi, G., Zhu, H., Yan, Z.G.: A priori subcell limiting approach for the FR/CPR method on unstructured meshes. Commun. Comput. Phys. (2022). https://doi.org/10.4208/cicp.OA-2021-0165
18. Sonntag, M., Munz, C.: Efficient parallelization of a shock capturing for discontinuous Galerkin methods using finite volume sub-cells. J. Sci. Comput. **70**(3), 1262–1289 (2017)
19. Wang, Z., Gao, H.: A unifying lifting collocation penalty formulation including the discontinudis Galerkin, spectral volume/difference methods for conservation laws on mixed grids. J. Comput. Phys. **228**, 8161–8186 (2009)
20. Wang, Z., Li, Y., Jia, F., Laskowski, G., Kopriva, J., Paliath, U., Bhaskaran, R.: Towards industrial large eddy simulation using the FR/CPR method. Comput. Fluids **156**, 579–589 (2017)
21. You, H., Kim, C.: High-order multi-dimensional limiting strategy with subcell resolution I: two-dimensional mixed meshes. J. Comput. Phys. **375**, 1005–1032 (2018)
22. Zhong, X., Shu, C.W.: A simple weighted essentially nonoscillatory limiter for Runge-Kutta discontinuous Galerkin methods. J. Comput. Phys. **232**(1), 397–415 (2013)
23. Zhu, H., Yan, Z., Liu, H., Mao, M., Deng, X.: High-order hybrid WCNS-CPR schemes on hybrid meshes with curved edges for conservation law I : spatial accuracy and geometric conservation laws. Commun. Comput. Phys. **23**(5), 1355–1392 (2018)
24. Zhu, H., Liu, H., Yan, Z., Shi, G., Deng, X.: A priori subcell limiting based on compact nonuniform nonlinear weighted schemes of high-order cpr method for hyperbolic conservation laws. Comput. Fluids **241**, 105456 (2022)